# HPLC METHODS FOR RECENTLY APPROVED PHARMACEUTICALS

# HPLC METHODS FOR RECENTLY APPROVED PHARMACEUTICALS

George Lunn

WILEY-INTERSCIENCE
A JOHN WILEY & SONS, INC., PUBLICATION

Published by John Wiley & Sons, Inc., Hoboken, New Jersey.
Published simultaneously in Canada.



***Library of Congress Cataloging-in-Publication Data:***

Lunn, George.
HPLC methods for recently approved pharmaceuticals / George Lunn.
p. cm.
Includes index.
ISBN 0-471-66941-5 (cloth)
1. High performance liquid chromatography. 2. Drugs–Analysis. I. Title.
RS189.5.H54L773 2005
615′.1–dc22

2004016046

Printed in the United States of America.

10 9 8 7 6 5 4 3 2 1

# CONTENTS

Preface / xi
Acknowledgements / xiii
About This Book / xv

Abacavir / 1
Acarbose / 5
Acetyl sulfisoxazole / 6
Acrivastine / 7
Adapalene / 10
Adefovir dipivoxil / 11
Adrenocorticotropic hormone / 13
Afloqualone / 15
Alacepril / 16
Alclometasone 17,21-dipropionate / 18
Alitretinoin / 21
Allethrin / 24
Almotriptan / 27
Alosetron / 29
Amcinonide / 30
Aminolevulinic acid / 33
Amprenavir / 36
Anagrelide / 42
Anakinra / 43
Apraclonidine / 45
Aprepitant / 46
Aranidipine / 48
Arotinolol / 49
Arteether / 52
Articaine / 54
Asparaginase / 57
Atazanavir sulfate / 58
Atipamezole / 60
Atomoxetine hydrochloride / 62
Atorvastatin / 64
Atosiban / 66
Balofloxacin / 67
Bambermycins / 69
Befunolol / 70
Benzalkonium chloride / 71
Betaine / 72
Bethanechol chloride / 74
Bexarotene / 75
Biapenem / 77
Bimatoprost / 79
Bioresmethrin / 80
Bivalirudin / 81
Boldenone / 82
Bosentan / 83
$\beta$-Boswellic acid / 86
Brimonidine / 88
Bromfenac / 90
Brovincamine / 92
Bucillamine / 93
Budipine / 94
Bulaquine / 95
Butacaine / 97
Butamben / 99
Butoconazole / 100
Butyl flufenamate / 101
Cambendazole / 102
Candesartan cilexetil / 104
Capecitabine / 106
Casanthranol / 108
Caspofungin / 109

Castor oil / 112
Cefbuperazone / 113
Cefditoren / 114
Cefoselis / 116
Cefozopran / 117
Cefuzonam / 118
Celecoxib / 119
Cerivastatin / 123
Cetrorelix / 125
Cetyl alcohol / 128
Cevimeline hydrochloride / 130
Chlorobutanol / 132
Chloroprocaine / 133
Chorionic gonadotropin / 134
Cilnidipine / 135
Cimetropium bromide / 136
Cisatracurium besylate / 137
Citric acid / 139
Clioquinol / 142
Clobetasol 17-propionate / 143
Clopidogrel / 147
Clopidol / 149
Cloricromen / 151
Clorsulon / 152
Colistin / 153
Cypermethrin / 155
Dalfopristin / 156
Dalteparin / 158
Daptomycin / 159
Deferiprone / 161
Deflazacort / 162
Desloratadine / 164
Desogestrel / 166
Desoximetasone / 167
Desoxycorticosterone / 169
Dexrazoxane / 172
Dextran / 174
Diacerein / 176
Dichloroacetic acid / 177
Dichlorophen / 178
Diclazuril / 179
Dihydrotachysterol / 181
Dimethyl sulfoxide / 183
Dinitolmide / 185
Dipivefrin / 186
Dithiazanine iodide / 187
Docarpamine / 188
Dofetilide / 189
Dolasetron / 191
Donepezil / 193
Doxefazepam / 195
Doxercalciferol / 196
Dropropizine / 198
Drospirenone / 199
Droxicam / 200
Droxidopa / 201
Ebrotidine / 202
Edaravone / 204
EDTA / 206
Efavirenz / 208
Efrotomycin / 212
Egualen / 213
Eletriptan / 214
Emtricitabine / 215
Enoxaparin sodium / 217
Entacapone / 218
Eperisone / 220
Eplerenone / 222
Epoprostenol / 224
Eprosartan / 225
Eptazocine / 227
Eptifibatide / 228
Erdosteine / 229
Ergotamine / 230
Ertapenem / 234
Ethopabate / 236
Ethyl icosapentate / 237
Etonogestrel / 238
Etoricoxib / 240
Etorphine / 242
Exemestane / 243
Ezetimibe / 245
Fadrozole / 247
Falecalcitriol / 248
Fenoxycarb / 250
Fenticonazole / 251
Fexofenadine / 253
Flomoxef / 257
Florfenicol / 258
Fludrocortisone / 260
Fluprostenol / 262
Flurandrenolide / 264
Flurithromycin / 267
Flurogestone acetate / 268
Fluticasone propionate / 270
Flutrimazole / 273
Fomepizole / 274
Fomivirsen / 276
Fondaparinux / 277
Formestane / 278

Formoterol / 279
Fosamprenavir calcium / 281
Fosinopril / 283
Fosphenytoin / 284
Frovatriptan / 286
Fumagillin / 288
Galantamine / 290
Ganirelix / 292
Gatifloxacin / 293
Gefitinib / 295
Gemcitabine / 296
Gemifloxacin / 298
Gestodene / 300
Gestrinone / 301
Glycerin / 302
Guanabenz / 304
Guanadrel / 305
Halobetasol propionate / 306
Halofuginone / 308
Hetastarch / 310
Hydroquinone / 311
Hygromycin B / 312
Iloprost / 313
Imatinib / 314
Imidocarb / 316
Iobenguane / 318
Iodixanol / 320
Iopanoic acid / 322
Iopromide / 324
Ioversol / 326
Ipratropium bromide / 327
Ipriflavone / 328
Isoflupredone / 329
Isopropamide iodide / 330
Itopride / 332
Kinetin / 333
Lafutidine / 334
Lamivudine / 335
Latanoprost / 339
Leflunomide / 341
Lercanidipine / 343
Letrozole / 345
Levetiracetam / 346
Levonordefrin / 348
Levosimendan / 349
Lidamidine / 351
Lincomycin / 352
Lindane / 354
Linezolid / 355
Liothyronine / 357
Lomerizine / 358
Lopinavir / 359
Loteprednol etabonate / 362
Marbofloxacin / 364
Masoprocol / 367
Maxacalcitol / 368
Medetomidine / 369
Meglutol / 371
Melatonin / 372
Melengestrol acetate / 375
Memantine / 378
Menthol / 380
Mepenzolate bromide / 381
Mepixanox / 383
Mequinol / 384
Methenamine / 385
Methoprene / 386
Methoxychlor / 387
Methyltestosterone / 392
Metrizamide / 393
Metyrosine / 394
Micafungin / 396
Milnacipran / 398
Mirtazapine / 400
Misoprostol / 405
Mizolastine / 406
Moexipril / 411
Mofezolac / 412
Mometasone furoate / 413
Monensin / 415
Morantel / 416
Mosapride / 417
Moxifloxacin / 420
Moxonidine / 423
Nadifloxacin / 424
Naftopidil / 425
Nandrolone / 427
Narasin / 429
Nartograstim / 430
Nateglinide / 431
Nebivolol / 433
Nelfinavir / 435
Nequinate / 440
Neridronic acid / 441
Nevirapine / 443
Nicarbazin / 447
Nilutamide / 448
Nipradilol / 449
Nitazoxanide / 450
Nitenpyram / 452

Nomegestrol / 453
Nonoxynol-9 / 454
Nystatin / 455
Octocrylene / 456
Oleic acid / 457
Olmesartan / 469
Olopatadine / 470
Orbifloxacin / 472
Orlistat / 475
Oseltamivir / 477
Oxaliplatin / 479
Oxiconazole / 481
Panipenem / 483
Parecoxib / 484
Paricalcitol / 485
Pazufloxacin / 487
Penciclovir / 488
Pentaerythritol tetranitrate / 490
Pentosan polysulfate / 491
Perflubron / 492
Perospirone / 493
Phenazopyridine hydrochloride / 494
Phentermine / 497
Phosphatidylcholine / 501
Phosphatidylglycerol / 504
Piketoprofen / 508
Pilsicainide / 509
Pioglitazone / 511
Pipercuronium bromide / 514
Pirlimycin / 515
Poloxalene / 518
Pramipexole / 519
Pranlukast / 521
Prednicarbate / 523
Propionylpromazine / 524
Propoxycaine hydrochloride / 527
Propylene glycol / 528
Propylhexedrine / 529
Protirelin / 530
Prulifloxacin / 532
Pyrethrins / 533
Quetiapine / 536
Quinfamide / 539
Quinupristin / 540
Rabeprazole / 542
Ractopamine / 544
Raloxifene / 547
Ramosetron / 549
Rapacuronium bromide / 551
Remifentanil / 553
Repaglinide / 555
Ricinoleic acid / 557
Rifaximin / 558
Rilmazafone / 559
Risedronate sodium / 560
Rizatriptan / 561
Rofecoxib / 562
Ropinirole / 565
Rosiglitazone / 566
Rosuvastatin calcium / 568
Sarafloxacin / 569
Selamectin / 571
Sermorelin / 572
Sibutramine / 574
Sildenafil / 576
Simethicone / 579
Sivelestat / 580
Sodium oxybate / 582
Somatropin / 583
Squalane / 584
Squalene / 585
Stanozolol / 587
Succimer / 588
Succinylcholine chloride / 589
Sulfabromomethazine / 591
Sulfachlorpyridazine / 592
Sulfaethoxypyridazine / 596
Sulfamerazine / 598
Sulfanitran / 600
Sultamicillin / 602
Tacalcitol / 604
Talipexole / 605
Taltirelin / 607
Technetium Tc 99m bicisate / 608
Tegaserod / 609
Telithromycin / 610
Telmesteine / 611
Telmisartan / 612
Temocapril / 614
Temozolomide / 615
Tenofovir disoproxil fumarate / 617
Teprenone / 620
Teriparatide / 621
Tetrachlorvinphos / 622
Tetrahydrozoline / 623
Thalidomide / 626
Thialbarbital / 628
Thyrotropin / 629
Tiagabine / 630
Tiletamine hydrochloride / 631

Tilmicosin / 633
Tiludronic acid / 634
Tirilazad / 635
Tirofiban / 637
Tiropramide / 639
Tizanidine / 641
Tolcapone / 643
Topiramate / 645
Topotecan / 647
Tosufloxacin / 649
Travoprost / 651
Trenbolone / 652
Treprostinil / 653
Trichlorfon / 655
Triethanolamine / 656
Trifluridine / 657
Triptorelin / 658
Troleandomycin / 660
Troxipide / 661
Ubenimex / 662
Unoprostone isopropyl ester / 663
Valacyclovir / 664
Valdecoxib / 666
Valganciclovir / 668
Valrubicin / 670
Valsartan / 671
Zaleplon / 673
Zaltoprofen / 676
Zanamivir / 678
Zinostatin / 680
Zofenopril calcium / 681
Zolazepam hydrochloride / 683

Cumulative Index / 685
Cross-Index to Other Substances / 703

# PREFACE

This book is a collection of procedures for the analysis of more than 390 pharmaceuticals using high-performance liquid chromatography (HPLC) and covers the literature up to the end of 2003. The current volume is a continuation of *HPLC Methods for Pharmaceutical Analysis*, published in four volumes from 1997 to 2000. The previous volumes described methods published in the literature through the middle of 1998.

The current work lists procedures for the analysis of drugs in three broad categories:

- Drugs that have been approved since the previous volumes were published.
- Drugs that were approved when the previous volumes were published but for which analytical methods were not then available in the literature.
- Drugs for which procedures allowing determination in a blood matrix have only become available since the previous volumes were published.

Please note that mention of a drug does not necessarily mean that it is currently approved for use in the United States or indeed in any country.

Despite the ready availability of computer-aided literature, searching this resource is not exploited as much as it might be. One reason for this reluctance is, of course, that a computer search merely produces a listing of possibly relevant references. Tedious and time-consuming searches in the library are necessary to find the most relevant reference that can be turned into a practical analytical procedure in the searcher's own laboratory. The reference finally chosen will, naturally, depend on the individual circumstances, such as the matrix in which the drug is present, availability of equipment, and so on. This book circumvents this lengthy process by providing a number of abstracted and evaluated procedures for the analysis of each drug. The analyst can rapidly identify a relevant procedure and put it into practice.

In addition to the analytical matrix, other factors may be important when choosing an analytical procedure. Accordingly, we have noted such features of the analytical procedures as sensitivity, mode of detection, other compounds that interfere with the analysis, other drugs that may be determined at the same time, and so on.

Readers familiar with our previous publications, *HPLC Methods for Pharmaceutical Analysis, Volumes 1–4* (George Lunn and Norman R. Schmuff, John Wiley, New York, 1997–2000) and *Handbook of Derivatization Reactions for HPLC* (George Lunn and Louise C. Hellwig, John Wiley, New York, 1998), will notice many similarities. The abstract structure is very similar, and the philosophy that the procedures should be reproducible without reference to the original literature is unchanged. A new feature is that the retention times (in minutes) of other drugs that may be determined using the same system have been added in parentheses after the drug name. Other data, such as the limit of detection (LOD), may also be added. The retention time is the number without units. Unlike the previous volumes, this book is not available on a CD in an electronic form.

At the end of the book a Cumulative Index and a Cross-Index to Other Substances are provided. The Cumulative Index provides a comprehensive listing of the drugs covered in this book and the previous volumes. The Cross-Index lists the other compounds that may also be chromatographed under the conditions described in the monographs in this book. Using the information in the monographs it may be possible to develop chromatographic procedures for these compounds.

George Lunn

# ACKNOWLEDGEMENTS

I am grateful for the use of the National Institutes of Health Library, the FDA Medical Library, and the National Library of Medicine and I would like to express my appreciation for the hard work of the staff of these libraries, particularly those diligent workers who reshelve the journal volumes after one of my forays. Although many people have helped with the preparation of this work the mistakes are my own. I would appreciate hearing from anyone who has corrections, comments, or suggestions. I can be reached at lunng@cder.fda.gov.

The content of this volume does not necessarily reflect the views or policies of the Food and Drug Administration, nor does the mention of trade names, commercial products, or organizations imply endorsement by the U.S. Government. Also, mention of a drug does not necessarily mean that it is currently approved for use in the United States or indeed in any country.

G.L.

# ABOUT THIS BOOK

## SCOPE

Newly approved drugs were identified from a variety of sources including the FDA's annual lists of drug approvals (available at www.fda.gov/cder) and *Annual Reports in Medicinal Chemistry* published by Elsevier/Academic Press.

The journals routinely surveyed for relevant articles are:

*American Journal of Health-System Pharmacy*
*Analyst*
*Analytica Chimica Acta*
*Analytical Chemistry*
*Analytical Letters*
*Analytical Sciences*
*Antimicrobial Agents and Chemotherapy*
*Arzneimittelforschung*
*Biological and Pharmaceutical Bulletin*
*Biomedical Chromatography*
*Biopharmaceutics and Drug Disposition*
*Chemical and Pharmaceutical Bulletin*
*Chromatographia*
*Clinical Chemistry*
*Clinical Pharmacology and Therapeutics*
*Drug Metabolism and Disposition*
*Farmaco*
*Food Additives and Contaminants*
*Journal of Analytical Toxicology*
*Journal of AOAC International*
*Journal of Chromatographic Science*
*Journal of Chromatography, Part A and Part B*
*Journal of Clinical Pharmacology*
*Journal of Forensic Sciences*

*Journal of Liquid Chromatography & Related Technology*
*Journal of Pharmaceutical and Biomedical Analysis*
*Journal of Pharmaceutical Sciences*
*Journal of Pharmacology and Experimental Therapeutics*
*Pharmaceutical Research*
*Pharmazie*
*Therapeutic Drug Monitoring*
*Xenobiotica*

Other journals were consulted when relevant articles were identified by computer searches.

The literature was surveyed from 1998 through the end of 2003, although methods from some older articles (and a few from 2004) are included.

## NOMENCLATURE

Each chapter is headed by the name and structure of the target compound as well as other useful data such as the CAS Registry Number, molecular formula, molecular weight, and Merck Index number (from the 13$^{th}$ edition).[1] More useful information such as melting point, solubility, optical rotation, references to reviews, and so on can be found in the Merck Index.

In general, the United States Adopted Name (USAN)[2] is used throughout to identify each drug. Names of derivatives, such as esters, which would have different chromatographic properties, are identified by placing the derivative name in parentheses after the retention time.

Increasingly, drugs previously marketed as racemates are being marketed as a single enantiomer with the name changed to reflect the enantiomer. For example, levofloxacin is the levorotatory form of ofloxacin. For an achiral HPLC method, the chromatography of a single enantiomer is no different from that of the racemate. In general, in this work and the preceding works, we have listed HPLC procedures under the name of the racemate rather than the single enantiomer. The interested reader is referred to the USP Dictionary[2] (page 1208) for the naming conventions used. Generally:

| | | |
|---|---|---|
| Levo rotatory | S isomer | Prefix lev/levo- |
| Levo rotatory | R isomer | Prefix ar- |
| Dextro rotatory | R isomer | Prefix dex/dextro- |
| Dextro rotatory | S isomer | Prefix es- |

For racemates, the rac- prefix is used.

In some cases, the chiral prefix is used. Thus, the following list shows the prefixes that are used in the different volumes:

Dexrazoxane in this volume
Dextromethorphan in Volume 2
Dextromoramide in Volume 2
Dextrothyroxine in Volume 2

Levallorphan in Volume 3
Levamisole in Volume 3
Levobunolol in Volume 3
Levodopa in Volume 3
Levonordefrin in Volume 3 and this volume
Levorphanol in Volume 3
Levosimendan in this volume
Levothyroxine in Volumes 1 and 3.

More generally, the name of the racemic compound is used. Thus,

| For | Consult | Volume |
|---|---|---|
| Arformoterol | Formoterol | 3, this volume |
| Dexamisole | Levamisole | 3 |
| Dexamphetamine | Amphetamine | 2 |
| Dexbrompheniramine | Brompheniramine | 2 |
| Dexbudesonide | Budesonide | 2 |
| Dexchlorpheniramine | Chlorpheniramine | 2 |
| Dexfenfluramine | Fenfluramine | 3 |
| Dexibuprofen | Ibuprofen | 1, 4 |
| Dexketoprofen | Ketoprofen | 1, 4 |
| Dexmedetomidine | Medetomidine | This volume |
| Dexmethylphenidate | Methylphenidate | 1 |
| Dexpropranolol | Propranolol | 4 |
| Dexsotalol | Sotalol | 4 |
| Dextroamphetamine | Amphetamine | 2 |
| Dextropropoxyphene | Propoxyphene | 1, 4 |
| Dexverapamil | Verapamil | 1, 4 |
| Esatenolol | Atenolol | 1, 2 |
| Escitalopram | Citalopram | 2 |
| Esflurbiprofen | Flurbiprofen | 3 |
| Esketamine | Ketamine | 3 |
| Esomeprazole | Omeprazole | 1, 3 |
| Esoxybutynin chloride | Oxybutynin chloride | 3 |
| Eszopiclone | Zopiclone | 4 |
| Levalbuterol | Albuterol | 1, 2 |
| Levamfetamine | Amphetamine | 2 |
| Levamphetamine | Amphetamine | 2 |
| Levcycloserine | Cycloserine | 2 |
| Levdobutamine | Dobutamine | 2 |
| Levmetamfetamine | Methamphetamine | 3 |
| Levobetaxolol | Betaxolol | 2 |
| Levobupivacaine | Bupivacaine | 2 |
| Levocarnitine | Carnitine | 2 |
| Levocetirizine | Cetirizine | 2 |
| Levodropropizine | Dropropizine | 2, this volume |

| | | |
|---|---|---|
| Levofenfluramine | Fenfluramine | 3 |
| Levofloxacin | Ofloxacin | 1, 3 |
| Levofuraltadone | Furaltadone | 3 |
| Levoleucovorin | Leucovorin | 3 |
| Levomenthol | Menthol | 3, this volume |
| Levomethadone | Methadone | 3 |
| Levomoprolol | Moprolol | 3 |
| Levonorgestrel | Norgestrel | 1 |
| Levopropoxyphene | Propoxyphene | 1, 4 |
| Levopropylhexedrine | Propylhexedrine | 4, this volume |
| Levosalbutamol | Albuterol | 1, 2 |
| Levosulpiride | Sulpiride | 4 |
| Racementhol | Menthol | 3, this volume |
| Racemethorphan | Dextromethorphan | 2 |
| Racemetirosine | Metyrosine | This volume |
| Racemorphan | Levorphanol | 3 |
| Racephedrine | Ephedrine | 3 |
| Racepinephrine | Epinephrine | 3 |

## BIBLIOGRAPHIES

For reasons of space, it is not possible to abstract every relevant paper, and so at the end of some chapters an Annotated Bibliography lists other relevant papers. After the citation, a few features of the method that are not obvious from the title of the paper may be briefly mentioned to help the reader decide if this paper may be of use. For example, the limit of quantitation of the method may be cited. Unless otherwise mentioned, it may be assumed that a method involves liquid–liquid extraction of a biological fluid from a human and uses reversed-phase HPLC with UV detection. Thus, if a method uses solid-phase extraction (SPE) or fluorescence detection, this will be mentioned.

## ABSTRACT STRUCTURE

The detailed procedures given normally contain the following sections. Of course, not all papers give full details, so some sections may be missing.

Matrix
Sample preparation
Guard column
Column
Mobile phase
Flow rate
Injection volume
Retention time
Detector

Internal standard
Column temperature
Extracted
Simultaneous
Also
Noninterfering
Interfering
Limit of detection
Limit of quantitation
Key words
Reference

## ABSTRACT CONVENTIONS

| | |
|---|---|
| Also | Compounds that can be analyzed at the same time. It is not specified whether they interfere, but they can be extracted. See also Extracted, Simultaneous. |
| Column | Dimensions are length (mm) × internal diameter (mm), and the material is stainless steel unless otherwise indicated. |
| Column temperature | If other than ambient (all temperatures are in degrees C). |
| Derivatization | Pre-column unless otherwise mentioned (in Key Words). |
| Detector | Wavelengths in nanometers |
| Extracted | Compounds that can be extracted from the matrix in question and analyzed at the same time and do not interfere. See also Also, Simultaneous. |
| Flow rates | In milliliters per minute. |
| Guard column | Dimensions are length (mm) × internal diameter (mm). |
| Injection volume | In microliters (μL). Injection volume may be either the volume actually injected or the volume of the injection loop. If it is the volume actually injected, this value is also given in the Sample preparation section. If the actual injection volume is not given in the Sample preparation section, the Injection volume given is that of the injection loop. |
| Interfering | Compounds that interfere with the analysis of the target compound. Compounds that interfere with the chromatography of the internal standard (IS) are listed under simultaneous (another IS can always be selected or an external standard procedure can be used). |
| Matrix | A controlled vocabulary is used (see below) |
| Mobile phase | Ratios are v/v and gradients are linear, unless otherwise noted. Times given when describing gradient elution and other procedures such as column switching are the times for each step, e.g., "MeOH:water 15:85 for 4 min, to 50:50 over 2 min, maintain at 50:50 for 4 min." If we were to include the cumulative times (t) in the example above it would read: "MeOH:water 15:85 for 4 min ($t = 4$), to 50:50 over 2 min ($t = 6$), maintain at 50:50 for 4 min ($t = 10$)." |

| | |
|---|---|
| Noninterfering | Compounds that do not interfere with the analysis for various reasons, e.g., they are not extracted, they are not detected. |
| Retention time | In minutes. This is frequently estimated from a reproduced chromatogram, and so the accuracy may not be great. When available, retention times are given for the analyte, the internal standard, and other compounds that may be chromatographed under the same conditions. For the internal standard and other compounds that may be chromatographed under the same conditions, the retention times are given in parentheses after the compound name. |
| Simultaneous | Compounds that can be analyzed at the same time and do not interfere. Note that the compound cannot necessarily be extracted from the matrix in question (although it may be). See also Also, Extracted. |
| SPE | For the sake of consistency, conditioning procedures for solid-phase extraction (SPE) cartridges are always described at the beginning of the sample preparation sections. Bear in mind, however, that the conditioning procedure should be carried out just prior to use. Thus, if sample preparation is a lengthy procedure, it may be necessary to delay SPE cartridge conditioning until the step requiring the cartridge. |
| Species | If other than human, noun is used instead of adjective, e.g., cow not bovine. In some cases, human may be specified. For example, if *both* human blood and rat blood are analyzed, *both* human and rat will be indicated (in Key Words). |

## MATRIX

To help with searching, a controlled vocabulary is used to limit the number of terms in the matrix section. For example, the terms raw material, drug substance, or API (active pharmaceutical ingredient) are not used; the term bulk is used instead. In a number of cases, the matrix is associated with various key words, which can be used to narrow the search. For example, the matrix term blood has the key words plasma, serum, and whole blood associated with it. Thus, if you are interested in the determination of the drug in blood in general, you should look in the matrix field for blood. If, however, you are specifically interested in finding the drug in plasma, you should look in the key words field for plasma.

| Matrix | Associated Key Words |
|---|---|
| Bile | |
| Blood | Plasma, serum, whole blood |
| Bulk | |
| CSF | |
| Formulations | Capsules, creams, injections, ointment, tablets, etc. |
| Microsomal incubations | |

| | |
|---|---|
| Milk | |
| Perfusate | |
| Reaction mixtures | |
| Saliva | |
| Tissue | Brain, heart, kidney, liver, muscle, spleen, etc. |
| Urine | |

## ABBREVIATIONS

| | |
|---|---|
| BHT | 2,6-Di-tert-butyl-4-methylphenol, butylated hydroxytoluene |
| DMSO | Dimethyl sulfoxide |
| E | Electrochemical detection |
| em | Emission wavelength |
| EtOH | Ethanol |
| ex | Excitation wavelength |
| F | Fluorescence detection |
| GPC | Gel permeation chromatography |
| h | Hour |
| HPLC | High-performance liquid chromatography |
| ID | Internal diameter |
| IS | Internal standard |
| L | Liter |
| LOD | Limit of detection or some other description indicating that this is the smallest concentration or quantity that can be detected or analyzed for |
| LOQ | Lower limit of quantitation, either given as such in the paper or taken as the lower limit of the linear quantitation range |
| M | Molar (i.e., moles/L) |
| MeCN | Acetonitrile |
| MeOH | Methanol |
| min | Minutes |
| mL | Milliliter |
| mM | Millimolar (i.e., millimoles/L) |
| MS | Mass spectrometric detection |
| MSPD | Matrix solid phase dispersion |
| MTBE | Methyl tert-butyl ether |
| nM | Nanomolar (i.e., nanomoles/L) |
| psi | Pounds/sq. in. (1 psi = 6.89476 kPa) |
| s | Seconds |
| SEC | Size Exclusion Chromatography |
| SFC | Supercritical fluid chromatography |
| SFE | Supercritical fluid extraction |
| SPE | Solid phase extraction |
| Temp | Temperature |
| U | Units |
| UV | Ultraviolet detection |
| vol | Volume |

## PIC REAGENTS

These reagents are offered by Waters as buffered solutions containing the following compounds:

PIC A is tetrabutylammonium sulfate
PIC B5 is pentanesulfonic acid
PIC B7 is heptanesulfonic acid.

## WORKING PRACTICES

In general, good working practice, for example, using high-grade materials is assumed. Solutions should be protected from light, and silanized glassware should be used unless you have good reason to believe that these precautions are not necessary. Details of solution preparation are generally not given. It should be remembered that the preparation of a dilute aqueous solution of a relatively water-insoluble compound can frequently be made by dissolving the compound in a small volume of a water-miscible organic solvent and diluting this solution with water. A number of excellent texts[3–9] discuss good working practices and procedures in HPLC. Please note that all the temperatures are in degrees C.

It is also assumed that safe working practices are observed. Organic solvents should only be evaporated in a properly functioning chemical fume hood, correct protective equipment should be worn when dealing with potentially hazardous biological materials, and waste solutions should be disposed of in accordance with all applicable regulations.

A number of solvents are particularly hazardous. For example, benzene is a human carcinogen;[10] chloroform,[11] dichloromethane,[12] dioxane,[13] and carbon tetrachloride [14] are carcinogenic in experimental animals; and DMF [15] and MTBE [16,17] may be carcinogenic. Organic solvents are, in general, flammable and toxic by inhalation, ingestion, and skin absorption. Sodium azide is carcinogenic and toxic and liberates explosive, volatile, toxic hydrazoic acid when mixed with acid. Sodium azide can form explosive heavy metal azides, for example, with plumbing fixtures, and so should not be discharged down the drain.[18] Disposal procedures have been described for a number of hazardous drugs and reagents,[18] and a procedure for the hydrolysis of acetonitrile in waste solvent to the much less toxic acetic acid and ammonia[19,20] has been described. *n*-Hexane is surprisingly toxic.[21]

## REFERENCES


1. O'Neil, M.J., Ed., *The Merck Index*, 13th edition, Merck & Co. Inc, Whitehouse Station, NJ, **2001**.
2. United States Pharmacopeial Convention. *USP Dictionary of USAN and International Drug Names*, United States Pharmacopeial Convention, Rockville, MD, **2004**.
3. Snyder, L.R.; Kirkland, J.J. *Introduction to Modern Liquid Chromatography*, 2nd edition, John Wiley & Sons, New York, **1979**.
4. Lawrence, J.F. *Organic Trace Analysis by Liquid Chromatography*, Academic Press, New York, **1981**.
5. Sadek, P.C. *The HPLC Solvent Guide*, John Wiley & Sons, New York, **1996**.

6. Snyder, L.R.; Kirkland, J.J.; Glajch, J.L. *Practical HPLC Method Development*, 2nd edition, John Wiley & Sons, New York, **1997**.
7. Meyer, V.R. *Pitfalls and Errors of HPLC in Pictures*, Hüthig, Heidelberg, Germany, **1997**.
8. Meyer, V.R. *Practical High-Performance Liquid Chromatography*, 3rd edition, John Wiley & Sons, Chichester, UK, **2000**.
9. Sadek, P.C. *Troubleshooting HPLC Systems. A Bench Manual*, John Wiley & Sons, New York, **2000**.
10. Lewis, R.J. Sr. *Sax's Dangerous Properties of Industrial Materials*, 8th edition, Van Nostrand Reinhold, New York, **1992**, pp. 356–358.
11. Lewis, R.J. Sr. *Sax's Dangerous Properties of Industrial Materials*, 8th edition, Van Nostrand Reinhold, New York, **1992**, pp. 815–816.
12. Lewis, R.J. Sr. *Sax's Dangerous Properties of Industrial Materials*, 8th edition, Van Nostrand Reinhold, New York, **1992**, pp. 2311–2312.
13. Lewis, R.J. Sr. *Sax's Dangerous Properties of Industrial Materials*, 8th edition, Van Nostrand Reinhold, New York, **1992**, pp. 1449–1450.
14. Lewis, R.J. Sr. *Sax's Dangerous Properties of Industrial Materials*, 8th edition, Van Nostrand Reinhold, New York, **1992**, pp. 701–702.
15. Lewis, R.J. Sr. *Sax's Dangerous Properties of Industrial Materials*, 8th edition, Van Nostrand Reinhold, New York, **1992**, p. 1378.
16. Belpoggi, F.; Soffritti, M.; Maltoni, C. Methyl-tertiary-butyl ether (MTBE) – a gasoline additive – causes testicular and lympho-haematopoietic cancers in rats, *Toxicol.Ind.Health.*, **1995**, *11*, 119–149.
17. Mehlman, M.A. Dangerous and cancer-causing properties of products and chemicals in the oil refining and petrochemical industry: Part XV. Health hazards and health risks from oxygenated automobile fuels (MTBE): lessons not heeded, *Int.J.Occup.Med.Toxicol.*, **1995**, *4*, 219–236.
18. Lunn, G.; Sansone, E.B. *Destruction of Hazardous Chemicals in the Laboratory*, 2nd edition, John Wiley & Sons, New York, **1994**.
19. Gilomen, K.; Stauffer, H.P.; Meyer, V.R. Detoxification of acetonitrile – water wastes from liquid chromatography, *Chromatographia*, **1995**, *41*, 488–491.
20. Gilomen, K.; Stauffer, H.P.; Meyer, V.R. Management and detoxification of acetonitrile wastes from liquid chromatography, *LC.GC*, **1996**, *14*, 56–58.
21. Meyer, V. A safer solvent, *Anal.Chem.*, **1997**, *69*, 18A.

# Abacavir

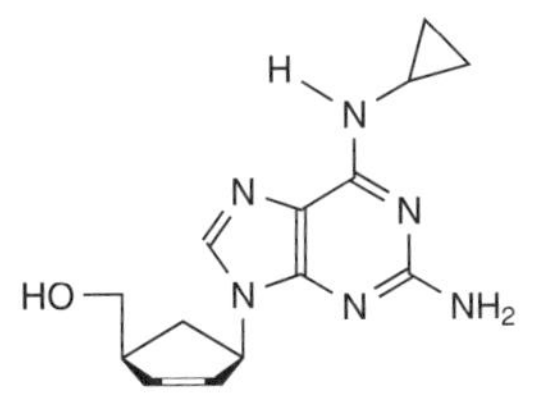

**Molecular formula:** $C_{14}H_{18}N_6O$
**Molecular weight:** 286.33
**CAS Registry No:** 136470-78-5 (base), 188062-50-2 (sulfate)
**Merck Index:** 13,1

## SAMPLE

**Matrix:** blood
**Sample preparation:** Condition a 1 mL 100 mg Bond Elut-C SPE cartridge with 1 mL MeOH and 1 mL 100 mM pH 7.0 ammonium acetate buffer. Heat plasma at 58° for 1 h to inactivate HIV. Vortex 800 μL plasma with 300 μL 2 μg/mL hexobarbital in 25 mM pH 7.0 ammonium acetate buffer for 30 s and centrifuge at 18 000 g for 5 min. Add 1 mL of the supernatant to the SPE cartridge, wash with 1 mL 100 mM pH 7.0 ammonium acetate buffer, suck dry for 1 min, elute with 800 μL MeOH. Evaporate the eluate to dryness under a stream of nitrogen at 40° and reconstitute the residue with 100 μL mobile phase. Vortex for 30 s, centrifuge at 18 000 g for 3 min, and inject an 80 μL aliquot.

## HPLC VARIABLES

**Guard column:** 20 × 3.9 5 μm Polarity dC18 (Waters)
**Column:** 150 × 3.9 5 μm Polarity dC18 (Waters)
**Column temperature:** 40
**Mobile phase:** Gradient. A was 10 mM pH 6.5 ammonium acetate buffer. B was 10 mM pH 6.5 ammonium acetate buffer:MeCN:MeOH 20:50:30. A:B 96:4 for 15 min, to 36:64 over 15 min, maintain at 36:64 for 3 min, re-equilibrate at initial conditions for 7 min.
**Flow rate:** 1.1
**Injection volume:** 80
**Detector:** UV 269 for 11 min, UV 250 for 3 min, UV 271 for 10 min, UV 230 for 9 min

## CHROMATOGRAM

**Retention time:** 25.1
**Internal standard:** hexobarbital (30.6)
**Limit of quantitation:** 10.0 ng/mL

## OTHER SUBSTANCES

**Extracted:** didanosine (13.6), lamivudine (8.6), nevirapine (27.3), stavudine (15.7), zalcitabine (5.9), zidovudine (23.8)
**Noninterfering:** tenofovir

## KEY WORDS

plasma; SPE

## REFERENCE


Rezk, N.L.; Tidwell, R.R.; Kashuba, A.D.M. Simultaneous determination of six HIV nucleoside analogue reverse transcriptase inhibitors and nevirapine by liquid chromatography with ultraviolet absorbance detection, *J.Chromatogr.B*, **2003**, *791*, 137–147.


## SAMPLE

**Matrix:** blood
**Sample preparation:** Condition a 100 mg Dual Zone C18 SPE cartridge (Diazem) with 2 mL MeOH and 2 mL water. Dilute 500 μL serum with 1 mL water, add to the SPE cartridge, wash with 500 μL water, elute with 1 mL MeOH. Evaporate the eluate to

dryness with vortexing under reduced pressure at 40° and reconstitute the residue with 300 μL MeOH, inject a 10 μL aliquot.

## HPLC VARIABLES

**Column:** two 150 × 4.6 3 μm Luna C18 columns in series
**Column temperature:** 60
**Mobile phase:** Gradient. MeCN:water from 5:95 to 45:55 over 20 min.
**Flow rate:** 0.85
**Injection volume:** 10
**Detector:** UV 250

## CHROMATOGRAM

**Retention time:** 17
**Limit of detection:** 75 ng/mL

## OTHER SUBSTANCES

**Extracted:** didanosine (10.5, LOD 120 ng/mL), lamivudine (9.5, LOD 260 ng/mL), stavudine (11.5, LOD 40 ng/mL), zalcitabine (7.5, LOD 440 ng/mL), zidovudine (16, LOD 30 ng/mL)

## KEY WORDS

SPE; serum

## REFERENCE

Simon, V.A.; Thiam, M.D.; Lipford, L.C. Determination of serum levels of thirteen human immunodeficiency virus-suppressing drugs by high-performance liquid chromatography, *J.Chromatogr.A*, **2001**, *913*, 447–453.

## SAMPLE

**Matrix:** blood
**Sample preparation:** Mix 300 μL plasma with 75 μL 20% perchloric acid for 30 s, centrifuge at 1300 g for 15 min, inject a 100 μL aliquot.

## HPLC VARIABLES

**Guard column:** 20 × 3.8 Symmetry C18 (Waters)
**Column:** 100 × 4.6 3.5 μm Symmetry C18 (Waters)
**Column temperature:** 41 ± 2
**Mobile phase:** MeCN:25 mM pH 7.0 phosphate buffer 15:85
**Flow rate:** 1
**Injection volume:** 100
**Detector:** UV 285

## CHROMATOGRAM

**Retention time:** 4.8
**Limit of quantitation:** 20 ng/mL

## OTHER SUBSTANCES

**Simultaneous:** didanosine, folic acid, ganciclovir, lamivudine, nevirapine, pyrazinamide, ranitidine, rifampin, stavudine, sulfamethoxazole, trimethoprim, zidovudine
**Noninterfering:** adefovir, amprenavir, delavirdine, efavirenz, fluconazole, indinavir, itraconazole, methadone, nelfinavir, oxazepam, pyrimethamine, rifampin, ritonavir, saquinavir, zalcitabine

## KEY WORDS

plasma

## REFERENCE

Veldkamp, A.I.; Sparidans, R.W.; Hoetelmans, R.M.W.; Beijnen, J.H. Quantitative determination of abacavir (1592U89), a novel nucleoside reverse transcriptase inhibitor, in human plasma using isocratic reversed-phase high-performance liquid chromatography with ultraviolet detection, *J.Chromatogr.B*, **1999**, *736*, 123–128.

## SAMPLE

**Matrix:** blood

**Sample preparation:** Centrifuge plasma at 4000 g for 20 min using a Centrifree micropartition device (Amicon), inject a 100 μL aliquot of the ultrafiltrate.

## HPLC VARIABLES

**Column:** 250 × 4.6 Adsorbsphere C18

**Mobile phase:** Gradient. A was MeCN:water 80:20. B was 50 mM ammonium acetate containing 0.1% triethylamine adjusted to pH 5.5. A:B from 0:100 to 50:50 over 30 min, re-equilibrate at initial conditions for 10 min.

**Flow rate:** 1

**Injection volume:** 100

**Detector:** UV 260, UV 285

## CHROMATOGRAM

**Retention time:** 23

## OTHER SUBSTANCES

**Extracted:** carbovir (20)

## KEY WORDS

rat; pharmacokinetics; plasma

## REFERENCE

Daluge, S.M.; Good, S.S.; Faletto, M.B.; Miller, W.H.; St.Clair, M.H.; Boone, L.R.; Tisdale, M.; Parry, N.R.; Reardon, J.E.; Dornsife, R.E.; Averett, D.R.; Krenitsky, T.A. 1592U89, a novel carbocyclic nucleoside analog with potent, selective anti-human immunodeficiency virus activity, *Antimicrob.Agents Chemother.*, **1997**, *41*, 1082–1093.

## SAMPLE

**Matrix:** CSF, urine

**Sample preparation:** Centrifuge CSF or urine at 12 000 g for 5 min, dilute a 75 μL aliquot to 750 μL with mobile phase, inject an aliquot.

## HPLC VARIABLES

**Column:** 150 × 3.2 5 μm Kromasil C18 (Phenomenex)

**Mobile phase:** Gradient. MeOH:25 mM pH 4.0 ammonium acetate buffer from 5:95 to 50:50 over 30 min, re-equilibrate at initial conditions for 10 min.

**Flow rate:** 0.7

**Detector:** UV 295

## CHROMATOGRAM

**Retention time:** 25.5

**Limit of quantitation:** 62 ng/mL (CSF), 629 ng/mL (urine)

## OTHER SUBSTANCES

**Extracted:** metabolites, abacavir 5′-glucuronide, abacavir 5′-carboxylate

## REFERENCE

Ravitch, J.R.; Moseley, C.G. High-performance liquid chromatographic assay for abacavir and its two major metabolites in human urine and cerebrospinal fluid, *J.Chromatogr.*, **2001**, *762*, 165–173.

## ANNOTATED BIBLIOGRAPHY

Fung, E.N.; Cai, Z.; Burnette, T.C.; Sinhababu, A.K. Simultaneous determination of Ziagen and its phosphorylated metabolites by ion-pairing high-performance liquid chromatography-tandem mass spectrometry, *J.Chromatogr.B*, **2001**, *754*, 285–295. [LC-MS]

Sparidans, R.W.; Hoetelmans, R.M.W.; Beijnen, J.H. Liquid chromatographic assay for simultaneous determination of abacavir and mycophenolic acid in human plasma using dual spectrophotometric detection, *J.Chromatogr.B*, **2001**, *750*, 155–161.

Thomas, S.A.; Bye, E.; Segal, M.B. Transport characteristics of the anti-human immunodeficiency virus nucleoside analog, abacavir, into brain and cerebrospinal fluid, *J.Pharmacol.Exp.Ther.*, **2001**, *298*, 947–953.

Yuen, G.J.; Lou, Y.; Thompson, N.F.; Otto, V.R.; Allsup, T.L.; Mahony, W.B.; Hutman, H.W. Abacavir/lamivudine/zidovudine as a combined formulation tablet: Bioequivalence compared with each component administered concurrently and the effect of food on absorption, *J.Clin.Pharmacol.*, **2001**, *41*, 277–288.

Aymard, G.; Legrand, M.; Trichereau, N.; Diquet, B. Determination of twelve antiretroviral agents in human plasma sample using reversed-phase high-performance liquid chromatography, *J.Chromatogr.B*, **2000**, *744*, 227–240. [for amprenavir; efavirenz; indinavir; nelfinavir; ritonavir; saquinavir; abacavir; didanosine; lamivudine; stavudine; nevirapine; zidovudine]

McDowell, J.A.; Lou, Y.; Symonds, W.S.; Stein, D.S. Multiple-dose pharmacokinetics and pharmacodynamics of abacavir alone and in combination with zidovudine in human immunodeficiency virus-infected adults, *Antimicrob.Agents Chemother.*, **2000**, *44*, 2061–2067.

Kumar, P.N.; Sweet, D.E.; McDowell, J.A.; Symonds, W.; Lou, Y.; Hetherington, S.; LaFon, S. Safety and pharmacokinetics of abacavir (1592U89) following oral administration of escalating single doses in human immunodeficiency virus type 1-infected adults, *Antimicrob.Agents Chemother.*, **1999**, *43*, 603–608.

McDowell, J.A.; Chittick, G.E.; Ravitch, J.R.; Polk, R.E.; Kerkering, T.M.; Stein, D.S. Pharmacokinetics of [$^{14}$C]abacavir, a human immunodeficiency virus type 1 (HIV-1) reverse transcriptase inhibitor, administered in a single oral dose to HIV-1-infected adults: a mass balance study, *Antimicrob.Agents Chemother.*, **1999**, *43*, 2855–2861.

Wang, L.H.; Chittick, G.E.; McDowell, J.A. Single-dose pharmacokinetics and safety of abacavir (1592U89), zidovudine, and lamivudine administered alone and in combination in adults with human immunodeficiency virus infection, *Antimicrob.Agents Chemother.*, **1999**, *43*, 1708–1715.

# Acarbose

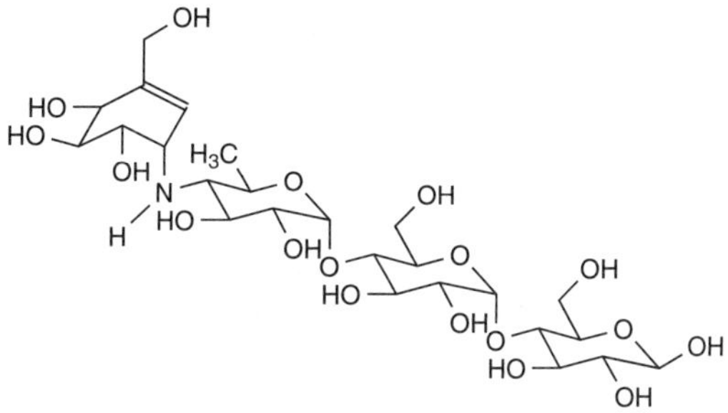


**Molecular formula:** $C_{25}H_{43}NO_{18}$
**Molecular weight:** 645.60
**CAS Registry No:** 56180-94-0
**Merck Index:** 13, 18

## SAMPLE

**Matrix:** formulations
**Sample preparation:** Powder tablet, extract 3 times with 5 mL aliquots of water with sonication for 15 min with vortexing at 5 min intervals each time, centrifuge at 2750 g for 5 min, combine supernatants, make up to 20 mL with water. Dilute a 50 μL aliquot to 1 mL with MeOH, filter (0.2 μM), inject a 20 μL aliquot.

## HPLC VARIABLES

**Column:** 250 × 4.6 5 μm Nucleosil-NH2
**Mobile phase:** MeOH:dichloromethane 65:35
**Flow rate:** 1
**Injection volume:** 20
**Detector:** ELSD, nebulizing gas air at 2.5 bar and 4 L/min, solvent evaporated at 40°

## CHROMATOGRAM

**Retention time:** 4.1
**Limit of detection:** 5 μg/mL
**Limit of quantitation:** 15 μg/mL

## OTHER SUBSTANCES

**Simultaneous:** sucrose (3.5)

## KEY WORDS

comparison with capillary electrophoresis; tablets

## REFERENCE

Cherkaoui, S.; Daali, Y.; Christen, P.; Veuthey, J.-L. Development and validation of liquid chromatography and capillary electrophoresis methods for acarbose determination in pharmaceutical tablets, *J.Pharm.Biomed.Anal.*, **1998**, *18*, 729–735.

# Acetyl sulfisoxazole

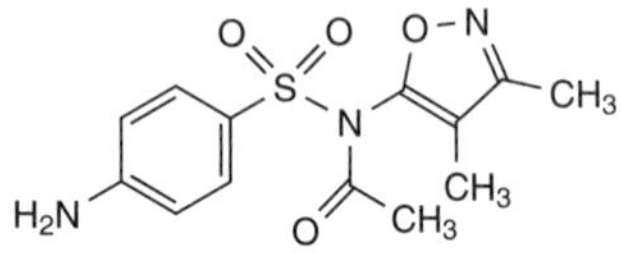


**Molecular formula:** $C_{13}H_{15}N_3O_4S$
**Molecular weight:** 309.35
**CAS Registry No:** 80-74-0
**Merck Index:** 13, 9041

## SAMPLE
**Matrix:** formulations
**Sample preparation:** Extract 1 mL suspension with three 15 mL aliquots of chloroform (Caution! Chloroform is a carcinogen!), combine the organic layers and make up to 50 mL with chloroform, filter (0.45 μm silver membrane, Selas Corp.). Evaporate a 2 mL aliquot of the filtrate to dryness under a stream of nitrogen, reconstitute with 5 mL 330 μg/mL benzanilide in MeCN, inject a 5 μL aliquot.

## HPLC VARIABLES
**Column:** 300 × 4 10 μm μBondapak C18
**Mobile phase:** MeCN:water 40:60
**Flow rate:** 1.5
**Injection volume:** 5
**Detector:** UV 254

## CHROMATOGRAM
**Retention time:** 7
**Internal standard:** benzanilide (11)

## OTHER SUBSTANCES
**Simultaneous:** sulfanilamide (2.5), sulfisoxazole (3)
**Noninterfering:** erythromycin ethylsuccinate

## KEY WORDS
oral suspensions

## REFERENCE
Elrod, L. Jr.; Cox, R.D.; Plasz, A.C. Analysis of oral suspensions containing sulfonamides in combination with erythromycin ethylsuccinate, *J.Pharm.Sci.*, **1982**, *71*, 161–166.

## ANNOTATED BIBLIOGRAPHY
Suber, R.L.; Edds, G.T. High performance liquid chromatographic determinations of sulfonamides by ionic suppression, *J.Liq.Chromatogr.*, **1980**, *3*, 257–268. [for sulfanilamide; sulfaguanidine; sulfamerazine; sulfamethazine; sulfapyridine; sulfisoxazole; *N*-acetylsulfisoxazole; sulfathiazole; in plasma]

# Acrivastine

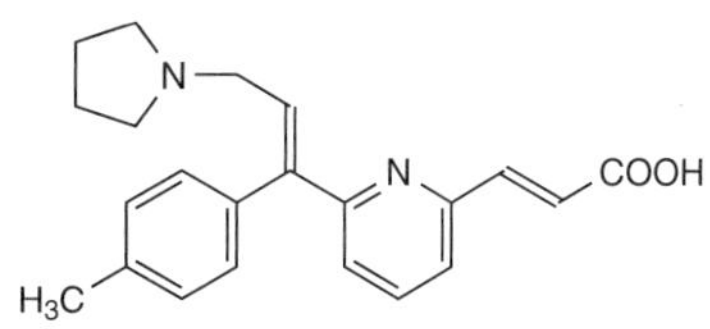


**Molecular formula:** $C_{22}H_{24}N_2O_2$
**Molecular weight:** 348.44
**CAS Registry No:** 87848-99-5
**Merck Index:** 13, 129

## SAMPLE

**Matrix:** blood

**Sample preparation:** Mix 1 mL whole blood with 20 μL 1 μg/mL dibenzepin in MeOH:water 50:50, add 300 μL pH 11 tris buffer, mix, add 500 μL butyl acetate, vortex for 2 min, centrifuge. Preserve the aqueous layer (A). Remove the organic layer and add it to 75 μL 10 mM ammonium acetate buffer containing 0.1% formic acid (pH 3.2), evaporate (?). Add 75 μL MeCN, sonicate for 5 min, centrifuge at 5000 rpm for 5 min, keep the extract as B. Add 20 μL 1 μg/mL enalapril in MeOH:water 50:50 and 120 mg NaCl to the aqueous layer (A), mix, add 500 μL pH 3 phosphate buffer, add 600 μL 8.5% phosphoric acid, add 5 mL dichloromethane:isopropanol 95:5, shake at 250 cycles/min in a bench-top shaker for 30 min, centrifuge at 5000 rpm for 5 min. Remove the lower organic layer and evaporate it to dryness under a stream of air at 45°. Reconstitute the residue with 150 μL initial mobile phase, sonicate, centrifuge, combine with extract B, inject a 30 μL aliquot. (Sample preparation from Gergov,M.; Robson,J.N.; Ojanperä,I.; Heinonen,O.P.; Vuori,E. Simultaneous screening and quantitation of 18 antihistamine drugs in blood by liquid chromatography ionspray tandem mass spectrometry. *Forensic Sci.Inter.* **2001**, *121*, 108–115.)

## HPLC VARIABLES

**Guard column:** 40 mm long 4 μm Purospher RP-18 LiChro Cart 4-4
**Column:** 100 × 2.1 4 μm Genesis C18 (Jones Chromatography)
**Column temperature:** 35
**Mobile phase:** Gradient. MeCN:buffer from 20:80 to 100:0 over 10 min, maintain at 0:100 for 3 min, re-equilibrate at initial conditions for 5 min. (Buffer was 10 mM ammonium acetate containing 0.1% formic acid (pH 3.2).
**Flow rate:** 0.2
**Injection volume:** 30
**Detector:** MS, PE Sciex API 365 triple stage quadrupole LC-MS-MS, PE Sciex Turbo Ion Spray interface, positive ion mode, needle voltage 5.2 kV, nebulizer gas air at 60 psi, curtain gas nitrogen at 40 psi, collision cell gas nitrogen at 40 psi, turbo ionspray heater 375°, heater gas flow 7 L/min

## CHROMATOGRAM

**Retention time:** 5.7
**Internal standard:** dibenzepin, enalapril
**Limit of detection:** <20 ng/mL

## OTHER SUBSTANCES

**Extracted:** acebutolol (3.8, LOD 0.1 μg/mL), acetaminophen (2.5, LOD <5 μg/mL), alprazolam (6.1, LOD <0.02 μg/mL), alprenolol (5.4, LOD 0.01 μg/mL), amantadine (3.4, LOD 0.1 μg/mL), amiloride (2.0, LOD 0.1 μg/mL), aminophenazone (2.8, LOD <5 μg/mL), amiodarone (10.2, LOD 0.05 μg/mL), amitriptyline (6.6, LOD <0.02 μg/mL), astemizole (5.8, LOD <0.02 μg/mL), atenolol (1.7, LOD 0.30 μg/mL), azacyclonol (5.1, LOD 0.02 μg/mL), benzhexol (6.6, LOD <0.02 μg/mL), benzoylecgonine (3.3, LOD 0.01 μg/mL), betaxolol (5.5, LOD 0.01 μg/mL), biperidine (6.2, LOD <0.02 μg/mL), bisoprolol (5.0, LOD <0.02 μg/mL), brompheniramine (5.3, LOD 0.002 μg/mL), bupivacaine (5.1, LOD <0.02 μg/mL), buprenorphine (5.9, LOD 0.01 μg/mL), buspirone (5.1, LOD 0.002 μg/mL), caffeine (2.8, LOD 1 μg/mL), carbamazepine

(6.1, LOD <0.02 μg/mL), carbinoxamine (5.1, LOD 0.002 μg/mL), carisoprodol (6.7, LOD <5 μg/mL), carvedilol (6.2, LOD <0.02 μg/mL), celiprolol (4.3, LOD 0.05 μg/mL), cetirizine (6.3, LOD 0.05 μg/mL), chlorcyclizine (6.6, LOD <0.02 μg/mL), chlordiazepoxide (5.7, LOD <0.02 μg/mL), chlormezanone (5.8, LOD <5 μg/mL), chloroquine (2.7, LOD 0.02 μg/mL), chlorpheniramine (5.1, LOD 0.002 μg/mL), chlorpromazine (7.0, LOD 0.02 μg/mL), chlorpropamide (6.7, LOD <5 μg/mL), chlorprothixene (7.0, LOD <0.02 μg/mL), cinnarizine (7.9, LOD <0.02 μg/mL), citalopram (5.7, LOD <0.02 μg/mL), clemastine (7.7, LOD 0.02 μg/mL), clobazam (7.3, LOD <0.02 μg/mL), clobutinol (5.3, LOD 0.02 μg/mL), clomethiazole (6.2, LOD 0.5 μg/mL), clomipramine (7.1, LOD <0.02 μg/mL), clonazepam (6.6, LOD <0.02 μg/mL), clonidine (2.8, LOD 0.1 μg/mL), clozapine (5.6, LOD <0.02 μg/mL), cocaine (4.6, LOD <0.02 μg/mL), codeine (2.5, LOD 0.1 μg/mL), coumatetralyl (8.4, LOD 0.05 μg/mL), cyclizine (5.8, LOD <0.02 μg/mL), dextropropoxyphene (6.6, LOD 0.05 μg/mL), demoxepam (5.8, LOD 0.02 μg/mL), dextromethorphan (5.5, LOD <0.02 μg/mL), diazepam (8.1, LOD 0.02 μg/mL), diltiazem (5.8, LOD <0.02 μg/mL), diphenhydramine (5.7, LOD <0.02 μg/mL), dipyridamole (5.4, LOD 0.005 μg/mL), disopyramine (4.4, LOD <0.02 μg/mL), dixyrazine (6.8, LOD 0.005 μg/mL), doxapram (4.8, LOD <0.02 μg/mL), doxepin (5.9, LOD <0.02 μg/mL), dronabinol (12.3, LOD 0.05 μg/mL), ebastine (9.6, LOD 0.005 μg/mL), embutramide (6.7, LOD 0.005 μg/mL), ergotamine (5.5, LOD 0.005 μg/mL), ethenzamide (5.0, LOD 0.05 μg/mL), ethylmorphine (3.2, LOD 0.05 μg/mL), ethylparathion (9.7, LOD <5 μg/mL), etodroxizine (6.4, LOD <0.02 μg/mL), felodipine (9.6, LOD 0.02 μg/mL), fenazepam (7.5, LOD <0.02 μg/mL), fenfluramine (5.3, LOD <0.02 μg/mL), fenkamfamine (5.1, LOD <0.02 μg/mL), fentanyl (5.5, LOD <0.02 μg/mL), fexofenadine (6.3, LOD <0.02 μg/mL), flecainide (5.9, LOD <0.02 μg/mL), fluconazole (4.0, LOD 0.1 μg/mL), flumazenil (5.2, LOD <0.02 μg/mL), flunitrazepam (7.1, LOD 0.002 μg/mL), fluoxetine (6.8, LOD 0.1 μg/mL), flupentixol (7.5, LOD 0.18 μg/mL), fluvoxamine (6.3, LOD 0.02 μg/mL), glibenclamide (8.5, LOD <0.02 μg/mL), glipizide (6.8, LOD <0.05 μg/mL), haloperidol (6.1, LOD <0.02 μg/mL), histapyrrodine (6.3, LOD 0.02 μg/mL), hydrocodone (3.0, LOD 0.05 μg/mL), hydroxychloroquine (2.4, LOD <0.3 μg/mL), hydroxyzine (6.3, LOD <0.02 μg/mL), imipramine (6.4, LOD 0.05 μg/mL), indomethacin (8.6, LOD 0.05 μg/mL), isoniazid (2.2, LOD 3 μg/mL), isradipine (8.6, LOD 0.05 μg/mL), ketamine (3.6, LOD <0.05 μg/mL), ketobemidone (3.3, LOD <0.05 μg/mL), ketoprofen (7.3, LOD 0.1 μg/mL), ketorolac (6.2, LOD 0.05 μg/mL), labetalol (4.9, LOD 0.05 μg/mL), lamotrigine (4.0, LOD 0.1 μg/mL), levocabastine (5.8, LOD 0.01 μg/mL), levomepromazine (6.5, LOD 0.02 μg/mL), lidocaine (3.7, LOD <0.05 μg/mL), loratadine (9.3, LOD 0.002 μg/mL), lorazepam (6.6, LOD 0.02 μg/mL), lormetazepam (7.4, LOD <0.02 μg/mL), LSD (4.7, LOD <0.02 μg/mL), malathion (8.9, LOD 10 μg/mL), maprotiline (6.4, LOD <0.02 μg/mL), MDMA (3.3, LOD 0.02 μg/mL), meclozine (8.5, LOD <0.02 μg/mL), medazepam (6.3, LOD <0.02 μg/mL), meloxicam (7.1, LOD 0.01 μg/mL), melperone (5.0, LOD <0.02 μg/mL), meperidine (4.7, LOD <0.02 μg/mL), mepivacaine (3.7, LOD <0.02 μg/mL), meprobamate (4.9, LOD 0.1 μg/mL), mesoridazine (5.4, LOD <0.02 μg/mL), methamphetamine (3.3, LOD 0.05 μg/mL), methadone (6.7, LOD <0.02 μg/mL), methylparathion (8.6, LOD 10 μg/mL), methylphenidate (4.2, LOD <0.02 μg/mL), metoclopramide (3.8, LOD <0.02 μg/mL), metoprolol (4.1, LOD 0.02 μg/mL), metronidazole (2.6, LOD 1 μg/mL), mexiletine (4.4, LOD 0.05 μg/mL), mianserin (5.7, LOD <0.02 μg/mL), midazolam (5.9, LOD <0.02 μg/mL), mirtazapine (4.4, LOD <0.02 μg/mL), mizolastine (5.5, LOD 0.01 μg/mL), moclobemide (3.7, LOD 0.05 μg/mL), molindone (4.0, LOD <0.02 μg/mL), monoacetylmorphine (2.7, LOD 0.1 μg/mL), morphine (2.0, LOD 0.1 μg/mL), nicotine (2.2, LOD 0.05 μg/mL), nifedipine (7.5, LOD 0.02 μg/mL), nikethamide (3.6, LOD <0.02 μg/mL), nitrazepam (6.5, LOD <0.02 μg/mL), nizatidine (1.7, LOD 1 μg/mL), nomifensine (4.6, LOD <0.02 μg/mL), nortriptyline (6.4, LOD <0.02 μg/mL), norverapamil (6.2, LOD 1 μg/mL), noscapine (5.0, LOD <0.02 μg/mL), olanzapine (3.0, LOD 0.05 μg/mL), ondansetron (4.6, LOD <0.02 μg/mL), orphenadrine (6.1, LOD <0.02 μg/mL), oxazepam (6.3, LOD <0.02 μg/mL), oxcarbazepine (5.3, LOD 0.02 μg/mL), oxprenolol (4.7, LOD 0.02 μg/mL), oxycodone (2.8, LOD 0.05 μg/mL), papaverine (4.8, LOD <0.02 μg/mL), paroxetine (6.2, LOD 0.02 μg/mL), pemoline (3.3, LOD 0.05 μg/mL), pentazocine (5.0, LOD <0.02 μg/mL), pentifylline (7.3, LOD <5 μg/mL), pentoxyverine (6.6, LOD <0.02 μg/mL), perphenazine (6.9, LOD 0.002 μg/mL), phenazone (3.9, LOD

0.05 μg/mL), phencyclidine (5.3, LOD 0.05 μg/mL), pheniramine (4.1, LOD 0.02 μg/mL), phenylbutazone (9.0, LOD <5 μg/mL), phenylpropanolamine (2.5, LOD 0.3 μg/mL), phenytoin (6.1, LOD 0.05 μg/mL), pindolol (3.3, LOD 0.05 μg/mL), piroxicam (6.6, LOD 0.02 μg/mL), pitofenone (5.4, LOD <0.02 μg/mL), pizotifen (6.5, LOD <0.02 μg/mL), practolol (1.8, LOD 0.1 μg/mL), prazosin (4.1, LOD 0.05 μg/mL), prilocaine (3.8, LOD <0.02 μg/mL), primidone (4.0, LOD <5 μg/mL), procainamide (2.2, LOD 0.05 μg/mL), prochlorperazine (7.5, LOD 0.02 μg/mL), promazine (6.2, LOD <0.02 μg/mL), promethazine (6.0, LOD 0.05 μg/mL), propafenone (6.3, LOD <0.02 μg/mL), propranolol (5.4, LOD 0.02 μg/mL), propyphenazone (6.6, LOD 0.50 μg/mL), pseudoephedrine (2.6, LOD 1 μg/mL), quinine (4.2, LOD 0.02 μg/mL), ranitidine (1.8, LOD 0.1 μg/mL), risperidone (4.9, LOD <0.02 μg/mL), rocurone (3.8, LOD 0.1 μg/mL), ropivacaine (4.6, LOD <0.02 μg/mL), salicylamide (4.2, LOD <5 μg/mL), selegiline (4.1, LOD 0.05 μg/mL), sertindole (7.2, LOD <0.02 μg/mL), sertraline (6.8, LOD 0.02 μg/mL), sulindac (6.5, LOD 0.02 μg/mL), simazine (6.0, LOD 0.1 μg/mL), sincocaine (6.5, LOD <0.02 μg/mL), sisapride (5.9, LOD <0.02 μg/mL), sotalol (2.1, LOD 0.1 μg/mL), strychnine (5.3, LOD 0.05 μg/mL), sulpiride (1.9, LOD 0.1 μg/mL), sulthiame (4.1, LOD 0.05 μg/mL), temazepam (7.2, LOD <0.02 μg/mL), terbutaline (2.3, LOD 0.1 μg/mL), terfenadine (8.1, LOD 0.002 μg/mL), terodiline (6.7, LOD <0.02 μg/mL), tetracaine (5.7, LOD <0.02 μg/mL), tetrahydrozoline (3.6, LOD 0.1 μg/mL), theobromine (2.3, LOD <5 μg/mL), theophylline (2.4, LOD <5 μg/mL), thioridazine (7.5, LOD 0.02 μg/mL), timolol (3.8, LOD 0.05 μg/mL), thiothixene (6.7, LOD 0.02 μg/mL), tolbutamide (7.1, LOD <5 μg/mL), toremifene (8.7, LOD 0.02 μg/mL), tramadol (4.2, LOD 0.02 μg/mL), trazodone (5.2, LOD <0.02 μg/mL), triamterene (3.2, LOD 0.1 μg/mL), triazolam (6.7, LOD 0.002 μg/mL), trimeprazine (6.4, LOD <0.02 μg/mL), trimethoprim (3.1, LOD 0.05 μg/mL), trimipramine (6.7, LOD <0.02 μg/mL), venlafaxine (4.9, LOD 0.02 μg/mL), verapamil (6.5, LOD <0.02 μg/mL), warfarin (7.9, LOD <0.02 μg/mL), yohimbine (4.5, LOD <0.02 μg/mL), zolpidem (4.7, LOD <0.02 μg/mL), zopiclone (4.0, LOD 0.1 μg/mL)

## KEY WORDS

whole blood

## REFERENCE

Gergov, M.; Ojanperä, I.; Vuori, E. Simultaneous screening for 238 drugs in blood by liquid chromatography-ionspray tandem mass spectrometry with multiple-reaction monitoring, *J.Chromatogr.B*, **2003**, *795*, 41–53.

# Adapalene

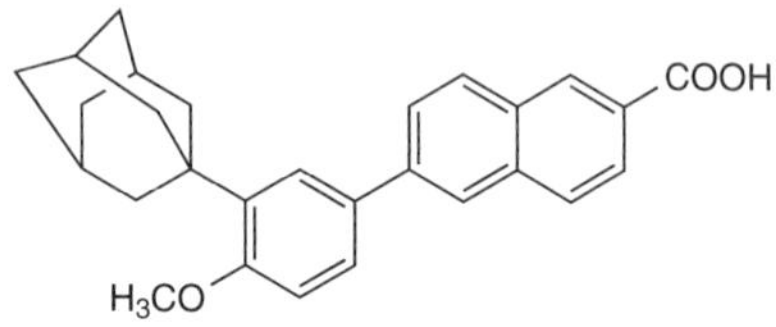


**Molecular formula:** $C_{28}H_{28}O_3$
**Molecular weight:** 412.52
**CAS Registry No:** 106685-40-9
**Merck Index:** 13, 150

## SAMPLE
**Matrix:** formulations
**Sample preparation:** Inject an aliquot of a 0.1% gel.

## HPLC VARIABLES
**Column:** 250 × 4 ODS-RP18 (Merck)
**Mobile phase:** MeCN:THF:water:trifluoroacetic acid 43:36:21:0.02
**Flow rate:** 1
**Detector:** UV 270

## CHROMATOGRAM
**Retention time:** 6.1

## OTHER SUBSTANCES
**Noninterfering:** tretinoin

## KEY WORDS
gel

## REFERENCE
Martin, B.; Meunier, C.; Montels, D.; Watts, O. Chemical stability of adapalene and tretinoin when combined with benzoyl peroxide in presence and in absence of visible light and ultraviolet radiation, *Br.J.Dermatol.*, **1998**, *139* (Suppl. 52), 8–11.

# Adefovir dipivoxil

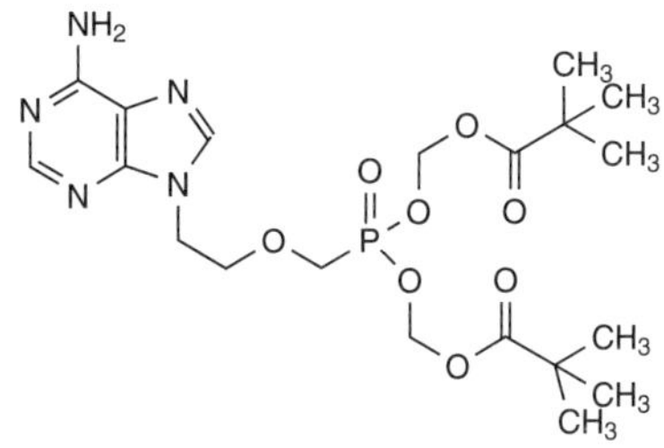


**Molecular formula:** $C_{20}H_{32}N_5O_8P$
**Molecular weight:** 501.47
**CAS Registry No:** 142340-99-6
**Merck Index:** 13, 151

## SAMPLE

**Matrix:** blood
**Sample preparation:** Mix 100 μL plasma with 200 μL 0.1% trifluoroacetic acid in MeCN. Evaporate the supernatant to dryness under reduced pressure at room temperature. Reconstitute with 0.34% chloroacetaldehyde in 100 mM pH 4.5 sodium acetate, vortex, centrifuge. Heat the supernatant at 95° for 40 min, evaporate to dryness, reconstitute with 100 μL 25 mM pH 6.0 potassium phosphate buffer containing 5 mM tetrabutylammonium hydrogen phosphate, inject a 50 μL aliquot.

## HPLC VARIABLES

**Guard column:** 15 × 3.2 Brownlee RP-18 Newguard
**Column:** 150 × 4.6 Zorbax RX-C18
**Column temperature:** 35
**Mobile phase:** Gradient. A was MeCN:25 mM pH 6.0 potassium phosphate buffer containing 5 mM tetrabutylammonium hydrogen phosphate 2:98. B was MeCN:25 mM pH 6.0 potassium phosphate buffer containing 5 mM tetrabutylammonium hydrogen phosphate 65:35. A:B 100:0 for 2 min, to 0:100 over 13 min, re-equilibrate at initial conditions for 10 min. (Only adefovir is detected in blood. However, the method is reported to distinguish between adefovir and adefovir dipivoxil.)
**Flow rate:** 1.5
**Injection volume:** 50
**Detector:** F ex 236 em 420

## KEY WORDS

derivatization; dog; pharmacokinetics; plasma

## REFERENCE

Cundy, K.C.; Sue, I.-L.; Visor, G.C.; Marshburn, J.; Nakamura, C.; Lee, W.A.; Shaw, J.P. Oral formulations of adefovir dipivoxil: In vitro dissolution and in vivo bioavailability in dogs, *J.Pharm.Sci.*, **1997**, *86*, 1334–1338.

## SAMPLE

**Matrix:** blood
**Sample preparation:** Vortex 200 μL plasma with 50 μL 20% trichloroacetic acid in water, centrifuge at 1300 g for 15 min. Remove 150 μL of the supernatant and mix it with 50 μL 160 mM chloroacetaldehyde in water containing 2 M sodium acetate, vortex, close the tube, heat at 98° for 30 min, cool to 2°, vortex, inject a 20 μL aliquot.

## HPLC VARIABLES

**Guard column:** 10 × 3 R3 (Chrompack)
**Column:** 150 × 4.6 5 μm Chromspher C8
**Column temperature:** 40 ± 2
**Mobile phase:** MeCN:buffer 10:90 (Buffer was 10 mM pH 7.0 sodium phosphate buffer containing 2 mM tetrabutylammonium hydrogen sulfate.)
**Flow rate:** 1.5

**Injection volume:** 20
**Detector:** F ex 254 em 425

---

## CHROMATOGRAM
**Retention time:** 4.5
**Limit of quantitation:** 10 ng/mL

---

## OTHER SUBSTANCES
**Extracted:** adefovir (4)

---

## KEY WORDS
derivatization; plasma

---

## REFERENCE
Sparidans, R.W.; Veldkamp, A.; Hoetelmans, R.M.W.; Beijnen, J.H. Improved and simplified liquid chromatographic assay for adefovir, a novel antiviral drug, in human plasma using derivatization with chloroacetaldehyde, *J.Chromatogr.B*, **1999**, *736*, 115–121.

# Adrenocorticotropic hormone

**CAS Registry No:** 9002-60-2
**Merck Index:** 13, 136

## SAMPLE

**Matrix:** blood
**Sample preparation:** Condition a 1 mL Analytichem weak cation-exchange (carboxymethylhydrogen form, CBA) SPE cartridge with 1 mL 1% trifluoroacetic acid in MeOH, 1 mL MeOH, and 2 mL water. Add 1 mL plasma to the SPE cartridge, rinse the tube with 1 mL water, add the rinse to the SPE cartridge, wash with 1 mL 1% trifluoroacetic acid in water, wash with 2 mL water, wash with 2 mL MeOH, elute with 2 mL 1% trifluoroacetic acid in MeOH. Evaporate the eluate to dryness under a stream of nitrogen, reconstitute the residue in 100 μL MeOH:buffer 50:50, inject a 5–75 μL aliquot. (Buffer was 5.7 g monochloroacetic acid, 2.0 g NaOH, and 0.2 g disodium EDTA in 1 L water, pH 3.2.) (The procedure was not necessarily validated for this compound.)

## HPLC VARIABLES

**Column:** 250 × 2 5 μm Ultrasphere octyl
**Column temperature:** 60
**Mobile phase:** Gradient. A was MeOH containing 10 mM sodium octanesulfonate. B was buffer containing 10 mM sodium octanesulfonate. A:B from 45:55 to 70:30 over 30 min, maintain at 70:30 for 1 h. (The buffer was 5.7 g monochloroacetic acid, 2.0 g NaOH, and 0.2 g disodium EDTA in 1 L water, pH 3.2.)
**Flow rate:** 0.3
**Injection volume:** 5–75
**Detector:** F ex 390 em 470 following post-column reaction. The column effluent mixed with 400 mM NaOH pumped at 0.15 mL/min and 0.05% ninhydrin pumped at 0.05 mL/min and the mixture flowed through a 12 m ×0.33 mm ID reaction coil at 70° to the detector.

## CHROMATOGRAM

**Retention time:** 45
**Limit of detection:** 100 fmole

## OTHER SUBSTANCES

**Simultaneous:** angiotensin I, angiotensin II, angiotensin III, atrial natriuretic peptide, bombesin, bradykinin, gonadorelin (LHRH), somatoliberin, vasopressin

## KEY WORDS

plasma; post-column reaction; SPE

## REFERENCE

Rhodes, G.R.; Boppana, V.K. High-performance liquid chromatographic analysis of arginine-containing peptides in biological fluids by means of a selective post-column reaction with fluorescence detection, *J.Chromatogr.*, **1988**, *444*, 123–131.

## SAMPLE

**Matrix:** solutions

## HPLC VARIABLES

**Column:** 300 × 3.9 10 μm μBondapak C18
**Mobile phase:** Gradient. A was 0.08% trifluoroacetic acid. B was MeCN:0.08% trifluoroacetic acid 70:30. A:B from 70:30 to 50:50 over 30 min.

**Flow rate:** 1
**Detector:** UV 206

## CHROMATOGRAM
**Retention time:** 25

## OTHER SUBSTANCES
**Simultaneous:** adrenocorticotropic hormone fragments, melanotropin

## KEY WORDS
human

## REFERENCE
McDermott, J.R.; Smith, A.I.; Biggins, J.A.; Al-Noaemi, M.C.; Edwardson, J.A. Characterization and determination of neuropeptides by high-performance liquid chromatography and radioimmunoassay, *J.Chromatogr.*, **1981**, *222*, 371–379.

---

## SAMPLE
**Matrix:** solutions
**Sample preparation:** Dissolve in 100 mM $NaH_2PO_4$ adjusted to pH 2.1 with orthophosphoric acid, inject a 100 μL aliquot.

## HPLC VARIABLES
**Column:** 250 × 4 Aquapore RP 300 (Kontron)
**Mobile phase:** Gradient. A was 100 mM $NaH_2PO_4$ adjusted to pH 2.1 with orthophosphoric acid. B was MeOH. A:B from 90:10 to 35:65 over 180 min.
**Flow rate:** 1
**Injection volume:** 100
**Detector:** UV 225

## CHROMATOGRAM
**Retention time:** 145

## OTHER SUBSTANCES
**Simultaneous:** adrenocorticotropin hormone fragments, lipotropic hormone and fragments, melanotropin, endorphins, prolactin, somatropin, menotropins

## KEY WORDS
pig

## REFERENCE
Richter, W.O.; Schwandt, P. Separation of neuropeptides by HPLC: evaluation of different supports, with analytical and preparative applications to human and porcine neurophysins, β-lipotropin, adrenocorticotropic hormone, and β-endorphin, *J.Neurochem.*, **1985**, *44*, 1697–1703.

---

## ANNOTATED BIBLIOGRAPHY
Capp, M.W.; Simonian, M.H. Separation of the major adrenal steroids by reversed-phase high-performance liquid chromatography, *Anal.Biochem.*, **1985**, *147*, 374–381.

Janssen, P.S.; van Nispen, J.W.; Hamelinck, R.L.; Melgers, P.A.; Goverde, B.C. Application of reversed-phase HPLC in some critical peptide separations, *J.Chromatogr.Sci.*, **1984**, *22*, 234–238.

Smith, A.I.; McDermott, J.R. High-performance liquid chromatography of neuropeptides using radially compressed polythene cartridges, *J.Chromatogr.*, **1984**, *306*, 99–108.

# Afloqualone

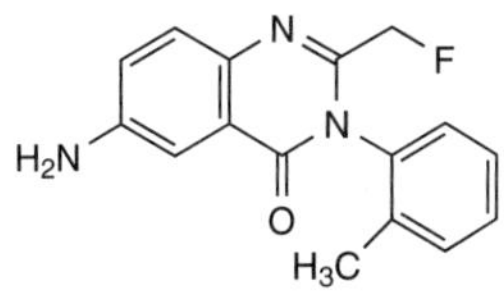


**Molecular formula:** $C_{16}H_{14}FN_3O$
**Molecular weight:** 283.30
**CAS Registry No:** 56287-74-2
**Merck Index:** 13, 183

## SAMPLE
**Matrix:** solutions

## HPLC VARIABLES
**Column:** Chiralpak AS
**Column temperature:** 50
**Mobile phase:** Hexane:EtOH 95:5
**Flow rate:** 1.3
**Detector:** UV 254

## CHROMATOGRAM
**Retention time:** 30, 35 (enantiomers)

## KEY WORDS
chiral

## REFERENCE
*Application Guide for Chiral Column Selection*, Second Edition; Chiral Technologies: Exton PA, **1995**, p. 43.

# Alacepril

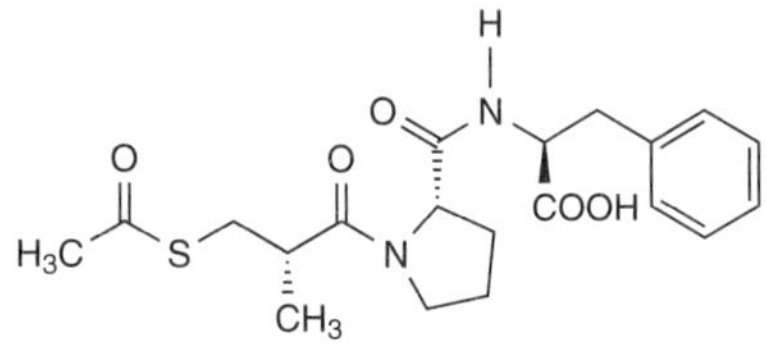


**Molecular formula:** $C_{20}H_{26}N_2O_5S$
**Molecular weight:** 406.50
**CAS Registry No:** 74258-86-9
**Merck Index:** 13, 200

## SAMPLE

**Matrix:** solutions

## HPLC VARIABLES

**Column:** 250 × 4.6 10 μm Cosmosil 5C18-MS
**Column temperature:** 50
**Mobile phase:** Gradient. MeCN:10 mM pH 2.5 potassium phosphate buffer 0:100 for 2 min, to 75:25 over 7.5 min, maintain at 75:25 for 5.5 min. Isocratic. MeCN:buffer 40:60
**Flow rate:** 1.5
**Detector:** UV (wavelength not specified)

## CHROMATOGRAM

**Retention time:** 10.9 (gradient) or 4.1 (isocratic)

## OTHER SUBSTANCES

**Simultaneous:** acetaminophen (7.9), ampicillin (7.9), aspirin (10.0), caffeine (8.5), carbenicillin (9.5), cefotiam (7.2), chlorpromazine (10.8), cromolyn (8.9), enalapril (9.9), loperamide (11.6), ofloxacin (8.3), procainamide (7.4), procaine (7.9), propranolol (9.6), sultamicillin tosylate (8.3), tegafur (8.4), temocapril (12.3), theophylline (8.0), tulobuterol (8.9) (gradient retention times; isocratic conditions may differ)

## REFERENCE

Sugiyama, T.; Matsuyama, R.; Usui, S.; Katagiri, Y.; Hirano, K. Selection of mobile phases in high-performance liquid chromatographic determination for medicines, *Biol.Pharm.Bull.*, **2000**, *23*, 274–278.

## SAMPLE

**Matrix:** enzyme reactions
**Sample preparation:** Mix 40 μL enzyme reaction mixture with 200 μL MeCN, add 200 μL of a 20 μg/mL solution of *n*-propyl paraben, centrifuge, inject a 30 μL aliquot of the supernatant.

## HPLC VARIABLES

**Column:** 250 × 4.6 Cosmosil 5-C18 MS
**Column temperature:** 50
**Mobile phase:** MeCN:10 mM pH 2.5 potassium phosphate buffer 40:60
**Flow rate:** 1.5
**Injection volume:** 30
**Detector:** UV 220

## CHROMATOGRAM

**Internal standard:** *n*-propyl paraben

## OTHER SUBSTANCES

**Extracted:** deacetylalacepril

## REFERENCE

Usui, S.; Kubota, M.; Iguchi, K.; Kiho, T.; Sugiyama, T.; Katagiri, Y.; Hirano, K. Sialic acid 9-*O*-acetylesterase catalyzes the hydrolyzing reaction from alacepril to deacetylalacepril, *Pharm.Res.*, **2003**, *20*, 1309–1316.

# Alclometasone 17,21-dipropionate

**Molecular formula:** $C_{28}H_{37}ClO_7$
**Molecular weight:** 521.05
**CAS Registry No:** 66734-13-2
**Merck Index:** 13, 219

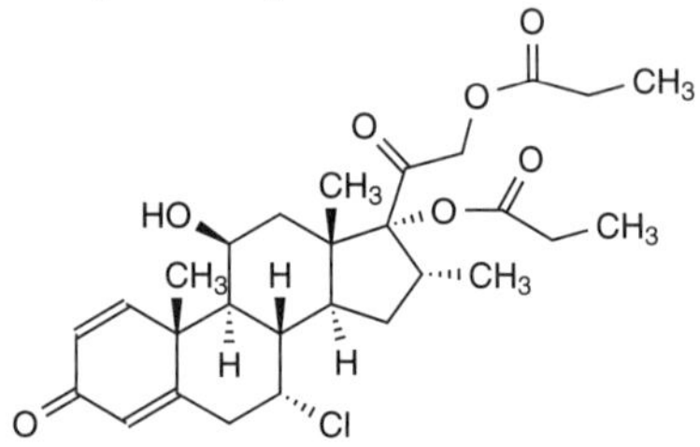

## SAMPLE

**Matrix:** formulations

**Sample preparation:** Condition a 3 mL 500 mg Megabond MF C18 SPE cartridge (Varian) with 3 mL MeOH and 3 mL water. Sonicate 1 g cosmetic with 10 mL MeOH or MeOH:dichloromethane 10:90 (depending on what appears visually to give best solubility) at 40° for 10 min, centrifuge, collect the clear supernatant. Add 5 mL of the supernatant to the SPE cartridge, wash with 4 mL acetone:water 20:80, wash with 1 mL *n*-hexane, elute with 4 mL diethyl ether. Evaporate the eluate to dryness under reduced pressure, reconstitute the residue with 5 mL (or more) MeOH, inject a 10 μL aliquot.

## HPLC VARIABLES

**Column:** 250 × 4.6 5 μm endcapped Purospher RP-18
**Column temperature:** 25
**Mobile phase:** Isocratic. MeCN:water 60:40. Gradient. MeCN:water from 25:75 to 90:10 over 30 min, maintain at 90:10 for 10 min.
**Flow rate:** 1
**Injection volume:** 10
**Detector:** UV 239

## CHROMATOGRAM

**Retention time:** k′ 2.55 (isocratic); 21.0 min (gradient)
**Limit of detection:** 300 ng/mL

## OTHER SUBSTANCES

**Simultaneous:** amcinonide (isocratic k′ 3.18; gradient retention time (min) 22.6; LOD 0.1 μg/mL), betamethasone (isocratic k′ 0.18; gradient retention time (min) 11.8; LOD 0.1 μg/mL), betamethasone-17-acetate (isocratic k′ 0.73; gradient retention time (min) 15.4; LOD 0.3 μg/mL), betamethasone-17-benzoate (isocratic k′ 2.04; gradient retention time (min) 20.6; LOD 0.3 μg/mL), betamethasone-17-propionate-21-stearate (isocratic k′ >13; gradient retention time (min) >35; LOD 0.5 μg/mL), betamethasone-17-propionate-21-butyrate (isocratic k′ 5.91; gradient retention time (min) 26.1; LOD 0.4 μg/mL), betamethasone-17-valerate-21-acetate (isocratic k′ 4.41; gradient retention time (min) 23.1; LOD 0.4 μg/mL), betamethasone-17-valerate (isocratic k′ 2.32; gradient retention time (min) 21.4; LOD 0.3 μg/mL), betamethasone-17,21-dipropionate (isocratic k′ 4.00; gradient retention time (min) 24.2; LOD 0.4 μg/mL), betamethasone-17,21-diacetate (isocratic k′ 1.81; gradient retention time (min) 20.5; LOD 0.3 μg/mL), betamethasone-17,21-divalerate (isocratic k′ 10.82; gradient retention time (min) 28.0; LOD 0.4 μg/mL), betamethasone-21-acetate (isocratic k′ 0.77; gradient retention time (min) 15.6; LOD 0.3 μg/mL), betamethasone propionate (isocratic k′ 0.82; gradient retention time (min) 17.1; LOD 0.3 μg/mL), clobetasol propionate (isocratic k′ 3.41; gradient retention time (min) 23.4; LOD 0.1 μg/mL), clobetasone butyrate (isocratic k′ 5.45; gradient retention time (min) 26.3; LOD 0.1 μg/mL), cortisone (isocratic k′ 0.18; gradient retention time (min) 11.1; LOD 0.6 μg/mL), cortisone acetate (isocratic k′ 0.73; gradient retention time (min) 15.2; LOD 0.6 μg/mL), dehydrocorticosterone (isocratic k′ 4.27; gradient retention time (min) 22.3; LOD 0.5 μg/mL), deoxymethasone (isocratic k′ 0.64; gradient retention time (min) 14.2; LOD 0.2 μg/mL), dexamethasone (isocratic k′ 0.27; gradient retention time (min) 11.9; LOD 0.1 μg/mL), dexamethasone-21-acetate (isocratic k′ 0.91; gradient retention time (min) 16.1; LOD 0.2 μg/mL), dexamethasone isonicotinate (isocratic k′ 1.05; gradient retention time (min) 17.7; LOD 0.4 μg/mL), dexamethasone pivalate (isocratic k′ 3.45; gradient retention time (min) 24.1; LOD 0.3 μg/mL), dexamethasone valerate (isocratic k′ 3.00; gradient retention time (min) 21.6; LOD 0.3 μg/mL), diflucortolone valerate (isocratic k′ 4.73; gradient retention time (min) 23.3; LOD 0.3 μg/mL), fludrocortisone acetate (isocratic k′ 0.59; gradient retention time (min) 14.1; LOD 0.3 μg/mL), flumethasone pivalate (isocratic k′ 2.68; gradient retention time (min) 21.2; LOD 0.3 μg/mL), fluocinolone acetonide (isocratic k′ 0.91; gradient retention time (min) 13.4; LOD 0.3 μg/mL), fluocinonide (isocratic k′ 1.45; gradient retention time (min) 20.5; LOD 0.1 μg/mL), fluocortin butyl ester (isocratic k′ 5.59; gradient retention time (min) 24.6; LOD 0.3 μg/mL), fluocortolone caproate (isocratic k′ 6.59; gradient retention time (min) 25.1; LOD 0.3 μg/mL), fluocortolone pivalate (isocratic k′ 4.50; gradient retention time (min) 23.6; LOD 0.3 μg/mL), fluorometholone (isocratic k′ 0.59; gradient retention time (min) 14.4; LOD 0.1 μg/mL), 9-$\alpha$-fluoroprednisolone (isocratic k′ 0.18; gradient retention time (min) 10.0; LOD 0.1 μg/mL), 9-$\alpha$-fluoroprednisolone acetate (isocratic k′ 0.50; gradient retention time (min) 13.9; LOD 0.2 μg/mL), flurandrenolide (isocratic k′ 0.50; gradient retention time (min) 13.5; LOD 0.1 μg/mL), halcinonide (isocratic k′ 1.64; gradient retention time (min) 20.6; LOD 0.1 μg/mL), hydrocortisone (isocratic k′ 0.18; gradient retention time (min) 10.0; LOD 0.4 μg/mL), hydrocortisone-17-butyrate (isocratic k′ 1.09; gradient retention time (min) 17.7; LOD 0.6 μg/mL), hydrocortisone-21-acetate (isocratic k′ 0.77; gradient retention time (min) 15.3; LOD 0.6 μg/mL), hydrocortisone pivalate (isocratic k′ 2.27; gradient retention time (min) 20.4; LOD 0.8 μg/mL), methylprednisolone (isocratic k′ 0.55; gradient retention time (min) 13.5; LOD 0.1 μg/mL), mometasone furoate (isocratic k′ 3.05; gradient retention time (min) 22.0; LOD 0.2 μg/mL), prednisolone-21-acetate (isocratic k′ 0.60; gradient retention time (min) 13.6; LOD 0.2 μg/mL), prednisolone acetonide (isocratic k′ 0.50; gradient retention time (min) 13.0; LOD 0.3 μg/mL), prednisolone pivalate (isocratic k′ 2.05; gradient retention time (min) 19.7; LOD 0.3 μg/mL), triamcinolone (isocratic k′ 0.14; gradient retention time (min) 7.2; LOD 0.1 μg/mL), triamcinolone acetonide (isocratic k′ 0.50; gradient retention time (min) 13.9; LOD 0.2 μg/mL), triamcinolone diacetate (isocratic k′ 0.45; gradient retention time (min) 13.9; LOD 0.3 μg/mL)

## KEY WORDS

cosmetics; SPE

## REFERENCE

Gagliardi, L.; De Orsi, D.; Del Giudice, M.R.; Gatta, F.; Porrà, R.; Chimenti, P.; Tonelli, D. Development of a tandem thin-layer chromatography-high-performance liquid chromatography method for the identification and determination of corticosteroids in cosmetic products, *Anal.Chim.Acta*, **2002**, *457*, 187–198.

## SAMPLE

**Matrix:** formulations

**Sample preparation:** Mix 1 g of a topical product, 5 mL 100 mM pH 4 citrate buffer saturated with NaCl, and 10 mL ethyl acetate, shake vigorously by hand to ensure that no large clumps stick to the tube, mix with the tube on its side on an oscillating shaker for 15 min. Remove the upper ethyl acetate layer and wash with citrate buffer as before. Dry the ethyl acetate over anhydrous sodium sulfate and evaporate to dryness. Dissolve the residue in a mixture of 10 mL heptane and 5 mL MeCN:water 90:10. Remove the upper heptane layer and extract it with 2 mL MeCN:water 90:10. Combine the MeCN/water layers and evaporate them to dryness, dissolve the residue in 300 μL MeOH, filter (0.45 μm nylon), inject a 5 μL aliquot.

## HPLC VARIABLES

**Guard column:** 15 × 3.2 7 μm Brownlee NewGuard C18
**Column:** 75 × 4.6 3.5 μm Symmetry C18 (Waters)
**Mobile phase:** Gradient. MeCN:water 18:82 for 2 min, to 82:18 over 12 min, maintain at 82:18 for 3 min, re-equilibrate at initial conditions for 12 min.
**Flow rate:** 1
**Injection volume:** 5
**Detector:** UV 240

## CHROMATOGRAM

**Retention time:** 10.93
**Limit of detection:** 0.001%

## OTHER SUBSTANCES

**Simultaneous:** amcinonide (10.90), beclomethasone 17,21-dipropionate (11.90), betamethasone (6.52), betamethasone 21-acetate (8.47), betamethasone 17-benzoate (10.28), betamethasone 17,21-dipropionate (11.42), betamethasone 17-valerate (10.25), budesonide (8.55, 8.69 (epimers)), clobetasol 17-propionate (11.06), cortisone (5.62), cortisone 21-acetate (8.07), dexamethasone (6.57), dexamethasone 21-acetate (8.68), desonide (6.99), desoximetasone (7.60), desoxycorticosterone acetate (10.90), desoxycorticosterone pivalate (14.45), diflorasone 17,21-diacetate (9.81), fluocinolone acetonide (7.38), fluocinonide (9.79), flurandrenolide (7.36), fludrocortisone 21-acetate (7.77), fluorometholone (7.67), flumethasone 21-pivalate (11.20), flunisolide (7.14), fluprednisolone (5.46), fluticasone 17-propionate (11.19), halcinonide (10.72), halobetasol propionate (10.98), hydrocortisone (5.50), hydrocortisone 21-acetate (7.65), hydrocortisone 17-butyrate (8.66), hydrocortisone 21-cypionate (12.54), hydrocortisone 17-valerate (9.53), mometasone 17-furoate (11.24), methylprednisolone (6.31), methylprednisolone 21-acetate (8.34), meprednisone (6.55), prednisolone (5.37), prednisolone 21-acetate (7.47), prednisolone 21-tebuate (11.01), paramethasone 21-acetate (8.64), prednicarbate (10.72), prednisone (5.46), triamcinolone (4.15), triamcinolone acetonide (7.04), triamcinolone 16,21-diacetate (7.49), triamcinolone hexacetonide (13.12)

## KEY WORDS

body wash, cream, gel, lotion, shampoo, spray

## REFERENCE

Reepmeyer, J.C. Screening for corticosteroids in topical pharmaceuticals by HPLC with a scanning ultraviolet detector, *J.Liq.Chromatogr.Rel.Technol.*, **2001**, *24*, 693–709.

# Alitretinoin

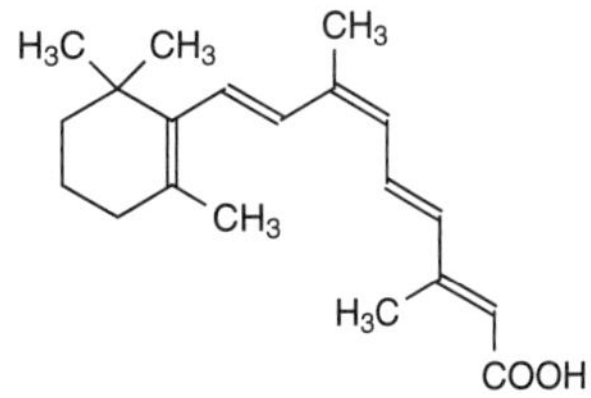


**Molecular formula:** $C_{20}H_{28}O_2$
**Molecular weight:** 300.43
**CAS Registry No:** 5300-03-8
**Merck Index:** 13, 244
[9-cis-retinoic acid]

## SAMPLE
**Matrix:** blood
**Sample preparation:** 1 mL plasma + 50 μL 500 μg/mL IS in MeOH:MeCN 50:50 + 1 mL 1 M pH 6.0 phosphate buffer, mix, add 6 mL MTBE, shake on a horizontal shaker for 10 min, freeze the aqueous layer in a dry ice/acetone bath. Decant the organic layer and evaporate it to dryness under nitrogen at 25°, reconstitute the residue with 200 μL MeOH, add 100 μL 5 mM ammonium acetate, centrifuge at 13 000 g for 3 min, inject a 100 μL aliquot. (Use silanized glassware. Process under yellow light.)

## HPLC VARIABLES
**Guard column:** 10 × 2 5 μm Hypersil BDS C18
**Column:** 100 × 4.6 3 μm Microsorb Short One C18 (Rainin)
**Column temperature:** 36
**Mobile phase:** Gradient. A was 5 mM pH 2.7 ammonium acetate/acetic acid buffer. B was 1% acetic acid in MeOH. A:B 30:70 for 6.5 min, to 20:80 over 0.5 min, to 11:80 over 14.4 min, to 30:70 over 0.5 min, maintain at 30:70 for 10 min.
**Flow rate:** 1
**Injection volume:** 100
**Detector:** UV 348

## CHROMATOGRAM
**Retention time:** 21
**Internal standard:** all-*trans*-3,7-dimethyl-9-(2,4,6-trimethylphenyl)-2,4,6,8-nonatetraenoic acid (Ro 11–5036) (19)
**Limit of detection:** 2.5 ng/mL

## OTHER SUBSTANCES
**Extracted:** isotretinoin (19.5), tretinoin (21.5), vitamin A (20.5)

## KEY WORDS
plasma

## REFERENCE
Dzerk, A.M.; Carlson, A.; Loewen, G.R.; Shirley, M.A.; Lee, J.W. A HPLC method for the determination of 9-cis retinoic acid (ALRT1057) and its 4-oxo metabolite in human plasma, *J.Pharm.Biomed.Anal.*, **1998**, *16*, 1013–1019.

## SAMPLE
**Matrix:** blood, food, formulations, tissue
**Sample preparation:** Serum. Extract one volume (20–100 μL) serum with three volumes isopropanol:dichloromethane 2:1 containing about 6 nM IS and 1 mM butylated hydroxytoluene (BHT, antioxidant), add glacial acetic acid (1 μL/20 μL serum). Vortex for 30 s, centrifuge for 1 min, inject a 20–70 μL aliquot of the supernatant. Tissue, food. Homogenize 100–200 mg human or rat liver, 200 mg–2 g other tissues, or 2–5 g pulp of fruits and fresh vegetables with 3–5 mL isopropanol:dichloromethane 2:1, make up to 10 mL with isopropanol:dichloromethane 2:1. Vortex for 1 min, keep under argon

at −20° overnight, vortex for 1 min, return to the freezer. On the third day, vortex the mixture, centrifuge or filter. Evaporate the supernatant or filtrate to dryness in a rotary evaporator. Dissolve the residue in 200 μL isopropanol:dichloromethane 2:1, inject a 20–40 μL aliquot. Multivitamin tablets. Grind tablet to a powder, add 10 mL isopropanol:dichloromethane 2:1. Vortex for 1 min, keep under argon at −20° overnight, vortex for 1 min, return to the freezer. On the third day, vortex the mixture, centrifuge about 500 μL solution, inject a 50 μL aliquot of the supernatant.

---

## HPLC VARIABLES

**Guard column:** C18 (Upchurch)
**Column:** 100 × 4.6 3 μm Microsorb MV
**Mobile phase:** Gradient. A was MeOH:water 75:25 containing 10 mM ammonium acetate. B was MeOH:dichloromethane 80:20. A:B from 100:0 to 0:100 over 15 min, maintain at 0:100 for 15–20 min, to 100:0 over 5 min, re-equilibrate at initial conditions for 10 min.
**Flow rate:** 0.8
**Injection volume:** 50
**Detector:** UV 340

---

## CHROMATOGRAM

**Retention time:** 10.2
**Internal standard:** retinyl acetate (13.8)

---

## OTHER SUBSTANCES

**Extracted:** β-carotene (27.1), isotretinoin (9.9), all-trans retinal (13.8), all-trans retinyl palmitate (24.1), all-trans retinyl stearate (26.4), tretinoin (10.5), vitamin A (12.9), vitamin E (18.7),

---

## KEY WORDS

human, ketchup, liver, mango, multivitamin tablets, papaya, rat, serum, spinach, tomato

---

## REFERENCE

Barua, A.B.; Olson, J.A. Reversed-phase gradient high-performance liquid chromatographic procedure for simultaneous analysis of very polar to nonpolar retinoids, carotenoids and tocopherols in animal and plant samples, *J.Chromatogr.B*, **1998**, *707*, 69–79.

---

## SAMPLE

**Matrix:** formulations
**Sample preparation:** Capsules. Cut open 10 capsules, sonicate three times at 30° for 5 min with 40 mL portions of MeCN:EtOH:1% acetic acid 70:20:10, centrifuge at 3500 rpm for 6 min. Filter the supernatants, combine, make up to 250 mL. Dilute a 1 mL aliquot to 10 mL with mobile phase, filter (nylon 0.45 μm), inject an aliquot. Gel. Sonicate a portion with 8 mL mobile phase for 1 min, centrifuge at 3500 rpm for 10 min. Filter the supernatant and make up to 10 mL. Dilute a 1 mL aliquot to 5 mL with mobile phase, filter (nylon 0.45 μm), inject an aliquot. Cream. Sonicate an aliquot twice for 5 min with 4 mL portions of MeCN:EtOH:1% acetic acid 70:20:10, centrifuge at 3500 rpm for 6 min. Filter the supernatants, combine, make up to 10 mL with MeCN:EtOH:1% acetic acid 70:20:10. Dilute a 2 mL aliquot to 5 mL with mobile phase, filter (nylon 0.45 μm), inject an aliquot.

---

## HPLC VARIABLES

**Column:** 250 × 3.2 Phenomenex Prodigy 5ODS
**Column temperature:** 32 ± 2
**Mobile phase:** MeCN:EtOH:1% acetic acid 68:8:24
**Flow rate:** 0.4
**Injection volume:** 20
**Detector:** F ex 350 em 520

## CHROMATOGRAM

**Retention time:** 31
**Limit of detection:** 11.09 pmole (S/N = 3)

## OTHER SUBSTANCES

**Simultaneous:** isotretinoin (28.5), tretinoin (33)

## KEY WORDS

avoid exposure to light, use amber-colored glassware, capsules, cream, gel

## REFERENCE

Gatti, R.; Gioia, M.G.; Cavrini, V. Analysis and stability study of retinoids in pharmaceuticals by LC with fluorescence detection, *J.Pharm.Biomed.Anal.*, **2000**, *23*, 147–159.

## ANNOTATED BIBLIOGRAPHY

Disdier, B.; Bun, H.; Placidi, M.; Durand, A. Excretion of oral 9-cis-retinoic acid in the rat, *Drug Metab.Dispos.*, **1996**, *24*, 1279–1281.

Disdier, B.; Bun, H.; Catalin, J.; Durand, A. Simultaneous determination of all-trans-, 13-cis, 9-cis-retinoic acid and their 4-oxometabolites in plasma by high-performance liquid chromatography, *J.Chromatogr.B*, **1996**, *683*, 143–154.

Gatti, R.; Gioia, M.G.; Di Pietra, A.M.; Cini, M. Determination of retinoids in galenicals by column liquid chromatography with fluorescence and diode-array detection, *J.Chromatogr.A*, **2001**, *905*, 345–350.

Marchetti, M.-N.; Sampol, E.; Bun, H.; Scoma, H.; Lacarelle, B.; Durand, A. In vitro metabolism of three major isomers of retinoic acid in rats. Intersex and interstrain comparison, *Drug Metab.Dispos.*, **1997**, *25*, 637–646.

Miyagi, M.; Yokoyama, H.; Shiraishi, H.; Matsumoto, M.; Ishii, H. Simultaneous quantification of retinol, retinal, and retinoic acid isomers by high-performance liquid chromatography with a simple gradiation, *J.Chromatogr.B*, **2001**, *757*, 365–368.

Rühl, R.; Schweigert, F.J. Automated solid-phase extraction and liquid chromatographic method for retinoid determination in biological samples, *J.Chromatogr.B*, **2003**, *798*, 309–316.

Shih, T.-W.; Lin, T.-H.; Shealy, Y.F.; Hill, D.L. Nonenzymatic isomerization of 9-cis-retinoic acid catalyzed by sulfhydryl compounds, *Drug Metab.Dispos.*, **1997**, *25*, 27–32.

Van Merris, V.; Meyer, E.; De Wasch, K.; Burvenich, C. Simple quantification of endogenous retinoids in bovine serum by high-performance liquid chromatography – diode-array detection, *Anal.Chim.Acta*, **2002**, *468*, 237–244.

Wyss, R.; Bucheli, F. Determination of endogenous levels of 13-cis-retinoic acid (isotretinoin), all-trans-retinoic acid (tretinoin) and their 4-oxo metabolites in human and animal plasma by high-performance liquid chromatography with automated column switching and ultraviolet detection, *J.Chromatogr.B*, **1997**, *700*, 31–47.

Yamakoshi, Y.; Fukasawa, H.; Yamauchi, T.; Waki, H.; Kadowaki, T.; Shudo, K.; Kagechika, H. Determination of endogenous levels of retinoic acid isomers in type II diabetes mellitus patients. Possible correlation with HbA1c values, *Biol.Pharm.Bull.*, **2002**, *25*, 1268–1271.

# Allethrin

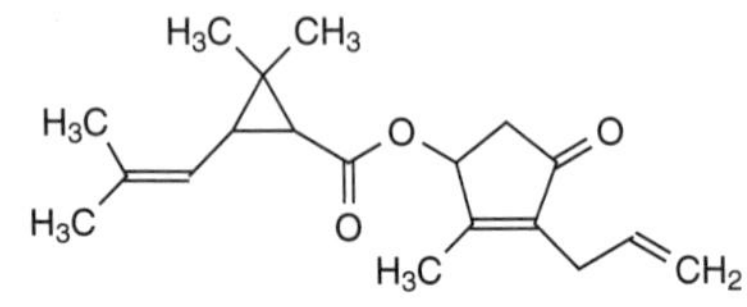


**Molecular formula:** $C_{19}H_{26}O_3$
**Molecular weight:** 302.41
**CAS Registry No:** 584-79-2
**Merck Index:** 13, 256

## SAMPLE
**Matrix:** fruit, vegetables
**Sample preparation:** Prepare a cleanup column by placing 4 g Florisil, 1 g activated charcoal, and a 20 mm layer of anhydrous sodium sulfate in a 400 × 10 glass column, wash with 40 mL toluene, wash with 40 mL toluene:MeCN 99:1. Homogenize 25 g chopped fruit or vegetable with 70 mL MeOH at high speed for 3 min, filter, homogenize solid with 30 mL MeOH, and filter. Combine the filtrates and add them to 60 mL toluene and 300 mL 10% NaCl in water, shake well for 3 min, let layers separate. Dry the organic layer by passing it through 20 g anhydrous sodium sulfate in a 20 mm diameter column, concentrate to about 5 mL under reduced pressure at 80°, add to the cleanup column, elute with 40 mL toluene:MeCN 99:1. Evaporate the eluate just to dryness under reduced pressure at 80°, reconstitute with 1 mL MeOH, inject an aliquot. (Reflux activated charcoal (20–40 mesh) with 1 M HCl for 4 h, wash with water until the washings are neutral, dry at 95–100° (*J.Assoc.Off.Anal.Chem.* **1983**, *66*, 1013). Heat 60–100 mesh Florisil at 200° for 24 h, cool, add 4% water, mix thoroughly, store in a sealed jar (*J.Assoc.Off.Anal.Chem.* **1983**, *66*, 1003).)

## HPLC VARIABLES
**Column:** 300 × 3.9 10 μm μBondapak C18
**Column temperature:** 50
**Mobile phase:** Gradient. MeCN:water from 62:38 to 78:22 over 32 min (Waters curve 6).
**Flow rate:** 1.5
**Detector:** UV 206

## CHROMATOGRAM
**Retention time:** 11.55
**Limit of detection:** 50 ng/g

## OTHER SUBSTANCES
**Simultaneous:** cypermethrin (21.05–22.08), permethrin (24.60, 27.01), tetramethrin (13.08)

## KEY WORDS
apple, cabbage, cucumber; peach, pear, tomato, SPE

## REFERENCE
Pang, G.-F.; Chao, Y.-Z.; Liu, X.-S.; Fan, C.-L. Multiresidue liquid chromatographic method for simultaneous determination of pyrethroid insecticides in fruits and vegetables, *J.AOAC Int.*, **1995**, *78*, 1474–1480.

## SAMPLE
**Matrix:** solutions

## HPLC VARIABLES
**Column:** Two 250 × 4 Phase 3019 columns in series (Phenomenex)
**Mobile phase:** Hexane:1,2-dichloroethane:EtOH 500:30:0.15

**Flow rate:** 0.8
**Detector:** UV 230

## CHROMATOGRAM
**Retention time:** 34, 36, 37, 39, 40, 42, 44, 46 (isomers)

## REFERENCE
*Phenomenex Catalogue*, Phenomenex: Torrance CA, **1994**, p. 1035.

## SAMPLE
**Matrix:** solutions

## HPLC VARIABLES
**Guard column:** 50 × 4 40 μm pellicular material
**Column:** 250 × 4.6 5 μm Ultrasphere octadecylsilica
**Mobile phase:** MeOH:water 80:20
**Flow rate:** 1
**Injection volume:** 10
**Detector:** UV 254

## CHROMATOGRAM
**Retention time:** k′ 4.68 (cis), k′ 5.32 (trans)

## OTHER SUBSTANCES
**Also analyzed**: cyfluthrin (baythroid) (k′ 7.41 (cis, S), k′ 7.77 (trans, R), k′ 8.01 (cis, S), k′ 8.73 (trans, R)), permethrin (k′ 14.9 (trans), k′ 19.5 (cis)), resmethrin (k′ 13.5 (cis), k′ 15.0 (trans)), tetramethrin (k′ 4.05 (cis), k′ 4.68 (trans))

## REFERENCE
Abidi, S.L. Column selectivity in high-performance liquid chromatography of substituted *gem*-dimethylcyclopropanes, *J.Chromatogr.*, **1986**, *368*, 59–76.

## SAMPLE
**Matrix:** solutions

## HPLC VARIABLES
**Column:** 250 × 4.6 5 μm Cyclobond I cyclodextrin-modified silica (Astec)
**Mobile phase:** MeCN:water 22:78
**Flow rate:** 1
**Detector:** UV 220

## CHROMATOGRAM
**Retention time:** 7 (cis isomers), 9.5 (1R,trans, αS), 10.5 (1S,trans, αR), 13 (1R,trans, αR), 15 (1S,trans, αS)

## KEY WORDS
comparison with GC

## REFERENCE
Kutter, J.P.; Class, T.J. Diastereoselective and enantioselective chromatography of the pyrethroid insecticides allethrin and cypermethrin, *Chromatographia*, **1992**, *33*, 103–112.

## SAMPLE
**Matrix:** solutions
**Sample preparation:** Inject an aliquot of a 0.1–1 mg/mL solution in hexane.

## HPLC VARIABLES

**Guard column:** 5 μm Spherisorb NH2
**Column:** 250 × 4.6 Pirkle ionic type 1-A column (Technicol)
**Mobile phase:** Hexane:isopropanol 99.85:0.15
**Flow rate:** 0.8
**Detector:** UV 230

## OTHER SUBSTANCES

**Also analyzed**: cypermethrin, fenpropathrin, fenvalerate, tetramethrin

## KEY WORDS

chiral

## REFERENCE

Lisseter, S.G.; Hambling, S.G. Chiral high-performance liquid chromatography of synthetic pyrethroid insecticides, *J.Chromatogr.*, **1991**, *539*, 207–210.

---

## SAMPLE

**Matrix:** urine
**Sample preparation:** Add 4 g solid NaCl, 3.5 mL MeCN, and 5 mL saturated NaCl solution to 5 mL MeCN, shake for 1 min. Remove the MeCN layer and extract the aqueous layer with 1 mL MeCN. Combine the MeCN layers and adjust to a known volume (0.5–1 mL), mix, filter (0.45 μm), inject a 40 μL aliquot.

## HPLC VARIABLES

**Column:** 150 × 3 3 μm Luna C18(2) (Phenomenex)
**Column temperature:** 30
**Mobile phase:** Gradient. MeCN:water 10:90 for 1 min, to 90:10 over 30 min, maintain at 90:10 for 4 min, to 100:0 over 1 min, maintain at 100:0 for 10 min, return to initial conditions over 1 min.
**Flow rate:** 0.5
**Injection volume:** 40
**Detector:** UV 235

## CHROMATOGRAM

**Retention time:** 31.8
**Limit of detection:** 5 ng/mL

## OTHER SUBSTANCES

**Extracted:** bifenthrin (37, LOD 5 ng/mL), cyfluthrin (34.3, LOD 5 ng/mL), fenvalerate (35.3, LOD 2 ng/mL), *cis*-permethrin (35.7, LOD 5 ng/mL), *trans*-permethrin (36.3, LOD 5 ng/mL), phenothrin (36.4, LOD 5 ng/mL), *m*-phenoxybenzyl alcohol (21, LOD 5 ng/mL), pyrethrin I (29.6, LOD 4 ng/mL), pyrethrin II (33.7, LOD 40 ng/mL), resmethrin (35.2, LOD 5 ng/mL), tetramethrin (31.4, LOD 5 ng/mL)

## REFERENCE

Loper, B.L.; Anderson, K.A. Determination of pyrethrin and pyrethroid pesticides in urine and water matrices by liquid chromatography with diode array detection, *J.AOAC Int.*, **2003**, *86*, 1236–1240.

# Almotriptan

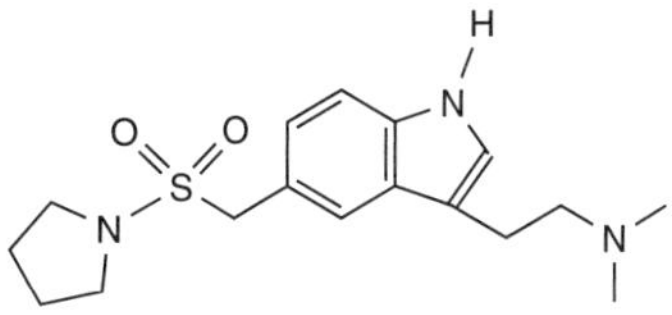

**Molecular formula:** $C_{17}H_{25}N_3O_2S$
**Molecular weight:** 335.47
**CAS Registry No:** 154323-57-6
**Merck Index:** 13, 301

## SAMPLE

**Matrix:** blood, urine
**Sample preparation:** Condition a C2 SPE cartridge (Baker) with 2 mL MeCN and 2 mL water. Dilute 500 μL plasma or 100 μL urine with 1 mL water containing IS, mix, add to the SPE cartridge, wash with 750 μL MeCN:water 30:70, wash with 250 μL water, elute with mobile phase over 1 min (straight onto column (?)).

## HPLC VARIABLES

**Guard column:** Guardpak μBondapak CN
**Column:** 150 × 4 5 μm Spherisorb ODS-2
**Mobile phase:** MeCN:50 mM pH 4.0 sodium phosphate buffer:triethylamine 20:80:0.2
**Flow rate:** 1
**Detector:** UV 227

## CHROMATOGRAM

**Retention time:** 6.5
**Internal standard:** 4-[3-(2-aminoethyl)-1H-indol-5-ylmethylsulfonyl]piperazine-1-carboxylic acid ethyl ester (10)
**Limit of quantitation:** 1 ng/mL (plasma), 50 ng/mL urine

## KEY WORDS

plasma, SPE

## REFERENCE

Jansat, J.M.; Costa, J.; Salvà, P.; Fernandez, F.J.; Martinez-Tobed, A. Absolute bioavailability, pharmacokinetics, and urinary excretion of the novel antimigraine agent almotriptan in healthy male volunteers, *J.Clin.Pharmacol.*, **2002**, *42*, 1303–1310.

## SAMPLE

**Matrix:** microsomal incubations
**Sample preparation:** Mix 500 μL microsomal incubation with 1 mL 200 mM pH 4 sodium acetate buffer, centrifuge, inject an aliquot.

## HPLC VARIABLES

**Guard column:** GuardPak μBondapak CN
**Column:** 300 × 3.9 10 μm μBondapak
**Mobile phase:** Gradient. A:B from 80:20 to 40:60 over 30 min. A was buffer. B was MeCN:buffer 80:20. Buffer was 10 mM orthophosphoric acid containing 0.1% triethylamine, adjusted to pH 6.5 with NaOH.
**Flow rate:** 1
**Detector:** UV 227

## CHROMATOGRAM

**Retention time:** 25

## OTHER SUBSTANCES

**Extracted:** metabolites

## KEY WORDS

human, liver

## REFERENCE

Salva, M.; Jansat, J.M.; Martinez-Tobed, A.; Palacios, J.M. Identification of the human liver enzymes involved in the metabolism of the antimigraine agent almotriptan, *Drug Metab.Dispos.*, **2003**, *31*, 404–411.

# Alosetron

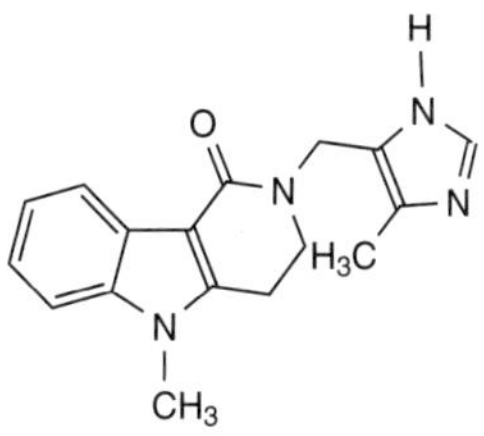

**Molecular formula:** $C_{17}H_{18}N_4O$
**Molecular weight:** 294.35
**CAS Registry No:** 122852-42-0, 122852-69-1 (HCl)
**Merck Index:** 13, 305

## SAMPLE

**Matrix:** blood
**Sample preparation:** Condition a 100 mg LRC Bond Elut ethyl (C2) SPE cartridge with 1 mL isopropanol and 1 mL buffer. Mix 1.1 mL plasma or serum with 1 mL buffer containing 10 ng/mL IS, vortex, add 2 mL to the SPE cartridge, wash with 2 mL buffer, dry with nitrogen for 30 s, wash with 2 mL MeCN, elute with two 2 mL aliquots of MeCN:buffer 90:10. Evaporate the eluate to dryness under a stream of nitrogen at 40°, reconstitute the residue with 300 μL mobile phase, vortex, inject a 200 μL aliquot. (The buffer was 10 mM ammonium acetate adjusted to pH 4.0 with glacial acetic acid.)

## HPLC VARIABLES

**Guard column:** 15 × 4.6 7 μm Spherisorb cyanopropyl
**Column:** 100 × 4.6 5 μm Spheri cyanopropyl (Brownlee)
**Column temperature:** 45
**Mobile phase:** MeOH:THF:10 mM pH 4.0 ammonium acetate buffer 24:6:70
**Flow rate:** 0.5
**Injection volume:** 200
**Detector:** F ex 295 em 370

## CHROMATOGRAM

**Retention time:** 10.1
**Internal standard:** GR87442, 6-fluoroalosetron (Glaxo) (13.7)
**Limit of quantitation:** 0.1 ng/mL

## OTHER SUBSTANCES

**Noninterfering:** amitriptyline, carbamazepine, carmustine, chlorpromazine, cimetidine, cisplatin, cyclophosphamide, dexamethasone, diazepam, digoxin, etoposide, furosemide, haloperidol, ibuprofen, imipramine, indomethacin, methotrexate, phenobarbital, phenytoin, propranolol, ranitidine, theophylline, triazolam, warfarin

## KEY WORDS

plasma; serum; SPE

## REFERENCE

Lloyd, T.L.; Gupta, S.K.; Gooding, A.E.; Alianti, J.R. Determination of alosetron in human plasma or serum by high-performance liquid chromatography with robotic sample preparation, *J.Chromatogr.B*, **1996**, *678*, 261–267.

# Amcinonide

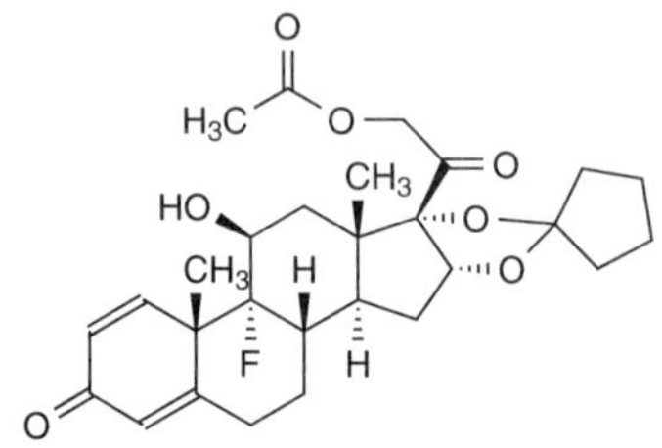

**Molecular formula:** $C_{28}H_{35}FO_7$
**Molecular weight:** 502.57
**CAS Registry No:** 51022-69-6
**Merck Index:** 13, 387

## SAMPLE

**Matrix:** formulations
**Sample preparation:** Condition a 3 mL 500 mg Megabond MF C18 SPE cartridge (Varian) with 3 mL MeOH and 3 mL water. Sonicate 1 g cosmetic with 10 mL MeOH or MeOH:dichloromethane 10:90 (depending on what appears visually to give the best solubility) at 40° for 10 min, centrifuge, collect the clear supernatant. Add 5 mL of the supernatant to the SPE cartridge, wash with 4 mL acetone:water 20:80, wash with 1 mL *n*-hexane, elute with 4 mL diethyl ether. Evaporate the eluate to dryness under reduced pressure, reconstitute the residue with 5 mL (or more) MeOH, inject a 10 μL aliquot.

## HPLC VARIABLES

**Column:** 250 × 4.6 5 μm endcapped Purospher RP-18
**Column temperature:** 25
**Mobile phase:** Isocratic.MeCN:water 60:40. Gradient. MeCN:water from 25:75 to 90:10 over 30 min, maintain at 90:10 for 10 min.
**Flow rate:** 1
**Injection volume:** 10
**Detector:** UV 239

## CHROMATOGRAM

**Retention time:** k′ 3.18 (isocratic); 22.6 min (gradient)
**Limit of detection:** 100 ng/mL

## OTHER SUBSTANCES

**Simultaneous:** alclometasone dipropionate (isocratic k′ 2.55; gradient retention time (min) 21.0; LOD 0.3 μg/mL), betamethasone (isocratic k′ 0.18; gradient retention time (min) 11.8; LOD 0.1 μg/mL), betamethasone-17-acetate (isocratic k′ 0.73; gradient retention time (min) 15.4; LOD 0.3 μg/mL), betamethasone-17-benzoate (isocratic k′ 2.04; gradient retention time (min) 20.6; LOD 0.3 μg/mL), betamethasone-17-propionate-21-stearate (isocratic k′ >13; gradient retention time (min) >35; LOD 0.5 μg/mL), betamethasone-17-propionate-21-butyrate (isocratic k′ 5.91; gradient retention time (min) 26.1; LOD 0.4 μg/mL), betamethasone-17-valerate-21-acetate (isocratic k′ 4.41; gradient retention time (min) 23.1; LOD 0.4 μg/mL), betamethasone-17-valerate (isocratic k′ 2.32; gradient retention time (min) 21.4; LOD 0.3 μg/mL), betamethasone-17,21-dipropionate (isocratic k′ 4.00; gradient retention time (min) 24.2; LOD 0.4 μg/mL), betamethasone-17,21-diacetate (isocratic k′ 1.81; gradient retention time (min) 20.5; LOD 0.3 μg/mL), betamethasone-17,21-divalerate (isocratic k′ 10.82; gradient retention time (min) 28.0; LOD 0.4 μg/mL), betamethasone-21-acetate (isocratic k′ 0.77; gradient retention time (min) 15.6; LOD 0.3 μg/mL), betamethasone propionate (isocratic k′ 0.82; gradient retention time (min) 17.1; LOD 0.3 μg/mL), clobetasol propionate (isocratic k′ 3.41; gradient retention time (min) 23.4; LOD 0.1 μg/mL), clobethasone butyrate (isocratic k′ 5.45; gradient retention time (min) 26.3; LOD 0.1 μg/mL), cortisone (isocratic k′ 0.18; gradient retention time (min) 11.1; LOD 0.6 μg/mL), cortisone acetate (isocratic k′ 0.73; gradient retention time (min) 15.2; LOD 0.6 μg/mL), dehydrocorticosterone (isocratic k′ 4.27; gradient retention time (min) 22.3; LOD 0.5 μg/mL), deoxymethasone (isocratic k′ 0.64; gradient retention time (min) 14.2; LOD 0.2 μg/mL), dexamethasone (isocratic k′ 0.27; gradient retention time (min) 11.9; LOD 0.1 μg/mL), dexamethasone-21-acetate (isocratic k′ 0.91; gradient retention time (min) 16.1; LOD 0.2 μg/mL), dexamethasone isonicotinate (isocratic k′ 1.05; gradient retention time (min) 17.7; LOD 0.4 μg/mL), dexamethasone pivalate (isocratic k′ 3.45; gradient retention time (min) 24.1; LOD 0.3 μg/mL), dexamethasone valerate (isocratic k′ 3.00; gradient retention time (min) 21.6; LOD 0.3 μg/mL), diflucortolone valerate (isocratic k′ 4.73; gradient retention time (min) 23.3; LOD 0.3 μg/mL), fludrocortisone acetate (isocratic k′ 0.59; gradient retention time (min) 14.1; LOD 0.3 μg/mL), flumethasone pivalate (isocratic k′ 2.68; gradient retention time (min) 21.2; LOD 0.3 μg/mL), fluocinolone acetonide (isocratic k′ 0.91; gradient retention time (min) 13.4; LOD 0.3 μg/mL), fluocinonide (isocratic k′ 1.45; gradient retention time (min) 20.5; LOD 0.1 μg/mL), fluocortin butyl ester (isocratic k′ 5.59; gradient retention time (min) 24.6; LOD 0.3 μg/mL), fluocortolone caproate (isocratic k′ 6.59; gradient retention time (min) 25.1; LOD 0.3 μg/mL), fluocortolone pivalate (isocratic k′ 4.50; gradient retention time (min) 23.6; LOD 0.3 μg/mL), fluorometholone (isocratic k′ 0.59; gradient retention time (min) 14.4; LOD 0.1 μg/mL), 9-$\alpha$-fluoroprednisolone (isocratic k′ 0.18; gradient retention time (min) 10.0; LOD 0.1 μg/mL), 9-$\alpha$-fluoroprednisolone acetate (isocratic k′ 0.50; gradient retention time (min) 13.9; LOD 0.2 μg/mL), flurandrenolide (isocratic k′ 0.50; gradient retention time (min) 13.5; LOD 0.1 μg/mL), halcinonide (isocratic k′ 1.64; gradient retention time (min) 20.6; LOD 0.1 μg/mL), hydrocortisone (isocratic k′ 0.18; gradient retention time (min) 10.0; LOD 0.4 μg/mL), hydrocortisone-17-butyrate (isocratic k′ 1.09; gradient retention time (min) 17.7; LOD 0.6 μg/mL), hydrocortisone-21-acetate (isocratic k′ 0.77; gradient retention time (min) 15.3; LOD 0.6 μg/mL), hydrocortisone pivalate (isocratic k′ 2.27; gradient retention time (min) 20.4; LOD 0.8 μg/mL), methylprednisolone (isocratic k′ 0.55; gradient retention time (min) 13.5; LOD 0.1 μg/mL), mometasone furoate (isocratic k′ 3.05; gradient retention time (min) 22.0; LOD 0.2 μg/mL), prednisolone-21-acetate (isocratic k′ 0.60; gradient retention time (min) 13.6; LOD 0.2 μg/mL), prednisolone acetonide (isocratic k′ 0.50; gradient retention time (min) 13.0; LOD 0.3 μg/mL), prednisolone pivalate (isocratic k′ 2.05; gradient retention time (min) 19.7; LOD 0.3 μg/mL), triamcinolone (isocratic k′ 0.14; gradient retention time (min) 7.2; LOD 0.1 μg/mL), triamcinolone acetonide (isocratic k′ 0.50; gradient retention time (min) 13.9; LOD 0.2 μg/mL), triamcinolone diacetate (isocratic k′ 0.45; gradient retention time (min) 13.9; LOD 0.3 μg/mL).

## KEY WORDS

cosmetics; SPE

## REFERENCE

Gagliardi, L.; De Orsi, D.; Del Giudice, M.R.; Gatta, F.; Porrà, R.; Chimenti, P.; Tonelli, D. Development of a tandem thin-layer chromatography-high-performance liquid chromatography method for the identification and determination of corticosteroids in cosmetic products, *Anal.Chim.Acta*, **2002**, *457*, 187–198.

## SAMPLE

**Matrix:** formulations

**Sample preparation:** Mix 1 g of a topical product, 5 mL 100 mM pH 4 citrate buffer saturated with NaCl, and 10 mL ethyl acetate, shake vigorously by hand to ensure that no large clumps stick to the tube, mix with the tube on its side on an oscillating shaker for 15 min. Remove the upper ethyl acetate layer and wash with citrate buffer as before. Dry the ethyl acetate over anhydrous sodium sulfate and evaporate to dryness. Dissolve the residue in a mixture of 10 mL heptane and 5 mL MeCN:water 90:10. Remove the upper heptane layer and extract it with 2 mL MeCN:water 90:10. Combine the MeCN/water layers and evaporate them to dryness, dissolve the residue in 300 μL MeOH, filter (0.45 μm nylon), inject a 5 μL aliquot.

## HPLC VARIABLES

**Guard column:** 15 × 3.2 7 μm Brownlee NewGuard C18

**Column:** 75 × 4.6 3.5 μm Symmetry C18 (Waters)

**Mobile phase:** Gradient. MeCN:water 18:82 for 2 min, to 82:18 over 12 min, maintain at 82:18 for 3 min, re-equilibrate at initial conditions for 12 min.
**Flow rate:** 1
**Injection volume:** 5
**Detector:** UV 240

## CHROMATOGRAM

**Retention time:** 10.90
**Limit of detection:** 0.001%

## OTHER SUBSTANCES

**Extracted:**
**Simultaneous:** alclometasone 17,21-dipropionate (10.93), beclomethasone 17,21-dipropionate (11.90), betamethasone (6.52), betamethasone 21-acetate (8.47), betamethasone 17-benzoate (10.28), betamethasone 17,21-dipropionate (11.42), betamethasone 17-valerate (10.25), budesonide (8.55, 8.69 (epimers)), clobetasol 17-propionate (11.06), cortisone (5.62), cortisone 21-acetate (8.07), dexamethasone (6.57), dexamethasone 21-acetate (8.68), desonide (6.99), desoximetasone (7.60), desoxycorticosterone acetate (10.90), desoxycorticosterone pivalate (14.45), diflorasone 17,21-diacetate (9.81), fluocinolone acetonide (7.38), fluocinonide (9.79), flurandrenolide (7.36), fludrocortisone 21-acetate (7.77), fluorometholone (7.67), flumethasone 21-pivalate (11.20), flunisolide (7.14), fluprednisolone (5.46), fluticasone 17-propionate (11.19), halcinonide (10.72), halobetasol propionate (10.98), hydrocortisone (5.50), hydrocortisone 21-acetate (7.65), hydrocortisone 17-butyrate (8.66), hydrocortisone 21-cypionate (12.54), hydrocortisone 17-valerate (9.53), mometasone 17-furoate (11.24), methylprednisolone (6.31), methylprednisolone 21-acetate (8.34), meprednisone (6.55), prednisolone (5.37), prednisolone 21-acetate (7.47), prednisolone 21-tebuate (11.01), paramethasone 21-acetate (8.64), prednicarbate (10.72), prednisone (5.46), triamcinolone (4.15), triamcinolone acetonide (7.04), triamcinolone 16,21-diacetate (7.49), triamcinolone hexacetonide (13.12)

## KEY WORDS

body wash, cream, gel, lotion, shampoo, spray

## REFERENCE

Reepmeyer, J.C. Screening for corticosteroids in topical pharmaceuticals by HPLC with a scanning ultraviolet detector, *J.Liq.Chromatogr.Rel.Technol.*, **2001**, *24*, 693–709.

# Aminolevulinic acid

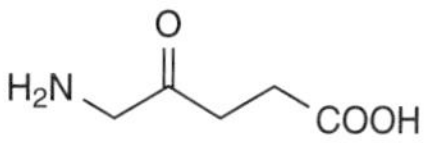

**Molecular formula:** $C_5H_9NO_3$
**Molecular weight:** 131.13
**CAS Registry No:** 106-60-5
**Merck Index:** 13, 445

## SAMPLE

**Matrix:** blood, tissue
**Sample preparation:** Deproteinize plasma by adding perchloric acid to a final concentration of 800 mM. Neutralize the supernatant by adding solid sodium bicarbonate until a pH of ca. 7.6 is reached. Homogenize tissue with 3 volumes of 10 mM pH 7.2 HEPES buffer containing 250 mM sucrose and 500 mM EDTA, centrifuge at 800 g for 5 min. Mix 10 μL sample with 5 μL reagent and 35 μL water, let stand at room temperature for 1 min, inject a 20 μL aliquot. (Prepare the reagent by dissolving 27 mg *o*-phthalaldehyde in 500 μL MeOH, add 5 mL 100 mM sodium tetraborate, add 20 μL mercaptoethanol, mix.)

## HPLC VARIABLES

**Column:** 150 × 3.9 4 μm C18 (Waters)
**Mobile phase:** MeCN:50 mM pH 7.0 phosphate buffer 10:90 containing 2.4 mM EDTA
**Flow rate:** 1
**Injection volume:** 20
**Detector:** E, Shimadzu LECD 6A, glassy carbon working electrode at +0.45 V, Ag/AgCl reference electrode

## CHROMATOGRAM

**Retention time:** 44.6
**Limit of detection:** 50 nM
**Limit of quantitation:** 100 nM

## KEY WORDS

brain; derivatization; human; liver; plasma; rat

## REFERENCE

Costa, C.A.; Trivelato, G.C.; Demasi, M.; Bechara, E.J.H. Determination of 5-aminolevulinic acid in blood plasma, tissues and cell cultures by high-performance liquid chromatography with electrochemical detection, *J.Chromatogr.B*, **1997**, *695*, 245–250.

## SAMPLE

**Matrix:** blood, urine
**Sample preparation:** 50 μL Plasma or urine + 3.5 mL reagent + 450 μL 10% formaldehyde, vortex for 3 s, heat at 100° for 10 min, cool in an ice bath, filter (0.8 μm, plasma samples only), inject a 10 (urine) or 20 (plasma) μL aliquot. (Prepare the reagent by mixing 15 mL acetylacetone, 10 mL EtOH, and 75 mL water.)

## HPLC VARIABLES

**Column:** 150 × 4.6 Shim-pack CLC-ODS (Shimadzu)
**Column temperature:** 40
**Mobile phase:** MeOH:water:acetic acid 50:50:1
**Flow rate:** 0.7
**Injection volume:** 10–20
**Detector:** F ex 370 em 460

## CHROMATOGRAM

**Retention time:** 6.1
**Limit of detection:** 3 ng/mL

## KEY WORDS

derivatization; plasma; protect from light

## REFERENCE

Oishi, H.; Nomiyama, H.; Nomiyama, K.; Tomokuni, K. Fluorometric HPLC determination of delta-aminolevulinic acid (ALA) in the plasma and urine of lead workers: biological indicators of lead exposure, *J.Anal.Toxicol.*, **1996**, *20*, 106–110.

## SAMPLE

**Matrix:** urine
**Sample preparation:** Centrifuge urine at 1000 g and store at −20°. 20 μL Urine + 5 mL acetylacetone:EtOH:4 g/L NaCl in water 15:10:75 + 450 μL 9.3% formaldehyde solution, mix, boil for 15 min, cool with water, store sample in the dark at 15° until injection, inject a 50 μL aliquot.

## HPLC VARIABLES

**Column:** 150 × 4.6 5 μm TSK-80 TM (Tosoh)
**Column temperature:** 40
**Mobile phase:** Gradient. A was MeCN:MeOH:water:acetic acid 10:35:54:1. B was MeCN. A:B 100:0 for 7.5 min, to 50:50 over 1.5 min, return to initial conditions over 2 min, re-equilibrate for 2 min.
**Flow rate:** 0.8
**Injection volume:** 50
**Detector:** F ex 246 em 458

## CHROMATOGRAM

**Retention time:** 7.3
**Limit of detection:** 10 ng/mL

## KEY WORDS

derivatization; protect from light; improved version of A. Okayama et al. *Clin.Chem.* **1990**, *36*, 1494.

## REFERENCE

Endo, Y.; Okayama, A.; Endo, G.; Ueda, T.; Nakazono, N.; Horiguchi, S. Improvement of urinary δ-aminolevulinic acid determination by HPLC and fluorescence detection using condensing reaction with acetylacetone and formaldehyde, *Jap.J.Ind.Health*, **1994**, *36*, 49–56.

## ANNOTATED BIBLIOGRAPHY

Dalton, J.T.; Meyer, M.C.; Golub, A.L. Pharmacokinetics of aminolevulinic acid after oral and intravenous administration to dogs, *Drug Metab.Dispos.*, **1999**, *27*, 432–435. [derivatization]

Ho, J.; Guthrie, R.; Tieckelmann, H. Detection of δ-aminolevulinic acid, porphobilinogen and porphyrins related to heme biosynthesis by high-performance liquid chromatography, *J.Chromatogr.*, **1986**, *375*, 57–63. [derivatization]

Ho, J.W. Micro assay for urinary δ-aminolevulinic acid and porphobilinogen by high-performance liquid chromatography with pre-column derivatization, *J.Chromatogr.*, **1990**, *527*, 134–139.

Kondo, M.; Kimura, H.; Maekubo, T.; Tomita, T.; Senda, M.; Urata, G.; Kajiwara, M. Direct injection method for quantitation of δ-aminolevulinic acid in urine by high-performance liquid chromatography, *Chem.Pharm.Bull.*, **1992**, *40*, 1948–1950. [derivatization]

Lim, C.K.; Rideout, J.M.; Samson, D.M. Determination of 5-aminolaevulinic acid and porphobilinogen by high-performance liquid chromatography, *J.Chromatogr.*, **1979**, *185*, 605–611.

Meisch, H.U.; Reinle, W.; Wolf, U. Determination of 5-aminolevulinic acid in biological samples by high-performance liquid chromatography, *Anal.Biochem.*, **1985**, *149*, 29–34.

Minder, E.I. Measurement of 5-aminolevulinic acid by reversed phase HPLC and fluorescence detection, *Clin.Chim.Acta*, **1986**, *161*, 11–18. [derivatization]

Miyajima, K.; Hirata, M.; Yoshida, T.; Kosaka, H.; Okayama, A. Study on measurement of delta-aminolevulinic acid in plasma by high-performance liquid chromatography, *J.Chromatogr.B*, **1994**, *654*, 165–169.

Okayama, A. Fluorimetric determination of urinary δ-aminolevulinic acid by high-performance liquid chromatography and post-column derivatization, *J.Chromatogr.*, **1988**, *426*, 365–369.

Okayama, A.; Fujii, S.; Miura, R. Optimized fluorometric determination of urinary delta-aminolevulinic acid by using pre-column derivatization, and identification of the derivative, *Clin.Chem.*, **1990**, *36*, 1494–1497.

Tomokuni, K.; Ichiba, M.; Hirai, Y.; Hasegawa, T. Optimized liquid-chromatographic method for fluorometric determination of urinary delta-aminolevulinic acid in workers exposed to lead, *Clin.Chem.*, **1987**, *33*, 1665–1667.

# Amprenavir

**Molecular formula:** $C_{25}H_{35}N_3O_6S$
**Molecular weight:** 505.64
**CAS Registry No:** 161814-49-9
**Merck Index:** 13, 594

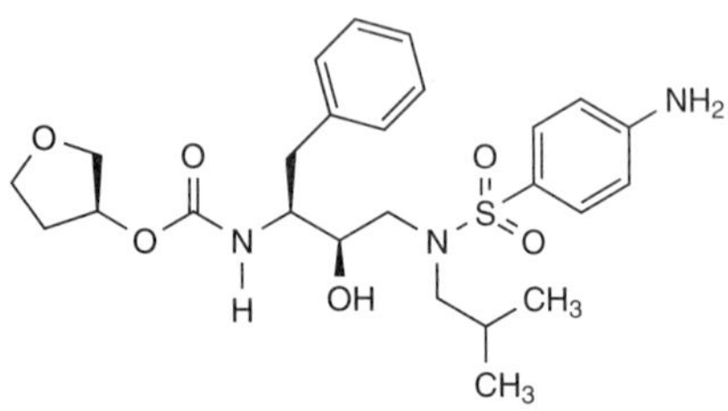


## SAMPLE

**Matrix:** blood
**Sample preparation:** Condition a C18 SPE cartridge (Baker) with 3 mL MeOH and 3 mL water. Do not allow to run dry. Add 1 mL plasma to the SPE cartridge, wash with 2 mL water, suck dry for 1 min, elute with 2.6 mL MeOH. Evaporate a 1 mL aliquot of the eluate to dryness under a stream of nitrogen at 40°, reconstitute the residue with 200 μL mobile phase, inject a 100 μL aliquot.

## HPLC VARIABLES

**Guard column:** GuardPak μBondapak C18
**Column:** 250 × 4.6 5 μm Symmetry C18
**Column temperature:** 37
**Mobile phase:** MeCN:40 mM disodium hydrogen phosphate containing 4% octanesulfonic acid 50:50. (At the end of each session, wash column with MeOH:water 50:50 and MeCN:water 80:20.)
**Flow rate:** 1.3
**Injection volume:** 100
**Detector:** UV 261 for 9 min, UV 241 for 11 min, UV 254 for 12 min

## CHROMATOGRAM

**Retention time:** 5.6
**Limit of quantitation:** 25 ng/mL

## OTHER SUBSTANCES

**Extracted:** efavirenz (15.2, LOQ 50 ng/mL), indinavir (4.8, LOQ 50 ng/mL), nelfinavir (19.2, LOQ 50 ng/mL), ritonavir (12.8, LOQ 50 ng/mL), saquinavir (16.8, LOQ 5 ng/mL)
**Noninterfering:** abacavir, acebutolol, acetaminophen, acetylcysteine, acyclovir, albendazole, alimemazine, alizapride, amikacin, amiodarone, amphotericin B, ampicillin, aspirin, bepridil, buprenorphine, butobarbital, caffeine, calcium folinate, captopril, carbamazepine, carbutamide, chloroquine, ciprofloxacin, clindamycin, clofazimine, clofibrate, clonazepam, clonidine, cloxacillin, clozapine, cocaine, codeine, cyamemazine, dantrolene, dexamethasone, dextropropoxyphene, diazepam, diclofenac, didanosine, digoxin, dihydroergotamine, diltiazem, doxycycline, ethambutol, flecainide, fluconazole, fluoxetine, fluvoxamine, foscarnet, furosemide, ganciclovir, gentamicin, glibenclamide, granisetron, halofantrine, haloperidol, hydrocortisone, imipramine, indomethacin, interferon alfa, isoniazid, itraconazole, josamycin, ketoconazole, lamivudine, levomepromazine, lidocaine, loperamide, loratadine, losartan, mefloquine, meprobamate, methadone, methylprednisolone, metoclopramide, metronidazole, mianserin, moclobemide, morphine, nevirapine, nifedipine, niflumic acid, nitrofurantoin, omeprazole, paroxetine, pentamidine, phenobarbital, phenytoin, piracetam, prazosin, prednisolone, prednisone, primidone, propranolol, quinidine, quinine, ranitidine, ribavirin, rifabutin, rifampin, roxithromycin, salicylic acid, simvastatin, stavudine, sulfadiazine, sulfamethoxazole, sulpiride, thalidomide, theophylline, trimethoprim, valproic acid, venlafaxine, vigabatrin, viloxazine, zidovudine, zolpidem, zopiclone
**Interfering:** delavirdine, flunitrazepam

## KEY WORDS
plasma; SPE

## REFERENCE
Aymard, G.; Legrand, M.; Trichereau, N.; Diquet, B. Determination of twelve antiretroviral agents in human plasma sample using reversed-phase high-performance liquid chromatography, *J.Chromatogr. B*, **2000**, *744*, 227–240.

## SAMPLE
**Matrix:** blood
**Sample preparation:** Mix 250 μL plasma with 50 μL MeOH, add 100 μL 2 μg/mL IS in MeOH, add 250 μL 1 M NaOH, add 3 mL hexane:ethyl acetate 50:50, shake at high speed for 25 min, centrifuge at 3000 g for 15 min. Evaporate the organic layer to dryness under a stream of air, reconstitute the residue with 1 mL initial mobile phase, inject a 20 μL aliquot.

## HPLC VARIABLES
**Guard column:** 10 × 2.1 Symmetry Shield
**Column:** 30 × 2.1 3.5 μm Symmetry C18
**Mobile phase:** Gradient. MeCN:5 mM pH 3.25 acetate buffer from 25:75 to 80:20 over 4 min using a nonlinear gradient (not specified).
**Flow rate:** 0.35
**Injection volume:** 20
**Detector:** MS, PE Sciex API 3000, turbo ionspray source, column effluent split 1:1 before entering source

## CHROMATOGRAM
**Retention time:** 2.7
**Internal standard:** Abbott A-86093 (3.2)
**Limit of detection:** 380 pg/mL
**Limit of quantitation:** 16.3 ng/mL

## OTHER SUBSTANCES
**Extracted:** indinavir (2.0, LOQ 16.3 ng/mL, LOD 1.5 ng/mL), lopinavir (3.1, LOQ 16.3 ng/mL, LOD 750 pg/mL), nelfinavir (2.5, LOQ 16.3 ng/mL, LOD 330 pg/mL), ritonavir (2.9, LOQ 51.2 ng/mL, LOD 650 pg/mL), saquinavir (2.4, LOQ 16.3 ng/mL, LOD 780 pg/mL)

## KEY WORDS
plasma

## REFERENCE
Frerichs, V.A.; DiFrancesco, R.; Morse, G.D. Determination of protease inhibitors using liquid chromatography-tandem mass spectrometry, *J.Chromatogr.B*, **2003**, *787*, 393–403.

## SAMPLE
**Matrix:** blood
**Sample preparation:** Mix 1 mL plasma with 200 μL 10 μg/mL IS in water, add 200 μL 100 mM NaOH, mix, add 4 mL diethyl ether, shake for 5 min, centrifuge at 2500 rpm for 5 min. Evaporate the organic layer to dryness under a stream of nitrogen, reconstitute the residue with 200 μL initial mobile phase, inject a 100 μL aliquot.

## HPLC VARIABLES
**Column:** 250 × 4.6 Stability RP18 (CIL, France)

**Mobile phase:** Gradient. MeCN:50 mM pH 5.65 phosphate buffer from 36:64 to 64:36 over 25 min, to 80:20 (step gradient), maintain at 80:20 for 10 min, re-equilibrate at initial conditions for 5 min.
**Flow rate:** 1.5
**Injection volume:** 100
**Detector:** UV 240 for 5 min, UV 215 for 22 min, UV 260 for rest of the run

---

### CHROMATOGRAM
**Retention time:** 11.2
**Internal standard:** JR051012 (Janssen Cilag) (28.2)
**Limit of quantitation:** 50 ng/mL

---

### OTHER SUBSTANCES
**Extracted:** efavirenz (19.9), indinavir (8.5), lopinavir (18.9), nelfinavir (24.1), nevirapine (3.3), ritonavir (17.6), saquinavir (16.7)
**Noninterfering:** acetaminophen, amineptine, amphotericin B, aspirin, bromazepam, buspirone, citalopram, clobazam, diazepam, didanosine, fluconazole, flunitrazepam, fluvoxamine, hydroxyitraconazole, isoniazid, itraconazole, lamivudine, loprazolam, lorazepam, metronidazole, minalcipram, nordiazepam, omeprazole, paroxetine, pyrimethamine, rifampin, sertraline, stavudine, sulfadiazine, trimethoprim, venlafaxine, zalcitabine, zidovudine, zolpidem, zopiclone

---

### KEY WORDS
plasma

---

### REFERENCE
Titier, K.; Lagrange, F.; Péhourcq, F.; Edno-Mcheik, L.; Moore, N.; Molimard, M. High-performance liquid chromatographic method for the simultaneous determination of the six HIV-protease inhibitors and two non-nucleoside reverse transcriptase inhibitors in human plasma, *Ther.Drug Monit.*, **2002**, *24*, 417–424.

---

### SAMPLE
**Matrix:** microsomal incubations
**Sample preparation:** Extract 100 μL incubation mixture twice with 5 mL MTBE. Evaporate the organic layer to dryness, reconstitute the residue with 100 μL MeCN, inject a 30 μL aliquot.

---

### HPLC VARIABLES
**Column:** 150 × 4.6 Beckman ODS Ultrasphere
**Column temperature:** 45
**Mobile phase:** Gradient. A was 0.1% formic acid in water. B was 0.1% formic acid in MeCN. A:B 100:0 for 1 min, to 30:70 over 3 min, to 5:95 over 3 min, maintain at 5:95 for 3 min, to 100:0 over 1 min.
**Flow rate:** 0.35
**Injection volume:** 30
**Detector:** MS, Hewlett-Packard 5989B, electrospray ionization, selected ion monitoring, m/z 506.6

---

### CHROMATOGRAM
**Limit of quantitation:** 25 ng/mL

---

### OTHER SUBSTANCES
**Also analyzed:** astemizole, indinavir, ketoconazole, methadone, nelfinavir, rifabutin, rifampin, ritonavir, saquinavir, terfenadine, trimethoprim

---

### KEY WORDS
human; liver; rat

## REFERENCE

Decker, C.J.; Laitinen, L.M.; Bridson, G.W.; Raybuck, S.A.; Tung, R.D.; Chaturvedi, P.R. Metabolism of amprenavir in liver microsomes: role of CYP3A4 inhibition for drug interactions, *J.Pharm.Sci.*, **1998**, *87*, 803–807.

## ANNOTATED BIBLIOGRAPHY

Brophy, D.F.; Israel, D.S.; Pastor, A.; Gillotin, C.; Chittick, G.E.; Symonds, W.T.; Lou, Y.; Sadler, B.M.; Polk, R.E. Pharmacokinetic interaction between amprenavir and clarithromycin in healthy male volunteers, *Antimicrob.Agents Chemother.*, **2000**, *44*, 978–984. [no HPLC of clarithromycin]

Chi, J.; Jayewardene, A.L.; Stone, J.A.; Motoya, T.; Aweeka, F.T. Simultaneous determination of five HIV protease inhibitors nelfinavir, indinavir, ritonavir, saquinavir and amprenavir in human plasma by LC/MS/MS, *J.Pharm.Biomed.Anal.*, **2002**, *30*, 675–684.

Cociglio, M.; Hillaire-Buys, D.; Peyrière, H.; Alric, R. Performance analysis of a rapid HPLC determination with the solvent demixing extraction of HIV antiproteases and efavirenz in plasma, *J.Chromatogr.Sci.*, **2003**, *41*, 80–86. [efavirenz; indinavir; amprenavir; ritonavir; saquinavir; nelfinavir]

Dailly, E.; Thomas, L.; Kergueris, M.F.; Jolliet, P.; Bourin, M. High-performance liquid chromatographic assay to determine the plasma levels of HIV-protease inhibitors (amprenavir, indinavir, nelfinavir, ritonavir and saquinavir) and the non-nucleoside reverse transcriptase inhibitor (nevirapine) after liquid-liquid extraction, *J.Chromatogr.B*, **2001**, *758*, 129–135.

Droste, J.A.H.; Verweij-van Wissen, C.P.W.G.M.; Burger, D.M. Simultaneous determination of the HIV drugs indinavir, amprenavir, saquinavir, ritonavir, lopinavir, nelfinavir, the nelfinavir hydroxymetabolite M8, and nevirapine in human plasma by reversed-phase high-performance liquid chromatography, *Ther.Drug Monit.*, **2003**, *25*, 393–399.

Edwards, J.E.; Brouwer, K.R.; McNamara, P.J. GF120918, a P-glycoprotein modulator, increases the concentration of unbound amprenavir in the central nervous system in rats, *Antimicrob.Agents Chemother.*, **2002**, *46*, 2284–2286.

Faux, J.; Venisse, N.; Le Moal, G.; Dupuis, A.; Bouquet, S. Simultaneous determination of six HIV protease inhibitors, one metabolite, and two non-nucleoside reverse transcriptase inhibitors in human plasma by isocratic reversed-phase liquid chromatography after solid-phase extraction, *Chromatographia* **2003**, *58*, 421–426. [amprenavir; indinavir; lopinavir; nelfinavir; ritonavir; saquinavir; efavirenz; nevirapine; prazepam]

Gao, W.; Kishida, T.; Kimura, K.; Kageyama, M.; Sumi, M.; Yoshikawa, Y.; Shibata, N.; Takada, K. Sensitive and simultaneous determination of HIV protease inhibitors in rat biological samples by liquid chromatography-mass spectrometry, *Biomed.Chromatogr.*, **2002**, *16*, 267–273. [indinavir; amprenavir; saquinavir; nelfinavir]

Goujard, C.; Vincent, I.; Meynard, J.-L.; Choudet, N.; Bollens, D.; Rousseau, C.; Demarles, D.; Gillotin, C.; Bidault, R.; Taburet, A.-M. Steady-state pharmacokinetics of amprenavir coadministered with ritonavir in human immunodeficiency virus type 1-infected patients, *Antimicrob.Agents Chemother.*, **2003**, *47*, 118–123.

Gunawan, S.; Griswold, M.P.; Kahn, D.G. Liquid chromatographic-tandem mass spectrometric determination of amprenavir (agenerase) in serum/plasma of human immunodeficiency virus type-1 infected patients receiving combination antiretroviral therapy, *J.Chromatogr.A*, **2001**, *914*, 1–4.

Huang, L.; Wring, S.A.; Woolley, J.L.; Brouwer, K.R.; Serabjit-Singh, C.; Polli, J.W. Induction of P-glycoprotein and cytochrome P450 3A by HIV protease inhibitors, *Drug Metab.Dispos.*, **2001**, *29*, 754–760. [fosamprenavir; amprenavir; nelfinavir]

Justesen, U.S.; Pedersen, C.; Klitgaard, N.A. Simultaneous quantitative determination of the HIV protease inhibitors indinavir, amprenavir, ritonavir, lopinavir, saquinavir, nelfinavir and the nelfinavir active metabolite M8 in plasma by liquid chromatography, *J.Chromatogr.B*, **2003**, *783*, 491–500.

Keil, K.; Frerichs, V.A.; DiFrancesco, R.; Morse, G. Reverse phase high-performance liquid chromatography method for the analysis of amprenavir, efavirenz, indinavir, lopinavir, nelfinavir and its active metabolite (M8), ritonavir, and saquinavir in heparinized human plasma, *Ther.Drug Monit.*, **2003**, *25*, 340–346.

Leibenguth, P.; Le Guellec, C.; Besnier, J.-M.; Bastides, F.; Macé, M.; Gaudet, M.-L.; Autret-Leca, E.; Paintaud, G. Therapeutic drug monitoring of HIV protease inhibitors using high-performance liquid chromatography with ultraviolet or photodiode array detection, *Ther.Drug Monit.*, **2001**, *23*, 679–688. [indinavir; saquinavir; lopinavir; ritonavir; nelfinavir; amprenavir; carbamazepine]

Marzolini, C.; Telenti, A.; Buclin, T.; Biollaz, J.; Decosterd, L.A. Simultaneous determination of the HIV protease inhibitors indinavir, amprenavir, saquinavir, ritonavir, nelfinavir and the non-nucleoside reverse transcriptase inhibitor efavirenz by high-performance liquid chromatography after solid-phase extraction, *J.Chromatogr.B*, **2000**, *740*, 43–58.

Marzolini, C.; Béguin, A.; Telenti, A.; Schreyer, A.; Buclin, T.; Biollaz, J.; Decosterd, L.A. Determination of lopinavir and nevirapine by high-performance liquid chromatography after solid-phase extraction: application for the assessment of their transplacental passage at delivery, *J.Chromatogr.B*, **2002**, *774*, 127–140. [lopinavir; nevirapine; amprenavir; efavirenz; clozapine]

Pereira, A.S.; Kenney, K.B.; Cohen, M.S.; Eron, J.J.; Tidwell, R.R.; Dunn, J.A. Determination of amprenavir, a HIV-1 protease inhibitor, in human seminal plasma using high-performance liquid chromatography-tandem mass spectrometry, *J.Chromatogr.B*, **2002**, *766*, 307–317.

Poirier, J.-M.; Radembino, N.; Robidou, P.; Jaillon, P. Simultaneous determination of the five HIV-protease inhibitors: Amprenavir, indinavir, nelfinavir, ritonavir, and saquinavir in human plasma by solid-phase extraction and column liquid chromatography, *Ther.Drug Monit.*, **2000**, *22*, 465–473.

Poirier, J.M.; Robidou, P.; Jaillon, P. Simultaneous determination of the six HIV protease inhibitors (amprenavir, indinavir, lopinavir, nelfinavir, ritonavir, and saquinavir) plus M8 nelfinavir metabolite and the nonnucleoside reverse transcription inhibitor efavirenz in human plasma by solid-phase extraction and column liquid chromatography, *Ther.Drug Monit.*, **2002**, *24*, 302–309.

Polk, R.E.; Brophy, D.F.; Israel, D.S.; Patron, R.; Sadler, B.M.; Chittick, G.E.; Symonds, W.T.; Lou, Y.; Kristoff, D.; Stein, D.S. Pharmacokinetic interaction between amprenavir and rifabutin or rifampin in healthy males, *Antimicrob.Agents Chemother.*, **2001**, *45*, 502–508.

Proust, V.; Toth, K.; Hulin, A.; Taburet, A.-M.; Gimenez, F.; Singlas, E. Simultaneous high-performance liquid chromatographic determination of the antiretroviral agents amprenavir, nelfinavir, ritonavir, saquinavir, delavirdine and efavirenz in human plasma, *J.Chromatogr.B*, **2000**, *742*, 453–458.

Rentsch, K.M. Sensitive and specific determination of eight antiretroviral agents in plasma by high-performance liquid chromatography-mass spectrometry, *J.Chromatogr.B*, **2003**, *788*, 339–350. [SPE; amprenavir; efavirenz; indinavir; lopinavir; nelfinavir; nevirapine; ritonavir; saquinavir]

Sadler, B.M.; Hanson, C.D.; Chittick, G.E.; Symonds, W.T.; Roskell, N.S. Safety and pharmacokinetics of amprenavir (141 W94), a human immunodeficiency virus (HIV) type I protease inhibitor, following oral administration of single doses to HIV-infected adults, *Antimicrob.Agents Chemother.*, **1999**, *43*, 1686–1692.

Sadler, B.M.; Chittick, G.E.; Polk, R.E.; Slain, D.; Kerkering, T.M.; Studenberg, S.D.; Lou, Y.; Moore, K.H.P.; Woolley, J.L.; Stein, D.S. Metabolic disposition and pharmacokinetics of [14C]-amprenavir, a human immunodeficiency virus Type 1 (HIV-1) protease inhibitor, administered as a single oral dose to healthy male subjects, *J.Clin.Pharmacol.*, **2001**, *41*, 386–396.

Sarasa-Nacenta, M.; López-Púa, Y.; Mallolas, J.; Blanco, J.L.; Gatell, J.M.; Carné, X. Simultaneous determination of the HIV-protease inhibitors indinavir, amprenavir, ritonavir, saquinavir and nelfinavir in human plasma by reversed-phase high-performance liquid chromatography, *J.Chromatogr.B*, **2001**, *757*, 325–332.

Sparidans, R.W.; Hoetelmans, R.M.W.; Beijnen, J.H. Sensitive liquid chromatographic assay for amprenavir, a human immunodeficiency virus protease inhibitor, in human plasma, cerebrospinal fluid and semen, *J.Chromatogr.B*, **2000**, *742*, 185–192.

Tran, J.Q.; Petersen, C.; Garrett, M.; Hee, B.; Kerr, B.M. Pharmacokinetic interaction between amprenavir and delavirdine: evidence of induced clearance by amprenavir, *Clin.Pharmacol.Ther.*, **2002**, *72*, 615–626.

Tréluyer, J.M.; Bowers, G.; Cazali, N.; Sonnier, M.; Rey, E.; Pons, G.; Cresteil, T. Oxidative metabolism of amprenavir in the human liver. Effect of the CYP3A maturation, *Drug Metab.Dispos.*, **2003**, *31*, 275–281.

Tribut, O.; Arvieux, C.; Michelet, C.; Chapplain, J.-M.; Allain, H.; Bentué-Ferrer, D. Simultaneous quantitative assay of six HIV protease inhibitors, one metabolite, and two non-nucleoside reverse transcriptase inhibitors in human plasma by isocratic reversed-phase liquid chromatography, *Ther.Drug Monit.*, **2002**, *24*, 554–562. [nevirapine; efavirenz; indinavir; amprenavir; nelfinavir; ritonavir; lopinavir; saquinavir]

Turner, M.L.; Reed-Walker, K.; King, J.R.; Acosta, E.P. Simultaneous determination of nine antiretroviral compounds in human plasma using liquid chromatography, *J.Chromatogr.B*, **2003**, *784*, 331–341. [indinavir; nelfinavir; saquinavir; ritonavir; amprenavir; delavirdine; efavirenz; lopinavir]

van Heeswijk, R.P.G.; Hoetelmans, R.M.W.; Harms, R.; Meenhorst, P.L.; Mulder, J.W.; Lange, J.M.A.; Beijnen, J.H. Simultaneous quantitative determination of the HIV protease inhibitors amprenavir, indinavir, nelfinavir, ritonavir and saquinavir in human plasma by ion-pair high-performance liquid chromatography with ultraviolet detection, *J.Chromatogr.B*, **1998**, *719*, 159–168.

Veronese, L.; Rautaureau, J.; Sadler, B.M.; Gillotin, C.; Petite, J.-P.; Pillegand, B.; Delvaux, M.; Masliah, C.; Fosse, S.; Lou, Y.; Stein, D.S. Single-dose pharmacokinetics of amprenavir, a human immunodeficiency virus type 1 protease inhibitor, in subjects with normal or impaired hepatic function, *Antimicrob.Agents Chemother.*, **2000**, *44*, 821–826.

Villani, P.; Feroggio, M.; Gianelli, L.; Bartoli, A.; Montagna, M.; Maserati, R.; Regazzi, M.B. Antiretrovirals: simultaneous determination of five protease inhibitors and three nonnucleoside transcriptase inhibitors in human plasma by a rapid high-performance liquid chromatography-mass spectrometry assay, *Ther.Drug Monit.*, **2001**, *23*, 380–388. [saquinavir; indinavir; ritonavir; nelfinavir; amprenavir; nevirapine; delavirdine; efavirenz]

Yamada, H.; Kotaki, H.; Nakamura, T.; Iwamoto, A. Simultaneous determination of the HIV protease inhibitors indinavir, amprenavir, saquinavir, ritonavir and nelfinavir in human plasma by high-performance liquid chromatography, *J.Chromatogr.B*, **2001**, *755*, 85–89.

Yu, L.; Bridgers, A.; Polli, J.; Vikers, A.; Long, S.; Roy, A.; Winnike, R.; Coffin, M. Vitamin E-TPGS increases absorption flux of an HIV protease inhibitor by enhancing its solubility and permeability, *Pharm.Res.*, **1999**, *16*, 1812–1817.

# Anagrelide

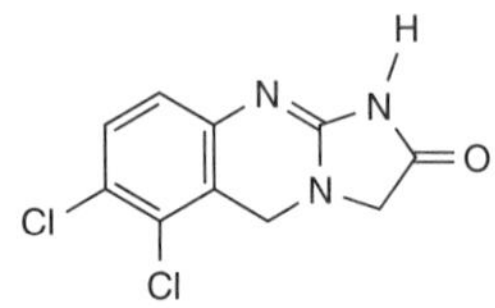


**Molecular formula:** $C_{10}H_7Cl_2N_3O$
**Molecular weight:** 256.09
**CAS Registry No:** 68475-42-3
**Merck Index:** 13, 629

## SAMPLE

**Matrix:** blood, urine

**Sample preparation:** Mix 2 mL plasma or urine with 2 mL 200 mM pH 7.0 phosphate buffer, extract twice with 10 mL portions of ethyl acetate. Evaporate the organic layer to dryness under a stream of nitrogen, reconstitute the residue with 60 μL DMSO, mix, sonicate, inject a 40 μL aliquot.

## HPLC VARIABLES

**Guard column:** 40 mm long μBondapak phenyl corasil
**Column:** 300 × 3.9 μBondapak phenyl
**Mobile phase:** MeCN:10 mM pH 4 sodium acetate buffer 25:75 for 10 min, DMSO for 8 min, return to original mobile phase
**Flow rate:** 2.5 for 13 min, 1 for 5 min, 2.5 for rest of the run
**Injection volume:** 40
**Detector:** UV 254; Radioactivity ($^{14}C$)

## CHROMATOGRAM

**Retention time:** 6–8

## KEY WORDS

plasma; radiolabeled

## REFERENCE

Gaver, R.C.; Deeb, G.; Pittman, K.A.; Smyth, R.D. Disposition of anagrelide, an inhibitor of platelet aggregation, *Clin.Pharmacol.Ther.*, **1981**, *29*, 381–386.

# Anakinra

**Molecular weight:** 17 000
**CAS Registry No:** 143090-92-0
**Merck Index:** 13, 5022

## SAMPLE
**Matrix:** blood, tissue
**Sample preparation:** Inject a 50 μL aliquot of plasma or tissue homogenate supernatant.

## HPLC VARIABLES
**Guard column:** 40 × 6 Spherogel TSK PWHR (Beckman)
**Column:** 300 × 7.8 5 μm Progel-TSK G2000 SWXL (Supelco)
**Mobile phase:** 10 mM pH 6.5 citrate buffer containing 140 mM NaCl and 0.5 mM EDTA
**Flow rate:** 0.5
**Injection volume:** 50
**Detector:** UV; Radioactivity ($^{35}S$); ELISA

## CHROMATOGRAM
**Retention time:** 20

## KEY WORDS
brain; gut; heart; kidney; liver; lung; muscle; plasma; rat; spleen

## REFERENCE
Kim, D.C.; Reitz, B.; Carmichael, D.F.; Bloedow, D.C. Kidney as a major clearance organ for recombinant human interleukin-1 receptor antagonist, *J.Pharm.Sci.*, **1995**, *84*, 575–580.

## SAMPLE
**Matrix:** solutions
**Sample preparation:** Inject a 100 μL aliquot of a 2–5 mg/mL solution in 10 mM pH 6.5 citrate buffer containing 140 mM NaCl and 0.5 mM EDTA.

## HPLC VARIABLES
**Column:** 75 × 7.5 Bio-Gel SP-5-PW (Bio-Rad)
**Mobile phase:** Gradient. A:B from 99:1 to 40:60 over 60 min. A was 20 mM pH 5.5 2-(*N*-morpholino)ethanesulfonic acid monohydrate. B was 20 mM pH 5.5 2-(*N*-morpholino)ethanesulfonic acid monohydrate containing 1.0 M NaCl.
**Flow rate:** 0.5
**Injection volume:** 100
**Detector:** UV 280

## REFERENCE
Nahata, M.C.; Morosco, R.S.; Sabados, B.K.; Weber, T.R. Stability and compatibility of anakinra with intravenous cimetidine hydrochloride or famotidine in 0.9% sodium chloride injection, *J.Clin.Pharm. Ther.*, **1995**, *20*, 97–99.

## ANNOTATED BIBLIOGRAPHY
Chang, B.S.; Beauvais, R.M.; Arakawa, T.; Narhi, L.O.; Dong, A.; Aparisio, D.I.; Carpenter, J.F. Formation of an active dimer during storage of interleukin-1 receptor antagonist in aqueous solution, *Biophys.J.*, **1996**, *71*, 3399–3406.

Chang, B.S.; Reeder, G.; Carpenter, J.F. Development of a stable freeze-dried formulation of recombinant human interleukin-1 receptor antagonist, *Pharm.Res.*, **1996**, *13*, 243–249.
Nahata, M.C.; Morosco, R.S.; Sabados, B.K.; Weber, T.R. Stability and compatibility of anakinra with ceftriaxone sodium injection in 0.9% sodium chloride or 5% dextrose injection, *J.Clin.Pharm.Ther.*, **1997**, *22*, 167–169.

# Apraclonidine

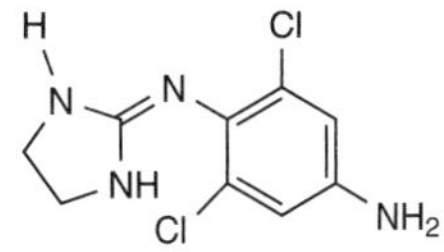


**Molecular formula:** $C_9H_{10}Cl_2N_4$
**Molecular weight:** 245.11
**CAS Registry No:** 66711-21-5
**Merck Index:** 13, 756

## SAMPLE

**Matrix:** solutions
**Sample preparation:** Inject a 50 μL aliquot of a solution in glutathione bicarbonated Ringer's solution (pH 7.4).

## HPLC VARIABLES

**Column:** 150 × 4 5 μm Ultrasphere ODS
**Mobile phase:** MeCN:water 20:80 to 60:40 (?) containing 5 mM sodium heptanesulfonic acid at pH 3.5
**Flow rate:** 1–1.5
**Injection volume:** 50
**Detector:** UV 254

## OTHER SUBSTANCES

**Simultaneous:** clonidine

## REFERENCE

Chien, D.S.; Homsy, J.J.; Gluchowski, C.; Tang-Liu, D.D.-S. Corneal and conjunctival/scleral penetration of *p*-aminoclonidine, AGN 190342, and clonidine in rabbit eyes, *Current Eye Res.*, **1990**, *9*, 1051–1059.

# Aprepitant

**Molecular formula:** $C_{23}H_{21}F_7N_4O_3$
**Molecular weight:** 534.43
**CAS Registry No:** 170729-80-3

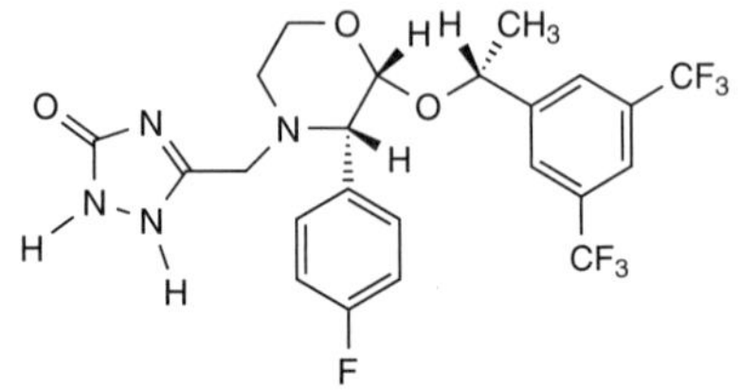

## SAMPLE

**Matrix:** blood, tissue
**Sample preparation:** Mix 200 μL plasma with 20 ng IS and 1.7 mL water, add 500 μL MeCN, add to a 500 mg Bond Elut C18 SPE cartridge, wash with 6 mL water, elute with 3 mL MeOH. Evaporate the eluate to dryness under a stream of nitrogen, reconstitute the residue with 300 μL mobile phase, inject an aliquot. Alternatively, mix 50 μL plasma or brain homogenate with 5 ng IS and 100 μL MeCN, vortex, centrifuge at 3000 g for 10 min, inject a 5–25 μL aliquot of the supernatant.

## HPLC VARIABLES

**Column:** 50 × 4.6 5 μm Spherisorb C8
**Mobile phase:** MeCN:10 mM ammonium acetate:formic acid 55:45:0.1
**Flow rate:** 1
**Injection volume:** 5–25
**Detector:** MS, Sciex API III+, heated nebulizer interface, dwell time 450 ms, m/z 535 to 179

## CHROMATOGRAM

**Retention time:** 1.5
**Internal standard:** desfluoroaprepitant (m/z 535 to 161) (1.5)

## KEY WORDS

brain; ferret; plasma; SPE

## REFERENCE

Huskey, S.-E.W.; Dean, B.J.; Bakhtiar, R.; Sanchez, R.I.; Tattersall, F.D.; Rycroft, W.; Hargreaves, R.; Watt, A.P.; Chicchi, G.G.; Keohane, C.; Hora, D.F.; Chiu, S.-H.L. Brain penetration of aprepitant, a substance P receptor antagonist, in ferrets, *Drug Metab.Dispos.*, **2003**, *31*, 785–791.

## SAMPLE

**Matrix:** blood, tissue
**Sample preparation:** Mix 3 mL plasma with 6 mL MeCN, centrifuge at 3000 g, evaporate the supernatant to dryness under a stream of nitrogen, reconstitute the residue with 1 mL MeOH:water 40:60, inject a 250–400 μL aliquot of the supernatant. Homogenize the brain with 3 volumes of water. Vortex 10 mL homogenate with 90 mL MeCN, sonicate for 5 min, centrifuge at 3000 g for 10 min, re-extract the pellet with 10 mL MeOH. Combine the organic layers and add to a Bond Elut C18 SPE cartridge equipped with an Acrodisc glass filter, elute with 5 mL MeOH:MeCN:water 50:25:25. Collect all the cartridge effluent and evaporate to dryness under a stream of nitrogen, reconstitute the residue with 5 mL MeOH, vortex, sonicate, centrifuge. Evaporate the supernatant to dryness under a stream of nitrogen, reconstitute the residue with 1 mL MeOH:water 40:60, inject a 400 μL aliquot of the supernatant.

## HPLC VARIABLES

**Column:** 250 × 4.6 Zorbax RX-C8

**Mobile phase:** Gradient. A:B 65:35 to 20:80 over 40 min. A was 10 mM ammonium acetate. B was MeCN:MeOH 92.8:7.2 containing 7.2 mM ammonium acetate. (Alternatively, A 10 mM ammonium acetate in water containing 0.1% trifluoroacetic acid and B MeCN:MeOH 92.8:7.2 containing 7.2 mM ammonium acetate and 0.1% trifluoroacetic acid with the same gradient.)
**Flow rate:** 1
**Injection volume:** 250–400
**Detector:** Radioactivity ($^{14}C$)

---

**CHROMATOGRAM**
**Retention time:** 26
**Internal standard:** desfluoroaprepitant (m/z 535 to 161) (1.5)

---

**OTHER SUBSTANCES**
**Extracted:** metabolites

---

**KEY WORDS**
brain; ferret; plasma; SPE

---

**REFERENCE**
Huskey, S.E.W.; Dean, B.J.; Bakhtiar, R.; Sanchez, R.I.; Tattersall, F.D.; Rycroft, W.; Hargreaves, R.; Watt, A.P.; Chicchi, G.G.; Keohane, C.; Hora, D.F.; Chiu, S.H.L. Brain penetration of aprepitant, a substance P receptor antagonist, in ferrets, *Drug Metab.Dispos.*, **2003**, *31*, 785–791.

---

**SAMPLE**
**Matrix:** solutions
**Sample preparation:** Inject a 1 μL aliquot of a solution in MeOH:water 10:90.

---

**HPLC VARIABLES**
**Column:** 20 × 2 5 μm DASH BetaBasic C8 (ThermoHypersil Keystone)
**Mobile phase:** Gradient. A was MeCN:water:formic acid 5:95:0.1. B was MeCN:water: formic acid 95:5:0.1. A:B 100:0 for 0.2 min, to 0:100 over 1.5 min.
**Flow rate:** 1.5
**Injection volume:** 1
**Detector:** MS, PE Sciex API-3000, turbo ionspray, electrospray 4500 V, ring 290 V, orifice 60 V, drying gas 400°, 20% of column effluent entered the detector, m/z 535.3–277

---

**CHROMATOGRAM**
**Retention time:** 1.4

---

**OTHER SUBSTANCES**
**Simultaneous:** amitriptyline (m/z 278.3–233) (1.1), diclofenac (m/z 296.1–215) (1.35), enoxacin (m/z 321.2–234) (0.7), fenofibrate (m/z 360.9–233) (1.6), finasteride (m/z 373.2–317) (1.2), indinavir (m/z 614.4–421) (0.93), pioglitazone (357.2–134) (0.87), raloxifene (m/z 474.1–112) (0.97)

---

**REFERENCE**
Romanyshyn, L.A.; Tiller, P.R. Ultra-short columns and ballistic gradients: considerations for ultra-fast chromatographic liquid chromatographic-tandem mass spectrometric analysis, *J.Chromatogr.A*, **2001**, *928*, 41–51.

# Aranidipine

**Molecular formula:** $C_{19}H_{20}N_2O_7$
**Molecular weight:** 388.37
**CAS Registry No:** 86780-90-7
**Merck Index:** 13, 772

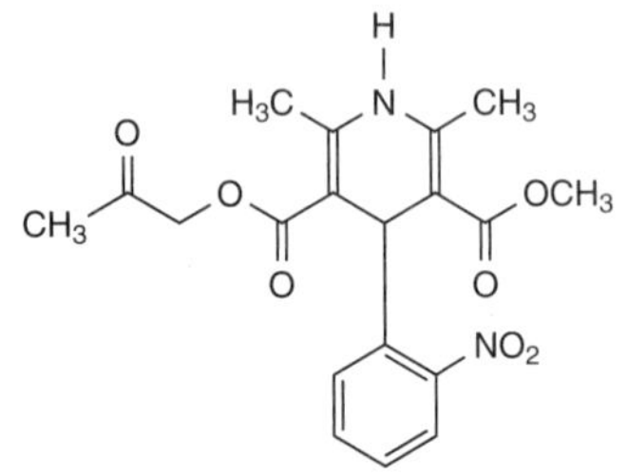


## SAMPLE

**Matrix:** blood

**Sample preparation:** Add 20 ng nifedipine and 500 μL 100 mM pH 9.0 borate buffer to 1 mL plasma, vortex for 10 s, add 6 mL toluene, shake mechanically for 10 min, centrifuge at 1000 g for 15 min. Evaporate the organic layer to dryness under a stream of nitrogen at 40°, reconstitute the residue with 200 μL mobile phase, inject a 50 μL aliquot. (Carry out all steps under yellow fluorescent lighting.)

## HPLC VARIABLES

**Column:** 150 × 4.6 5 μm Inertsil ODS-2
**Column temperature:** 40
**Mobile phase:** MeOH:360 mM sodium perchlorate 45:55
**Flow rate:** 0.8
**Injection volume:** 50
**Detector:** E, BAS LC-4B/17AT, +0.92 V versus Ag/AgCl

## CHROMATOGRAM

**Retention time:** 16
**Internal standard:** nifedipine (26)
**Limit of quantitation:** 2.5 ng/mL

## OTHER SUBSTANCES

**Extracted:** metabolites

## KEY WORDS

dog; pharmacokinetics; plasma

## REFERENCE

Iida, Y.; Kinouchi, Y.; Takeichi, Y.; Imai, T.; Otagiri, M. Simultaneous determination of a new dihydropyridine calcium antagonist (MPC-1304) and its metabolite in dog plasma by high-performance liquid chromatography with electrochemical detection, *J.Chromatogr.*, **1991**, *571*, 277–282.

# Arotinolol

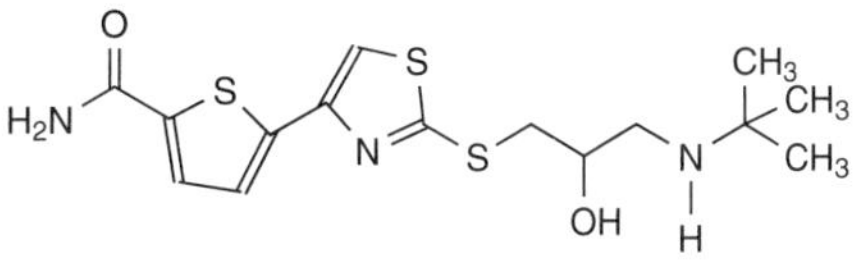

**Molecular formula:** $C_{15}H_{21}N_3O_2S_3$
**Molecular weight:** 371.55
**CAS Registry No:** 68377-92-4, 68377-91-3 (HCl)
**Merck Index:** 13, 797

## SAMPLE
**Matrix:** blood
**Sample preparation:** Add 100 μL 10 mg/mL IS in water to 500 μL plasma, make up to 1 mL with water, add 100 μL 3 M pH 9 ammonium acetate, vortex vigorously for 2 min, centrifuge at 3000 g for 10 min. Extract the aqueous layer three times with 1 mL portions of ether, evaporate the extracts to dryness under reduced pressure, reconstitute the residue with 100 μL 100 mM HCl, inject an aliquot.

## HPLC VARIABLES
**Column:** 250 × 4.6 Chirobiotic T (Advanced Separation Technologies)
**Mobile phase:** MeOH:acetic acid:triethylamine 100:0.1:0.1
**Flow rate:** 0.8
**Detector:** UV 317

## CHROMATOGRAM
**Retention time:** 17.25 (S-(+)), 20.06 (R-(−) )
**Internal standard:** labetalol hydrochloride (21.98, 23.43 (enantiomers))
**Limit of detection:** 50 ng/mL
**Limit of quantitation:** 100 ng/mL

## KEY WORDS
chiral; plasma

## REFERENCE
Aboul-Enein, H.Y.; Hefnawy, M.M. Enantioselective determination of arotinolol in human plasma by HPLC using teicoplanin chiral stationary phase, *Biomed.Chromatogr.*, **2003**, *17*, 453–457.

## SAMPLE
**Matrix:** blood, urine
**Sample preparation:** Condition a 1 mL C18 Bakerbond SPE cartridge with MeOH. Mix 1 mL plasma with 50 μL 5 μg/mL alpiropride in water. Mix 100 μL pure or diluted urine with 250 μL blank plasma and 100 μL 5 μg/mL alpiropride in water. Add the sample to the SPE cartridge, wash three times with 1 mL portions of water, wash three times with 1 mL portions of *n*-hexane:diethyl ether 50:50, elute with two 1 mL portions of chloroform:triethylamine 90:10 (Caution! Chloroform is a carcinogen!). Evaporate the eluate to dryness under reduced pressure, reconstitute the residue with 150 μL mobile phase, inject a 100 μL aliquot.

## HPLC VARIABLES
**Guard column:** 15 × 4.6 7 μm C18
**Column:** 250 × 4.6 5 μm ODS Hypersil
**Column temperature:** 25
**Mobile phase:** MeCN:MeOH:buffer 12.5:12.5:75 (The buffer was 67 mM pH 5.6 phosphate buffer containing 0.6 mM tetrabutylammonium chloride.)
**Flow rate:** 1.2
**Injection volume:** 100
**Detector:** F ex 310 em 395; UV 310

---

## CHROMATOGRAM

**Retention time:** 10.0
**Internal standard:** alpiropride (4.7)
**Limit of detection:** 0.11 ng/mL (plasma), 11 ng/mL (urine)

---

## OTHER SUBSTANCES

**Extracted:** metabolites

---

## KEY WORDS

pharmacokinetics; plasma; SPE

---

## REFERENCE

Moulin, A.; Maillet, E.; Truffer, D.; Dufour, A. High performance liquid chromatographic determination of arotinolol and AC 623, its main metabolite in biological samples, *J.Liq.Chromatogr.*, **1992**, *15*, 151–164.

---

## SAMPLE

**Matrix:** blood, urine
**Sample preparation:** Condition a 1 mL C18 Bakerbond SPE cartridge with MeOH. Mix 1 mL plasma with 50 μL 5 μg/mL alpiropride in water. Mix 100 μL pure or diluted urine with 250 μL blank plasma and 100 μL 5 μg/mL alpiropride in water. Add the sample to the SPE cartridge, wash three times with 1 mL portions of water, wash three times with 1 mL portions of *n*-hexane:diethyl ether 50:50, elute with two 1 mL portions of chloroform:triethylamine 90:10 (Caution! Chloroform is a carcinogen!). Evaporate the eluate to dryness under reduced pressure, reconstitute the residue with 150 μL mobile phase, inject a 100 μL aliquot.

---

## HPLC VARIABLES

**Guard column:** 15 × 4.6 7 μm diol
**Column:** 200 × 4.6 5 μm Lichrosorb diol
**Column temperature:** 25
**Mobile phase:** Dichloromethane containing 10 mM Z-glycyl-L-proline:MeOH 100:1
**Flow rate:** 2
**Injection volume:** 100
**Detector:** F ex 320 em 425

---

## CHROMATOGRAM

**Retention time:** 12 (R-(−) ), 15 (S-(+))
**Internal standard:** alpiropride (7.5)
**Limit of detection:** 2 ng/mL

---

## OTHER SUBSTANCES

**Extracted:** metabolites

---

## KEY WORDS

chiral; plasma; SPE

---

## REFERENCE

Moulin, A.; Truffer, D.; Maillet, E.; Dufour, A. High performance liquid chromatographic determination of the optical isomers of arotinolol and AC 623, its main metabolite, in biological samples, *J.Liq.Chromatogr.*, **1992**, *15*, 165–181.

## ANNOTATED BIBLIOGRAPHY

Nakamura, K.; Fujima, H.; Kitagawa, H.; Wada, H.; Makino, K. Preparation and chromatographic characteristics of a chiral-recognizing perphenylated cyclodextrin column, *J.Chromatogr.A*, **1995**, *694*, 111–118. [ibuprofen; chlorpheniramine; acetylpheneturide; alprenolol; arotinolol; atenolol; benzoin; biperiden; bunitrolol; chlormezanone; chlorphenesin; eperisone; flavanone; oxprenolol; phenylethyl alcohol; phenylethylamine; pindolol; proglumide; propranolol; trihexyphenidyl]

# Arteether

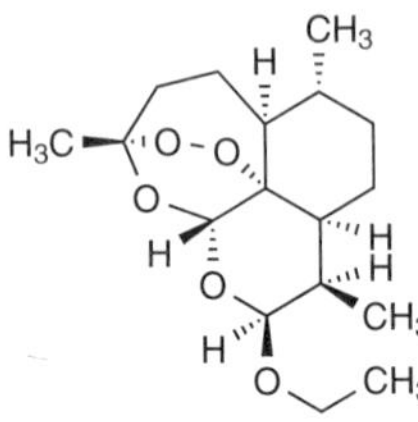


**Molecular formula:** $C_{17}H_{28}O_5$
**Molecular weight:** 312.40
**CAS Registry No:** 75887-54-6
**Merck Index:** 13, 822

## SAMPLE
**Matrix:** blood
**Sample preparation:** Add 5 μL 10 μg/mL artemisinin in MeOH to 200 μL serum, vortex, add 2 mL hexane, vortex for 1 min, centrifuge at 1000 g for 5 min, freeze in liquid nitrogen. Repeat the extraction. Combine the organic layers and evaporate to dryness, reconstitute the residue with 40 μL MeOH, inject a 20 μL aliquot.

## HPLC VARIABLES
**Column:** 300 × 4.6 5 μm Ultracarb 5 ODS 20 (Phenomenex)
**Mobile phase:** MeOH:100 mM sodium acetate 80:20
**Flow rate:** 1
**Injection volume:** 20
**Detector:** MS, Quattro II triple quadrupole, electrospray, nebulizing gas nitrogen 10 L/h, curtain gas nitrogen 250 L/h, ESI capillary at 3.5 kV, cone voltage 52 V, positive mode, m/z 335 [M + Na]+, one tenth of column effluent was allowed into MS

## CHROMATOGRAM
**Retention time:** 1.73 ($\alpha$), 2.81 ($\beta$)
**Internal standard:** artemisinin (m/z 305) (1.02)
**Limit of detection:** 5 ng/mL
**Limit of quantitation:** 20 ng/mL

## KEY WORDS
rat; serum

## REFERENCE
Rajanikanth, M.; Madhusudanan, K.P.; Gupta, R.C. Liquid chromatographic-mass spectrometric method for the determination of $\alpha$-, $\beta$-arteether in rat serum, *J.Chromatogr.B*, **2003**, *783*, 391–399.

## SAMPLE
**Matrix:** solutions

## HPLC VARIABLES
**Column:** 250 × 4.6 5 μm IBO-SIL C18 (Phenomenex)
**Mobile phase:** MeOH:water 80:20
**Flow rate:** 0.9
**Detector:** UV 260

## REFERENCE
Illapakurthy, A.C.; Sabnis, Y.A.; Avery, B.A.; Avery, M.A.; Wyandt, C.M. Interaction of artemisinin and its related compounds with hydroxypropyl-$\beta$-cyclodextrin in solution state: experimental and molecular-modeling studies, *J.Pharm.Sci.*, **2003**, *92*, 649–655.

## ANNOTATED BIBLIOGRAPHY

Baker, J.K.; Yarber, R.H.; Hufford, C.D.; Lee, I.-S.; ElSohly, H.N.; McChesney, J.D. Thermospray mass spectroscopy/high performance liquid chromatographic identification of the metabolites formed from arteether using a rat liver microsome preparation, *Biomed.Environ.Mass Spectrom.*, **1989**, *18, 337–351.*

Baker, J.K.; McChesney, J.D.; Chi, H.T. Decomposition of arteether in simulated stomach acid yielding compounds retaining antimalarial activity, *Pharm.Res.*, **1993**, *10*, 662–666.

Benakis, A.; Schopfer, C.; Paris, M.; Plessas, C.; Karayannakos, P.E.; Dondas, I.; Kotsarelis, D.; Plessas, S.T.; Skalkeas, G. Pharmacokinetics of arteether in dog, *Eur.J.Drug Metab.Pharmacokinet.*, **1991**, *16*, 325–328.

Chi, H.T.; Ramu, K.; Baker, J.K.; Hufford, C.D.; Lee, I.S.; Zeng, Y.-L.; McChesney, J.D. Identification of the in vivo metabolites of the antimalarial arteether by thermospray high-performance liquid chromatography/mass spectrometry, *Biol.Mass Spectrom.*, **1991**, *20*, 609–628.

Idowu, O.R.; Edwards, G.; Ward, S.A.; Orme, M.L'E.; Breckenridge, A.M. Determination of arteether in blood plasma by high-performance liquid chromatography with ultraviolet detection after hydrolysis with acid, *J.Chromatogr.*, **1989**, *493*, 125–136.

Leo, K.U.; Grace, J.M.; Li, Q.; Peggins, J.; Mitchell, A.L.; Aguilar, T.; Brewer, T.G. Effects of *Plasmodium berghei* infection on arteether metabolism and disposition, *Pharmacology*, **1997**, *54*, 276–284.

Leskovac, V.; Theoharides, A.D. Hepatic metabolism of artemisinin drugs – I. Drug metabolism in rat liver microsomes, *Comp.Biochem.Physiol.C*, **1991**, *99*, 383–390. [arteether; dihydroartemisinin]

Li, Q.G.; Brueckner, R.P.; Peggins, J.O.; Trotman, K.M.; Brewer, T.G. Arteether toxicokinetics and pharmacokinetics in rats after 25 mg/kg/day single and multiple doses, *Eur.J.Drug Metab.Pharmacokinet.*, **1999**, *24*, 213–223.

Melendez, V.; Peggins, J.O.; Brewer, T.G.; Theoharides, A.D. Determination of the antimalarial arteether and its deethylated metabolite dihydroartemisinin in plasma by high-performance liquid chromatography with reductive electrochemical detection, *J.Pharm.Sci.*, **1991**, *80*, 132–138.

Rajanikanth, M.; Madhusudanan, K.P.; Gupta, R.C. An HPLC-MS method for simultaneous estimation of $\alpha,\beta$-arteether and its metabolite dihydroartemisinin, in rat plasma for application to pharmacokinetic study, *Biomed.Chromatogr.*, **2003**, *17*, 440–446.

Ramu, K.; Baker, J.K. Identification of the glucuronides of the hydroxylated metabolites of the antimalarial arteether in rat plasma and urine by thermospray high-performance liquid chromatography/mass spectrometry, *J.Pharm.Sci.*, **1997**, *86*, 915–920.

# Articaine

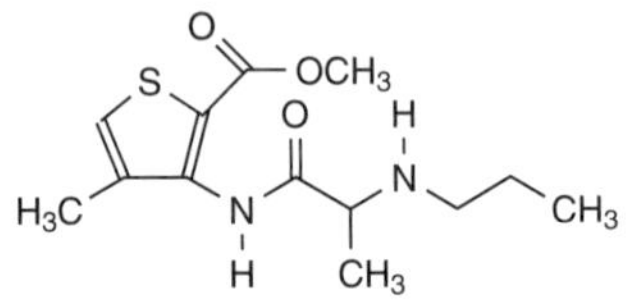


**Molecular formula:** $C_{13}H_{20}N_2O_3S$
**Molecular weight:** 284.38
**CAS Registry No:** 23964-58-1, 23964-57-0 (HCl)
**Merck Index:** 13, 1884

## SAMPLE
**Matrix:** blood
**Sample preparation:** 1 mL Plasma + 50 μL 2 μg/mL etidocaine in water + 100 μL 1 M NaOH + 3 mL heptane:ethyl acetate 90:10, shake for 2 min, centrifuge at 1200 g for 10 min. Remove the organic phase and add it to 50 μL 50 mM sulfuric acid, shake for 2 min, centrifuge at 1200 g for 5 min. Remove the aqueous phase and add it to 820 μg sodium acetate, inject a 40 μL aliquot. (The sodium acetate was measured out by adding 50 μL 200 mM sodium acetate in MeOH to the tube and evaporating the MeOH.)

## HPLC VARIABLES
**Column:** 250 × 4 10 μm μBondapak C18
**Column temperature:** 30
**Mobile phase:** MeCN:10 mM sodium dihydrogen phosphate 7:93, adjusted to pH 2.1
**Flow rate:** 1
**Injection volume:** 40
**Detector:** UV 205

## CHROMATOGRAM
**Retention time:** 19
**Internal standard:** etidocaine (10)
**Limit of detection:** 5 ng/mL

## OTHER SUBSTANCES
**Extracted:** mepivacaine (15)

## KEY WORDS
plasma; rabbit

## REFERENCE
Le Guévello, P.; Le Corre, P.; Chevanne, P.; Le Verge, R. High-performance liquid chromatographic determination of bupivacaine in plasma samples for biopharmaceutical studies and application to seven other local anaesthetics, *J.Chromatogr.*, **1993**, *622*, 284–290.

## SAMPLE
**Matrix:** blood
**Sample preparation:** Condition a 1 mL 4 mm SDB-RPS SPE disk cartridge (3M Empore) with 500 μL MeOH, 500 μL air, 500 μL water, and 1 mL air. Mix 1 mL serum with 50 μL perchloric acid, let stand for 10 min, mix, centrifuge at 16 000 g for 10 min. Add 800 μL to the cartridge followed by 2 mL air. Wash with 800 μL 0.5% phosphoric acid in MeOH:water 20:80, push through 1.5 mL air, wash with 700 μL water, push through 2 mL air, elute with 500 μL MeOH containing 1% ammonia, push through 1.2 mL air. Evaporate the eluate to dryness under a stream of air at 70° and reconstitute the residue with 50 μL mobile phase, inject a 40 μL aliquot.

## HPLC VARIABLES
**Column:** 125 × 3 5 μm Nucleosil 50-5 endcapped RP-8

**Column temperature:** 35
**Mobile phase:** MeCN:buffer 12:88 (Buffer was 880 mL 20 mM potassium dihydrogen phosphate containing 500 μL phosphoric acid, pH 3.)
**Flow rate:** 1
**Injection volume:** 40
**Detector:** UV 274

---

## CHROMATOGRAM
**Retention time:** 9.5
**Limit of quantitation:** 10 ng/mL

---

## OTHER SUBSTANCES
**Extracted:** articainic acid (3.5)

---

## KEY WORDS
serum; SPE

---

## REFERENCE
Richter, K.; Oertel, R. Solid-phase extraction and high-performance liquid chromatographic determination of articaine and its metabolite articainic acid in human serum, *J.Chromatogr.B*, **1999**, *724*, 109–115.

---

## SAMPLE
**Matrix:** solutions

---

## HPLC VARIABLES
**Column:** 250 × 4.6 10 μm Chiralcel OD
**Mobile phase:** *n*-Hexane:isopropanol 80:20
**Flow rate:** 0.4
**Injection volume:** 5
**Detector:** UV 274

---

## CHROMATOGRAM
**Retention time:** 12, 14 (enantiomers)

---

## KEY WORDS
chiral

---

## REFERENCE
Rustichelli, C.; Ferioli, V.; Gamberini, G.; Stancanelli, R. Enantiomeric separation of local anaesthetic drug by HPLC on chiral stationary phases, *Chromatographia*, **2001**, *54*, 731–736.

---

## ANNOTATED BIBLIOGRAPHY
Chankvetadze, B.; Chankvetadze, L.; Sidamonidze, S.; Yashima, E.; Okamoto, Y. High performance liquid chromatography enantioseparation of chiral pharmaceuticals using tris(chloro-methylphenylcarbamate)s of cellulose, *J.Pharm.Biomed.Anal.*, **1996**, *14*, 1295–1303. [oxazepam; oxazolam; hexobarbital; phenobarbital; thiopental; lorazepam; lopirazepam; camazepam; cloxazolam; ketazolam; pindolol; propranolol; sotalol; alprenolol; bupranolol; acebutolol; penbutolol; toliprolol; enilconazole; econazole; miconazole; bifonazole; ornidazole; bayleton; metomidate; nisoldipine; nimodipine; isradipine; nicardipine; triadimefon; pheniramine; doxylamine; chlorphenoxamine; carbinoxamine; azelastine; mequitazine; paramethadione; norgestrel; tesicam; mesuximide; metofoline; tramadol; clofedanol; lofexidine; clenbuterol; piprozolin; etozolin; doxapram; chlormezanone; aminoglutethimide; articaine; etidocaine; nefopam]

Oertel, R.; Richter, K.; Weile, K.; Gramatte, T.; Berndt, A.; Feller, K. A simple method for the determination of articaine and its metabolite articainic acid in dentistry: application to a comparison of articaine and lidocaine concentrations in alveolus blood, *Methods Find.Exp.Clin.Pharmacol.*, **1993**, *15*, 541–547.

Oertel, R.; Richter, K.; Gramatté, T.; Kirch, W. Determination of drugs in biological fluids by high-performance liquid chromatography with on-line sample processing, *J.Chromatogr.A*, **1998**, *797*, 203–209. [metoprolol; talinolol; celiprolol; tiracizine; triamterene; ajmaline; articaine; lamotrigine]

Rop, P.P.; Grimaldi, F.; Bresson, M.; Fornaris, M.; Viala, A. Liquid chromatographic analysis of cocaine, benzoylecgonine, local anaesthetic agents and some of their metabolites in biological fluids, *J.Liq.Chromatogr.*, **1993**, *16*, 2797–2811. [cocaine; benzoylecgonine; procaine; *p*-aminobenzoic acid; butacaine; tetracaine; articaine; prilocaine; *o*-toluidine; lidocaine; monoethylglycine xylidide; bupivacaine; pipecolylxylidene; etidocaine; dibucaine; caffeine; amphetamine; ephedrine; epinephrine; morphine; monoacetylmorphine; diamorphine; ethylmorphine; codeine; acetylcodeine; fluorescence detection; UV detection; SPE]

Vree, T.B.; Baars, A.M.; van Oss, G.E.; Booij, L.H. High-performance liquid chromatography and preliminary pharmacokinetics of articaine and its 2-carboxy metabolite in human serum and urine, *J.Chromatogr.*, **1988**, *424*, 440–444.

# Asparaginase

**Molecular weight:** ca. 136 000
**CAS Registry No:** 9015-68-3
**Merck Index:** 13, 841

## SAMPLE
**Matrix:** reaction mixtures
**Sample preparation:** Adjust to pH 7 with 2 M NaOH, inject a 20 μL aliquot.

## HPLC VARIABLES
**Column:** 250 × 4 10 μm HEMA-BIO 1000 (Tessek, Prague) (hydroxyethyl methacrylate–type column)
**Mobile phase:** 100 mM Potassium dihydrogen phosphate adjusted to pH 6.9 with 2 M NaOH
**Flow rate:** 0.8
**Injection volume:** 20
**Detector:** UV 210

## CHROMATOGRAM
**Retention time:** 2.1
**Limit of detection:** 0.51 U/mL

## REFERENCE
Barek, J.; Cvacka, J.; Zima, J.; De Méo, M.; Laget, M.; Michelon, J.; Castegnaro, M. Chemical degradation of wastes of antineoplastic agents amsacrine, azathioprine, asparaginase and thiotepa, *Ann.Occup.Hyg.*, **1998**, *42*, 259–266.

# Atazanavir sulfate

**Molecular formula:** $C_{38}H_{52}N_6O_7.H_2O_4S$
**Molecular weight:** 802.93
**CAS Registry No:** 229975-97-7

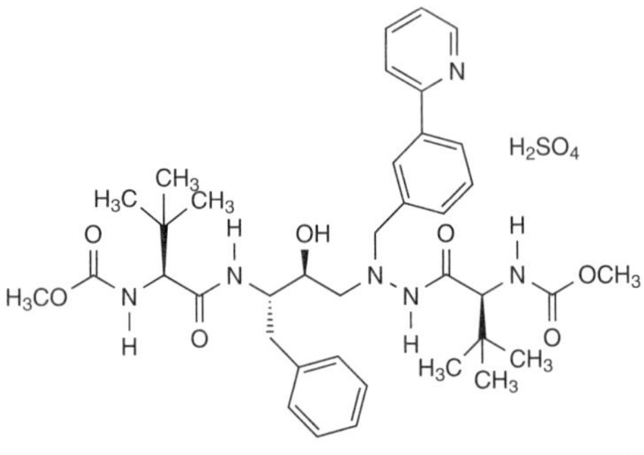

### SAMPLE

**Matrix:** blood
**Sample preparation:** Condition each well of a 3M Empore C2-SD 96 well plate with 250 μL MeOH and 500 μL 0.1% acetic acid, do not allow to go dry. Add 50 μL 200 ng/mL IS in MeOH:water 60:40 to 200 μL MeOH:water 60:40 containing 5 million cells, sonicate for 10 min, centrifuge at 2600 g for 10 min. Evaporate the supernatant to dryness, reconstitute with 50 μL MeOH, add 200 μL water, add 250 μL 0.1% acetic acid, mix, add to a well on the SPE plate, allow to pass through under vacuum over 2 min, wash with 500 μL 0.1% acetic acid, dry under vacuum for 2 min, elute twice with 200 μL portions of MeCN:MeOH 50:50, pulling to dryness after each portion. Evaporate the eluate to dryness under a stream of nitrogen at 60° over ca. 40 min, reconstitute the residue with 200 μL mobile phase, vortex, inject a 20 μL aliquot.

### HPLC VARIABLES

**Guard column:** 10 × 2 5 μm YMC Basic
**Column:** 50 × 2 5 μm YMC Basic
**Mobile phase:** MeCN:MeOH:water:88% formic acid 30:30:40:0.025
**Flow rate:** 0.25
**Injection volume:** 20
**Detector:** MS, Sciex API 3000 turbo ionspray, electrospray, positive mode at 400°, m/z 705 to 335, IonSpray 4600 V, declustering potential 56 V, entrance potential −10 V, focusing potential 220 V, TurboIon gas nitrogen 8 L/min, collision energy 42 V, collision cell exit potential 24 V, dwell time 500 ms, pause time 5 ms

### CHROMATOGRAM

**Retention time:** <4
**Internal standard:** $^{13}C_6$-atazanavir
**Limit of quantitation:** 5 fmole/million cells

### KEY WORDS

peripheral blood mononuclear cells; SPE

### REFERENCE

Jemal, M.; Rao, S.; Gatz, M.; Whigan, D. Liquid chromatography-tandem mass spectrometric quantitative determination of the HIV protease inhibitor atazanavir (BMS-232632) in human peripheral blood mononuclear cells (PBMC): practical approaches to PBMC preparation and PBMC assay design for high-throughput analysis, *J.Chromatogr.B*, **2003**, *795*, 273–289.

### SAMPLE

**Matrix:** blood
**Sample preparation:** Condition each well of a10 mg Oasis HLB 96 well SPE plate with 1 mL MeOH and 1 mL 0.1% acetic acid. Add 40 μL 5 μg/mL IS in water and 300 μL 0.1%

acetic acid to 250 μL plasma, mix, add to a well of the SPE plate, wash with 500 μL 0.1% acetic acid, wash with 500 μL MeOH:water 20:80, elute with 300 μL MeOH. Evaporate the eluate to dryness under a stream of nitrogen at 60°, reconstitute the residue with 500 μL MeCN:MeOH:10 mM pH 5.5 ammonium acetate 30:30:40, inject a 15 μL aliquot.

## HPLC VARIABLES

**Column:** 33 × 4.6 3 μm Uptisphere HDO C18 (Interchim)
**Mobile phase:** Gradient. MeCN:5 mM ammonium acetate 50:50 for 0.5 min, to 60:40 over 0.1 min, maintain at 60:40 for 1.7 min, return to initial conditions over 0.1 min, re-equilibrate for 2.1 min.
**Flow rate:** 0.8
**Injection volume:** 15
**Detector:** MS, Micromass Quattro Ultima, atmospheric pressure electrospray ionization, column effluent split 1:20 before entering MS, positive ion mode, capillary sprayer voltage 3.2 kV, sample cone voltage 80 V, source temperature 100°, desolvation temperature 350°, nebulizing gas nitrogen, cone gas nitrogen at 37 L/h, desolvation gas nitrogen at 500 L/h, collision gas argon at 2.6 μbar, collision energy was set at 40 eV, resolution set at 0.7 mass units at half height for the first and third quadrupoles.

## CHROMATOGRAM

**Retention time:** 2.3
**Internal standard:** $^{13}C_6$-atazanavir
**Limit of quantitation:** 1 ng/mL

## KEY WORDS

plasma; SPE

## REFERENCE

Schuster, A.; Burzawa, S.; Jemal, M.; Loizillon, E.; Couerbe, P.; Whigan, D. Quantitative determination of the HIV protease inhibitor atazanavir (BMS-232632) in human plasma by liquid chromatography-tandem mass spectrometry following automated solid-phase extraction, *J.Chromatogr.B*, **2003**, *788*, 377–386.

# Atipamezole

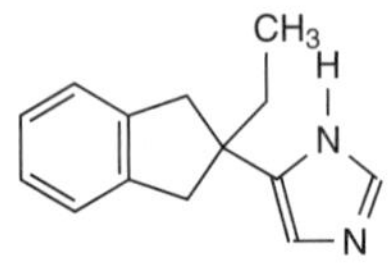

**Molecular formula:** $C_{14}H_{16}N_2$
**Molecular weight:** 212.29
**CAS Registry No:** 104054-27-5
**Merck Index:** 13, 866

### SAMPLE
**Matrix:** blood
**Sample preparation:** Condition a Sep-Pak C18 SPE cartridge with water, MeOH, and 100 mM ammonium acetate. 5 mL Plasma + 250 ng detomidine, add to the SPE cartridge, wash with 100 mM ammonium acetate, elute with MeOH:100 mM ammonium acetate 75:25. Evaporate the eluate to dryness under reduced pressure, reconstitute the residue in 200 μL mobile phase, inject a 50 μL aliquot.

### HPLC VARIABLES
**Column:** 150 × 4.6 5 μm Hitachi gel #3056
**Mobile phase:** MeOH:100 mM ammonium acetate 65:35
**Flow rate:** 1
**Injection volume:** 50
**Detector:** MS, Hitachi M-1000, APCI interface, drift voltage 21 V, nebulizer 260°, vaporizer 399°, multiplier voltage 1500 VF, m/z 213

### CHROMATOGRAM
**Retention time:** 8
**Internal standard:** detomidine (m/z 187) (6.5)
**Limit of quantitation:** 1–2 ng/mL

### OTHER SUBSTANCES
**Extracted:** medetomidine (7.5, m/z 201), midazolam (10.5, m/z 326)

### KEY WORDS
pharmacokinetics; pig; plasma; SPE

### REFERENCE
Kanazawa, H.; Nishimura, R.; Sasaki, N.; Takeuchi, A.; Takai, N.; Nagata, Y.; Matsushima, Y. Determination of medetomidine, atipamazole and midazolam by liquid chromatography-mass spectrometry, *Biomed.Chromatogr.*, **1995**, *9*, 188–191.

### SAMPLE
**Matrix:** blood
**Sample preparation:** Mix 200 μL 250 mM NaOH with 500 μL plasma, add 6 mL dichloromethane, mix gently for 10 min, centrifuge at 1700 g for 10 min. Evaporate 4 mL of the lower organic layer to dryness under a stream of nitrogen, reconstitute the residue with 100 μL 50 mM pH 3.2 phosphate buffer, vortex for 1.5 min, centrifuge at 1700 g for 10 min, inject a 50 μL aliquot.

### HPLC VARIABLES
**Guard column:** 20 × 4.6 5 μm Supelguard LC-DP
**Column:** 250 × 4.6 5 μm Supelcosil LC-DP
**Mobile phase:** MeCN:50 mM phosphate buffer:triethylamine 27:73:0.05, adjusted to pH 3.2
**Flow rate:** 1
**Injection volume:** 50

**Detector:** UV 215

## CHROMATOGRAM

**Retention time:** 16.2
**Limit of detection:** 5 ng/mL

## OTHER SUBSTANCES

**Extracted:** medetomidine (14.6)

## KEY WORDS

pharmacokinetics; plasma; reindeer

## REFERENCE

Ranheim, B.; Horsberg, T.E.; Nymoen, U.; Soli, N.E.; Tyler, N.J.; Arnemo, J.M. Reversal of medetomidine-induced sedation in reindeer (*Rangifer tarandus tarandus*) with atipamezole increases the medetomidine concentration in plasma, *J.Vet.Pharmacol.Ther.*, **1997**, *20*, 350–354.

## ANNOTATED BIBLIOGRAPHY

Kanazawa, H.; Nagata, Y.; Matsushima, Y.; Takai, N.; Uchiyama, H.; Nishimura, R.; Takeuchi, A. Liquid chromatography-mass spectrometry for the determination of medetomidine and other anaesthetics in plasma, *J.Chromatogr.*, **1993**, *631*, 215–220.

# Atomoxetine hydrochloride

**Molecular formula:** $C_{17}H_{21}NO.HCl$
**Molecular weight:** 291.82
**CAS Registry No:** 82248-59-7

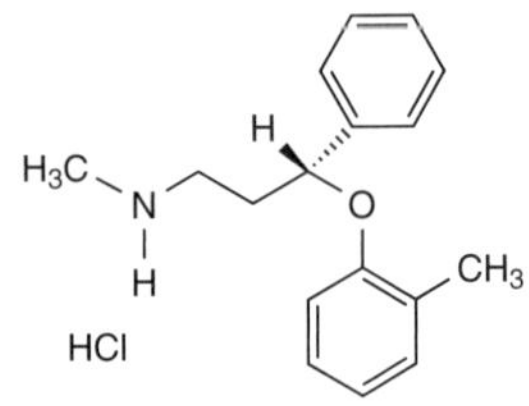

## SAMPLE

**Matrix:** blood
**Sample preparation:** Add 500 μL plasma to a Varian SDB-XC SPE cartridge, wash with 1 mL MeOH:water 15:85, elute with 750 μL MeCN containing 0.1% trifluoroacetic acid. Evaporate the eluate to dryness under a stream of nitrogen at 45°, reconstitute the residue with 100 μL MeCN, mix with 25 μL water, inject an aliquot.

## HPLC VARIABLES

**Column:** 100 × 4.6 5 μm Brownlee Spheri-5 ODS
**Mobile phase:** MeCN:water 85:15 containing 5 mM ammonium acetate, 0.2% formic acid, and 0.03% trifluoroacetic acid
**Flow rate:** 1
**Detector:** MS, PE Sciex API III, MS/MS, positive atmospheric pressure chemical ionization, heated nebulizer interface, m/z 256 to 44

## CHROMATOGRAM

**Limit of quantitation:** 0.25 ng/mL

## OTHER SUBSTANCES

**Extracted:** metabolites

## KEY WORDS

dog; plasma; rat; SPE

## REFERENCE

Mattiuz, E.L.; Ponsler, G.D.; Barbuch, R.J.; Wood, P.G.; Mullen, J.H.; Shugert, R.L.; Li, Q.; Wheeler, W.J.; Kuo, F.; Conrad, P.C.; Sauer, J.-M. Disposition and metabolic fate of atomoxetine hydrochloride: Pharmacokinetics, metabolism, and excretion in the Fischer 344 rat and beagle dog, *Drug Metab.Dispos.*, **2003**, *31*, 88–97.

## SAMPLE

**Matrix:** blood, urine
**Sample preparation:** Plasma. Mix 3 mL MeCN with 1.5 mL plasma, centrifuge, evaporate the supernatant to dryness under a stream of nitrogen, reconstitute the residue with 200 μL MeCN:water 10:90, inject an aliquot. Urine. Lyophilize urine, reconstitute with MeCN:water 10:90 to one-tenth original volume, vortex, filter (0.45 μm), inject an aliquot.

## HPLC VARIABLES

**Column:** 150 × 4.6 5 μm Zorbax Eclipse XDB-C18
**Column temperature:** 30
**Mobile phase:** Gradient. MeCN:50 mM ammonium acetate from 10:90 to 60:40 over 30 min. (Use 25 mM ammonium acetate for MS detector.)
**Flow rate:** 1

**Detector:** Radioactivity ($^{14}C$); MS, Finnigan TSQ 700 or TSQ 7000, positive electrospray, collision gas argon, 0.2 mL/min of column effluent entered MS

---

## CHROMATOGRAM
**Retention time:** 20

---

## OTHER SUBSTANCES
**Extracted:** metabolites

---

## KEY WORDS
plasma

---

## REFERENCE
Sauer, J.-M.; Ponsler, G.D.; Mattiuz, E.L.; Long, A.J.; Witcher, J.W.; Thomasson, H.R.; Desante, K.A. Disposition and metabolic fate of atomoxetine hydrochloride: the role of CYP2D6 in human disposition and metabolism, *Drug Metab.Dispos.*, **2003**, *31*, 98–107.

# Atorvastatin

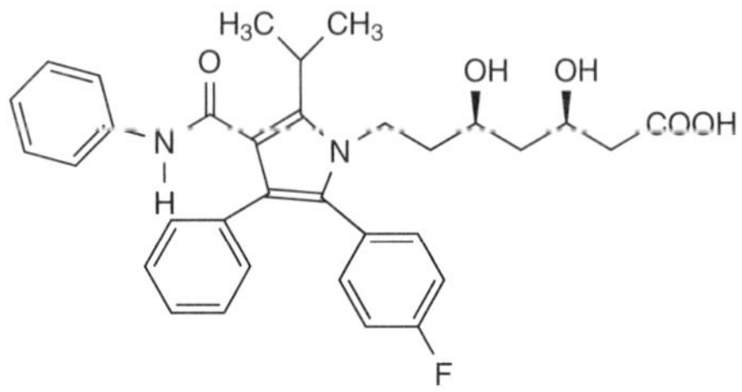


**Molecular formula:** $C_{33}H_{35}FN_2O_5$
**Molecular weight:** 558.64
**CAS Registry No:** 134523-00-5
**Merck Index:** 13, 868

## SAMPLE
**Matrix:** blood
**Sample preparation:** Mix 500 μL serum with IS, acidify to pH 6 with sodium acetate buffer, extract with MTBE, centrifuge. Remove the organic layer and evaporate it to dryness, reconstitute the residue, inject an aliquot.

## HPLC VARIABLES
**Column:** YMC Basic
**Mobile phase:** Gradient. A:B from 70:30 to 45:55 over 1 min, maintain at 45:55 for 0.5 min, return to initial conditions over 0.1 min, maintain at 70:30 for 1.9 min. A was MeOH:water:88% formic acid 5:95:0.0043. B was MeCN:MeOH:88% formic acid 95:5:0.0043.
**Detector:** MS, Finnigan TSQ-7000, electrospray, m/z 559-440

## CHROMATOGRAM
**Internal standard:** deuterated atorvastatin
**Limit of quantitation:** 500 pg/mL

## OTHER SUBSTANCES
**Extracted:** metabolites

## KEY WORDS
pharmacokinetics; serum

## REFERENCE
Kantola, T.; Kivistö, K.T.; Neuvonen, P.J. Effect of itraconazole on the pharmacokinetics of atorvastatin, *Clin.Pharmacol.Ther.*, **1998**, *64*, 58–65.

## SAMPLE
**Matrix:** formulations
**Sample preparation:** Shake 10 tablets with 50 mL MeCN:THF:50 mM pH 4 ammonium citrate buffer 27:20:53 at 450 rpm for 1 h, make up to 100 mL with the same solution, filter, dilute a 2 mL aliquot to 10 mL, inject an aliquot.

## HPLC VARIABLES
**Guard column:** 4 × 3 5 μm C18 Luna (Phenomenex)
**Column:** 250 × 4.6 5 μm C18 Luna (Phenomenex)
**Mobile phase:** Gradient. MeCN:THF:20 mM pH 4.0 ammonium acetate buffer from 25:5:70 to 70:5:25 over 50 min, maintain at 70:5:25 for 10 min.
**Flow rate:** 1
**Injection volume:** 100
**Detector:** UV 248

## CHROMATOGRAM
**Retention time:** 30

**Limit of detection:** 13 ng/mL
**Limit of quantitation:** 130 ng/mL

## OTHER SUBSTANCES

**Simultaneous:** impurities

## KEY WORDS

tablets

## REFERENCE

Ertürk, S.; Aktas, E.S.; Ersoy, L.; Fiçicioglu, S. An HPLC method for the determination of atorvastatin and its impurities in bulk drug and tablets, *J.Pharm.Biomed.Anal.*, **2003**, *33*, 1017–1023.

## ANNOTATED BIBLIOGRAPHY

Al-Rawithi, S.; Hussein, R.F.; Alzahrani, A. Sensitive assay for the determination of fluvastatin in plasma utilizing high-performance liquid chromatography with fluorescence detection, *Ther.Drug Monit.*, **2003**, *25*, 88–92. [atorvastatin is IS]

Jacobsen, W.; Kuhn, B.; Soldner, A.; Kirchner, G.; Sewing, K.-F.; Kollman, P.A.; Benet, L.Z.; Christians, U. Lactonization is the critical first step in the disposition of the 3-hydroxy-3-methylglutaryl-CoA reductase inhibitor atorvastatin, *Drug Metab.Dispos.*, **2000**, *28*, 1369–1378. [microsomal incubations]

Mazzu, A.L.; Lasseter, K.C.; Shamblen, E.C.; Agarwal, V.; Lettieri, J.; Sundaresen, P. Itraconazole alters the pharmacokinetics of atorvastatin to a greater extent than either cerivastatin or pravastatin, *Clin.Pharmacol.Ther.*, **2000**, *68*, 391–400. [SPE]

Miao, X.-S.; Metcalfe, C.D. Determination of cholesterol-lowering statin drugs in aqueous samples using liquid chromatography-electrospray ionization tandem mass spectrometry, *J.Chromatogr.A*, **2003**, *998*, 133–141. [atorvastatin; lovastatin; pravastatin; simvastatin]

Prueksaritanont, T.; Tang, C.; Qiu, Y.; Mu, L.; Subramanian, R.; Lin, J.H. Effects of fibrates on metabolism of statins in human hepatocytes, *Drug Metab.Dispos.*, **2002**, *30*, 1280–1287. [cerivastatin; simvastatin; atorvastatin; rosuvastatin; pravastatin]

# Atosiban

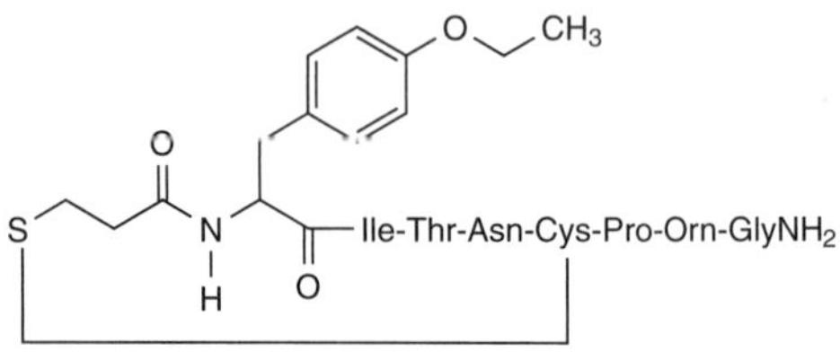

**Molecular formula:** $C_{43}H_{67}N_{11}O_{12}S_2$
**Molecular weight:** 994.20
**CAS Registry No:** 90779-69-4
**Merck Index:** 13, 869

## SAMPLE
**Matrix:** solutions

## HPLC VARIABLES
**Guard column:** Kromasil C8 pre-column
**Column:** 100 × 2.1 KR 100-5 C8 1572 (Hichrom)
**Mobile phase:** MeCN:water:triethylammonium phosphate 27:72.9:0.1
**Flow rate:** 0.2
**Detector:** UV 190

## REFERENCE
Lundin, S.; Svedman, P.; Höglund, P.; Jönsson, K.; Broeders, A.; Melin, P. Absorption of an oxytocin antagonist (antocin) and a vasopressin analogue (dDAVP) through a standardized skin erosion in volunteers, *Pharm.Res.*, **1995**, *12*, 2024–2029.

## SAMPLE
**Matrix:** solutions
**Sample preparation:** Inject an aliquot of a solution in mobile phase.

## HPLC VARIABLES
**Column:** 250 × 4.6 5 μm Spherisorb $NH_2$ (Before use flush column with isopropanol at 60° (to remove hexane-shipping solvent) and then with aqueous trifluoroacetic acid (pH 2.0) at 75° (to protonate amino groups).)
**Column temperature:** 40
**Mobile phase:** MeCN:water 92.35:7.65 containing 2.5 mM ammonium acetate and 250 mM sodium perchlorate
**Flow rate:** 0.5–1.2
**Detector:** UV 210

## CHROMATOGRAM
**Retention time:** 230 (0.5 mL/min), 130 (1.2 mL/min)

## OTHER SUBSTANCES
**Simultaneous:** impurities

## REFERENCE
Oyler, A.R.; Armstrong, B.L.; Cha, J.Y.; Zhou, M.X.; Yang, Q.; Robinson, R.I.; Dunphy, R.; Burinsky, D.J. Hydrophilic interaction chromatography on amino-silica phases complements reversed-phase high-performance liquid chromatography and capillary electrophoresis for peptide analysis, *J.Chromatogr.A*, **1996**, *724*, 378–383.

# Balofloxacin

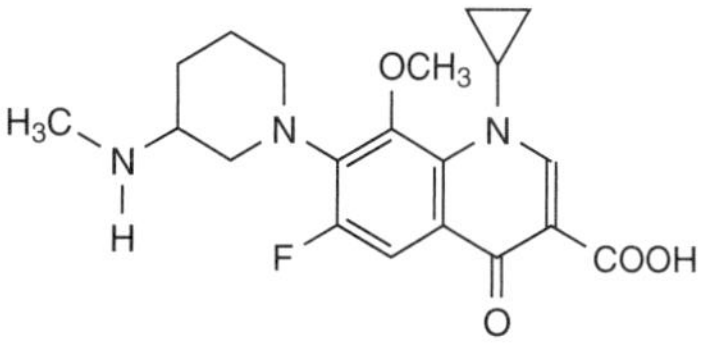


**Molecular formula:** $C_{20}H_{24}FN_3O_4$
**Molecular weight:** 389.42
**CAS Registry No:** 127294-70-6
**Merck Index:** 13, 946

### SAMPLE
**Matrix:** bile, blood, urine
**Sample preparation:** Dilute urine and bile 100-fold with 100 mM pH 7.4 sodium phosphate buffer. Mix 100 μL plasma or diluted urine or bile with 100 μL IS solution, 100 μL 100 mM pH 7.4 sodium phosphate buffer, and 5 mL dichloromethane, shake for 10 min, centrifuge at 3000 rpm for 3 min. Remove the organic layer and evaporate it to dryness under a stream of nitrogen, reconstitute the residue with 100 μL mobile phase, inject a 20 μL aliquot.

### HPLC VARIABLES
**Column:** 250 × 4.6 Capcell Pak C18
**Mobile phase:** MeCN:10 mM pH 2.5 potassium phosphate buffer containing 5 mM PIC B-5 22:78
**Flow rate:** 1
**Injection volume:** 20
**Detector:** F ex 295 em 500

### CHROMATOGRAM
**Internal standard:** l-cyclopropyl-6,8-difluoro-l,4-dihydro-7-(3-methylaminopiperidin-l-yl)-4-oxoquinoline-3-carboxylic acid
**Limit of quantitation:** 10 ng/mL

### OTHER SUBSTANCES
**Extracted:** metabolites

### KEY WORDS
plasma; rat

### REFERENCE
Nakagawa, T.; Ishigai, M.; Hiramatsu, Y.; Kinoshita, H.; Ishitani, Y.; Ohkubo, K.; Okazaki, A. Determination of the new fluoroquinolone balofloxacin and its metabolites in biological fluids by high performance liquid chromatography, *Arzneimittelforschung*, **1995**, *45*, 716–718.

### SAMPLE
**Matrix:** tissue
**Sample preparation:** Homogenize 100 mg mouse ears with 100 mM pH 7.2 phosphate buffered saline.

### HPLC VARIABLES
**Column:** Capcell Pak C18
**Mobile phase:** MeCN:10 mM potassium dihydrogen phosphate 22:78 containing 5 mM 1-pentanesulfonic acid
**Detector:** F ex 295 em 500

### CHROMATOGRAM
**Retention time:** 7

## KEY WORDS
ear; mouse

## REFERENCE
Marutani, K.; Matsumoto, M.; Otabe, Y.; Nagamuta, M.; Tanaka, K.; Miyoshi, A.; Hasegawa, T.; Nagano, H.; Matsubara, S.; Kamide, R.; Yokota, T.; Matsumoto, F.; Ueda, Y. Reduced phototoxicity of a fluoroquinolone antibacterial agent with a methoxy group at the 8 position in mice irradiated with long-wavelength UV light, *Antimicrob.Agents Chemother.*, **1993**, *37*, 2217–2223.

## ANNOTATED BIBLIOGRAPHY
Kozawa, O.; Uematsu, T.; Matsuno, H.; Niwa, M.; Nagashima, S.; Kanamaru, M. Comparative study of pharmacokinetics of two new fluoroquinolones, balofloxacin and grepafloxacin, in elderly subjects, *Antimicrob.Agents Chemother.*, **1996**, *40*, 2824–2828.

Uematsu, T.; Ohsawa, Y.; Mizuno, A.; Nakashima, M. Analysis of a new fluoroquinolone derivative (Q-35) in human scalp hair as an index of drug exposure and as a time marker in hair, *Int.J.Legal Med.*, **1994**, *106*, 237–243.

# Bambermycins

**CAS Registry No:** 11015-37-5
**Merck Index:** 13, 954

## SAMPLE
**Matrix:** solutions

## HPLC VARIABLES
**Column:** 250 × 4 Superspher-100 RP18 endcapped
**Mobile phase:** MeCN:MeOH:buffer 25:35:40 (The buffer was 10 mg potassium dihydrogen phosphate in 1 L water, adjusted to pH 7.8 with dipotassium hydrogen phosphate.)
**Flow rate:** 0.4
**Detector:** UV 258

## CHROMATOGRAM
**Retention time:** 24.6 (moenomycins A and C3)

## KEY WORDS
According to the paper, moenomycins A and C3 can be separated using 100 × 2 5 μm Nucleosil C18 with a 10 min gradient from 0.09% trifluoroacetic acid in MeCN to 0.1% trifluoroacetic acid (sic) at 0.2 mL/min using electrospray MS triple quadrupole detection (8.2 and 8.5 min, respectively).

## REFERENCE
Subramaniam-Niehaus, B.; Schneider, T.; Metzger, J.W.; Wohlleben, W. Isolation and analysis of moenomycin and its biosynthetic intermediates from Streptomyces ghanaensis (ATCC 14672) wildtype and selected mutants, *Z.Naturforsch.C*, **1997**, *52*, 217–226.

# Befunolol

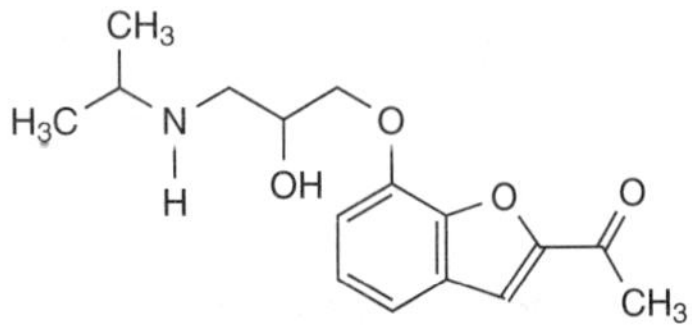

**Molecular formula:** $C_{16}H_{21}NO_4$
**Molecular weight:** 291.34
**CAS Registry No:** 39552-01-7, 39543-79-8 (HCl), 66717-59-7 ((S)-(−) HCl), 66685-79-8 ((R)-(+) HCl)
**Merck Index:** 13, 1023

## SAMPLE

**Matrix:** blood
**Sample preparation:** Condition a Sep-Pak C18 SPE cartridge with 5 mL MeOH, 5 mL water, and 5 mL MeOH:water 30:70. Mix 1 mL plasma with 450 μL 3 μg/mL IS in MeOH, centrifuge at 1700 g for 10 min, add the supernatant to the SPE cartridge, wash with 5 mL MeOH:water 30:70, wash with 50 μL MeOH, suck dry under reduced pressure for 1 min, elute with 5 mL MeOH. Evaporate the eluate to dryness under a stream of nitrogen at 50°, reconstitute the residue with 200 μL mobile phase, inject a 150 μL aliquot.

## HPLC VARIABLES

**Column:** 250 × 4 LiChrosorb RP-Select B
**Mobile phase:** MeCN:20 mM potassium dihydrogen phosphate 20:80
**Flow rate:** 1
**Injection volume:** 150
**Detector:** UV 290

## CHROMATOGRAM

**Retention time:** 11
**Internal standard:** levobunolol (13)
**Limit of quantitation:** 1 ng/mL

## OTHER SUBSTANCES

**Extracted:** dihydrobefunolol (UV 245, LOQ 10 ng/mL) (10)

## KEY WORDS

pharmacokinetics; plasma; rabbit

## REFERENCE

Nozaki, Y.; Imai, T.; Goto, K.; Imamura, Y.; Underberg, W.J.M.; Otagiri, M. Simultaneous determination of befunolol and its major metabolite, dihydrobefunolol, in plasma by high performance liquid chromatography, *Anal.Sci.*, **1989**, *5*, 395–397.

# Benzalkonium chloride

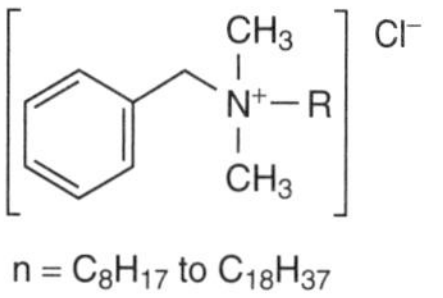


**Molecular formula:** $C_{21}H_{38}ClN$ (C12), $C_{23}H_{42}ClN$ (C14)
**Molecular weight:** 339.99 (C12), 368.04 (C14)
**CAS Registry No:** 8001-54-5
**Merck Index:** 13, 1058

## SAMPLE

**Matrix:** blood
**Sample preparation:** Mix 2 mL plasma with 4 mL water, add to a 3 mL Baker C18 SPE cartridge, wash with three 3 mL portions of water, wash with two 3 mL portions of MeOH, wash with two 3 mL portions of ethyl acetate, elute with 4 mL MeOH:ethyl acetate 50:50 containing 0.01% ammonium chloride, and evaporate the eluate to dryness under a stream of nitrogen. Reconstitute the residue with 1 mL ethyl acetate, 1 mL 10% sodium carbonate, and 200 μL 0.1% bromophenol blue. Evaporate the organic layer to dryness under a stream of nitrogen, reconstitute the residue with 50 μL mobile phase, and inject an aliquot. (The ammonium chloride is eliminated in the aqueous layer and the benzalkonium forms an ion pair with the bromophenol blue and goes into the organic layer.)

## HPLC VARIABLES

**Column:** Spherisorb CN
**Mobile phase:** MeCN:buffer 90:10 (The buffer was a 161 mM pH 5.4 propionate buffer made by mixing 75 mL 10% sodium carbonate, 1.5 L water, and 12 mL propionic acid. Make up to 2 L with water.)
**Flow rate:** 2
**Detector:** UV 214, UV 254

## CHROMATOGRAM

**Retention time:** 5.8 (C14), 6.2 (C12)
**Limit of detection:** 5 ng/mL (UV 214)

## KEY WORDS

dog; human; plasma; SPE

## REFERENCE

Bleau, G.; Desaulniers, M. High-performance liquid chromatographic assay of benzalkonium in plasma, *J.Chromatogr.*, **1989**, *487*, 221–227.

# Betaine

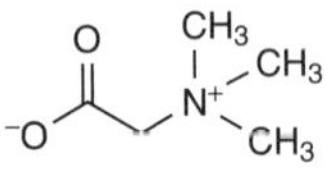

**Molecular formula:** $C_5H_{11}NO_2$
**Molecular weight:** 117.15
**CAS Registry No:** 107-43-7
**Merck Index:** 13, 1182

## SAMPLE

**Matrix:** blood, urine
**Sample preparation:** Dilute urine 10-fold with water. Mix 50 μL plasma or diluted urine with 50 μL 100 mM potassium dihydrogen phosphate, vortex, add 900 μL reagent, vortex, heat at 80° for 1 h, cool to room temperature, vortex, centrifuge at 1000 g, inject a 15 μL aliquot of the supernatant. (Prepare the reagent by dissolving 66 mg 18-crown-6 and 1.39 g 4-bromophenacyl bromide in 100 mL MeCN.)

## HPLC VARIABLES

**Column:** 250 × 4.6 5 μm Supelcosil LC-SCX
**Mobile phase:** MeCN:water 90:10 containing 22 mM choline
**Flow rate:** 1.5
**Injection volume:** 15
**Detector:** UV 254

## CHROMATOGRAM

**Retention time:** 14.8
**Limit of detection:** 5 μM

## OTHER SUBSTANCES

**Extracted:** *N,N*-dimethylglycine (12.7)

## KEY WORDS

derivatization; plasma

## REFERENCE

Laryea, M.D.; Steinhagen, F.; Pawliczek, S.; Wendel, U. Simple method for the routine determination of betaine and *N,N*-dimethylglycine in blood and urine, *Clin.Chem.*, **1998**, *44*, 1937–1941.

## SAMPLE

**Matrix:** blood, urine
**Sample preparation:** Vortex 100 μL plasma or urine with 1 mL MeCN:MeOH 90:10, add 300–400 mg drying mixture, mix thoroughly with occasional inversion for 1 h, centrifuge at 1000 g for 2 min. Remove a 200 μL portion of the supernatant and mix with 20 μL of a 10% suspension of magnesium oxide in water, vortex, add 50 μL 100 mM 4-bromophenacyl triflate in MeCN, continue mixing for 5 min, centrifuge at 1000 g for 2 min (Lever,M. et al. *Anal.Biochem.* **1992**, *205*, 14–21), add 10 μL of a suspension of Dowex 1 in the dichloroacetate form (to eliminate excess derivatization reagent), inject a 20 μL aliquot. (Drying mixture was 90% anhydrous disodium hydrogen phosphate and 10% argentous oxide. Prepare 4′-bromophenacyl trifluoromethanesulfonate as follows. Add 8.8 g *p*-bromobenzoyl chloride in 40 mL dry ether over 20–30 min to 100 mmol diazomethane stirred in an ice bath, stir in an ice bath for 8–9 h, let stand at room temperature for 3 h, evaporate the solvent under reduced pressure, recrystallize 4′-bromo-2-diazoacetophenone from ether/hexane (mp 123.5–124° d) (*J.Am.Chem.Soc.* **1951**, *73*, 5301). Condense 50 mL anhydrous sulfur dioxide in a flask fitted with a calcium sulfate drying tube, cool in a dry ice/acetone bath, add 2.25 g 4′-bromo-2-diazoacetophenone, stir for 5 min, add 900 μL anhydrous trifluoromethanesulfonic acid

from a freshly opened bottle in one portion, stir for 15 min, remove the cooling bath, after 30 min use an ice/water bath to evaporate the solvent. Dissolve the residue in 100 mL boiling dichloromethane, treat twice with 5 g portions of decolorizing carbon, filter, evaporate the filtrate, recrystallize the residue from pentane:dichloromethane 80:20 to give 4′-bromophenacyl trifluoromethanesulfonate as colorless plates (mp 137–8°) (*J.Chromatogr.* **1984**, *299*, 365.)

## HPLC VARIABLES

**Column:** 250 × 4 5 μm silica (Merck)
**Mobile phase:** MeCN:dichloromethane:water 80:10:10 containing 9 mM triethylamine and 6 mM citric acid
**Flow rate:** 1
**Injection volume:** 20

## KEY WORDS

normal phase; plasma

## REFERENCE

Lever, M.; Sizeland, P.C.; Bason, L.M.; Hayman, C.M.; Chambers, S.T. Glycine betaine and proline betaine in human blood and urine, *Biochim.Biophys.Acta*, **1994**, *1200*, 259–264.

# Bethanechol chloride

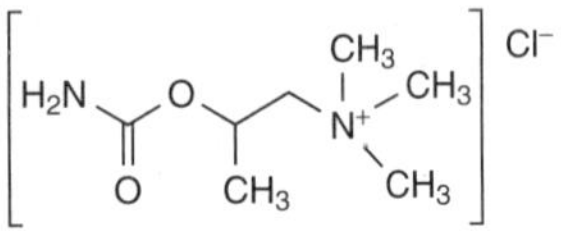


**Molecular formula:** $C_7H_{17}ClN_2O_2$
**Molecular weight::** 196.68
**CAS Registry No:** 590-63-6
**Merck Index:** 13, 1189

## SAMPLE
**Matrix:** solutions

## HPLC VARIABLES
**Column:** 250 × 4.6 5 μm Bakerbond phenylethyl
**Mobile phase:** MeCN:water 33:67
**Flow rate:** 0.7
**Detector:** UV 200

## CHROMATOGRAM
**Retention time:** 3.0

## OTHER SUBSTANCES
**Simultaneous:** degradants

## KEY WORDS
stability-indicating

## REFERENCE
Allen, L.V. Jr.; Erickson,M.A. III. Stability of bethanechol chloride, pyrazinamide, quinidine sulfate, rifampin, and tetracycline hydrochloride in extemporaneously compounded oral liquids, *Am.J.Health Syst.Pharm.*, **1998**, *55*, 1804–1809.

# Bexarotene

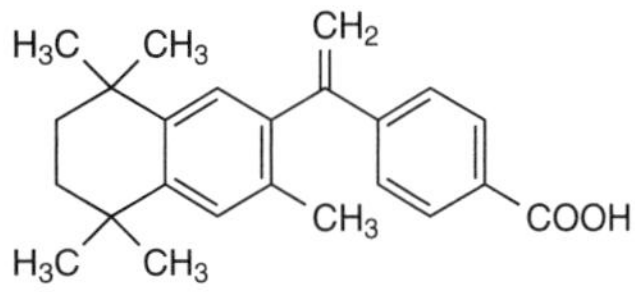


**Molecular formula:** $C_{24}H_{28}O_2$
**Molecular weight:** 348.48
**CAS Registry No:** 153559-49-0
**Merck Index:** 13, 1196

## SAMPLE

**Matrix:** bile, blood, microsomal incubation
**Sample preparation:** Mix plasma with 5 vol of MeOH, cool to −20°, centrifuge at 4°, evaporate the supernatant to dryness under reduced pressure, reconstitute the residue with MeCN:10 mM ammonium acetate 40:60 containing 1% acetic acid, inject an aliquot. Mix 1 mL (?) microsomal incubation with 1.5 mL ice-cold EtOH, cool at <5° for 1 h, centrifuge, evaporate the supernatant to dryness under reduced pressure, reconstitute the residue with MeCN:10 mM ammonium acetate 40:60 containing 1% acetic acid, inject an aliquot. Dilute bile twofold with 10 mM ammonium acetate, centrifuge, inject an aliquot.

## HPLC VARIABLES

**Column:** 250 × 4.6 5 μm Microsorb MV C18
**Column temperature:** 40
**Mobile phase:** Gradient. A:B from 20:80 to 80:20 over 20 min, maintain at 80:20 for 15 min. A was MeCN:acetic acid 100:1. B was 10 mM ammonium acetate:acetic acid 100:1.
**Detector:** UV 262

## CHROMATOGRAM

**Retention time:** 32

## OTHER SUBSTANCES

**Extracted:** metabolites

## KEY WORDS

dog; plasma; rat

## REFERENCE

Howell, S.R.; Shirley, M.A.; Grese, T.A.; Neel, D.A.; Wells, K.E.; Ulm, E.H. Bexarotene metabolism in rat, dog, and human, synthesis of oxidative metabolites, and in vitro activity at retinoid receptors, *Drug Metab.Dispos.*, **2001**, *29*, 990–998.

## SAMPLE

**Matrix:** blood
**Sample preparation:** Mix 1 mL plasma with 20 μL 2.5 μg/mL IS, vortex for 10 s, add five 250 μL portions of MeCN and one 1 mL portion of 500 mM HCl with each addition followed by vortexing for a short period. Add 5 mL *n*-hexane:isoamyl alcohol 98:2, rotate at 45 rpm for 20 min, centrifuge at 3200 g for 10 min, freeze, remove the organic layer. Evaporate the organic layer to dryness under a stream of nitrogen at 40°, reconstitute the residue with 400 μL MeCN, vortex for 30 s, sonicate for 2 min, add 100 μL 10 mM ammonium acetate, centrifuge at 13 000 g for 5 min, inject a 50 μL aliquot.

## HPLC VARIABLES

**Column:** 150 × 4.6 5 μm Inertsil ODS-2
**Column temperature:** 35
**Mobile phase:** MeCN:10 mM ammonium acetate:acetic acid 80:20:0.8

**Injection volume:** 50
**Detector:** F ex 260 em 430

## CHROMATOGRAM

**Retention time:** 8.5
**Internal standard:** LG100130 (4-[1-(5,6,7,8-tetrahydro-3-(1-methyl)ethyl-5,5,8,8-tetramethyl-2-naphthalenyl)ethenyl]benzoic acid, Ligand Pharmaceuticals, San Diego) (11)
**Limit of detection:** 0.3 ng/mL

## OTHER SUBSTANCES

**Extracted:** metabolites

## KEY WORDS

pharmacokinetics; plasma

## REFERENCE

Van de Merbel, N.C.; van Veen, J.H.; Wilkens, G.; Loewen, G. Validated liquid chromatographic method for the determination of bexarotene in human plasma, *J.Chromatogr.B*, **2002**, *775*, 189–195.

# Biapenem

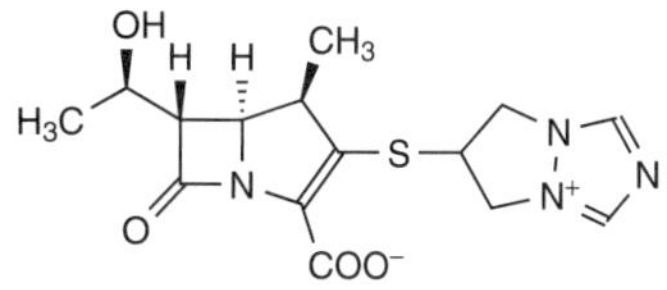


**Molecular formula:** $C_{15}H_{18}N_4O_4S$
**Molecular weight:** 350.40
**CAS Registry No:** 120410-24-4
**Merck Index:** 13, 1200

## SAMPLE

**Matrix:** blood, urine
**Sample preparation:** Mix plasma or urine with an equal volume of 1 M pH 7.0 MOPS (3-(*N*-morpholino)propanesulfonic acid) buffer. Dilute plasma with 30% ammonium sulfate solution, centrifuge at 3000 rpm.

## HPLC VARIABLES

**Column:** 150 × 4.6 TSK gel ODS 80TM (Tosoh)
**Mobile phase:** MeCN:100 mM acetate buffer 1.5:98.5 (plasma) or MeCN:MeOH:sodium 1-octanesulfonate solution:acetic acid 110:12:480:3 (urine)
**Flow rate:** 1.2 (plasma), 1.1 (urine)
**Detector:** UV 300 (plasma), UV 310 (urine)

## CHROMATOGRAM

**Internal standard:** 5-hydroxyindole-3-acetic acid (plasma), *o*-nitroacetanilide (urine)
**Limit of detection:** 100 ng/mL (plasma), 1 μg/mL (urine)

## KEY WORDS

plasma

## REFERENCE

Kozawa, O.; Uematsu, T.; Matsuno, H.; Niwa, M.; Takiguchi, Y.; Matsumoto, S.; Minamoto, M.; Niida, Y.; Yokokawa, M.; Nagashima, S.; Kanamaru, M. Pharmacokinetics and safety of a new parenteral carbapenem antibiotic, biapenem (L-627), in elderly subjects, *Antimicrob.Agents Chemother.*, **1998**, *42*, 1433–1436.

## SAMPLE

**Matrix:** solutions
**Sample preparation:** Mix an aliquot of a solution in 50 mM pH 7.0 MOPS (3-(*N*-morpholino)propanesulfonic acid) buffer with 2 aliquots of MeCN, vortex. Add a volume of chloroform equal to 50% of the total volume (Caution! Chloroform is a carcinogen!), mix well, centrifuge at 5000 g for 10 min, inject a 20 μL aliquot of the supernatant.

## HPLC VARIABLES

**Guard column:** 100 × 4.6 Inertsil ODS-2
**Column:** 150 × 4.6 Inertsil ODS-2
**Mobile phase:** MeCN:50 mM pH 5.5 acetate buffer 2:100
**Flow rate:** 0.9
**Injection volume:** 20
**Detector:** UV 295

## CHROMATOGRAM

**Retention time:** 6.3

## REFERENCE

Hikida, M.; Kawashima, K.; Nishiki, K.; Furukawa, Y.; Nishizawa, K.; Saito, I.; Kuwao, S. Renal dehydropeptidase-I stability of LJC 10,627, a new carbapenem antibiotic, *Antimicrob.Agents Chemother.*, **1992**, *36*, 481–483.

# Bimatoprost

**Molecular formula:** $C_{25}H_{37}NO_4$
**Molecular weight:** 415.57
**CAS Registry No:** 155206-00-1

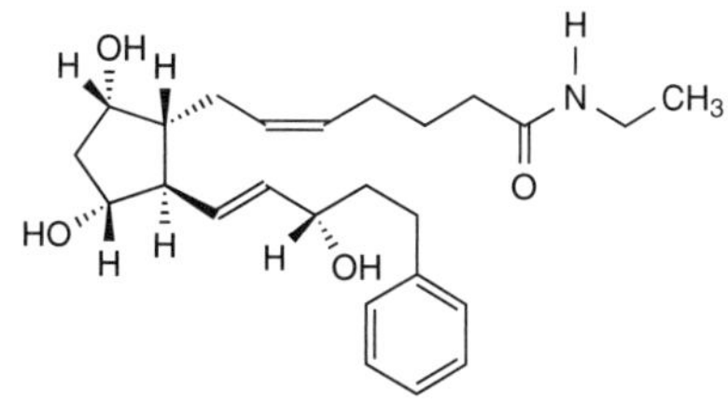


## SAMPLE
**Matrix:** solutions

## HPLC VARIABLES
**Column:** 150 × 4.6 5 μm Ultrasphere IP
**Mobile phase:** Gradient. A:B 100:0 for 1 min, to 40:60 over 16 min (+ curved), maintain at 40:60 for 4 min, return to initial conditions over 1 min, re-equilibrate at initial conditions for 8 min. A was MeCN:20 mM pH 2.8 potassium phosphate buffer 20:80. B was MeCN:20 mM pH 2.8 potassium phosphate buffer 50:50.
**Flow rate:** 1
**Injection volume:** 20–100
**Detector:** UV, Radioactivity ($^{3}H$)

## CHROMATOGRAM
**Retention time:** 16

## OTHER SUBSTANCES
**Simultaneous:** metabolites

## REFERENCE
Woodward, D.F.; Krauss, A.H.-P.; Chen, J.; Liang, Y.; Li, C.; Protzman, C.E.; Bogardus, A.; Chen, R.; Kedzie, K.M.; Krauss, H.A.; Gil, D.W.; Kharlamb, A.; Wheeler, L.A.; Babusis, D.; Welty, D.; Tang-Liu, D.D.-S.; Cherukury, M.; Andrews, S.W.; Burk, R.M.; Garst, M.E. Pharmacological characterization of a novel antiglaucoma agent, bimatoprost (AGN 192024), *J.Pharmacol.Exp.Ther.*, **2003**, *305*, 772–785.

# Bioresmethrin

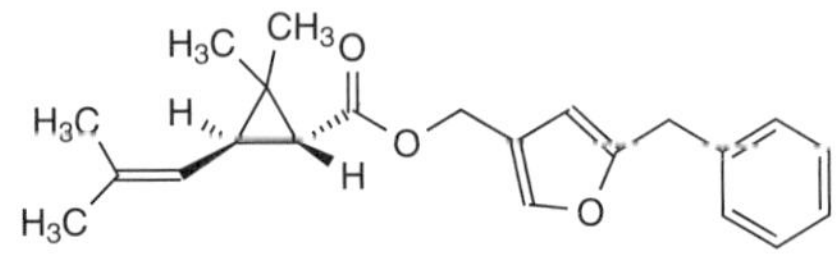

**Molecular formula:** $C_{22}H_{26}O_3$
**Molecular weight:** 338.44
**CAS Registry No:** 28434-01-7
**Merck Index:** 13, 1230

## SAMPLE

**Matrix:** urine
**Sample preparation:** Add 4 g solid NaCl, 3.5 mL MeCN, and 5 mL saturated NaCl solution to 5 mL MeCN, shake for 1 min. Remove the MeCN layer and extract the aqueous layer with 1 mL MeCN. Combine the MeCN layers and adjust to a known volume (0.5–1 mL), mix, filter (0.45 μm), inject a 40 μL aliquot.

## HPLC VARIABLES

**Column:** 150 × 3 3 μm Luna C18(2) (Phenomenex)
**Column temperature:** 30
**Mobile phase:** Gradient. MeCN:water 10:90 for 1 min, to 90:10 over 30 min, maintain at 90:10 for 4 min, to 100:0 over 1 min, maintain at 100:0 for 10 min, return to initial conditions over 1 min.
**Flow rate:** 0.5
**Injection volume:** 40
**Detector:** UV 235

## CHROMATOGRAM

**Retention time:** 35.2
**Limit of detection:** 5 ng/mL

## OTHER SUBSTANCES

**Extracted:** allethrin (31.8, LOD 5 ng/mL), bifenthrin (37, LOD 5 ng/mL), cyfluthrin (34.3, LOD 5 ng/mL), fenvalerate (35.3, LOD 2 ng/mL), *cis*-permethrin (35.7, LO 5 ng/mL), *trans*-permethrin (36.3, LOD 5 ng/mL), phenothrin (36.4, LOD 5 ng/mL), *m*-phenoxybenzyl alcohol (21, LOD 5 ng/mL), pyrethrin I (29.6, LOD 4 ng/mL), pyrethrin II (33.7, LOD 40 ng/mL), tetramethrin (31.4, LOD 5 ng/mL)

## REFERENCE

Loper, B.L.; Anderson, K.A. Determination of pyrethrin and pyrethroid pesticides in urine and water matrices by liquid chromatography with diode array detection, *J.AOAC Int.*, **2003**, *86*, 1236–1240.

# Bivalirudin

**Molecular formula:** $C_{98}H_{138}N_{24}O_{33}$
**Molecular weight:** 2180.28
**CAS Registry No:** 128270-60-0
**Merck Index:** 13, 1306

## SAMPLE

**Matrix:** blood, urine
**Sample preparation:** Condition a 1 mL 30 mg Oasis HLB SPE cartridge with 1 mL MeCN:trifluoroacetic acid 98:2 and 1 mL water. Plasma. Mix 200 μL plasma with 100 μL of a solution of IS in water, make up to 1100 μL with water, add 1 mL to the SPE cartridge, wash with 500 μL water, wash with 500 μL MeOH:water 30:70, elute with 1 mL MeCN:trifluoroacetic acid 98:2, evaporate the eluate to dryness under a stream of nitrogen at 40°, reconstitute the residue with 100 μL MeOH:water 50:50, centrifuge at 10 000 rpm at 4° for 10 min, inject a 30 μL aliquot. Urine. Mix 50 μL urine with 400 μL of a solution of IS in water, make up to 500 μL with water, inject a 10 μL aliquot.

## HPLC VARIABLES

**Column:** 30 × 4.6 5 μm Hypersil BDS C18
**Mobile phase:** Gradient. A:B from 30:70 to 100:0 over 0.5 min, maintain at 100:0 for 2 min, return to initial conditions over 0.2 min, re-equilibrate at initial conditions for 1.5 min. A was MeCN:0.1% formic acid 60:40. B was 0.1% formic acid.
**Injection volume:** 10–30
**Detector:** MS, Perkin Elmer, positive ion mode, TurboIonSpray heater 400°, m/z 1090.9 to 227.1, dwell time 400 ms, ionspray voltage 5.2 kV, ring voltage 230 V, nebulizer gas nitrogen at 12 units, TurboIonSpray gas nitrogen at 7.0 L/min, collision gas nitrogen at 3 units, curtain gas nitrogen at 10 units, deflector −100 V, electron multiplier 2400–2700 V, collision energy −83 V

## CHROMATOGRAM

**Internal standard:** Gln-9-hirulog (Polypeptide Laboratories, Torrance CA) (m/z 1097.9 to 199.1)

## KEY WORDS

pharmacokinetics; plasma; SPE

## REFERENCE

Robson, R.; White, H.; Aylward, P.; Frampton, C. Bivalirudin pharmacokinetics and pharmacodynamics: effect of renal function, dose, and gender, *Clin.Pharmacol.Ther.*, **2002**, *71*, 433–439.

# Boldenone

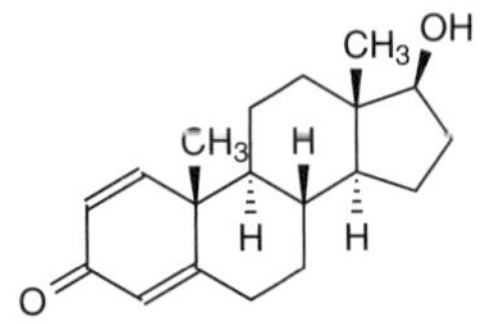

**Molecular formula:** $C_{19}H_{26}O_2$
**Molecular weight::** 286.41
**CAS Registry No:** 846-48-0
**Merck Index:** 13, 1315

## SAMPLE

**Matrix:** blood, urine
**Sample preparation:** Adjust the pH of 3 mL serum to 9.6, add to an Extrelut 3 SPE column, extract with 15 mL diethyl ether, evaporate the ether to dryness, reconstitute the residue with 50 μL mobile phase containing 10 ng IS, inject an aliquot. Adjust the pH of 3 mL urine to 9.6 with sodium bicarbonate:sodium carbonate 2:1, add to an Extrelut 3 SPE column, extract with 15 mL diethyl ether, evaporate the ether to dryness, reconstitute the residue with 50 μL mobile phase containing 10 ng IS, inject the whole amount of aliquot.

## HPLC VARIABLES

**Column:** 250 × 4.6 5 μm Ultrasphere ODS
**Mobile phase:** MeCN:water 37:63 (A) or THF:water 30:70 (B)
**Flow rate:** 1
**Detector:** ELISA (collect 300 μL fractions)

## CHROMATOGRAM

**Retention time:** 13 (A), 10 (B)
**Internal standard:** methandienone (17 (A), 12 (B))

## OTHER SUBSTANCES

**Extracted:** testosterone (20 (A), 15 (B))

## KEY WORDS

horse; serum; SPE

## REFERENCE

Hagedorn, H.-W.; Schulz, R.; Jaeschke, G. Identification and verification of the anabolic steroid boldenone in equine blood and urine by HPLC/ELISA, *Biomed.Chromatogr.*, **1994**, *8*, 63–68.

# Bosentan

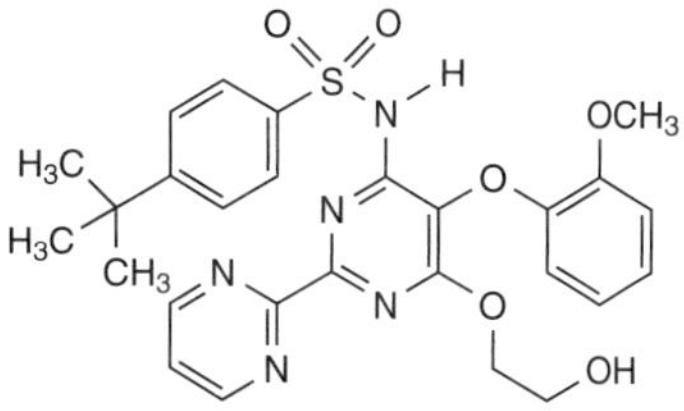

**Molecular formula:** $C_{27}H_{29}N_5O_6S$
**Molecular weight:** 551.62
**CAS Registry No:** 147536-97-8, 150726-52-6 (Na salt)
**Merck Index:** 13, 1341

## SAMPLE

**Matrix:** blood
**Sample preparation:** Mix 500 μL plasma with 10 μL 5 μg/mL IS in EtOH, add 750 μL MeCN, vortex, store at 5° for 10 min. Add the supernatant at pH 11 to 7 mL dichloromethane, shake on a rotating shaker for 20 min, centrifuge for 5 min. Evaporate the organic layer to dryness under reduced pressure, reconstitute the residue with 1 mL MeCN, centrifuge for 5 min. Evaporate the supernatant to dryness under reduced pressure, reconstitute the residue with 100 μL MeCN:5 mM ammonium acetate:acetic acid 30:70:1, inject a 30–90 μL aliquot.

## HPLC VARIABLES

**Guard column:** 10 × 2 4 μm Superspher RP-18
**Column:** 125 × 2 4 μm Superspher RP-18
**Column temperature:** 35
**Mobile phase:** MeCN:5 mM ammonium acetate:acetic acid 75:25:1 or 70:30:1
**Injection volume:** 30–90
**Detector:** UV 270; MS, Perkin-Elmer SCIEX API III plus triple quadrupole, ionspray, collision gas argon, collision energy 50 eV

## CHROMATOGRAM

**Retention time:** 3.2
**Internal standard:** tetradeutero bosentan; Ro 47–8761 (4-cyclopropyl-*N*-[6-(2-hydroxyethoxy)-5-(2-methoxyphenoxy)[2,2′-bipyrimidin]-4-yl]benzenesulfonamide)
**Limit of quantitation:** 0.5 ng/mL (MS)

## KEY WORDS

plasma

## REFERENCE

Lausecker, B.; Hopfgartner, G. Determination of an endothelin receptor antagonist in human plasma by narrow-bore liquid chromatography and ionspray tandem mass spectrometry, *J.Chromatogr.A*, **1995**, *712*, 75–83.

## SAMPLE

**Matrix:** blood, tissue
**Sample preparation:** Homogenize (Polytron) liver with 2 vol of 100 mM phosphate buffer at 8000 rpm for 10–15 s. Mix 250 μL plasma or liver homogenate with 50 μL 2 μg/mL IS in MeCN:5 mM ammonium acetate:acetic acid 10:90:1, add 1 mL MeCN:EtOH 50:50, mix, centrifuge at 14 000 g for 10 min. Remove the supernatant and add it to 1 mL pH 4.0 buffer (Merck Tetrisol citrate/HCl), mix, add 6 mL *n*-chlorobutane:dichloromethane 80:20, rotate at 40 rpm for 20 min. Evaporate the organic layer to dryness under reduced pressure, reconstitute the residue with 1 mL MeCN:5 mM ammonium acetate:acetic acid 10:90:1, inject a 950 μL aliquot onto column A and elute to waste with mobile phase A. After 2 min, elute the contents of column A onto column B with mobile phase B. (After the elution of the analytes, flush column B with MeCN:acetic acid 99:1 at 0.35 mL/min for 1 min and then re-equilibrate with initial mobile phase B at 0.35 mL/min for 30 s. Wash column A in backflush and forward flush

mode with MeCN:MeOH:5 mM ammonium acetate:acetic acid 45:45:10:1 at 1 mL/min for 3.5 min, then re-equilibrate with mobile phase A at 1 mL/min for 2.5 min.)

---

## HPLC VARIABLES

**Column:** A 25 × 4 5 μm Superspher RP-Select B; B 10 × 2 Superspher RP−18 + 150 × 2.1 Symmetry RP-18

**Mobile phase:** A 5 mM ammonium acetate containing 1% acetic acid; B Gradient. C:D from 50:50 to 10:90 over 4.5 min. C was MeCN:MeOH:5 mM ammonium acetate:acetic acid 25:25:50:1. D was MeCN:MeOH:5 mM ammonium acetate:acetic acid 45:45:10:1.

**Flow rate:** A 1; B 0.25

**Injection volume:** 950

**Detector:** MS, Perkin-Elmer SCIEX API 300 triple quadrupole, ionspray, collision gas argon, collision energy 50 eV, m/z 552 to 202

---

## CHROMATOGRAM

**Retention time:** 6.8

**Internal standard:** tetradeutero bosentan

**Limit of quantitation:** 1 ng/mL

---

## OTHER SUBSTANCES

**Extracted:** metabolites

---

## KEY WORDS

column-switching; dog; human; liver; plasma; rat

---

## REFERENCE

Lausecker, B.; Hess, B.; Fischer, G.; Mueller, M.; Hopfgartner, G. Simultaneous determination of bosentan and its three major metabolites in various biological matrices and species using narrow bore liquid chromatography with ion spray tandem mass spectrometric detection, *J.Chromatogr.B*, **2000**, *749*, 67–83.

---

## SAMPLE

**Matrix:** blood

**Sample preparation:** Mix 250 μL plasma with 50 μL 2 μg/mL IS in MeCN:5 mM ammonium acetate:acetic acid 10:90:1, add 750 μL MeOH. Mix the supernatant with 2 mL 50 mM pH 10 ammonium acetate buffer, add to a 30 mg Oasis SPE cartridge, wash with 1 mL water, wash with 2 mL 20 mM phosphoric acid, wash with 2.1 mL MeOH:water 20:80, wash with 1 mL water, elute with MeCN:MeOH:triethylamine 30:70:2. Evaporate the eluate to dryness, reconstitute the residue with 150 μL MeCN:5 mM ammonium acetate:acetic acid 10:90:1, inject an aliquot.

---

## HPLC VARIABLES

**Column:** 125 × 2 Superspher

**Mobile phase:** MeCN:MeOH:5 mM ammonium acetate:acetic acid 37.5:37.5:25:1

**Flow rate:** 0.25

**Detector:** MS, Perkin-Elmer SCIEX API 300 triple quadrupole, ionspray, collision gas argon, collision energy 50 eV, m/z 552 to 202

---

## CHROMATOGRAM

**Retention time:** 3

**Internal standard:** tetradeutero bosentan

**Limit of quantitation:** 1 ng/mL

---

## OTHER SUBSTANCES

**Extracted:** metabolites

## KEY WORDS

human; plasma; rat; SPE

## REFERENCE

Lausecker, B.; Hess, B.; Fischer, G.; Mueller, M.; Hopfgartner, G. Simultaneous determination of bosentan and its three major metabolites in various biological matrices and species using narrow bore liquid chromatography with ion spray tandem mass spectrometric detection, *J.Chromatogr.B*, **2000**, *749*, 67–83.

## ANNOTATED BIBLIOGRAPHY

Dell, D.; Lausecker, B.; Hopfgartner, G.; van Giersbergen, P.L.M.; Dingemanse, J. Evolving bioanalytical methods for the cardiovascular drug bosentan, *Chromatographia*, **2002**, *55*, S115–S119. [review]

Ubeaud, G.; Schmitt, C.; Jaeck, D.; Lave, T.; Coassolo, P. Bosentan, a new endothelin receptor antagonist: prediction of the systemic plasma clearance in man from combined in vivo and in vitro data, *Xenobiotica*, **1995**, *25*, 1381–1390. [gradient; microsomal incubations; rat; mouse; dog; rabbit; metabolites]

Weber, C.; Schmitt, R.; Birnboeck, H.; Hopfgartner, G.; van Marle, S.P.; Peeters, P.A.M.; Jonkman, J.H.G.; Jones, C.-R. Pharmacokinetics and pharmacodynamics of the endothelin-receptor antagonist bosentan in healthy human subjects, *Clin.Pharmacol.Ther.*, **1996**, *60*, 124–137. [LC-MS; UV detection; urine; plasma; LOQ 50 ng/mL]

# β-Boswellic acid

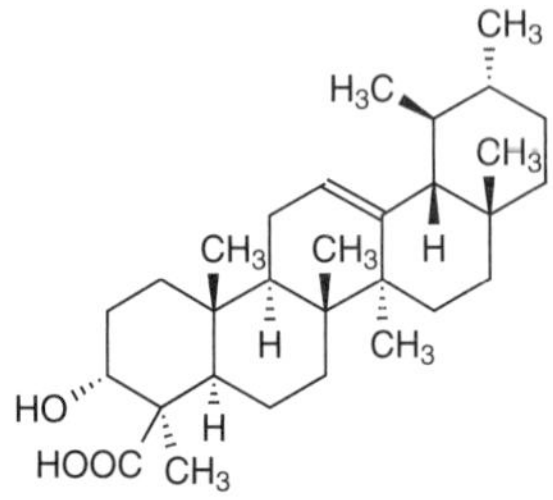


**Molecular formula:** $C_{30}H_{48}O_3$
**Molecular weight:** 456.70
**CAS Registry No:** 631-69-6
**Merck Index:** 13, 1343

## SAMPLE

**Matrix:** blood
**Sample preparation:** Add 1 mL plasma to an Extrelut NT SPE cartridge, let stand for 15 min, elute with THF:hexane:ethyl acetate:isopropanol 32:32:32:3. Evaporate the eluate to dryness under a stream of nitrogen at 60°, reconstitute the residue with 80 μL DMSO, add this solution to a 4 mL 150 mg Carbograph SPE cartridge (Alltech, preconditioned with 6 mL MeOH). Rinse the vial twice with 1 mL portions of MeOH:isopropanol 95:5 and add the rinses to the Carbograph SPE cartridge, wash with 3 mL MeOH:isopropanol 95:5, apply a vacuum for 5 s, elute with 2.5 mL cyclohexane:acetone 50:50. Evaporate the eluate to dryness under a stream of nitrogen at 60°, reconstitute the residue with 80 μL DMSO, inject an aliquot.

## HPLC VARIABLES

**Column:** 250 × 3 5 μm ReproSil-Pur 120 ODS-3 (Dr Maisch, Ammerbuch, Germany)
**Column temperature:** 28
**Mobile phase:** Gradient. A:B from 62:38 to 51:49 over 20 min, to 39:61 over 15 min, to 32:68 over 5 min, to 31:69 over 5 min, to 0:100 over 5 min, maintain at 0:100 for 10 min, re-equilibrate at initial conditions for 6 min. A was MeOH:water:acetic acid 80:20:0.2. B was MeOH:acetic acid 100:0.2.
**Flow rate:** 0.56 for 50 min, 0.9 for 10 min, 0.56 for 6 min
**Detector:** UV 210

## CHROMATOGRAM

**Retention time:** 32

## OTHER SUBSTANCES

**Extracted:** similar triterpenic acids

## KEY WORDS

plasma; SPE

## REFERENCE

Büchele, B.; Simmet, T. Analysis of 12 different pentacyclic triterpenic acids from frankincense in human plasma by high-performance liquid chromatography and photodiode array detection, *J.Chromatogr.B*, **2003**, *795*, 355–362.

## SAMPLE

**Matrix:** plant resin
**Sample preparation:** Sonicate 100–500 mg resin in 3 mL MeOH for 10 min, centrifuge, repeat extraction twice more. Combine the supernatants and make up to 10 mL with MeOH, filter (0.45 μm), inject a 10 μL aliquot.

## HPLC VARIABLES

**Column:** 150 × 4.6 4 μm Synergi MAX-RP 80 A (Phenomenex)
**Column temperature:** 60

**Mobile phase:** Gradient. A:B 65:35 for 7 min, to 90:10 over 21 min, maintain at 90:10 for 5 min, to 100:0 (step gradient), maintain at 100:0 for 5 min, re-equilibrate at initial conditions for 10 min. A was MeCN containing 0.05% phosphoric acid. B was water containing 0.05% phosphoric acid.
**Flow rate:** 1
**Injection volume:** 10
**Detector:** UV 210

## CHROMATOGRAM
**Retention time:** 25
**Limit of detection:** 270 ng/mL

## OTHER SUBSTANCES
**Simultaneous:** similar triterpenic acids

## REFERENCE
Ganzera, M.; Khan, I.A. A reversed phase high-performance liquid chromatography method for the analysis of boswellic acids in *Boswellia serrata*, *Planta Med.*, **2001**, *67*, 778–780.

# Brimonidine

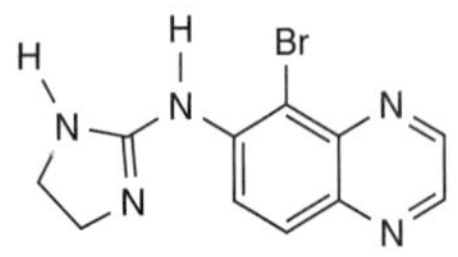

**Molecular formula:** $C_{11}H_{10}BrN_5$
**Molecular weight:** 292.14
**CAS Registry No:** 59803-98-4
**Merck Index:** 13, 1361

## SAMPLE

**Matrix:** aqueous humor, blood
**Sample preparation:** Mix 100–300 μL serum or aqueous humor with 2 vol buffer, centrifuge at 11 000 g for 10 min,, add the supernatant to a Sep-Pak C18 SPE cartridge, wash with 5 mL buffer, elute with 5 mL MeCN:buffer 50:50. Evaporate the eluate to dryness under a stream of nitrogen, reconstitute the residue with 100 μL mobile phase, inject a 10–20 μL aliquot. (The buffer was 10 mM triethylamine adjusted to pH 3.2 with phosphoric acid.)

## HPLC VARIABLES

**Guard column:** 20 × 4.6 RP-18 (Supelco)
**Column:** 250 × 4.6 5 μm Supelcosil LC-18
**Mobile phase:** MeCN:buffer 10:90 (The buffer was 10 mM triethylamine adjusted to pH 3.2 with phosphoric acid.)
**Flow rate:** 1
**Injection volume:** 10–20
**Detector:** UV 248

## CHROMATOGRAM

**Retention time:** 7
**Limit of detection:** 30 pg/mL

## OTHER SUBSTANCES

**Noninterfering:** benzalkonium chloride, cyclopentolate, phenylephrine, tropicamide

## KEY WORDS

serum; SPE

## REFERENCE

Karamanos, N.K.; Lamari, F.; Katsimpris, J.; Gartaganis, S. Development of an HPLC method for determining the alpha2-adrenergic receptor agonist brimonidine in blood serum and aqueous humor of the eye, *Biomed.Chromatogr.*, **1999**, *13*, 86–88.

## SAMPLE

**Matrix:** tissue
**Sample preparation:** Vortex <3–25 mg ocular tissue with 200 μL MeOH for 3 min, centrifuge at 3000 g for 10 min. Mix a 35 μL aliquot of the supernatant with 65 μL 0.01% trifluoroacetic acid in water, inject the whole amount.

## HPLC VARIABLES

**Column:** 300 × 3.9 15 μm Delta PAK C18 (Waters)
**Mobile phase:** MeOH:water:trifluoroacetic acid 35:65:0.01
**Flow rate:** 1
**Injection volume:** 100
**Detector:** UV 360

## CHROMATOGRAM

**Retention time:** 3.5
**Limit of detection:** 20 ng/mL

## OTHER SUBSTANCES

**Extracted:** mitomycin C (6, LOD 10 ng/mL), timolol (10, LOD 50 ng/mL)

## KEY WORDS

eye

## REFERENCE

Xiong, X.; Lim, B.A.; Lat-Luna, M.; Chew, P.; Tan, D. Quantitation of mitomycin C in human ocular tissues by high-performance liquid chromatography-photo-diode array detection, *J.Chromatogr.B*, **2001**, *755*, 65–72.

## ANNOTATED BIBLIOGRAPHY

Acheampong, A.A.; Shackleton, M.; Tang-Liu, D.D.-S. Comparative ocular pharmacokinetics of brimonidine after a single dose application to the eyes of albino and pigmented rabbits, *Drug Metab.Dispos.*, **1995**, *23*, 708–712. [UV detection; radioactivity detection]

Acheampong, A.A.; Chien, D.-S.; Lam, S.; Vekich, S.; Breau, A.; Usansky, J.; Harcourt, D.; Munk, S.A.; Nguyen, H.; Garst, M.; Tang-Liu, D. Characterization of brimonidine metabolism with rat, rabbit, dog, monkey and human liver fractions and rabbit liver aldehyde oxidase, *Xenobiotica*, **1996**, *26*, 1035–1055.

Acheampong, A.A.; Shackleton, M.; John, B.; Burke, J.; Wheeler, L.; Tang-Liu, D. Distribution of brimonidine into anterior and posterior tissues of monkey, rabbit, and rat eyes, *Drug Metab.Dispos.*, **2002**, *30*, 421–429.

# Bromfenac

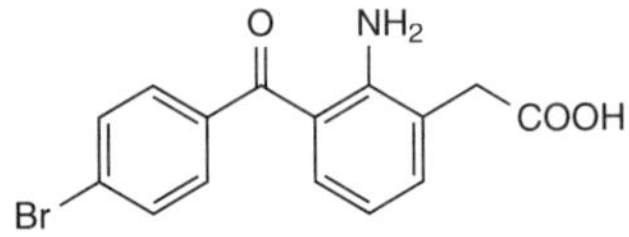


**Molecular formula:** $C_{15}H_{12}BrNO_3$
**Molecular weight:** 334.17
**CAS Registry No:** 91714-94-2
**Merck Index:** 13, 1374

## SAMPLE

**Matrix:** blood, urine
**Sample preparation:** Mix 1 mL plasma with 1 mL MeCN, filter (0.2 μm) by centrifuging at 800 g for 15 min, inject a 500 μL aliquot of the filtrate. Mix 1 mL urine with 1 mL 100 mM pH 5 ammonium acetate buffer and 100 μL Glusulase (40 000 units), heat at 37° for 16 h, centrifuge at 800 g for 15 min, inject a 500 μL aliquot of the supernatant.

## HPLC VARIABLES

**Column:** 150 × 4.6 5 μm HiChrom (Regis) (for electrospray negative ion MS (Finnigan-MAT TSQ 700) detection use 250 × 2 C18 DB (Supelco) column with the gradient below at 0.2 mL/min)
**Mobile phase:** Gradient. MeCN:100 mM pH 4.9 ammonium acetate buffer from 10:90 to 45:55 over 50 min, maintain at 45:55 for 10 min, re-equilibrate at initial conditions for 15 min.
**Flow rate:** 1
**Injection volume:** 500
**Detector:** UV 270

## CHROMATOGRAM

**Retention time:** 42

## OTHER SUBSTANCES

**Extracted:** metabolites

## KEY WORDS

plasma

## REFERENCE

Osman, M.; Chandrasekaran, A.; Chan, K.; Scatina, J.; Ermer, J.; Cevallos, W.; Sisenwine, S.F. Metabolic disposition of $^{14}$C-bromfenac in healthy male volunteers, *J.Clin.Pharmacol.*, **1998**, *38*, 744–752.

## SAMPLE

**Matrix:** blood
**Sample preparation:** Mix 500 μL plasma with 200 μL 1 μg/mL IS in water, add 7 mL hexane, add 500 μL 5% ammonium hydroxide, shake rapidly on a reciprocating shaker for 10 min, centrifuge at 550 g for 5 min, discard the organic layer. Add 1 mL 2 M HCl to the aqueous layer, mix briefly, let stand for 20 min. Add 7 mL hexane, shake rapidly for 10 min, centrifuge at 550 g for 5 min. Evaporate the organic layer to dryness under a stream of nitrogen at 60°, reconstitute the residue with 200 μL mobile phase, inject a 100 μL aliquot.

## HPLC VARIABLES

**Column:** 300 × 3.9 10 μm μBondapak C18
**Mobile phase:** MeCN:THF:50 mM pH 6.5 sodium acetate buffer 39:6:55
**Flow rate:** 1.5
**Injection volume:** 100
**Detector:** UV 270

---

## CHROMATOGRAM

**Retention time:** 9.5
**Internal standard:** 2-amino-3-(4-chlorobenzoyl)benzeneacetic acid (chloro analogue) (8.5)
**Limit of quantitation:** 30 ng/mL

---

## KEY WORDS

pharmacokinetics; plasma; rat

---

## REFERENCE

Osman, M.A.; Dunning, L.K.; Cheng, L.K.; Wright, G.J. Determination of bromfenac in plasma by high-performance liquid chromatography, *J.Chromatogr.*, **1989**, *489*, 452–458.

# Brovincamine

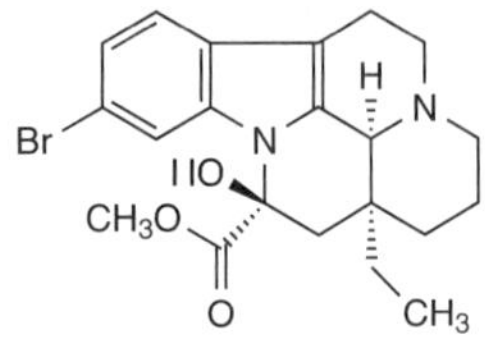


**Molecular formula:** $C_{21}H_{25}BrN_2O_3$
**Molecular weight:** 433.35
**CAS Registry No:** 57475-17-9
**Merck Index:** 13, 1436

## SAMPLE

**Matrix:** blood
**Sample preparation:** Mix 1 mL plasma with 40 μL 10 μg/mL IS in MeCN, add 500 μL ammonia (0.88):water 1:10, mix, add 5 mL diethyl ether, rotate for 10 min, centrifuge at 2000 g for 10 min. Evaporate the organic layer to dryness under a stream of nitrogen at 37°, rinse the walls of the tube with ether, again evaporate to dryness, reconstitute the residue with 50 μL mobile phase, centrifuge at 2000 g for 10 min, inject an aliquot.

## HPLC VARIABLES

**Guard column:** 70 × 2 25–37 μm Co:Pell ODS (Whatman)
**Column:** 300 × 4 10 μm μBondapak C18
**Mobile phase:** MeCN:buffer 35:65 (The buffer was 0.1% sodium dihydrogen phosphate adjusted to pH 3.5 with phosphoric acid.)
**Flow rate:** 2
**Detector:** UV 232

## CHROMATOGRAM

**Retention time:** 6.8
**Internal standard:** vincamine (3.6)
**Limit of detection:** 10 ng/mL

## KEY WORDS

plasma

## REFERENCE

Brodie, R.R.; Chasseaud, L.F. Determination of 11-bromovincamine in human plasma by high-performance liquid chromatography, *J.Chromatogr.*, **1982**, *228*, 413–417.

# Bucillamine

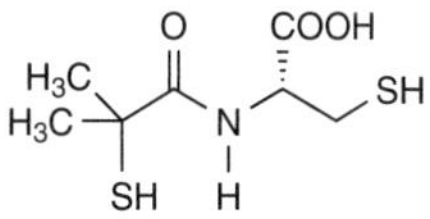


**Molecular formula:** $C_7H_{13}NO_3S_2$
**Molecular weight:** 223.32
**CAS Registry No:** 65002-17-7
**Merck Index:** 13, 1445

## SAMPLE

**Matrix:** blood
**Sample preparation:** Mix 50 μL blood with 50 μL buffer, immediately add 300 μL 0.167 mM *N*-(1-pyrenyl)maleimide in MeCN, vortex, let stand at room temperature for 15 min, add 5 μL 167 mM HCl, centrifuge at 12 000 rpm for 2 min, inject an aliquot of the supernatant. (The buffer was 10 mM pH 8.0 Tris-HCl containing 1 mM EDTA.)

## HPLC VARIABLES

**Column:** 150 × 4.6 5 μm C18 (Kanto, Japan)
**Mobile phase:** MeCN:EtOH:water:orthophosphoric acid:acetic acid 210:250:290:0.5:0.5
**Flow rate:** 0.5
**Injection volume:** 20
**Detector:** F ex 330 em 380

## CHROMATOGRAM

**Retention time:** 10
**Limit of detection:** 2.5 nM (S/N 3)
**Limit of quantitation:** 3 nM

## KEY WORDS

derivatization; whole blood

## REFERENCE

Higashi, Y.; Yamashiro, M.; Yamamoto, R.; Fujii, Y. HPLC analysis of bucillamine by derivatization with *N*-(1-pyrenyl)maleimide in human blood, *J.Liq.Chromatogr.Rel.Technol.*, **2003**, *26*, 3265–3275.

# Budipine

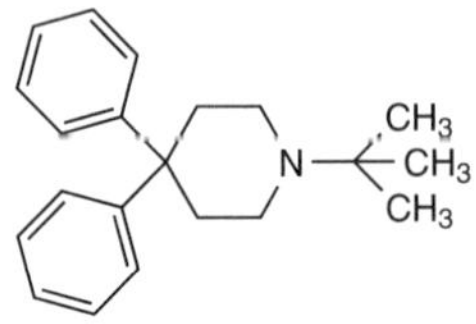


**Molecular formula:** $C_{21}H_{27}N$
**Molecular weight:** 293.44
**CAS Registry No:** 57982-78-2
**Merck Index:** 13, 1455

## SAMPLE

**Matrix:** urine

**Sample preparation:** Adjust pH of urine to 5.0 with acetic acid, add 1300 U/mL glusulase and 50 U/mL sulfatase, heat at 37° for 24 h, adjust pH to 9, pass through a column of Amberlite XAD-2 resin, wash with 3 vol of water, elute with MeOH. Evaporate the eluate to dryness, reconstitute the residue with water, adjust to pH 8, extract with ethyl acetate. Wash the ethyl acetate extract with one-tenth the volume of saturated NaCl, dry over anhydrous sodium sulfate, reconstitute with MeOH, inject an aliquot. Alternatively, adjust pH of urine to 5.0 with acetic acid, add glusulase/sulfatase, heat at 37°, adjust pH to 9, extract with dichloromethane:isopropanol 90:10. Evaporate the extract to dryness under reduced pressure, reconstitute, inject an aliquot. (Both the procedures are given.)

## HPLC VARIABLES

**Column:** 10 μm Viosfer C8
**Mobile phase:** MeCN:water:isopropylamine 79.96:20:0.04
**Detector:** UV 220

## OTHER SUBSTANCES

**Extracted:** metabolites

## KEY WORDS

rat; SPE

## REFERENCE

Caputo, O.; Grosa, G.; Ceruti, M.; Rocco, F.; Biglino, G. The metabolic fate of the anti-parkinsonian drug budipine in rats, *Eur.J.Drug Metab.Pharmacokinet.*, **1991**, *16*, 113–118.

# Bulaquine

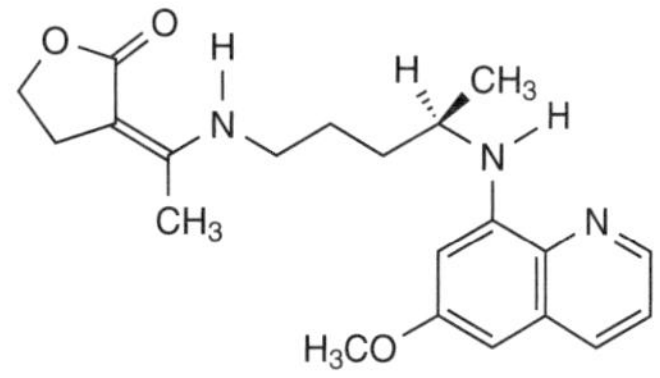


**Molecular formula:** $C_{21}H_{27}N_3O_3$
**Molecular weight:** 369.46
**CAS Registry No:** 223661-25-4

## SAMPLE

**Matrix:** formulations
**Sample preparation:** Extract capsule fill containing 5 mg bulaquine three times with 3 mL MeOH:dimethyloctylamine 99:1. Combine the extracts and make up to 10 mL with the same solvent, filter, dilute 500 μL of the filtrate to 25 mL with MeOH, inject an aliquot.

## HPLC VARIABLES

**Column:** 250 × 4 5 μm Lichrospher RP select-B C8
**Mobile phase:** MeCN:10 mM pH 5.6 sodium acetate buffer 55:45
**Flow rate:** 1
**Injection volume:** 20
**Detector:** UV 265

## CHROMATOGRAM

**Retention time:** 17.26
**Limit of detection:** 400 ng/mL
**Limit of quantitation:** 2 μg/mL

## OTHER SUBSTANCES

**Simultaneous:** chloroquine (4.22), primaquine (5.72)

## KEY WORDS

capsules; comparison with HPTLC

## REFERENCE

Dwivedi, A.K.; Saxena, D.; Singh, S. HPLC and HPTLC assays for the antimalarial agents chloroquine, primaquine and bulaquine, *J.Pharm.Biomed.Anal.*, **2003**, *33*, 851–858.

## SAMPLE

**Matrix:** blood
**Sample preparation:** Mix 500 μL plasma with 50 μL 1 M KOH, add 25 μL 20 μg/mL IS in MeCN, add 3 mL extraction solvent, vortex for 1 min, centrifuge at 1000 g for 10 min, freeze the aqueous layer, remove the organic layer, repeat the extraction. Evaporate the combined organic layers to dryness under reduced pressure, reconstitute the residue with 200 μL MeCN:50 mM pH 7.0 ammonium acetate buffer 50:50, centrifuge, inject an aliquot of the supernatant. (Extraction solvent was *n*-hexane:isopropanol:dimethyloctylamine 98:2:0.1.)

## HPLC VARIABLES

**Guard column:** 30 × 4.6 5 μm Spheri-5 cyano
**Column:** 220 × 4.6 5 μm Spheri-5 cyano
**Mobile phase:** Gradient. A:B from 55:45 to 90:10 over 15 min, maintain at 90:10 for 2 min, return to initial conditions over 2 min. A was MeCN:50 mM pH 6.0 ammonium acetate buffer 65:35. B was 50 mM pH 6.0 ammonium acetate buffer.

**Flow rate:** 1
**Injection volume:** 50
**Detector:** UV 261

## CHROMATOGRAM

**Retention time:** 8
**Internal standard:** 3-bromoprimaquine diphosphate (13.5)
**Limit of detection:** 10 ng/mL
**Limit of quantitation:** 20 ng/mL

## OTHER SUBSTANCES

**Extracted:** primaquine (11)

## KEY WORDS

pharmacokinetics; plasma; rabbit

## REFERENCE

Lal, J.; Mehrotra, N.; Gupta, R.C. Analysis and pharmacokinetics of bulaquine and its major metabolite primaquine in rabbits using an LC-UV method – a pilot study, *J.Pharm.Biomed.Anal.*, **2003**, *32*, 141–150.

## ANNOTATED BIBLIOGRAPHY

Nitin, M.; Rajanikanth, M.; Lal, J.; Madhusudanan, K.P.; Gupta, R.C. Liquid chromatography-tandem mass spectrometric assay with a novel method of quantitation for the simultaneous determination of bulaquine and its metabolite, primaquine, in monkey plasma, *J.Chromatogr.B*, **2003**, *793*, 253–263.

# Butacaine

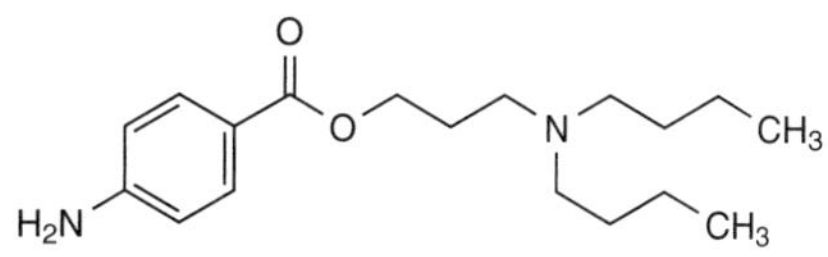

**Molecular formula:** $C_{18}H_{30}N_2O_2$
**Molecular weight:** 306.44
**CAS Registry No:** 149-16-6
**Merck Index:** 13, 1495

## SAMPLE
**Matrix:** blood
**Sample preparation:** Buffer plasma with 100 mM pH 7.0 phosphate buffer, inject a 300 µL aliquot onto column A and elute to waste with mobile phase A; after 11 min, backflush the contents of column A onto column B with mobile phase B. (Re-equilibrate column A with mobile phase A at 1.8 mL/min for 3.3 min.)

## HPLC VARIABLES
**Column:** A 25 × 4 LiChrospher RP-18 ADS; B 4 × 4 5 µm LiChrospher 100 RP−18 + 125 × 3 5 µm LiChrospher 100 RP-18
**Column temperature:** 35
**Mobile phase:** A 100 mM pH 7.0 phosphate buffer; B MeCN:50 mM pH 7.0 phosphate buffer 50:50
**Flow rate:** 0.6
**Injection volume:** 300
**Detector:** UV 210

## KEY WORDS
column-switching; plasma

## REFERENCE
Baeyens, W.R.G.; Van Der Weken, G.; Haustraete, J.; Smet, E. Application of the ADS precolumn-switching technique to the separation of some drugs from plasma, *Biomed.Chromatogr.*, **2000**, *14*, 61–63.

## SAMPLE
**Matrix:** blood, urine
**Sample preparation:** 2 mL Whole blood, plasma, or urine + 1 mL saturated sodium carbonate + 20 µL 100 µg/mL tetracaine, add to a 3 mL Extrelut SPE cartridge, elute with 15 mL dichloromethane. Evaporate eluate to dryness under a stream of nitrogen at 40°, reconstitute in 100 µL mobile phase, inject a 40 µL aliquot.

## HPLC VARIABLES
**Guard column:** 5 × 6 µBondapak Guard Pak
**Column:** 300 × 3.9 10 µm µBondapak C18
**Mobile phase:** MeCN:100 mM ammonium acetate 50:50
**Flow rate:** 1.5
**Injection volume:** 40
**Detector:** UV 280; F ex 280 em 350

## CHROMATOGRAM
**Retention time:** 10
**Internal standard:** tetracaine (14)
**Limit of detection:** 20 ng/mL

## OTHER SUBSTANCES
**Extracted:** *p*-aminobenzoic acid (4), procaine (5)

**Also analyzed:** articaine, prilocaine, *o*-toluidine, lidocaine, bupivacaine, etidocaine, dibucaine, caffeine, amphetamine, ephedrine, epinephrine, morphine, monoacetylmorphine, diamorphine, ethylmorphine, codeine, acetylcodeine

---

## KEY WORDS

plasma; SPE; whole blood

---

## REFERENCE

Rop, P.P.; Grimaldi, F.; Bresson, M.; Fornaris, M.; Viala, A. Liquid chromatographic analysis of cocaine, benzoylecgonine, local anaesthetic agents and some of their metabolites in biological fluids, *J.Liq.Chromatogr.*, **1993**, *16*, 2797–2811.

# Butamben

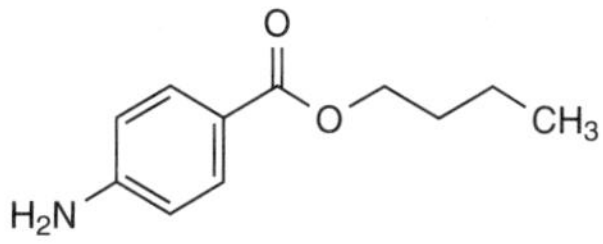


**Molecular formula:** $C_{11}H_{15}NO_2$
**Molecular weight:** 193.24
**CAS Registry No:** 94-25-7
**Merck Index:** 13, 1502

## SAMPLE

**Matrix:** blood
**Sample preparation:** Mix 500 μL plasma with 600 μL 100 mM KOH, add 1 mL 8.0 M urea, vortex gently, add 4 mL ethyl acetate, vortex for 1 min, centrifuge at 3000 g for 10 min. Evaporate the organic layer to dryness under a stream of nitrogen, reconstitute the residue with 200 μL MeOH, inject a 20 μL aliquot. (Perform the analyses under yellow light.)

## HPLC VARIABLES

**Guard column:** Corasil
**Column:** 250 × 4.6 5 μm Spherisorb ODS 2
**Mobile phase:** MeOH:10 mM pH 6.1 ammonium acetate buffer 62:38
**Flow rate:** 1
**Injection volume:** 20
**Detector:** UV 247

## CHROMATOGRAM

**Retention time:** 8.45

## OTHER SUBSTANCES

**Extracted:** nifedipine (7.14)

## KEY WORDS

butamben is IS; plasma

## REFERENCE

Thongnopnua, P.; Viwatwongsa, K. Quantitative analysis of nifedipine in plasma by high-performance liquid chromatography, *J.Pharm.Biomed.Anal.*, **1994**, *12*, 119–125.

# Butoconazole

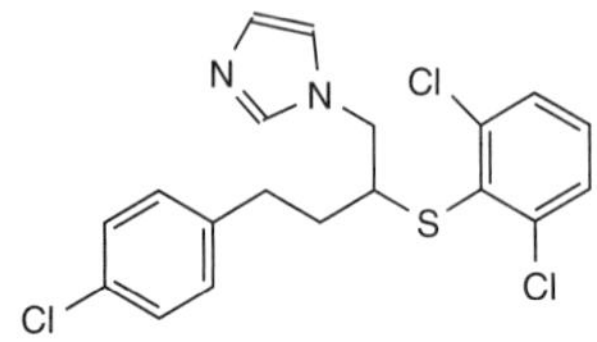

**Molecular formula:** $C_{19}H_{17}Cl_3N_2S$
**Molecular weight:** 411.78
**CAS Registry No:** 64872-76-0; 64872-77-1 (nitrate)
**Merck Index:** 13, 1525

## SAMPLE
**Matrix:** solutions
**Sample preparation:** Extract swab with MeOH or MeOH:THF 95:5, add IS, filter, inject an aliquot.

## HPLC VARIABLES
**Column:** 250 × 4.6 Partisil 10 SCX (Whatman)
**Mobile phase:** MeOH:135 mM pH 4.35 potassium acetate buffer 61:39
**Flow rate:** 1
**Detector:** UV 230

## CHROMATOGRAM
**Internal standard:** 1-benzylimidazole

## REFERENCE
Weinstein, L.; Henzel, M.R.; Tsina, I.W. Vaginal retention of 2% butoconazole nitrate cream: comparison of a standard and a sustained-release preparation, *Clin.Ther.*, **1994**, *16*, 930–934.

# Butyl flufenamate

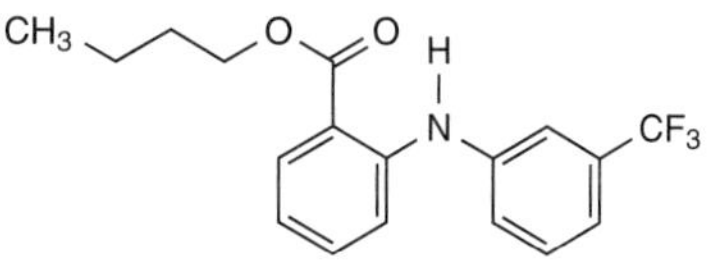

**Molecular formula:** $C_{18}H_{18}F_3NO_2$
**Molecular weight:** 337.34
**CAS Registry No:** 67330-25-0
**Merck Index:** 13, 4158

## SAMPLE

**Matrix:** formulations
**Sample preparation:** Add 100–300 mg gel ointment to 3 mL MeOH, mix vigorously, filter (0.2 μm), inject a 10 μL aliquot.

## HPLC VARIABLES

**Column:** 300 × 3.9 10 μm μBondapak C18
**Mobile phase:** MeOH
**Flow rate:** 1
**Injection volume:** 10
**Detector:** UV 280

## CHROMATOGRAM

**Retention time:** 4.3

## KEY WORDS

ointment

## REFERENCE

Yamamura, K.; Yamada, J.-I.; Yotsuyanagi, T. High-performance liquid chromatographic assay of anti-inflammatory drugs incorporated in gel ointments. Separation and stability testing, *J.Chromatogr.*, **1985**, *331*, 383–388.

# Cambendazole

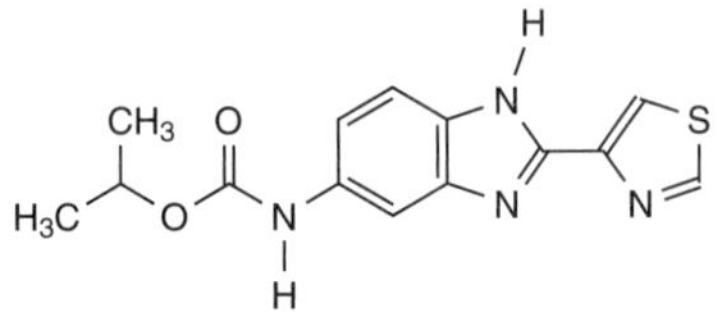

**Molecular formula:** $C_{14}H_{14}N_4O_2S$
**Molecular weight:** 302.36
**CAS Registry No:** 26097-80-3
**Merck Index:** 13, 1733

## SAMPLE

**Matrix:** abomasal fluid, blood, duodenal fluid, rumen fluid
**Sample preparation:** Shake 4 mL plasma, rumen fluid, abomasal fluid, or duodenal fluid with 4 mL pH 7.4 phosphate buffer and 20 mL ether on a rotary mixer for 10 min, remove 16 mL of the ether layer, add 20 mL ether, shake on a rotary mixer for 10 min, remove 20 mL of the ether layer. Combine the ether layers and evaporate them under a stream of nitrogen at 60° to dryness, reconstitute in 50 µL MeOH, sonicate, inject a 5 µL aliquot.

## HPLC VARIABLES

**Column:** 100 × 8 ODS Hypersil
**Mobile phase:** MeOH:50 mM ammonium carbonate 65:35
**Flow rate:** 1.5
**Injection volume:** 5
**Detector:** UV 292

## CHROMATOGRAM

**Retention time:** 4.3
**Limit of detection:** 20 ng/mL

## OTHER SUBSTANCES

**Extracted:** albendazole (7.5), fenbendazole (10), mebendazole (5), oxfendazole (3), oxibendazole (5.2), parbendazole (11), thiabendazole (3.7)

## KEY WORDS

plasma; sheep

## REFERENCE

Bogan, J.A.; Marriner, S. Analysis of benzimidazoles in body fluids by high-performance liquid chromatography, *J.Pharm.Sci.*, **1980**, *69*, 422–423.

## SAMPLE

**Matrix:** egg, tissue
**Sample preparation:** Condition an SDB (styrol-divinyl-benzene) SPE cartridge (Baker) with 3 mL MeOH and 3 mL water. Vortex 3 g muscle or liver with ? µL 10 µg/mL IS in MeOH, 1.5 g sodium sulfate, 500 µL 4 M potassium carbonate, and 5 mL ethyl acetate. Centrifuge at 2500 g for 5 min and remove the organic layer. Repeat the extraction. Combine the organic layers and evaporate them to dryness under a stream of nitrogen at 50° or under reduced pressure. Add 5 mL *n*-hexane to the residue, shake to get a good mixture, add 1 mL EtOH/HCl, vortex, centrifuge at 1000 g for 2 min, discard the upper layer. Evaporate the lower layer to dryness under reduced pressure, reconstitute the residue with 500 µL 10 mM pH 5.5 ammonium acetate and 500 µL MeOH, add to the SPE cartridge, wash with 3 mL water, wash with two 3 mL portions of MeOH:water 50:50, dry under vacuum, elute with 3 mL MeOH:ethyl acetate 20:80. Evaporate the eluate to dryness under reduced pressure, reconstitute the residue with 250 µL mobile phase, inject a 4 (ion spray) or 15 (turbo ion spray) µL aliquot. Preparation of egg samples is similar except that the mixture is vortexed for 10 s and sonicated for 15 min

after the addition of the ethyl acetate. (EtOH/HCl was made by mixing 66 mL EtOH with 33 mL 200 mM HCl.)

## HPLC VARIABLES

**Column:** 150 × 2.1 5 μm Zorbax RX
**Mobile phase:** MeCN:10 mM ammonium acetate containing 0.5% acetic acid 60:40
**Flow rate:** 0.04 (ion spray), 0.2 (turbo ion spray)
**Injection volume:** 4–15
**Detector:** MS, Perkin-Elmer SCIEX API 365 tandem, ion spray or turbo ion spray, m/z 303–261

## CHROMATOGRAM

**Retention time:** 1.9
**Internal standard:** 2-*n*-butylmercaptobenzimidazole (2.7) (m/z 207–151)
**Limit of detection:** 4 ng/g
**Limit of quantitation:** 6 ng/g

## OTHER SUBSTANCES

**Extracted:** albendazole (2.5, m/z 266–233, LOD 19 ng/g, LOQ 305 ng/g), febantel (4, m/z 447–382, LOD 10 ng/g, LOQ 15 ng/g), fenbendazole (2.9, m/z 300–268, LOD 5 ng/g, LOQ 7 ng/g), flubendazole (2.1, m/z 314–281, LOD 3 ng/g, LOQ 5 ng/g), mebendazole (2, m/z 296–264, LOD 3 ng/g, LOQ 5 ng/g), oxfendazole (1.7, m/z 316–159, LOD 5 ng/g, LOQ 8 ng/g), oxibendazole (2.2, m/z 250–176, LOD 4 ng/g, LOQ 6 ng/g), thiabendazole (2, m/z 202–175, LOD 3 ng/g, LOQ 5 ng/g), triclabendazole (5.5, m/z 359–343, LOD 6 ng/g, LOQ 9 ng/g)

## KEY WORDS

cow; egg; liver; muscle; pig; sheep; SPE

## REFERENCE

Balizs, G. Determination of benzimidazole residues using liquid chromatography and tandem mass spectrometry, *J.Chromatogr.B*, **1999**, *727*, 167–177.

## ANNOTATED BIBLIOGRAPHY

Danaher, M.; O'Keeffe, M.; Glennon, J.D. Development and optimisation of a method for the extraction of benzimidazoles from animal liver using supercritical carbon dioxide, *Anal.Chim.Acta*, **2003**, *483*, 313–324. [albendazole; fenbendazole; mebendazole; thiabendazole; oxibendazole; flubendazole; oxfendazole; netobimin; triclabendazole; cambendazole]

# Candesartan cilexetil

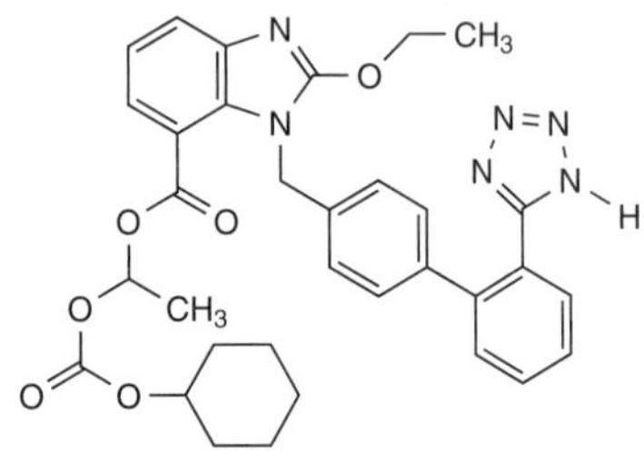

**Molecular formula:** $C_{33}H_{34}N_6O_6$
**Molecular weight:** 610.66
**CAS Registry No:** 145040-37-5, 139481-59-7 (candesartan only)
**Merck Index:** 13, 1747

## SAMPLE
**Matrix:** blood, formulations
**Sample preparation:** Mix 1 mL plasma with 2 mL MeCN, vortex briefly, let stand at room temperature for 5 min, centrifuge at 4000 g for 20 min, inject a 20 μL aliquot of the supernatant. Mechanically shake a pulverized tablet with 100 mL MeOH for 10 min, centrifuge an aliquot at 4000 rpm for 5 min, filter (0.45 μm), dilute, inject a 20 μL aliquot.

## HPLC VARIABLES
**Column:** 150 × 4.6 5 μm Supelcosil C18
**Mobile phase:** MeCN:MeOH:10 mM potassium dihydrogen phosphate 18:80:2
**Flow rate:** 1
**Injection volume:** 20
**Detector:** UV 260

## CHROMATOGRAM
**Retention time:** 3.5
**Limit of detection:** 2 ng/mL
**Limit of quantitation:** 11 ng/mL

## OTHER SUBSTANCES
**Extracted:** hydrochlorothiazide (6.5, LOD 3.58 ng/mL, LOQ 6.75 ng/mL)
**Simultaneous:** benazepril (4.5), cilazapril (2.5), hydroflumethiazide (11), lisinopril (7.3)
**Noninterfering:** amiloride, losartan, valsartan

## KEY WORDS
plasma; tablets

## REFERENCE
Erk, N. Simultaneous analysis of candesartan cilexetil and hydrochlorothiazide in human plasma and dosage forms using HPLC with a photodiode array detector, *J.Liq.Chromatogr.Rel.Technol.*, **2003**, *26*, 2581–2591.

## SAMPLE
**Matrix:** blood
**Sample preparation:** Condition a 1 mL 100 mg Bond Elut C8 SPE cartridge with 2 mL MeOH and 1 mL 100 mM pH 2 phosphate buffer. Mix 250 μL plasma with IS, add 250 μL 1 M phosphoric acid, shake, centrifuge at 10000 g at 4° for 5 min, add the supernatant to the SPE cartridge, wash with 500 μL MeOH:100 mM pH 2 phosphate buffer 50:50, dry at full vacuum for 20 min, elute with 500 μL MeOH. Add 100 μL MeOH:ethylene glycol 90:10 to the eluate (to prevent adsorption of the drug), vortex, evaporate to dryness under a stream of nitrogen at 40°, reconstitute the residue with 250 μL of the initial mobile phase, inject a 20 μL aliquot.

## HPLC VARIABLES

**Guard column:** 20 × 3.9 4 μm Novapak C18 (Waters)
**Column:** 300 × 3.9 10 μm μBondapak C18
**Mobile phase:** Gradient. MeCN:5 mM pH 4 acetate buffer from 30:70 to 60:40 over 15 min, to 95:5 over 6 min, return to initial conditions over 3 min, re-equilibrate at initial conditions for 1 min.
**Flow rate:** From 1 to 1.2 over 15 min, maintain at 1.2 for 6 min, to 1 over 3 min, maintain at 1 for 1 min
**Injection volume:** 20
**Detector:** F ex 250 em 375

## CHROMATOGRAM

**Retention time:** 22.6
**Internal standard:** bumetanide (13.5)
**Limit of quantitation:** 3 ng/mL

## OTHER SUBSTANCES

**Extracted:** metabolites, irbesartan (12.6, LOQ 50 ng/mL), losartan (11.5, LOQ 16 ng/mL), valsartan (14.4, LOQ 50 ng/mL)

## KEY WORDS

plasma; SPE

## REFERENCE

González, L.; López, J.A.; Alonso, R.M.; Jiménez, R.M. Fast screening method for the determination of angiotensin II receptor antagonists in human plasma by high-performance liquid chromatography with fluorimetric detection, *J.Chromatogr.A*, **2002**, *949*, 49–60.

## ANNOTATED BIBLIOGRAPHY

González, L.; Alonso, R.M.; Jiménez, R.M. A high-performance liquid chromatographic method for screening angiotensin II receptor antagonists in human urine, *Chromatographia*, **2000**, *52*, 735–740. [losartan; irbesartan; valsartan; candesartan; SPE]

Stenhoff, H.; Lagerström, P.O.; Andersen, C. Determination of candesartan cilexetil, candesartan and a metabolite in human plasma and urine by liquid chromatography and fluorometric detection, *J.Chromatogr.B*, **1999**, *731*, 411–417.

# Capecitabine

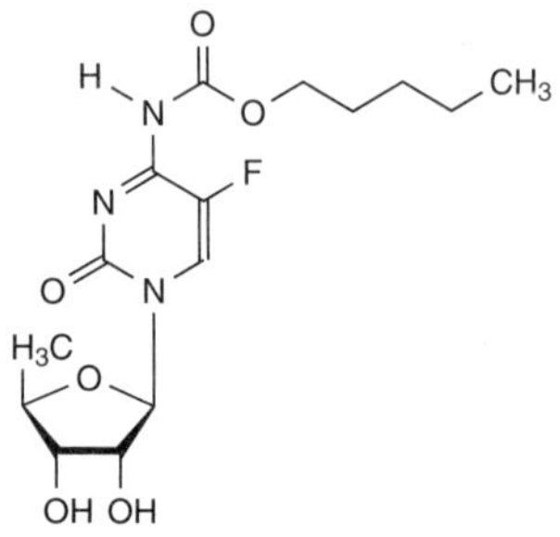

**Molecular formula:** $C_{15}H_{22}FN_3O_6$
**Molecular weight:** 359.35
**CAS Registry No:** 154361-50-9
**Merck Index:** 13, 1759

## SAMPLE
**Matrix:** blood
**Sample preparation:** Add 50 μL 50 μg/mL IS in MeOH to a tube, evaporate to dryness under reduced pressure at 30° for 20 min, add 500 μL plasma, add 500 μL 5 mM pH 6.8 ammonium acetate, vortex, let stand on ice for 15 min, centrifuge at 3000 g at 41° (sic) for 10 min. Inject a 20 μL aliquot of the supernatant onto column A and elute to waste with the mobile phase; after 0.8 min, divert the effluent from column A onto column B; after another 19.1 min, elute only column A to waste.

## HPLC VARIABLES
**Column:** A 20 × 2.1 Oasis HLB (Waters); B SecurityGuard C18 (Phenomenex) + 150 × 2 5 μm YMC ODS-AQ
**Column temperature:** 30 (column B only)
**Mobile phase:** Gradient. MeCN:5 mM pH 6.8 ammonium acetate 0:100 for 2 min, to 10:90 over 0.2 min, to 30:70 over 7.8 min, to 70:30 over 2 min, maintain at 70:30 for 3 min, to 0:100 over 0.2 min, maintain at 0:100 for 4.8 min, to 95:5 over 0.2 min, maintain at 95:5 for 1 min, return to initial conditions over 0.2 min, re-equilibrate for 2.6 min.
**Flow rate:** 3 for 0.8 min, 0.2 for 19.2 min, then 3
**Injection volume:** 20
**Detector:** MS, electrospray, m/z 360, cone voltage 20 V, 33% of column effluent went to MS

## CHROMATOGRAM
**Retention time:** 18
**Internal standard:** 5-chloro-2′-deoxyuridine (11.5, m/z 261, cone voltage 35 V)
**Limit of quantitation:** 1.4 ng/mL

## OTHER SUBSTANCES
**Extracted:** metabolites

## KEY WORDS
column-switching; pharmacokinetics; plasma

## REFERENCE
Xu, Y.; Grem, J.L. Liquid chromatography-mass spectrometry method for the analysis of the anti-cancer agent capecitabine and its nucleoside metabolites in human plasma, *J.Chromatogr.B*, **2003**, *783*, 273–285.

## SAMPLE
**Matrix:** blood
**Sample preparation:** Mix plasma and IS with MeCN, centrifuge, add the supernatant to a Bond Elut C18 SPE cartridge, elute with MeOH.

---

## HPLC VARIABLES

**Column:** 150 × 6 5 μm YMC-Pack C8-AM
**Mobile phase:** MeCN:buffer 40:60 (The buffer was 50 mL pH 4.0 citrate buffer (Titrisol, Merck) diluted to 1200 mL with water.)
**Flow rate:** 1
**Detector:** UV 310

---

## CHROMATOGRAM

**Retention time:** 6.2
**Internal standard:** Ro 09–1977 (8.8)
**Limit of quantitation:** 50 ng/mL

---

## KEY WORDS

pharmacokinetics; plasma; SPE

---

## REFERENCE

Reigner, B.; Verweij, J.; Dirix, L.; Cassidy, J.; Twelves, C.; Allman, D.; Weidekamm, E.; Roos, B.; Banken, L.; Utoh, M.; Osterwalder, B. Effect of food on the pharmacokinetics of capecitabine and its metabolites following oral administration in cancer patients, *Clin.Cancer Res.*, **1998**, *4*, 941–948.

# Casanthranol

**CAS Registry No:** 8024-48-4
**Merck Index:** 13, 1893

## SAMPLE
**Matrix:** solutions
**Sample preparation:** Inject a 20 μL aliquot of a solution in MeOH.

## HPLC VARIABLES
**Guard column:** 4 × 4 5 μm LiChrospher 100 RP18
**Column:** 125 × 4 5 μm LiChrospher 100 RP18
**Column temperature:** 40
**Mobile phase:** MeCN:MeOH:water:acetic acid 5:55:39:1
**Flow rate:** 1
**Injection volume:** 20
**Detector:** UV 254

## CHROMATOGRAM
**Retention time:** 6.1 (aloe-emodin), 20.8 (emodin)

## OTHER SUBSTANCES
**Extracted:** rhein (9.1), sennidin A (16.4), sennidin B (24.8)

## REFERENCE
Grimminger, W.; Witthohn, K. Analytics of senna drugs with regard to the toxicological discussion of anthranoids, *Pharmacology*, **1993**, *47*(Suppl. 1), 98–109.

## ANNOTATED BIBLIOGRAPHY
Matthees, D. Determination of emodin in feeds, *J.Agric.Food Chem.*, **1983**, *31*, 453–454.
Koch, A. Metabolism of aloin – the influence of nutrition, *J.Pharm.Biomed.Anal.*, **1996**, *14*, 1335–1338.

# Caspofungin

**Molecular formula:** $C_{52}H_{88}N_{10}O_{15}$
**Molecular weight:** 1093.31
**CAS Registry No:** 162808-62-0
**Merck Index:** 13, 1899

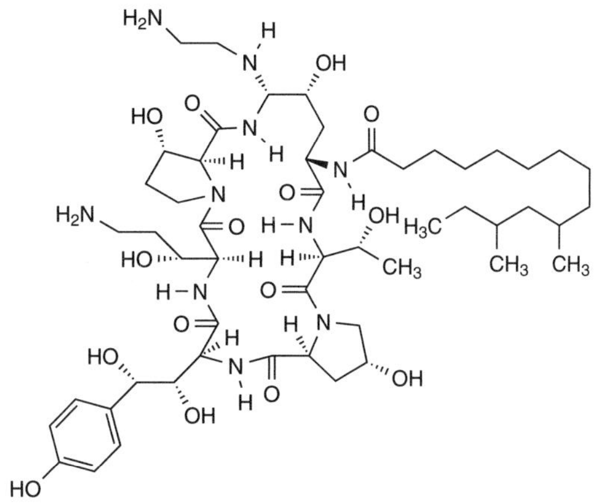

## SAMPLE

**Matrix:** blood

**Sample preparation:** Condition a 3 mL diol SPE cartridge (Baker) with 3 mL MeOH and 3 mL water. Mix 1 mL plasma with 10 μL 50 μg/mL IS in MeCN:0.1% trifluoroacetic acid adjusted to pH 3.0 with triethylamine 35:65 and 250 μL 1 M pH 4.9 potassium acetate buffer, add to SPE cartridge, wash with 3 mL water, wash with 3 mL MeOH, elute with 1 mL MeOH containing 250 mM ammonium hydroxide and 0.1% trifluoroacetic acid. Evaporate the eluate to dryness under reduced pressure at 50°, reconstitute the residue with 100 μL mobile phase, inject a 50 μL aliquot. (See also Schwartz,M. et al. *Anal.Chim.Acta* **1997**, *352*, 299–307.)

## HPLC VARIABLES

**Column:** 150 × 4.6 5 μm Zorbax SB C80

**Mobile phase:** MeCN:buffer 35:65. After 5.5 min, wash column with MeCN:buffer 90:10 for 0.9 min, re-equilibrate at initial conditions for 5 min. (The buffer was 675 μL trifluoroacetic acid in 1 L water, adjusted to pH 3.0 with ammonium hydroxide.)

**Flow rate:** 1.2

**Injection volume:** 50

**Detector:** MS, PE Sciex API III plus triple quadrupole, ion spray, 5% of column effluent entered MS, m/z 1093.7

## CHROMATOGRAM

**Retention time:** 5.5

**Internal standard:** isostere (oxygen replaces nitrogen at 5 position, m/z 1094.7) (6.8)

**Limit of quantitation:** 10 ng/mL (LOQ was 2.5 ng/mL using a 75 μL injection and turbo ion spray, but precision and accuracy was less good)

## KEY WORDS

plasma; SPE

## REFERENCE

Chavez-Eng, C.M.; Schwartz, M.; Constanzer, M.L.; Matuszewski, B.K. Determination of a cyclic hexapeptide, a novel antifungal agent, in human plasma by high-performance liquid chromatography with ion spray and turbo ion spray tandem mass spectrometric detection, *J.Chromatogr.B*, **1999**, *721*, 229–238.

## SAMPLE

**Matrix:** blood

**Sample preparation:** Mix plasma with MeCN/MeOH containing 0.1% trifluoroacetic acid, vortex, centrifuge. Evaporate the supernatant to dryness under reduced pressure, reconstitute the residue with 0.1% trifluoroacetic acid, vortex, centrifuge, inject an aliquot.

---

## HPLC VARIABLES

**Column:** 250 × 4.6 5 μm Zorbax RX C8
**Mobile phase:** Gradient. MeCN:0.1% trifluoroacetic acid 18:82 for 4 min, to 50:50 over 36 min.
**Detector:** Radioactivity ($^3$H)

---

## CHROMATOGRAM

**Retention time:** 32

---

## OTHER SUBSTANCES

**Extracted:** metabolites

---

## KEY WORDS

plasma

---

## REFERENCE

Balani, S.K.; Xu, X.; Arison, B.H.; Silva, M.V.; Gries, A.; deLuna, F.A.; Cui, D.; Kari, P.H.; Ly, T.; Hop, C.E.C.A.; Singh, R.; Wallace, M.A.; Dean, D.C.; Lin, J.H.; Pearson, P.G.; Baillie, T.A. Metabolites of caspofungin acetate, a potent antifungal agent, in human plasma and urine, *Drug Metab.Dispos.*, **2000**, *28*, 1274–1278.

---

## SAMPLE

**Matrix:** urine
**Sample preparation:** Evaporate to dryness under reduced pressure, reconstitute, inject an aliquot.

---

## HPLC VARIABLES

**Column:** 250 × 4.6 5 μm Luna C18 (Phenomenex) + 250 × 4.6 5 μm Zorbax SAX in series
**Mobile phase:** Gradient. MeCN:0.1% trifluoroacetic acid 0:100 for 15 min, to 80:20 over 25 min
**Detector:** Radioactivity ($^3$H)

---

## CHROMATOGRAM

**Retention time:** 35

---

## OTHER SUBSTANCES

**Extracted:** metabolites

---

## REFERENCE

Balani, S.K.; Xu, X.; Arison, B.H.; Silva, M.V.; Gries, A.; deLuna, F.A.; Cui, D.; Kari, P.H.; Ly, T.; Hop, C.E.C.A.; Singh, R.; Wallace, M.A.; Dean, D.C.; Lin, J.H.; Pearson, P.G.; Baillie, T.A. Metabolites of caspofungin acetate, a potent antifungal agent, in human plasma and urine, *Drug Metab.Dispos.*, **2000**, *28*, 1274–1278.

---

## ANNOTATED BIBLIOGRAPHY

Groll, A.H.; Gullick, B.M.; Petraitiene, R.; Petraitis, V.; Candelario, M.; Piscitelli, S.C.; Walsh, T.J. Compartmental pharmacokinetics of the antifungal echinocandin caspofungin (MK-0991) in rabbits, *Antimicrob.Agents Chemother.*, **2001**, *45*, 596–600.

Petraitiene, R.; Petraitis, V.; Groll, A.H.; Sein, T.; Schaufele, R.L.; Francesconi, A.; Bacher, J.; Avila, N.A.; Walsh, T.J. Antifungal efficacy of caspofungin (MK-0991) in experimental pulmonary aspergillosis in persistently neutropenic rabbits: pharmacokinetics, drug disposition, and relationship to galactomannan antigenemia, *Antimicrob.Agents Chemother.*, **2002**, *46*, 12–23.

Schwartz, M.; Kline, W.; Matuszewski, B. Determination of a cyclic hexapeptide (L-743 872), a novel pneumocandin antifungal agent in human plasma and urine by high-performance liquid chromatography with fluorescence detection, *Anal.Chim.Acta*, **1997**, *352*, 299–307.

# Castor oil

**CAS Registry No:** 8001-79-4
**Merck Index:** 13, 1908

## SAMPLE

**Matrix:** cosmetics
**Sample preparation:** Inject a 5–50 μL aliquot of a solution of the lipstick in MeCN.

## HPLC VARIABLES

**Column:** 300 × 3.9 Bondapak C18
**Mobile phase:** Gradient. MeCN:water 50:50 for 5 min, to 100:0 (step gradient)
**Flow rate:** 2
**Injection volume:** 5–50
**Detector:** UV 254

## CHROMATOGRAM

**Retention time:** 8, 20, 27 (multiple peaks)

## OTHER SUBSTANCES

**Simultaneous:** lanolin (13, 14), propyl paraben (4)

## KEY WORDS

lipstick

## REFERENCE

Reuland, D.J.; Trinler, W.A. A comparison of lipstick smears by high performance liquid chromatography, *J.Forensic Sci.Soc.*, **1980**, *20*, 111–120.

# Cefbuperazone

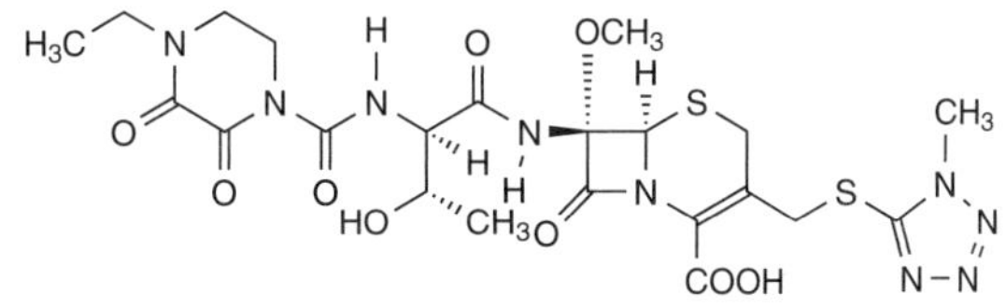

**Molecular formula:** $C_{22}H_{29}N_9O_9S_2$
**Molecular weight:** 627.66
**CAS Registry No:** 76610-84-9
**Merck Index:** 13, 1930

## SAMPLE

**Matrix:** blood

**Sample preparation:** Condition a 3 mL 200 mg Bond Elut C18 SPE cartridge with two 3 mL portions of MeOH and two 3 mL portions of 5% acetic acid. Mix 500 μL serum with 500 μL 10% acetic acid, add to the SPE cartridge, wash with 3 mL water, elute with four 500 μL portions of MeOH:water 60:40. Evaporate the combined eluates to dryness under a stream of nitrogen, reconstitute the residue with 250 μL water, filter (0.22 μm). Inject a 100 μL aliquot onto column A and elute to waste with mobile phase A. After an unspecified time, backflush the contents of column A onto column B using mobile phase B. Monitor the effluent from column B. (Sato,K.; Kobayashi,K.; Moore,C.M.; Mizuno,Y.; Katsumata,Y. Semi-quantitative analysis of cefaclor in human serum by capillary high performance liquid chromatography/fast atom bombardment mass spectrometry. *Forensic Sci.Int.* **1993**, *59*, 71–77.)

## HPLC VARIABLES

**Column:** A 30 × 0.5 10 μm Develosil PhA (phenethyl); B 150 × 0.5 5 μm Develosil PhA (phenethyl)

**Mobile phase:** A 10 mM ammonium acetate:glycerol 99.5:0.5, adjusted to pH 5 with acetic acid; B MeOH:water:acetic acid:glycerol 40:59:0.5:0.5 (pH ca. 3)

**Flow rate:** A 0.05; B 0.004

**Injection volume:** 100

**Detector:** MS, JMS DX303 double focusing, xenon FAB ion source, gun current 10 mA, voltage 3 kV, positive ion mode

## CHROMATOGRAM

**Retention time:** 20.8

**Limit of detection:** 50–200 ng

## OTHER SUBSTANCES

**Extracted:** cefaclor (14.4), cefamandole (31.0), cefazolin (19.7), cefixime (19.2), cefmenoxime (18.8), cefmetazole (21.4), cefoperazone (24.2), cefotaxime (16.1), cefotetan (16.8), cefotiam (11.7), cefpiramide (20.9), cefsulodin (12.2), ceftazidime (12.0), ceftizoxime (15.6), ceftriaxone (16.7), cefuroxime (16.9), cefuzonam (32.0), cephalexin (14.8), cephaloglycine (15.7), cephaloridine (15.6), cephalothin (34.7), flomoxef (18.3), latamoxef (15.1)

## KEY WORDS

column-switching; serum; SPE

## REFERENCE

Kobayashi, K.; Sato, K.; Mizuno, Y.; Katsumata, Y. Capillary high-performance liquid chromatography-fast atom bombardment mass spectrometry of 24 cephem antibiotics, *J.Chromatogr.B*, **1996**, *677*, 275–290.

# Cefditoren

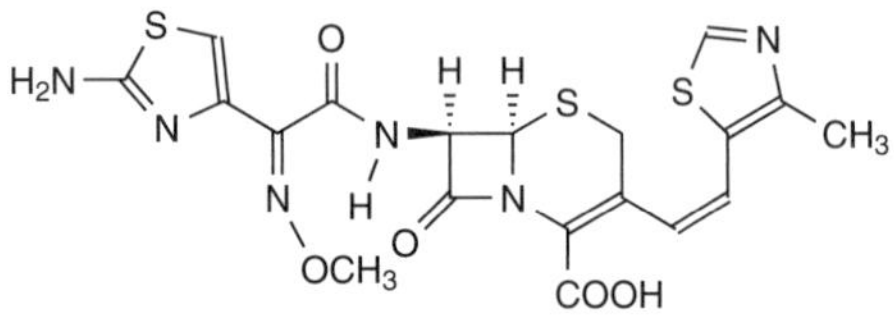

**Molecular formula:** $C_{19}H_{18}N_6O_5S_3$
**Molecular weight:** 506.59
**CAS Registry No:** 104145-95-1, 117467-28-4 (pivoxil)
**Merck Index:** 13, 1934

## SAMPLE
**Matrix:** blood
**Sample preparation:** Mix 200 μL serum or urine with 200 μL 5 μg/mL IS in MeCN, centrifuge at 3500 rpm for 10 min, inject a 20 μL aliquot.

## HPLC VARIABLES
**Column:** 150 × 4.6 ODS C18
**Column temperature:** 30
**Mobile phase:** MeOH:5% amine acetate (sic) 30:70
**Flow rate:** 1
**Injection volume:** 20
**Detector:** UV 295

## CHROMATOGRAM
**Retention time:** 11.5
**Internal standard:** not specified (15.8)

## KEY WORDS
pharmacokinetics; serum

## REFERENCE
Li, J.T.; Hou, F.; Lu, H.; Li, T.Y.; Li, H. Phase I clinical trial of cefditoren pivoxil (ME 1207): pharmacokinetics in healthy volunteers, *Drugs Exp.Clin.Res.*, **1997**, *23*, 145–150.

## SAMPLE
**Matrix:** blood
**Sample preparation:** Mix 500 μL MeCN with ? μL plasma and ? μL 2 μg/mL IS in MeCN:water 50:50, vortex at high speed for 30 s, centrifuge at 1180 g for 5 min. Evaporate the supernatant to dryness, reconstitute the residue with 100–200 μL mobile phase (?), inject a 10–25 μL aliquot.

## HPLC VARIABLES
**Column:** 100 × 2 4 μm YMC J'sphere M80
**Mobile phase:** MeOH:10 mM pH 4.5 ammonium acetate 40:60
**Flow rate:** 0.2
**Injection volume:** 10–25
**Detector:** MS, PE-Sciex API-365 triple quadrupole, turbo ion spray source 450°, nitrogen 7 L/min, m/z 506.9–240.8

## CHROMATOGRAM
**Retention time:** 6.2
**Internal standard:** cefotaxime (m/z 455.9–324) (2.2)
**Limit of quantitation:** 10 ng/mL

## KEY WORDS
plasma

## REFERENCE

Zhu, T.; Cheung, B.W.Y.; Cartier, L.L.; Giebink, G.S.; Sawchuk, R.J. Simultaneous intravenous and intramiddle-ear dosing to determine cefditoren influx and efflux clearances in middle ear fluid in freely moving chinchillas, *J.Pharm.Sci.*, **2003**, *92*, 1947–1956.

## SAMPLE

**Matrix:** dialysate
**Sample preparation:** Inject a 10 μL aliquot of artificial middle ear fluid (pH 7.45 phosphate-buffered saline-containing 15 mM phosphate and 150 mM sodium) containing 200 ng/mL IS.

## HPLC VARIABLES

**Column:** 100 × 4.6 YMC ODS-A S-5 C18
**Mobile phase:** MeOH:50 mM pH 4.0 ammonium acetate buffer 35:65
**Flow rate:** 0.5
**Injection volume:** 10
**Detector:** UV 295

## CHROMATOGRAM

**Retention time:** 9.6
**Internal standard:** (+)-(6*R*,7*R*)-7-[(*Z*)-2-(2-aminothiazol-4-yl)-2-methoxyiminoacetamido]-3-[(*Z*)-2-(4-methylthiazol-5-yl)ethenyl]-8-oxo-5-thia-1-azabicyclo[4.2.0]oct-3-ene-2-carboxylic acid (Meiji Seika Kaisha, Tokyo) (8.2)
**Limit of quantitation:** 5 ng/mL

## OTHER SUBSTANCES

**Extracted:** metabolites

## REFERENCE

Zhu, T.; Cheung, B.W.Y.; Cartier, L.L.; Giebink, G.S.; Sawchuk, R.J. Simultaneous intravenous and intramiddle-ear dosing to determine cefditoren influx and efflux clearances in middle ear fluid in freely moving chinchillas, *J.Pharm.Sci.*, **2003**, *92*, 1947–1956.

# Cefoselis

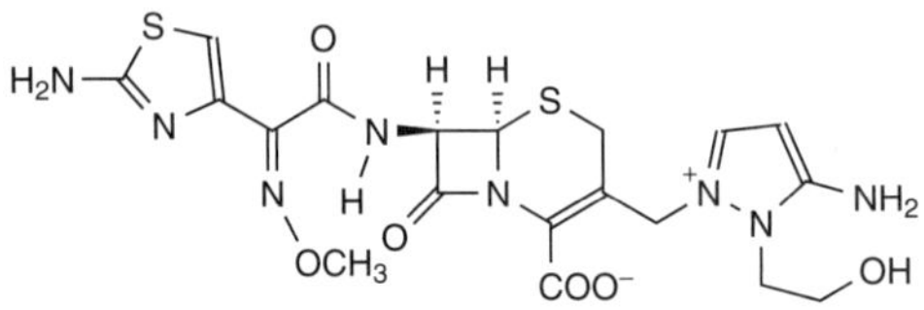

**Molecular formula:** $C_{19}H_{22}N_8O_6S_2$
**Molecular weight:** 522.57
**CAS Registry No:** 122841-10-5, 122841-12-7 (sulfate)
**Merck Index:** 13, 1945

## SAMPLE

**Matrix:** blood, CSF, tissue
**Sample preparation:** Mix 100 μL serum with 10 μL 100 mg/mL (sic) IS in 100 mM pH 7.0 phosphate buffer and 100 μL MeCN, vortex for 10 s, centrifuge at 13000 rpm for 2 min. Remove 100 μL of the supernatant and mix it with 400 μL 20 mM pH 2.5 phosphate buffer, vortex for 10 s, inject a 5 μL aliquot. Mix 30 μL CSF with 1 mg/mL IS in 100 mM pH 7.0 phosphate buffer (?), inject a 50 μL aliquot. Homogenize brain tissue with 5 vol of saline. Mix 200 μL homogenate with 20 μL 1 mg/mL IS in 100 mM pH 7.0 phosphate buffer and 200 μL MeCN, vortex for 10 s, inject a 50 μL aliquot.

## HPLC VARIABLES

**Column:** 150 × 4.6 5 μm TSKgel ODS-80TM
**Mobile phase:** MeCN:20 mM pH 2.5 phosphate buffer 0.065:99.935
**Flow rate:** 1
**Injection volume:** 5–50
**Detector:** UV 254

## CHROMATOGRAM

**Internal standard:** cefpirome sulfate

## KEY WORDS

brain; rat; serum

## REFERENCE

Nagata, M.; Yasuhara, M. Effect of experimental renal failure on the pharmacodynamics of cefoselis-induced seizures in rats, *Biol.Pharm.Bull.*, **2001**, *24*, 1049–1052.

# Cefozopran

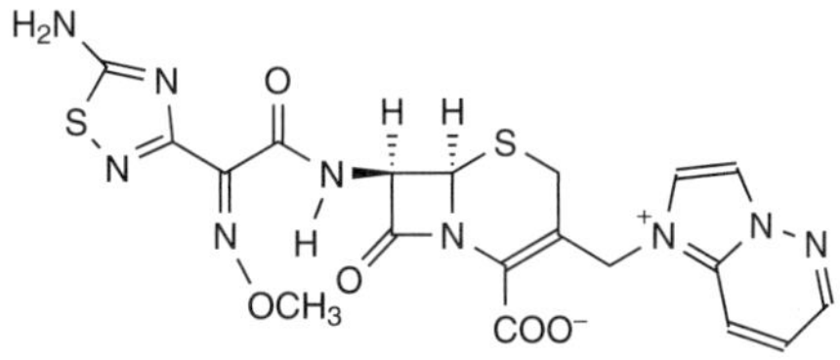


**Molecular formula:** $C_{19}H_{17}N_9O_5S_2$
**Molecular weight:** 515.53
**CAS Registry No:** 113359-04-9
**Merck Index:** 13, 1950

## SAMPLE

**Matrix:** blood, urine
**Sample preparation:** Serum. Mix 300 μL serum with 300 μL 5 mM pH 3.2 sodium phosphate buffer and 800 μL MeCN, shake mechanically for 2 min, centrifuge at 13000 g for 5 min. Remove 1.4 mL of the supernatant and add it to 2 mL dichloromethane, agitate for 1 min, centrifuge at 2000 g for 5 min. Remove a 300 μL aliquot of the supernatant and add it to 600 μL mobile phase, inject a 20 μL aliquot. Urine. Mix 100 μL urine with 900 μL of the aqueous portion of the mobile phase, inject a 5 μL aliquot.

## HPLC VARIABLES

**Guard column:** 30 × 4 30–40 μm Perisorb RP18
**Column:** 125 × 4 5 μm Nucleosil 5C18
**Mobile phase:** MeCN:buffer 4.5:95.5 (Make mobile phase by adding 45 mL MeCN and 3 bottles of PIC B7 solution (15 M heptanesulfonic acid, pH 3.2 (Waters)) to water and making up to 1 L with water.)
**Flow rate:** 1.5
**Injection volume:** 5 (urine), 20 (serum)
**Detector:** UV 235

## CHROMATOGRAM

**Retention time:** 5.3
**Limit of detection:** 600 ng/mL (serum), 3.5 μg/mL (urine)
**Limit of quantitation:** 1 μg/mL (serum), 5 μg/mL (urine)

## KEY WORDS

serum

## REFERENCE

Borner, K.; Borner, E.; Lode, H. Determination of a new cephalosporin, SCE-2787, in serum and urine by high-performance liquid chromatography, *J.Chromatogr.*, **1993**, *615*, 174–179.

# Cefuzonam

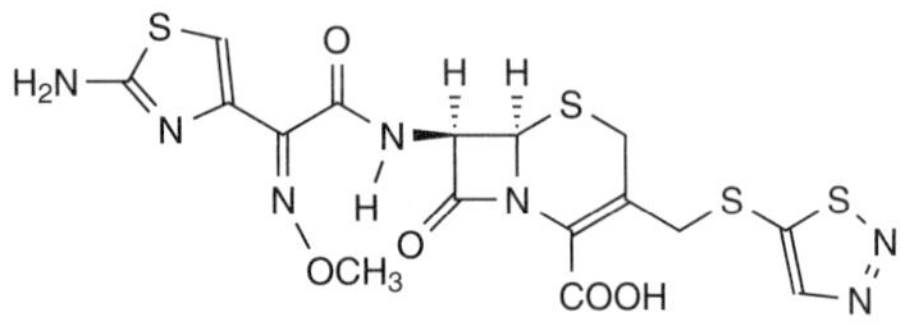


**Molecular formula:** $C_{16}H_{15}N_7O_5S_4$
**Molecular weight:** 513.60
**CAS Registry No:** 82219-78-1
**Merck Index:** 13, 1967

## SAMPLE

**Matrix:** blood
**Sample preparation:** Condition a 3 mL 200 mg Bond Elut C18 SPE cartridge with two 3 mL portions of MeOH and two 3 mL portions of 5% acetic acid. Mix 500 μL serum with 500 μL 10% acetic acid, add to the SPE cartridge, wash with 3 mL water, elute with four 500 μL portions of MeOH:water 60:40. Evaporate the combined eluates to dryness under a stream of nitrogen, reconstitute the residue with 250 μL water, filter (0.22 μm). Inject a 100 μL aliquot onto column A and elute to waste with mobile phase A. After an unspecified time, backflush the contents of column A onto column B using mobile phase B. Monitor the effluent from column B. (Sato,K.; Kobayashi,K.; Moore,C.M.; Mizuno,Y.; Katsumata,Y. Semi-quantitative analysis of cefaclor in human serum by capillary high performance liquid chromatography/fast atom bombardment mass spectrometry. *Forensic Sci.Int.* **1993**, *59*, 71–77.)

## HPLC VARIABLES

**Column:** A 30 × 0.5 10 μm Develosil PhA (phenethyl); B 150 × 0.5 5 μm Develosil PhA (phenethyl)
**Mobile phase:** A 10 mM ammonium acetate:glycerol 99.5:0.5, adjusted to pH 5 with acetic acid; B MeOH:water:acetic acid:glycerol 40:59:0.5:0.5 (pH ca .3)
**Flow rate:** A 0.05; B 0.004
**Injection volume:** 100
**Detector:** MS, JMS DX303 double focusing, xenon FAB ion source, gun current 10 mA, voltage 3 kV, positive ion mode

## CHROMATOGRAM

**Retention time:** 32.0
**Limit of detection:** 200–1000 ng

## OTHER SUBSTANCES

**Extracted:** cefaclor (14.4), cefamandole (31.0), cefazolin (19.7), cefbuperazone (20.8), cefixime (19.2), cefmenoxime (18.8), cefmetazole (21.4), cefoperazone (24.2), cefotaxime (16.1), cefotetan (16.8), cefotiam (11.7), cefpiramide (20.9), cefsulodin (12.2), ceftazidime (12.0), ceftizoxime (15.6), ceftriaxone (16.7), cefuroxime (16.9), cephalexin (14.8), cephaloglycine (15.7), cephaloridine (15.6), cephalothin (34.7), flomoxef (18.3), latamoxef (15.1)

## KEY WORDS

column-switching; serum; SPE

## REFERENCE

Kobayashi, K.; Sato, K.; Mizuno, Y.; Katsumata, Y. Capillary high-performance liquid chromatography-fast atom bombardment mass spectrometry of 24 cephem antibiotics, *J.Chromatogr.B*, **1996**, *677*, 275–290.

# Celecoxib

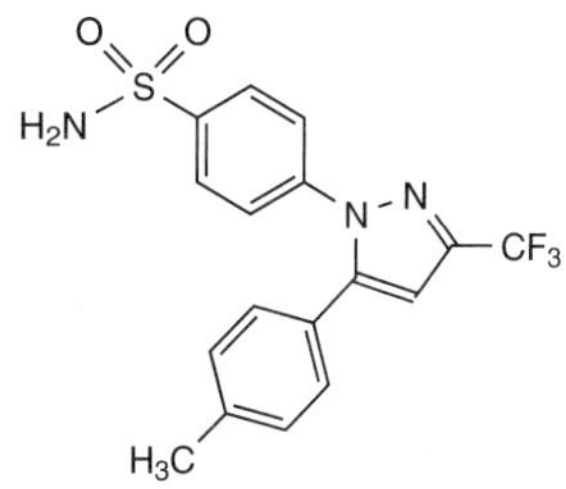

**Molecular formula:** $C_{17}H_{14}F_3N_3O_2S$
**Molecular weight:** 381.38
**CAS Registry No:** 169590-42-5
**Merck Index:** 13, 1968

## SAMPLE

**Matrix:** blood
**Sample preparation:** Condition a 1 mL 100 mg Chromabond C18 SPE cartridge with two 1 mL portions of MeCN and two 1 mL portions of water. Mix 200 μL plasma with 200 μL 100 mM pH 4.0 phosphate buffer and 25 μL 450 ng/mL IS in MeCN:water 50:50, add to the SPE cartridge, wash with two 1 mL portions of water, dry under vacuum for at least 5 min, elute with two 1 mL portions of dichloromethane. Evaporate the eluate to dryness under a stream of nitrogen at 40°, reconstitute the residue with 200 μL mobile phase, inject a 10 μL aliquot.

## HPLC VARIABLES

**Column:** 30 × 2 5 μm Nucleosil C18
**Mobile phase:** MeCN:water:25% ammonium hydroxide 85:15:0.1
**Flow rate:** 0.2
**Injection volume:** 10
**Detector:** MS, PE Sciex API 3000 triple quadrupole, turbo ion spray interface, negative ion mode at -3700 V and 400°, auxiliary gas nitrogen 4.5 L/min, nebulizer gas nitrogen 1.23 L/min, curtain gas nitrogen 1.08 L/min, collision gas nitrogen $2.92 \times 10^{15}$ molecules/cm$^2$, m/z 380–316 (−32 eV)

## CHROMATOGRAM

**Retention time:** 0.8
**Internal standard:** 4-[5-phenyl-3-(trifluoromethyl)-1H-pyrazol-1-yl]benzenesulfonamide (The IS is celecoxib lacking the methyl group. Preparation is as follows. Reflux 200 mL EtOH containing 16 mmol 4,4,4-trifluoro-1-phenyl-1,3-butadiene and 17.6 mmol 4-sulfonamidophenylhydrazine hydrochloride with stirring for 22 h, evaporate under reduced pressure. Take up the residue in ethyl acetate, wash with water, wash with saline, evaporate the organic solvent, recrystallize from *n*-hexane:ethyl acetate 50:50 to obtain the IS as white needles.) (m/z 366–302 (−30 eV)) (0.8)
**Limit of quantitation:** 0.25 ng/mL

## KEY WORDS

pharmacokinetics; plasma; SPE

## REFERENCE

Bräutigam, L.; Vetter, G.; Tegeder, I.; Heinkele, G.; Geisslinger, G. Determination of celecoxib in human plasma and rat microdialysis samples by liquid chromatography tandem mass spectrometry, *J.Chromatogr.B*, **2001**, *761*, 203–212.

## SAMPLE

**Matrix:** blood
**Sample preparation:** Mix 500 μL serum with 25 μL 30 μg/mL IS in MeCN, add 500 μL saturated NaCl, add 1 mL MeCN, mix, add 8 mL chloroform (Caution! Chloroform is a carcinogen!), extract for 15 min, centrifuge at 1500 g for 15 min. Evaporate the organic

layer to dryness under a stream of nitrogen, reconstitute the residue with 100 μL mobile phase, sonicate for 5 min, vortex, centrifuge for 15 min, inject a 10 μL aliquot.

---

## HPLC VARIABLES

**Column:** 150 × 3 3 μm Prontosil C18 AQ
**Column temperature:** 15
**Mobile phase:** MeCN:water 60:40
**Flow rate:** 0.35
**Injection volume:** 10
**Detector:** F ex 240 em 380

---

## CHROMATOGRAM

**Retention time:** 9.1
**Internal standard:** 4-[5-phenyl-3-(trifluoromethyl)-1H-pyrazol-1-yl]benzenesulfonamide (The IS is celecoxib lacking the methyl group. Preparation is as follows. Reflux 200 mL EtOH containing 16 mmol 4,4,4-trifluoro-1-phenyl-1,3-butadiene and 17.6 mmol 4-sulfonamidophenylhydrazine hydrochloride with stirring for 22 h, evaporate under reduced pressure. Take up the residue in ethyl acetate, wash with water, wash with saline, evaporate the organic solvent, recrystallize from *n*-hexane:ethyl acetate 50:50 to obtain the IS as white needles.) (11.7)
**Limit of quantitation:** 12.5 ng/mL

---

## KEY WORDS

pharmacokinetics; serum

---

## REFERENCE

Schönberger, F.; Heinkele, G.; Mürdter, T.E.; Brenner, S.; Klotz, U.; Hofmann, U. Simple and sensitive method for the determination of celecoxib in human serum by high-performance liquid chromatography with fluorescence detection, *J.Chromatogr.B*, **2002**, *768*, 255–260.

---

## SAMPLE

**Matrix:** blood
**Sample preparation:** Condition a 1 mL Oasis HLB SPE cartridge with 1 mL MeOH and 1 mL water. Mix 500 μL plasma with 15 μL 10 μg/mL IS in MeOH and 300 μL 200 mM pH 5.0 sodium acetate buffer, add 1.8 mL MeCN, vortex, centrifuge at 15000 g for 5 min. Evaporate the organic part of the supernatant under a stream of nitrogen at 50°, dilute the remaining aqueous phase with 2 mL water, add to the SPE cartridge at 0.07 mL/min, wash with two 1.5 mL portions of MeOH:water 5:95, elute with 2 mL MeOH. Evaporate the eluate to dryness under a stream of nitrogen at 50°, reconstitute the residue with 75 μL MeOH, inject a 50 μL aliquot.

---

## HPLC VARIABLES

**Column:** 150 × 4.6 5 μm Luna C18 (Phenomenex)
**Column temperature:** 60
**Mobile phase:** Gradient. A:B 90:10 for 10 min, to 62:38 over 37 min, to 27:73 over 16 min, maintain at 27:73 for 5 min, return to initial conditions over 0.1 min, re-equilibrate for 6.9 min. A was MeCN:10 mM pH 5.4 sodium phosphate buffer 10:90. B was MeCN:10 mM pH 5.4 sodium phosphate buffer 80:20.
**Flow rate:** 1.5
**Injection volume:** 50
**Detector:** UV 254

---

## CHROMATOGRAM

**Retention time:** 61
**Internal standard:** phenacetin (12)
**Limit of quantitation:** 10 ng/mL

## OTHER SUBSTANCES

**Extracted:** metabolites

## KEY WORDS

pharmacokinetics; plasma; SPE

## REFERENCE

Störmer, E.; Bauer, S.; Kirchheiner, J.; Brockmöller, J.; Roots, I. Simultaneous determination of celecoxib, hydroxycelecoxib, and carboxycelecoxib in human plasma using gradient reversed-phase liquid chromatography with ultraviolet absorbance detection, *J.Chromatogr.B*, **2003**, *783*, 207–212.

## SAMPLE

**Matrix:** formulations

**Sample preparation:** Weigh out capsule contents containing 5 mg celecoxib, dissolve in 25 mL mobile phase, filter (Whatman No 1 paper), dilute with mobile phase, inject an aliquot.

## HPLC VARIABLES

**Column:** 250 × 4.6 5 μm Inertsil C8

**Mobile phase:** MeCN:water 65:35

**Flow rate:** 1.25

**Injection volume:** 20

**Detector:** UV 230

## CHROMATOGRAM

**Retention time:** 8.05

**Limit of detection:** 25 ng/mL

**Limit of quantitation:** 75 ng/mL

## KEY WORDS

capsules

## REFERENCE

Saha, R.N.; Sajeev, C.; Jadhav, P.R.; Patil, S.P.; Srinivasan, N. Determination of celecoxib in pharmaceutical formulations using UV spectrophotometry and liquid chromatography, *J.Pharm.Biomed.Anal.*, **2002**, *28*, 741–751.

## ANNOTATED BIBLIOGRAPHY


Abdel-Hamid, M.; Novotny, L.; Hamza, H. Liquid chromatographic-mass spectrometric determination of celecoxib in plasma using single-ion monitoring and its use in clinical pharmacokinetics, *J.Chromatogr.B*, **2001**, *753*, 401–408.

Abdel-Hamid, M.E. LC-MS analysis of selected sulfur-containing non-steroid antiinflammatory agents: Applications to pharmaceutical products, *J.Liq.Chromatogr.Rel.Technol.*, **2000**, *23*, 3095–3107. [rofecoxib; sulindac; celecoxib; piroxicam; tenoxicam]

Chow, H.-H.S.; Anavy, N.; Salazar, D.; Frank, D.H.; Alberts, D.S. Determination of celecoxib in human plasma using solid-phase extraction and high-performance liquid chromatography, *J.Pharm.Biomed.Anal.*, **2004**, *34*, 167–174.

Jayasagar, G.; Kumar, M.K.; Chandrasekhar, K.; SivaPrasad, P.; Madhusudan Rao, Y. Validated HPLC method for the determination of celecoxib in human serum and its application in a clinical pharmacokinetic study, *Pharmazie*, **2002**, *57*, 619–621. [tolbutamide is internal standard]

Mamidi, R.N.V.S.; Mullangi, R.; Kota, J.; Bhamidipati, R.; Khan, A.A.; Katneni, K.; Datla, S.; Singh, S.K.; Rao, K.Y.; Rao, C.S.; Srinivas, N.R.; Rajagopalan, R. Pharmacological and pharmacokinetic evaluation of celecoxib prodrugs in rats, *Biopharm.Drug Dispos.*, **2002**, *23*, 273–282.

Mamidi, R.N.V.S.; Benjamin, B.; Ramesh, M.; Srinivas, N.R. Simple method for the determination of rosiglitazone in human plasma using a commercially available internal standard, *Biomed.Chromatogr.*, **2003**, *17*, 417–420.

Paulson, S.K.; Engel, L.; Reitz, B.; Bolten, S.; Burton, E.G.; Maziasz, T.J.; Yan, B.; Schoenhard, G.L. Evidence for polymorphism in the canine metabolism of the cyclooxygenase 2 inhibitor, celecoxib, *Drug Metab.Dispos.*, **1999**, *27*, 1133–1142.

Paulson, S.K.; Kaprak, T.A.; Gresk, C.J.; Fast, D.M.; Baratta, M.T.; Burton, E.G.; Breau, A.P.; Karim, A. Plasma protein binding of celecoxib in mice, rat, rabbit, dog and human, *Biopharm.Drug Dispos.*, **1999**, *20*, 293–299.

Paulson, S.K.; Hribar, J.D.; Liu, N.W.K.; Hajdu, E.; Bible, R.H. Jr.; Piergies, A.; Karim, A. Metabolism and excretion of [$^{14}$C]celecoxib in healthy male volunteers, *Drug Metab.Dispos.*, **2000**, *28*, 308–314. [SPE]

Paulson, S.K.; Zhang, J.Y.; Breau, A.P.; Hribar, J.D.; Liu, N.W.K.; Jessen, S.M.; Lawal, Y.M.; Cogburn, J.N.; Gresk, C.J.; Markos, C.S.; Maziasz, T.J.; Schoenhard, G.L.; Burton, E.G. Pharmacokinetics, tissue distribution, metabolism, and excretion of celecoxib in rats, *Drug Metab.Dispos.*, **2000**, *28*, 514–521.

Rao, D.S.; Srinivasu, M.K.; Narayana, C.L.; Reddy, G.O. LC separation of ortho and meta isomers of celecoxib in bulk and formulations using a chiral column, *J.Pharm.Biomed.Anal.*, **2001**, *25*, 21–30.

Rose, M.J.; Woolf, E.J.; Matuszewski, B.K. Determination of celecoxib in human plasma by normal-phase high-performance liquid chromatography with column switching and ultraviolet absorbance detection, *J.Chromatogr.B*, **2000**, *738*, 377–385.

Srinivasu, M.K.; Narayana, C.L.; Rao, D.S.; Reddy, G.O. A validated LC method for the quantitative determination of celecoxib in pharmaceutical dosage forms and purity evaluation in bulk drugs, *J.Pharm.Biomed.Anal.*, **2000**, *22*, 949–956.

Stempak, D.; Gammon, J.; Klein, J.; Koren, G.; Baruchel, S. Single-dose and steady-state pharmacokinetics of celecoxib in children, *Clin.Pharmacol.Ther.*, **2002**, *72*, 490–497.

Tang, C.; Shou, M.; Mei, Q.; Rushmore, T.H.; Rodrigues, A.D. Major role of human liver microsomal cytochrome P450 2C9 (CYP2C9) in the oxidative metabolism of celecoxib, a novel cyclooxygenase-II inhibitor, *J.Pharmacol.Exp.Ther.*, **2000**, *293*, 453–459.

Tang, C.; Shou, M.; Rodrigues, A.D. Substrate-dependent effect of acetonitrile on human liver microsomal cytochrome P450 2C9 (CYP2C9) activity, *Drug Metab.Dispos.*, **2000**, *28*, 567–572. [diclofenac; phenytoin; celecoxib; tolbutamide; flurbiprofen; chlorpropamide]

Werner, U.; Werner, D.; Mundkowski, R.; Gillich, M.; Brune, K. Selective and rapid liquid chromatography-mass spectrometry method for the quantification of rofecoxib in pharmacokinetic studies with humans, *J.Chromatogr.B*, **2001**, *760*, 83–90.

Werner, U.; Werner, D.; Pahl, A.; Mundkowski, R.; Gillich, M.; Brune, K. Investigation of the pharmacokinetics of celecoxib by liquid chromatography-mass spectrometry, *Biomed.Chromatogr.*, **2002**, *16*, 56–60. [rofecoxib is internal standard]

Yener, G.; Gonullu, U.; Uner, M.; Degim, T.; Araman, A. Effect of vehicles and penetration enhancers on the in vitro percutaneous absorption of celecoxib through human skin, *Pharmazie*, **2003**, *58*, 330–333.

# Cerivastatin

**Molecular formula:** $C_{26}H_{34}FNO_5$
**Molecular weight:** 459.55
**CAS Registry No:** 145599-86-6
**Merck Index:** 13, 2004

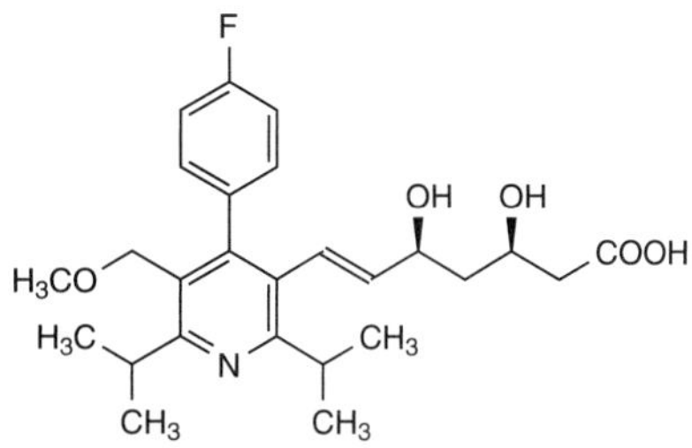

## SAMPLE

**Matrix:** blood
**Sample preparation:** Add 50 μL serum to 500 μL chilled (4°) 100 mM pH 5.0 sodium acetate buffer, mix, add 50 μL IS in MeCN:water 50:50, vortex, add 5 mL MTBE, shake mechanically for 15 min, centrifuge for 5 min, freeze in dry ice/MeOH. Remove the organic layer, evaporate to dryness under reduced pressure with nitrogen at 35°, reconstitute the residue with 50 μL MeCN:10 mM pH 4.0 ammonium formate buffer 50:50, centrifuge, inject a 25 μL aliquot.

## HPLC VARIABLES

**Guard column:** 10 × 2 YMC Basic
**Column:** 50 × 2 5 μm YMC Basic
**Mobile phase:** Gradient. MeCN:10 mM pH 4.0 ammonium formate buffer from 60:40 to 100:0 over 0.3 min, maintain at 100:0 for 0.7 min, return to initial conditions over 0.1 min, re-equilibrate for 2.4 min.
**Flow rate:** 0.3
**Injection volume:** 25
**Detector:** MS, Sciex API 365, positive turbo ion spray at 400°, ion spray +5.5 kV, orifice voltage 45 V, nebulizer gas nitrogen 5.5 bar, turbo ion spray gas at 7 L/min, dwell time 200 ms, collision cell energy 45 eV, m/z 460.4-356.0 (acid), m/z 442.2–354.0 (lactone)

## CHROMATOGRAM

**Retention time:** 1.52 (acid), 1.75 (lactone)
**Internal standard:** $d_3$-cerivastatin acid (1.49), $d_3$-cerivastatin lactone (1.72)
**Limit of quantitation:** 0.01 ng/mL

## OTHER SUBSTANCES

**Extracted:** metabolites

## KEY WORDS

serum

## REFERENCE

Jemal, M.; Rao, S.; Salahudeen, I.; Chen, B.-C.; Kates, R. Quantitation of cerivastatin and its seven acid and lactone biotransformation products in human serum by liquid chromatography-electrospray tandem mass spectrometry, *J.Chromatogr.B*, **1999**, *736*, 19–41.

## SAMPLE

**Matrix:** blood, formulations
**Sample preparation:** Vortex 2 mL serum and 500 μL MeCN for 2 min, centrifuge at 4000 g for 10 min, inject a 50 μL aliquot of the supernatant. Crush tablets and weigh out an amount corresponding to 10 mg cerivastatin, sonicate with MeOH for 10 min, make up to 10 mL with MeOH, filter, add IS to the filtrate, dilute with mobile phase, inject a 50 μL aliquot.

## HPLC VARIABLES

**Column:** 250 × 4.6 5 μm LC18 (Waters)
**Mobile phase:** MeCN:buffer 35:65 (Buffer was 10 mM potassium dihydrogen phosphate adjusted to pH 3.1 with phosphoric acid.)
**Flow rate:** 1
**Injection volume:** 50
**Detector:** UV 232

## CHROMATOGRAM

**Retention time:** 6.90
**Internal standard:** losartan (8.78)
**Limit of detection:** 0.62 ng/mL
**Limit of quantitation:** 207 ng/mL

## KEY WORDS

serum; tablets

## REFERENCE

Ozkan, S.A.; Ozkan, Y.; Aboul-Enein, H.Y. Quality control and drug dissolution studies of pharmaceutical preparations containing cerivastatin sodium by means of RP-HPLC, *J.Liq.Chromatogr.Rel.Technol.*, **2002**, *25*, 251–262.

## ANNOTATED BIBLIOGRAPHY

Boberg, M.; Angerbauer, R.; Fey, P.; Kanhai, W.K.; Karl, W.; Kern, A.; Ploschke, J.; Radtke, M. Metabolism of cerivastatin by human liver microsomes in vitro. Characterization of primary metabolic pathways and of cytochrome P450 isozymes involved, *Drug Metab.Dispos.*, **1997**, *25*, 321–331. [LC-MS]

Krol, G.J.; Beck, G.W.; Ritter, W.; Lettieri, J.T. LC separation and induced fluorometric detection of rivastatin in blood plasma, *J.Pharm.Biomed.Anal.*, **1993**, *11*, 1269–1275.

Mazzu, A.L.; Lasseter, K.C.; Shamblen, E.C.; Agarwal, V.; Lettieri, J.; Sundaresen, P. Itraconazole alters the pharmacokinetics of atorvastatin to a greater extent than either cerivastatin or pravastatin, *Clin.Pharmacol.Ther.*, **2000**, *68*, 391–400. [SPE]

Prueksaritanont, T.; Tang, C.; Qiu, Y.; Mu, L.; Subramanian, R.; Lin, J.H. Effects of fibrates on metabolism of statins in human hepatocytes, *Drug Metab.Dispos.*, **2002**, *30*, 1280–1287. [cerivastatin; simvastatin; atorvastatin; rosuvastatin; pravastatin]

# Cetrorelix

**Molecular formula:** $C_{70}H_{92}ClN_{17}O_{14}$
**Molecular weight:** 1431.06
**CAS Registry No:** 120287-85-6
**Merck Index:** 13, 2036

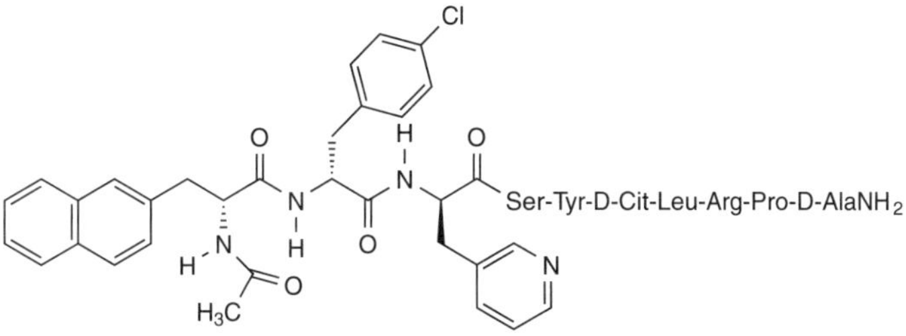


## SAMPLE

**Matrix:** bile, feces, urine

**Sample preparation:** Bile. Dilute bile with 2 vol of 30% acetic acid, inject a 50–100 μL aliquot. After each run, regenerate column with a water injection. Feces. Homogenize feces with 1.5 vol of water. Extract 5 g homogenized feces three times with 5 mL aliquots of MeOH:ethyl acetate 2:1 using an ultra turrax (TP 18–10; IKA, Staufen, Germany) at about 10000 rpm with three 30 s applications, centrifuge at 10000 g for 15 min. Combine the supernatants, evaporate to dryness under reduced pressure at 40°, reconstitute the residue with 2 mL 30% acetic acid in water, filter (minisart GF and minisart NML, 0.8 μm, Sartorius), inject a 50–100 μL aliquot. Urine. Directly inject a 50–100 μL aliquot of urine.

## HPLC VARIABLES

**Guard column:** 30 × 3 Merck
**Column:** 250 × 3 5 μm LiChrospher WP 300 RP-18
**Mobile phase:** Gradient. A:B from 95:5 to 20:80 over 180 min, to 0:100 over 2 min, maintain at 0:100 for 10 min, return to initial conditions over 3 min. A was 0.1% trifluoroacetic acid in water, adjusted to pH 2.0 with 1 M NaOH. B was MeCN:water:trifluoroacetic acid 90:10:0.1, adjusted to pH 2.0 with 1 M NaOH.
**Flow rate:** 0.5
**Injection volume:** 50–100
**Detector:** Radioactivity ($^{14}C$); UV 226

## CHROMATOGRAM

**Retention time:** 82

## OTHER SUBSTANCES

**Extracted:** metabolites

## KEY WORDS

dog; rat

## REFERENCE

Schwahn, M.; Schupke, H.; Gasparic, A.; Krone, D.; Peter, G.; Hempel, R.; Kronbach, T.; Locher, M.; Jahn, W.; Engel, J. Disposition and metabolism of cetrorelix, a potent luteinizing hormone-releasing hormone antagonist, in rats and dogs, *Drug Metab.Dispos.*, **2000**, *28*, 10–20.

## SAMPLE

**Matrix:** blood, urine

**Sample preparation:** Plasma. Condition a 500 mg Sep-Pak Vac C8 with 20 mL MeOH and 20 mL water. Vortex 1 mL plasma with 10 μL 20 μg/mL IS in mobile phase, add 10 mL MeOH:water 10:90, mix, add to the SPE cartridge, wash with 10 mL MeOH:water 10:90, wash with 10 mL MeOH:water 50:50, elute with 10 mL MeOH:water:trifluoroacetic acid 90:10:0.1. Evaporate the eluate to dryness under reduced pressure, reconstitute the residue with 100 μL mobile phase, vortex, add to an Ultrafree-MC/LCR filter (0.2 μm, area 0.2 $cm^2$), filter while centrifuging at 3500 g at 4° for 20 min, inject a 40 μL aliquot of the filtrate. Urine. Condition a 500 mg Sep-Pak Vac C8 with 20 mL MeOH and 20 mL water. Mix 2 mL urine with 200 μL blank human plasma, add 10 μL 20 μg/mL IS in mobile phase, mix, add 8 mL MeOH:water 10:70, mix, add to the SPE cartridge, wash with 10 mL MeOH:water 10:90, wash with 10 mL MeOH:water 50:50, elute with 10 mL MeOH:water:trifluoroacetic acid 90:10:0.1. Evaporate the eluate to dryness under reduced pressure, reconstitute the residue with 100 μL mobile phase, vortex, add to an Ultrafree-MC/LCR filter (0.2 μm, area 0.2 $cm^2$), filter while centrifuging at 3500 g at 4° for 20 min, inject a 40 μL aliquot of the filtrate.

## HPLC VARIABLES

**Column:** 150 × 2.1 5 μm μ-Bondasphere C18
**Mobile phase:** MeCN:water:trifluoroacetic acid 35:65:0.1
**Flow rate:** 0.2
**Injection volume:** 40
**Detector:** MS, Micromass Platform single-stage quadrupole, electrospray, negative ion detection, nebulizer gas nitrogen at 30 L/h, drying gas nitrogen at 300 L/h, source temperature 120°, capillary voltage 3 kV, m/z 1429, 1543, 1657

## CHROMATOGRAM

**Retention time:** 6
**Internal standard:** brominated cetrorelix (dissolve 10 mg cetrorelix in 1 mL glacial acetic acid, slowly add 15 μL 10% bromine in acetic acid while slowly stirring at room temperature, stir for 1 h, evaporate to dryness under a stream of nitrogen, reconstitute the residue with glacial acetic acid, store at 4°.) (m/z 1586, 1700, 1813) (8)
**Limit of quantitation:** 1 ng/mL

## KEY WORDS

pharmacokinetics; plasma; SPE

## REFERENCE

Niwa, M.; Enomoto, K.; Yamashita, K. Measurement of the novel decapeptide cetrorelix in human plasma and urine by liquid chromatography-electrospray ionization mass spectrometry, *J.Chromatogr.B*, **1999**, *729*, 245–253.

## SAMPLE

**Matrix:** blood
**Sample preparation:** Mix 4 mL plasma with 200 μL 1 M HCl, add 100 μL 1 μg/mL IS in 10 mM acetic acid, add 9 mL ethyl acetate:1-butanol 90:10, extract on a Reax mixer for 15 min, centrifuge at 1000 g for 3 min, remove 5 mL of the organic layer, add 5 mL ethyl acetate:1-butanol 90:10, extract for 15 min, centrifuge at 1000 g for 3 min, remove 5 mL of the organic layer. Combine the organic layers and extract with 150 μL 10 mM HCl, centrifuge at 1000 g for 3 min, inject a 150 μL aliquot of the aqueous layer.

## HPLC VARIABLES

**Column:** 250 × 4 5 μm LiChrospher 60 RP-Select B
**Mobile phase:** MeCN:MeOH:50 mM pH 4 ammonium acetate buffer 25.5:25.5:49
**Flow rate:** 0.4
**Injection volume:** 150
**Detector:** F ex 227 em 340

## CHROMATOGRAM

**Retention time:** 25
**Internal standard:** D-21740 (Ac-D-$Nal^1$-(*p*-Cl)-D-$Phe^2$-D-$Pal^3$-$Ser^4$-$Tyr^5$-D-$Cit^6$-$Leu^7$-$Arg^8$-$Ala^9$-D-$Ala^{10}$-$NH_2$ trifluoroacetate) (23)
**Limit of quantitation:** 2 ng/mL

## KEY WORDS

plasma

## REFERENCE

Raffel, H.H.; Locher, M.; Borbe, H.O. High-performance liquid chromatographic assay for the determination of the decapeptide cetrorelix, a novel luteinizing hormone-releasing hormone antagonist, in human plasma, *J.Chromatogr.B*, **1994**, *653*, 102–105.

# Cetyl alcohol

$CH_3(CH_2)_{14}CH_2OH$

**Molecular formula:** $C_{16}H_{34}O$
**Molecular weight:** 242.44
**CAS Registry No:** 36653-82-4
**Merck Index:** 13, 2037

## SAMPLE

**Matrix:** solutions

**Sample preparation:** Prepare a solution of the alcohol by dissolving 10 mg alcohol in 200 μL pyridine and diluting to 10 mL in MeCN. Mix 1 mL solution with 100 μL reagent and 100 μL 2% 1-isopropyl-3-(3-dimethylaminopropyl)carbodiimide perchlorate in MeCN, heat at 80° for 20 min, cool to room temperature, add 1 mL water, add 2 mL isopropanol:water 50:50, add to a Sep-Pak ODS SPE cartridge, rinse the tube with 3 mL isopropanol:water 50:50, add the rinse to the SPE cartridge, wash with 3 mL isopropanol:water 50:50, elute with 2 mL isopropanol, inject a 20 μL aliquot of the eluate. (The reagent was 10 mg 2-(4-carboxyphenyl)-5,6-dimethylbenzimidazole in 1 mL pyridine, add 700 mg 4-piperidinopyridine, dilute to 10 mL with MeCN. Prepare 2-(4-carboxyphenyl)-5,6-dimethylbenzimidazole as follows. Add 13 g 4-carboxybenzaldehyde (terephthalaldehydic acid) in 400 mL EtOH dropwise to 4,5-dimethyl-1,2-phenylenediamine in 400 mL EtOH in an ice bath, after 1 h reflux for 8 h, cool to room temperature, collect the precipitate, recrystallize three times from MeOH:water 50:50 to give 2-(4-carboxyphenyl)-5,6-dimethylbenzimidazole as a white amorphous product (mp >300°). 4-Piperidinopyridine is not commercially available but 4-dimethylaminopyridine or 4-pyrrolidinopyridine can be used instead although interferences are greater. Alternatively, 4-piperidinopyridine can be synthesized as follows. Add 200 mmol piperidine dropwise with stirring to 15 g phosphorus pentoxide and 9.51 g 4-hydroxypyridine, heat at 250° for 7 h, cautiously pour onto 200 g ice, add 400 mL 1 M NaOH, add 200 mL ether. Remove the ether layer and extract the aqueous layer three times with 100 mL portions of ether. Combine the organic layers and dry them over anhydrous potassium carbonate, evaporate, distill the residue, recrystallize from petroleum ether (bp 80–100°) to give 4-piperidinopyridine (bp 167–170°/11 mm Hg; mp 79–80°) (*Synthesis* **1978**, 844). Alternatively, add 1.94 g 4-bromopyridine hydrochloride to 5 mL 50% NaOH, add 5 mL piperidine, add 2.72 g benzyltriethylammonium bromide, heat at 100° for 5 h, remove excess piperidine by distillation, add 25 mL water, extract four times with 25 mL portions of benzene. Combine the organic layers and dry them over anhydrous sodium sulfate, boil the residue with petroleum ether to give 4-piperidinopyridine (mp 80°) (*Syn.Commun.* **1979**, *9*, 251). Prepare 1-isopropyl-3-(3-dimethylaminopropyl)carbodiimide perchlorate as follows. Stir 1.41 mol isopropylisocyanate in 750 mL dichloromethane at 5°, add 144 g 3-dimethylaminopropylamine (*N*, *N*-dimethyl-1,3-propanediamine) in 250 mL dichloromethane at such a rate that the temperature does not exceed 10°, add 500 mL triethylamine, add 300 g *p*-toluenesulfonyl chloride in 300 mL dichloromethane at such a rate that the temperature does not exceed 10°, reflux for 3 h, add 400 g anhydrous sodium carbonate, add 3.5 L ice water, stir vigorously for 30 min, remove the organic phase. Extract the aqueous phase three times with 500 mL portions of dichloromethane. Combine the organic layers and dry them over anhydrous sodium sulfate, evaporate under reduced pressure, distil the residue to give 1-isopropyl-3-(3-dimethylaminopropyl)carbodiimide (bp 91–92°/10 mm Hg (*Ber.* **1941**, *74B*, 1285)) (cf. *Org.Syn.* **1973**, *Coll Vol. V*, 555). Prepare pyridine perchlorate from pyridine and 20% perchloric acid, crystallize from EtOH (*Ber.* **1926**, *59*, 446). Add 18 g pyridine perchlorate in portions to 100 mmol 1-isopropyl-3-(3-dimethylaminopropyl)carbodiimide stirred in 200 mL dichloromethane at 0°, let stand for 30 min, filter, add 200 mL anhydrous diethyl ether to the filtrate. Filter off the precipitate and recrystallize it from dichloromethane/diethyl ether to give 1-isopropyl-

3-(3-dimethylaminopropyl)carbodiimide perchlorate (mp 88–90°) (*Chem.Pharm.Bull.* **1985**, 33, 5375).)

---

**HPLC VARIABLES**
**Guard column:** 50 × 4.6 7 μm Zorbax ODS
**Column:** 250 × 4.6 7 μm Zorbax ODS
**Mobile phase:** MeOH:isopropanol 85:15
**Flow rate:** 1
**Injection volume:** 20
**Detector:** F ex 338 em 428

---

**CHROMATOGRAM**
**Retention time:** 9.4
**Limit of detection:** 10–20 pg/mL

---

**OTHER SUBSTANCES**
**Simultaneous:** dodecyl alcohol (6.2), tetradecyl alcohol (7.2), stearyl alcohol (12.5), eicosyl alcohol (16.5)

---

**KEY WORDS**
derivatization; SPE

---

**REFERENCE**
Katayama, M.; Masuda, Y.; Taniguchi, H. Determination of alcohols by high-performance liquid chromatography after pre-column derivatization with 2-(4-carboxyphenyl)-5,6-dimethylbenzimidazole, *J.Chromatogr.*, **1991**, *585*, 219–224.

# Cevimeline hydrochloride

**Molecular formula:** $C_{10}H_{17}NOS.HCl.1/2H_2O$

**Molecular weight:** 244.78

**CAS Registry No:** 153504-70-2, 107233-08-9 (free base)

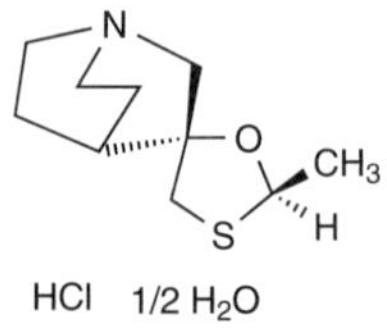


## SAMPLE

**Matrix:** blood

**Sample preparation:** Condition a Bond Elut Certify SPE cartridge with 3 mL MeOH and 3 mL 100 mM pH 6.0 phosphate buffer. Mix 1 mL plasma with 100 μL water, 100 μL 1 μg/mL IS, and 2 mL 100 mM pH 6.0 phosphate buffer, add to the SPE cartridge, wash with 4 mL 10 μM acetic acid, wash with 5 mL MeOH, elute with 6 mL dichloromethane:isopropanol 80:20 containing 2% ammonia water. Concentrate the eluate under a stream of nitrogen at 40° for 10 min, add 200 μL 0.02% DL-tartaric acid in MeOH, evaporate to dryness under a stream of nitrogen at 40°, reconstitute the residue with 200 μL MeOH, evaporate to dryness again, reconstitute the residue with 50 μL MeOH:water 20:80, inject an aliquot.

## HPLC VARIABLES

**Column:** 250 × 4.6 5 μm Capcell pak C18 UG120

**Mobile phase:** Gradient. MeCN:0.075% heptafluorobutyric acid in water from 5:95 to 25:75 over 25 min.

**Detector:** MS, JEOL AX-505 W, 1% glycerol in MeOH is added post-column at 0.3 mL/min, chamber temperature 60°, emission current 5 mA, collision gas xenon, FAB positive ion

## CHROMATOGRAM

**Limit of quantitation:** 5 ng/mL

## KEY WORDS

dog; pharmacokinetics; plasma; rat

## REFERENCE

Washio, T.; Kohsaka, K.; Arisawa, H.; Masunaga, H. Pharmacokinetics and metabolism of the novel muscarinic receptor agonist SNI-2011 in rats and dogs, *Arzneimittelforschung*, **2003**, *53*, 26–33.

## SAMPLE

**Matrix:** urine

**Sample preparation:** Freeze-dry urine and reconstitute with 0.05% trifluoroacetic acid, inject an aliquot.

## HPLC VARIABLES

**Column:** 250 × 4.6 Inertsil ODS-2

**Mobile phase:** Gradient. MeCN:0.05% trifluoroacetic acid from 0:100 to 20:80 over 40 min.

**Flow rate:** 1

**Detector:** Radioactivity ($^{14}C$); Refractive Index

## CHROMATOGRAM

**Retention time:** 29

## OTHER SUBSTANCES

**Extracted:** metabolites

## REFERENCE

Washio, T.; Kohsaka, K.; Arisawa, H.; Masunaga, H.; Nagatsuka, S.-i.; Satoh, Y. Pharmacokinetics and metabolism of radiolabelled SNI-2011, a novel muscarinic receptor agonist, in healthy volunteers, *Arzneimittelforschung*, **2003**, *53*, 80–86.

# Chlorobutanol

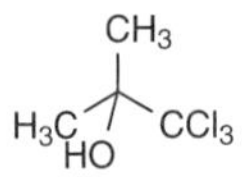


**Molecular formula:** $C_4H_7Cl_3O$
**Molecular weight:** 177.46
**CAS Registry No:** 57-15-8
**Merck Index:** 13, 2148

## SAMPLE

**Matrix:** formulations
**Sample preparation:** Dissolve ophthalmic ointment containing 20 mg chlorobutanol in 50 mL hexane, extract 3 times with 15 mL portions of MeOH:water 75:25. Combine the extracts, make up to 50 mL with MeOH:water 75:25, inject a 100 μL aliquot.

## HPLC VARIABLES

**Guard column:** 10 × 2 47 μm Bondapak C18/Corasil
**Column:** 250 × 4 10 μm ODS-10 (Bio-Rad)
**Mobile phase:** MeOH:water 50:50
**Flow rate:** 1.8
**Injection volume:** 100
**Detector:** UV 210

## CHROMATOGRAM

**Retention time:** 6

## KEY WORDS

ophthalmic ointment

## REFERENCE

Dunn, D.L.; Jones, W.J.; Dorsey, E.D. Analysis of chlorobutanol in ophthalmic ointments and aqueous solutions by reverse-phase high-performance liquid chromatography, *J.Pharm.Sci.*, **1983**, *72*, 277–280.

# Chloroprocaine

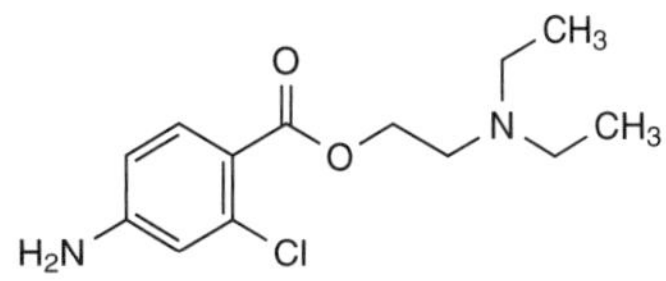

**Molecular formula:** $C_{13}H_{19}ClN_2O_2$
**Molecular weight:** 270.76
**CAS Registry No:** 3858-89-7 (HCl)
**Merck Index:** 13, 2177

## SAMPLE
**Matrix:** blood
**Sample preparation:** Mix 1 mL plasma with 50 μL 50 μg/mL lidocaine in water and 200 μL 2 M NaOH, extract with diethyl ether. Evaporate the organic layer to dryness under a stream of nitrogen, reconstitute with 150 μL mobile phase, inject a 50 μL aliquot.

## HPLC VARIABLES
**Column:** 150 × 3.9 μBondapak C18
**Mobile phase:** MeCN:50 mM pH 5.8 sodium phosphate buffer 30:70
**Flow rate:** 1
**Injection volume:** 50
**Detector:** UV 210

## CHROMATOGRAM
**Retention time:** 4
**Internal standard:** lidocaine (5)
**Limit of detection:** 100 ng/mL

## OTHER SUBSTANCES
**Extracted:** metabolites

## KEY WORDS
dog; pharmacokinetics; plasma

## REFERENCE
Janicki, P.K.; Johnson, R.; Kambam, J.R. Rapid determination of chloroprocaine and its major metabolite, 2-chloroaminobenzoic acid, in plasma by high-performance liquid chromatography, *J.Chromatogr.B*, **1996**, *675*, 336–341.

# Chorionic gonadotropin

**CAS Registry No:** 9002-61-3 (human), 9002-70-4 (horse)
**Merck Index:** 13, 2237

## SAMPLE
**Matrix:** solutions
**Sample preparation:** Inject a 20 μL aliquot of a 1 mg/mL solution.

## HPLC VARIABLES
**Column:** 250 × 4.6 10 μm Vydac 214TP
**Mobile phase:** Gradient. A:B from 30:70 to 0:100 over 40 min. A was 100 mM pH 6.8 sodium phosphate buffer. B was MeCN:100 mM pH 6.8 sodium phosphate buffer 50:50.
**Flow rate:** 2
**Injection volume:** 20
**Detector:** UV 220

## CHROMATOGRAM
**Retention time:** 19

## OTHER SUBSTANCES
**Simultaneous:** luteinizing hormone (20), thyroid-stimulating hormone (16)

## REFERENCE
Chlenov, M.A.; Kandyba, E.I.; Nagornaya, L.V.; Orlova, I.L.; Volgin, Y.V. High-performance liquid chromatography of human glycoprotein hormones, *J.Chromatogr.*, **1993**, *631*, 261–267.

## SAMPLE
**Matrix:** solutions

## HPLC VARIABLES
**Guard column:** 40 × 4.1 SynChropak RSC (SynChrom)
**Column:** 250 × 4.1 SynChropak RP-P (SynChrom)
**Mobile phase:** Gradient. A:B 85:15 for 6 min, to 75:25 over 1 min, maintain at 75:25 for 5 min, return to initial conditions over 1 min, re-equilibrate for 13 min. A was water:ethylene glycol dimethyl ether:trifluoroacetic acid 93:7:0.012. B was isopropanol.
**Flow rate:** 1
**Detector:** UV 220

## CHROMATOGRAM
**Retention time:** 11

## REFERENCE
Wilks, J.W.; Butler, S.S. Biologic activity of human chorionic gonadotropin following reversed-phase high-performance liquid chromatography, *J.Chromatogr.*, **1984**, *298*, 123–130.

## ANNOTATED BIBLIOGRAPHY
Pask-Hughes, R.A. Characterisation and purification of some glycoproteins by high-performance liquid chromatography, *J.Chromatogr.*, **1987**, *393*, 273–284.

# Cilnidipine

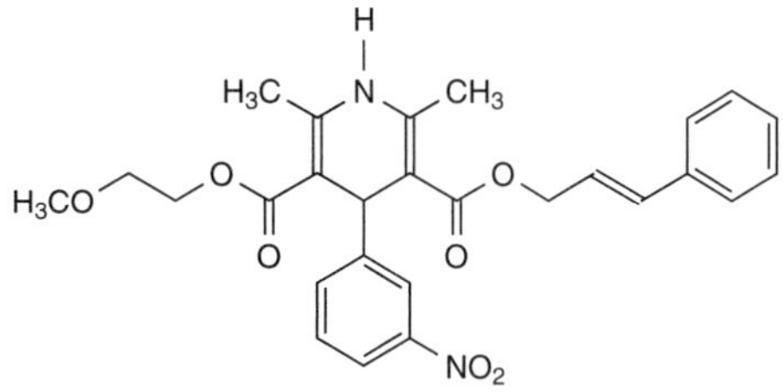


**Molecular formula:** $C_{27}H_{28}N_2O_7$
**Molecular weight:** 492.52
**CAS Registry No:** 132203-70-4
**Merck Index:** 13, 2297

## SAMPLE

**Matrix:** blood
**Sample preparation:** Condition a Bond Elut C18 SPE cartridge with 5 mL MeCN and 10 mL water. Mix 10 μL 10 μg/mL IS in MeCN with 1 mL plasma, add 2 mL MeCN, vortex for 30 s, centrifuge at 1670 g at 4° for 10 min, remove the supernatant, repeat the extraction with 1 mL MeCN. Add the supernatants to 6 mL water, vortex for 30 s, add to the SPE cartridge, wash with 2 mL MeCN:water 40:60, elute with 2 mL MeCN. Evaporate the eluate to dryness under reduced pressure at 40°, reconstitute the residue with 100 μL MeCN, inject a 10 μL aliquot.

## HPLC VARIABLES

**Column:** 60 × 4.6 3 μm Hypersil ODS
**Mobile phase:** MeCN:100 mM ammonium acetate 60:40
**Flow rate:** 1
**Injection volume:** 10
**Detector:** MS, Hewlett-Packard Model 5988A thermospray quadrupole, vaporizer stem 95°, ion source 275°, negative ion filament on mode, m/z 375

## CHROMATOGRAM

**Retention time:** 3.7
**Internal standard:** $d_3$-cilnidipine (m/z 378) (3.7)
**Limit of quantitation:** 0.5 ng/mL

## OTHER SUBSTANCES

**Extracted:** metabolites

## KEY WORDS

plasma; SPE

## REFERENCE

Hatada, K.; Kimura, M.; Ono, I.; Ozaki, M. Determination of a new calcium antagonist and its main metabolite in plasma by thermospray liquid chromatography-mass spectrometry, *J.Chromatogr.*, **1992**, *583*, 116–121.

# Cimetropium bromide

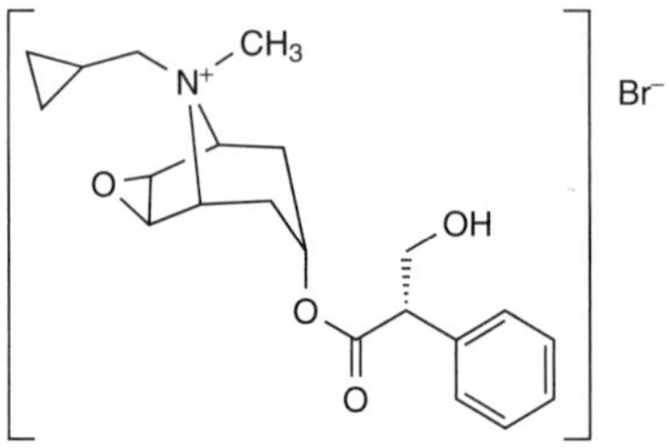


**Molecular formula:** $C_{21}H_{28}BrNO_4$
**Molecular weight:** 438.36
**CAS Registry No:** 51598-60-8
**Merck Index:** 13, 2301

## SAMPLE

**Matrix:** microsomal incubations

**Sample preparation:** Condition a Sep-Pak C18 SPE cartridge (unspecified). Mix 3.5 mL microsomal incubation with 1 mL 1% zinc sulfate solution, centrifuge at 3000 g for 15 min, add to the SPE cartridge, wash with 500 μL water, elute with 5 mL MeOH. Evaporate the eluate to dryness under a stream of nitrogen at 37°, reconstitute the residue with 100 μL MeOH, inject an aliquot.

## HPLC VARIABLES

**Column:** 250 × 4.6 10 μm Spherisorb ODS-2

**Mobile phase:** MeCN:MeOH:water 20:20:55 or 10:5:55 containing 20 or 35 mM sodium heptanesulfonate and 100 mM triethylamine, pH 3.7

**Flow rate:** 1

**Detector:** MS, VG 70-SEQ, positive ion FAB, collision gas xenon at 8.5 keV

## CHROMATOGRAM

**Internal standard:** glycopyrrolate or benzyl alcohol

## REFERENCE

Kajbaf, M.; Jahanshahi, M.; Pattichis, K.; Gorrod, J.W.; Naylor, S. Rapid and efficient purification of cimetropium bromide and mifentidine drug metabolite mixtures derived from microsomal incubates for analysis by mass spectrometry, *J.Chromatogr.*, **1992**, *575*, 75–85.

# Cisatracurium besylate

**Molecular formula:** $C_{65}H_{82}N_2O_{18}S_2$
**Molecular weight:** 1243.50
**CAS Registry No:** 96946-42-8
**Merck Index:** 13, 872
**(See also atracurium besylate in Volume 2)**

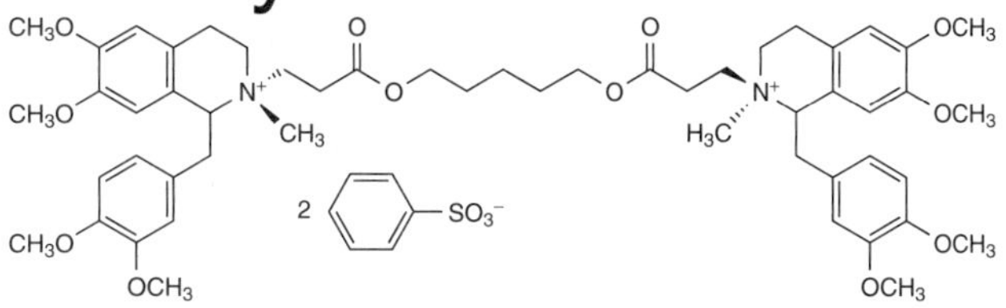

## SAMPLE
**Matrix:** blood, gastric contents
**Sample preparation:** Immediately vortex 500 μL plasma, serum, whole blood, or gastric contents with 20 μL 50 mM sulfuric acid, store at -20° (if required), add 20 μL 10 μg/mL IS in MeOH, add 1 mL MeCN, vortex, let stand for 10 min, centrifuge at 2000 g for 5 min. Evaporate the supernatant to 100–150 μL under a stream of nitrogen, inject a 3 μL aliquot.

## HPLC VARIABLES
**Column:** 150 × 1 3.5 μm X-Terra MS C18
**Mobile phase:** Gradient. A was 2 mM pH 3 ammonium formate buffer. B was MeCN:2 mM pH 3 ammonium formate buffer 90:10. A:B 85:15 for 2 min, to 57:43 over 8 min, re-equilibrate at initial conditions for 1 min.
**Flow rate:** 0.05
**Injection volume:** 3
**Detector:** MS, PE Sciex API-100, electrospray, ion spray 5800 V, positive ion mode, nebulizer gas nitrogen, curtain gas nitrogen, collision gas argon, m/z 464.6, 358.4

## CHROMATOGRAM
**Retention time:** 9.1
**Internal standard:** ambenonium besylate (m/z 125.0, 411.6)
**Limit of detection:** 5 ng/mL
**Limit of quantitation:** 10 ng/mL

## OTHER SUBSTANCES
**Extracted:** laudanosine (m/z 206.1, 358.4) (7.0), mivacurium (m/z 446.6, 402.5) (6.3), pancuronium (m/z 430.5, 472.5) (6.4), rocuronium (m/z 529.4, 358.4) (5.3), vecuronium (m/z 557.4, 398.4) (6.8)

## KEY WORDS
plasma; serum; whole blood

## REFERENCE
Sayer, H.; Quintela, O.; Marquet, P.; Dupuy, J.-L.; Gaulier, J.M.; Lachâtre, G. Identification and quantitation of six non-depolarizing neuromuscular blocking agents by LC-MS in biological fluids, *J.Anal.Toxicol.*, **2004**, *28*, 105–110.

## SAMPLE
**Matrix:** solutions

## HPLC VARIABLES
**Column:** 250 × 4.6 5 μm Spherisorb ODS-2
**Column temperature:** 40

**Mobile phase:** Gradient. A was MeCN:buffer 50:50. B was MeOH. A:B 100:0 for 9 min, to 0:100 over 2 min, maintain at 0:100 for 4 min. (The buffer was 300 mM ammonium formate adjusted to pH 5.6 with formic acid.)
**Flow rate:** 1
**Injection volume:** 20
**Detector:** UV 280; MS, Micromass Quattro II, triple quadrupole, ESI, source 150°, source capillary voltage 2500–3500 V, cone voltage 25–50 V or APCI, positive ion mode to negative ion mode at 11 min, source 150°, probe 550°, corona voltage 2500 V, cone voltage 35 V

## CHROMATOGRAM
**Retention time:** 9.3
**Limit of detection:** 100 pg/mL (positive ion mode)

## OTHER SUBSTANCES
**Simultaneous:** degradation products, propofol (negative ion mode) (13)

## REFERENCE
Wang, P.; Zhang, H.; Stewart, J.T.; Bartlett, M.G. Simultaneous detection of cisatracurium, its degradation products and propofol using positive ion detection followed by negative ion detection in a single LC/MS run, *J.Pharm.Biomed.Anal.*, **1998**, *17*, 547–554.

## SAMPLE
**Matrix:** solutions

## HPLC VARIABLES
**Column:** 250 × 4.6 5 μm Spherisorb ODS-2
**Mobile phase:** MeCN:buffer 50:50 (The buffer was 300 mM ammonium formate adjusted to pH 5.2 with formic acid.)
**Flow rate:** 1
**Injection volume:** 20
**Detector:** UV 280; MS, Micromass Quattro II, EI positive ion mode or APCI, scan m/z 150–600, m/z 464

## CHROMATOGRAM
**Retention time:** 8.8

## OTHER SUBSTANCES
**Simultaneous:** degradation products, laudanosine (5, m/z 358), propofol (13, m/z 178)

## KEY WORDS
stability-indicating

## REFERENCE
Zhang, H.; Wang, P.; Bartlett, M.G.; Stewart, J.T. HPLC determination of cisatracurium besylate and propofol mixtures with LC-MS identification of degradation products, *J.Pharm.Biomed.Anal.*, **1998**, *16*, 1241–1249.

## ANNOTATED BIBLIOGRAPHY
Bryant, B.J.; James, C.D. Jr.; Cook, D.R.; Harrelson, J.C. High performance liquid chromatographic assay for cisatracurium and its metabolites in human urine, *J.Liq.Chromatogr.Rel.Technol.*, **1997**, *20*, 2041–2051.

Xu, Q.A.; Zhang, Y.-P.; Trissel, L.A.; Gilbert, D.L.; Martinez, J.F.; Fox, J.L. Stability of cisatracurium besylate in vials, syringes, and infusion admixtures, *Am.J.Health-Syst.Pharm.*, **1998**, *55*, 1037–1041.

# Citric acid

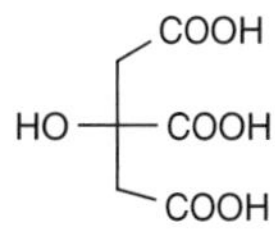

**Molecular formula:** $C_6H_8O_7$
**Molecular weight:** 192.12
**CAS Registry No:** 77-92-9
**Merck Index:** 13, 2350

## SAMPLE
**Matrix:** blood
**Sample preparation:** Condition a 30 × 10 column of DEAE-cellulose (Serva) with 10 vol of 100 mM pH 7.0 sodium perchlorate and 15 vol of water. Mix 6 mL plasma with 6 mL MeCN, centrifuge at 4000 g for 10 min, suspend the pellet in 6 mL MeCN:water 50:50, centrifuge at 4000 g for 10 min. Combine the supernatants, adjust the pH of a 10 mL aliquot to 7.0 with 100 mM NaOH, add to the DEAE-cellulose column, wash with 10 mL water, elute with 10 mL 100 mM perchloric acid. Freeze-dry the eluate overnight at 0°, reconstitute the residue with 500 μL mobile phase, centrifuge at 12000 g for 2 min, inject a 100 μL aliquot.

## HPLC VARIABLES
**Guard column:** 50 × 4 micro-Guard-NH2
**Column:** 300 × 7.8 Aminex HPX-87 cation-exchange (Bio-Rad)
**Mobile phase:** 6.5 mM sulfuric acid
**Flow rate:** 0.5
**Injection volume:** 100
**Detector:** UV 210

## CHROMATOGRAM
**Retention time:** 5

## OTHER SUBSTANCES
**Extracted:** adipic acid (12), meglutol (8), 2-oxoglutaric acid (6.4),

## KEY WORDS
plasma; SPE

## REFERENCE
Lippe, G.; Trevisan, R.; Nosadini, R.; Fabris, R.; Deana, R. 3-Hydroxy-3-methylglutaric, adipic, and 2-oxoglutaric acids measured by HPLC in the plasma from diabetic patients, *Clin.Biochem.*, **1987**, *20*, 275–279.

## SAMPLE
**Matrix:** plants
**Sample preparation:** Extract 20 g leaves with 250 mL water at 15 psi for 20 min, filter, repeat extraction and filtration twice more. Combine the extracts, concentrate to 50 mL under reduced pressure, add 200 mL EtOH, centrifuge, concentrate the supernatant to 25 mL under reduced pressure, inject a 20 μL aliquot.

## HPLC VARIABLES
**Column:** 250 × 4.6 μBondapak C18
**Mobile phase:** 8 mM sulfuric acid
**Flow rate:** 1
**Injection volume:** 20
**Detector:** UV 210

## CHROMATOGRAM
**Retention time:** 5.7
**Limit of quantitation:** 1.4 μg

## KEY WORDS
leaves

## REFERENCE
Jayaprakasha, G.K.; Sakariah, K.K. Determination of organic acids in leaves and rinds of *Garcinia indica* (Desr.) by LC, *J.Pharm.Biomed.Anal.*, **2002**, *28*, 379–384.

## SAMPLE
**Matrix:** solutions
**Sample preparation:** Inject a 5 μL aliquot.

## HPLC VARIABLES
**Column:** 150 × 4.6 5 μm LC-18
**Mobile phase:** 0.5% Ammonium dihydrogen phosphate adjusted to pH 2.80 with 1 M phosphoric acid
**Flow rate:** 1
**Injection volume:** 5
**Detector:** UV 214

## CHROMATOGRAM
**Retention time:** 5
**Limit of detection:** 30 ng

## OTHER SUBSTANCES
**Simultaneous:** ascorbic acid (3.3, LOD 10 ng), malic acid (2.5, LOD 50 ng), oxalic acid (1, LOD 10 ng), succinic acid (5.7, 100 ng), tartaric acid (1.7, LOD 10 ng)

## REFERENCE
Zhanguo, C.; Jiuru, L. Simultaneous and direct determination of oxalic acid, tartaric acid, malic acid, vitamin C, citric acid, and succinic acid in *Fructus mume* by reversed-phase high-performance liquid chromatography, *J.Chromatogr.Sci.*, **2002**, *40*, 35–39.

## ANNOTATED BIBLIOGRAPHY
Jayaprakasha, G.K.; Jena, B.S.; Sakariah, K.K. Improved liquid chromatographic method for determination of organic acids in leaves, pulp, fruits, and rinds of *Garcinia*, *J.AOAC Int.*, **2003**, *86*, 1063–1068. [SPE]

Nozal, M.J.; Bernal, J.L.; Diego, J.C.; Gómez, L.A.; Higes, M. HPLC determination of low molecular weight organic acids in honey with series-coupled ion-exclusion columns, *J.Liq.Chromatogr.Rel. Technol.*, **2003**, *26*, 1231–1253. [oxalic acid; glucuronic acid; citric acid; galacturonic acid; propionic acid; pyruvic acid; malic acid; citramalic acid; quinic acid; gluconic acid; lactic acid; formic acid; glutaric acid; fumaric acid; succinic acid; butyric acid]

Pérez-Ruiz, T.; Martínez-Lozano, C.; Tomás, V.; Martín, J. High-performance liquid chromatographic separation and quantification of citric, lactic, malic, oxalic and tartaric acids using a post-column photochemical reaction and chemiluminescence detection, *J.Chromatogr.A*, **2004**, *1026*, 57–64.

Podgornik, A.; Barut, M.; Jaksa, S.; Jancar, J.; Strancar, A. Application of very short monolithic columns for separation of low and high molecular mass substances, *J.Liq.Chromatogr.Rel.Technol.*, **2002**, *25*, 3099–3116. [citric acid; malic acid; ketoglutaric acid]

Qiu, J.; Jin, X. Development and optimization of organic acid analysis in tobacco with ion chromatography and suppressed conductivity detection, *J.Chromatogr.A*, **2002**, *950*, 81–88. [citric acid; malonic acid; malic acid; succinic acid; lactic acid; formic acid; acetic acid; pyroglutamic acid]

Sanarico, D.; Motta, S.; Bertolini, L.; Antonelli, A. HPLC determination of organic acids in traditional balsamic vinegar of Reggio Emilia, *J.Liq.Chromatogr.Rel.Technol.*, **2003**, *26*, 2177–2187. [dextrose; fructose; citric acid; tartaric acid; gluconic acid; malic acid; succinic acid; lactic acid; acetic acid]

Shui, G.; Leong, L.P. Separation and determination of organic acids and phenolic compounds in fruit juices and drinks by high-performance liquid chromatography, *J.Chromatogr.A*, **2002**, *977*, 89–96. [vitamin C; catechin; epicatechin; myricetin; quercetin; eugenol; kaempferol; tartaric acid; oxalic acid; malic acid; ascorbic acid; malonic acid; lactic acid; acetic acid; citric acid; fumaric acid; gallic acid; hydroxybenzoic acid; *p*-hydroxybenzoic acid; chlorogenic acid; caffeic acid; syringic acid; ferulic acid; benzoic acid; ellagic acid; salicylic acid; cinnamic acid]

Skelly, N.E. Separation of aliphatic and aromatic acids, aromatic sulfonates, quaternary ammonium compounds, and chelating agents on a reversed-phase column without ion pairing, *J.Chromatogr.Sci.*, **2003**, *41*, 22–25. [nonoxynol-9; citric acid; benzenesulfonic acid; phthalic acid; hydrobromic acid; nitrilotriacetic acid; oxalic acid; nitric acid; hydriodic acid; glycolic acid; formic acid; nitrous acid; cyanuric acid; lactic acid; acetic acid; NTA; benzalkonium; EDTA]

Suárez-Luque, S.; Mato, I.; Huidobro, J.F.; Simal-Lozano, J. Solid-phase extraction procedure to remove organic acids from honey, *J.Chromatogr.B*, **2002**, *770*, 77–82. [malic acid; maleic acid; citric acid; succinic acid; fumaric acid]

Suárez-Luque, S.; Mato, I.; Huidobro, J.F.; Simal-Lozano, J.; Sancho, M.T. Rapid determination of minority organic acids in honey by high-performance liquid chromatography, *J.Chromatogr.A*, **2002**, *955*, 207–214. [malic acid; maleic acid; citric acid; succinic acid; fumaric acid; SPE]

Wei, M.-C.; Chang, C.-T.; Jen, J.-F. Determination of organic acids in fermentation products of milk with high performance liquid chromatography/on-lined micro-dialysis, *Chromatographia*, **2001**, *54*, 601–605. [citric acid; acetic acid; lactic acid; formic acid]

Yamamoto, A.; Kodama, S.; Matsunaga, A.; Inoue, Y.; Aoyama, T.; Kumagai, Y. Characteristics of a column suitable for capacity gradient chromatography with a borate eluent, *The Analyst*, **2001**, *126*, 465–468.

# Clioquinol

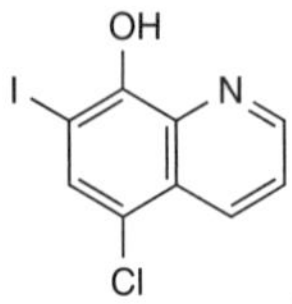


**Molecular formula:** $C_9H_5ClINO$
**Molecular weight:** 305.50
**CAS Registry No:** 130-26-7
**Merck Index:** 13, 5053

## SAMPLE

**Matrix:** bile, blood, tissue, urine
**Sample preparation:** Mix 200 μL plasma, bile, urine, or kidney with 20 μL IS in MeOH, 200 μL 200 mM disodium EDTA, and 800 μL water. Add 4 mL benzene:pyridine 90:10 (Caution! Benzene is a carcinogen!), shake vigorously for 1 min, centrifuge at 3500 g for 5 min. Evaporate the organic layer to dryness under reduced pressure, reconstitute the residue with 200 μL mobile phase, inject a 2–20 μL aliquot. This sample measures free clioquinol. To measure the glucuronide conjugate, wash the aqueous phase left over from the previous extraction twice with 6 mL portions of benzene. Discard the benzene. Add IS in MeOH, β-glucuronidase (final concentration 200 U/mL), and 150 μL 1 M pH 5 acetate buffer to the aqueous phase. Shake at 37° for 2 h. Add 4 mL benzene:pyridine 90:10, shake vigorously for 1 min, centrifuge at 3500 g for 5 min. Evaporate the organic layer to dryness under reduced pressure, reconstitute the residue with 200 μL mobile phase, inject a 2–20 μL aliquot. This sample measures clioquinol glucuronide. To measure the sulfate conjugate, wash the aqueous phase left over from the previous extraction twice with 6 mL portions of benzene. Discard the benzene. Add 6 M HCl to achieve an HCl concentration of 1 M, add IS, heat at 40° for 2 h, almost (sic) neutralize with 3 M NaOH, add 4 mL benzene:pyridine 90:10, shake vigorously for 1 min, centrifuge at 3500 g for 5 min. Evaporate the organic layer to dryness under reduced pressure, reconstitute the residue with 200 μL mobile phase, inject a 2–20 μL aliquot. This sample measures clioquinol sulfate.

## HPLC VARIABLES

**Column:** 300 × 3 10 μm Iatrobeads 6cp-2010 (polystyrene-type porous polymer) in a Pyrex glass column
**Column temperature:** 37 ± 0.5
**Mobile phase:** MeOH:*n*-hexane:100 mM citric acid 86:6:8
**Flow rate:** 0.75
**Injection volume:** 2–20
**Detector:** UV 254

## CHROMATOGRAM

**Retention time:** 9.6
**Internal standard:** 5,7-dichloro-8-hydroxyquinoline (6.8)
**Limit of quantitation:** 600 ng/mL

## KEY WORDS

kidney; plasma; rabbit

## REFERENCE

Hayakawa, K.; Kitada, K.; Hamaki, M.; Miyazaki, M. High-performance liquid chromatographic determination of clioquinol and its conjugates in biological materials, *J.Chromatogr.*, **1982**, *229*, 159–165.

# Clobetasol 17-propionate

**Molecular formula:** $C_{25}H_{32}ClFO_5$
**Molecular weight:** 466.98
**CAS Registry No:** 25122-46-7
**Merck Index:** 13, 2387

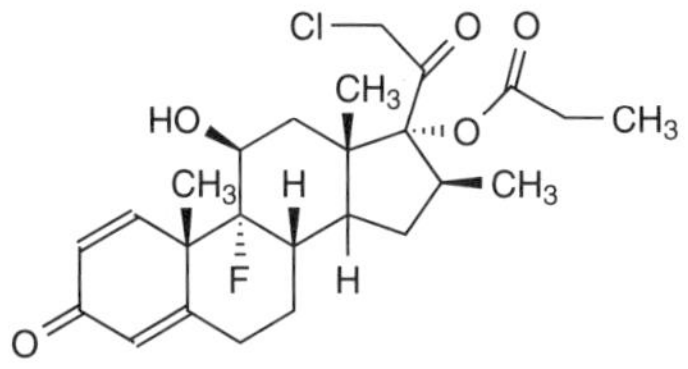

## SAMPLE

**Matrix:** formulations
**Sample preparation:** Condition a 3 mL 500 mg Megabond MF C18 SPE cartridge (Varian) with 3 mL MeOH and 3 mL water. Sonicate 1 g cosmetic with 10 mL MeOH or MeOH:dichloromethane 10:90 (depending on what appears visually to give the best solubility) at 40° for 10 min, centrifuge, collect the clear supernatant. Add 5 mL of the supernatant to the SPE cartridge, wash with 4 mL acetone:water 20:80, wash with 1 mL *n*-hexane, elute with 4 mL diethyl ether. Evaporate the eluate to dryness under reduced pressure, reconstitute the residue with 5 mL (or more) MeOH, inject a 10 μL aliquot.

## HPLC VARIABLES

**Column:** 250 × 4.6 5 μm endcapped Purospher RP-18
**Column temperature:** 25
**Mobile phase:** Isocratic.MeCN:water 60:40. Gradient. MeCN:water from 25:75 to 90:10 over 30 min, maintain at 90:10 for 10 min.
**Flow rate:** 1
**Injection volume:** 10
**Detector:** UV 239

## CHROMATOGRAM

**Retention time:** k′ 3.41 (isocratic); 23.4 min (gradient)
**Limit of detection:** 100 ng/mL

## OTHER SUBSTANCES

**Simultaneous:** alclometasone dipropionate (isocratic k′ 2.55; gradient retention time (min) 21.0; LOD 0.3 μg/mL), amcinonide (isocratic k′ 3.18; gradient retention time (min) 22.6; LOD 0.1 μg/mL), betamethasone (isocratic k′ 0.18; gradient retention time (min) 11.8; LOD 0.1 μg/mL), betamethasone-17-acetate (isocratic k′ 0.73; gradient retention time (min) 15.4; LOD 0.3 μg/mL), betamethasone-17-benzoate (isocratic k′ 2.04; gradient retention time (min) 20.6; LOD 0.3 μg/mL), betamethasone-17-propionate-21-stearate (isocratic k′ >13; gradient retention time (min) >35; LOD 0.5 μg/mL), betamethasone-17-propionate-21-butyrate (isocratic k′ 5.91; gradient retention time (min) 26.1; LOD 0.4 μg/mL), betamethasone-17-valerate-21-acetate (isocratic k′ 4.41; gradient retention time (min) 23.1; LOD 0.4 μg/mL), betamethasone-17-valerate (isocratic k′ 2.32; gradient retention time (min) 21.4; LOD 0.3 μg/mL), betamethasone-17,21-dipropionate (isocratic k′ 4.00; gradient retention time (min) 24.2; LOD 0.4 μg/mL), betamethasone-17,21-diacetate (isocratic k′ 1.81; gradient retention time (min) 20.5; LOD 0.3 μg/mL), betamethasone-17,21-divalerate (isocratic k′ 10.82; gradient retention time (min) 28.0; LOD 0.4 μg/mL), betamethasone-21-acetate (isocratic k′ 0.77; gradient retention time (min) 15.6; LOD 0.3 μg/mL), betamethasone propionate (isocratic k′ 0.82; gradient retention time (min) 17.1; LOD 0.3 μg/mL), clobetasone butyrate (isocratic k′ 5.45; gradient retention time (min) 26.3; LOD 0.1 μg/mL), cortisone (isocratic k′ 0.18; gradient retention time (min) 11.1; LOD 0.6 μg/mL), cortisone acetate (isocratic k′ 0.73; gradient retention time (min) 15.2; LOD 0.6 μg/mL), dehydrocorticosterone (isocratic k′ 4.27; gradient retention time (min) 22.3; LOD 0.5 μg/mL), deoxymethasone (isocratic k′ 0.64; gradient retention time (min) 14.2; LOD 0.2 μg/mL), dexamethasone (isocratic k′ 0.27; gradient retention time (min) 11.9; LOD 0.1 μg/mL), dexamethasone-21-acetate (isocratic k′ 0.91; gradient retention time (min) 16.1; LOD 0.2 μg/mL), dexamethasone isonicotinate (isocratic k′ 1.05; gradient retention time (min) 17.7; LOD 0.4 μg/mL), dexamethasone pivalate (isocratic k′ 3.45; gradient retention time (min) 24.1; LOD 0.3 μg/mL), dexamethasone valerate (isocratic k′ 3.00; gradient retention time (min) 21.6; LOD 0.3 μg/mL), diflucortolone valerate (isocratic k′ 4.73; gradient retention time (min) 23.3; LOD 0.3 μg/mL), fludrocortisone acetate (isocratic k′ 0.59; gradient retention time (min) 14.1; LOD 0.3 μg/mL), flumethasone pivalate (isocratic k′ 2.68; gradient retention time (min) 21.2; LOD 0.3 μg/mL), fluocinolone acetonide (isocratic k′ 0.91; gradient retention time (min) 13.4; LOD 0.3 μg/mL), fluocinonide (isocratic k′ 1.45; gradient retention time (min) 20.5; LOD 0.1 μg/mL), fluocortin butyl ester (isocratic k′ 5.59; gradient retention time (min) 24.6; LOD 0.3 μg/mL), fluocortolone caproate (isocratic k′ 6.59; gradient retention time (min) 25.1; LOD 0.3 μg/mL), fluocortolone pivalate (isocratic k′ 4.50; gradient retention time (min) 23.6; LOD 0.3 μg/mL), fluorometholone (isocratic k′ 0.59; gradient retention time (min) 14.4; LOD 0.1 μg/mL), 9-α-fluoroprednisolone (isocratic k′ 0.18; gradient retention time (min) 10.0; LOD 0.1 μg/mL), 9-α-fluoroprednisolone acetate (isocratic k′ 0.50; gradient retention time (min) 13.9; LOD 0.2 μg/mL), flurandrenolide (isocratic k′ 0.50; gradient retention time (min) 13.5; LOD 0.1 μg/mL), halcinonide (isocratic k′ 1.64; gradient retention time (min) 20.6; LOD 0.1 μg/mL), hydrocortisone (isocratic k′ 0.18; gradient retention time (min) 10.0; LOD 0.4 μg/mL), hydrocortisone-17-butyrate (isocratic k′ 1.09; gradient retention time (min) 17.7; LOD 0.6 μg/mL), hydrocortisone-21-acetate (isocratic k′ 0.77; gradient retention time (min) 15.3; LOD 0.6 μg/mL), hydrocortisone pivalate (isocratic k′ 2.27; gradient retention time (min) 20.4; LOD 0.8 μg/mL), methylprednisolone (isocratic k′ 0.55; gradient retention time (min) 13.5; LOD 0.1 μg/mL), mometasone furoate (isocratic k′ 3.05; gradient retention time (min) 22.0; LOD 0.2 μg/mL), prednisolone-21-acetate (isocratic k′ 0.60; gradient retention time (min) 13.6; LOD 0.2 μg/mL), prednisolone acetonide (isocratic k′ 0.50; gradient retention time (min) 13.0; LOD 0.3 μg/mL), prednisolone pivalate (isocratic k′ 2.05; gradient retention time (min) 19.7; LOD 0.3 μg/mL), triamcinolone (isocratic k′ 0.14; gradient retention time (min) 7.2; LOD 0.1 μg/mL), triamcinolone acetonide (isocratic k′ 0.50; gradient retention time (min) 13.9; LOD 0.2 μg/mL), triamcinolone diacetate (isocratic k′ 0.45; gradient retention time (min) 13.9; LOD 0.3 μg/mL)

## KEY WORDS

cosmetics; SPE

## REFERENCE

Gagliardi, L.; De Orsi, D.; Del Giudice, M.R.; Gatta, F.; Porrà, R.; Chimenti, P.; Tonelli, D. Development of a tandem thin-layer chromatography-high-performance liquid chromatography method for the identification and determination of corticosteroids in cosmetic products, *Anal.Chim.Acta*, **2002**, *457*, 187–198.

## SAMPLE

**Matrix:** formulations

**Sample preparation:** Mix 1 g of a topical product, 5 mL 100 mM pH 4 citrate buffer saturated with NaCl, and 10 mL ethyl acetate, shake vigorously by hand to ensure that no large clumps stick to the tube, mix with the tube on its side on an oscillating shaker for 15 min. Remove the upper ethyl acetate layer and wash with citrate buffer as before. Dry the ethyl acetate over anhydrous sodium sulfate and evaporate to dryness. Dissolve the residue in a mixture of 10 mL heptane and 5 mL MeCN:water 90:10. Remove the upper heptane layer and extract it with 2 mL MeCN:water 90:10. Combine the MeCN/water layers and evaporate them to dryness, dissolve the residue in 300 μL MeOH, filter (0.45 μm nylon), inject a 5 μL aliquot.

## HPLC VARIABLES

**Guard column:** 15 × 3.2 7 μm Brownlee NewGuard C18

**Column:** 75 × 4.6 3.5 μm Symmetry C18 (Waters)
**Mobile phase:** Gradient. MeCN:water 18:82 for 2 min, to 82:18 over 12 min, maintain at 82:18 for 3 min, re-equilibrate at initial conditions for 12 min.
**Flow rate:** 1
**Injection volume:** 5
**Detector:** UV 240

---

## CHROMATOGRAM

**Retention time:** 11.06
**Limit of detection:** 0.001%

---

## OTHER SUBSTANCES

**Simultaneous:** alclometasone 17,21-dipropionate (10.93), amcinonide (10.90), beclomethasone 17,21-dipropionate (11.90), betamethasone (6.52), betamethasone 21-acetate (8.47), betamethasone 17-benzoate (10.28), betamethasone 17,21-dipropionate (11.42), betamethasone 17-valerate (10.25), budesonide (8.55, 8.69 (epimers)), cortisone (5.62), cortisone 21-acetate (8.07), dexamethasone (6.57), dexamethasone 21-acetate (8.68), desonide (6.99), desoximetasone (7.60), desoxycorticosterone acetate (10.90), desoxycorticosterone pivalate (14.45), diflorasone 17,21-diacetate (9.81), fluocinolone acetonide (7.38), fluocinonide (9.79), flurandrenolide (7.36), fludrocortisone 21-acetate (7.77), fluorometholone (7.67), flumethasone 21-pivalate (11.20), flunisolide (7.14), fluprednisolone (5.46), fluticasone 17-propionate (11.19), halcinonide (10.72), halobetasol propionate (10.98), hydrocortisone (5.50), hydrocortisone 21-acetate (7.65), hydrocortisone 17-butyrate (8.66), hydrocortisone 21-cypionate (12.54), hydrocortisone 17-valerate (9.53), mometasone 17-furoate (11.24), methylprednisolone (6.31), methylprednisolone 21-acetate (8.34), meprednisone (6.55), prednisolone (5.37), prednisolone 21-acetate (7.47), prednisolone 21-tebuate (11.01), paramethasone 21-acetate (8.64), prednicarbate (10.72), prednisone (5.46), triamcinolone (4.15), triamcinolone acetonide (7.04), triamcinolone 16,21-diacetate (7.49), triamcinolone hexacetonide (13.12)

---

## KEY WORDS

body wash, cream, gel, lotion, shampoo, spray

---

## REFERENCE

Reepmeyer, J.C. Screening for corticosteroids in topical pharmaceuticals by HPLC with a scanning ultraviolet detector, *J.Liq.Chromatogr.Rel.Technol.*, **2001**, *24*, 693–709.

---

## SAMPLE

**Matrix:** solutions
**Sample preparation:** The receptor solution for skin diffusion experiments consisted of 4% bovine serum albumin (BSA) in phosphate-buffered saline. Mix receptor solution with an equal volume of 2 μg/mL IS in MeCN, let stand for 3 h, centrifuge for 20 min, inject a 50 μL aliquot of the supernatant.

---

## HPLC VARIABLES

**Guard column:** 4 × 3 C18 (ODS) (Phenomenex)
**Column:** 150 × 4.6 5 μm Luna C18 (2) (Phenomenex)
**Mobile phase:** MeCN:MeOH:100 mM pH 3 potassium dihydrogen phosphate 50:10:40
**Flow rate:** 1
**Injection volume:** 50
**Detector:** UV 239

---

## CHROMATOGRAM

**Internal standard:** propyl paraben

## REFERENCE

Mueller, B.; Anissimov, Y.G.; Roberts, M.S. Unexpected clobetasol propionate profile in human stratum corneum after topical application in vitro, *Pharm.Res.*, **2003**, *20*, 1835–1837.

## ANNOTATED BIBLIOGRAPHY

Marini, R.D.; Pantella, A.; Bimazubute, M.A.; Chiap, P.; Hubert, P.; Crommen, J. Optimisation and validation of a generic method for the LC assay of six corticosteroids and salicylic acid in dermopharmaceutical forms, *Chromatographia*, **2002**, *55*, 263–269. [salicylic acid; methyl paraben; propyl paraben; triamcinolone acetonide; hydrocortisone acetate; betamethasone valerate; clobetasol propionate; clobetasone butyrate; betamethasone dipropionate]

Reepmeyer, J.C.; Revelle, L.K.; Vidavsky, I. Detection of clobetasol propionate as an undeclared steroid in zinc pyrithione formulations by high-performance liquid chromatography with rapid-scanning ultraviolet spectroscopy and mass spectrometry, *J.Chromatogr.A*, **1998**, *828*, 239–246.

Tsai, J.-C.; Cheng, C.-L.; Tsai, Y.-F.; Sheu, H.-M.; Chou, C.-H. Evaluation of in vivo bioequivalence methodology for topical clobetasol 17-propionate based on pharmacodynamic modeling using Chinese skin, *J.Pharm.Sci.*, **2004**, *93*, 207–217. [betamethasone dipropionate is internal standard]

# Clopidogrel

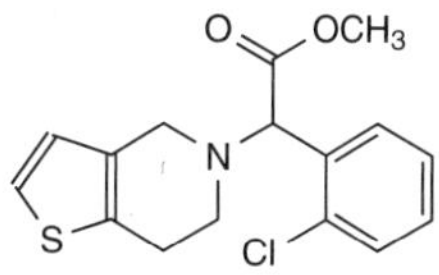

**Molecular formula:** $C_{16}H_{16}ClNO_2S$
**Molecular weight:** 321.83
**CAS Registry No:** 113665-84-2
**Merck Index:** 13, 2421

## SAMPLE
**Matrix:** microsomal incubations
**Sample preparation:** Add 200 μL microsomal incubation to 100 μL ice-cold MeCN, centrifuge, inject a 200 μL aliquot of the supernatant.

## HPLC VARIABLES
**Column:** 150 × 4.6 5 μm Resolve C18 (Waters)
**Mobile phase:** Gradient. A:B from 17:83 to 43:57 over 30 min, to 100:0 over 15 min. A was MeCN. B was MeCN:10 mM ammonium acetate 10:90.
**Flow rate:** 1
**Injection volume:** 200
**Detector:** UV 220; MS, Micromass VG Fisons Platform single quadrupole electrospray, m/z 322.2

## CHROMATOGRAM
**Retention time:** 26

## REFERENCE
Clarke, T.A.; Waskell, L.A. The metabolism of clopidogrel is catalyzed by human cytochrome P450 3A and is inhibited by atorvastatin, *Drug Metab.Dispos.*, **2003**, *31*, 53–59.

## SAMPLE
**Matrix:** formulations
**Sample preparation:** Weigh out an amount of pulverized tablet containing 75 mg of clopidogrel, sonicate with 40 mL MeCN for 20 min, make up to 50 mL with MeCN, centrifuge at 2890 g for 5 min, dilute a 1 mL aliquot to 100 mL with water. Remove a 1 mL aliquot of this solution and dilute to 10 mL with 100 ng/mL IS in mobile phase, inject a 5 μL aliquot.

## HPLC VARIABLES
**Column:** 250 × 2.1 5 μm BDS C8
**Mobile phase:** MeCN:10 mM pH 3.0 sodium phosphate buffer 65:35
**Flow rate:** 0.3
**Injection volume:** 5
**Detector:** UV 235

## CHROMATOGRAM
**Retention time:** 3.08
**Internal standard:** naproxen (6.28)
**Limit of detection:** 120 ng/mL
**Limit of quantitation:** 390 ng/mL

## KEY WORDS
stability-indicating; tablets

## REFERENCE

Mitakos, A.; Panderi, I. A validated LC method for the determination of clopidogrel in pharmaceutical preparations, *J.Pharm.Biomed.Anal.*, **2002**, *28*, 431–438.

## ANNOTATED BIBLIOGRAPHY

Reist, M.; Roy-De Vos, M.; Montseny, J.P.; Mayer, J.M.; Carrupt, P.-A.; Berger, Y.; Testa, B. Very slow chiral inversion of clopidogrel in rats: a pharmacokinetic and mechanistic investigation, *Drug Metab.Dispos.*, **2000**, *28*, 1405–1410. [The metabolite clopidogrel acid is measured in rat plasma by chiral HPLC with derivatization.]

# Clopidol

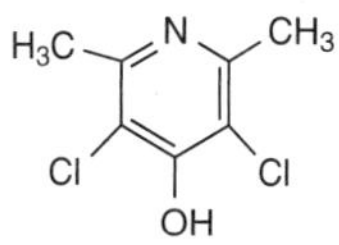

**Molecular formula:** $C_7H_7Cl_2NO$
**Molecular weight:** 192.05
**CAS Registry No:** 2971-90-6
**Merck Index:** 13, 2422

## SAMPLE

**Matrix:** feed
**Sample preparation:** Moisten 10 g pulverized feed with 5 mL water for 1 min, add 45 mL DMF:water 95:5, shake horizontally for 1 h. Remove a 10 mL aliquot of the supernatant and centrifuge at 3000 rpm for 5 min. Add 5 mL of the supernatant to a non-pretreated 6 mL 1 g Isolute A1-B SPE cartridge, discard the first 1 mL, collect the next 2 mL, inject a 10 μL aliquot.

## HPLC VARIABLES

**Guard column:** 20 × 4.6 5 μm Supelguard LC-18
**Column:** 250 × 4.6 5 μm Supelcosil LC-18
**Mobile phase:** Gradient. MeCN:10 mM pH 4.6 acetate buffer from 0:100 to 30:70 over 10 min, to 80:20 over 5 min, maintain at 80:20 for 4 min, return to initial conditions over 1 min, re-equilibrate for 4 min.
**Flow rate:** 1.2
**Injection volume:** 10
**Detector:** UV 265 for 12 min, UV 345 for rest of the run

## CHROMATOGRAM

**Retention time:** 9
**Limit of quantitation:** 2 μg/g

## OTHER SUBSTANCES

**Extracted:** nicarbazin (17.5)

## KEY WORDS

SPE

## REFERENCE

Dusi, G.; Faggionato, E.; Gamba, V.; Baiguera, A. Determination of nicarbazin and clopidol in poultry feeds by liquid chromatography, *J.Chromatogr.A*, **2000**, *882*, 79–84.

## SAMPLE

**Matrix:** tissue
**Sample preparation:** Fill a 400 × 20 glass tube fitted with a glass wool plug and a PTFE stopcock about a third full with MeOH, add 15 g neutral alumina (70–230 mesh), drain the solvent to the top of the column. Fill a 200 × 12 glass tube fitted with a glass wool plug and a PTFE stopcock with a slurry of Dowex 1-X8 anion-exchange resin (acetate form, 100–200 mesh) in MeOH to a bed height of 15 mm after settling, drain the solvent to the top of the bed. Place the alumina column on top of the Dowex column. Homogenize 10 g minced tissue, 20 g anhydrous sodium sulfate, and 50 mL MeCN at 10000 rpm for 2 min, add the supernatant to the alumina tube. Re-extract the residue with 50 mL MeCN, centrifuge, add to the columns, rinse the centrifuge tube with 20 mL MeOH, and add the rinse to the columns. Allow the liquid to flow through the columns at 3 mL/min. When the various solutions have passed through, discard the alumina column and elute the Dowex column with 20 mL 0.5% acetic acid in MeOH. Evaporate

the eluate to dryness under reduced pressure at 60°, reconstitute the residue with 1 mL MeOH, inject an aliquot.

---

## HPLC VARIABLES

**Column:** 150 × 4.6 Inertsil ODS
**Mobile phase:** MeCN:water 20:80
**Flow rate:** 0.8
**Injection volume:**
**Detector:** MS, Finnigan MAT TSQ-7000 quadrupole, atmospheric pressure chemical ionization, m/z 190, corona voltage 1.8 kV, corona current 5.0 μA, capillary temperature 220°, electron multiplier 1400 V, vaporizer temperature 400°

---

## CHROMATOGRAM

**Retention time:** 3.47
**Limit of detection:** 5 ng/g
**Limit of quantitation:** 10 ng/g

---

## KEY WORDS

chicken; muscle; SPE

---

## REFERENCE

Pang, G.-F.; Cao, Y.-Z.; Fan, C.-L.; Zhang, J.-J.; Li, X.-M.; Wang, C. Determination of clopidol residues in chicken tissues by high-performance liquid chromatography-mass spectrometry, *J.Chromatogr.A*, **2000**, *882*, 85–88.

---

## ANNOTATED BIBLIOGRAPHY

Malisch, R. Multimethode zur Bestimmung der Ruckstande von Chemotherapeutica, Antiparasitica und Wachstumsforderern in Lebensmitt tierischer Herkunft. 3. Mitteilung: Bestimmung von Chloramphenicol und Meticlorpindol [Method for the simultaneous determination of residues of chemotherapeutic agents, antiparasitic agents and growth promoters in food of animal origin. 3. Determination of chloramphenicol and meticlorpindol], *Z.Lebensm.-Unters.-Forsch.*, **1987**, *184*, 467–477. [clopidol; chloramphenicol]

Mattern, E.M.; Kan, C.A.; van Gend, H.W. An automated HPLC determination of meticlorpindol in eggs with UV absorbance detection, using on-line dialysis and pre-concentration as sample clean-up; occurrence in and carry over to eggs, *Z.Lebensm.-Unters.-Forsch.*, **1990**, *190*, 25–30.

Pang, G.-F.; Cao, Y.-Z.; Fan, C.-L.; Zhang, J.-J.; Li, X.-M.; Jia, X.; Song, W.-B. Determination of clopidol residues in chicken tissues by liquid chromatography: Part I. Optimization of analytical conditions and comparison with AOAC gas chromatography method, *J.AOAC Int.*, **2001**, *84*, 1337–1342. [SPE]

Pang, G.-F.; Cao, Y.-Z.; Fan, C.-L.; Zhang, J.-J.; Li, X.-M.; Zhang, Z.-Y. Determination of clopidol residues in chicken tissues by liquid chromatography: Part II. Distribution and depletion of clopidol in chicken tissues, *J.AOAC Int.*, **2001**, *84*, 1343–1346. [SPE]

Pang, G.-F.; Cao, Y.-Z.; Fan, C.-L.; Zhang, J.-J.; Li, X.-M.; MacNeil, J.D. Determination of clopidol residues in chicken tissues by liquid chromatography : collaborative study, *J.AOAC Int.*, **2003**, *86*, 685–693.

# Cloricromen

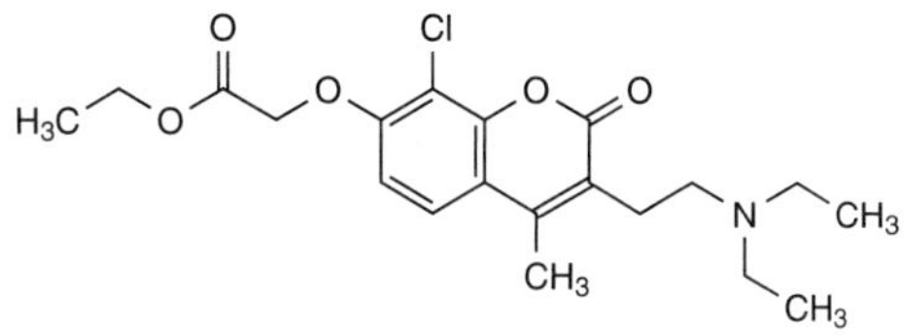

**Molecular formula:** $C_{20}H_{26}ClNO_5$
**Molecular weight:** 395.88
**CAS Registry No:** 68206-94-0
**Merck Index:** 13, 2431

## SAMPLE

**Matrix:** aqueous humor
**Sample preparation:** Vortex 100 μL aqueous humor with 100 μL 4.8 μg/mL IS in MeCN containing 0.6% perchloric acid for 1 min, centrifuge at 10000 g for 5 min, filter (0.2 μm) the supernatant, inject a 20 μL aliquot of the supernatant.

## HPLC VARIABLES

**Guard column:** 7.5 × 4.6 5 μm Hypersil ODS C18
**Column:** 150 × 4.6 5 μm Hypersil ODS C18
**Column temperature:** 25
**Mobile phase:** Gradient. MeCN:buffer from 10:90 to 53:47 over 13 min, return to initial conditions over 5 min, re-equilibrate for 3 min. (The buffer was 1% triethylamine adjusted to pH 3.5 with phosphoric acid.)
**Flow rate:** 1
**Injection volume:** 20
**Detector:** UV 318

## CHROMATOGRAM

**Retention time:** 11.25
**Internal standard:** timolol maleate (6.48)
**Limit of quantitation:** 10 ng/mL

## OTHER SUBSTANCES

**Extracted:** metabolite

## KEY WORDS

pharmacokinetics; rabbit

## REFERENCE

Maltese, A.; Bucolo, C. Simultaneous determination of cloricromene and its active metabolite in rabbit aqueous humor by high-performance liquid chromatography, *J.Chromatogr.B*, **2002**, *767*, 153–158.

# Clorsulon

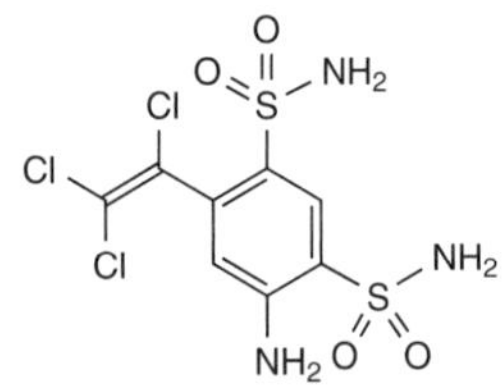

**Molecular formula:** $C_8H_8Cl_3N_3O_4S_2$
**Molecular weight:** 380.66
**CAS Registry No:** 60200-06-8
**Merck Index:** 13, 2435

## SAMPLE

**Matrix:** milk

**Sample preparation:** Matrix Solid-Phase Dispersion (MSPD) – Blend 500 μL milk with 2 g 40 μm Bondesil octadecylsilyl silica (Analytichem) in a mortar and pestle until a homogeneous mixture is obtained, place the mixture in the barrel of an 8 mL syringe fitted with a frit, compress the matrix using the syringe plunger, wash with 3 mL hexane; when all the hexane has eluted, place under vacuum for 5 s. Place this column on top of a 6 mL 1 g Supelclean LC-Florisil SPE cartridge (prewashed with diethyl ether), elute with three 3 mL portions of diethyl ether. Remove the C18 column and elute the Florisil with 2 mL ether. Combine all the ether eluates, evaporate to dryness under a stream of nitrogen at 40°, reconstitute the residue with 1 mL mobile phase, vortex, filter, inject a 200 μL aliquot. Solid-Phase Extraction (SPE) – Condition a 6 mL 1 g Mega Bond Elut cyclohexyl SPE cartridge with two 3 mL portions of MeOH and two 3 mL portions of water. Vortex 6 g milk with 6 mL 200 mg/mL hydroxylamine hydrochloride in water, add 4 mL MeOH, shake on a rotating shaker at 180 rpm for 30 min, centrifuge at 8000 rpm for 30 min. Add 10 mL of the supernatant to the SPE cartridge, elute at not more than 2 drops/s, discard the effluent, wash with two 4 mL portions of water, elute with two 4 mL portions of MeCN. Evaporate the eluate to dryness under a stream of nitrogen at 60°, reconstitute the residue with 1 mL mobile phase, vortex for 5 s, filter (0.2 μm PTFE), inject a 50 μL aliquot. (SPE method found in Schenck,F.J.; Wagner,R.; Bargo,W. Determination of clorsulon residues in milk using a solid-phase extraction cleanup and liquid chromatographic determination. *J.Liq.Chromatogr.* **1993**, *16*, 513–520, other details are identical.)

## HPLC VARIABLES

**Column:** 150 × 4.6 3 μm Econosphere ODS
**Mobile phase:** MeCN:10 mM pH 7.0 potassium phosphate buffer 25:75
**Flow rate:** 1
**Injection volume:** 50–200
**Detector:** UV 265

## CHROMATOGRAM

**Retention time:** 11.5
**Limit of detection:** 5 ng/g (SPE, S/N = 4)
**Limit of quantitation:** 50 ng/g (MSPD), 15 ng/g (SPE)

## KEY WORDS

matrix solid-phase dispersion; SPE

## REFERENCE

Schenck, F.J.; Barker, S.A.; Long, A.R. Matrix solid phase dispersion (MSPD) isolation and liquid chromatographic determination of clorsulon in milk, *J.Liq.Chromatogr.*, **1991**, *14*, 2827–2834.

# Colistin

**Molecular formula:** $C_{58}H_{105}N_{16}Na_5O_{28}S_5$ (colistin A sodium methanesulfonate)

**Molecular weight:** 1749.82

**CAS Registry No:** 1066-17-7, 8068-28-8 (sodium methanesulfonate), 1264-72-8 (sulfate)

**Merck Index:** 13, 2503

## SAMPLE

**Matrix:** blood

**Sample preparation:** Condition a 100 mg C18 SPE cartridge (Baker) with 1 mL MeOH and 1 mL carbonate buffer. Mix 20 μL 5 μg/mL IS in water with 250 μL plasma, add 50 μL MeOH:10% trichloroacetic acid 50:50, vortex for 1 min, centrifuge at 1000 g at 4° for 10 min. Remove the supernatant, add 10 μL 1 M NaOH, vortex for 1 min, add 250 μL MeOH:10 mM HCl 50:50, vortex for 1 min, add to the SPE cartridge, wash with 1 mL carbonate buffer, add 30 μL 100 mM 9-fluorenylmethyl chloroformate in MeCN, let stand for 10 min, dry by drawing air through the cartridge, elute with 900 μL acetone. Add the eluate to 600 μL 200 mM boric acid solution, vortex for 2 min, inject a 50 μL aliquot. (Prepare the carbonate buffer by adjusting the pH of 1% sodium bicarbonate solution to 10 with 10% NaOH.)

## HPLC VARIABLES

**Column:** 250 × 4.6 5 μm Ultrasphere C18
**Mobile phase:** MeCN:THF:water 87:4:13
**Flow rate:** 1
**Injection volume:** 50
**Detector:** F ex 260 em 315

## CHROMATOGRAM

**Retention time:** 21.8 (colistin B), 26.1 (colistin A)
**Internal standard:** netilmicin sulfate (17.1)
**Limit of quantitation:** 100 ng/mL

## OTHER SUBSTANCES

**Noninterfering:** aztreonam, ceftazidime, ciprofloxacin, meropenem, piperacillin, ticarcillin, tobramycin

## KEY WORDS

derivatization; plasma; SPE

## REFERENCE

Li, J.; Milne, R.W.; Nation, R.L.; Turnidge, J.D.; Coulthard, K.; Johnson, D.W. A simple method for the assay of colistin in human plasma, using pre-column derivatization with 9-fluorenylmethyl chloroformate in solid-phase extraction cartridges and reversed-phase high-performance liquid chromatography, *J.Chromatogr.B*, **2001**, *761*, 167–175.

## SAMPLE

**Matrix:** blood, sputum, urine

**Sample preparation:** Plasma, sputum. Add 1 mL plasma or sonicated sputum to 1.1 mL water and 200 μL 3 M perchloric acid, vortex for 10 s, centrifuge at 3000 g at 4° for 25 min. Remove 1.4 mL of the supernatant and add it to 200 μL 1.5 M KOH, vortex, add 200 μL 100 mM HCl, vortex, add 100 μL 9% sodium carbonate, vortex, add 400 μL 20 mg/mL dansyl chloride, vortex, heat at 54°; after 5 min, add 100 μL 150 mg/mL proline, heat at 54° for another 55 min, vortex, heat at 54° for 1 h, cool. Remove a 2 mL aliquot and add it to 1 mL ethyl acetate, rotate for 10 min (Roto-torque Model 7637 setting #5), centrifuge at 3000 rpm for 10 min, inject an aliquot of the top layer. Urine. Dilute urine 0–20 fold with 100 mM pH 6.8 sodium phosphate buffer. Add 1 mL diluted urine to 1 mL water and 200 μL 3 M perchloric acid, vortex for 10 s. Remove 1.4 mL and add it to 200 μL 1.5 M KOH, vortex, add 200 μL 100 mM HCl, vortex, add 100 μL 9% sodium carbonate, vortex, add 400 μL 20 mg/mL dansyl chloride, vortex, heat at 57°; after 5 min, add 100 μL 300 mg/mL proline, heat at 57° for another 55 min, vortex, heat at 57° for 1 h, cool. Remove a 2 mL aliquot and add it to 1 mL ethyl acetate, rotate for 10 min (Roto-torque Model 7637 setting #5), centrifuge at 3000 rpm for 2 min, inject an aliquot of the top layer.

## HPLC VARIABLES

**Guard column:** 40 μm Vydac pellicular resin
**Column:** 150 × 4.6 Eclipse XDB-C8
**Mobile phase:** Gradient. A:B from 40:60 to 80:20 over 15 min, maintain at 80:20 for 3 min. A was 0.1% trifluoroacetic acid in MeCN. B was 0.1% trifluoroacetic acid in water.
**Flow rate:** 1
**Injection volume:** 50
**Detector:** F ex 350 em 500

## CHROMATOGRAM

**Limit of quantitation:** 5 μg/mL

## KEY WORDS

derivatization; pharmacokinetics; plasma

## REFERENCE

Reed, M.D.; Stern, R.C.; O'Riordan, M.A.; Blumer, J.L. The pharmacokinetics of colistin in patients with cystic fibrosis, *J.Clin.Pharmacol.*, **2001**, *41*, 645–654.

## ANNOTATED BIBLIOGRAPHY

Cabanes, A.; Cajal, Y.; Haro, I.; Garcia Anton, J.M.; Arboix, M.; Reig, F. Gentamicin determination in biological fluids by HPLC, using tobramycin as internal standard, *J.Liq.Chromatogr.*, **1991**, *14*, 1989–2010. [gentamicin; tobramycin is internal standard; plasma; urine; rabbit; derivatization with *o*-phthalaldehyde; fluorescence detection; LOD 300 ng/mL; kanamycin; colistin]

Gmur, D.J.; Bredl, C.R.; Steele, S.J.; Cai, S.; VanDevanter, D.R.; Nardella, P.A. Determination of polymyxin E1 in rat plasma by high-performance liquid chromatography, *J.Chromatogr.B*, **2003**, *789*, 365–372. [SPE; derivatization; fluorescence detection; LOQ 50 ng/mL; rat; dog]

Le Brun, P.P.H.; de Graaf, A.I.; Vinks, A.A.T.M.M. High-performance liquid chromatographic method for the determination of colistin in serum, *Ther.Drug Monit.*, **2000**, *22*, 589–593. [colistin; derivatization; polymyxin]

# Cypermethrin

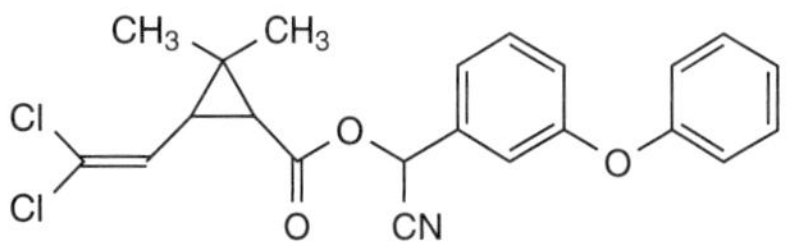


**Molecular formula:** $C_{22}H_{19}Cl_2NO_3$
**Molecular weight:** 416.30
**CAS Registry No:** 52315-07-8
**Merck Index:** 13, 2796

## SAMPLE

**Matrix:** blood, milk
**Sample preparation:** Activate silica gel 60 in air at 130° for 5 h, hydrate by adding water equivalent to 10% of its weight, keep in a sealed container for 30 min before use. Prepare a chromatographic column with 4 g of this material and wash with 1 mL *n*-hexane:diethyl ether 90:10. Acidify 10 mL or 10 g whole blood or whole milk with 1.0 M HCl to approximately pH 4.0, add 50 mL MeCN, shake mechanically for 30 min, filter (Whatman paper No. 42 or 44). Add 25 mL MeCN to the residue, shake for 15 min, filter (Whatman paper No. 42 or 44). Combine the filtrates and add them to 15 mL *n*-hexane, shake for 1 min, repeat twice. Retain the acetonitrile phases. Add 45 mL MeCN to the hexane phases, shake for 1 min. Combine all the acetonitrile layers, evaporate to dryness under a stream of nitrogen at 50°. Dissolve the residue in 10 mL *n*-hexane. Add to the silica column, elute with 7 mL *n*-hexane:diethyl ether 90:10, evaporate the eluate to dryness at room temperature, dissolve the residue in 1.0 mL MeCN, filter (0.45 μm cellulose membrane), inject an aliquot of the filtrate. (All diethyl ether should be peroxide free.)

## HPLC VARIABLES

**Column:** 250 × 4.6 Nucleosil 120-5 C18
**Mobile phase:** MeCN:water 80:20
**Flow rate:** 1
**Injection volume:** 20
**Detector:** UV 266

## CHROMATOGRAM

**Retention time:** 8.6, 8.9
**Limit of detection:** 1 ng/g

## OTHER SUBSTANCES

**Extracted:** cyhalothrin (7.7), deltamethrin (9.8), flumethrin (3.8, 4.1)

## KEY WORDS

cow; SPE; whole blood

## REFERENCE

Bissacot, D.Z.; Vassilieff, I. HPLC determination of flumethrin, deltamethrin, cypermethrin, and cyhalothrin residues in the milk and blood of lactating dairy cows, *J.Anal.Toxicol.*, **1997**, *21*, 397–402.

# Dalfopristin

**Molecular formula:** $C_{34}H_{50}N_4O_9S$
**Molecular weight:** 690.86
**CAS Registry No:** 112362-50-2
**Merck Index:** 13, 2831

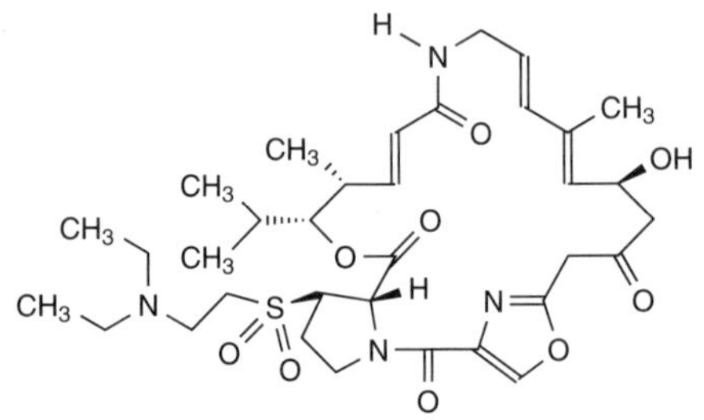

## SAMPLE

**Matrix:** blood
**Sample preparation:** Condition a CN SPE cartridge (Lida-Interchim) with 1 mL MeOH, 1 mL water, and 1 mL buffer. Add 1 mL 3.8% sodium citrate and 2.5 mL 250 mM HCl to 10 mL whole blood. Shake gently by hand and centrifuge at 2000 g at 4° for 15 min. Add 1 mL buffer and 50 μL 100 μg/mL IS in MeOH to 1.35 mL acidified plasma. Vortex for a few seconds, centrifuge at 4000 g at 4° for 5 min. Add either supernatant to the SPE cartridge and dry the SPE cartridge with 3 mL air. Elute with 500 μL MeOH:water 70:30 containing 3.5 mM pentane sulfonic acid, inject an aliquot. (The buffer was a mixture of 85 mM pH 3.0 citric acid monohydrate containing 81 mM NaOH and 60 mM HCl.)

## HPLC VARIABLES

**Guard column:** 10 μm μBondapak C18
**Column:** 125 × 4.6 5 μm Kromasil C18 (Higgins Analytical)
**Mobile phase:** Gradient. MeCN:buffer 30:70 for 11 min, then 32:68 for 4 min (step gradient), to 40:60 over 0.6 min, maintain at 40:60 for 0.4 min, at 38:62 for 18 min (step gradient), at 80:20 for 2 min (step gradient), re-equilibrate at 30:70 for 9 min. (Prepare buffer by adding 800 μL 70% perchloric acid to 1 L water.)
**Flow rate:** 0.5 for 11 min, 1 for 25 min, 0.5 for 9 min
**Injection volume:** 500
**Detector:** UV 235

## CHROMATOGRAM

**Retention time:** 12.7
**Internal standard:** dimethylamino-3-propyl thiomethylene-5 virginiamycin S (31.0)
**Limit of quantitation:** 25 pg/mL

## OTHER SUBSTANCES

**Extracted:** metabolites, pristinamycin II A (24.1), quinupristin (22.1)

## KEY WORDS

plasma; SPE; whole blood

## REFERENCE

Le Liboux, A.; Pasquier, O.; Montay, G. Simultaneous high-performance liquid chromatographic determination of quinupristin, dalfopristin and their main metabolites in human plasma, *J.Chromatogr.B*, **1998**, *708*, 161–168.

## SAMPLE

**Matrix:** formulations
**Sample preparation:** Inject a 5–10 μL aliquot of the infusion solution.

## HPLC VARIABLES

**Column:** 125 × 4 5 μm LiChrospher-100 RP18
**Column Temperature:** 40 ± 1
**Mobile phase:** Gradient. A:B from 0:100 to 66:34 over 42.5 min, return to initial conditions over 1.5 min, re-equilibrate for 5 min. A was MeCN:buffer 65:35. B was MeCN:buffer 20:80. The buffer was 30 mM potassium dihydrogen phosphate adjusted to pH 2.9 with phosphoric acid.
**Flow rate:** 1.1
**Injection volume:** 5–10
**Detector:** UV 254

## CHROMATOGRAM

**Retention time:** 8.5
**Limit of quantitation:** 0.05%

## OTHER SUBSTANCES

**Simultaneous:** impurities, quinupristin (LOQ 0.12%) (23.9, 23.1, 27.0)

## KEY WORDS

infusion; injection; stability-indicating

## REFERENCE

Vasselle, B.; Gousset, G.; Bounine, J.-P. Development and validation of a high-performance liquid chromatographic stability-indicating method for the analysis of Synercid in quality control, stability and compatibility studies, *J.Pharm.Biomed.Anal.*, **1999**, *19*, 641–657.

## ANNOTATED BIBLIOGRAPHY

Abdel-Hamid, M.E.; Phillips, O.A. LC-MS/MS determination of Synercid injections, *J.Pharm.Biomed. Anal.*, **2003**, *32*, 1167–1174. [pristinamycin; quinupristin; dalfopristin]

# Dalteparin

**Molecular weight:** 4000–6000
**CAS Registry No:** 9041-08-1 (Na salt)
**Merck Index:** 13, 2832

## SAMPLE

**Matrix:** solutions
**Sample preparation:** Prepare a 10 mg/mL solution in mobile phase, stir for 4 h, filter (0.2 μm) while centrifuging, inject a 200 μL aliquot.

## HPLC VARIABLES

**Guard column:** Ohpak SB-G
**Column:** 300 × 8 Shodex Ohpak SB-803 HQ
**Mobile phase:** 100 mM pH 7 ammonium acetate buffer containing 0.05% sodium azide (Caution! Sodium azide is highly toxic and can form shock-sensitive and highly explosive azides when it comes in contact with heavy metals! Sodium azide should not be discharged to the plumbing system!)
**Flow rate:** 0.8
**Injection volume:** 200
**Detector:** Refractive Index; Light Scattering Detector (miniDAWN, Wyatt Technology, Santa Barbara CA) (detector measures scattered light intensity (690 nm) at three angles (45°, 90°, 135°)

## CHROMATOGRAM

**Retention time:** 10

## REFERENCE

Knobloch, J.E.; Shaklee, P.N. Absolute molecular weight distribution of low-molecular-weight heparins by size-exclusion chromatography with multiangle laser light scattering detection, *Anal.Biochem.*, **1997**, *245*, 231–241.

# Daptomycin

**Molecular formula:** $C_{72}H_{101}N_{17}O_{26}$
**Molecular weight:** 1620.67
**CAS Registry No:** 103060-53-3

---

### SAMPLE
**Matrix:** solutions

### HPLC VARIABLES
**Column:** 250 mm long DuPont C8
**Mobile phase:** MeCN:0.5% pH 3.5 ammonium dihydrogen phosphate 38:62
**Flow rate:** 1.5
**Detector:** UV 214

### CHROMATOGRAM
**Retention time:** 25

### OTHER SUBSTANCES
**Simultaneous:** anhydrodaptomycin (35), $\beta$-asp daptomycin (23)

### REFERENCE
Kirsch, L.E.; Molloy, R.M.; Debono, M.; Baker, P.; Farid, K.Z. Kinetics of the aspartyl transpeptidation of daptomycin, a novel lipopeptide antibiotic, *Pharm.Res.*, **1989**, *6*, 387–393.

---

### SAMPLE
**Matrix:** solutions

### HPLC VARIABLES
**Column:** 250 × 4.6 5 μm Zorbax 300SB-C8
**Mobile phase:** MeCN:50 mM pH 5 phosphate buffer 29:71
**Flow rate:** 1
**Injection volume:** 10
**Detector:** UV 214

### CHROMATOGRAM
**Retention time:** <30

### OTHER SUBSTANCES
**Simultaneous:** degradation products

### REFERENCE
Muangsiri, W.; Kirsch, L.E. The kinetics of the alkaline degradation of daptomycin, *J.Pharm.Sci.*, **2001**, *90*, 1066–1075.

---

### SAMPLE
**Matrix:** blood
**Sample preparation:** Precipitate proteins before analysis (no other details)

### HPLC VARIABLES
**Guard column:** Xterra RP18 (Waters)
**Column:** Hypersil C8
**Mobile phase:** MeCN:0.5% ammonium phosphate 32.6:67.4

**Flow rate:** 1.5
**Detector:** UV 214

## CHROMATOGRAM
**Retention time:** 14–16
**Internal standard:** ethyl paraben
**Limit of quantitation:** 7.5 μg/mL

## KEY WORDS
plasma

## REFERENCE
Sakoulas, G.; Eliopoulos, G.M.; Alder, J. Thauvin-Eliopoulos,C. Efficacy of daptomycin in experimental endocarditis due to methicillin-resistant staphylococcus aureus, *Antimicrob.Agents Chemother.*, **2003**, *47*, 1714–1718.

# Deferiprone

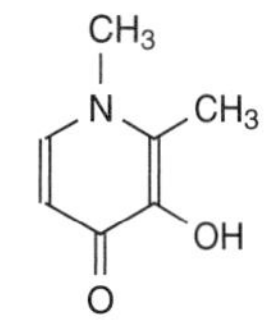

**Molecular formula:** $C_7H_9NO_2$
**Molecular weight:** 139.15
**CAS Registry No:** 30652-11-0
**Merck Index:** 13, 2878

## SAMPLE

**Matrix:** blood, urine
**Sample preparation:** Add IS to a final concentration of 100 μM and perchloric acid to a final concentration of 500 mM to plasma or serum, centrifuge at 11 600 g for 10 min, inject a 50 μL aliquot. Centrifuge urine at 960 g for 20 min, filter (0.45 μm), dilute 10-fold with 50 mM pH 2.0 potassium phosphate buffer, inject a 50 μL aliquot.

## HPLC VARIABLES

**Guard column:** LiChrosorb RP Select B
**Column:** 5 μm LiChrosorb RP Select B
**Mobile phase:** Gradient MeOH:50 mM pH 2.0 potassium phosphate buffer 27:73 containing 10 mM octanesulfonic acid for 16 min, to 70:30 over 2 min, maintain at 70:30 for 2 min, return to initial conditions over 2 min, re-equilibrate for 8 min. (Methanol wash not necessary for urine samples.)
**Flow rate:** 1.6
**Injection volume:** 50
**Detector:** UV 280

## CHROMATOGRAM

**Retention time:** 8.2
**Internal standard:** 1-ethyl-2-methyl-3-hydroxypyrid-4-one (13.2)
**Limit of quantitation:** 25 μM

## KEY WORDS

plasma; serum

## REFERENCE

Goddard, J.G.; Kontoghlorghes, G.J. Development of an HPLC method for measuring orally administered 1-substituted 2-alkyl-3-hydroxypyrid-4-one iron chelators in biological fluids, *Clin.Chem.*, **1990**, *36*, 5–8.

# Deflazacort

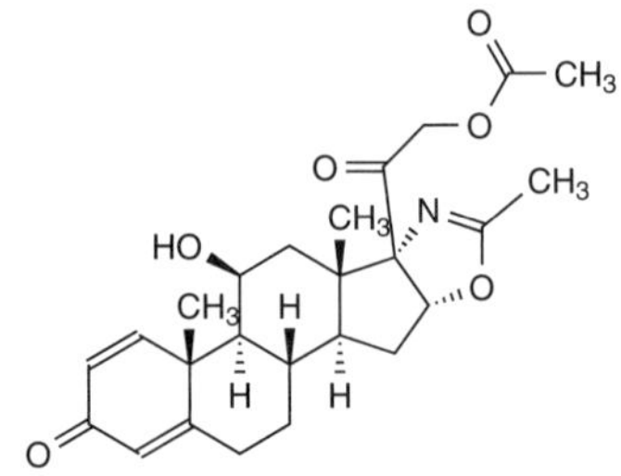


**Molecular formula:** $C_{25}H_{31}NO_6$
**Molecular weight:** 441.52
**CAS Registry No:** 14484-47-0
**Merck Index:** 13, 2881

## SAMPLE
**Matrix:** blood, formulations
**Sample preparation:** Mix 1 mL serum with 1 μg IS and 1 mL MeOH, vortex for 5 min, centrifuge at 5000 g for 5 min, inject a 20 μL aliquot of the supernatant. Weigh out amount of crushed tablets containing 10 mg deflazacort, add 7 mL mobile phase, sonicate for 10 min, make up to 10 mL with mobile phase, centrifuge. Mix an aliquot of the supernatant with an aliquot of 2 μg/mL IS in mobile phase, dilute with mobile phase, inject a 20 μL aliquot.

## HPLC VARIABLES
**Column:** 250 × 4.6 5 μm Spherisorb C18
**Mobile phase:** MeCN:MeOH:67 mM potassium dihydrogen phosphate 27:20:53, adjusted to pH 6.5 with 3 M NaOH
**Flow rate:** 0.75
**Injection volume:** 20
**Detector:** UV 244

## CHROMATOGRAM
**Retention time:** 12.3
**Internal standard:** etodolac (5.5)
**Limit of detection:** 4 ng/mL
**Limit of quantitation:** 13.6 ng/mL

## KEY WORDS
plasma; tablets

## REFERENCE
Ozkan, Y.; Savaser, A.; Tas, C.; Uslu, B.; Ozkan, S.A. Drug dissolution studies and determination of deflazacort in pharmaceutical formulations and human serum samples by RP-HPLC, *J.Liq.Chromatogr. Rel.Technol.*, **2003**, *26*, 2141–2156.

## SAMPLE
**Matrix:** blood
**Sample preparation:** Condition a 3 mL 200 mg Clean-Up C18 SPE cartridge (Worldwide Monitoring) with two 2 mL portions of MeOH and two 2 mL portions of water. Centrifuge plasma at 2000 rpm for 5 min prior to analysis. Mix 2 mL plasma with 20 μL 5 μg/mL IS in MeCN and 20 μL MeCN, add 1 mL water, vortex until homogeneous, add to the SPE cartridge, wash with two 2 mL portions of acetone:water 20:80, wash with 1 mL 50 mM pH 2.7 phosphate buffer, place under vacuum for 3 min, elute with two 2.5 mL portions of ethyl acetate. Evaporate the eluate to dryness under reduced pressure at 40°, reconstitute the residue with 75 μL mobile phase B, vortex, add 100 μL 50 mM pH 3 phosphate buffer, centrifuge at 5000 rpm for 5 min, inject a 145 μL aliquot.

## HPLC VARIABLES
**Column:** 100 × 2 5 μm YMC Basic

**Column temperature:** 50
**Mobile phase:** Gradient. A:B 20:80 for 1 min, to 50:50 over 7 min, to 65:35 over 3 min, to 100:0 (step gradient), maintain at 100:0 for 1 min, return to initial conditions, re-equilibrate for 4 min. A was MeCN:50 mM pH 3.0 potassium dihydrogen phosphate buffer 50:50. B was MeOH:50 mM pH 3.0 potassium dihydrogen phosphate buffer 20:80.
**Flow rate:** 0.3
**Injection volume:** 145
**Detector:** UV 246

## CHROMATOGRAM
**Retention time:** 13.8
**Internal standard:** fludrocortisone acetate (12.4)

## OTHER SUBSTANCES
**Extracted:** 21-hydroxydeflazacort (10)
**Simultaneous:** cortisone, hydrocortisone

## KEY WORDS
plasma; SPE; method validated for 21-hydroxydeflazacort rather than deflazacort

## REFERENCE
Reynolds, D.L.; Burmaster, S.D.; Eichmeier, L.S. Quantitative determination of 21-hydroxy-deflazacort in human plasma using gradient semi-microbore liquid chromatography, *Biomed.Chromatogr.*, **1994**, *8*, 230–235.

## ANNOTATED BIBLIOGRAPHY
Ozkan, S.A.; Uslu, B. Rapid HPLC assay for lamivudine in pharmaceuticals and human serum, *J.Liq.Chromatogr.Rel.Technol.*, **2002**, *25*, 1447–1456. [deflazacort is internal standard]

# Desloratadine

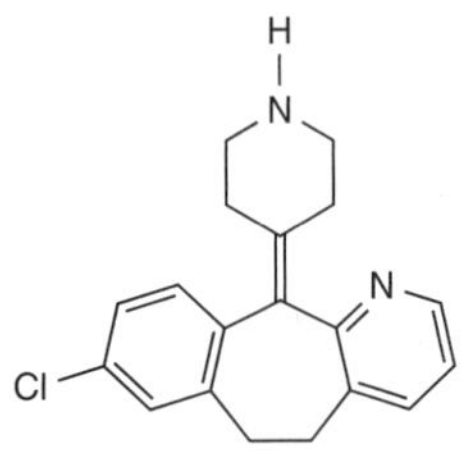


**Molecular formula:** $C_{19}H_{19}ClN_2$
**Molecular weight:** 310.83
**CAS Registry No:** 100643-71-8
**Merck Index:** 13, 2939

## SAMPLE

**Matrix:** blood
**Sample preparation:** Mix 1 mL plasma, 100 μL MeOH, 50 μL MeOH containing 20 ng/mL $d_3$-loratadine and 100 ng/mL $d_3$-desloratadine, and 50 μL 0.1% ammonium hydroxide, vortex for 30 s, add 5 mL hexane, vortex for 2 min, centrifuge at 1200 g for 5 min, freeze in dry ice acetone. Evaporate the organic layer to dryness under a stream of nitrogen at 40°, reconstitute the residue with 200 μL 0.1% trifluoroacetic acid in MeCN, vortex for 2 min, inject a 35 μL aliquot.

## HPLC VARIABLES

**Column:** 50 × 3 5 μm Betasil silica (Keystone)
**Mobile phase:** MeCN:water:trifluoroacetic acid 90:10:0.1
**Flow rate:** 0.5
**Injection volume:** 35
**Detector:** MS, PE Sciex API 3000 turbo ionspray, positive ion mode, ionspray needle 5 kV, turbo gas temperature 300°, auxiliary gas flow 8 L/min, m/z 311–259

## CHROMATOGRAM

**Retention time:** 2
**Internal standard:** $d_3$-loratadine (m/z 388–342), $d_3$-desloratadine (m/z 316–262)
**Limit of quantitation:** 25 pg/mL

## OTHER SUBSTANCES

**Extracted:** loratadine (LOQ 10 pg/mL, m/z 383–337) (1.2)
**Noninterfering:** acetaminophen, albuterol, aspirin, caffeine, clonidine, fentanyl, ibuprofen, naltrexone, ritonavir

## KEY WORDS

plasma

## REFERENCE

Naidong, W.; Addison, T.; Schneider, T.; Jiang, X.; Halls, T.D.J. A sensitive LC/MS/MS method using silica column and aqueous-organic mobile phase for the analysis of loratadine and descarboethoxy-loratadine in human plasma, *J.Pharm.Biomed.Anal.*, **2003**, *32*, 609–617.

## SAMPLE

**Matrix:** blood
**Sample preparation:** Mix 1 mL plasma with 20 μL 7.5 μg/mL IS in MeOH:50 mM HCl 20:80 and 200 μL 1 M NaOH, vortex, add 3 mL toluene, shake for 20 min, centrifuge at 2000 g for 10 min, freeze, remove organic layer, repeat extraction. Combine the organic layers and evaporate to dryness under a stream of nitrogen at 50°, reconstitute the residue with 200 μL MeOH:50 mM HCl 20:80, vortex, mobile phase, inject a 100 μL aliquot.

## HPLC VARIABLES

**Column:** 150 × 3.9 5 μm Symmetry C18 (Waters)

**Column temperature:** 35
**Mobile phase:** Gradient. MeCN:MeOH:50 mM pH 2.0 potassium phosphate buffer from 14:5:81 to 24:3:7.3 over 5.5 min, to 40:3:57 over 4.5 min, return to initial conditions over 4 min, re-equilibrate for 6 min.
**Flow rate:** 1.2
**Injection volume:** 100
**Detector:** F ex 290 em 480

## CHROMATOGRAM
**Retention time:** 3.4
**Internal standard:** propranolol hydrochloride (8.6)
**Limit of detection:** 0.25 ng/mL
**Limit of quantitation:** 0.5 ng/mL

## OTHER SUBSTANCES
**Extracted:** loratadine (11.2)

## KEY WORDS
pharmacokinetics; plasma

## REFERENCE
Yin, O.Q.P.; Shi, X.; Chow, M.S.S. Reliable and specific high-performance liquid chromatographic method for simultaneous determination of loratadine and its metabolite in human plasma, *J.Chromatogr.B*, **2003**, *796*, 165–172.

## ANNOTATED BIBLIOGRAPHY
Rupérez, F.J.; Fernández, H.; Barbas, C. LC determination of loratadine and related impurities, *J.Pharm. Biomed.Anal.*, **2002**, *29*, 35–41.

Sutherland, F.C.W.; de Jager, A.D.; Badenhorst, D.; Scanes, T.; Hundt, H.K.L.; Swart, K.J.; Hundt, A.F.. Sensitive liquid chromatography-tandem mass spectrometry method for the determination of loratadine and its major active metabolite descarboethoxyloratadine in human plasma. *J.Chromatogr.A* **2001**, *914*, 37–43.

# Desogestrel

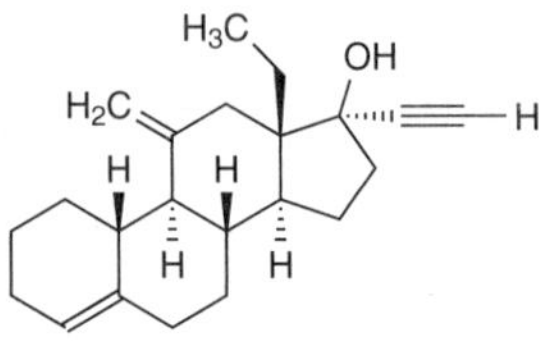


**Molecular formula:** $C_{22}H_{30}O$
**Molecular weight:** 310.47
**CAS Registry No:** 54024-22-5
**Merck Index:** 13, 2943

## SAMPLE
**Matrix:** blood
**Sample preparation:** Add plasma to a Bakerbond C18 SPE cartridge, wash with 100 mM pH 4.2 ammonium acetate buffer, elute with MeOH. Evaporate the eluate to dryness under reduced pressure, reconstitute the residue with MeOH, inject an aliquot.

## HPLC VARIABLES
**Column:** 300 × 3.9 μBondapak C18
**Column temperature:** 50
**Mobile phase:** Gradient. MeOH:100 mM pH 4.2 ammonium acetate from 10:90 to 90:10 over 30 min
**Flow rate:** 1.7
**Detector:** Radioactivity ($^3H$); UV 254

## CHROMATOGRAM
**Retention time:** 32

## OTHER SUBSTANCES
**Extracted:** metabolites

## KEY WORDS
dog; plasma; SPE

## REFERENCE
Verhoeven, C.H.J.; Krebbers, S.F.M.; Wagenaars, G.N.; Vos, R.M.E. In vitro and in vivo metabolism of desogestrel in several species, *Drug Metab.Dispos.*, **1998**, *26*, 927–936.

# Desoximetasone

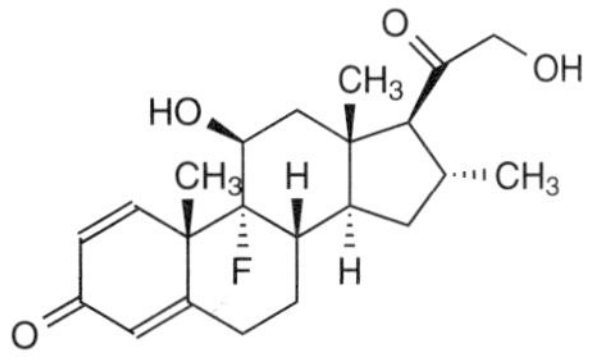


**Molecular formula:** $C_{22}H_{29}FO_4$
**Molecular weight:** 376.46
**CAS Registry No:** 382-67-2
**Merck Index:** 13, 2947

## SAMPLE

**Matrix:** formulations
**Sample preparation:** Mix 1 g of a topical product, 5 mL 100 mM pH 4 citrate buffer saturated with NaCl, and 10 mL ethyl acetate, shake vigorously by hand to ensure that no large clumps stick to the tube, mix with the tube on its side on an oscillating shaker for 15 min. Remove the upper ethyl acetate layer and wash with citrate buffer as before. Dry the ethyl acetate over anhydrous sodium sulfate and evaporate to dryness. Dissolve the residue in a mixture of 10 mL heptane and 5 mL MeCN:water 90:10. Remove the upper heptane layer and extract it with 2 mL MeCN:water 90:10. Combine the MeCN/water layers and evaporate them to dryness, dissolve the residue in 300 μL MeOH, filter (0.45 μm nylon), inject a 5 μL aliquot.

## HPLC VARIABLES

**Guard column:** 15 × 3.2 7 μm Brownlee NewGuard C18
**Column:** 75 × 4.6 3.5 μm Symmetry C18 (Waters)
**Mobile phase:** Gradient. MeCN:water 18:82 for 2 min, to 82:18 over 12 min, maintain at 82:18 for 3 min, re-equilibrate at initial conditions for 12 min.
**Flow rate:** 1
**Injection volume:** 5
**Detector:** UV 240

## CHROMATOGRAM

**Retention time:** 7.60
**Limit of detection:** 0.001%

## OTHER SUBSTANCES

**Simultaneous:** alclometasone 17,21-dipropionate (10.93), amcinonide (10.90), beclomethasone 17,21-dipropionate (11.90), betamethasone (6.52), betamethasone 21-acetate (8.47), betamethasone 17-benzoate (10.28), betamethasone 17,21-dipropionate (11.42), betamethasone 17-valerate (10.25), budesonide (8.55, 8.69 (epimers)), clobetasol 17-propionate (11.06), cortisone (5.62), cortisone 21-acetate (8.07), dexamethasone (6.57), dexamethasone 21-acetate (8.68), desonide (6.99), desoximetasone (7.60), desoxycorticosterone acetate (10.90), desoxycorticosterone pivalate (14.45), diflorasone 17,21-diacetate (9.81), fluocinolone acetonide (7.38), fluocinonide (9.79), flurandrenolide (7.36), fludrocortisone 21-acetate (7.77), fluorometholone (7.67), flumethasone 21-pivalate (11.20), flunisolide (7.14), fluprednisolone (5.46), fluticasone 17-propionate (11.19), halcinonide (10.72), halobetasol propionate (10.98), hydrocortisone (5.50), hydrocortisone 21-acetate (7.65), hydrocortisone 17-butyrate (8.66), hydrocortisone 21-cypionate (12.54), hydrocortisone 17-valerate (9.53), mometasone 17-furoate (11.24), methylprednisolone (6.31), methylprednisolone 21-acetate (8.34), meprednisone (6.55), prednisolone (5.37), prednisolone 21-acetate (7.47), prednisolone 21-tebuate (11.01), paramethasone 21-acetate (8.64), prednicarbate (10.72), prednisone (5.46), triamcinolone (4.15), triamcinolone acetonide (7.04), triamcinolone 16,21-diacetate (7.49), triamcinolone hexacetonide (13.12)

## KEY WORDS

body wash, cream, gel, lotion, shampoo, spray

## REFERENCE

Reepmeyer, J.C. Screening for corticosteroids in topical pharmaceuticals by HPLC with a scanning ultraviolet detector, *J.Liq.Chromatogr.Rel.Technol.*, **2001**, *24*, 693–709.

## SAMPLE

**Matrix:** urine

**Sample preparation:** Mix 5 mL urine with 100 ng IS and 400 μL 1 M pH 4.1 sodium acetate buffer, adjust pH to 5.0, add 600 μL β-glucuronidase solution (10800 U), heat at 65° for 3.5 h or at 37° overnight, cool, add 6 mL ethyl acetate, rotate for 10 min. Remove the organic layer and wash it with 3 mL 1 M NaOH containing 150 mM NaCl by rotating for 5 min, centrifuge at 1900 g for 30 s. Remove the organic layer and pass it through an anhydrous sodium sulfate drying column. Evaporate the eluate to dryness under a stream of nitrogen at 60°, reconstitute the residue with 30 μL MeOH, vortex, inject a 10 μL aliquot.

## HPLC VARIABLES

**Column:** 75 × 4.6 3 μm DB-8 (Supelco)

**Column temperature:** 25

**Mobile phase:** Gradient. MeOH:1% acetic acid from 0:100 to 100:0 over 15 min, maintain at 100:0 for 3 min.

**Flow rate:** 1

**Injection volume:** 10

**Detector:** MS, Finnigan MAT LCQ Classic, APCI, source 450°, capillary 150°, source +5 kV for positive ions and – 5 kV for negative ions, collision gas helium, m/z 377

## CHROMATOGRAM

**Retention time:** 27.6

**Internal standard:** $d_4$-hydrocortisone (24.4)

**Limit of detection:** 5 ng/mL

## OTHER SUBSTANCES

**Extracted:** beclomethasone (m/z 409) (26.4), betamethasone (m/z 393) (25.9), deoxycortone (m/z 331) (28.1), 21-deoxydexamethasone (m/z 377) (26.8), dexamethasone (m/z 393) (25.9), dichlorisone (m/z 413) (26.4), fluclorolone acetonide (m/z 487) (27.8), fludrocortisone (m/z 381) (24.2), flumethasone (m/z 411) (25.4), fluocinolone acetonide (m/z 453) (26.4), fluocinonide (m/z 495) (28.5), fluocortolone (m/z 377) (26.8), fluorometholone (m/z 3.77) (26.6), fluprednisolone (m/z 379) (23.9), flurandrenolide (m/z 437) (26.8), hydrocortisone (m/z 363) (24.4), isoflupredone (m/z 379) (23.9), methylprednisolone (m/z 375) (26.1), prednisolone (m/z 361) (24.4), prednisone (m/z 359) (23.4), triamcinolone (m/z 395) (19), triamcinolone acetonide (m/z 435) (29.3)

## KEY WORDS

horse

## REFERENCE

Tang, P.W.; Law, W.C.; Wan, T.S.M. Analysis of corticosteroids in equine urine by liquid chromatography-mass spectrometry, *J.Chromatogr.B*, **2001**, *754*, 229–244.

# Desoxycorticosterone

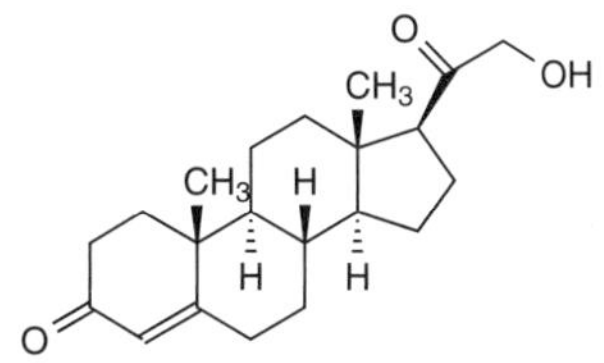


**Molecular formula:** $C_{21}H_{30}O_3$
**Molecular weight:** 330.46
**CAS Registry No:** 64-85-7
**Merck Index:** 13, 2917

### SAMPLE
**Matrix:** blood
**Sample preparation:** Vortex 500 μL serum with 5 mL diethyl ether for 3 min, centrifuge at 3500 rpm for 5 min. Evaporate the organic layer to dryness at 50°, reconstitute the residue with 50 μL MeOH:water 60:40, vortex for 1 min, sonicate for 1 min, centrifuge at 4000 rpm for 2 min, inject a 20 μL aliquot of the supernatant.

### HPLC VARIABLES
**Guard column:** 50 × 2.1 unspecified
**Column:** 150 × 6 Shimpack CLC-ODS
**Column temperature:** 48
**Mobile phase:** MeOH:THF:water 26:18:56
**Flow rate:** 1
**Injection volume:** 20
**Detector:** UV 254

### CHROMATOGRAM
**Retention time:** 10
**Limit of detection:** 1 pmol

### OTHER SUBSTANCES
**Extracted:** androstenedione (9.5), corticosterone (7), cortisone (5), 11-deoxycortisol (7.5), dexamethasone acetate (13), estradiol (F ex 285 em 310) (15), estriol (F ex 285 em 310) (5), estrone (F ex 285 em 310) (17), hydrocortisone (6), hydroxyprogesterone (14), prednisone acetate (8), progesterone (23), testosterone (11)

### KEY WORDS
serum; SPE

### REFERENCE
Wei, J.-q.; Wei, J.-l.; Zhou, X.-t.; Cheng, J.-p. Isocratic reversed phase high performance liquid chromatography determination of twelve natural corticosteroids in serum with on-line ultraviolet and fluorescence detection, *Biomed.Chromatogr.*, **1990**, *4*, 161–164.

### SAMPLE
**Matrix:** formulations
**Sample preparation:** Mix 1 g of a topical product, 5 mL 100 mM pH 4 citrate buffer saturated with NaCl, and 10 mL ethyl acetate, shake vigorously by hand to ensure that no large clumps stick to the tube, mix with the tube on its side on an oscillating shaker for 15 min. Remove the upper ethyl acetate layer and wash with citrate buffer as before. Dry the ethyl acetate over anhydrous sodium sulfate and evaporate to dryness. Dissolve the residue in a mixture of 10 mL heptane and 5 mL MeCN:water 90:10. Remove the upper heptane layer and extract it with 2 mL MeCN:water 90:10. Combine the MeCN/water layers and evaporate them to dryness, dissolve the residue in 300 μL MeOH, filter (0.45 μm nylon), inject a 5 μL aliquot.

## HPLC VARIABLES

**Guard column:** 15 × 3.2 7 μm Brownlee NewGuard C18
**Column:** 75 × 4.6 3.5 μm Symmetry C18 (Waters)
**Mobile phase:** Gradient. MeCN:water 18:82 for 2 min, to 82:18 over 12 min, maintain at 82:18 for 3 min, re-equilibrate at initial conditions for 12 min.
**Flow rate:** 1
**Injection volume:** 5
**Detector:** UV 240

## CHROMATOGRAM

**Retention time:** 10.90 (acetate), 14.45 (pivalate)
**Limit of detection:** 0.001%

## OTHER SUBSTANCES

**Simultaneous:** alclometasone 17,21-dipropionate (10.93), amcinonide (10.90), beclomethasone 17,21-dipropionate (11.90), betamethasone (6.52), betamethasone 21-acetate (8.47), betamethasone 17-benzoate (10.28), betamethasone 17,21-dipropionate (11.42), betamethasone 17-valerate (10.25), budesonide (8.55, 8.69 (epimers)), clobetasol 17-propionate (11.06), cortisone (5.62), cortisone 21-acetate (8.07), dexamethasone (6.57), dexamethasone 21-acetate (8.68), desonide (6.99), desoximetasone (7.60), diflorasone 17,21-diacetate (9.81), fluocinolone acetonide (7.38), fluocinonide (9.79), flurandrenolide (7.36), fludrocortisone 21-acetate (7.77), fluorometholone (7.67), flumethasone 21-pivalate (11.20), flunisolide (7.14), fluprednisolone (5.46), fluticasone 17-propionate (11.19), halcinonide (10.72), halobetasol propionate (10.98), hydrocortisone (5.50), hydrocortisone 21-acetate (7.65), hydrocortisone 17-butyrate (8.66), hydrocortisone 21-cypionate (12.54), hydrocortisone 17-valerate (9.53), mometasone 17-furoate (11.24), methylprednisolone (6.31), methylprednisolone 21-acetate (8.34), meprednisone (6.55), prednisolone (5.37), prednisolone 21-acetate (7.47), prednisolone 21-tebuate (11.01), paramethasone 21-acetate (8.64), prednicarbate (10.72), prednisone (5.46), triamcinolone (4.15), triamcinolone acetonide (7.04), triamcinolone 16,21-diacetate (7.49), triamcinolone hexacetonide (13.12)

## KEY WORDS

body wash, cream, gel, lotion, shampoo, spray

## REFERENCE

Reepmeyer, J.C. Screening for corticosteroids in topical pharmaceuticals by HPLC with a scanning ultraviolet detector, *J.Liq.Chromatogr.Rel.Technol.*, **2001**, *24*, 693–709.

---

## SAMPLE

**Matrix:** cell cultures
**Sample preparation:** Extract cell medium twice with two volumes of ethyl acetate for 5 min. Evaporate the combined extracts to dryness under a stream of nitrogen, reconstitute the residue with 100 μL MeCN, inject an aliquot.

## HPLC VARIABLES

**Guard column:** 15 × 3.2 Newguard 7 μm RP-C18
**Column:** 220 × 4.6 5 μm RP-C18 Spheri (Kontron)
**Column temperature:** 30
**Mobile phase:** Gradient. MeCN:water 30:70 for 12 min, to 70:30 over 33 min.
**Flow rate:** 1.2
**Injection volume:** 20
**Detector:** UV 254

## CHROMATOGRAM

**Retention time:** 16.9
**Limit of quantitation:** 50 ng

## OTHER SUBSTANCES

**Extracted:** corticosterone (8.3), 18-hydroxydeoxycorticosterone (6.9), progesterone (29.6)

## REFERENCE

Matilla, M.J.; Jimenez, M.M.; Montiel, M. Deoxycorticosterone, 18-OH-deoxycorticosterone and corticosterone determination by high performance liquid chromatography in monolayer adrenal cell culture, *Biochem.Int.*, **1991**, *24*, 951–957.

## ANNOTATED BIBLIOGRAPHY

Sheikh, S.U.; Touchstone, J. HPLC of steroids in non-aqueous mobile phase at subambient temperature, *J.Liq.Chromatogr.*, **1987**, *10*, 2489–2496. [column temperature −50°; desoxycorticosterone; estrone; estradiol; cortisone; hydrocortisone]

Smith, E. Liquid chromatographic determination of desoxycorticosterone acetate in oil injections, *J.Assoc.Off.Anal.Chem.*, **1979**, *62*, 812–817.

Wei, J.Q.; Wei, J.L.; Lucarelli, C.; Zhou, X.T.; Wang, D.Q.; Dai, W.J.; Li, S.; Li, S.M.; Liu, R.T. Serum steroid hormonal profiles by reversed-phase liquid chromatography in patients with 17-hydroxylase deficiency and in an affected family, *Clin.Chem.*, **1992**, *38*, 76–82. [cortisone; hydrocortisone; deoxycortisol; androstenedione; testosterone; progesterone; hydroxyprogesterone; desoxycorticosterone]

# Dexrazoxane

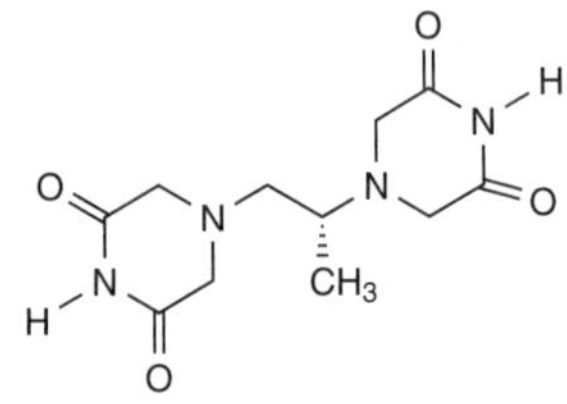

**Molecular formula:** $C_{11}H_{16}N_4O_4$
**Molecular weight:** 268.27
**CAS Registry No:** 24584-09-6, 21416-87-5 (racemic)
**Merck Index:** 13, 8208

## SAMPLE

**Matrix:** blood
**Sample preparation:** Condition a Sep-Pak Plus C18 SPE cartridge with MeCN and water. Acidify 1 mL plasma with 50 μL 43% phosphoric acid. Add 70 μL 6 M HCl to 260 μL acidified plasma, centrifuge, remove the supernatant, wash the pellet three times with 250 μL portions of 10 mM HCl. Combine the supernatant and the washings and adjust the pH to 6.0 with 80 μL 5 M NaOH, add to the SPE cartridge, wash with 5 mL water, wash with 1 mL hexane, elute with 7.5 mL MeCN. Evaporate the eluate to dryness under a stream of argon, reconstitute the residue with 130 μL mobile phase EtOH:MeOH:isopropanol 90:5:5, inject a 50 μL aliquot.

## HPLC VARIABLES

**Guard column:** 3.2 × 1.5 NewGuard silica (Brownlee)
**Column:** 250 × 4.6 10 μm Chiralcel OD
**Mobile phase:** EtOH:MeOH:isopropanol:*n*-hexane 76.5:4.25:4.25:15
**Flow rate:** 0.5
**Injection volume:** 50
**Detector:** UV 207

## CHROMATOGRAM

**Retention time:** 18.5
**Limit of quantitation:** 2 μM

## OTHER SUBSTANCES

**Extracted:** levrazoxane (16.5)

## KEY WORDS

chiral; plasma; pharmacokinetics; rat; SPE

## REFERENCE

Hasinoff, B.B.; Aoyama, R.G. Stereoselective metabolism of dexrazoxane (ICRF-187) and levrazoxane (ICRF-186), *Chirality*, **1999**, *11*, 286–290.

## SAMPLE

**Matrix:** blood, urine
**Sample preparation:** Plasma. Mix 1 mL plasma with 25 μg/mL IS, 100 μL water, and 2 mL MeCN, shake for 10 min, centrifuge at 2500 g for 10 min. Remove the supernatant, add 20 mL chloroform:2-methyl-2-propanol 90:10 (Caution! Chloroform is a carcinogen!), shake for 30 min, centrifuge at 2500 g for 10 min. Remove a 15 mL portion of the organic layer and evaporate it to dryness under a stream of nitrogen at 40°, reconstitute the residue with 500 μL MeOH:100 mM HCl 20:80, inject a 50 μL aliquot. Urine. Mix 1 mL urine with 250 μg/mL IS, make up to 100 mL with water. Remove a 1 mL portion of this mixture, add 20 mL chloroform:2-methyl-2-propanol 90:10, shake for 30 min, centrifuge at 2500 g for 10 min. Remove 15 mL of the organic layer and evaporate it to dryness under a stream of nitrogen at 40°, reconstitute the residue with 500 μL MeOH:100 mM HCl 20:80, inject a 50 μL aliquot.

## HPLC VARIABLES

**Guard column:** 10 × 3 reverse phase (Chrompack)
**Column:** 300 × 4.6 10 μm μBondapak Phenyl
**Mobile phase:** MeOH:10 mM pH 4.7 potassium phosphate buffer 20:80
**Flow rate:** 1
**Injection volume:** 50
**Detector:** UV 208

## CHROMATOGRAM

**Retention time:** 6.2
**Internal standard:** 8-chlorotheophylline (16)
**Limit of quantitation:** 100 ng/mL (plasma), 10 μg/mL (urine)

## OTHER SUBSTANCES

**Simultaneous:** acetaminophen, 5-bromouracil, caffeine, 5-chlorouracil, 5-fluorocytosine, 5-fluorouracil, phenylethyleneglycol, theophylline
**Noninterfering:** codeine, cyclophosphamide, doxorubicin, indomethacin, methylparaben, phenacetin, phenazone, prednisone, triamcinolone acetate, triamcinolone

## KEY WORDS

plasma; pharmacokinetics

## REFERENCE

Rosing, H.; van Gijn, R.; ten Bokkel Huinink, W.W.; Beijnen, J.H. High performance liquid chromatographic analysis of the cardioprotective agent dexrazoxane in human plasma and urine, *J.Liq.Chromatogr.Rel.Technol.*, **1997**, *20*, 583–601.

## ANNOTATED BIBLIOGRAPHY

Hasinoff, B.B.; Aoyama, R.G. Relative plasma levels of the cardioprotective drug dexrazoxane and its two active ring-opened metabolites in the rat, *Drug Metab.Dispos.*, **1999**, *27*, 265–268.

# Dextran

**CAS Registry No:** 9004-54-0
**Merck Index:** 13, 2965

## SAMPLE

**Matrix:** blood
**Sample preparation:** Dilute plasma 4 times with 100 mM pH 12 sodium glycinate buffer, filter (Amicon Centrifree) while centrifuging. Lyophilize the ultrafiltrate and reconstitute the residue with 125 μL pH 7 ammonium phosphate buffer, inject a 100 μL aliquot.

## HPLC VARIABLES

**Column:** Tosohaas 2000SWXL + Tosohaas 2500PWXL in series
**Mobile phase:** 100 mM pH 6 sodium acetate buffer
**Flow rate:** 0.8
**Injection volume:** 100
**Detector:** Radioactivity ($^3H$)

## CHROMATOGRAM

**Retention time:** 18

## KEY WORDS

plasma; rat

## REFERENCE

Hartman, N.R.; Johns, D.G.; Mitsuya, H. Pharmacokinetic analysis of dextran sulfate in rats as pertains to its clinical usefulness for therapy of HIV infection, *AIDS Res.Hum.Retroviruses*, **1990**, *6*, 805–812.

## SAMPLE

**Matrix:** blood
**Sample preparation:** Filter serum (0.45 μm), inject an aliquot.

## HPLC VARIABLES

**Guard column:** 48 × 4.6 (ES Industries)
**Column:** 300 × 7.8 5 μm Chromega diol size-exclusion (ES Industries)
**Column temperature:** 40
**Mobile phase:** Buffer containing 25 mM potassium dihydrogen phosphate, 25 mM dipotassium hydrogen phosphate, and 50 mM potassium chloride, pH ca. 6.8
**Flow rate:** 1
**Injection volume:** 10
**Detector:** UV 525 following post-column reaction with an 18.5 μg/mL solution of 1,9-dimethylmethylene blue in water pumped at 0.5 mL/min. The column effluent mixed with the dye solution in a 3.1 μL mixing tee and flowed through a 28.8 cm length of 0.254 mm ID tubing to the detector.

## CHROMATOGRAM

**Retention time:** 7
**Limit of detection:** 300 ng/mL
**Limit of quantitation:** 8 μg/mL

## KEY WORDS

post-column reaction; rat; serum

## REFERENCE

Maderich, A.B.; Sugita, E.T. Size-exclusion chromatographic determination of dextran sulfate in rat serum, *J.Chromatogr.*, **1993**, *620*, 137–142.

## ANNOTATED BIBLIOGRAPHY

Hemmelder, M.H.; de Jong, P.E.; De Zeeuw, D. A comparison of analytic procedures for measurement of fractional dextran clearances, *J.Lab.Clin.Med.*, **1998**, *132*, 390–403. [RI detection; detection by reaction of fractions with anthrone reagent]

# Diacerein

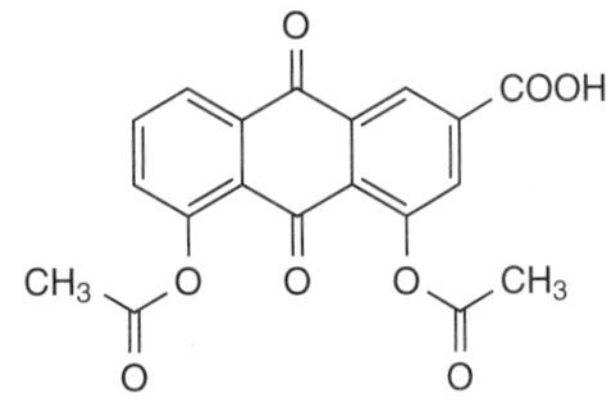


**Molecular formula:** $C_{19}H_{12}O_8$
**Molecular weight:** 368.29
**CAS Registry No:** 13739-02-1
**Merck Index:** 13, 2979

## SAMPLE

**Matrix:** blood, urine
**Sample preparation:** Mix 1 mL plasma or urine with 1 mL MeCN, centrifuge, inject a 40 μL aliquot of the supernatant.

## HPLC VARIABLES

**Column:** Nucleosil C18
**Mobile phase:** MeCN:pH 2.2 McIlvaine buffer 47:53
**Flow rate:** 1
**Injection volume:** 40
**Detector:** UV 432

## CHROMATOGRAM

**Limit of detection:** 100 ng/mL

## KEY WORDS

determined as rhein, the active metabolite; pharmacokinetics; plasma

## REFERENCE

Debord, P.; Louchahi, K.; Tod, M.; Cournot, A.; Perret, G.; Petitjean, O. Influence of renal function on the pharmacokinetics of diacerein after a single oral dose, *Eur.J.Drug Metab.Pharmacokinet.*, **1994**, *19*, 13–19.

# Dichloroacetic acid

$Cl_2CHCOOH$

**Molecular formula:** $C_2H_2Cl_2O_2$
**Molecular weight:** 128.94
**CAS Registry No:** 79-43-6
**Merck Index:** 13, 3075

## SAMPLE
**Matrix:** blood
**Sample preparation:** Inject 15–25 μL serum directly.

## HPLC VARIABLES
**Column:** Asahipak GS-320 gel permeation
**Mobile phase:** MeCN:10 mM pH 4.0 ammonium acetate buffer 10:90
**Flow rate:** 2
**Injection volume:** 15–25
**Detector:** UV 220

## CHROMATOGRAM
**Retention time:** 12
**Limit of detection:** 5 μg/mL

## KEY WORDS
pharmacokinetics; serum

## REFERENCE
Sakakihara, Y.; Nakamura, G.; Tokoeda, Y.; Abe, T.; Kamoshita, S. A rapid microassay for dichloroacetate in serum by gel-permeation chromatography, *Eur.J.Clin.Chem.Clin.Biochem.*, **1994**, *32*, 79–83.

# Dichlorophen

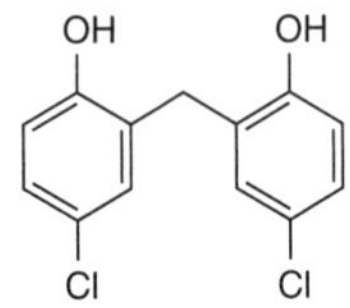


**Molecular formula:** $C_{13}H_{10}Cl_2O_2$
**Molecular weight:** 269.13
**CAS Registry No:** 97-23-4
**Merck Index:** 13, 3096

## SAMPLE

**Matrix:** solutions
**Sample preparation:** Inject a 10–40 μL aliquot of a 500 μg/mL solution in MeOH.

## HPLC VARIABLES

**Column:** Zorbax C8
**Column temperature:** 31
**Mobile phase:** Gradient. A:B from 0:100 to 100:0 over 30 min. A was MeCN:water:phosphoric acid 89.9:10:0.1. B was 0.1% phosphoric acid.
**Flow rate:** 2
**Injection volume:** 10–40
**Detector:** UV 230

## CHROMATOGRAM

**Retention time:** 21.9

## OTHER SUBSTANCES

**Simultaneous:** acetaminophen (8.2), acetophenetidine (14.6), aminophylline (8.8), amobarbital (15.5), antipyrine (12.9), aprobarbital (12.3), aspirin (13.6), barbital (9.4), benzoic acid (13.2), butabarbital (13.1), butethal (14.2), caffeine (10.2), chloramphenicol (14.4), chlorothiazide (9.1), chlorpropamide (17.7), colchicine (15.3), cortisone (15.0), coumarin (15.1), cyclothiazide (18.6), cyheptamide (17.8), danazol (19.3), danthron (22.3), dapsone (13.1), diethylstilbestrol (21.5), dronabinol (26.3), estradiol (19.4), estriol, 15.0), estrone (20.3), ethosuximide (9.8), eugenol (18.8), fenoprofen (19.2), fluorouracil, 3.0), fluoxymesterone (15.0), furosemide (17.0), gitoxigenin (16.0), glutethimide (16.3), guaiacol (11.4), hexobarbital (15.1), hippuric acid (9.6), hydrocortisone (14.4), hydroquinone (3.3), ibuprofen (22.4), indomethacin (22.00, isocarbostyril (11.8), mefenamic acid (23.7), methocarbamol (12.6), methyl salicylate (19.3), methyldopa (6.4), methylparaben (12.8), methyltestosterone (19.6), niacin (3.9), nitrofurantoin (11.2), normethsuximide (14.5), oxyphenbutazone (19.8), paraxanthine (8.7), pentobarbital (15.2), phenylbutazone (23.4), piperonyl butoxide (26.1), prednisolone (15.2), prednisone (15.5), primidone (11.3), probenecid (20.5), progesterone (21.8), propylparaben (18.3), pyrithydione (11.2), pyrocatechol (14.6), reserpine (22.5), resorcinol (4.3), saccharin (8.0), salicylamide (11.0), salicylic acid (15.0), secobarbital (15.7), stanozolol (19.2), sulfacetamide (8.4), sulfadimethoxine (7.7), sulfaethidole (13.4), sulfamerazine (10.1), sulfamethazine (7.0), sulfamethizole (11.5), sulfamethoxazole (13.6), sulfanilamide (4.0), sulfapyridine (9.6), sulfisoxazole (14.2), sulindac (19.1), testosterone acetate (20.5), testosterone 17 β-cypionate (25.2), testosterone enanthate (25.0), testosterone propionate, 24.1), theobromine (8.3), thiobarbituric acid (2.4), thiosalicylic acid (15.4), tolbutamide, 18.8), tolmetin (18.7), triamcinolone (13.8), triamcinolone acetonide (17.8), warfarin (20.0)

## REFERENCE

Hill, D.W.; Langner, K.J. HPLC photodiode array UV detection for toxicological drug analysis, *J.Liq. Chromatogr.*, **1987**, *10*, 377–409.

# Diclazuril

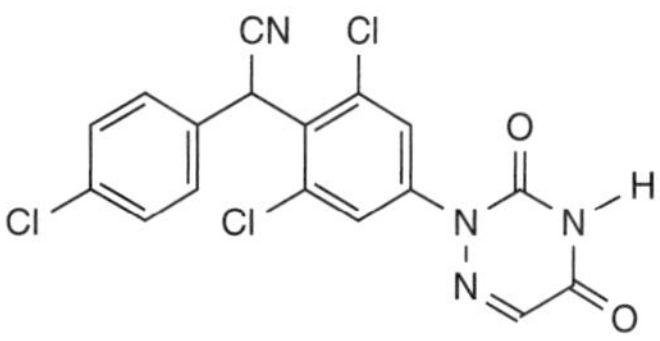


**Molecular formula:** $C_{17}H_9Cl_3N_4O_2$
**Molecular weight:** 407.64
**CAS Registry No:** 101831-37-2
**Merck Index:** 13, 3106

## SAMPLE

**Matrix:** blood
**Sample preparation:** Condition a Mega Bond Elut C18 SPE cartridge with 2 mL MeOH and 2 mL 100 mM pH 6.0 phosphate buffer. Mix 20 μL 100 μg/mL IS in DMF:water 50:50 with ? mL plasma, add 2 mL 100 mM pH 6.0 phosphate buffer, mix, adjust pH to 6.0, add to the SPE cartridge, allow to pass through taking at least 2 min, wash with 2 mL 100 mM pH 6.0 phosphate buffer, wash with 2 mL 1 M acetic acid, wash with 2 mL hexane (allow cartridge to dry for 5–10 min after each wash). Elute with 4 mL MeOH:conc. HCl 95:5. Evaporate the eluate to dryness under a stream of nitrogen at 40°, reconstitute the residue with 100 μL DMF, vortex, sonicate, add 100 μL water, vortex, sonicate, inject a 20 μL aliquot.

## HPLC VARIABLES

**Guard column:** Alltech C18
**Column:** 150 × 4.6 5 μm Ultrasphere ODS
**Mobile phase:** MeCN:MeOH:buffer 20:43.2:36.8 (The buffer was 0.5% ammonium acetate containing 10 mM tetrabutylammonium hydrogen sulfate.)
**Flow rate:** 1
**Injection volume:** 20
**Detector:** UV 280

## CHROMATOGRAM

**Retention time:** 13
**Internal standard:** R062646 (diclazuril with methyl $\alpha$ to cyano group, Janssen) (14.5)
**Limit of detection:** 10 ng/mL

## KEY WORDS

horse; pharmacokinetics; plasma; SPE; use silanized glassware

## REFERENCE

Dirikolu, L.; Lehner, F.; Nattrass, C.; Bentz, B.G.; Woods, W.E.; Carter, W.G.; Karpiesiuk, W.; Jacobs, J.; Boyles, J.; Harkins, J.D.; Granstrom, D.E.; Tobin, T. Diclazuril in the horse: its identification and detection and preliminary pharmacokinetics, *J.Vet.Pharmacol.Ther.*, **1999**, *22*, 374–379.

## SAMPLE

**Matrix:** feed
**Sample preparation:** Mix 50 g ground feed with 1 mL 50 μg/mL IS in DMF and 200 mL acidified MeOH, stir overnight. Remove 20 mL of the supernatant and dilute with 20 mL water, add to a Mega Bond C18 SPE cartridge, wash with 25 mL acidified MeOH:water 65:35, elute with 25 mL acidified MeOH:water 80:20. Evaporate the eluate to dryness at 60°, reconstitute the residue with 1 mL DMF and 1.5 mL water, filter (0.45 μm), inject a 20 μL aliquot. (Acidified MeOH was 5 mL conc. HCl in 1 L MeOH.)

## HPLC VARIABLES

**Column:** 100 × 4.6 3 μm Hypersil ODS or BDS
**Mobile phase:** Gradient. MeCN:MeOH:10 mM tetrabutylammonium hydrogen sulfate 20:20:60 for 10 min, to 20:35:45 over 30 min, flush column with MeCN for 10 min.

**Flow rate:** 2
**Injection volume:** 20
**Detector:** UV 280

## CHROMATOGRAM

**Retention time:** 24
**Internal standard:** R062646 (diclazuril with methyl $\alpha$ to cyano group, Janssen) (26)
**Limit of quantitation:** 1 μg/g

## OTHER SUBSTANCES

**Simultaneous:** degradant

## KEY WORDS

SPE

## REFERENCE

De Kock, J.; de Smet, M.; Sneyers, R. Determination of diclazuril in animal feed by liquid chromatography, *J.Chromatogr.*, **1992**, *606*, 141–146.

## ANNOTATED BIBLIOGRAPHY

Fontaine, A.; Haustraete, K. Liquid-chromatographic determination of diclazuril in premix and supplemented feed: Interlaboratory study, *J.AOAC Int.*, **1994**, *77*, 1359–1361. [SPE]

Mortier, L.; Daeseleire, E.; Delahaut, P. Simultaneous detection of five coccidiostats in eggs by liquid chromatography-tandem mass spectrometry, *Anal.Chim.Acta*, **2003**, *483*, 27–37. [diclazuril; dimetridazole; halofuginone; nicarbazin; robenidine]

# Dihydrotachysterol

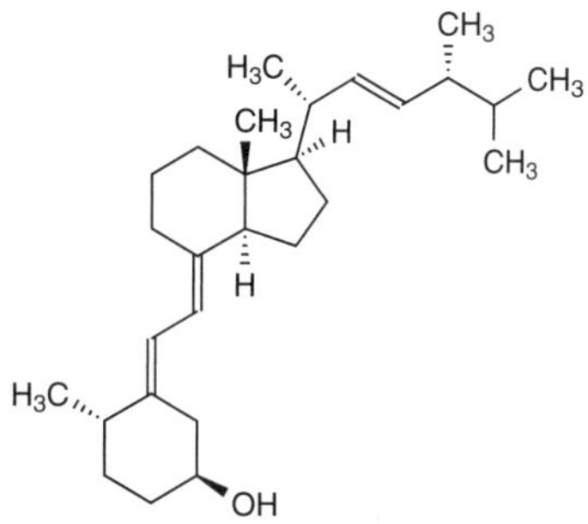


**Molecular formula:** $C_{28}H_{46}O$
**Molecular weight:** 398.66
**CAS Registry No:** 67-96-9
**Merck Index:** 13, 3202

## SAMPLE

**Matrix:** blood
**Sample preparation:** Mix 1 mL serum with 2 mL 50 ng/mL IS in EtOH, centrifuge at 1500 g for 15 min, add to an activated (by an unspecified method) Chromabond C18 ec SPE cartridge, wash with 3 mL EtOH:500 mM ammonium acetate 2:1, wash with 1 mL water, elute with 5 mL MeCN. Evaporate the eluate to dryness under a stream of nitrogen at 45°, reconstitute the residue with 90 μL MeCN, inject a 50 μL aliquot.

## HPLC VARIABLES

**Column:** 250 × 3 Nucleosil 100-5 C18 AB
**Mobile phase:** MeCN:water:acetic acid 95:5:5
**Flow rate:** 0.75
**Injection volume:** 50
**Detector:** UV 252

## CHROMATOGRAM

**Retention time:** 15.4
**Internal standard:** vitamin $D_2$ (17)
**Limit of quantitation:** 0.5 ng/mL

## KEY WORDS

pharmacokinetics; serum

## REFERENCE

Koytchev, R.; Alken, R.G.; Vagaday, M.; Kunter, U.; Kirkov, V. Differences in the bioavailability of dihydrotachysterol preparations, *Eur.J.Clin.Pharmacol.*, **1994**, *47*, 81–84.

## SAMPLE

**Matrix:** blood
**Sample preparation:** Condition a Sep-Pak C18 SPE cartridge with 20 mL MeOH and 10 mL water. Mix 5 mL plasma with 5 mL freshly charcoal-washed MeCN, let stand for 1 h with occasional vortexing, centrifuge at 7000 g for 15 min. Add the supernatant to 2.5 mL 200 mM pH 5.6 acetate buffer, add to the SPE cartridge, wash with 3 mL MeOH:water 60:40, elute with 6 mL MeOH. Evaporate the eluate to dryness under a stream of nitrogen or under reduced pressure, reconstitute the residue with hexane:isopropanol:MeOH 91:7:2, inject an aliquot. (SPE procedure from: Coldwell,R.D.; Trafford,D.J.H.; Makin,H.L.J.; Varley,M.J.; Kirk,D.N. Specific mass fragmentographic assay for 25,26-dihydroxyvitamin D in human plasma using a deuterated internal standard. *J.Chromatogr.* **1985**, *338*, 289–302.)

## HPLC VARIABLES

**Column:** 250 × 6.2 Zorbax-SIL

**Mobile phase:** Hexane:isopropanol:MeOH 91:7:2
**Flow rate:** 1.8–2
**Detector:** UV 254

---

## KEY WORDS

normal phase; plasma; rat; SPE

---

## REFERENCE

Qaw, F.; Calverley, M.J.; Schroeder, N.J.; Trafford, D.J.; Makin, H.L.; Jones, G. In vivo metabolism of the vitamin D analog, dihydrotachysterol. Evidence for formation of 1α,25- and 1β,25-dihydroxy-dihydrotachysterol metabolites and studies of their biological activity, *J.Biol.Chem.*, **1993**, *268*, 282–292.

---

## ANNOTATED BIBLIOGRAPHY

Taylor, A.; Bikle, D.D.; Norman, M.E. Serum dihydrotachysterol levels and biological action in normal man, *J.Clin.Endocrinol.Metab.*, **1988**, *67*, 198–202. [normal phase]

# Dimethyl sulfoxide

$(CH_3)_2SO$

**Molecular formula:** $C_2H_6OS$
**Molecular weight:** 78.13
**CAS Registry No:** 67-68-5
**Merck Index:** 13, 3285

## SAMPLE
**Matrix:** bulk
**Sample preparation:** Dissolve 40 mg bisnafide drug substance in 20 mL water with sonication and gentle heating, inject a 200 μL aliquot. (Dimethyl sulfoxide is determined as an impurity in bisnafide drug substance.)

## HPLC VARIABLES
**Column:** Zorbax Rx C8
**Column temperature:** 45 ± 2
**Mobile phase:** Gradient. A:B 100:0 for 10 min, to 0:100 (step gradient) maintain at 0:100 for 5 min, 100:0 for 15 min. A was 10 mM NaCl containing 0.10% phosphoric acid. B was MeCN:10 mM NaCl containing 0.10% phosphoric acid 90:10.
**Flow rate:** 0.8
**Injection volume:** 200
**Detector:** UV 215

## CHROMATOGRAM
**Retention time:** 7
**Limit of detection:** 51 ng/mL
**Limit of quantitation:** 219 ng/mL

## REFERENCE
Walker, J.T.; Paolini, D.L.; Segretario, J. Quantitation of residual dimethylsulfoxide in a drug substance (bisnafide) by reversed-phase high-performance liquid chromatography, *J.Chromatogr.Sci.*, **1996**, *34*, 513–516.

## SAMPLE
**Matrix:** tissue
**Sample preparation:** Extract 1 g tissue with 10 mL MeOH:water 10:90 for at least 6 h (no other details), dilute an aliquot 10 fold with MeOH:water 10:90, centrifuge for 5 min, inject an aliquot.

## HPLC VARIABLES
**Guard column:** 5 μm Spheri-18 C18 (Brownlee)
**Column:** 250 × 4.6 5 μm Spherisorb ODS2 C18
**Mobile phase:** MeOH:water 10:90
**Flow rate:** 1
**Injection volume:** 20
**Detector:** UV 214

## CHROMATOGRAM
**Retention time:** 3.5
**Limit of quantitation:** 500 ng/mL

## KEY WORDS

myocardium; pig

## REFERENCE

Carpenter, J.F.; Dawson, P.E. Quantitation of dimethyl sulfoxide in solutions and tissues by high-performance liquid chromatography, *Cryobiology*, **1991**, *28*, 210–215.

# Dinitolmide

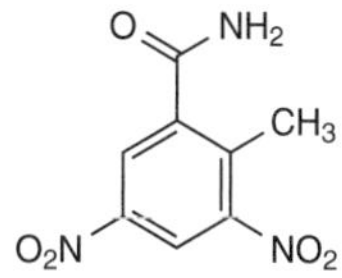

**Molecular formula:** $C_8H_7N_3O_5$
**Molecular weight:** 225.16
**CAS Registry No:** 148-01-6
**Merck Index:** 13, 3297

## SAMPLE

**Matrix:** feed
**Sample preparation:** Mix 5 g ground feed with 50 mL MeCN:water 85:15, heat at 50° with frequent shaking for 10 min, shake on a wrist-action shaker at room temperature for 10 min, filter, inject a 20 μL aliquot of the filtrate.

## HPLC VARIABLES

**Column:** 250 × 4.9 5 μm Partisil
**Mobile phase:** MeCN:dichloromethane 50:50
**Flow rate:** 2
**Injection volume:** 20
**Detector:** UV 270

## CHROMATOGRAM

**Retention time:** 3.5
**Limit of quantitation:** 20 μg/g

## OTHER SUBSTANCES

**Noninterfering:** amprolium, avoparcin, decoquinate, dimetridazole, ethopabate, furazolidone, halquinol, monensin, nifursol, nitrovin, robenidine, vitamin A, vitamin $D_3$, vitamin E, zinc bacitracin
**Interfering:** sulfaquinoxaline (can be resolved using MeCN:dichloromethane 30:70)

## KEY WORDS

normal phase

## REFERENCE

Burns, I.W.; Jones, A.D. Determination of 3,5-dinitro-o-toluamide in poultry feedstuffs and pre-mixes by high-performance liquid chromatography, *The Analyst*, **1980**, *105*, 509–512.

# Dipivefrin

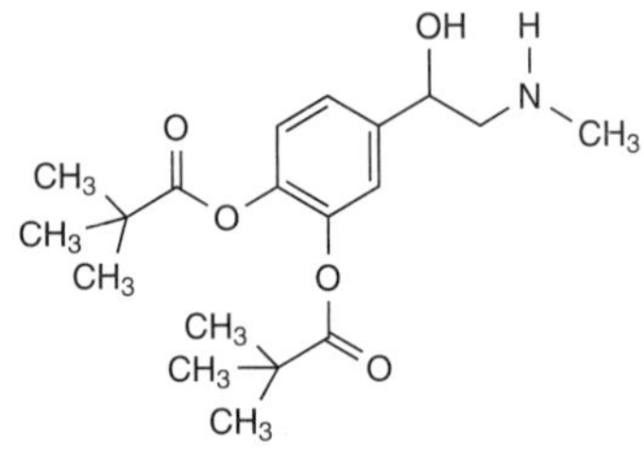


**Molecular formula:** $C_{19}H_{29}NO_5$
**Molecular weight:** 351.44
**CAS Registry No:** 52365-63-6
**Merck Index:** 13, 3372

## SAMPLE
**Matrix:** solutions

## HPLC VARIABLES
**Column:** reversed phase
**Mobile phase:** MeOH:20 mM pH 7.9 Tris buffer 80:20
**Flow rate:** 1.6
**Injection volume:** 30
**Detector:** UV ?

## CHROMATOGRAM
**Retention time:** 12–16

## REFERENCE
Tamaru, R.D.; Davis, W.L.; Anderson, J.A. Comparison of ocular disposition of free pivalic acid and pivalic acid esterified in dipivefrin, *Arch.Ophthalmol.*, **1983**, *101*, 1127–1129.

# Dithiazanine iodide

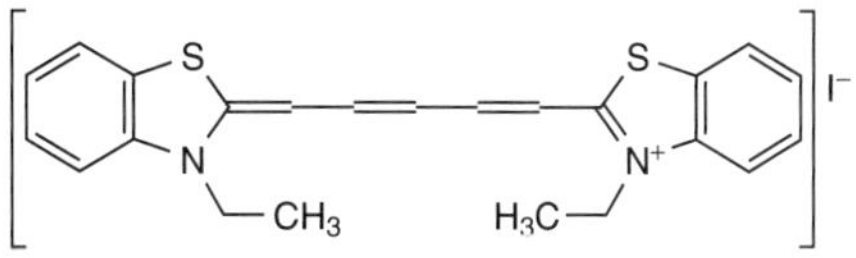


**Molecular formula:** $C_{23}H_{23}IN_2S_2$
**Molecular weight:** 518.49
**CAS Registry No:** 514-73-8
**Merck Index:** 13, 3405

## SAMPLE
**Matrix:** solutions

## HPLC VARIABLES
**Column:** 150 × 4.6 5 μm TSKgel ODS 80TM
**Mobile phase:** MeCN:25 mM pH 6.5 imidazole buffer 60:40 containing 10 mM sodium 1-propanesulfonate
**Flow rate:** 1
**Injection volume:**
**Detector:** Chemiluminescence, the column effluent mixed with MeCN containing 0.25 mM bis(4-nitro-2-(3,6,9-trioxadecyloxycarbonyl)phenyl) oxalate (TDPO) and 25 mM hydrogen peroxide pumped at 1.3 mL/min and the mixture flowed into the flow cell and the chemiluminescence was measured using a red-sensitive photomultiplier (Hamamatsu R2228)

## CHROMATOGRAM
**Retention time:** 14
**Limit of detection:** 0.19 fmole

## OTHER SUBSTANCES
**Simultaneous:** methylene blue (3.5)

## REFERENCE
Kimoto, K.; Gohda, R.; Murayama, K.; Santa, T.; Fukushima, T.; Homma, H.; Imai, K. Sensitive detection of near-infrared fluorescent dyes using high-performance liquid chromatography with peroxalate chemiluminescence detection system, *Biomed.Chromatogr.*, **1996**, *10*, 189–190.

# Docarpamine

**Molecular formula:** $C_{21}H_{30}N_2O_8S$
**Molecular weight:** 470.54
**CAS Registry No:** 74639-40-0
**Merck Index:** 13, 3430

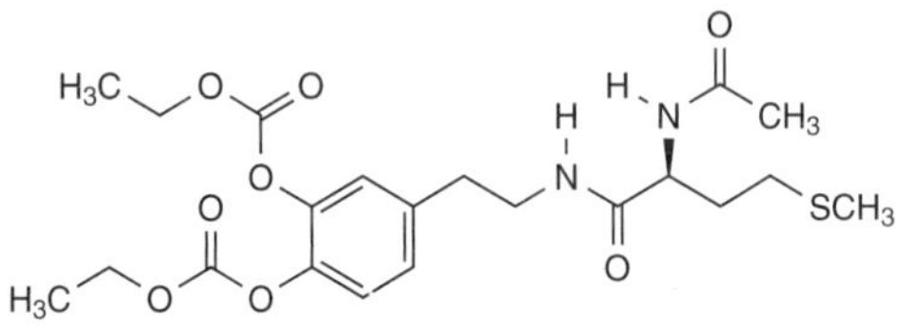

## SAMPLE
**Matrix:** solutions

## HPLC VARIABLES
**Column:** 250 × 4.6 10 μm Cosmosil 5C18-MS
**Column temperature:** 50
**Mobile phase:** Gradient. MeCN:10 mM pH 4.2 potassium phosphate buffer 0:100 for 2 min, to 75:25 over 7.5 min, maintain at 75:25 for 5.5 min. Isocratic. MeCN:buffer 40:60
**Flow rate:** 1.5
**Detector:** UV (wavelength not specified)

## CHROMATOGRAM
**Retention time:** 11.4 (gradient) or 6.5 (isocratic)

## OTHER SUBSTANCES
**Simultaneous:** acetazolamide (7.9), acyclovir (7.0), allopurinol (7.3), aniracetam (8.6), benazepril (11.5), betamethasone (10.8), camostat mesylate (8.8), carbamazepine (10.8), cilazapril (10.7), cimetidine (7.5), clofibrate (15.0), clonazepam (11.3), cyclophosphamide (9.9), delapril (11.9), dexamethasone (10.9), diazepam (12.5), digoxin (9.0), dilazep (10.6), diltiazem (10.2), eperisone (8.7), ethosuximide (9.0), fenbufen (11.8), fluconazole (9.2), flutamide (12.7), fominoben (12.6), hydrocortisone acetate (12.1), imidapril (10.1), indomethacin (12.6), irinotecan (9.2), maprotiline (10.5), methotrexate (8.0), nefiracetam (9.7), nifedipine (11.9), nitrazepam (11.2), pentobarbital (10.9), phenobarbital (10.0), phenytoin (10.8), pindolol (8.3), pranlukast (13.4), pranoprofen (10.4), prednisolone (10.3), primidone (8.9), quinapril (10.5), spironolactone (12.4), sulpiride (7.6), sulthiame (9.3), tolbutamide (11.6), tranilast (10.5), triamcinolone (11.2), warfarin (12.0), zonisamide (9.5) (gradient retention times; isocratic conditions may differ)

## REFERENCE
Sugiyama, T.; Matsuyama, R.; Usui, S.; Katagiri, Y.; Hirano, K. Selection of mobile phases in high-performance liquid chromatographic determination for medicines, *Biol.Pharm.Bull.*, **2000**, *23*, 274–278.

# Dofetilide

**Molecular formula:** $C_{19}H_{27}N_3O_5S_2$
**Molecular weight:** 441.57
**CAS Registry No:** 115256-11-6
**Merck Index:** 13, 3443

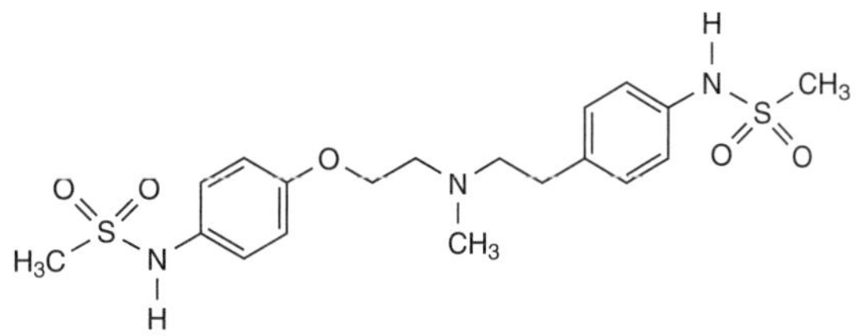


## SAMPLE

**Matrix:** blood
**Sample preparation:** Dilute plasma with an equal volume of MeOH:water 10:90 containing 1 M monochloroacetic acid, centrifuge at 4° at 3000 rpm for 1 h, inject a 200 μL aliquot onto column A and elute to waste with mobile phase A, backflush column A with mobile phase A, backflush the contents of column A onto column B with mobile phase C, elute column B with mobile phase C. Wash column A with mobile phase B and wash column B with mobile phase D

## HPLC VARIABLES

**Column:** A 50 × 1 50 μm HTLC Turbo C18 (Cohesive Technologies); B 33 × 2.1 HTLC HiRes C18 (Cohesive Technologies) (Fit a 5 μm filter before column A and a 2 μm filter before column B.)
**Mobile phase:** A 0.01% trifluoroacetic acid in water; B MeCN:0.1% aqueous ammonia 90:10; C MeCN:MeOH:20 mM ammonium acetate 45:45:10; D MeCN:THF 50:50
**Flow rate:** A 5; B 1
**Injection volume:** 200
**Detector:** MS, PE Sciex API 2000 triple quadrupole, positive ion mode, TurboIonspray, nebulizer gas nitrogen at 25 psi, applied voltage 5200 V, collision energy 50 V, turbo gas 100° and 40 psi, a flow splitter was used to deliver 50 μL/mL to the detector, m/z 442–198

## CHROMATOGRAM

**Retention time:** 1.1
**Limit of quantitation:** 5 ng/mL

## OTHER SUBSTANCES

**Extracted:** doxazosin (m/z 452–344) (1.1)

## KEY WORDS

column-switching; dog; plasma

## REFERENCE

Chassaing, C.; Luckwell, J.; Macrae, P.; Saunders, K.; Wright, P.; Venn, R. Direct analysis of crude plasma samples by turbulent flow chromatography/tandem mass spectrometry, *Chromatographia*, **2001**, *53*, 122–130.

## SAMPLE

**Matrix:** urine
**Sample preparation:** Mix 1 mL urine with 20 μL 5 μg/mL IS in MeOH, add 1 mL 100 mM pH 7.4 phosphate buffer, mix, add 5 mL MTBE, rotate at 30 rpm for 10 min, centrifuge at 1730 g for 5 min. Remove the organic layer and extract it with 1 mL 20 mM phosphoric acid, centrifuge at 1730 g for 5 min. Add the aqueous layer to 2 mL 100 mM pH 7.4 phosphate buffer, extract with 5 mL MTBE, centrifuge at 1730 g for 5 min. Evaporate the organic layer to dryness under a stream of nitrogen, reconstitute the residue with 100 μL mobile phase, inject an 80 μL aliquot.

## HPLC VARIABLES

**Column:** 250 × 4.6 5 μm Spherisorb silica
**Mobile phase:** MeCN:20 mM pH 7.0 ammonium phosphate buffer 35:65
**Flow rate:** 1
**Injection volume:** 80
**Detector:** UV 230

## CHROMATOGRAM

**Retention time:** 8.5
**Internal standard:** UK-69,308 (*N*,*N*-bis(4-methanesulfonamidophenethyl)methylamine) (10)
**Limit of quantitation:** 2.5 ng/mL

## OTHER SUBSTANCES

**Extracted:** metabolite

## REFERENCE

Walker, D.K.; Smith, D.A.; Stopher, D.A. Liquid-liquid extraction and high-performance liquid chromatography for the determination of a novel antidysrhythmic agent (UK-68,798) in human urine, *J.Chromatogr.*, **1991**, *568*, 475–480.

# Dolasetron

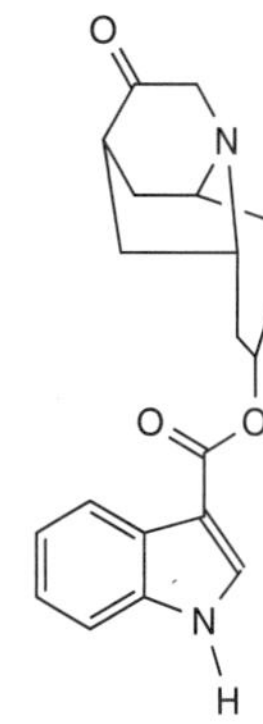

**Molecular formula:** $C_{19}H_{20}N_2O_3$
**Molecular weight:** 324.37
**CAS Registry No:** 115956-12-2
**Merck Index:** 13, 3445

## SAMPLE

**Matrix:** blood, urine
**Sample preparation:** Plasma. Mix 1 mL plasma with IS and 100 μL 5 M citric acid, add 1 mL 2 M sodium carbonate, extract with 4 mL ethyl acetate:*n*-hexane 75:25 for 30 min. Remove the organic layer and extract it with 1 mL 100 mM HCL for 15 min, centrifuge, discard the organic layer. Add 1 mL 2 M sodium carbonate to the aqueous layer, extract with 4 mL ethyl acetate:*n*-hexane 75:25, evaporate the organic layer to dryness under a stream of nitrogen, reconstitute with 150 μL mobile phase, inject a 100 μL aliquot. Urine. Dilute urine 1:100 with water. Mix 1 mL diluted urine with IS and 1 mL 2 M sodium carbonate, extract with 4 mL ethyl acetate:*n*-hexane 75:25 for 30 min. Remove the organic layer and extract it with 1 mL 100 mM HCL for 15 min, centrifuge, discard the organic layer. Add 1 mL 2 M sodium carbonate to the aqueous layer, extract with 4 mL ethyl acetate:*n*-hexane 75:25, evaporate the organic layer to dryness under a stream of nitrogen, reconstitute with 150 μL mobile phase, inject a 100 μL aliquot.

## HPLC VARIABLES

**Guard column:** 4 × 6 10 μm μBondapak C18
**Column:** 150 × 4.6 5 μm Ultrasphere IP C18
**Column temperature:** 30
**Mobile phase:** MeCN:*n*-butanol:buffer 5:6:89 (The buffer was 50 mM sodium dihydrogen phosphate adjusted to pH 2.5 with orthophosphoric acid.)
**Flow rate:** 0.7
**Injection volume:** 100
**Detector:** F ex 285 em 345

## CHROMATOGRAM

**Retention time:** 7.6
**Internal standard:** trans-octahydro-3-hydroxy-2,6-methano-2H-quinolizin-8-yl-1H-5-methyl-indole-3-carboxylate (MDL 101,8588; Marion Merrell Dow Research Institute, Strasbourg) (17.0)
**Limit of quantitation:** 10 nM (plasma), 50 nM (urine)

## OTHER SUBSTANCES

**Extracted:** metabolite

## KEY WORDS

pharmacokinetics; plasma

## REFERENCE

Huebert, N.D.; Schwartz, J.J.; Zeidler, L.; Schwach, V.; Haegele, K.D. Simultaneous measurement of dolasetron and its major metabolite, MDL 74,156, in human plasma and urine, *J.Chromatogr.B*, **1996**, *685*, 291–297.

## SAMPLE

**Matrix:** formulations

**Sample preparation:** Prepare a liquid suspension by crushing twelve 50 mg tablets, slowly add 60 mL of Ora-Plus:Ora-Sweet SF 50:50, dilute to 10 μg/mL with MeCN:water 24:76, shake for 15 s, centrifuge at 1000 rpm for 2 min, inject a 5 μL aliquot.

## HPLC VARIABLES

**Column:** 150 × 4.6 3 μm Spherisorb CN

**Column temperature:** 30

**Mobile phase:** MeCN:buffer 24:76 (The buffer was 50 mM ammonium acetate adjusted to pH 7.5 with dilute ammonium hydroxide.)

**Flow rate:** 0.8

**Injection volume:** 5

**Detector:** UV 280

## CHROMATOGRAM

**Retention time:** 6.9

## KEY WORDS

stability-indicating; suspension

## REFERENCE

Johnson, C.E.; Wagner, D.S.; Bussard, W.E. Stability of dolasetron in two oral liquid vehicles, *Am.J. Health-Syst.Pharm.*, **2003**, *60*, 2242–2244.

## ANNOTATED BIBLIOGRAPHY

McElvain, J.S.; Vandiver, V.J.; Eichemeier, L.S. Validation of a reversed-phase HPLC method for directly quantifying the enantiomers of MDL 74,156, the primary metabolite of dolasetron mesylate, in human plasma, *J.Pharm.Biomed.Anal.*, **1997**, *15*, 513–521. [HPLC of metabolites only]

Reith, M.K.; Sproles, G.D.; Cheng, L.K. Human metabolism of dolasetron mesylate, a 5-HT3 receptor antagonist, *Drug Metab.Dispos.*, **1995**, *23*, 806–812. [column temp 30; metabolites; urine]

Sanwald, P.; Huebert, N.D.; Haegele, K.D. Simultaneous measurement of the major metabolites of dolasetron mesilate in human urine using solid-phase extraction and high-performance liquid chromatography, *J.Chromatogr.B*, **1994**, *661*, 101–107.

# Donepezil

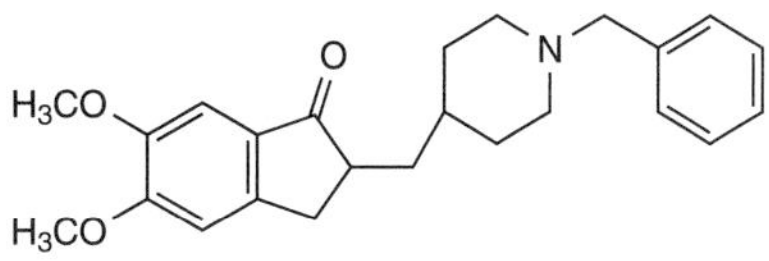

**Molecular formula:** $C_{24}H_{29}NO_3$
**Molecular weight:** 379.49
**CAS Registry No:** 120014-06-4
**Merck Index:** 13, 3453

## SAMPLE

**Matrix:** blood
**Sample preparation:** Add 100 μL 200 ng/mL IS in 1 mM HCl and 5 mL *n*-hexane:isopropanol 97:3 to 1 mL plasma, shake for 5 min, centrifuge at 1800 g for 1 min. Remove the organic layer and add it to 200 μL 1 mM HCl, shake for 1 min, centrifuge at 1800 g for 1 min, inject a 75 μL aliquot of the aqueous layer.

## HPLC VARIABLES

**Column:** 150 × 2.1 5 μm Bioptick AV-1 (GL Sciences)
**Mobile phase:** MeOH:10 mM formic acid 25:75
**Flow rate:** 0.2
**Injection volume:** 75
**Detector:** MS, Finnigan MAT TSQ7000, source 4.5 kV, capillary 200°, sheath gas nitrogen 70 psi, auxiliary gas nitrogen, collision gas argon 1.5 mtorr 40 eV, m/z 380–91

## CHROMATOGRAM

**Retention time:** 8 (R-enantiomer), 13 (S-enantiomer)
**Internal standard:** $d_7$-donepezil (m/z 387–98)
**Limit of detection:** 7.7 pg
**Limit of quantitation:** 20 pg/mL

## KEY WORDS

chiral; pharmacokinetics; plasma

## REFERENCE

Matsui, K.; Oda, Y.; Nakata, H.; Yoshimura, T. Simultaneous determination of donepezil (aricept) enantiomers in human plasma by liquid chromatography-electrospray tandem mass spectrometry, *J.Chromatogr.B*, **1999**, *729*, 147–155.

## SAMPLE

**Matrix:** blood
**Sample preparation:** Add 100 μL 3 μg/mL IS in 1 mM HCl and 500 μL 100 mM NaOH to 1 mL plasma, vortex for 10 s, add 4 mL *n*-hexane:isopropanol 97:3, shake for 10 min, centrifuge at 620 g at 4° for 10 min. Remove the organic layer and add it to 75 μL 100 mM HCl, mix at 100 cycles/min for 10 min, centrifuge at 1710 g at 4° for 5 min. Carefully remove the upper organic layer, then remove traces of organic solvent with a stream of air, inject a 50 μL aliquot of the aqueous layer.

## HPLC VARIABLES

**Column:** 150 × 4.6 5 μm STR ODS-II (Shinwa)
**Column temperature:** 30 (in body of paper) or 40 (in abstract)
**Mobile phase:** MeCN:20 mM pH 4.6 phosphate buffer:6 M perchloric acid 40:59.5:0.5
**Flow rate:** 1
**Injection volume:** 50
**Detector:** UV 315

## CHROMATOGRAM

**Retention time:** 5.1
**Internal standard:** cisapride (7.1)
**Limit of quantitation:** 3 ng/mL

## OTHER SUBSTANCES

**Simultaneous:** alprazolam, chlorpromazine, diazepam, levomepromazine, nitrazepam
**Noninterfering:** haloperidol, risperidone

## KEY WORDS

pharmacokinetics; plasma

## REFERENCE

Yasui-Furukori, N.; Furuya, R.; Takahata, T.; Tateishi, T. Determination of donepezil, an acetylcholinesterase inhibitor, in human plasma by high-performance liquid chromatography with ultraviolet absorbance detection, *J.Chromatogr.B*, **2002**, *768*, 261–265.

## ANNOTATED BIBLIOGRAPHY

Kaddoumi, A.; Mori, M.; Nanashima, K.; Kono, M.; Nakashima, K. High performance liquid chromatographic determination of mazindol in human plasma, *The Analyst*, **2001**, *126*, 1963–1968. [donepezil is internal standard]

Matsui, K.; Mishima, M.; Nagai, Y.; Yuzuriha, T.; Yoshimura, T. Absorption, distribution, metabolism, and excretion of donepezil (Aricept) after a single oral administration to rat, *Drug Metab.Dispos.*, **1999**, *27*, 1406–1414.

Pappa, H.; Farrú, R.; Vilanova, P.O.; Palacios, M.; Pizzorno, M.T. A new HPLC method to determine Donepezil hydrochloride in tablets, *J.Pharm.Biomed.Anal.*, **2002**, *27*, 177–182.

# Doxefazepam

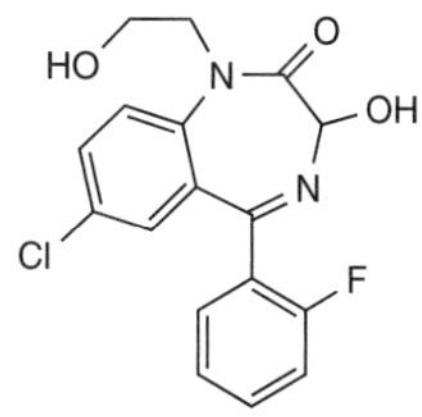


**Molecular formula:** $C_{17}H_{14}ClFN_2O_3$
**Molecular weight:** 348.76
**CAS Registry No:** 40762-15-0
**Merck Index:** 13, 3467

## SAMPLE

**Matrix:** blood

**Sample preparation:** Condition a Supelclean LC-18 SPE cartridge with 1 mL MeOH and 2 mL water. Mix 1 mL plasma with 100 μL 2 μg/mL IS in MeOH, add to the SPE cartridge, wash with 2 column volumes of water, wash with 100 μL MeOH, elute with 400 μL MeOH. Evaporate the eluate to dryness under a stream of nitrogen, reconstitute the residue with 200 μL mobile phase, inject a 25 μL aliquot.

## HPLC VARIABLES

**Guard column:** 20 × 4 Supelguard C18
**Column:** 300 × 4 μBondapak C18
**Mobile phase:** MeOH:water 20:80
**Flow rate:** 2
**Injection volume:** 25
**Detector:** UV 280

## CHROMATOGRAM

**Retention time:** 5.2
**Internal standard:** diazepam (3.1)
**Limit of quantitation:** 100 ng/mL

## KEY WORDS

plasma; SPE

## REFERENCE

Carlucci, G. A high-performance liquid chromatographic method for the determination of doxefazepam in human plasma using a solid-phase extraction column, *J.Liq.Chromatogr.*, **1988**, *11*, 1559–1568.

# Doxercalciferol

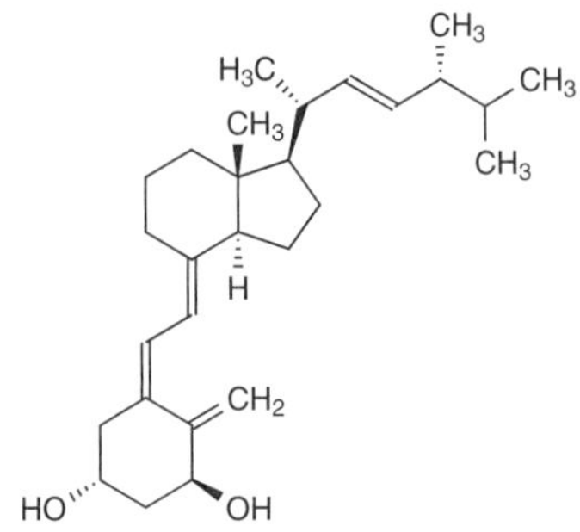

**Molecular formula:** $C_{28}H_{44}O_2$
**Molecular weight:** 412.65
**CAS Registry No:** 54573-75-0
**Merck Index:** 13, 3470
**[1α-Hydroxyergocalciferol]**

## SAMPLE

**Matrix:** blood

**Sample preparation:** Condition a 60 mg Oasis HLB SPE cartridge with 2 mL ethyl acetate, 2 mL MeOH, and 2 mL water. Condition a 500 mg Bond Elut silica SPE cartridge with 4 mL chloroform:MeOH 30:1 (Caution! Chloroform is a carcinogen!) and 4 mL chloroform. Mix 10 μL 40 ng/mL IS in EtOH with 1 mL plasma, let stand for 15 min, add this mixture to 1 mL MeCN in another tube, rinse the first tube with 250 μL MeCN and add it to the plasma/MeCN mixture. Vortex for 30 s, centrifuge at 1500 g for 10 min. Add 2 mL water to the supernatant and pass the mixture through the Oasis SPE cartridge. Wash with 2 mL water, wash with 2 mL MeOH:water 70:30, wash with 1 mL hexane, elute with 1 mL ethyl acetate. Evaporate the eluate to dryness under a stream of nitrogen, reconstitute the residue with two 200 μL portions of chloroform. Add the chloroform layers to the silica SPE cartridge, wash with 3 mL chloroform, wash with 2.2 mL chloroform:MeOH 40:1, elute with 2 mL chloroform:MeOH 30:1. Evaporate the eluate to dryness, place the residue under reduced pressure for 10 min, reconstitute the residue with 25 μL 100 μg/mL DMEQTAD in ethyl acetate, let stand at room temperature for 30 min, add another 25 μL 100 μg/mL DMEQTAD in ethyl acetate, let stand at room temperature for 1 h, add 40 μL EtOH, evaporate to dryness, reconstitute with 10 μL acetic anhydride and 20 μL pyridine, heat at 50° for 1 h, add 40 μL EtOH, evaporate the solvent, dissolve the residue in 40 μL mobile phase, inject a 15 μL aliquot. (DMEQTAD is 4-[2-(6,7-dimethoxy-4-methyl-3-oxo-3,4-dihydroquinoxalyl)ethyl]-1,2,4-triazoline-3,5-dione and can be purchased from Wako.)

## HPLC VARIABLES

**Column:** 150 × 4.6 4 μm YMC J'sphere ODS H-80
**Column temperature:** 40
**Mobile phase:** MeCN:water 92:8
**Flow rate:** 1
**Injection volume:** 15
**Detector:** MS, ThermoQuest LCQ, APCI, positive ion mode, collision gas helium, source current 5 μA, heated capillary 225°, vaporizer 475°, capillary 3 V, tube lens offset 15 V

## CHROMATOGRAM

**Retention time:** 6.8
**Internal standard:** $d_4$-doxercalciferol (6.7)
**Limit of detection:** 6.3 fmol
**Limit of quantitation:** 25 pg/mL

## KEY WORDS

derivatization; plasma; SPE; Doxercalciferol can be determined without derivatization using MeOH:water 95:5, retention time 5.4 min, LOD 200 pg/injection.

## REFERENCE

Higashi, T.; Awada, D.; Shimada, K. Liquid chromatography-mass spectrometric method combined with derivatization for determination of 1α-hydroxyvitamin $D_3$ in human plasma, *J.Chromatogr.B*, **2002**, *772*, 229–238.

## SAMPLE

**Matrix:** cell cultures

**Sample preparation:** Precipitate cells with MeOH. Homogenize 1 vol of cells with 2 vol of MeOH, add 1 vol of chloroform (Caution! Chloroform is a carcinogen!), homogenize (blender or Polytron) for 2 min, add 1 vol of chloroform, homogenize for 30 s, add 1 vol of water, homogenize for 30 s, centrifuge, extract solids with 0.4 vol chloroform:MeOH 50:50. Combine the liquids and separate the chloroform layer. Pass the chloroform layer through anhydrous sodium sulfate and evaporate under reduced pressure at 40°, reconstitute, inject an aliquot. (*Current Protocols in Food Analytical Chemistry*, Wrolstad,R.E. (ed), Wiley, New York, 2003, page D.1.1.5; after Bligh,E.G.; Dyer,W.J. A rapid method of total lipid extraction and purification. *Can.J.Biochem.Physiol.* **1959**, *37*, 911–917.))

## HPLC VARIABLES

**Column:** 80 × 6.2 3 μm Zorbax SIL
**Mobile phase:** Hexane:isopropanol:MeOH 91:7:2
**Flow rate:** 1.5
**Detector:** UV 265

## CHROMATOGRAM

**Retention time:** 5.6

## OTHER SUBSTANCES

**Simultaneous:** metabolites

## KEY WORDS

normal phase

## REFERENCE

Strugnell, S.; Byford, V.; Makin, H.L.; Moriarty, R.M.; Gilardi, R.; LeVan, L.W.; Knutson, J.C.; Bishop, C.W.; Jones, G. 1α,24(S)-dihydroxyvitamin $D_2$: a biologically active product of 1α-hydroxyvitamin $D_2$ made in the human hepatoma, Hep3B, *Biochem.J.*, **1995**, *310*, 233–241.

# Dropropizine

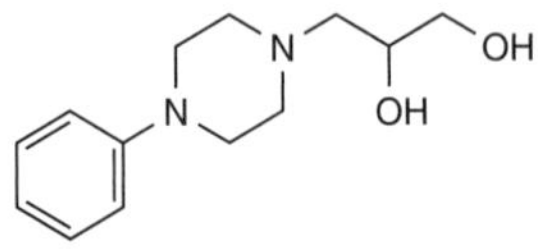


**Molecular formula:** $C_{13}H_{20}N_2O_2$
**Molecular weight:** 236.31
**CAS Registry No:** 17692-31-8, 99291-24-4 (levodropropizine)
**Merck Index:** 13, 3486

## SAMPLE

**Matrix:** blood
**Sample preparation:** Mix 200 μL plasma or serum with 10 ng IS and 200 μL 100 mM pH 8.9 disodium hydrogen phosphate, add 5 mL chloroform:isopropanol 90:10 (Caution! Chloroform is a carcinogen!), vortex for 2 min, centrifuge at 700 g for 10 min. Evaporate 4 mL of the organic layer to dryness under a stream of nitrogen, reconstitute the residue with 500 μL water, inject a 100 μL aliquot.

## HPLC VARIABLES

**Column:** 150 × 4.6 5 μm Bio-Gel PRP 70-5 poly(styrene-divinylbenzene) (Bio-Rad)
**Mobile phase:** MeOH:THF:100 mM pH 3 potassium phosphate buffer 30:0.5:70
**Flow rate:** 0.5
**Injection volume:** 100
**Detector:** F ex 240 em 350 (slits 18 and 30 nm, respectively)

## CHROMATOGRAM

**Retention time:** 10
**Internal standard:** *p*-methoxylevodropropizine (15)
**Limit of detection:** 1–2 ng/mL
**Limit of quantitation:** 3.1 ng/mL

## OTHER SUBSTANCES

**Noninterfering:** acetaminophen, amitriptyline, amobarbital, amphetamine, aprobarbital, atropine, barbital, benzoylecgonine, benztropine, butabarbital, caffeine, carbamazepine, carisoprodol, chlorpheniramine, chlorpromazine, chlorprothixene, cimetidine, cocaine, codeine, dextromethorphan, diazepam, dihydrocodeine, diphenhydramine, diphenoxylate, diphenylhydantoin, disopyramide, doxepin, doxylamine, emetine, erythromycin, ethinamate, ethylmorphine, flurazepam, glutethimide, hydrocodone, hydrocortisone, hydromorphone, hydroxyzine, imipramine, lidocaine, loxapine, meperidine, meprobamate, methadone, methamphetamine, methapyrilene, methaqualone, methocarbamol, methylphenidate, morphine, naloxone, nicotine, nordiazepam, nortriptyline, orphenadrine, oxycodone, papaverine, pentazocine, pentobarbital, phenacetin, phencyclidine, phenmetrazine, phenobarbital, phenolphthalein, phentermine, phenylpropanolamine, phenytoin, phetidine, prazepam, procainamide, procaine, propoxyphene, propranolol, protriptyline, pseudoephedrine, pyrilamine, quinine, salicylamide, secobarbital, spironolactone, strychnine, terpin hydrate, thioridazine, thiothixene, triamterene, trifluperazine, triflupromazine, trihexyphenidyl, trimeprazine, trimethoprim, trimetobenzamide

## KEY WORDS

plasma; serum

## REFERENCE

Tagliaro, F.; Moffa, M.; De Battisti, Z.; Smith, F.P.; Gentile, M. High-performance liquid chromatographic determination of levodropropizine in human plasma with fluorometric detection, *J.Chromatogr.B*, **1996**, *685*, 165–170.

# Drospirenone

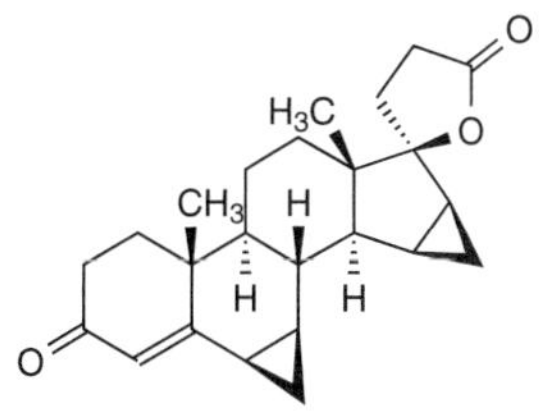


**Molecular formula:** $C_{24}H_{30}O_3$
**Molecular weight:** 366.49
**CAS Registry No:** 67392-87-4
**Merck Index:** 13, 3488

## SAMPLE

**Matrix:** blood
**Sample preparation:** Vortex 3 mL plasma with 3 mL *n*-hexane:toluene 50:50 for 1 min, centrifuge at 1200 g for 5 min, repeat the extraction. Combine the organic layers, evaporate them to dryness under a stream of nitrogen, reconstitute the residue with 200 μL mobile phase, inject a 150 μL aliquot. To increase sensitivity, extract 5 mL plasma twice with 2 mL aliquots of *n*-hexane:toluene 50:50.

## HPLC VARIABLES

**Column:** 250 × 4.6 10 μm LiChrosorb RP-18
**Mobile phase:** MeOH:water 60:40
**Flow rate:** 2
**Injection volume:** 150
**Detector:** UV 254

## CHROMATOGRAM

**Retention time:** 13.4
**Limit of detection:** <5 ng/mL

## OTHER SUBSTANCES

**Extracted:** spirenone (9.2)

## KEY WORDS

pharmacokinetics; plasma

## REFERENCE

Krause, W.; Jakobs, U. Determination of plasma levels of spirorenone, a new aldosterone antagonist, and one of its metabolites by high-performance liquid chromatography, *J.Chromatogr.*, **1982**, *230*, 37–45.

# Droxicam

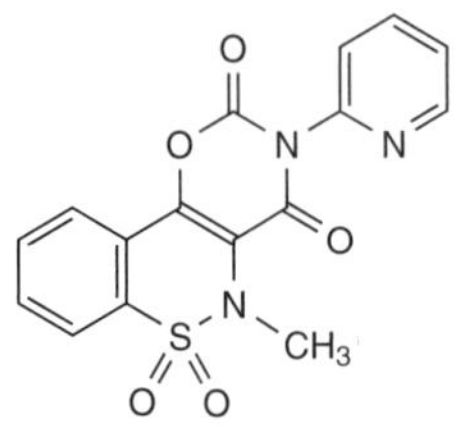

**Molecular formula:** $C_{16}H_{11}N_3O_5S$
**Molecular weight:** 357.35
**CAS Registry No:** 90101-16-9
**Merck Index:** 13, 3491

## SAMPLE
**Matrix:** solutions

## HPLC VARIABLES
**Column:** 250 × 4 5 μm LiChrospher 60 RP-Select B
**Mobile phase:** MeOH:water:acetic acid 48:45:7
**Flow rate:** 1.1
**Injection volume:** 20
**Detector:** UV 340

## OTHER SUBSTANCES
**Simultaneous:** isoxicam, piroxicam

## KEY WORDS
Droxicam cannot be detected in plasma, but it can be chromatographed under the above conditions.

## REFERENCE
Maya, M.T.; Pais, J.P.; Morais, J.A. A rapid method for the determination of piroxicam in plasma by high-performance liquid chromatography, *J.Pharm.Biomed.Anal.*, **1995**, *13*, 319–322.

# Droxidopa

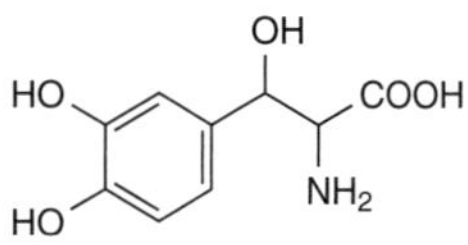

**Molecular formula:** $C_9H_{11}NO_5$
**Molecular weight:** 213.19
**CAS Registry No:** 23561-95-8
**Merck Index:** 13, 3492

## SAMPLE
**Matrix:** solutions

## HPLC VARIABLES
**Guard column:** 50 × 4 Develosil ODS-10
**Column:** 250 × 4 Develosil ODS-5
**Mobile phase:** MeCN:buffer 10:90 (The buffer was 50 mM pH 3.4 sodium dihydrogen phosphate containing 50 mM trichloroacetic acid, 200 μg/mL sodium dodecylsulfate, and 10 μg/mL disodium EDTA.)
**Flow rate:** 0.6
**Detector:** E, Coulochem 5100A, conditioning cell Model 5021 + 400 mV, analytical cell Model 5011, first electrode −100 mV, second electrode (recording electrode) −400 mV

## CHROMATOGRAM
**Retention time:** 4

## REFERENCE
Naoi, M.; Nagatsu, T. Uptake of L-threo-dihydroxyphenylserine into human brain synaptosomes, *J.Neural Transm.*, **1987**, *70*, 51–61.

# Ebrotidine

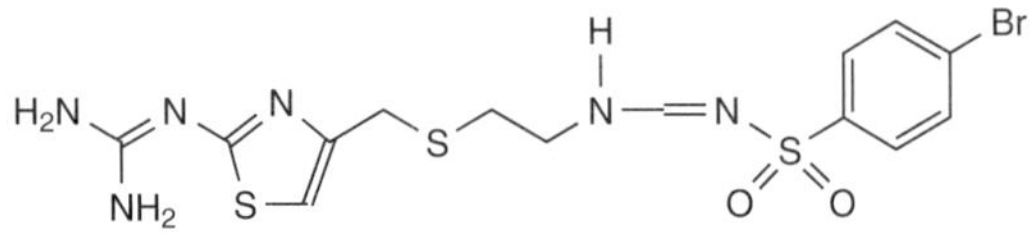

**Molecular formula:** $C_{14}H_{17}BrN_6O_2S_3$
**Molecular weight:** 477.43
**CAS Registry No:** 100981-43-9
**Merck Index:** 13, 3518

## SAMPLE
**Matrix:** bulk
**Sample preparation:** Inject a 20 μL aliquot of an 0.2% solution in MeOH.

## HPLC VARIABLES
**Column:** 250 × 4 10 μm Spherisorb CN
**Column temperature:** 30
**Mobile phase:** MeCN:10 mM sodium dihydrogen phosphate 35:65
**Flow rate:** 1.5
**Injection volume:** 20
**Detector:** UV 245

## CHROMATOGRAM
**Retention time:** 11

## OTHER SUBSTANCES
**Simultaneous:** impurities

## REFERENCE
Albet, C.; Fernandez, J.M.; Castello, J.M.; Sacristan, A.; Ortiz, J.A. Physicochemical properties, analytical determinations and stability of ebrotidine, *Arzneimittelforschung*, **1997**, *47*, 435–438.

## SAMPLE
**Matrix:** urine
**Sample preparation:** Vortex 1 mL urine with 100 μL 1 M NaOH, add 5 mL dichloromethane:isopropanol 90:10, shake mechanically for 15 min, centrifuge at 2700 g for 5 min. Evaporate 4 mL of the organic layer to dryness under a stream of nitrogen at 40°, reconstitute the residue with 200 μL MeOH, vortex, filter (0.45 μm), inject a 25 μL aliquot.

## HPLC VARIABLES
**Column:** 300 × 3.9 10 μm μBondapak C18
**Mobile phase:** Gradient. MeCN:buffer from 20:80 to 35:65 over 30 min (The buffer was 5 mM 1-hexanesulfonic acid adjusted to pH 3 with glacial acetic acid.)
**Flow rate:** 1.5
**Injection volume:** 25
**Detector:** UV 235

## CHROMATOGRAM
**Retention time:** 24
**Limit of detection:** 97 ng/mL

## OTHER SUBSTANCES
**Extracted:** metabolites
**Simultaneous:** acetaminophen, amoxicillin, aspirin, caffeine, cimetidine, codeine, diazepam, omeprazole, ranitidine, theophylline
**Interfering:** imipramine

## KEY WORDS

SPE

## REFERENCE

Rozman, E.; Galcerán, M.T.; Albet, C. Determination of ebrotidine and its metabolites in human urine by reversed-phase ion-pair high-performance liquid chromatography, *J.Chromatogr.B*, **1997**, *688*, 107–115.

## ANNOTATED BIBLIOGRAPHY

Rozman, E.; Galceran, M.T.; Anglada, L.; Albet, C. Investigation of the metabolism of ebrotidine in human urine by liquid chromatography-atmospheric pressure chemical ionization mass spectrometry, *Drug Metab.Dispos.*, **1995**, *23*, 976–981. [APCI].

Rozman, E.; Galcerán, M.T.; Albet, C. Ebrotidine and its metabolites studied by mass spectrometry with electrospray ionization. Comparison of tandem and in-source fragmentation, *Rapid Commun.Mass Spectrom.*, **1995**, *9*, 1492–1498.

# Edaravone

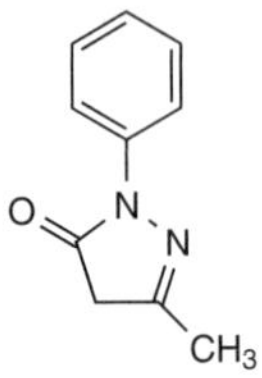


**Molecular formula:** $C_{10}H_{10}N_2O$
**Molecular weight:** 174.20
**CAS Registry No:** 89-25-8
**Merck Index:** 13, 6746

## SAMPLE

**Matrix:** blood, urine

**Sample preparation:** Plasma. Add 25 μL 15 μg/mL IS in MeOH, 500 μL 750 mM pH 5.0 sodium acetate buffer containing 40 mg/mL sodium metabisulfite, and 10 mg $\beta$-glucuronidase/arylsulfatase (Limpet acetone powder type 1:*Platela vulgata* (Sigma)) to 500 μL plasma, heat at 37° for 2 h, add 100 mg NaCl, add 1.2 mL chloroform:EtOH 90:10 (Caution! Chloroform is a carcinogen!), vortex for 2 min, centrifuge at 1800 g for 20 min. Remove the organic layer and add it to 50 μL 100 mM HCl. Evaporate to dryness under a stream of nitrogen at room temperature, reconstitute the residue with 75 μL mobile phase, wash with 25 μL *n*-hexane, inject a 20 μL aliquot of the lower layer. Urine. Add 25 μL 500 μg/mL IS in MeOH, 500 μL 750 mM pH 5.0 sodium acetate buffer containing 40 mg/mL sodium metabisulfite, and 10 mg $\beta$-glucuronidase/arylsulfatase (Limpet acetone powder type 1: *Platela vulgata* (Sigma)) to 500 μL urine, heat at 37° for 2 h, add 200 mg NaCl, add 2 mL dichloromethane:isopropanol 90:10, vortex for 1.5 min, centrifuge at 1800 g for 30 min. Remove the organic layer, evaporate to dryness under a stream of nitrogen at room temperature, reconstitute the residue with 200 μL mobile phase (keep in a dry ice bath), vortex for 15 s, inject a 20 μL aliquot. (Wash NaCl with extraction mixture before use.)

## HPLC VARIABLES

**Column:** 150 × 3.9 Nova-Pak C18
**Mobile phase:** MeOH:250 mM pH 5.0 sodium acetate buffer 30:70
**Flow rate:** 1
**Injection volume:** 20
**Detector:** UV 254

## CHROMATOGRAM

**Retention time:** 7
**Internal standard:** phenacetin (10)
**Limit of quantitation:** 50 ng/mL

## OTHER SUBSTANCES

**Extracted:** antipyrine (5)
**Noninterfering:** albuterol, aspirin, chlorpropamide, chlorthalidone, codeine, diazepam, diclofenac, dipyrone, enalapril, furosemide, heparin, nifedipine, phenytoin, ranitidine, warfarin
**Interfering:** aminophylline, dapsone

## KEY WORDS

plasma

## REFERENCE

Lanchote, V.L.; Ping, W.C.; Santos, S.R.C.J. Determination of antipyrine and metabolites in plasma of a patient with mild renal failure, *Ther.Drug Monit.*, **1997**, *19*, 705–710.

## SAMPLE

**Matrix:** urine

**Sample preparation:** Centrifuge urine, mix a 100 μL aliquot with 1 mL acetate buffer and 50 μL water containing 5000–6000 IU β-glucuronidase and 150–200 IU sulfatase from *Helix pomatia*, Sigma) (hydrolysis conditions not specified but 3 h at 37° is recommended by St.Peter,J.V.; Awni,W.M. *J.Chromatogr.* **1989**, *494*, 424–427), add 8 mL dichloromethane, add 1 g NaCl, extract on a linear agitator for 10 min, centrifuge at 1500 g for 5 min, remove the organic layer and add it to 1 mL washing buffer, extract on a linear agitator for 5 min, centrifuge at 1500 g for 5 min, remove the organic layer and add it to 500 μL water. Evaporate the dichloromethane to dryness under a stream of nitrogen at 37°, add 500 μL MeOH, homogenize, inject a 20 μL aliquot. (Stock acetate buffer was 27 g/L sodium acetate trihydrate containing 3 mL/L glacial acetic acid. Store at 4°. Prepare the working acetate buffer (pH 4.9) fresh each day by adding 2% sodium metabisulfite to the stock solution. Washing buffer was 500 mM disodium hydrogen phosphate containing 35 g/L NaCl adjusted to pH 7.8 with 38% KOH.)

## HPLC VARIABLES

**Guard column:** 20 × 4.6 10 μm C8
**Column:** 150 × 4.6 5 μm Ultrasphere C8
**Mobile phase:** MeCN:buffer 15:85 (The buffer was 5 mL/L glacial acetic acid containing 3 g/L sodium acetate trihydrate and 0.1 mM (20 μL/L) *n*-decylamine, pH 3.8–3.9.)
**Flow rate:** 1
**Injection volume:** 20
**Detector:** UV 242

## CHROMATOGRAM

**Retention time:** 11
**Internal standard:** pyramidon (5)
**Limit of detection:** 1 μg/mL
**Limit of quantitation:** 15 μg/mL

## REFERENCE

Palette, C.; Cordonnier, P.; Naline, E.; Advenier, C.; Pays, M. High-performance liquid chromatographic method for the determination of the three main oxidative and 3-carboxylic antipyrine metabolites in human urine, *J.Chromatogr.*, **1991**, *563*, 103–113.

## ANNOTATED BIBLIOGRAPHY

Ali, H.A.; el-Yazigi, A.; Sieck, J.O.; Dossing, M.; Saour, J.; Raines, D.A.; Ernst, P. Elimination studies of antipyrine and its metabolites in healthy Saudi Arabians, *Hum.Exp.Toxicol.*, **1994**, *13*, 658–662.

Mikati, M.A.; Szabo, G.K.; Pylilo, R.J.; LeDuc, B.W.; Browne, T.R.; Greenblatt, D.J. Improved high-performance liquid chromatographic assay of antipyrine, hydroxymethylantipyrine, 4-hydroxyantipyrine and norantipyrine in urine, *J.Chromatogr.*, **1988**, *433*, 305–311.

Nakagawa, A.; Nakamura, K.; Ishizaki, T.; Chiba, K. Automated high-performance liquid chromatographic method for the determination of antipyrine and its metabolites in urine. Some preliminary results obtained from smokers and non-smokers, *J.Chromatogr.*, **1982**, *231*, 349–360.

St.Peter, J.V.; Awni, W.M. Modified high-performance liquid chromatographic assay for antipyrine and its three major metabolites in urine, *J.Chromatogr.*, **1989**, *494*, 424–427.

# EDTA

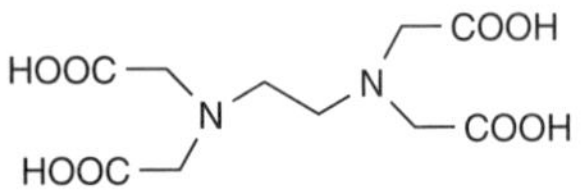

**Molecular formula:** $C_{10}H_{16}N_2O_8$
**Molecular weight:** 292.24
**CAS Registry No:** 60-00-4
**Merck Index:** 13, 3546

## SAMPLE

**Matrix:** blood
**Sample preparation:** Place 0.5 $cm^2$ of a dried blood stain in 50–100 μL 25 mM copper(II) sulfate, let stand for 3 h, vortex, centrifuge at 3000–9000 rpm for 10 min, filter (0.2 μm), inject a 25 μL aliquot. Alternatively, dilute 200 μL whole blood with 2 mL water, mix with an equal volume of 50 mM copper(II) sulfate, centrifuge for 7 min, inject an aliquot of the supernatant.

## HPLC VARIABLES

**Column:** Hamilton PRP X-100
**Mobile phase:** MeOH:3 mM sulfuric acid 5:95
**Flow rate:** 2
**Injection volume:** 25
**Detector:** UV 243

## CHROMATOGRAM

**Retention time:** 6
**Limit of detection:** 5 ppm

## KEY WORDS

complexation; derivatization; dried blood; whole blood

## REFERENCE

Miller, M.L.; McCord, B.R.; Martz, R.; Budowle, B. The analysis of EDTA in dried bloodstains by electrospray LC-MS-MS and ion chromatography, *J.Anal.Toxicol.*, **1997**, *21*, 521–528.

## SAMPLE

**Matrix:** blood
**Sample preparation:** Place 0.5 $cm^2$ of a dried blood stain with 25 μL water in an Ultrafree-MC centrifugal filter with a PTTK polysulfone membrane (cutoff 30 000 Da), let stand for 45 min, centrifuge for 10 min, inject an aliquot of the filtrate.

## HPLC VARIABLES

**Column:** 150 × 2.1 Hamilton PRP-1
**Mobile phase:** MeCN:water:ammonium hydroxide 80:20:0.03
**Flow rate:** 0.3
**Detector:** MS, Finnigan MAT TSQ700, triple-stage quadrupole, electrospray, collision gas argon, positive ion mode, spray voltage 4 kV, sheath gas 90 psi, interface capillary 200°, collision offset – 20 V, m/z 293

## CHROMATOGRAM

**Retention time:** 0.9

## KEY WORDS

dried blood

## REFERENCE

Miller, M.L.; McCord, B.R.; Martz, R.; Budowle, B. The analysis of EDTA in dried bloodstains by electrospray LC-MS-MS and ion chromatography, *J.Anal.Toxicol.*, **1997**, *21*, 521–528.

# Efavirenz

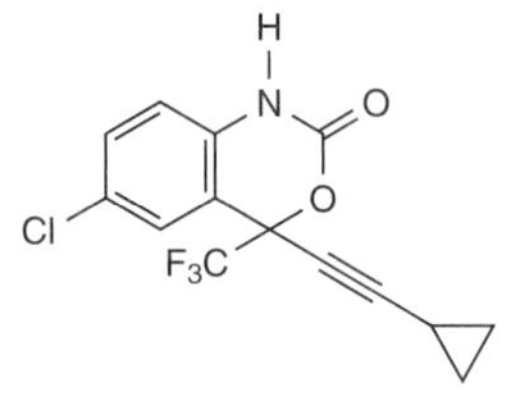


**Molecular formula:** $C_{14}H_9ClF_3NO_2$
**Molecular weight:** 315.68
**CAS Registry No:** 154598-52-4
**Merck Index:** 13, 3552

## SAMPLE

**Matrix:** blood
**Sample preparation:** Condition an Extrasep C18 SPE cartridge (Lida) with 2 mL MeOH and 2 mL water. Dilute 500 μL serum with 500 μL water, add to the SPE cartridge, wash with 500 μL water, elute with 1 mL MeOH. Evaporate the eluate to dryness with vortexing under reduced pressure at 40° and reconstitute the residue with 300 μL MeOH, inject a 10 μL aliquot.

## HPLC VARIABLES

**Column:** two 150 × 4.6 3 μm Luna C18 columns in series
**Column temperature:** 60
**Mobile phase:** Gradient. MeCN:4 mM sulfuric acid from 8:92 to 63:37 over 45 min, maintain at 63:37 for 5 min.
**Flow rate:** 0.85
**Injection volume:** 10
**Detector:** UV 265 for 31 min then UV 240

## CHROMATOGRAM

**Retention time:** 51
**Limit of detection:** 62 ng/mL

## OTHER SUBSTANCES

**Extracted:** delavirdine (25.5, LOD 110 ng/mL), indinavir (24.5, LOD 210 ng/mL), nelfinavir (33.5, LOD 400 ng/mL), nevirapine (23.5, LOD 84 ng/mL), ritonavir (50.5, LOD 510 ng/mL), saquinavir (35, LOD 100 ng/mL)

## KEY WORDS

serum; SPE

## REFERENCE

Simon, V.A.; Thiam, M.D.; Lipford, L.C. Determination of serum levels of thirteen human immunodeficiency virus-suppressing drugs by high-performance liquid chromatography, *J.Chromatogr.A*, **2001**, *913*, 447–453.

## SAMPLE

**Matrix:** blood
**Sample preparation:** Mix 1 mL plasma with 200 μL 10 μg/mL IS in water, add 200 μL 100 mM NaOH, mix, add 4 mL diethyl ether, shake for 5 min, centrifuge at 2500 rpm for 5 min. Evaporate the organic layer to dryness under a stream of nitrogen, reconstitute the residue with 200 μL initial mobile phase, inject a 100 μL aliquot.

## HPLC VARIABLES

**Column:** 250 × 4.6 Stability RP18 (CIL, France)
**Mobile phase:** Gradient. MeCN:50 mM pH 5.65 phosphate buffer from 36:64 to 64:36 over 25 min, to 80:20 (step gradient), maintain at 80:20 for 10 min, re-equilibrate at initial conditions for 5 min.
**Flow rate:** 1.5

**Injection volume:** 100
**Detector:** UV 240 for 5 min, UV 215 for 22 min, UV 260 for rest of run

## CHROMATOGRAM

**Retention time:** 19.9
**Internal standard:** JR051012 (Janssen Cilag) (28.2)
**Limit of quantitation:** 50 ng/mL

## OTHER SUBSTANCES

**Extracted:** amprenavir (11.2), indinavir (8.5), lopinavir (18.9), nelfinavir (24.1), nevirapine (3.3), ritonavir (17.6), saquinavir (16.7)

**Noninterfering:** acetaminophen, amineptine, amphotericin B, aspirin, bromazepam, buspirone, citalopram, clobazam, diazepam, didanosine, fluconazole, flunitrazepam, fluvoxamine, hydroxyitraconazole, isoniazid, itraconazole, lamivudine, loprazolam, lorazepam, metronidazole, minalcipram, nordiazepam, omeprazole, paroxetine, pyrimethamine, rifampin, sertraline, stavudine, sulfadiazine, trimethoprim, venlafaxine, zalcitabine, zidovudine, zolpidem, zopiclone

## KEY WORDS

plasma

## REFERENCE

Titier, K.; Lagrange, F.; Péhourcq, F.; Edno-Mcheik, L.; Moore, N.; Molimard, M. High-performance liquid chromatographic method for the simultaneous determination of the six HIV-protease inhibitors and two non-nucleoside reverse transcriptase inhibitors in human plasma, *Ther.Drug Monit.*, **2002**, *24*, 417–424.

---

## SAMPLE

**Matrix:** bulk, formulations
**Sample preparation:** Inject a 35 μL aliquot of a solution in MeCN:water 52.5:47.5.

## HPLC VARIABLES

**Column:** 150 × 4.6 Zorbax SB-CN
**Column temperature:** 40
**Mobile phase:** Gradient. A:B from 40:60 to 50:50 over 16 min, to 65:35 over 7 min, to 70:30 over 5 min, to 80:20 over 1 min, maintain at 80:20 for 2 min, return to initial conditions over 1 min, re-equilibrate for 8 min. A was MeOH:water:trifluoroacetic acid 90:10:0.05. B was MeOH:water:trifluoroacetic acid 10:90:0.05.
**Flow rate:** 1.5
**Injection volume:** 35
**Detector:** UV 250

## CHROMATOGRAM

**Retention time:** 15
**Limit of detection:** 0.01%
**Limit of quantitation:** 0.05%

## OTHER SUBSTANCES

**Simultaneous:** impurities

## KEY WORDS

capsules; stability-indicating

## REFERENCE

Montgomery, E.R.; Edmanson, A.L.; Cook, S.C.; Hovsepian, P.K. Development and validation of a reverse-phase HPLC method for analysis of efavirenz and its related substances in the drug substance and in a capsule formulation, *J.Pharm.Biomed.Anal.*, **2001**, *25*, 267–284.

## ANNOTATED BIBLIOGRAPHY

Aarnoutse, R.E.; Grintjes, K.J.T.; Telgt, D.S.C.; Stek, M. Jr.; Hugen, P.W.H.; Reiss, P.; Koopmans, P.P.; Hekster, Y.A.; Burger, D.M. The influence of efavirenz on the pharmacokinetics of a twice-daily combination of indinavir and low-dose ritonavir in healthy volunteers, *Clin.Pharmacol.Ther.*, **2002**, *71*, 57–67.

Aymard, G.; Legrand, M.; Trichereau, N.; Diquet, B. Determination of twelve antiretroviral agents in human plasma sample using reversed-phase high-performance liquid chromatography, *J.Chromatogr.B*, **2000**, *744*, 227–240. [amprenavir; efavirenz; indinavir; nelfinavir; ritonavir; saquinavir; abacavir; didanosine; lamivudine; stavudine; nevirapine; zidovudine]

Balani, S.K.; Kauffman, L.R.; deLuna, F.A.; Lin, J.H. Nonlinear pharmacokinetics of efavirenz (DMP-266), a potent HIV-1 reverse transcriptase inhibitor, in rats and monkeys, *Drug Metab.Dispos.*, **1999**, *27*, 41–45.

Cociglio, M.; Hillaire-Buys, D.; Peyrière, H.; Alric, R. Performance analysis of a rapid HPLC determination with the solvent demixing extraction of HIV antiproteases and efavirenz in plasma, *J.Chromatogr.Sci.*, **2003**, *41*, 80–86. [efavirenz; indinavir; amprenavir; ritonavir; saquinavir; nelfinavir]

Fan, B.; Bartlett, M.G.; Stewart, J.T. Determination of lamivudine/stavudine/efavirenz in human serum using liquid chromatography/electrospray tandem mass spectrometry with ionization polarity switch, *Biomed.Chromatogr.*, **2002**, *16*, 383–389. [aprobarbital is internal standard; SPE]

Faux, J.; Venisse, N.; Le Moal, G.; Dupuis, A.; Bouquet, S. Simultaneous determination of six HIV protease inhibitors, one metabolite, and two non-nucleoside reverse transcriptase inhibitors in human plasma by isocratic reversed-phase liquid chromatography after solid-phase extraction, *Chromatographia*, **2003**, *58*, 421–426. [amprenavir; indinavir; lopinavir; nelfinavir; ritonavir; saquinavir; efavirenz; nevirapine; prazepam is internal standard; SPE]

Keil, K.; Frerichs, V.A.; DiFrancesco, R.; Morse, G. Reverse phase high-performance liquid chromatography method for the analysis of amprenavir, efavirenz, indinavir, lopinavir, nelfinavir and its active metabolite (M8), ritonavir, and saquinavir in heparinized human plasma, *Ther.Drug Monit.*, **2003**, *25*, 340–346.

Langmann, P.; Schirmer, D.; Väth, T.; Zilly, M.; Klinker, H. High-performance liquid chromatographic method for the determination of HIV-1 non-nucleoside reverse transcriptase inhibitor efavirenz in plasma of patients during highly active antiretroviral therapy, *J.Chromatogr.B*, **2001**, *755*, 151–156.

Lavison, G.; Thiébaut, D. Evaluation of a ristocetin bonded stationary phase for subcritical fluid chromatography of enantiomers, *Chirality*, **2003**, *15*, 630–636. [warfarin; efavirenz; chiral; thalidomide]

Marzolini, C.; Telenti, A.; Buclin, T.; Biollaz, J.; Decosterd, L.A. Simultaneous determination of the HIV protease inhibitors indinavir, amprenavir, saquinavir, ritonavir, nelfinavir and the non-nucleoside reverse transcriptase inhibitor efavirenz by high-performance liquid chromatography after solid-phase extraction, *J.Chromatogr.B*, **2000**, *740*, 43–58.. [SPE]

Marzolini, C.; Béguin, A.; Telenti, A.; Schreyer, A.; Buclin, T.; Biollaz, J.; Decosterd, L.A. Determination of lopinavir and nevirapine by high-performance liquid chromatography after solid-phase extraction: application for the assessment of their transplacental passage at delivery, *J.Chromatogr.B*, **2002**, *774*, 127–140. [SPE; clozapine is internal standard]

Matthews, C.Z.; Woolf, E.J.; Mazenko, R.S.; Haddix-Wiener, H.; Chavez-Eng, C.M.; Constanzer, M.L.; Doss, G.A.; Matuszewski, B.K. Determination of efavirenz, a selective non-nucleoside reverse transcriptase inhibitor, in human plasma using HPLC with post-column photochemical derivatization and fluorescence detection, *J.Pharm.Biomed.Anal.*, **2002**, *28*, 925–934.

Maurin, M.B.; Rowe, S.M.; Blom, K.; Pierce, M.E. Kinetics and mechanism of hydrolysis of efavirenz, *Pharm.Res.*, **2002**, *19*, 517–521.

Mutlib, A.E.; Chen, H.; Nemeth, G.; Gan, L.; Christ, D.D. Liquid chromatography/mass spectrometry and high-field nuclear magnetic resonance characterization of novel mixed diconjugates of the non-nucleoside human immunodeficiency virus-1 reverse transcriptase inhibitor, efavirenz, *Drug Metab.Dispos.*, **1999**, *27*, 1045–1056.

Poirier, J.-M.; Robidou, P.; Jaillon, P. Simultaneous determination of the six HIV protease inhibitors (amprenavir, indinavir, lopinavir, nelfinavir, ritonavir, and saquinavir) plus M8 nelfinavir metabolite and the nonnucleoside reverse transcription inhibitor efavirenz in human plasma by solid-phase extraction and column liquid chromatography, *Ther.Drug Monit.*, **2002**, *24*, 302–309.

Proust, V.; Toth, K.; Hulin, A.; Taburet, A.-M.; Gimenez, F.; Singlas, E. Simultaneous high-performance liquid chromatographic determination of the antiretroviral agents amprenavir, nelfinavir, ritonavir, saquinavir, delavirdine and efavirenz in human plasma, *J.Chromatogr.B*, **2000**, *742*, 453–458.

Rentsch, K.M. Sensitive and specific determination of eight antiretroviral agents in plasma by high-performance liquid chromatography-mass spectrometry, *J.Chromatogr.B*, **2003**, *788*, 339–350. [SPE; amprenavir; efavirenz; indinavir; lopinavir; nelfinavir; nevirapine; ritonavir; saquinavir]

Rezk, N.L.; Tidwell, R.R.; Kashuba, A.D.M. Simple and rapid quantification of the non-nucleoside reverse transcriptase inhibitors nevirapine, delavirdine, and efavirenz in human blood plasma using high-performance liquid chromatography with ultraviolet absorbance detection, *J.Chromatogr.B*, **2002**, *774*, 79–88. [SPE]

Sarasa-Nacenta, M.; López-Púa, Y.; López-Cortés, L.F.; Mallolas, J.; Gatell, J.M.; Carné, X. Determination of efavirenz in human plasma by high-performance liquid chromatography with ultraviolet detection, *J.Chromatogr.B*, **2001**, *763*, 53–59.

Simon, V.A.; Thiam, M.D.; Lipford, L.C. Determination of serum levels of thirteen human immunodeficiency virus-suppressing drugs by high-performance liquid chromatography, *J.Chromatogr.A*, **2001**, *913*, 447–453. [zalcitabine; lamivudine; stavudine; didanosine; zidovudine; nevirapine; abacavir; indinavir; delavirdine; nelfinavir; saquinavir; ritonavir; efavirenz]

Störmer, E.; von Moltke, L.L.; Perloff, M.D.; Greenblatt, D.J. Differential modulation of P-glycoprotein expression and activity by non-nucleoside HIV-1 reverse transcriptase inhibitors in cell culture, *Pharm.Res.*, **2002**, *19*, 1038–1045. [verapamil; efavirenz; nevirapine; delavirdine]

Tribut, O.; Arvieux, C.; Michelet, C.; Chapplain, J.-M.; Allain, H.; Bentué-Ferrer, D. Simultaneous quantitative assay of six HIV protease inhibitors, one metabolite, and two non-nucleoside reverse transcriptase inhibitors in human plasma by isocratic reversed-phase liquid chromatography, *Ther.Drug Monit.*, **2002**, *24*, 554–562. [nevirapine; efavirenz; indinavir; amprenavir; nelfinavir; ritonavir; lopinavir; saquinavir]

Turner, M.L.; Reed-Walker, K.; King, J.R.; Acosta, E.P. Simultaneous determination of nine antiretroviral compounds in human plasma using liquid chromatography, *J.Chromatogr.B*, **2003**, *784*, 331–341. [indinavir; nelfinavir; saquinavir; ritonavir; amprenavir; delavirdine; efavirenz; lopinavir]

Usami, Y.; Oki, T.; Nakai, M.; Sagisaka, M.; Kaneda, T. A simple HPLC method for simultaneous determination of lopinavir, ritonavir and efavirenz, *Chem.Pharm.Bull.*, **2003**, *26*, 715–718.

Veldkamp, A.I.; van Heeswijk, R.P.G.; Meenhorst, P.L.; Mulder, J.W.; Lange, J.M.A.; Beijnen, J.H.; Hoetelmans, R.M.W. Quantitative determination of efavirenz (DMP 266), a novel non-nucleoside reverse transcriptase inhibitor, in human plasma using isocratic reversed-phase high-performance liquid chromatography with ultraviolet detection, *J.Chromatogr.B*, **1999**, *734*, 55–61.

Villani, P.; Pregnolato, M.; Banfo, S.; Rettani, M.; Burroni, D.; Seminari, E.; Maserati, R.; Regazzi, M.B. High-performance liquid chromatography method for analyzing the antiretroviral agent efavirenz in human plasma, *Ther.Drug Monit.*, **1999**, *21*, 346–350. [saquinavir is internal standard]

Villani, P.; Feroggio, M.; Gianelli, L.; Bartoli, A.; Montagna, M.; Maserati, R.; Regazzi, M.B. Antiretrovirals: simultaneous determination of five protease inhibitors and three nonnucleoside transcriptase inhibitors in human plasma by a rapid high-performance liquid chromatography-mass spectrometry assay, *Ther.Drug Monit.*, **2001**, *23*, 380–388. [saquinavir; indinavir; ritonavir; nelfinavir; amprenavir; nevirapine; delavirdine; efavirenz]

Ward, B.A.; Gorski, J.C.; Jones, D.R.; Hall, S.D.; Flockhart, D.A.; Desta, Z. The cytochrome P450 2B6 (CYP2B6) is the main catalyst of efavirenz primary and secondary metabolism: implication for HIV/AIDS therapy and utility of efavirenz as a substrate marker of CYP2B6 catalytic activity, *J.Pharmacol.Exp.Ther.*, **2003**, *306*, 287–300.

Weissburg, R.P.; Montgomery, E.R.; Junnier, L.A.; Segretario, J.; Cook, S.; Hovsepian, P.K. Investigation of critical factors for the resolution of SR695, a key impurity, from efavirenz in the reversed-phase assay of efavirenz dosage forms, *J.Pharm.Biomed.Anal.*, **2002**, *28*, 45–56.

Xu, J.Q.; Aubry, A.-F. Impurity profiling of non-nucleoside reverse transcriptase inhibitors by HPLC using a porous graphitic carbon stationary phase, *Chromatographia*, **2003**, *57*, 67–71.

# Efrotomycin

**Molecular formula:**
$C_{59}H_{88}N_2O_{20}$
**Molecular weight:**
1145.33
**CAS Registry No:**
56592-32-6
**Merck Index:**
13, 3556

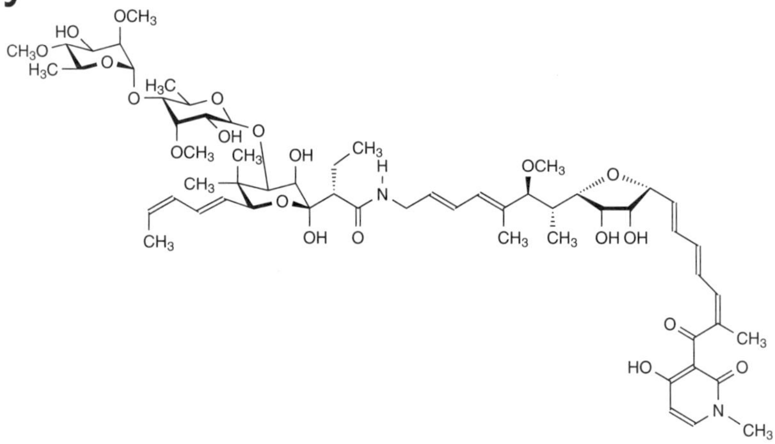

## SAMPLE

**Matrix:** feed

**Sample preparation:** Condition a 2.8 mL Bond-Elut $NH_2$ SPE cartridge with 1 mL MeCN:dichloromethane 50:50 saturated with 50 mM pH 7.5 potassium phosphate buffer. Shake 100 g blended homogenized feed with 320 mL 50 mM pH 7.5 potassium phosphate buffer in a 950 mL amber bottle for 30 min, add 280 mL MeCN, shake for 15 min, centrifuge a 30 mL aliquot at 2500 rpm for 10 min. Remove a 5 mL aliquot and add it to 10 mL dichloromethane and 20 mL MeCN:dichloromethane 50:50 saturated with 50 mM pH 7.5 potassium phosphate buffer, shake for 30 min, centrifuge at 2500 rpm for 10 min. Add 10 mL of the lower phase to the SPE cartridge, wash with 2 mL MeOH, wash with 2 mL ethyl acetate, wash with 2 mL hexane, draw air through the cartridge for 5 min, elute with 1 mL elution solvent, inject a 100 μL aliquot of the eluate. (The elution solvent was 250 mM pH 5 phosphate buffer containing 1.5 M LiCl saturated with MeCN. Prepare by mixing 500 mL of this aqueous phase with 250 mL MeCN.)

## HPLC VARIABLES

**Column:** 150 × 4.6 Zorbax C8
**Column temperature:** 55
**Mobile phase:** MeCN:MeOH:water:85% phosphoric acid 32.5:22.5:57.5:1 (Adjust with MeCN or water to give a k' of 19–22.)
**Flow rate:** 2
**Injection volume:** 100
**Detector:** UV 335

## CHROMATOGRAM

**Retention time:** 10
**Limit of quantitation:** 2 ppm

## KEY WORDS

SPE

## REFERENCE

Stong, J.D. Determination of efrotomycin in feeds by high-performance liquid chromatography, *Analyst*, **1986**, *111*, 853–855.

# Egualen

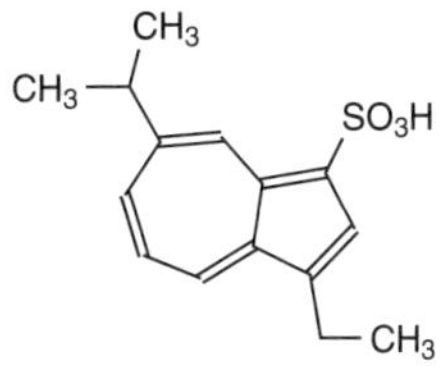

**Molecular formula:** $C_{15}H_{18}O_3S$
**Molecular weight:** 278.37
**CAS Registry No:** 99287-30-6

## SAMPLE

**Matrix:** blood
**Sample preparation:** Prepare plasma using a Sep-Pak C18 SPE cartridge (no further details).

## HPLC VARIABLES

**Column:** 250 × 4.6 TSK-gel ODS-80TM
**Mobile phase:** Gradient. MeCN:20 mM pH 6.0 phosphate buffer 20:80 for 10 min, to 30:70 over 10 min, maintain at 30:70 for 15 min.
**Flow rate:** 1
**Detector:** Radioactivity ($^{14}C$)

## OTHER SUBSTANCES

**Extracted:** metabolites

## KEY WORDS

pharmacokinetics; plasma; SPE

## REFERENCE

Sato, M.; Suzaka, H.; Miyazaki, H. Sex-related differences in urinary excretion of egualen sodium in rats, *Drug Metab.Dispos.*, **2000**, *28*, 21–27.

# Eletriptan

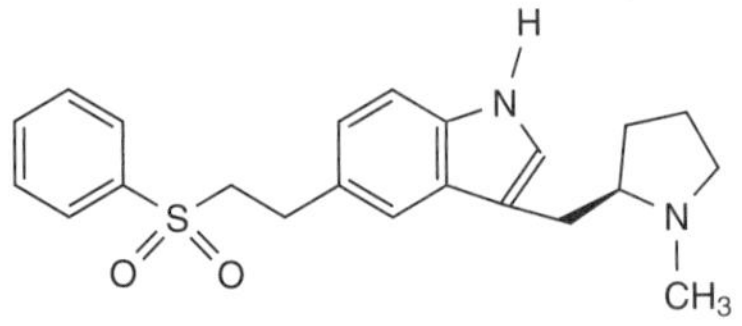


**Molecular formula:** $C_{22}H_{26}N_2O_2S$
**Molecular weight:** 382.53
**CAS Registry No:** 143322-58-1
**Merck Index:** 13, 3577

## SAMPLE

**Matrix:** blood, saliva

**Sample preparation:** Mix 570 μL plasma or saliva with 130 μL 1 (plasma) or 0.1 (saliva) M monochloroacetic acid containing 500 ng/mL IS. Place 690 μL of this mixture in the donor channel of a dialyzer fitted with a15 kDa cuprophan membrane (regenerated cellulose, Enka, Germany), pass 3500 μL recipient solvent in 500 μL pulsed bursts through the acceptor channel over 4 min, pass the recipient solvent through column A, wash column A with 200 μL MeCN:water 60:40, wash with 800 μL donor solvent, backflush the contents of column A onto column B with the mobile phase, monitor the effluent from column B. (The donor solvent was 10 mM pH 7.0 potassium phosphate buffer. Recipient solvent was MeOH:10 mM pH 7.0 potassium phosphate buffer 10:90. After each run, purge the donor channel with 1.5 mL donor solvent. Purge the HPLC system with 14 mL of donor and recipient solvent. Condition column A with 200 μL recipient solvent.)

## HPLC VARIABLES

**Column:** A 5 × 4.6 10 μm Hypersil C1; B 100 × 4.6 5 μm Kromasil C1

**Mobile phase:** MeCN:500 mM pH 3.5 potassium phosphate buffer:water 30:6:64 (Add 20 mM diethylamine hydrochloride to the buffer/water mixture before adding the MeCN.)

**Flow rate:** 1

**Detector:** UV 225

## CHROMATOGRAM

**Retention time:** 5

**Internal standard:** UK-136,509 (7)

**Limit of quantitation:** 0.5 ng/mL

## OTHER SUBSTANCES

**Extracted:** metabolite

## KEY WORDS

column-switching; dialysis; pharmacokinetics; plasma

## REFERENCE

Cooper, J.D.H.; Muirhead, D.C.; Taylor, J.E. Determination of eletriptan in plasma and saliva using automated sequential trace enrichment of dialysate and high-performance liquid chromatography, *J.Pharm.Biomed.Anal.*, **1999**, *21*, 787–796.

# Emtricitabine

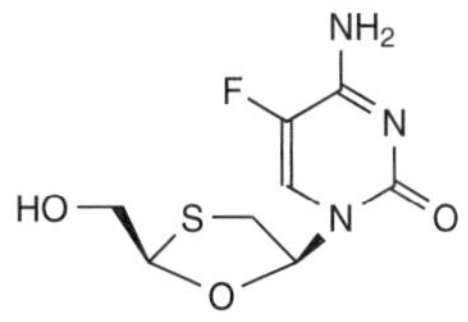


**Molecular formula:** $C_8H_{10}FN_3O_3S$
**Molecular weight:** 247.25
**CAS Registry No:** 143491-57-0
**Merck Index:** 13, 3597

## SAMPLE

**Matrix:** blood, CSF, urine
**Sample preparation:** Plasma. Mix 100 μL serum, 50 μL 20 μg/mL IS, and 50 μL 2 M perchloric acid, centrifuge at 5000 g for 5 min, add 50 μL 2 M KOH, mix well, centrifuge at 5000 g for 5 min, inject a 15–200 μL aliquot. CSF. Mix 100 μL CSF with 20 μL IS solution and 80 μL water, inject a 100 μL aliquot. Urine. Dilute 50 μL urine and 100 μL IS solution to 1 mL with water, inject a 15–100 μL aliquot.

## HPLC VARIABLES

**Column:** 150 × 4.6 5 μm Hypersil octadecanoylsulfate (sic) (Alltech)
**Mobile phase:** MeOH:40 mM pH 2.5 potassium phosphate buffer 5.5:94.5
**Flow rate:** 2
**Injection volume:** 15–200
**Detector:** UV 279

## CHROMATOGRAM

**Retention time:** 5.5
**Internal standard:** 3′deoxy-2′,3′-didehydrothymidine (9.5)
**Limit of quantitation:** 100 ng/mL

## OTHER SUBSTANCES

**Extracted:** 2,3′-dideoxy-5-fluoro-3′-thiauridine (FTU) (8.3)

## KEY WORDS

monkey; pharmacokinetics; plasma

## REFERENCE

Schinazi, R.F.; Boudinot, F.D.; Ibrahim, S.S.; Manning, C.; McClure, H.M.; Liotta, D.C. Pharmacokinetics and metabolism of racemic 2′,3′-dideoxy-5-fluoro-3′-thiacytidine in rhesus monkeys, *Antimicrob.Agents Chemother.*, **1992**, *36*, 2432–2438.

## SAMPLE

**Matrix:** cell cultures
**Sample preparation:** Extract cells twice with 4 mL portions of MeOH:water 60:40 at − 70° overnight, centrifuge at 2000 g for 10 min. Evaporate the methanol from a 500 μL aliquot under a stream of nitrogen, adjust the volume to 200 μL with water, inject a 180 μL aliquot.

## HPLC VARIABLES

**Column:** 250 × 4.6 10 μm Partisil SAX
**Mobile phase:** Gradient. A:B 100:0 for 10 min, to 0:100 over 65 min. A was 8 mM pH 3.5 potassium phosphate buffer. B was 1 M pH 3.5 potassium phosphate buffer.
**Flow rate:** 1
**Injection volume:** 180
**Detector:** Radioactivity ($^3$H)

## CHROMATOGRAM
**Retention time:** 5

## OTHER SUBSTANCES
**Extracted:** emtricitabine monophosphate (25), emtricitabine diphosphate (40), emtricitabine triphosphate (57)

## KEY WORDS
peripheral blood mononuclear cells

## REFERENCE
Darque, A.; Valette, G.; Rousseau, F.; Wang, L.H.; Sommadossi, J.-P.; Zhou, X.-J. Quantitation of intracellular triphosphate for emtricitabine in peripheral blood mononuclear cells from human immunodeficiency virus-infected patients, *Antimicrob.Agents Chemother.*, **1999**, *43*, 2245–2250.

## SAMPLE
**Matrix:** solutions

## HPLC VARIABLES
**Column:** 250 × 4.6 10 μm Chiralpak AS
**Mobile phase:** Isopropanol
**Flow rate:** 0.8
**Detector:** UV 270

## CHROMATOGRAM
**Retention time:** 5.9 (emtricitabine), 9.5 ((+)-enantiomer)

## KEY WORDS
chiral

## REFERENCE
Schinazi, R.F.; McMillan, A.; Cannon, D.; Mathis, R.; Lloyd, R.M.; Peck, A.; Sommadossi, J.-P.; St. Clair, M.; Wilson, J.; Furman, P.A.; Painter, G.; Choi, W.-B.; Liotta, D.C. Selective inhibition of human immunodeficiency viruses by racemates and enantiomers of *cis*-5-fluoro-1-[2-(hydroxymethyl)-1,3-oxathiolan-5-yl]cytosine, *Antimicrob.Agents Chemother.*, **1992**, *36*, 2423–2431.

## ANNOTATED BIBLIOGRAPHY
Cass, Q.B.; Watanabe, C.S.F.; Rabi, J.A.; Bottari, P.Q.; Costa, M.R.; Nascimento, R.M.; Cruz, J.E.D.; Ronald, R.C. Polysaccharide-based chiral phase under polar organic mode of elution in the determination of the enantiomeric purity of emtricitabine an anti-HIV analogue nucleoside, *J.Pharm.Biomed. Anal.*, **2003**, *33*, 581–587.

Paff, M.T.; Averett, D.R.; Prus, K.L.; Miller, W.H.; Nelson, D.J. Intracellular metabolism of (-) and (+)-*cis*-5-fluoro-1-[2-(hydroxymethyl)- 1,3-oxathiolan-5-yl]cytosine in HepG2 derivative 2.2.15 (Subclone P5A) cells, *Antimicrob.Agents Chemother.*, **1994**, *38*, 1230–1238.

# Enoxaparin sodium

**Molecular weight:** ca. 4500
**CAS Registry No:** 9041-08-1
**Merck Index:** 13, 3621

## SAMPLE

**Matrix:** solutions

## HPLC VARIABLES

**Column:** 300 × 3.2 Superdex (Pharmacia PC 3.2/30)
**Mobile phase:** 200 mM NaCl
**Flow rate:** 0.1
**Detector:** Refractive Index

## CHROMATOGRAM

**Retention time:** 12–16

## REFERENCE

Intes, O.; Renault, J.-H.; Sinquin, C.; Zèches-Hanrot, M.; Nuzillard, J.-M. Fractionation of low-molecular-mass heparin by centrifugal partition chromatography in the ion-exchange displacement mode, *J.Chromatogr.A*, **2001**, *918*, 47–57.

# Entacapone

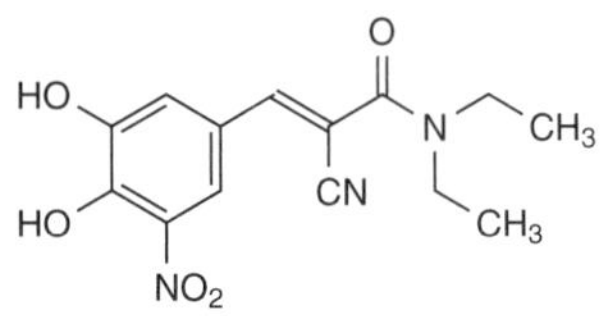


**Molecular formula:** $C_{14}H_{15}N_3O_5$
**Molecular weight:** 305.29
**CAS Registry No:** 130929-57-6
**Merck Index:** 13, 3626

## SAMPLE

**Matrix:** blood, urine
**Sample preparation:** Plasma. Add 50 μL 50 mM pH 7.2 phosphate buffer to 1 mL plasma, vortex, add 100 μL 2 M HCl, add 6 mL *n*-hexane:ethyl acetate 50:50, vortex for 2 min, centrifuge at 3500 g for 5 min. Remove 5 mL of the organic layer, evaporate it to dryness under a stream of nitrogen at 40°, reconstitute the residue with 500 μL DMSO, inject a 20 μL aliquot. Urine. Add 50 μL 50 mM pH 7.2 phosphate buffer to 1 mL urine, vortex, add 100 μL 2 M HCl, vortex, add 5 mL *n*-hexane:ethyl acetate 75:25, vortex for 2 min, centrifuge at 3500 g for 5 min. Remove 4 mL of the organic layer and add it to 1 mL 50 mM pH 7.2 phosphate buffer, vortex for 2 min, centrifuge at 3500 g for 5 min, inject a 30 μL aliquot of the aqueous layer. (Protect from light during preparation.)

## HPLC VARIABLES

**Guard column:** μBondapak Guard-PAC C18
**Column:** 250 × 4 10 μm Lichrosorb C18
**Column temperature:** 35 (plasma), 28 (urine)
**Mobile phase:** MeOH:THF:buffer A 50:5:63 (plasma) or MeCN:THF:buffer B 50:5:125 (urine) (Buffer A was 50 mM sodium dihydrogen phosphate containing 20 mM citric acid and 0.25 mM EDTA, adjusted to pH 2.0 with phosphoric acid. Buffer B was 50 mM sodium dihydrogen phosphate containing 20 mM citric acid and 0.25 mM EDTA, adjusted to pH 3.0 with 10 M NaOH.)
**Flow rate:** 1.5
**Injection volume:** 20 (plasma), 30 (urine)
**Detector:** E, Bioanalytical Systems LC-4B, glassy carbon electrode 700 mV, Ag/AgCl reference electrode

## CHROMATOGRAM

**Retention time:** 7.5 (plasma), 10.7 (urine)
**Limit of quantitation:** 10 ng/mL

## OTHER SUBSTANCES

**Extracted:** Z-isomer (5.3 (plasma), 9.1 (urine))
**Noninterfering:** carbidopa, 3,4-dihydroxyphenylacetic acid, homovanillic acid, levodopa, 3-O-methyldopa

## KEY WORDS

pharmacokinetics; plasma

## REFERENCE

Karlsson, M.; Wikberg, T. Liquid chromatographic determination of a new catechol-O-methyltransferase inhibitor, entacapone, and its *Z*-isomer in human plasma and urine, *J.Pharm.Biomed.Anal.*, **1992**, *10*, 593–600.

## ANNOTATED BIBLIOGRAPHY

Keski-Hynnilä, H.; Kurkela, M.; Elovaara, E.; Antonio, L.; Magdalou, J.; Luukkanen, L.; Taskinen, J.; Kostiainen, R. Comparison of electrospray, atmospheric pressure chemical ionization, and atmospheric pressure photoionization in the identification of apomorphine, dobutamine, and entacapone phase II metabolites in biological samples, *Anal.Chem.*, **2002**, *74*, 3449–3457. [urine; microsomal incubations; entacapone metabolites only]

Luukkanen, L.; Kilpelainen, I.; Kangas, H.; Ottoila, P.; Elovaara, E.; Taskinen, J. Enzyme-assisted synthesis and structural characterization of nitrocatechol glucuronides, *Bioconjug.Chem.*, **1999**, *10*, 150–154. [HPLC of metabolites only]

Wikberg, T.; Ottoila, P.; Taskinen, J. Identification of major urinary metabolites of the catechol-O-methyltransferase inhibitor entacapone in the dog, *Eur.J.Drug Metab.Pharmacokinet.*, **1993**, *18*, 359–367.

Wikberg, T.; Vuorela, A.; Ottoila, P.; Taskinen, J. Identification of major metabolites of the catechol-O-methyltransferase inhibitor entacapone in rats and humans, *Drug Metab.Dispos.*, **1993**, *21*, 81–92.

Wikberg, T.; Vuorela, A. Metabolite profiles of two [$^{14}$C]-labelled catechol O-methyltransferase inhibitors, nitecapone and entacapone, in rat and mouse urine and rat bile, *Eur.J.Drug Metab.Pharmacokinet.*, **1994**, *19*, 125–135. [HPLC of metabolites only]

# Eperisone

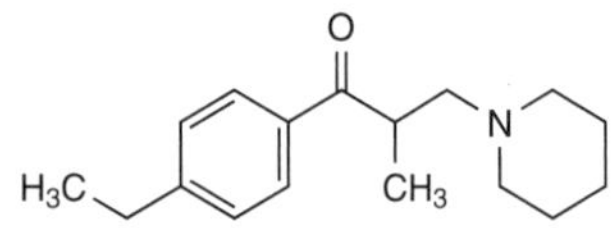


**Molecular formula:** $C_{17}H_{25}NO$
**Molecular weight:** 259.39
**CAS Registry No:** 64840-90-0
**Merck Index:** 13, 3637

## SAMPLE
**Matrix:** blood
**Sample preparation:** Mix 1 mL water, 100 μL 1 μg/mL IS in MeOH, and 1 mL 1 M pH 7.2 phosphate buffer with 100 μL plasma, add 4 mL diethyl ether, extract, repeat extraction. Combine the organic layers and add them to 200 μL 100 mM HCl, extract, inject a 70 μL aliquot of the aqueous layer.

## HPLC VARIABLES
**Column:** 250 × 4.6 YMC pack A-303
**Mobile phase:** MeCN:5 mM pH 2.5 sodium dodecyl sulfate 50:50
**Flow rate:** 1
**Injection volume:** 70
**Detector:** UV 256

## CHROMATOGRAM
**Internal standard:** tolperisone HCl
**Limit of quantitation:** 2 ng/mL

## KEY WORDS
plasma

## REFERENCE
Matsunaga, M.; Uemura, Y.; Yonemoto, Y.; Kanai, K.; Etoh, H.; Tanaka, S.; Atsuta, Y.; Nishizawa, Y.; Yamanishi, Y. Long-lasting muscle relaxant activity of eperisone hydrochloride after percutaneous administration in rats, *Jpn.J.Pharmacol.*, **1997**, *73*, 215–220.

## SAMPLE
**Matrix:** solutions

## HPLC VARIABLES
**Column:** 250 × 4.6 10 μm Cosmosil 5C18-MS
**Column temperature:** 50
**Mobile phase:** Gradient. MeCN:10 mM pH 4.2 potassium phosphate buffer 0:100 for 2 min, to 75:25 over 7.5 min, maintain at 75:25 for 5.5 min. Isocratic. MeCN:buffer 20:80
**Flow rate:** 1.5
**Detector:** UV (wavelength not specified)

## CHROMATOGRAM
**Retention time:** 8.7 (gradient) or 5.8 (isocratic)

## OTHER SUBSTANCES
**Simultaneous:** acetazolamide (7.9), acyclovir (7.0), allopurinol (7.3), aniracetam (8.6), benazepril (11.5), betamethasone (10.8), camostat mesylate (8.8), carbamazepine (10.8), cilazapril (10.7), cimetidine (7.5), clofibrate (15.0), clonazepam (11.3), cyclophosphamide (9.9), delapril (11.9), dexamethasone (10.9), diazepam (12.5), digoxin (9.0), dilazep (10.6), diltiazem (10.2), docarpamine (11.4), ethosuximide (9.0), fenbufen (11.8), fluconazole (9.2), flutamide (12.7), fominoben (12.6), hydrocortisone acetate (12.1), imidapril

(10.1), indomethacin (12.6), irinotecan (9.2), maprotiline (10.5), methotrexate (8.0), nefiracetam (9.7), nifedipine (11.9), nitrazepam (11.2), pentobarbital (10.9), phenobarbital (10.0), phenytoin (10.8), pindolol (8.3), pranlukast (13.4), pranoprofen (10.4), prednisolone (10.3), primidone (8.9), quinapril (10.5), spironolactone (12.4), sulpiride (7.6), sulthiame (9.3), tolbutamide (11.6), tranilast (10.5), triamcinolone (11.2), warfarin (12.0), zonisamide (9.5) (gradient retention times; isocratic conditions may differ)

## REFERENCE

Sugiyama, T.; Matsuyama, R.; Usui, S.; Katagiri, Y.; Hirano, K. Selection of mobile phases in high-performance liquid chromatographic determination for medicines, *Biol.Pharm.Bull.*, **2000**, *23*, 274–278.

## ANNOTATED BIBLIOGRAPHY

Nakamura, K.; Fujima, H.; Kitagawa, H.; Wada, H.; Makino, K. Preparation and chromatographic characteristics of a chiral-recognizing perphenylated cyclodextrin column, *J.Chromatogr.A*, **1995**, *694*, 111–118. [ibuprofen; chlorpheniramine; acetylpheneturide; alprenolol; arotinolol; atenolol; benzoin; biperiden; bunitrolol; chlormezanone; chlorphenesin; eperisone; flavanone; oxprenolol; phenylethyl alcohol; phenylethylamine; pindolol; proglumide; propranolol; trihexyphenidyl]

# Eplerenone

**Molecular formula:** $C_{24}H_{30}O_6$
**Molecular weight:** 414.49
**CAS Registry No:** 107724-20-9

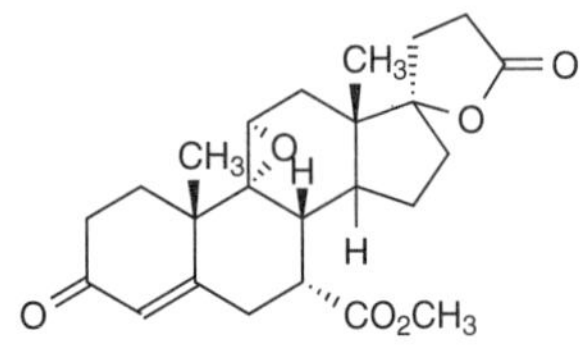

## SAMPLE

**Matrix:** urine

**Sample preparation:** Urine. Condition a 1 mL 100 mg Bond Elut C18 SPE cartridge with 2 mL MeCN and 2 mL water. Vortex 500 μL urine and 500 μL 1 μg/mL IS in 20 mM pH 7.4 ammonium acetate, add to the SPE cartridge, wash with 3 mL water, elute with 250 μL MeCN. Vortex the eluate with 250 μL 20 mM pH 7.4 ammonium acetate buffer, inject a 20 μL aliquot. Plasma. Condition a 1 mL 100 mg Bond Elut C18 SPE cartridge with 2 mL MeOH and 2 mL water. Centrifuge plasma at 2000 g at 4° for 10 min. Vortex 400 μL of the supernatant and 400 μL 500 ng/mL IS in water, add to the SPE cartridge, wash with 2 mL water, elute with 500 μL MeCN. Evaporate the eluate to dryness under a stream of nitrogen at room temperature, reconstitute the residue with 100 μL mobile phase, inject a 10 μL aliquot. (Plasma preparation from Zhang,J.Y.; Fast,D.M.; Breau,A.P. Development and validation of a liquid chromatography-tandem mass spectrometric assay for Eplerenone and its hydrolyzed metabolite in human plasma. *J.Chromatogr.B* **2003**, *787*, 333–344.)

## HPLC VARIABLES

**Column:** 50 × 2.1 5 μm Zorbax XDB-C8
**Mobile phase:** MeCN:water 40:60 containing 10 mM ammonium acetate, pH 7.4
**Flow rate:** 0.1
**Injection volume:** 20 (urine), 10 (plasma)
**Detector:** MS, PE Sciex API-III Plus quadrupole, ionspray, positive ionization, ionspray interface 4400 V, orifice 67 V, nebulizer gas nitrogen at 60 psi, curtain gas nitrogen at 1.8 L/min, collision gas argon, m/z 415–163

## CHROMATOGRAM

**Retention time:** 3.2
**Internal standard:** $^{13}C^{2}H_3$-eplerenone (m/z 419–163)
**Limit of quantitation:** 50 ng/mL (urine), 10 ng/mL (plasma)

## OTHER SUBSTANCES

**Extracted:** metabolite

## KEY WORDS

pharmacokinetics; SPE

## REFERENCE

Zhang, J.Y.; Fast, D.M.; Breau, A.P. A validated SPE-LC-MS/MS assay for Eplerenone and its hydrolyzed metabolite in human urine, *J.Pharm.Biomed.Anal.*, **2003**, *31*, 103–115.

## ANNOTATED BIBLIOGRAPHY

Cook, C.S.; Berry, L.M.; Kim, D.H.; Burton, E.G.; Hribar, J.D.; Zhang, L. Involvement of CYP3A in the metabolism of eplerenone in humans and dogs: Differential metabolism by CYP3A4 and CYP3A5, *Drug Metab.Dispos.*, **2002**, *30*, 1344–1351. [radioactivity detection ($^{14}C$); LC-MS]

Cook, C.S.; Zhang, L. Atypical dose-route-dependent food effects of eplerenone in the dog: Presence of food effects following intravenous dosing and lack of food effects of following oral dosing, *J.Pharm.Sci.*, **2002**, *91*, 607–614.

Cook, C.S.; Berry, L.M.; Bible, R.H.; Hribar, J.D.; Hajdu, E.; Liu, N.W. Pharmacokinetics and metabolism of [$^{14}$C]eplerenone after oral administration to humans, *Drug Metab.Dispos.*, **2003**, *31*, 1448–1455. [radioactivity detection; LC-MS]

Cook, C.S.; Zhang, L.; Ames, G.B.; Fischer, J.; Zhang, J.; Levin, S. Single-and repeated-dose pharmacokinetics of eplerenone, a selective aldosterone receptor blocker, in rats, *Xenobiotica*, **2003**, *33*, 305–321. [LC-MS; SPE]

# Epoprostenol

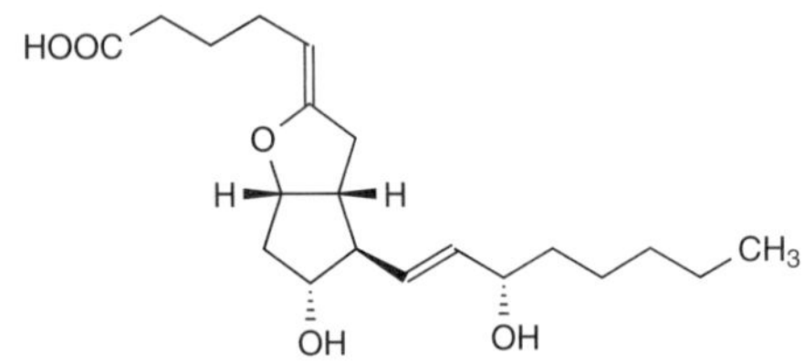


**Molecular formula:** $C_{20}H_{32}O_5$
**Molecular weight:** 352.46
**CAS Registry No:** 35121-78-9, 61849-14-7 (Na salt)
**Merck Index:** 13, 7967

## SAMPLE

**Matrix:** blood
**Sample preparation:** Pack a polypropylene column with 1 g Hi-Flosil (80/100) C18 material (Applied Science) in an MeCN slurry, add 5 mL 10% dimethyldichlorosilane in toluene, wash with 15 mL toluene, wash with 15 mL MeCN. Just prior to use, equilibrate with 15 mL 2 mM sodium borate adjusted to pH 10 with NaOH. Mix 1 mL whole blood with 10 mg/mL sodium carbonate solution, keep on ice, centrifuge at 2000 g at 4° for 10 min, add a 500 µL aliquot to the SPE column, wash with two 3 mL portions of 2 mM pH 10 sodium borate, elute with three 3 mL portions of MeCN:2 mM pH 10 sodium borate 40:60. Collect the eluate in a container containing 200 µL 200 mM NaOH. Evaporate the eluate to dryness under reduced pressure at 40°, reconstitute the residue with 1 mL 2 mM pH 9.1 sodium borate, inject a 100 µL aliquot.

## HPLC VARIABLES

**Column:** 250 × 4 MCH-10
**Mobile phase:** Gradient. MeCN:2 mM pH 9.1 sodium borate from 12:88 to 27:73 over 7 min.
**Flow rate:** 2
**Injection volume:** 100
**Detector:** UV 200; Radioactivity ($^3$H)

## CHROMATOGRAM

**Retention time:** 6

## OTHER SUBSTANCES

**Extracted:** 6-ketoprostaglandin $F_{1\alpha}$ (2)

## KEY WORDS

SPE; whole blood

## REFERENCE

Skrinska, V.; Lucas, F.V. Isolation of prostacyclin from whole blood, *Prostaglandins*, **1981**, *22*, 365–375.

# Eprosartan

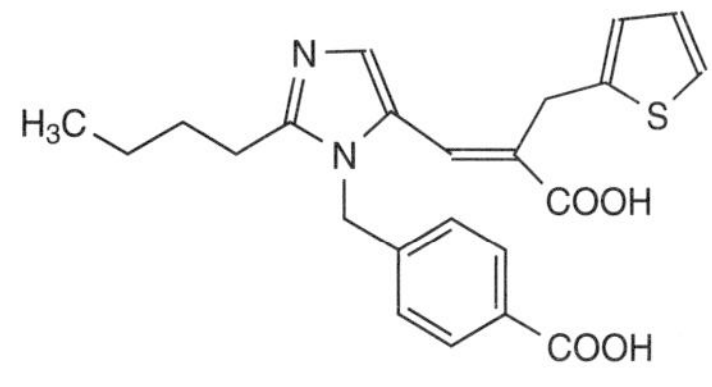


**Molecular formula:** $C_{23}H_{24}N_2O_4S$
**Molecular weight:** 424.52
**CAS Registry No:** 133040-01-4
**Merck Index::** 13, 3664

### SAMPLE

**Matrix:** blood
**Sample preparation:** Condition a 1 mL 100 mg phenyl SPE cartridge (Analytichem) with 2 mL MeOH and 2 mL water. Add 500 μL 100 mM pH 3.5 citrate buffer to 500 μL plasma containing 200 ng IS, vortex briefly, centrifuge at 8800 g for 5 min. Add the supernatant to the SPE cartridge, wash with 2 mL 50 mM acetic acid, dry the cartridge by passing air through it for 45 s, wash with 1 mL ethyl acetate containing 0.1% triethylamine, dry with air for 45 s, elute with 2 mL MeOH:50 mM acetic acid 90:10. Evaporate the eluate to dryness under nitrogen at 45°, reconstitute the residue with 125 μL mobile phase, vortex, centrifuge at 1875 g for 10 min, inject a 50 μL aliquot of the supernatant.

### HPLC VARIABLES

**Guard column:** 20 × 2 5 μm BDS-Hypersil C18
**Column:** 150 × 2 5 μm BDS-Hypersil C18
**Mobile phase:** THF:50 mM pH 3.5 citrate buffer 32:68
**Flow rate:** 0.25
**Injection volume:** 50
**Detector:** UV 300

### CHROMATOGRAM

**Retention time:** 9
**Internal standard:** SB-200062 [(E)-3-[2-butyl-1-[(4-carboxy-phenyl)methyl]-1H-imidazol-5-yl]-2-[(2-thienyl)-ethyl]propenoic acid] (12.3)
**Limit of quantitation:** 10 ng/mL

### KEY WORDS

pharmacokinetics; plasma; SPE

### REFERENCE

Lundberg, D.E. Jr.; Person, C.R.; Knox, S.; Cyronak, M.J. Determination of SK&F 108566 (Teveten) in human plasma by reversed-phase high-performance liquid chromatography, *J.Chromatogr.B*, **1998**, *707*, 328–333.

### SAMPLE

**Matrix:** urine
**Sample preparation:** Mix 100 μL urine with 200 μL 500 ng/mL IS in MeCN, vortex, centrifuge at 11 000 g, inject a 1 μL aliquot of the supernatant.

### HPLC VARIABLES

**Column:** 20 × 2 3 μm Hypersil APS-2 amino
**Mobile phase:** MeCN:10 mM pH 3.0 ammonium formate 80:20
**Flow rate:** 0.25
**Injection volume:** 1
**Detector:** MS, PE Sciex API-III Plus, ionspray, positive ion mode, split column effluent so 50 μL/min enters MS

## CHROMATOGRAM

**Retention time:** 0.42
**Internal standard:** unspecified (0.44)
**Limit of quantitation:** 50 ng/mL

## KEY WORDS

SPE

## REFERENCE

Martin, D.E.; Chapelsky, M.C.; Ilson, B.; Tenero, D.; Boike, S.C.; Zariffa, N.; Jorkasky, D.K. Pharmacokinetics and protein binding of eprosartan in healthy volunteers and in patients with varying degrees of renal impairment, *J.Clin.Pharmacol.*, **1998**, *38*, 129–137.

# Eptazocine

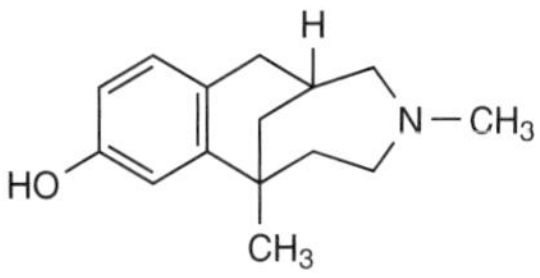


**Molecular formula:** $C_{15}H_{21}NO$
**Molecular weight::** 231.33
**CAS Registry No:** 72522-13-5
**Merck Index::** 13, 3668

## SAMPLE

**Matrix:** blood
**Sample preparation:** Add 100 μL 100 ng/mL IS in MeOH to 50 μL plasma, add 300 μL water, add 20 μL 25% ammonium hydroxide, vortex gently for 1 min, add 400 μL ethyl acetate, vortex for 5 min, centrifuge at 10 000 g for 10 min. Evaporate the organic layer to dryness under a stream of nitrogen at 38°, reconstitute the residue with 200 μL MeOH, inject a 20 μL aliquot.

## HPLC VARIABLES

**Column:** 150 × 4.6 5 μm Cosmosil 5C18-MS
**Column temperature:** 25
**Mobile phase:** MeOH:70 mM pH 3.0 sodium phosphate buffer containing 5 mM sodium heptylsulfonate 45:55
**Flow rate:** 0.8
**Injection volume:** 20
**Detector:** F ex 278 em 324

## CHROMATOGRAM

**Retention time:** 4.9
**Internal standard:** 7-methyleptazocine (7.2)
**Limit of detection:** 5 ng/mL
**Limit of quantitation:** 10 ng/mL

## KEY WORDS

pharmacokinetics; plasma; rat

## REFERENCE

Suzuki, T.; Shimizu, R.; Suganuma, T.; Nishino, J.; Tomono, K.; Hanano, M.; Watanabe, J. Pharmacokinetic/pharmacodynamic relationship of eptazocine, a narcotic-antagonist analgesic, in rats, *Biol.Pharm. Bull.*, **2000**, *23*, 1504–1510.

# Eptifibatide

**Molecular formula:** $C_{35}H_{49}N_{11}O_9S_2$
**Molecular weight:** 831.98
**CAS Registry No:** 188627-80-7
**Merck Index:** 13, 3670

## SAMPLE

**Matrix:** formulations
**Sample preparation:** Dilute oral solution with MeCN:water 70:30, inject a 100 μL aliquot.

## HPLC VARIABLES

**Column:** 150 × 4.6 5 μm Pinnacle octyl amine (C8) (Restek)
**Mobile phase:** Gradient. MeCN:buffer 17:83 for 15 min, to 100:0 over 5 min, return to initial conditions over 2 min. (The buffer was 0.1% trifluoroacetic acid in water containing 0.1% triethylamine.)
**Flow rate:** 1
**Injection volume:** 100
**Detector:** UV 220

## CHROMATOGRAM

**Retention time:** 10.1
**Limit of quantitation:** 500 ng/mL

## OTHER SUBSTANCES

**Simultaneous:** degradants

## KEY WORDS

oral solution

## REFERENCE

Zhao, L.; Yalkowsky, S.H. Stabilization of eptifibatide by cosolvents, *Int.J.Pharm.*, **2001**, *218*, 43–56.

## SAMPLE

**Matrix:** solutions

## HPLC VARIABLES

**Column:** 30 mm long C18 (Perkin-Elmer)
**Mobile phase:** Gradient. MeCN:water:trifluoroacetic acid from 80:20:0.1 to 75:25:0.1 (sic) over 5 min.
**Detector:** UV 280

## REFERENCE

Schachter, D.M.; Kohn, J. A synthetic polymer matrix for the delayed or pulsatile release of water-soluble peptides, *J.Control.Rel.*, **2002**, *78*, 143–153.

# Erdosteine

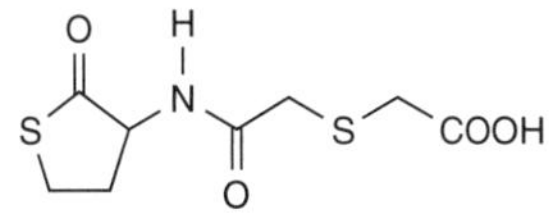

**Molecular formula:** $C_8H_{11}NO_4S_2$
**Molecular weight:** 249.31
**CAS Registry No:** 84611-23-4
**Merck Index:** 13, 3677

## SAMPLE

**Matrix:** blood
**Sample preparation:** Mix 100 μL plasma, 500 μL water, 560 μL 100 mM pH 9.3 borate buffer, 10 μL 100 μM IS in water, and 420 μL 762 μg/mL DBD-F in MeCN, let stand at room temperature for 30 min (to derivatize thiol in metabolites and IS), evaporate the MeCN under reduced pressure, add 2 mL ethyl acetate to the remaining solution, mix vigorously, centrifuge at 3000 rpm for 2 min. Wash the aqueous layer three more times with ethyl acetate. Adjust the pH of the aqueous layer to 1–2 with ca. 1.5 mL 100 mM HCl, extract three times with 2 mL portions of ethyl acetate. Combine the organic layers and evaporate them to dryness under reduced pressure, reconstitute the residue with 100 μL MeCN. Remove a 50 μL aliquot, add 5 μL DPPA, add 45 μL 17.8 mg/mL *R*-(−)-DBD-Apy in MeCN, let stand at room temperature for 2 h (to derivatize carboxylic acid), inject an aliquot. (DBD-F is 4-(*N*,*N*-dimethylaminosulfonyl)-7-fluoro-2,1,3-benzoxadiazole. *R*-(−)-DBD-Apy is *R*-(−)-4-(*N*,*N*-dimethylaminosulfonyl)-7-(3-aminopyrrolidin-1-yl)-2,1,3-benzoxadiazole. DPPA is diphenyl phosphoryl azide. They are all available from Tokyo Kasei or TCI America (www.tciamerica.com).)

## HPLC VARIABLES

**Column:** 150 × 4.6 5 μm Ultron VX-ODS
**Column temperature:** 40
**Mobile phase:** Gradient. MeCN:water:trifluoroacetic acid 30:70:0.1 for 10 min, to 37:63:0.1 over 25 min, to 43:57:0.1 over 30 min.
**Flow rate:** 1
**Detector:** F ex 468 em 563

## CHROMATOGRAM

**Retention time:** 11
**Internal standard:** captopril (41)
**Limit of detection:** 0.22 pmol

## OTHER SUBSTANCES

**Extracted:** metabolites

## KEY WORDS

derivatization; plasma; rat

## REFERENCE

Muramatsu, N.; Toyo'oka, T.; Yamaguchi, K.; Kobayashi, S. High-performance liquid chromatographic determination of erdosteine and its optical active metabolite utilizing a fluorescent chiral tagging reagent, *R*-(−)-4-(N,N-dimethylaminosulfonyl)-7-(3-aminopyrrolidin-1-yl)-2,1,3-benzoxadiazole, *J. Chromatogr.B*, **1998**, *719*, 177–189.

# Ergotamine

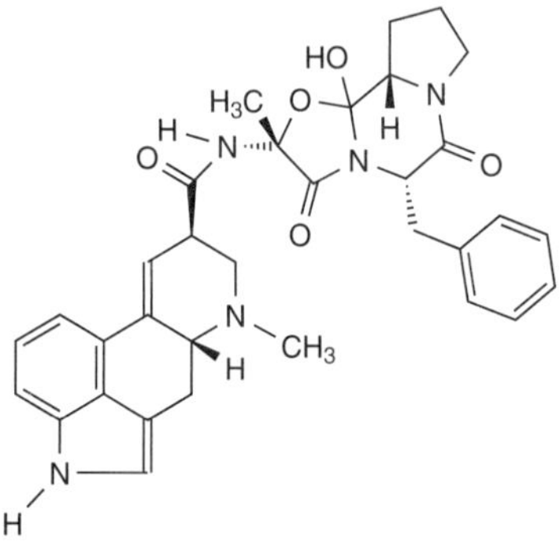

**Molecular formula:** $C_{33}H_{35}N_5O_5$
**Molecular weight:** 581.66
**CAS Registry No:** 113-15-5
**Merck Index:** 13, 3696

## SAMPLE

**Matrix:** blood
**Sample preparation:** Mix 1 mL whole blood with 20 μL 1 μg/mL dibenzepin in MeOH:water 50:50, add 300 μL pH 11 Tris buffer, mix, add 500 μL butyl acetate, vortex for 2 min, centrifuge. Preserve the aqueous layer (A). Remove the organic layer and add it to 75 μL 10 mM ammonium acetate buffer containing 0.1% formic acid (pH 3.2), evaporate (?). Add 75 μL MeCN, sonicate for 5 min, centrifuge at 5000 rpm for 5 min, keep the extract as B. Add 20 μL 1 μg/mL enalapril in MeOH:water 50:50 and 120 mg NaCl to the aqueous layer (A), mix, add 500 μL pH 3 phosphate buffer, add 600 μL 8.5% phosphoric acid, add 5 mL dichloromethane:isopropanol 95:5, shake at 250 cycles/min in a bench-top shaker for 30 min, centrifuge at 5000 rpm for 5 min. Remove the lower organic layer and evaporate it to dryness under a stream of air at 45°. Reconstitute the residue with 150 μL initial mobile phase, sonicate, centrifuge, combine with extract B, inject a 30 μL aliquot. (Sample preparation from Gergov,M.; Robson,J.N.; Ojanperä,I.; Heinonen,O.P.; Vuori,E. Simultaneous screening and quantitation of 18 antihistamine drugs in blood by liquid chromatography ionspray tandem mass spectrometry. *Forensic Sci.Inter.* **2001**, *121*, 108–115.)

## HPLC VARIABLES

**Guard column:** 40 mm long 4 μm Purospher RP-18 LiChro Cart 4-4
**Column:** 100 × 2.1 4 μm Genesis C18 (Jones Chromatography)
**Column temperature:** 35
**Mobile phase:** Gradient. MeCN:buffer from 20:80 to 100:0 over 10 min, maintain at 0:100 for 3 min, re-equilibrate at initial conditions for 5 min. (The buffer was 10 mM ammonium acetate containing 0.1% formic acid (pH 3.2).)
**Flow rate:** 0.2
**Injection volume:** 30
**Detector:** MS, PE Sciex API 365 triple-stage quadrupole LC-MS-MS, PE Sciex TurboIon-Spray interface, positive ion mode, needle voltage 5.2 kV, nebulizer gas air at 60 psi, curtain gas nitrogen at 40 psi, collision cell gas nitrogen at 40 psi, turbo ionspray heater 375°, heater gas flow 7 L/min

## CHROMATOGRAM

**Retention time:** 5.5
**Internal standard:** dibenzepin, enalapril
**Limit of detection:** 5 ng/mL

## OTHER SUBSTANCES

**Extracted:** acebutolol (3.8, LOD 0.1 μg/mL), acrivastine (5.7, LOD <0.02 μg/mL), alprazolam (6.1, LOD <0.02 μg/mL), alprenolol (5.4, LOD 0.01 μg/mL), amantadine (3.4, LOD 0.1 μg/mL), amiloride (2.0, LOD 0.1 μg/mL), aminophenazone (2.8, LOD <5 μg/mL), amiodarone (10.2, LOD 0.05 μg/mL), amitriptyline (6.6, LOD <0.02 μg/mL), astemizole (5.8, LOD <0.02 μg/mL), atenolol (1.7, LOD 0.30 μg/mL), azacyclonol (5.1, LOD 0.02 μg/mL), benzhexol (6.6, LOD <0.02 μg/mL), benzoylecgonine (3.3, LOD 0.01 μg/mL), betaxolol (5.5, LOD 0.01 μg/mL), biperidine (6.2, LOD <0.02 μg/mL), bisoprolol (5.0, LOD <0.02 μg/mL), brompheniramine (5.3, LOD 0.002 μg/mL), bupivacaine (5.1, LOD <0.02 μg/mL), buprenorphine (5.9, LOD 0.01 μg/mL), buspirone (5.1, LOD 0.002 μg/mL), caffeine (2.8, LOD 1 μg/mL), carbamazepine (6.1, LOD <0.02 μg/mL), carbinoxamine (5.1, LOD 0.002 μg/mL), carisoprodol (6.7, LOD <5 μg/mL), carvedilol (6.2, LOD <0.02 μg/mL), celiprolol (4.3, LOD 0.05 μg/mL), cetirizine (6.3, LOD 0.05 μg/mL), chlorcyclizine (6.6, LOD <0.02 μg/mL), chlordiazepoxide (5.7, LOD <0.02 μg/mL), chlormezanone (5.8, LOD <5 μg/mL), chloroquine (2.7, LOD 0.02 μg/mL), chlorpheniramine (5.1, LOD 0.002 μg/mL), chlorpromazine (7.0, LOD 0.02 μg/mL), chlorpropamide (6.7, LOD <5 μg/mL), chlorprothixene (7.0, LOD <0.02 μg/mL), cinnarizine (7.9, LOD <0.02 μg/mL), citalopram (5.7, LOD <0.02 μg/mL), clemastine (7.7, LOD 0.02 μg/mL), clobazam (7.3, LOD <0.02 μg/mL), clobutinol (5.3, LOD 0.02 μg/mL), clomethiazole (6.2, LOD 0.5 μg/mL), clomipramine (7.1, LOD <0.02 μg/mL), clonazepam (6.6, LOD <0.02 μg/mL), clonidine (2.8, LOD 0.1 μg/mL), clozapine (5.6, LOD <0.02 μg/mL), cocaine (4.6, LOD <0.02 μg/mL), codeine (2.5, LOD 0.1 μg/mL), coumatetralyl (8.4, LOD 0.05 μg/mL), cyclizine (5.8, LOD <0.02 μg/mL), dextropropoxyphene (6.6, LOD 0.05 μg/mL), demoxepam (5.8, LOD 0.02 μg/mL), dextromethorphan (5.5, LOD <0.02 μg/mL), diazepam (8.1, LOD 0.02 μg/mL), diltiazem (5.8, LOD <0.02 μg/mL), diphenhydramine (5.7, LOD <0.02 μg/mL), dipyridamole (5.4, LOD 0.005 μg/mL), disopyramine (4.4, LOD <0.02 μg/mL), dixyrazine (6.8, LOD 0.005 μg/mL), doxapram (4.8, LOD <0.02 μg/mL), doxepin (5.9, LOD <0.02 μg/mL), ebastine (9.6, LOD 0.005 μg/mL), embutramide (6.7, LOD 0.005 μg/mL), ethenzamide (5.0, LOD 0.05 μg/mL), ethylmorphine (3.2, LOD 0.05 μg/mL), ethylparathion (9.7, LOD <5 μg/mL), etodroxizine (6.4, LOD <0.02 μg/mL), felodipine (9.6, LOD 0.02 μg/mL), fenazepam (7.5, LOD <0.02 μg/mL), fenfluramine (5.3, LOD <0.02 μg/mL), fenkamfamine (5.1, LOD <0.02 μg/mL), fentanyl (5.5, LOD <0.02 μg/mL), fexofenadine (6.3, LOD <0.02 μg/mL), flecainide (5.9, LOD <0.02 μg/mL), fluconazole (4.0, LOD 0.1 μg/mL), flumazenil (5.2, LOD <0.02 μg/mL), flunitrazepam (7.1, LOD 0.002 μg/mL), fluoxetine (6.8, LOD 0.1 μg/mL), flupentixol (7.5, LOD 0.18 μg/mL), fluvoxamine (6.3, LOD 0.02 μg/mL), glibenclamide (8.5, LOD <0.02 μg/mL), glipizide (6.8, LOD <0.05 μg/mL), haloperidol (6.1, LOD <0.02 μg/mL), histapyrrodine (6.3, LOD 0.02 μg/mL), hydrocodone (3.0, LOD 0.05 μg/mL), hydroxychloroquine (2.4, LOD <0.3 μg/mL), hydroxyzine (6.3, LOD <0.02 μg/mL), imipramine (6.4, LOD 0.05 μg/mL), indomethacin (8.6, LOD 0.05 μg/mL), isoniazid (2.2, LOD 3 μg/mL), isradipine (8.6, LOD 0.05 μg/mL), ketamine (3.6, LOD <0.05 μg/mL), ketobemidone (3.3, LOD <0.05 μg/mL), ketoprofen (7.3, LOD 0.1 μg/mL), ketorolac (6.2, LOD 0.05 μg/mL), labetalol (4.9, LOD 0.05 μg/mL), lamotrigine (4.0, LOD 0.1 μg/mL), levocabastine (5.8, LOD 0.01 μg/mL), levomepromazine (6.5, LOD 0.02 μg/mL), lidocaine (3.7, LOD <0.05 μg/mL), loratadine (9.3, LOD 0.002 μg/mL), lorazepam (6.6, LOD 0.02 μg/mL), lormetazepam (7.4, LOD <0.02 μg/mL), LSD (4.7, LOD <0.02 μg/mL), malathion (8.9, LOD 10 μg/mL), maprotiline (6.4, LOD <0.02 μg/mL), MDMA (3.3, LOD 0.02 μg/mL), meclozine (8.5, LOD <0.02 μg/mL), medazepam (6.3, LOD <0.02 μg/mL), meloxicam (7.1, LOD 0.01 μg/mL), melperone (5.0, LOD <0.02 μg/mL), meperidine (4.7, LOD <0.02 μg/mL), mepivacaine (3.7, LOD <0.02 μg/mL), meprobamate (4.9, LOD 0.1 μg/mL), mesoridazine (5.4, LOD <0.02 μg/mL), methamphetamine (3.3, LOD 0.05 μg/mL), methadone (6.7, LOD <0.02 μg/mL), methylparathion (8.6, LOD 10 μg/mL), methylphenidate (4.2, LOD <0.02 μg/mL), metoclopramide (3.8, LOD <0.02 μg/mL), metoprolol (4.1, LOD 0.02 μg/mL), metronidazole (2.6, LOD 1 μg/mL), mexiletine (4.4, LOD 0.05 μg/mL), mianserin (5.7, LOD <0.02 μg/mL), midazolam (5.9, LOD <0.02 μg/mL), mirtazapine (4.4, LOD <0.02 μg/mL), mizolastine (5.5, LOD 0.01 μg/mL), moclobemide (3.7, LOD 0.05 μg/mL), molindone (4.0, LOD <0.02 μg/mL), monoacetylmorphine (2.7, LOD 0.1 μg/mL), morphine (2.0, LOD 0.1 μg/mL), nicotine (2.2, LOD 0.05 μg/mL), nifedipine (7.5, LOD 0.02 μg/mL), nikethamide (3.6, LOD <0.02 μg/mL), nitrazepam (6.5, LOD <0.02 μg/mL), nizatidine (1.7, LOD 1 μg/mL), nomifensine (4.6, LOD <0.02 μg/mL), nortriptyline (6.4, LOD <0.02 μg/mL), norverapamil (6.2, LOD 1 μg/mL), noscapine (5.0, LOD <0.02 μg/mL), olanzapine (3.0, LOD 0.05 μg/mL), ondansetron (4.6, LOD <0.02 μg/mL), orphenadrine (6.1, LOD <0.02 μg/mL), oxazepam (6.3, LOD

<0.02 μg/mL), oxcarbazepine (5.3, LOD 0.02 μg/mL), oxprenolol (4.7, LOD 0.02 μg/mL), oxycodone (2.8, LOD 0.05 μg/mL), papaverine (4.8, LOD <0.02 μg/mL), acetaminophen (2.5, LOD <5 μg/mL), paroxetine (6.2, LOD 0.02 μg/mL), pemoline (3.3, LOD 0.05 μg/mL), pentazocine (5.0, LOD <0.02 μg/mL), pentifylline (7.3, LOD <5 μg/mL), pentoxyverine (6.6, LOD <0.02 μg/mL), perphenazine (6.9, LOD 0.002 μg/mL), phenazone (3.9, LOD 0.05 μg/mL), phencyclidine (5.3, LOD 0.05 μg/mL), pheniramine (4.1, LOD 0.02 μg/mL), phenylbutazone (9.0, LOD <5 μg/mL), phenylpropanolamine (2.5, LOD 0.3 μg/mL), phenytoin (6.1, LOD 0.05 μg/mL), pindolol (3.3, LOD 0.05 μg/mL), piroxicam (6.6, LOD 0.02 μg/mL), pitofenone (5.4, LOD <0.02 μg/mL), pizotifen (6.5, LOD <0.02 μg/mL), practolol (1.8, LOD 0.1 μg/mL), prazosin (4.1, LOD 0.05 μg/mL), prilocaine (3.8, LOD <0.02 μg/mL), primidone (4.0, LOD <5 μg/mL), procainamide (2.2, LOD 0.05 μg/mL), prochlorperazine (7.5, LOD 0.02 μg/mL), promazine (6.2, LOD <0.02 μg/mL), promethazine (6.0, LOD 0.05 μg/mL), propafenone (6.3, LOD <0.02 μg/mL), propranolol (5.4, LOD 0.02 μg/mL), propyphenazone (6.6, LOD 0.50 μg/mL), pseudoephedrine (2.6, LOD 1 μg/mL), quinine (4.2, LOD 0.02 μg/mL), ranitidine (1.8, LOD 0.1 μg/mL), risperidone (4.9, LOD <0.02 μg/mL), rocurone (3.8, LOD 0.1 μg/mL), ropivacaine (4.6, LOD <0.02 μg/mL), salicylamide (4.2, LOD <5 μg/mL), selegiline (4.1, LOD 0.05 μg/mL), sertindole (7.2, LOD <0.02 μg/mL), sertraline (6.8, LOD 0.02 μg/mL), sulindac (6.5, LOD 0.02 μg/mL), simazine (6.0, LOD 0.1 μg/mL), sincocaine (6.5, LOD <0.02 μg/mL), sisapride (5.9, LOD <0.02 μg/mL), sotalol (2.1, LOD 0.1 μg/mL), strychnine (5.3, LOD 0.05 μg/mL), sulpiride (1.9, LOD 0.1 μg/mL), sulthiame (4.1, LOD 0.05 μg/mL), temazepam (7.2, LOD <0.02 μg/mL), terbutaline (2.3, LOD 0.1 μg/mL), terfenadine (8.1, LOD 0.002 μg/mL), terodiline (6.7, LOD <0.02 μg/mL), tetracaine (5.7, LOD <0.02 μg/mL), dronabinol (12.3, LOD 0.05 μg/mL), tetrahydrozoline (3.6, LOD 0.1 μg/mL), theobromine (2.3, LOD <5 μg/mL), theophylline (2.4, LOD <5 μg/mL), thioridazine (7.5, LOD 0.02 μg/mL), timolol (3.8, LOD 0.05 μg/mL), thiothixene (6.7, LOD 0.02 μg/mL), tolbutamide (7.1, LOD <5 μg/mL), toremifene (8.7, LOD 0.02 μg/mL), tramadol (4.2, LOD 0.02 μg/mL), trazodone (5.2, LOD <0.02 μg/mL), triamterene (3.2, LOD 0.1 μg/mL), triazolam (6.7, LOD 0.002 μg/mL), trimeprazine (6.4, LOD <0.02 μg/mL), trimethoprim (3.1, LOD 0.05 μg/mL), trimipramine (6.7, LOD <0.02 μg/mL), venlafaxine (4.9, LOD 0.02 μg/mL), verapamil (6.5, LOD <0.02 μg/mL), warfarin (7.9, LOD <0.02 μg/mL), yohimbine (4.5, LOD <0.02 μg/mL), zolpidem (4.7, LOD <0.02 μg/mL), zopiclone (4.0, LOD 0.1 μg/mL)

## KEY WORDS

whole blood

## REFERENCE

Gergov, M.; Ojanperä, I.; Vuori, E. Simultaneous screening for 238 drugs in blood by liquid chromatography-ionspray tandem mass spectrometry with multiple-reaction monitoring, *J.Chromatogr.B*, **2003**, *795*, 41–53.

## SAMPLE

**Matrix:** blood

**Sample preparation:** Rotate 2 mL plasma and 5 mL diethyl ether for 30 min, centrifuge, repeat the extraction. Combine the organic layers and evaporate them to dryness under a stream of nitrogen at room temperature, reconstitute the residue with 100 μL MeOH, inject a 50 μL aliquot.

## HPLC VARIABLES

**Guard column:** 15 × 3.2 7 μm silica Newguard

**Column:** 100 × 4.6 5 μm Spheri-5 silica

**Mobile phase:** MeOH:water:acetic acid:MeCN 40:2:10 (sic)

**Flow rate:** 1

**Injection volume:** 50

**Detector:** F ex 322 em 405

## CHROMATOGRAM
**Retention time:** 7.5
**Internal standard:** ergotamine

## OTHER SUBSTANCES
**Extracted:** lisuride (12)

## KEY WORDS
ergotamine is IS in original paper; plasma; protect from light

## REFERENCE
Wolthers, B.G.; Verhagen Kamerbeek, W.D.J.; van Beusekom, C.M.; Elshof, F.; de Ruyter Buitenhuis, A.W.; Brunt, E.P.R.; Lakke, J.P.W.F. Quantitative determination of the dopamine agonist lisuride in plasma using high-performance liquid chromatography with fluorescence detection, *J.Chromatogr.*, **1993**, *622*, 33–38.

---

## SAMPLE
**Matrix:** blood
**Sample preparation:** Vortex 500 μL plasma, 200 μL 2.5 M potassium carbonate, and 3 mL diethyl ether. Remove a 2 mL aliquot of the organic layer and add it to 100 μL 50 mM sulfuric acid, vortex for 1 min, inject a 50 μL aliquot of the aqueous layer.

## HPLC VARIABLES
**Column:** 250 × 4 5 μm Hypersil ODS RP-C18
**Mobile phase:** MeCN:MeOH:water 8:56:36 containing 200 mg (per 500 mL (?)) sodium heptanesulfonate
**Flow rate:** 1
**Injection volume:** 50
**Detector:** UV 254

## CHROMATOGRAM
**Retention time:** 7
**Internal standard:** ergotamine

## OTHER SUBSTANCES
**Extracted:** bromocriptine (13)

## KEY WORDS
ergotamine is IS in original paper; plasma; rabbit

## REFERENCE
Degim, I.T.; Acartürk, F.; Erdogan, D.; Lortlar, N.D. Transdermal administration of bromocriptine, *Biol.Pharm.Bull.*, **2003**, *26*, 501–505.

# Ertapenem

**Molecular formula:** $C_{22}H_{25}N_3O_7S$
**Molecular weight:** 475.52
**CAS Registry No:** 153832-46-3
**Merck Index:** 13, 3706

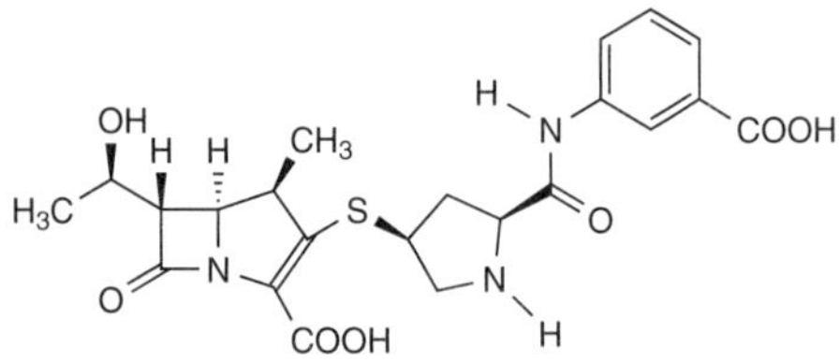


## SAMPLE

**Matrix:** blood
**Sample preparation:** Filter (Amicon Centrifree) 1 mL plasma while centrifuging at 1500 g for 15 min, mix an aliquot of the filtrate with an equal volume of buffer, inject a 50 μL aliquot. (Prepare the buffer by dissolving 15.5 g sodium 2-[*N*-morpholino]ethane sulfonate (MES) and 5.58 g 2-[*N*-morpholino]ethane sulfonic acid in 100 mL water. Dilute this stock solution 10-fold before use, pH 6.5.)

## HPLC VARIABLES

**Column:** 100 × 4.6 5 μm Hypersil C18
**Mobile phase:** MeOH:25 mM phosphate buffer 10:90, apparent pH 6.5
**Flow rate:** 2
**Injection volume:** 50
**Detector:** UV 300

## CHROMATOGRAM

**Retention time:** 5.6
**Limit of quantitation:** 250 ng/mL

## KEY WORDS

pharmacokinetics; plasma; ultrafiltrate

## REFERENCE

Musson, D.G.; Birk, K.L.; Kitchen, C.J.; Zhang, J.; Hsieh, J.Y.K.; Fang, W.; Majumdar, A.K.; Rogers, J.D. Assay methodology for the quantitation of unbound ertapenem, a new carbapenem antibiotic, in human plasma, *J.Chromatogr.B*, **2003**, *783*, 1–9.

## SAMPLE

**Matrix:** blood
**Sample preparation:** Mix 50 μL 50 μg/mL IS with 200 μL serum, add 800 μL MeCN, vortex briefly, centrifuge at 2600 g for 10 min. remove the supernatant and add it to 2.5 mL dichloromethane, vortex briefly, centrifuge, inject an aliquot of the aqueous layer.

## HPLC VARIABLES

**Column:** 100 × 4.6 5 μm C18 (Keystone)
**Mobile phase:** MeOH:25 mM pH 6.5 phosphate buffer 9.5:100
**Flow rate:** 1
**Detector:** UV 300

## CHROMATOGRAM

**Retention time:** 10.5
**Internal standard:** meropenem (14.9)
**Limit of quantitation:** 150 ng/mL

## KEY WORDS

mouse; pharmacokinetics; serum

## REFERENCE

Xuan, D.; Banevicius, M.; Capitano, B.; Kim, M.-K.; Nightingale, C.; Nicolau, D. Pharmacodynamic assessment of ertapenem (MK-0826) against streptococcus pneumoniae in a murine neutropenic thigh infection model, *Antimicrob.Agents Chemother.*, **2002**, *46*, 2990–2995.

## ANNOTATED BIBLIOGRAPHY

Kitchen, C.J.; Musson, D.G.; Fisher, A.L. Column-switching technique for the sensitive determination of ertapenem in human cerebrospinal fluid using liquid chromatography and ultraviolet absorbance detection, *J.Chromatogr.B*, **2004**, *799*, 9–14.

McQuade, M.S.; Van Nostrand, V.; Schariter, J.; Kanike, J.D.; Forsyth, R.J. Stability and compatibility of reconstituted ertapenem with commonly used i.v. infusion and coinfusion solutions, *Am.J.Health-Syst.Pharm.*, **2004**, *61*, 38–45.

Musson, D.G.; Kitchen, C.J.; Hsieh, J.Y.-K.; Birk, K.L. Modified high-performance liquid chromatographic method for the determination of ertapenem in human urine: enhanced selectivity and automation, *J.Chromatogr.B*, **2002**, *779*, 341–346.

Musson, D.G.; Majumdar, A.; Birk, K.; Holland, S.; Wickersham, P.; Li, S.X.; Mistry, G.; Fisher, A.; Waldman, S.; Greenberg, H.; Deutsch, P.; Rogers, J.D. Pharmacokinetics of intramuscularly administered ertapenem, *Antimicrob.Agents Chemother.*, **2003**, *47*, 1732–1735.

Sajonz, P.; Natishan, T.K.; Wu, Y.; Williams, J.M.; Pipik, B.; DiMichele, L.; Novak, T.; Pitzenberger, S.; Dubost, D.; Almarsson, O. Preparation, isolation, and characterization of dimeric degradation products of the 1$\beta$-methylcarbapenem antibiotic, ertapenem, *J.Liq.Chromatogr.Rel.Technol.*, **2001**, *24*, 2999–3015.

# Ethopabate

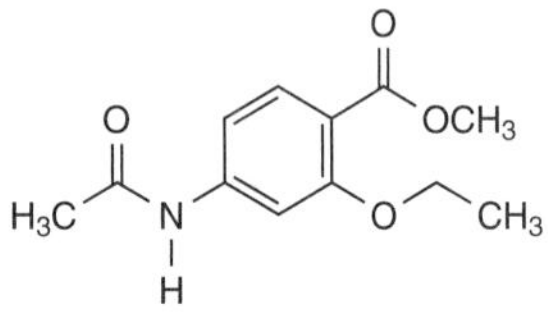

**Molecular formula:** $C_{12}H_{15}NO_4$
**Molecular weight:** 237.25
**CAS Registry No:** 59-06-3
**Merck Index:** 13, 3781

## SAMPLE

**Matrix:** blood
**Sample preparation:** Vortex 100 μL MeOH:water 80:20 and 400 μL trichloroacetic acid solution with 500 μL plasma, centrifuge at 4000 rpm for 4 min, filter (Spin-X) the supernatant. Mix the filtrate with an equal quantity of water and inject a 30 μL aliquot. (Prepare trichloroacetic acid solution as follows. Dissolve 85 g trichloroacetic acid in 15 mL water. Store this solution in the refrigerator. Dilute 150 μL of this solution with 100 mL acetone.)

## HPLC VARIABLES

**Column:** 250 × 4.6 5 μm Supelcosil LC-ABZ + Plus
**Mobile phase:** MeCN:10 mM ammonium acetate 35:65
**Flow rate:** 1
**Injection volume:** 30
**Detector:** MS, PE Sciex API 100 single quadrupole, turbo ionspray, 50 μL/min flowed into detector, m/z 238.2

## CHROMATOGRAM

**Retention time:** 7.5
**Limit of quantitation:** 15 ng/mL

## KEY WORDS

chicken; plasma

## REFERENCE

Hormazábal, V.; Yndestad, M. Determination of amprolium, ethopabate, lasalocid, monensin, narasin, and salinomycin in chicken tissues, plasma, and egg using liquid chromatography-mass spectrometry, *J.Liq.Chromatogr.Rel.Technol.*, **2000**, *23*, 1585–1598.

# Ethyl icosapentate

**Molecular formula:** $C_{22}H_{34}O_2$
**Molecular weight:** 330.51
**CAS Registry No:** 73310-10-8

## SAMPLE
**Matrix:** solutions
**Sample preparation:** Inject a 2 μL aliquot.

## HPLC VARIABLES
**Column:** 125 × 4 4 μm Supersphere Si 60
**Column temperature:** 27
**Mobile phase:** *n*-Hexane:diethyl ether 98:2
**Flow rate:** 1.8
**Injection volume:** 2
**Detector:** UV 213

## CHROMATOGRAM
**Retention time:** 2
**Internal standard:** 2,4-dinitrochlorobenzene (6)

## KEY WORDS
normal phase

## REFERENCE
Teraoka, R.; Otsuka, M.; Matsuda, Y. Chemical stability of ethyl icosapentate against autoxidation. I. Effect of temperature on oxidation kinetics, *Pharm.Res.*, **1992**, *9*, 1673–1676.

# Etonogestrel

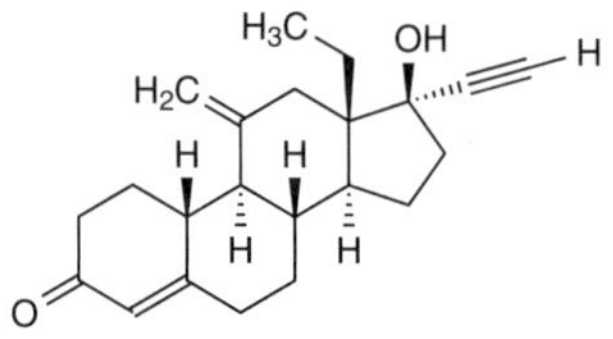


**Molecular formula:** $C_{22}H_{28}O_2$
**Molecular weight:** 324.46
**CAS Registry No:** 54048-10-1
**Merck Index:** 13, 3916

## SAMPLE
**Matrix:** cell culture
**Sample preparation:** Extract medium twice with 2 mL diethyl ether. Evaporate the extracts to dryness, reconstitute the residue with MeOH, inject an aliquot.

## HPLC VARIABLES
**Column:** 250 × 4.6 Ultratech 5-ODS (HPLC Technology)
**Mobile phase:** MeCN:MeOH:0.5% pH 3.0 ammonium dihydrogen phosphate 15:47:38
**Flow rate:** 1.2
**Detector:** UV 214

## CHROMATOGRAM
**Retention time:** 13

## OTHER SUBSTANCES
**Extracted:** metabolites

## REFERENCE
Wild, M.J.; Rudland, P.S.; Back, D.J. Metabolism of the oral contraceptive steroids ethynylestradiol, norgestimate and 3-ketodesogestrel by a human endometrial cancer cell line (HEC-1A) and endometrial tissue in vitro, *J.Steroid Biochem.Mol.Biol.*, **1993**, *45*, 407–420.

## SAMPLE
**Matrix:** microsomal incubations

## HPLC VARIABLES
**Guard column:** Novapak C18
**Column:** 150 × 3.9 Novapak C18
**Column temperature:** 50
**Mobile phase:** Gradient. MeCN:MeOH:water from 5:20:75 to 8:32:60 over 5 min, to 14:56:30 over 10 min, to 18:72:10 over 2 min.
**Flow rate:** 1
**Detector:** Radioactivity ($^3H$); UV 254

## CHROMATOGRAM
**Retention time:** 16

## OTHER SUBSTANCES
**Extracted:** metabolites

## KEY WORDS
dog; human; liver; rabbit, rat

## REFERENCE
Verhoeven, C.H.J.; Krebbers, S.F.M.; Wagenaars, G.N.; Vos, R.M.E. In vitro and in vivo metabolism of desogestrel in several species, *Drug Metab.Dispos.*, **1998**, *26*, 927–936.

## SAMPLE
**Matrix:** solutions
**Sample preparation:** Extract 15 mL aqueous solution with dichloromethane, evaporate the extract to dryness, take up the residue in 3 mL mobile phase, inject a 50 μL aliquot.

## HPLC VARIABLES
**Column:** C18
**Mobile phase:** MeOH:water 82:18
**Injection volume:** 50
**Detector:** UV 242

## CHROMATOGRAM
**Internal standard:** progesterone
**Limit of quantitation:** 50 ng/mL

## OTHER SUBSTANCES
**Simultaneous:** ethinyl estradiol (F ex 200 em 300), mestranol (F ex 200 em 300)

## REFERENCE
de Leede, L.G.; Govers, C.P.M.; de Nijs, H. A multi-compartment vaginal ring system for independently adjustable release of contraceptive steroids, *Contraception*, **1986**, *34*, 589–602.

## SAMPLE
**Matrix:** solutions

## HPLC VARIABLES
**Column:** 150 × 3.9 Novapak C18
**Column temperature:** 30
**Mobile phase:** MeCN:water 30:70
**Flow rate:** 1.5
**Injection volume:** 10
**Detector:** UV 205

## OTHER SUBSTANCES
**Simultaneous:** ethinyl estradiol

## REFERENCE
van Laarhoven, J.A.H.; Kruft, M.A.B.; Vromans, H. In vitro release properties of etonogestrel and ethinyl estradiol from a contraceptive vaginal ring, *Int.J.Pharm.*, **2002**, *232*, 163–173.

# Etoricoxib

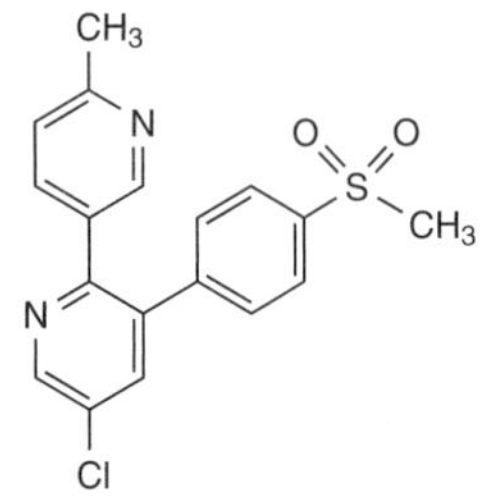

**Molecular formula:** $C_{18}H_{15}ClN_2O_2S$
**Molecular weight:** 358.84
**CAS Registry No:** 202409-33-4

## SAMPLE

**Matrix:** blood

**Sample preparation:** Condition a 1 mL 30 mg Oasis HLB SPE cartridge with 1 mL MeOH and 1 mL water. Mix 200 μL plasma with 20 μL 500 ng/mL IS in water, centrifuge at 12 000 g, add to the SPE cartridge, wash with 1 mL MeOH:water 5:95, dry under reduced pressure for 5 min, elute with 2 mL MeCN:ethyl acetate 50:50. Evaporate the eluate to dryness under reduced pressure at 45°, reconstitute the residue with 200 μL mobile phase, inject a 15 μL aliquot.

## HPLC VARIABLES

**Column:** 30 × 2 5 μm Nucleodur C18
**Mobile phase:** MeCN:water 90:10
**Flow rate:** 0.3
**Injection volume:** 15
**Detector:** MS, PE Sciex API 3000 triple quadrupole, turbo ionspray, positive ion mode 5400 V and 400°, nebulizer gas nitrogen at 1.49 L/min, curtain gas nitrogen at 1.25 L/min, collision gas nitrogen, collision energy 43 eV, m/z 359.0–280.1

## CHROMATOGRAM

**Retention time:** 0.5
**Internal standard:** phenazone (m/z 189.0–104.0, 33 eV) (0.5)
**Limit of quantitation:** 0.2 ng/mL

## KEY WORDS

pharmacokinetics; plasma

## REFERENCE

Bräutigam, L.; Nefflen, J.U.; Geisslinger, G. Determination of etoricoxib in human plasma by liquid chromatography-tandem mass spectrometry with electrospray ionisation, *J.Chromatogr.B*, **2003**, *788*, 309–315.

## SAMPLE

**Matrix:** solutions

**Sample preparation:** Inject a 10 μL aliquot of a 100 μg/mL solution in MeCN:water 50:50.

## HPLC VARIABLES

**Column:** 150 × 4.6 3 μm YMC AQ-ODS
**Column temperature:** 35
**Mobile phase:** Gradient. MeCN:buffer 28:72 for 11 min, to 70:30 over 19 min, to 90:10 over 5 min. (The buffer was 10 mM potassium dihydrogen phosphate adjusted to pH 3.1 with 2.2 mM phosphoric acid.)
**Flow rate:** 1
**Injection volume:** 10
**Detector:** UV 220

## CHROMATOGRAM

**Retention time:** 12
**Limit of detection:** 0.02%
**Limit of quantitation:** 0.04%

## OTHER SUBSTANCES

**Simultaneous:** degradants, impurities

## KEY WORDS

stability-indicating; robust

## REFERENCE

Hartman, R.; Abrahim, A.; Clausen, A.; Mao, B.; Crocker, L.S.; Ge, Z. Development and validation of an HPLC method for the impurity and quantitative analysis of etoricoxib, *J.Liq.Chromatogr.Rel.Technol.*, **2003**, *26*, 2551–2566.

## ANNOTATED BIBLIOGRAPHY

Kassahun, K.; McIntosh, I.S.; Shou, M.; Walsh, D.J.; Rodeheffer, C.; Slaughter, D.E.; Geer, L.A.; Halpin, R.A.; Agrawal, N.; Rodrigues, A.D. Role of human liver cytochrome P4503A in the metabolism of etoricoxib, a novel cyclooxygenase-2 selective inhibitor, *Drug Metab.Dispos.*, **2001**, *29*, 813–820.

Niederberger, E.; Tegeder, I.; Schäfer, C.; Seegel, M.; Grösch, S.; Geisslinger, G. Opposite effects of rofecoxib on nuclear factor-kappaB and activating protein-1 activation, *J.Pharmacol.Exp.Ther.*, **2003**, *304*, 1153–1160. [etoricoxib is internal standard; post-column photochemical derivatization]

Rodrigues, A.D.; Halpin, R.A.; Geer, L.A.; Cui, D.; Woolf, E.J.; Matthews, C.Z.; Gottesdiener, K.M.; Larson, P.J.; Lasseter, K.C.; Agrawal, N.G.B. Absorption, metabolism, and excretion of etoricoxib, a potent and selective cyclooxygenase-2 inhibitor, in healthy male volunteers, *Drug Metab.Dispos.*, **2003**, *31*, 224–232.

Rose, M.J.; Agrawal, N.; Woolf, E.J.; Matuszewski, B.K. Simultaneous determination of unlabeled and carbon-13-labeled etoricoxib, a new cyclooxygenase-2 inhibitor, in human plasma using HPLC-MS/MS, *J.Pharm.Sci.*, **2002**, *91*, 405–416. [SPE]

# Etorphine

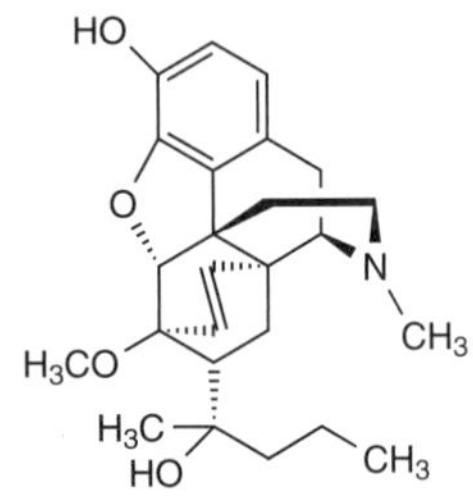


**Molecular formula:** $C_{25}H_{33}NO_4$
**Molecular weight:** 411.53
**CAS Registry No:** 14521-96-1
**Merck Index:** 13, 3919

## SAMPLE
**Matrix:** blood, urine
**Sample preparation:** Add 3 mL 200 mM sodium carbonate solution containing 100 ng/mL IS and 3 mL 1-chlorobutane to 3 mL blood or urine, shake for 3 min, centrifuge at 4500 rpm for 5 min, repeat the extraction. Combine the organic layers and add them to 100 μL 50 mM sulfuric acid, shake for 3 min, centrifuge at 3500 rpm for 3 min, inject a 20 μL aliquot of the aqueous layer.

## HPLC VARIABLES
**Guard column:** 10 × 4 Spherisorb S5ODS2
**Column:** 150 × 4.6 Spherisorb S50D/CN
**Mobile phase:** MeCN:water 20:80
**Flow rate:** 2
**Injection volume:** 20
**Detector:** UV 220

## CHROMATOGRAM
**Retention time:** 3.77
**Internal standard:** pentazocine (6.31)
**Limit of quantitation:** 2 ng/mL

## KEY WORDS
whole blood

## REFERENCE
Elliott, S.P.; Hale, K.A. Analysis of etorphine in postmortem samples by HPLC with UV diode-array detection, *Forensic Sci.Int.*, **1999**, *101*, 9–16.

# Exemestane

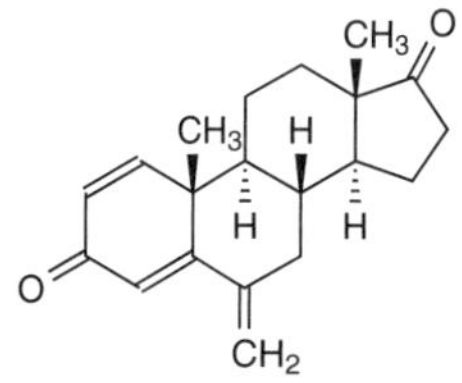


**Molecular formula:** $C_{20}H_{24}O_2$
**Molecular weight:** 296.40
**CAS Registry No:** 107868-30-4
**Merck Index:** 13, 3944

## SAMPLE

**Matrix:** blood
**Sample preparation:** Mix 1 mL plasma with 100 μL 1 μg/mL IS in water, add 600 μL 500 mM pH 7.4 potassium phosphate buffer, mix, add 3 mL dichloromethane:isooctane 40:60, shake for 10 min, centrifuge at 1200 g for 15 min, repeat the extraction. Combine the organic layers and evaporate them to dryness under a stream of nitrogen at 37°, reconstitute the residue with 200 μL MeCN:water 50:50, inject a 150 μL aliquot.

## HPLC VARIABLES

**Guard column:** 37–53 μm pellicular ODS
**Column:** 125 × 4.6 5 μm Lichrocart RP18
**Mobile phase:** MeCN:50 mM pH 4.5 potassium phosphate buffer 35:65
**Flow rate:** 1.5
**Injection volume:** 150
**Detector:** UV 247

## CHROMATOGRAM

**Retention time:** 14
**Internal standard:** norgestrel (17)
**Limit of detection:** 2 ng
**Limit of quantitation:** 10 ng/mL

## OTHER SUBSTANCES

**Extracted:** metabolite

## KEY WORDS

pharmacokinetics; plasma

## REFERENCE

Breda, M.; Pianezzola, E.; Strolin Benedetti, M.S. Determination of exemestane, a new aromatase inhibitor, in plasma by high-performance liquid chromatography with ultraviolet detection, *J.Chromatogr.*, **1993**, *620*, 225–231.

## SAMPLE

**Matrix:** blood
**Sample preparation:** Condition a 2 mL 50 mg C2 end-capped SPE cartridge (in a 96 well plate format) with two 1 mL portions of MeCN and two 1 mL portions of water. Vortex 500 μL plasma with 50 μL 1.11 μg/mL IS in MeOH:water 50:50 and 500 μL water, add to the SPE cartridge, wash with 1 mL MeCN:water 10:90, dry under vacuum for 30 min, elute with two 150 μL portions of MeCN containing 0.1% trifluoroacetic acid. Centrifuge the eluate at 1500 rpm for 2 min and inject an 80 μL aliquot of the supernatant.

## HPLC VARIABLES

**Guard column:** 15 × 3.2 7 μm New Guard RP8 Aquapore Octyl
**Column:** 150 × 4.6 5 μm Zorbax SB C8
**Column temperature:** 45

**Mobile phase:** MeCN
**Flow rate:** 1
**Injection volume:** 80
**Detector:** MS, PE Sciex API 300 triple quadrupole, APCI, nebulizer probe 375°, collision energy 30 eV, m/z 297–121

## CHROMATOGRAM
**Retention time:** 2.3
**Internal standard:** $^{13}C_3$-exemestane (m/z 300–123) (2.3)
**Limit of quantitation:** 50 pg/mL

## KEY WORDS
plasma; SPE

## REFERENCE
Cenacchi, V.; Barattè, S.; Cicioni, P.; Frigerio, E.; Long, J.; James, C. LC-MS-MS determination of exemestane in human plasma with heated nebulizer interface following solid-phase extraction in the 96 well plate format, *J.Pharm.Biomed.Anal.*, **2000**, *22*, 451–460.

## ANNOTATED BIBLIOGRAPHY
Del Nero, S.; Di Somma, M.; Vigevani, A. High-performance liquid chromatographic analysis of FCE 24304 (6-methylenandrosta-1,4-diene-3,17-dione) and FCE 24928 (4-aminoandrosta-1,4,6-triene-3,17-dione), two new aromatase inhibitors, *J.Chromatogr.*, **1992**, *593*, 25–28.

# Ezetimibe

**Molecular formula:** $C_{24}H_{21}F_2NO_3$
**Molecular weight:** 409.42
**CAS Registry No:** 163222-33-1
**Merck Index:** 13, 3949

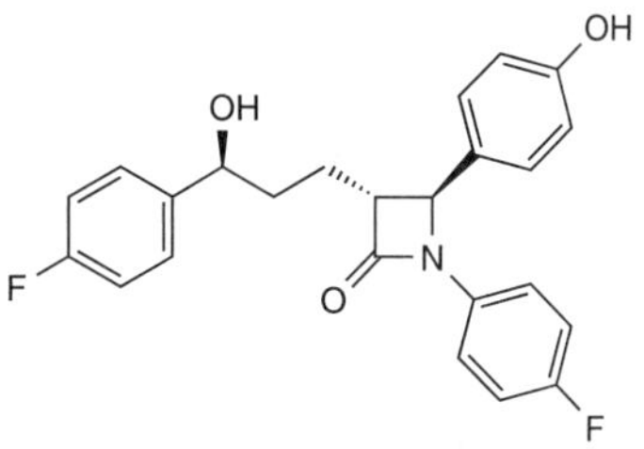

### SAMPLE
**Matrix:** bile
**Sample preparation:** Extract bile with 1.5 vol of MeCN.

### HPLC VARIABLES
**Column:** 150 × 4.6 5 μm Inertsil C8
**Mobile phase:** Gradient. MeCN:100 mM pH 6 ammonium acetate from 30:70 to 100:0 over 40 min (concave gradient (Waters Expert-Ease Curve #10)).
**Flow rate:** 1
**Detector:** UV 245; Radioactivity ($^3$H)

### CHROMATOGRAM
**Retention time:** 40

### OTHER SUBSTANCES
**Extracted:** metabolites

### KEY WORDS
rat

### REFERENCE
Van Heek, M.; Farley, C.; Compton, D.S.; Hoos, L.; Alton, K.B.; Sybertz, E.J.; Davis, H.R. Jr. Comparison of the activity and disposition of the novel cholesterol absorption inhibitor, SCH58235, and its glucuronide, SCH60663. *Br.J.Pharmacol.*, **2000**, *129*, 1748–1754.

### SAMPLE
**Matrix:** blood
**Sample preparation:** For unconjugated ezetimibe, add 100 μL 513 ng/mL IS in water and 1 mL water to 200 μL plasma, add 8 mL 1-chlorobutane. For total ezetimibe, add 100 μL 513 ng/mL IS in water, 500 μL 500 mM pH 5.0 sodium acetate buffer, and 50 μL β-glucuronidase (100 000 U/mL) to 200 μL plasma, heat at 50° for 1 h, add 500 μL 1 M sodium borate solution, add 8 mL 1-chlorobutane. Shake each mixture for 15 min and centrifuge at 491 g for 10 min. Evaporate the organic layer to dryness under reduced pressure, reconstitute the residue with 500 μL MeOH, evaporate to dryness, reconstitute with 50 μL MeOH, inject an aliquot.

### HPLC VARIABLES
**Column:** 100 × 4.6 10 μm Spherisorb ODS-2
**Mobile phase:** MeOH:25 mM ammonium acetate 90:10
**Flow rate:** 1.5
**Detector:** MS, PE Sciex API-III, positive ion mode, m/z 392.3–133.1

## CHROMATOGRAM

**Internal standard:** SCH 58053 ((+)-7-(4-chlorophenyl)-2-(4-fluorophenyl)-7-hydroxy-3(*R*)-4-hydroxyphenyl)-2-azaspiro[3.5]nonan-1-one (enantiomer A)) (m/z 434.2–216.1)
**Limit of quantitation:** 1 ng/mL (unconjugated), 5.02 ng/mL (total)

## KEY WORDS

pharmacokinetics; plasma

## REFERENCE

Patrick, J.E.; Kosoglou, T.; Stauber, K.L.; Alton, K.B.; Maxwell, S.E.; Zhu, Y.; Statkevich, P.; Iannucci, R.; Chowdhury, S.; Affrime, M.; Cayen, M.N. Disposition of the selective cholesterol absorption inhibitor ezetimibe in healthy male subjects, *Drug Metab.Dispos.*, **2002**, *30*, 430–437.

# Fadrozole

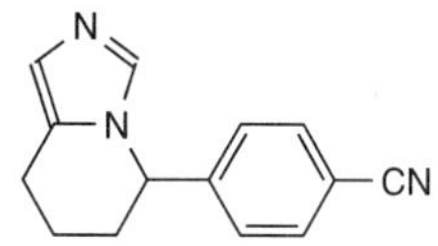


**Molecular formula:** $C_{14}H_{13}N_3$
**Molecular weight:** 223.27
**CAS Registry No:** 102676-47-1
**Merck Index:** 13, 3958

## SAMPLE
**Matrix:** solutions

## HPLC VARIABLES
**Column:** Chiralcel OD
**Mobile phase:** Hexane:isopropanol 70:30
**Detector:** UV

## CHROMATOGRAM
**Retention time:** (+)-enantiomer elutes first, $\alpha = 1.40$

## KEY WORDS
chiral

## REFERENCE
Furet, P.; Batzl, C.; Bhatnagar, A.; Francotte, E.; Rihs, G.; Lang, M. Aromatase inhibitors: synthesis, biological activity, and binding mode of azole-type compounds, *J.Med.Chem.*, **1993**, *36*, 1393–1400.

# Falecalcitriol

**Molecular formula:** $C_{27}H_{38}F_6O_3$
**Molecular weight:** 524.58
**CAS Registry No:** 83805-11-2

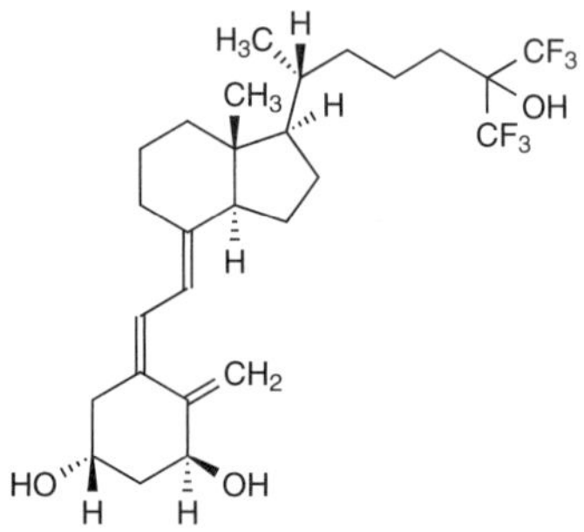


## SAMPLE

**Matrix:** cell suspensions
**Sample preparation:** Condition a Sep-Pak silica SPE cartridge with *n*-hexane:isopropanol 80:20. Sonicate 2 (?) mL cell suspension at 0° for 2 min, add 1 mL THF, add 4 mL ethyl acetate, vortex, centrifuge at 2200 rpm at 4° for 5 min, extract the aqueous layer four times with 4 mL portions of ethyl acetate. Dry the combined organic layers over anhydrous sodium sulfate, evaporate to dryness, reconstitute with *n*-hexane:isopropanol 96:4, pass through the SPE cartridge. Evaporate the eluate to dryness under a stream of nitrogen, reconstitute the residue with 200 μL toluene:EtOH 50:50, inject an aliquot.

## HPLC VARIABLES

**Column:** 150 × 6 Sumipax ODS A212
**Mobile phase:** MeCN:THF:water 45:5:50
**Flow rate:** 1.5
**Detector:** Radioactivity ($^3$H)

## CHROMATOGRAM

**Retention time:** 60

## OTHER SUBSTANCES

**Extracted:** metabolites

## KEY WORDS

SPE

## REFERENCE

Miyahara, T.; Gomyo, S.; Ueda, Y.; Ohyama, Y.; Sigeno, C.; Kozakai, A.; Takamura, T.; Yamazaki, R.; Higuchi, S.; Yamamoto, M.; Sakuma, T.; Nemoto, N. Metabolism of 26,27-hexafluoro-1$\alpha$,25-dihydroxyvitamin $D_3$ and 26,27-hexafluoro-1$\alpha$,23(*S*)25–trihydroxyvitamin $D_3$ in ROS17/2.8 cells transfected with a plasmid expressing CYP24, *Xenobiotica*, **2000**, *30*, 1055–1062.

## SAMPLE

**Matrix:** enzyme incubations
**Sample preparation:** Extract enzyme incubation with 3 vol dichloromethane, evaporate the extract to dryness, dissolve the residue in MeCN, inject an aliquot.

## HPLC VARIABLES

**Column:** 300 × 4 μBondapak C18
**Column temperature:** 50
**Mobile phase:** Gradient. MeCN:water from 50:50 to 100:0 over 15 (?) min.
**Flow rate:** 1
**Detector:** UV 265

## CHROMATOGRAM
**Retention time:** 12

## OTHER SUBSTANCES
**Extracted:** metabolites

## REFERENCE
Hayashi, K.; Akiyoshi-Shibata, M.; Sakaki, T.; Yabusaki, Y. Rat CYP24 catalyses 23S-hydroxylation of 26,26,26,27,27,27-hexafluorocalcitriol in vitro, *Xenobiotica*, **1998**, *28*, 457–463.

## SAMPLE
**Matrix:** solutions

## HPLC VARIABLES
**Column:** 250 × 4.6 Zorbax Sil
**Mobile phase:** *n*-Hexane:dichloromethane:methanol 49:48:3
**Flow rate:** 1.5
**Detector:** UV 254

## CHROMATOGRAM
**Retention time:** 12

## OTHER SUBSTANCES
**Simultaneous:** metabolites

## KEY WORDS
normal phase

## REFERENCE
Imanishi, Y.; Inaba, M.; Seki, H.; Koyama, H.; Nishizawa, Y.; Morii, H.; Otani, S. Increased biological potency of hexafluorinated analogs of 1,25-dihydroxyvitamin $D_3$ on bovine parathyroid cells, *J.Steroid Biochem.Mol.Biol.*, **1999**, *70*, 243–248.

## ANNOTATED BIBLIOGRAPHY
Harada, M.; Miyahara, T.; Kajita-Kondo, S.; Kozakai, A.; Higuchi, S.; Otomo, S.; Kozuka, H. Differences in metabolism between 26,26,26,27,27,27-hexafluoro-1$\alpha$,25-dihydroxyvitamin $D_3$ and 1$\alpha$,25-dihydroxyvitamin $D_3$ in cultured neonatal mouse calvaria, *Res.Commun.Mol.Pathol.Pharmacol.*, **1994**, *86*, 183–193.

Miyahara, T.; Harada, M.; Kozakai, A.; Matsumoto, M.; Hashimoto, K.; Inoue, H.; Yoda, K.; Nakatsu, T.; Kajita, S.; Yamazaki, R.; Higuchi, S.; Kozuka, H.; Nemoto, N. Comparison of 26,27-hexafluoro-1$\alpha$,25-dihydroxyvitamin $D_3$ and 1$\alpha$,25-dihydroxyvitamin $D_3$ on the resorption of bone explants ex vivo. *J.Nutr.Sci.Vitaminol.*, **1999**, *45*, 239–249.

# Fenoxycarb

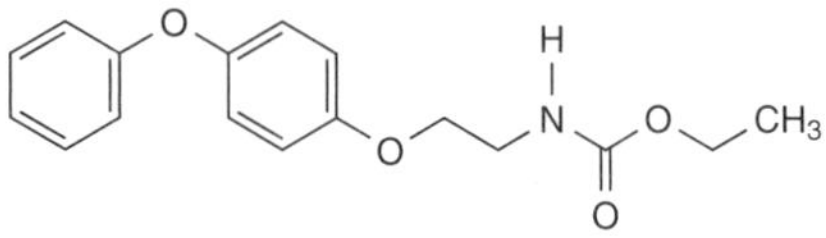

**Molecular formula:** $C_{17}H_{19}NO_4$
**Molecular weight:** 301.34
**CAS Registry No:** 72490-01-8
**Merck Index:** 13, 4013

## SAMPLE

**Matrix:** fruit, vegetables
**Sample preparation:** Blend 500 mg chopped fruit or vegetable with 500 mg 45–55 μm C8 reversed-phase material in a mortar and pestle for 5 min, place in a 100 × 9 glass column, elute with 10 mL MeCN:dichloromethane 40:60. Concentrate the eluate to 500 μL under a stream of nitrogen, inject a 5 μL aliquot.

## HPLC VARIABLES

**Guard column:** 10 × 4.6 5 μm LiChrosorb RP-8
**Column:** 150 × 4.6 3 μm Spherisorb C8
**Mobile phase:** Gradient. MeOH:water 50:50 for 15 min, to 70:30 over 5 min, maintain at 70:30 for 5 min, to 90:10 over 5 min, maintain at 90:10 for 5 min.
**Flow rate:** 0.5
**Injection volume:** 5
**Detector:** MS, HP, electrospray, positive mode, gas 350°, drying gas flow 13 L/min, nebulizer gas pressure 30 psi, capillary 4000 V, m/z 302.1

## CHROMATOGRAM

**Retention time:** 30
**Limit of detection:** 5 ng/g

## OTHER SUBSTANCES

**Extracted:** carbaryl (19), carbofuran (17), diethofencarb (24), ethiofencarb (20), fenobucarb (24), isoprocarb (21), methiocarb (26), metholcarb (13.5), oxamyl (4.5), pirimicarb (22), propoxur (16), thiobencarb (32.5)

## KEY WORDS

matrix solid phase dispersion

## REFERENCE

Fernández, M.; Picó, Y.; Mañes, J. Determination of carbamate residues in fruits and vegetables by matrix solid-phase dispersion and liquid chromatography-mass spectrometry, *J.Chromatogr.A*, **2000**, *871*, 43–56.

## ANNOTATED BIBLIOGRAPHY

Fernández, M.; Picó, Y.; Mañes, J. Simultaneous determination of carbamate and organophosphorus pesticides in honeybees by liquid chromatography-mass spectrometry, *Chromatographia*, **2003**, *58*, 151–158. [very similar to above method]

# Fenticonazole

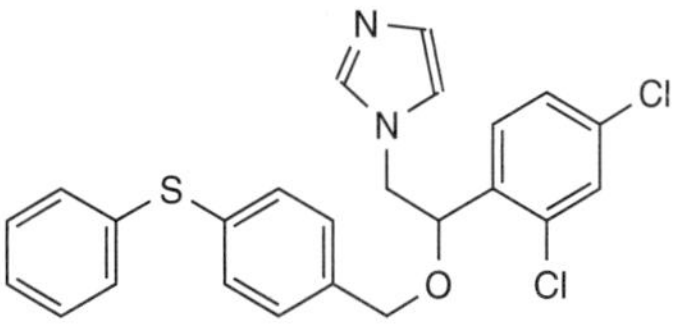


**Molecular formula:** $C_{24}H_{20}Cl_2N_2OS$
**Molecular weight:** 455.41
**CAS Registry No:** 72479-26-6
**Merck Index:** 13, 4033

## SAMPLE
**Matrix:** blood
**Sample preparation:** Vortex 50 μL plasma with 100 μL 2 ng/mL IS in MeCN, centrifuge at 12 000 rpm for 2 min, inject a 20 μL aliquot of the supernatant.

## HPLC VARIABLES
**Guard column:** 15 × 3.2 Brownlee C18
**Column:** 50 × 4.6 3 μm Shandon BDS C8
**Column temperature:** 45
**Mobile phase:** MeCN:buffer 70:30 (The buffer was 10 mM ammonium formate adjusted to pH 3.5 with formic acid.)
**Flow rate:** 1
**Injection volume:** 20
**Detector:** MS, PE Sciex API-III+ triple quadrupole, APCI, positive ion, heated nebulizer interface, nebulizer probe 500°, nebulizer gas nitrogen 0.6 L/min, curtain gas nitrogen at 0.8 L/min, auxiliary gas nitrogen 2.0 L/min, nebulizer gas pressure 70 psi, m/z 455–199

## CHROMATOGRAM
**Retention time:** 2.25
**Internal standard:** econazole (m/z 381–125) (1.5)
**Limit of quantitation:** 0.49 ng/mL

## OTHER SUBSTANCES
**Simultaneous:** clindamycin (1)

## KEY WORDS
plasma

## REFERENCE
Speed, W.; Long, J.M.; Simmonds, R.J.; Enos, T.A. The development and validation of a high performance liquid chromatography (HPLC)/tandem mass spectrometry assay for fenticonazole in human plasma and comparison with an HPLC-UV method, *Rapid Commun.Mass Spectrom.*, **1995**, *9*, 1452–1456.

## SAMPLE
**Matrix:** blood
**Sample preparation:** Liquid–liquid extraction of 1 mL basified plasma (not otherwise specified), inject an 80 μL aliquot of the supernatant.

## HPLC VARIABLES
**Guard column:** Guard Pak μBondapak C18
**Column:** 150 × 4.6 4 μm phenyl
**Column temperature:** 40
**Mobile phase:** MeCN:buffer 76:24 (The buffer was 120 mM acetic acid containing 0.02 mM sodium bisulfate.)
**Flow rate:** 1.5
**Injection volume:** 80
**Detector:** UV 254

## CHROMATOGRAM

**Retention time:** 12.9
**Internal standard:** miconazole (9.9)
**Limit of quantitation:** 5 ng/mL

## OTHER SUBSTANCES

**Simultaneous:** clindamycin (1)

## KEY WORDS

plasma

## REFERENCE

Speed, W.; Long, J.M.; Simmonds, R.J.; Enos, T.A. The development and validation of a high performance liquid chromatography (HPLC)/tandem mass spectrometry assay for fenticonazole in human plasma and comparison with an HPLC-UV method, *Rapid Commun.Mass Spectrom.*, **1995**, *9*, 1452–1456.

# Fexofenadine

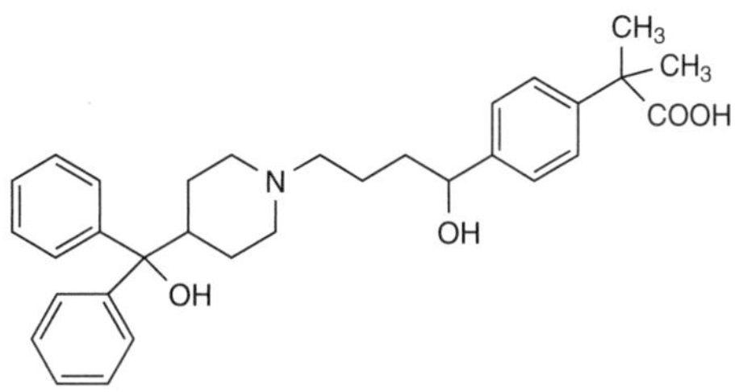

**Molecular formula:** $C_{32}H_{39}NO_4$
**Molecular weight:** 501.65
**CAS Registry No:** 83799-24-0
**Merck Index:** 13, 4101

## SAMPLE

**Matrix:** blood
**Sample preparation:** Mix 1 mL whole blood with 20 μL 1 μg/mL dibenzepin in MeOH:water 50:50, add 300 μL pH 11 Tris buffer, mix, add 500 μL butyl acetate, vortex for 2 min, centrifuge. Preserve the aqueous layer (A). Remove the organic layer and add it to 75 μL 10 mM ammonium acetate buffer containing 0.1% formic acid (pH 3.2), evaporate (?). Add 75 μL MeCN, sonicate for 5 min, centrifuge at 5000 rpm for 5 min, keep the extract as B. Add 20 μL 1 μg/mL enalapril in MeOH:water 50:50 and 120 mg NaCl to the aqueous layer (A), mix, add 500 μL pH 3 phosphate buffer, add 600 μL 8.5% phosphoric acid, add 5 mL dichloromethane:isopropanol 95:5, shake at 250 cycles/min in a bench-top shaker for 30 min, centrifuge at 5000 rpm for 5 min. Remove the lower organic layer and evaporate it to dryness under a stream of air at 45°. Reconstitute the residue with 150 μL initial mobile phase, sonicate, centrifuge, combine with extract B, inject a 30 μL aliquot. (Sample preparation from Gergov,M.; Robson,J.N.; Ojanperä,I.; Heinonen,O.P.; Vuori,E. Simultaneous screening and quantitation of 18 antihistamine drugs in blood by liquid chromatography ionspray tandem mass spectrometry. *Forensic Sci.Inter.* **2001**, *121*, 108–115.)

## HPLC VARIABLES

**Guard column:** 40 mm long 4 μm Purospher RP-18 LiChro Cart 4-4
**Column:** 100 × 2.1 4 μm Genesis C18 (Jones Chromatography)
**Column temperature:** 35
**Mobile phase:** Gradient. MeCN:buffer from 20:80 to 100:0 over 10 min, maintain at 0:100 for 3 min, re-equilibrate at initial conditions for 5 min. (The buffer was 10 mM ammonium acetate containing 0.1% formic acid (pH 3.2).)
**Flow rate:** 0.2
**Injection volume:** 30
**Detector:** MS, PE Sciex API 365 triple-stage quadrupole LC-MS-MS, PE Sciex Turbo Ion Spray interface, positive ion mode, needle voltage 5.2 kV, nebulizer gas air at 60 psi, curtain gas nitrogen at 40 psi, collision cell gas nitrogen at 40 psi, turbo ionspray heater 375°, heater gas flow 7 L/min

## CHROMATOGRAM

**Retention time:** 6.3
**Internal standard:** dibenzepin, enalapril
**Limit of detection:** <20 ng/mL

## OTHER SUBSTANCES

**Extracted:** acebutolol (3.8, LOD 0.1 μg/mL), acrivastine (5.7, LOD <0.02 μg/mL), alprazolam (6.1, LOD <0.02 μg/mL), alprenolol (5.4, LOD 0.01 μg/mL), amantadine (3.4, LOD 0.1 μg/mL), amiloride (2.0, LOD 0.1 μg/mL), aminophenazone (2.8, LOD <5 μg/mL), amiodarone (10.2, LOD 0.05 μg/mL), amitriptyline (6.6, LOD <0.02 μg/mL), astemizole (5.8, LOD <0.02 μg/mL), atenolol (1.7, LOD 0.30 μg/mL), azacyclonol (5.1, LOD 0.02 μg/mL), benzhexol (6.6, LOD <0.02 μg/mL), benzoylecgonine (3.3, LOD 0.01 μg/mL), betaxolol (5.5, LOD 0.01 μg/mL), biperidine (6.2, LOD <0.02 μg/mL), bisoprolol (5.0, LOD <0.02 μg/mL), brompheniramine (5.3, LOD 0.002 μg/mL), bupivacaine (5.1, LOD <0.02 μg/mL), buprenorphine (5.9, LOD 0.01 μg/mL), buspirone (5.1, LOD 0.002 μg/mL), caffeine (2.8, LOD 1 μg/mL), carbamazepine (6.1, LOD <0.02 μg/mL), carbinoxamine (5.1, LOD 0.002 μg/mL), carisoprodol (6.7, LOD <5 μg/mL), carvedilol (6.2, LOD <0.02 μg/mL), celiprolol (4.3, LOD 0.05 μg/mL), cetirizine (6.3, LOD 0.05 μg/mL), chlorcyclizine (6.6, LOD <0.02 μg/mL), chlordiazepoxide (5.7, LOD <0.02 μg/mL), chlormezanone (5.8, LOD <5 μg/mL), chloroquine (2.7, LOD 0.02 μg/mL), chlorpheniramine (5.1, LOD 0.002 μg/mL), chlorpromazine (7.0, LOD 0.02 μg/mL), chlorpropamide (6.7, LOD <5 μg/mL), chlorprothixene (7.0, LOD <0.02 μg/mL), cinnarizine (7.9, LOD <0.02 μg/mL), citalopram (5.7, LOD <0.02 μg/mL), clemastine (7.7, LOD 0.02 μg/mL), clobazam (7.3, LOD <0.02 μg/mL), clobutinol (5.3, LOD 0.02 μg/mL), clomethiazole (6.2, LOD 0.5 μg/mL), clomipramine (7.1, LOD <0.02 μg/mL), clonazepam (6.6, LOD <0.02 μg/mL), clonidine (2.8, LOD 0.1 μg/mL), clozapine (5.6, LOD <0.02 μg/mL), cocaine (4.6, LOD <0.02 μg/mL), codeine (2.5, LOD 0.1 μg/mL), coumatetralyl (8.4, LOD 0.05 μg/mL), cyclizine (5.8, LOD <0.02 μg/mL), dextropropoxyphene (6.6, LOD 0.05 μg/mL), demoxepam (5.8, LOD 0.02 μg/mL), dextromethorphan (5.5, LOD <0.02 μg/mL), diazepam (8.1, LOD 0.02 μg/mL), diltiazem (5.8, LOD <0.02 μg/mL), diphenhydramine (5.7, LOD <0.02 μg/mL), dipyridamole (5.4, LOD 0.005 μg/mL), disopyramine (4.4, LOD <0.02 μg/mL), dixyrazine (6.8, LOD 0.005 μg/mL), doxapram (4.8, LOD <0.02 μg/mL), doxepin (5.9, LOD <0.02 μg/mL), ebastine (9.6, LOD 0.005 μg/mL), embutramide (6.7, LOD 0.005 μg/mL), ergotamine (5.5, LOD 0.005 μg/mL), ethenzamide (5.0, LOD 0.05 μg/mL), ethylmorphine (3.2, LOD 0.05 μg/mL), ethylparathion (9.7, LOD <5 μg/mL), etodroxizine (6.4, LOD <0.02 μg/mL), felodipine (9.6, LOD 0.02 μg/mL), fenazepam (7.5, LOD <0.02 μg/mL), fenfluramine (5.3, LOD <0.02 μg/mL), fenkamfamine (5.1, LOD <0.02 μg/mL), fentanyl (5.5, LOD <0.02 μg/mL), flecainide (5.9, LOD <0.02 μg/mL), fluconazole (4.0, LOD 0.1 μg/mL), flumazenil (5.2, LOD <0.02 μg/mL), flunitrazepam (7.1, LOD 0.002 μg/mL), fluoxetine (6.8, LOD 0.1 μg/mL), flupentixol (7.5, LOD 0.18 μg/mL), fluvoxamine (6.3, LOD 0.02 μg/mL), glibenclamide (8.5, LOD <0.02 μg/mL), glipizide (6.8, LOD <0.05 μg/mL), haloperidol (6.1, LOD <0.02 μg/mL), histapyrrodine (6.3, LOD 0.02 μg/mL), hydrocodone (3.0, LOD 0.05 μg/mL), hydroxychloroquine (2.4, LOD <0.3 μg/mL), hydroxyzine (6.3, LOD <0.02 μg/mL), imipramine (6.4, LOD 0.05 μg/mL), indomethacin (8.6, LOD 0.05 μg/mL), isoniazid (2.2, LOD 3 μg/mL), isradipine (8.6, LOD 0.05 μg/mL), ketamine (3.6, LOD <0.05 μg/mL), ketobemidone (3.3, LOD <0.05 μg/mL), ketoprofen (7.3, LOD 0.1 μg/mL), ketorolac (6.2, LOD 0.05 μg/mL), labetalol (4.9, LOD 0.05 μg/mL), lamotrigine (4.0, LOD 0.1 μg/mL), levocabastine (5.8, LOD 0.01 μg/mL), levomepromazine (6.5, LOD 0.02 μg/mL), lidocaine (3.7, LOD <0.05 μg/mL), loratadine (9.3, LOD 0.002 μg/mL), lorazepam (6.6, LOD 0.02 μg/mL), lormetazepam (7.4, LOD <0.02 μg/mL), LSD (4.7, LOD <0.02 μg/mL), malathion (8.9, LOD 10 μg/mL), maprotiline (6.4, LOD <0.02 μg/mL), MDMA (3.3, LOD 0.02 μg/mL), meclozine (8.5, LOD <0.02 μg/mL), medazepam (6.3, LOD <0.02 μg/mL), meloxicam (7.1, LOD 0.01 μg/mL), melperone (5.0, LOD <0.02 μg/mL), meperidine (4.7, LOD <0.02 μg/mL), mepivacaine (3.7, LOD <0.02 μg/mL), meprobamate (4.9, LOD 0.1 μg/mL), mesoridazine (5.4, LOD <0.02 μg/mL), methamphetamine (3.3, LOD 0.05 μg/mL), methadone (6.7, LOD <0.02 μg/mL), methylparathion (8.6, LOD 10 μg/mL), methylphenidate (4.2, LOD <0.02 μg/mL), metoclopramide (3.8, LOD <0.02 μg/mL), metoprolol (4.1, LOD 0.02 μg/mL), metronidazole (2.6, LOD 1 μg/mL), mexiletine (4.4, LOD 0.05 μg/mL), mianserin (5.7, LOD <0.02 μg/mL), midazolam (5.9, LOD <0.02 μg/mL), mirtazapine (4.4, LOD <0.02 μg/mL), mizolastine (5.5, LOD 0.01 μg/mL), moclobemide (3.7, LOD 0.05 μg/mL), molindone (4.0, LOD <0.02 μg/mL), monoacetylmorphine (2.7, LOD 0.1 μg/mL), morphine (2.0, LOD 0.1 μg/mL), nicotine (2.2, LOD 0.05 μg/mL), nifedipine (7.5, LOD 0.02 μg/mL), nikethamide (3.6, LOD <0.02 μg/mL), nitrazepam (6.5, LOD <0.02 μg/mL), nizatidine (1.7, LOD 1 μg/mL), nomifensine (4.6, LOD <0.02 μg/mL), nortriptyline (6.4, LOD <0.02 μg/mL), norverapamil (6.2, LOD 1 μg/mL), noscapine (5.0, LOD <0.02 μg/mL), olanzapine (3.0, LOD 0.05 μg/mL), ondansetron (4.6, LOD <0.02 μg/mL), orphenadrine (6.1, LOD <0.02 μg/mL), oxazepam (6.3, LOD <0.02 μg/mL), oxcarbazepine (5.3, LOD 0.02 μg/mL), oxprenolol (4.7, LOD 0.02 μg/mL), oxycodone (2.8, LOD 0.05 μg/mL), papaverine (4.8, LOD <0.02 μg/mL), acetaminophen (2.5, LOD <5 μg/mL), paroxetine (6.2, LOD 0.02 μg/mL), pemoline (3.3, LOD 0.05 μg/mL), pentazocine (5.0, LOD <0.02 μg/mL), pentifylline (7.3, LOD <5 μg/mL), pentoxyverine (6.6, LOD <0.02 μg/mL), perphenazine (6.9, LOD 0.002 μg/mL), phenazone (3.9, LOD 0.05 μg/mL), phencyclidine (5.3, LOD 0.05 μg/mL), pheniramine (4.1, LOD 0.02 μg/mL), phenylbutazone (9.0, LOD <5 μg/mL), phenylpropanolamine (2.5, LOD 0.3 μg/mL), phenytoin (6.1, LOD 0.05 μg/mL), pindolol (3.3, LOD 0.05 μg/mL), piroxicam (6.6, LOD 0.02 μg/mL), pitofenone (5.4, LOD <0.02 μg/mL), pizotifen (6.5, LOD <0.02 μg/mL), practolol (1.8, LOD 0.1 μg/mL), prazosin (4.1, LOD 0.05 μg/mL), prilocaine (3.8, LOD <0.02 μg/mL), primidone (4.0, LOD <5 μg/mL), procainamide (2.2, LOD 0.05 μg/mL), prochlorperazine (7.5, LOD 0.02 μg/mL), promazine (6.2, LOD <0.02 μg/mL), promethazine (6.0, LOD 0.05 μg/mL), propafenone (6.3, LOD <0.02 μg/mL), propranolol (5.4, LOD 0.02 μg/mL), propyphenazone (6.6, LOD 0.50 μg/mL), pseudoephedrine (2.6, LOD 1 μg/mL), quinine (4.2, LOD 0.02 μg/mL), ranitidine (1.8, LOD 0.1 μg/mL), risperidone (4.9, LOD <0.02 μg/mL), rocurone (3.8, LOD 0.1 μg/mL), ropivacaine (4.6, LOD <0.02 μg/mL), salicylamide (4.2, LOD <5 μg/mL), selegiline (4.1, LOD 0.05 μg/mL), sertindole (7.2, LOD <0.02 μg/mL), sertraline (6.8, LOD 0.02 μg/mL), sulindac (6.5, LOD 0.02 μg/mL), simazine (6.0, LOD 0.1 μg/mL), sincocaine (6.5, LOD <0.02 μg/mL), sisapride (5.9, LOD <0.02 μg/mL), sotalol (2.1, LOD 0.1 μg/mL), strychnine (5.3, LOD 0.05 μg/mL), sulpiride (1.9, LOD 0.1 μg/mL), sulthiame (4.1, LOD 0.05 μg/mL), temazepam (7.2, LOD <0.02 μg/mL), terbutaline (2.3, LOD 0.1 μg/mL), terfenadine (8.1, LOD 0.002 μg/mL), terodiline (6.7, LOD <0.02 μg/mL), tetracaine (5.7, LOD <0.02 μg/mL), dronabinol (12.3, LOD 0.05 μg/mL), tetrahydrozoline (3.6, LOD 0.1 μg/mL), theobromine (2.3, LOD <5 μg/mL), theophylline (2.4, LOD <5 μg/mL), thioridazine (7.5, LOD 0.02 μg/mL), timolol (3.8, LOD 0.05 μg/mL), thiothixene (6.7, LOD 0.02 μg/mL), tolbutamide (7.1, LOD <5 μg/mL), toremifene (8.7, LOD 0.02 μg/mL), tramadol (4.2, LOD 0.02 μg/mL), trazodone (5.2, LOD <0.02 μg/mL), triamterene (3.2, LOD 0.1 μg/mL), triazolam (6.7, LOD 0.002 μg/mL), trimeprazine (6.4, LOD <0.02 μg/mL), trimethoprim (3.1, LOD 0.05 μg/mL), trimipramine (6.7, LOD <0.02 μg/mL), venlafaxine (4.9, LOD 0.02 μg/mL), verapamil (6.5, LOD <0.02 μg/mL), warfarin (7.9, LOD <0.02 μg/mL), yohimbine (4.5, LOD <0.02 μg/mL), zolpidem (4.7, LOD <0.02 μg/mL), zopiclone (4.0, LOD 0.1 μg/mL)

## KEY WORDS

whole blood

## REFERENCE

Gergov, M.; Ojanperä, I.; Vuori, E. Simultaneous screening for 238 drugs in blood by liquid chromatography-ion spray tandem mass spectrometry with multiple-reaction monitoring, *J.Chromatogr.B*, **2003**, *795*, 41–53.

## SAMPLE

**Matrix:** formulations

**Sample preparation:** Extract tablets with MeOH, centrifuge, dilute the supernatant with mobile phase, inject an aliquot.

## HPLC VARIABLES

**Column:** 150 × 4.6 5 μm Eclipse XDB C8

**Mobile phase:** MeCN:MeOH:buffer 20:20:60 (The buffer was 1% triethylamine adjusted to pH 3.7 with concentrated orthophosphoric acid.)

**Flow rate:** 1.2

**Injection volume:** 10

**Detector:** UV 210

## CHROMATOGRAM

**Retention time:** 14

**Internal standard:** 5-methyl-2-nitrophenol (12)

**Limit of detection:** 0.12–0.18 μg/mL

**Limit of quantitation:** 0.5–0.6 μg/mL

## OTHER SUBSTANCES

**Simultaneous:** impurities

## KEY WORDS

robust; tablets; validated

## REFERENCE

Radhakrishna, T.; Reddy, G.O. Simultaneous determination of fexofenadine and its related compounds by HPLC, *J.Pharm.Biomed.Anal.*, **2002**, *29*, 681–690.

## ANNOTATED BIBLIOGRAPHY

Dresser, G.K.; Bailey, D.G.; Leake, B.F.; Schwarz, U.I.; Dawson, P.A.; Freeman, D.J.; Kim, R.B. Fruit juices inhibit organic anion transporting polypeptide-mediated drug uptake to decrease the oral availability of fexofenadine, *Clin.Pharmacol.Ther.*, **2002**, *71*, 11–20.

Hamman, M.A.; Bruce, M.A.; Haehner-Daniels, D.; Hall, S.D. The effect of rifampin administration on the disposition of fexofenadine, *Clin.Pharmacol.Ther.*, **2001**, *69*, 114–121. [SPE]

Hofmann, U.; Seiler, M.; Drescher, S.; Fromm, M.F. Determination of fexofenadine in human plasma and urine by liquid chromatography-mass spectrometry, *J.Chromatogr.B*, **2002**, *766*, 227–233. [SPE]

Kim, R.B.; Leake, B.F.; Choo, E.F.; Dresser, G.K.; Kubba, S.V.; Schwarz, U.I.; Taylor, A.; Xie, H.-G.; McKinsey, J.; Zhou, S.; Lan, L.-B.; Schuetz, J.D.; Schuetz, E.G.; Wilkinson, G.R. Identification of functionally variant MDR1 alleles among European Americans and African Americans, *Clin.Pharmacol.Ther.*, **2001**, *70*, 189–199.

Milne, R.W.; Larsen, L.A.; Jorgensen, K.L.; Bastlund, J.; Stretch, G.R.; Evans, A.M. Hepatic disposition of fexofenadine: Influence of the transport inhibitors erythromycin and dibromosulphothalein, *Pharm.Res.*, **2000**, *17*, 1511–1515.

Putnam, W.S.; Ramanathan, S.; Pan, L.; Takahashi, L.H.; Benet, L.Z. Functional characterization of monocarboxylic acid, large neutral amino acid, bile acid and peptide transporters, and P-glycoprotein in MDCK and Caco-2 cells, *J.Pharm.Sci.*, **2002**, *91*, 2622–2635. [column-switching]

Robbins, D.K.; Castles, M.A.; Pack, D.J.; Bhargava, V.O.; Weir, S.J. Dose proportionality and comparison of single and multiple dose pharmacokinetics of fexofenadine (MDL 16455) and its enantiomers in healthy male volunteers, *Biopharm.Drug Dispos.*, **1998**, *19*, 455–463.

Soldner, A.; Christians, U.; Susanto, M.; Wacher, V.J.; Silverman, J.A.; Benet, L.Z. Grapefruit juice activates P-glycoprotein-mediated drug transport, *Pharm.Res.*, **1999**, *16*, 478–485. [losartan; felodipine; nifedipine; fexofenadine; column-switching]

Tannergren, C.; Petri, N.; Knutson, L.; Hedeland, M.; Bondesson, U.; Lennernäs, H. Multiple transport mechanisms involved in the intestinal absorption and first-pass extraction of fexofenadine, *Clin.Pharmacol.Ther.*, **2003**, *74*, 423–436. [SPE; LC-MS; UV detection]

# Flomoxef

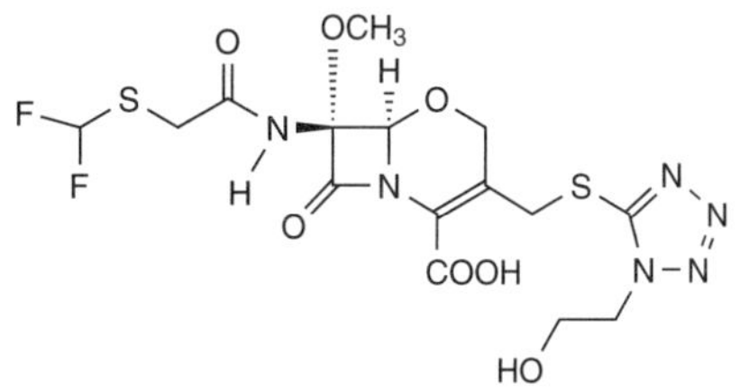


**Molecular formula:** $C_{15}H_{18}F_2N_6O_7S_2$
**Molecular weight:** 496.47
**CAS Registry No:** 99665-00-6
**Merck Index:** 13, 4132

## SAMPLE

**Matrix:** blood

**Sample preparation:** Condition a 3 mL 200 mg Bond-Elut C18 SPE cartridge with two 3 mL portions of MeOH and two 3 mL portions of 5% acetic acid. Mix 500 μL serum with 500 μL 10% acetic acid, add to the SPE cartridge, wash with 3 mL water, elute with four 500 μL portions of MeOH:water 60:40. Evaporate the combined eluates to dryness under a stream of nitrogen, reconstitute the residue with 250 μL water, filter (0.22 μm). Inject a 100 μL aliquot onto column A and elute to waste with mobile phase A. After an unspecified time, backflush the contents of column A onto column B using mobile phase B. Monitor the effluent from column B. (Sato,K.; Kobayashi,K.; Moore,C.M.; Mizuno,Y.; Katsumata,Y. Semi-quantitative analysis of cefaclor in human serum by capillary high performance liquid chromatography/fast atom bombardment mass spectrometry. *Forensic Sci.Int.* **1993**, *59*, 71–77.)

## HPLC VARIABLES

**Column:** A 30 × 0.5 10 μm Develosil PhA (phenethyl); B 150 × 0.5 5 μm Develosil PhA (phenethyl)

**Mobile phase:** A 10 mM ammonium acetate:glycerol 99.5:0.5, adjusted to pH 5 with acetic acid; B MeOH:water:acetic acid:glycerol 40:59:0.5:0.5 (pH ca. 3)

**Flow rate::** A 0.05; B 0.004

**Injection volume:** 100

**Detector:** MS, JMS DX303 double focusing, xenon FAB ion source, gun current 10 mA, voltage 3 kV, positive ion mode

## CHROMATOGRAM

**Retention time:** 18.3

## LIMIT OF DETECTION:

200–1000 ng

## OTHER SUBSTANCES

**Extracted:** cefaclor (14.4), cefamandole (31.0), cefazolin (19.7), cefbuperazone (20.8), cefixime (19.2), cefmenoxime (18.8), cefmetazole (21.4), cefoperazone (24.2), cefotaxime (16.1), cefotetan (16.8), cefotiam (11.7), cefpiramide (20.9), cefsulodin (12.2), ceftazidime (12.0), ceftizoxime (15.6), ceftriaxone (16.7), cefuroxime (16.9), cefuzonam (32.0), cephalexin (14.8), cephaloglycine (15.7), cephaloridine (15.6), cephalothin (34.7), latamoxef (15.1)

## KEY WORDS

column-switching; serum; SPE

## REFERENCE

Kobayashi, K.; Sato, K.; Mizuno, Y.; Katsumata, Y. Capillary high-performance liquid chromatography-fast atom bombardment mass spectrometry of 24 cephem antibiotics, *J.Chromatogr.B*, **1996**, *677*, 275–290.

# Florfenicol

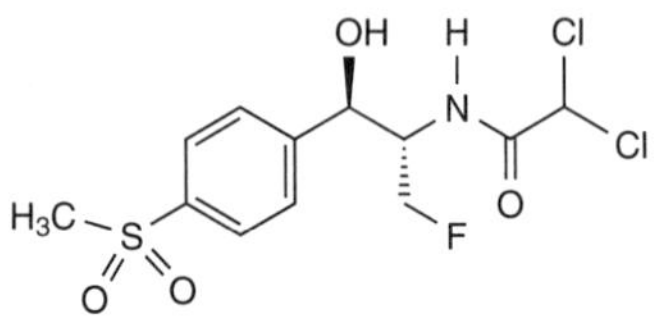

**Molecular formula:** $C_{12}H_{14}Cl_2FNO_4S$
**Molecular weight:** 358.22
**CAS Registry No:** 73231-34-2
**Merck Index:** 13, 4135

## SAMPLE
**Matrix:** blood, CSF
**Sample preparation:** Condition a 3 mL 500 mg Bond-Elut C18 SPE cartridge with 2 mL MeOH and 2 mL water. Mix 250 μL plasma or CSF with 750 μL water and 40 μL 8 μg/mL IS in MeOH:water 2:98 or 400 μg/mL IS in MeOH, vortex. Add to the SPE cartridge, wash with 2 mL MeCN:water 15:85, wash with 3 mL hexane, elute with 3 mL MeCN. Evaporate the eluate to dryness under reduced pressure. Dissolve the residue in 500 μL MeCN:water 22:78, filter (0.45 μm polyvinylidene difluoride), inject a 100 μL aliquot.

## HPLC VARIABLES
**Column:** 250 × 4.6 5 μm Hypersil ODS
**Column temperature:** 40
**Mobile phase:** MeCN:50 mM ammonium acetate 22:78
**Flow rate:** 1.5
**Injection volume:** 100
**Detector:** UV 224

## CHROMATOGRAM
**Retention time:** 6
**Internal standard:** chloramphenicol (7.50)
**Limit of detection:** 20 ng/mL
**Limit of quantitation:** 30 ng/mL

## KEY WORDS
calf; cow; pharmacokinetics; plasma; SPE

## REFERENCE
de Craene, B.A.; Deprez, P.; D'Haese, E.; Nelis, H.J.; Van den Bossche, W.; De Leenheer, P. Pharmacokinetics of florfenicol in cerebrospinal fluid and plasma of calves, *Antimicrob.Agents Chemother.*, **1997**, *41*, 1991–1995.

## SAMPLE
**Matrix:** water
**Sample preparation:** Inject a 25 μL aliquot.

## HPLC VARIABLES
**Column:** 150 × 3.9 4 μm Nova-Pak C18
**Column temperature:** 30
**Mobile phase:** MeCN:buffer 15:85 (The buffer was 3.28 g sodium acetate and 1.01 g sodium 1-heptanesulfonate in 1 L water, adjusted to pH 4.6 with phosphoric acid.)
**Flow rate:** 1
**Injection volume:** 25
**Detector:** UV 224

## CHROMATOGRAM
**Retention time:** 8

**Limit of detection:** 0.1 μg/mL
**Limit of quantitation:** 0.2 μg/mL

## OTHER SUBSTANCES

**Simultaneous:** degradants, thiamphenicol (3.3)

## REFERENCE

Hayes, J.M.; Eichman, J.; Katz, T.; Gilewicz, R. Stability of florfenicol in drinking water, *J.AOAC Int.*, **2003**, *86*, 22–29.

## ANNOTATED BIBLIOGRAPHY

De Wasch, K.; Van Hoof, N.; Poelmans, S.; Okerman, L.; Courtheyn, D.; Ermens, A.; Cornelis, M.; De Brabander, H.F. Identification of "unknown analytes" in injection sites: a semi-quantitative interpretation, *Anal.Chim.Acta*, **2003**, *483*, 387–399.

Hormazabal, V.; Steffenak, I.; Yndestad, M. Simultaneous determination of residues of florfenicol and the metabolite florfenicol amine in fish tissues by high-performance liquid chromatography, *J.Chromatogr.*, **1993**, *616*, 161–165.

Hormazabal, V.; Steffenak, I.; Yndestad, M. Simultaneous extraction and determination of florfenicol and the metabolite florfenicol amine in sediment by high-performance liquid chromatography, *J.Chromatogr.A*, **1996**, *724*, 364–366.

Liu, J.; Fung, K.-F.; Chen, Z.; Zeng, Z.; Zhang, J. Pharmacokinetics of florfenicol in healthy pigs and in pigs experimentally infected with Actinobacillus pleuropneumoniae, *Antimicrob.Agents Chemother.*, **2003**, *47*, 820–823.

Lobell, R.D.; Varma, K.J.; Johnson, J.C.; Sams, R.A.; Gerken, D.F.; Ashcraft, S.M. Pharmacokinetics of florfenicol following intravenous and intramuscular doses to cattle, *J.Vet.Pharmacol.Ther.*, **1994**, *17*, 253–258. [thiamphenicol is internal standard]

Nagata, T.; Saeki, M. Simultaneous determination of thiamphenicol, florfenicol and chloramphenicol residues in muscles of animals and cultured fish by liquid chromatography, *J.Liq.Chromatogr.*, **1992**, *15*, 2045–2056. [LOD 0.01 ppm; column temp 55°; SPE]

Vue, C.; Schmidt, L.J.; Stehly, G.R.; Gingerich, W.H. Liquid chromatographic determination of florfenicol in the plasma of multiple species of fish, *J.Chromatogr.B*, **2002**, *780*, 111–117.

# Fludrocortisone

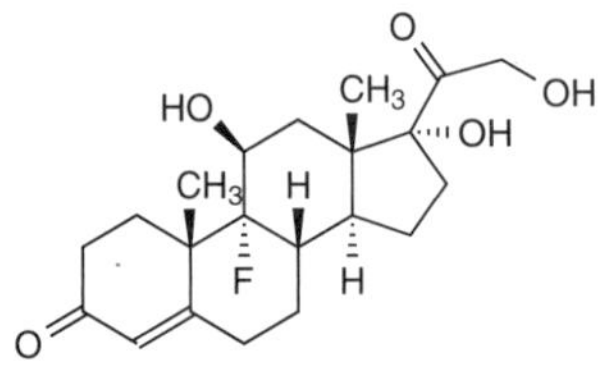


**Molecular formula:** $C_{21}H_{29}FO_5$
**Molecular weight:** 380.45
**CAS Registry No:** 127-31-1, 514-36-3 (21-acetate)
**Merck Index:** 13, 4156

## SAMPLE

**Matrix:** blood
**Sample preparation:** Condition a 3 mL 500 mg Sep-Pak Vac C18 SPE cartridge with 3 mL MeOH and 3 mL water. Mix 500 μL 200 mM pH 3.85 acetate buffer with 1 mL serum, add 400 μL 2.5 μM IS in mobile phase, centrifuge, add to the SPE cartridge, wash with 3 mL acetone:water 20:80, wash with 3 mL water, wash with 3 mL hexane, elute with 3 mL diethyl ether. Vortex the eluate with 1 mL 200 mM NaOH, centrifuge. Evaporate the organic layer to dryness under a stream of nitrogen, reconstitute the residue with 250 μL mobile phase, place on a rotary mixer for 5 min, inject a 60 μL aliquot. (Fludrocortisone is IS. Extraction from serum has not been strictly demonstrated.)

## HPLC VARIABLES

**Column:** 250 × 4.6 5 μm Spherex C18 (Phenomenex)
**Mobile phase:** MeOH:THF:water 3:25:72
**Flow rate:** 1
**Injection volume:** 60
**Detector:** UV 254

## CHROMATOGRAM

**Retention time:** 15.90
**Internal standard:** fludrocortisone (15.90)
**Limit of detection:** 5 nM
**Limit of quantitation:** 10 nM

## OTHER SUBSTANCES

**Extracted:** 11-deoxycortisol (22.10), dexamethasone (29.85), hydrocortisone (12.85), methylprednisolone (21.00), prednisolone (12.00)

## KEY WORDS

fludrocortisone is IS in original method; serum; SPE

## REFERENCE

McWhinney, B.C.; Ward, G.; Hickman, P.E. Improved HPLC method for simultaneous analysis of cortisol, 11-deoxycortisol, prednisolone, methylprednisolone, and dexamethasone in serum and urine, *Clin.Chem.*, **1996**, *42*, 979–981.

## SAMPLE

**Matrix:** blood
**Sample preparation:** Condition a 3 mL 200 mg Clean-Up C18 SPE cartridge (Worldwide Monitoring) with two 2 mL portions of MeOH and two 2 mL portions of water. Centrifuge plasma at 2000 rpm for 5 min prior to analysis. Mix 2 mL plasma with 40 μL MeCN, add 1 mL water, vortex until homogeneous, add to the SPE cartridge, wash with two 2 mL portions of acetone:water 20:80, wash with 1 mL 50 mM pH 2.7 phosphate buffer, place under vacuum for 3 min, elute with two 2.5 mL portions of ethyl acetate. Evaporate the eluate to dryness under reduced pressure at 40°, reconstitute the residue with 75 μL mobile phase B, vortex, add 100 μL 50 mM pH 3 phosphate buffer, centrifuge at 5000 rpm for 5 min, inject a 145 μL aliquot.

## HPLC VARIABLES

**Column:** 100 × 2 5 μm YMC Basic
**Column temperature:** 50
**Mobile phase:** Gradient. A:B 20:80 for 1 min, to 50:50 over 7 min, to 65:35 over 3 min, to 100:0 (step gradient), maintain at 100:0 for 1 min, return to initial conditions, re-equilibrate for 4 min. A was MeCN:50 mM pH 3.0 potassium dihydrogen phosphate buffer 50:50. B was MeOH:50 mM pH 3.0 potassium dihydrogen phosphate buffer 20:80.
**Flow rate:** 0.3
**Injection volume:** 145
**Detector:** UV 246

## CHROMATOGRAM

**Retention time:** 12.4
**Internal standard:** fludrocortisone acetate

## OTHER SUBSTANCES

**Extracted:** deflazacort (13.8), 21-hydroxydeflazacort (10)
**Simultaneous:** cortisone, hydrocortisone

## KEY WORDS

fludrocortisone acetate is internal standard; plasma; SPE

## REFERENCE

Reynolds, D.L.; Burmaster, S.D.; Eichmeier, L.S. Quantitative determination of 21-hydroxy-deflazacort in human plasma using gradient semi-microbore liquid chromatography, *Biomed.Chromatogr.*, **1994**, *8*, 230–235.

# Fluprostenol

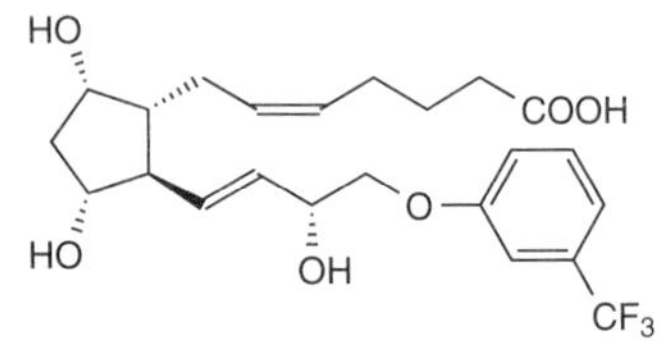


**Molecular formula:** $C_{23}H_{29}F_3O_6$
**Molecular weight:** 458.47
**CAS Registry No:** 40666-16-8
**Merck Index:** 13, 4220

## SAMPLE

**Matrix:** blood
**Sample preparation:** Condition a 1 mL 100 mg IST Isolute ODS SPE cartridge (Jones Chromatography) with 1 mL MeOH and 1 mL MeOH:water:formic acid 3:97:0.1. Vortex 200 μL plasma with 200 pg IS for 30 s, add to the SPE cartridge, wash with 1 mL MeOH:water:formic acid 3:97:0.1, wash with 1 mL MeOH:water 25:75, elute with 1 mL MeOH:water 90:10. Evaporate the eluate to dryness under a stream of nitrogen at 35°, reconstitute the residue with 120 μL MeOH:water 30:70, inject a 25 μL aliquot.

## HPLC VARIABLES

**Column:** 10 × 2.1 3 μm Symmetry Sentry Guard C18 (Waters)
**Mobile phase:** MeOH:water:formic acid 38:62:0.1 containing 70 μg/mL ammonium acetate
**Flow rate:** 0.4
**Injection volume:** 25
**Detector:** MS, PE Sciex API-III Plus triple quadrupole, TurboIonSpray, positive ionization, ESI 3800 V, orifice 40 V, nebulizer gas nitrogen at 62 psi, TurboProbe 475°, TurboProbe nitrogen at 8 L/min, collision gas argon, m/z 476–279

## CHROMATOGRAM

**Retention time:** 0.63
**Internal standard:** 3,3,4,4-$^{2}H_4$-fluprostenol (m/z 480–283) (0.63)
**Limit of quantitation:** 25 pg/mL

## KEY WORDS

pharmacokinetics; plasma; rat; SPE

## REFERENCE

Eichhold, T.H.; Kuhlenbeck, D.L.; Baker, T.R.; Stella, M.E.; Amburgey, J.S.; deLong, M.A.; Hartke, J.R.; Cruze, C.A.; Pierce, S.A.; Wehmeyer, K.R. Use of short high-performance liquid chromatography columns and tandem-mass spectrometry for the rapid analysis of a prostaglandin analog, fluprostenol, in rat plasma, *J.Chromatogr.B*, **2000**, *741*, 213–220.

## SAMPLE

**Matrix:** blood
**Sample preparation:** Condition a SPEC 3 mL 15 mg MP1 nonpolar reversed-phase/strong cation exchange SPE Cartridge (Ansys) with 500 μL MeOH and 500 μL 40 mM formic acid. Mix 1 mL plasma with 15 μL 20 ng/mL IS and 1 mL 100 mM formic acid, add to the SPE cartridge, rinse tube with 0.5–1 mL water, add rinse to the SPE cartridge, wash with 500 μL water, dry under vacuum, wash with two 500 μL portions of toluene:dichloromethane 60:40, dry under vacuum, elute with 600 μL toluene:methylformate 20:80. Evaporate the eluate to dryness under a stream of nitrogen at 30°, reconstitute the residue with 125 μL MeOH:water 50:50, inject a 35 μL aliquot.

## HPLC VARIABLES

**Column:** 150 × 2 5 μm Phenomenex Columbus C18

**Mobile phase:** MeOH:5 mM pH 6.3 ammonium formate 70:30
**Flow rate:** 0.2
**Injection volume:** 35
**Detector:** MS, Micromass Quattro LC, electrospray, capillary 3.0 kV, sample cone 40 V, extraction cone 2 V, RF lens 0.3 V, source temperature 125°, drying gas 250°, MS1 parameters LM resolution 14, HM resolution 14, ion energy, 1.2 V, entrance and exit set to 0 and 1, collision energy 30 eV, MS2 parameters LM resolution 15.0, HM resolution 15.0, ion energy 1.2 V, multiplier 650 V, nebulizing gas 75 L/h, drying gas 570 L/h, collision gas argon, m/z 457–161

## CHROMATOGRAM

**Retention time:** 5.3
**Internal standard:** AL-5848X (tetradeutero travoprost free acid) (m/z 461–161)
**Limit of quantitation:** 10 pg/mL

## KEY WORDS

plasma; SPE

## REFERENCE

McCue, B.A.; Cason, M.M.; Curtis, M.A.; Faulkner, R.D.; Dahlin, D.C. Determination of travoprost and travoprost free acid in human plasma by electrospray HPLC/MS/MS, *J.Pharm.Biomed.Anal.*, **2002**, *28*, 199–208.

# Flurandrenolide

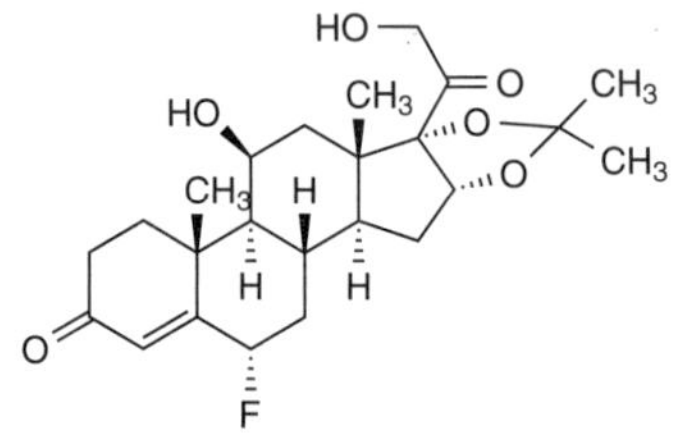


**Molecular formula:** $C_{24}H_{33}FO_6$
**Molecular weight:** 436.51
**CAS Registry No:** 1524-88-5
**Merck Index:** 13, 4223

## SAMPLE

**Matrix:** formulations
**Sample preparation:** Mix 1 g of a topical product, 5 mL 100 mM pH 4 citrate buffer saturated with NaCl, and 10 mL ethyl acetate, shake vigorously by hand to ensure that no large clumps stick to the tube, mix with the tube on its side on an oscillating shaker for 15 min. Remove the upper ethyl acetate layer and wash with citrate buffer as before. Dry the ethyl acetate over anhydrous sodium sulfate and evaporate to dryness. Dissolve the residue in a mixture of 10 mL heptane and 5 mL MeCN:water 90:10. Remove the upper heptane layer and extract it with 2 mL MeCN:water 90:10. Combine the MeCN/water layers and evaporate them to dryness, dissolve the residue in 300 μL MeOH, filter (0.45 μm nylon), inject a 5 μL aliquot.

## HPLC VARIABLES

**Guard column:** 15 × 3.2 7 μm Brownlee NewGuard C18
**Column:** 75 × 4.6 3.5 μm Symmetry C18 (Waters)
**Mobile phase:** Gradient. MeCN:water 18:82 for 2 min, to 82:18 over 12 min, maintain at 82:18 for 3 min, re-equilibrate at initial conditions for 12 min.
**Flow rate:** 1
**Injection volume:** 5
**Detector:** UV 240

## CHROMATOGRAM

**Retention time:** 7.36
**Limit of detection:** 0.001%

## OTHER SUBSTANCES

**Simultaneous:** alclometasone 17,21-dipropionate (10.93), amcinonide (10.90), beclomethasone 17,21-dipropionate (11.90), betamethasone (6.52), betamethasone 21-acetate (8.47), betamethasone 17-benzoate (10.28), betamethasone 17,21-dipropionate (11.42), betamethasone 17-valerate (10.25), budesonide (8.55, 8.69 (epimers)), clobetasol 17-propionate (11.06), cortisone (5.62), cortisone 21-acetate (8.07), dexamethasone (6.57), dexamethasone 21-acetate (8.68), desonide (6.99), desoximetasone (7.60), desoxycorticosterone acetate (10.90), desoxycorticosterone pivalate (14.45), diflorasone 17,21-diacetate (9.81), fluocinolone acetonide (7.38), fluocinonide (9.79), fludrocortisone 21-acetate (7.77), fluorometholone (7.67), flumethasone 21-pivalate (11.20), flunisolide (7.14), fluprednisolone (5.46), fluticasone 17-propionate (11.19), halcinonide (10.72), halobetasol propionate (10.98), hydrocortisone (5.50), hydrocortisone 21-acetate (7.65), hydrocortisone 17-butyrate (8.66), hydrocortisone 21-cypionate (12.54), hydrocortisone 17-valerate (9.53), mometasone 17-furoate (11.24), methylprednisolone (6.31), methylprednisolone 21-acetate (8.34), meprednisone (6.55), prednisolone (5.37), prednisolone 21-acetate (7.47), prednisolone 21-tebuate (11.01), paramethasone 21-acetate (8.64), prednicarbate (10.72), prednisone (5.46), triamcinolone (4.15), triamcinolone acetonide (7.04), triamcinolone 16,21-diacetate (7.49), triamcinolone hexacetonide (13.12)

## KEY WORDS

body wash, cream, gel, lotion, shampoo, spray

## REFERENCE

Reepmeyer, J.C. Screening for corticosteroids in topical pharmaceuticals by HPLC with a scanning ultraviolet detector, *J.Liq.Chromatogr.Rel.Technol.*, **2001**, *24*, 693–709.

## SAMPLE

**Matrix:** urine

**Sample preparation:** Mix 5 mL urine with 100 ng IS and 400 μL 1 M pH 4.1 sodium acetate buffer, adjust pH to 5.0, add 600 μL β-glucuronidase solution (10800 U), heat at 65° for 3.5 h or at 37° overnight, cool, add 6 mL ethyl acetate, rotate for 10 min. Remove the organic layer and wash it with 3 mL 1 M NaOH containing 150 mM NaCl by rotating for 5 min, centrifuge at 1900 g for 30 s. Remove the organic layer and pass it through an anhydrous sodium sulfate drying column. Evaporate the eluate to dryness under a stream of nitrogen at 60°, reconstitute the residue with 30 μL MeOH, vortex, inject a 10 μL aliquot.

## HPLC VARIABLES

**Column:** 75 × 4.6 3 μm DB-8 (Supelco)

**Column temperature:** 25

**Mobile phase:** Gradient. MeOH:1% acetic acid from 0:100 to 100:0 over 15 min, maintain at 100:0 for 3 min.

**Flow rate:** 1

**Injection volume:** 10

**Detector:** MS, Finnigan MAT LCQ Classic, APCI, source 450°, capillary 150°, source +5 kV for positive ions and – 5 kV for negative ions, collision gas helium, m/z 437

## CHROMATOGRAM

**Retention time:** 26.8

**Internal standard:** $d_4$-hydrocortisone (24.4)

**Limit of detection:** 5 ng/mL

## OTHER SUBSTANCES

**Extracted:** beclomethasone (m/z 409) (26.4), betamethasone (m/z 393) (25.9), deoxycortone (m/z 331) (28.1), 21-deoxydexamethasone (m/z 377) (26.8), desoximetasone (m/z 377) (27.6), dexamethasone (m/z 393) (25.9), dichlorisone (m/z 413) (26.4), fluclorolone acetonide (m/z 487) (27.8), fludrocortisone (m/z 381) (24.2), flumethasone (m/z 411) (25.4), fluocinolone acetonide (m/z 453) (26.4), fluocinonide (m/z 495) (28.5), fluocortolone (m/z 377) (26.8), fluorometholone (m/z 3.77) (26.6), fluprednisolone (m/z 379) (23.9), hydrocortisone (m/z 363) (24.4), isoflupredone (m/z 379) (23.9), methylprednisolone (m/z 375) (26.1), prednisolone (m/z 361) (24.4), prednisone (m/z 359) (23.4), triamcinolone (m/z 395) (19), triamcinolone acetonide (m/z 435) (29.3)

## KEY WORDS

horse

## REFERENCE

Tang, P.W.; Law, W.C.; Wan, T.S.M. Analysis of corticosteroids in equine urine by liquid chromatography-mass spectrometry, *J.Chromatogr.B*, **2001**, *754*, 229–244.

## ANNOTATED BIBLIOGRAPHY

Gagliardi, L.; De Orsi, D.; Del Giudice, M.R.; Gatta, F.; Porrà, R.; Chimenti, P.; Tonelli, D. Development of a tandem thin-layer chromatography-high-performance liquid chromatography method for the identification and determination of corticosteroids in cosmetic products, *Anal.Chim.Acta*, **2002**, *457*, 187–198. [hydrocortisone; hydrocortisone acetate; hydrocortisone butyrate; hydrocortisone pivalate; prednisolone pivalate; prednisolone acetonide; prednisolone acetate; fluorocortolone pivalate; clobetasone butyrate; flumethasone pivalate; fluorocortolone caproate; prednisolone; fluorocortolone; clobetasone; flumethasone; triamcinolone; triamcinolone diacetate; triamcinolone acetonide; clobetasol propionate; clobetasol; fluocortin; fluocortin butyl ester; fludrocortisone; fludrocortisone acetate; dexamethasone; dexamethasone acetate; dexamethasone pivalate; dexamethasone isonicotinate; dexamethasone valerate; betamethasone; betamethasone acetate; betamethasone benzoate; betamethasone propionate butyrate; betamethasone propionate stearate; betamethasone propionate; betamethasone dipropionate; betamethasone valerate; betamethasone divalerate; betamethasone valerate acetate; fluocinolone; fluocinolone acetonide; fluocinonide; cortisone; cortisone acetate; amcinonide; flurandrenolide; fluorometholone; methylprednisolone; halcinonide; deoxymethasone; diflucortolone; valerate; dehydrocorticosterone; fluoroprednisolone; beclomethasone; alclometasone; mometasone; fluoroprednisolone acetate; beclomethasone dipropionate; alclometasone dipropionate; mometasone furoate]

Tang, P.W.; Law, W.C.; Wan, T.S.M. Analysis of corticosteroids in equine urine by liquid chromatography-mass spectrometry, *J.Chromatogr.B*, **2001**, *754*, 229–244. [triamcinolone; prednisone; fluprednisolone; isoflupredone; fludrocortisone; hydrocortisone; prednisolone; flumethasone; betamethasone; dexamethasone; methylprednisolone; dichlorisone; beclomethasone; fluocinolone acetonide; fluocinolone; fluorometholone; fluocortolone; desoximetasone; triamcinolone acetonide; flurandrenolide; fluclorolone acetonide; fluclorolone; deoxycortone; fluocinonide; flucloronide]

# Flurithromycin

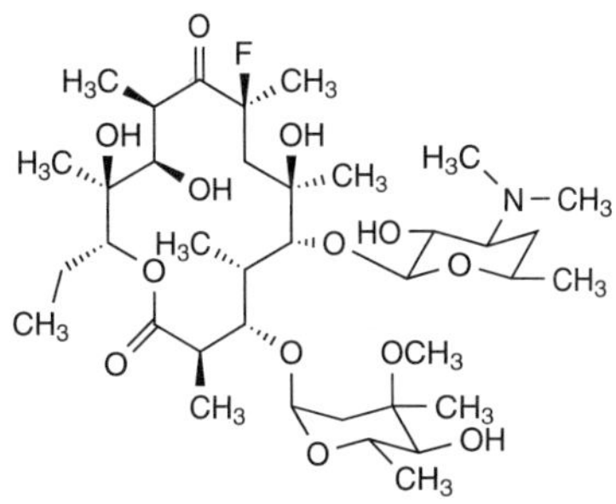


**Molecular formula:** $C_{37}H_{66}FNO_{13}$
**Molecular weight:** 751.92
**CAS Registry No:** 82664-20-8

### SAMPLE
**Matrix:** solutions

### HPLC VARIABLES
**Column:** 250 × 4.6 10 μm Lichrosorb RP18
**Column temperature:** 35
**Mobile phase:** MeCN:buffer 68:32 (The buffer was 10 mM potassium dihydrogen phosphate adjusted to pH 7 with 20% NaOH.)
**Flow rate:** 1.5
**Detector:** UV 214

### CHROMATOGRAM
**Retention time:** 3.9 (hemiketal), 5.6 (ketone)

### REFERENCE
Colombo, N.; Depaoli, A.; Gobetti, M.; Saorin, M.G. Analytical-physical profile of the novel macrolide antibiotic flurithromycin ethylsuccinate, *Arzneimittelforschung*, **1994**, *44*, 850–855.

### SAMPLE
**Matrix:** solutions

### HPLC VARIABLES
**Column:** 125 × 4 3 μm LKB Spherisorb ODS-2
**Mobile phase:** MeCN:MeOH:100 mM pH 7.0 ammonium acetate 65:20:15
**Flow rate:** 0.5
**Injection volume:** 10–50
**Detector:** UV 235

### CHROMATOGRAM
**Retention time:** 5.1

### OTHER SUBSTANCES
**Simultaneous:** azithromycin (4.5), clarithromycin (7.6), erythromycin (5.4), oleandomycin (4.4), roxithromycin (7.9)

### REFERENCE
Lingerfelt, B.; Champney, W.S. Macrolide and ketolide antibiotic separation by reversed phase high performance liquid chromatography, *J.Pharm.Biomed.Anal.*, **1999**, *20*, 459–469.

# Flurogestone acetate

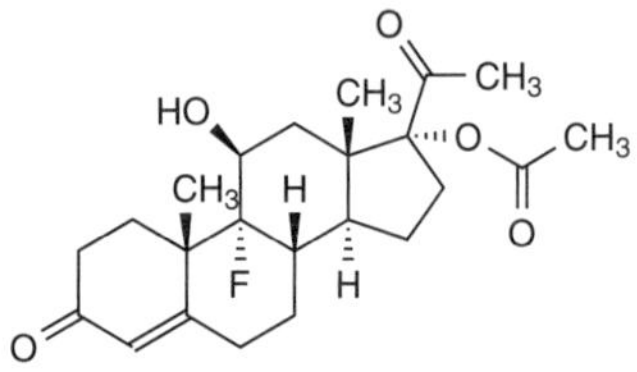

**Molecular formula:** $C_{23}H_{31}FO_5$
**Molecular weight:** 406.49
**CAS Registry No:** 2529-45-5
**Merck Index:** 13, 4226

## SAMPLE

**Matrix:** solutions

## HPLC VARIABLES

**Column:** 300 × 3.9 10 μm μBondapak C18
**Mobile phase:** MeOH:water 50:50
**Flow rate:** 2
**Detector:** UV 254

## CHROMATOGRAM

**Retention time:** 21
**Internal standard:** testosterone (32)

## OTHER SUBSTANCES

**Simultaneous:** degradants

## REFERENCE

Kabadi, M.B.; Valia, K.H.; Chien, Y.W. Intravaginal controlled administration of flurogestone acetate I: development of a stability-indicating liquid chromatographic method and stability kinetics of flurogestone acetate, *J.Pharm.Sci.*, **1984**, *73*, 1461–1464.

## SAMPLE

**Matrix:** tissue
**Sample preparation:** Condition a 100 mg C18 SPE cartridge (Varian) with MeOH and 10 mM pH 8.5 Tris buffer. Fill a 22 mL vessel (from bottom to top) with 5 g alumina (dried at 120° for 48 h before use), 6 g anhydrous sodium sulfate, and 2 g melted (microwave) kidney fat. There is filter paper between the alumina and sodium sulfate. Pass hexane down through the layers at 60° at 1500 psi followed by MeCN at 50° at 1500 psi. Store the MeCN at – 20° for 30 min to precipitate fat, filter through a plug of glass wool, evaporate to dryness under a stream of nitrogen at 40°, reconstitute the residue with 2 mL MeOH, mix with 5 mL water, add to the SPE cartridge, wash with 2 mL 20 mM pH 8.5 Tris buffer, wash with 2 mL MeOH:water 40:60, elute with 2 mL MeOH. Evaporate the eluate to dryness under a stream of nitrogen at 40°, reconstitute the residue with 75 μL MeCN, add 75 μL 0.5% formic acid, mix, inject a 75 μL aliquot. (The extraction is done in a Dionex ASE system (advanced solvent extraction).)

## HPLC VARIABLES

**Column:** 150 × 3 5 μm Symmetry (Waters)
**Column temperature:** 40
**Mobile phase:** Gradient. A:B from 45:55 to 100:0 over 12 min. A was MeCN:water:formic acid 10:90:0.5. B was MeCN:water:formic acid 90:10:0.5.
**Flow rate:** 0.4
**Injection volume:** 75
**Detector:** MS, Micromass Quattro Ultima, positive electrospray, capillary 2.5 kV, cone 40 V, source 120°, cone gas 189 L/h, desolvation gas 652 L/h, m/z 407.2–267.4–225.4

## CHROMATOGRAM
**Retention time:** 4.5
**Limit of detection:** <2 ng/g

## OTHER SUBSTANCES
**Extracted:** chloromadinone acetate (m/z 405.2–309.2–345.3) (8.6), chlorotestosterone acetate (m/z 365.2–305.2–323.3) (12.2), delmadinone acetate (m/z 403.2–205.1–181.1) (7.8), medroxyprogesterone acetate (m/z 387.2–327.3–285.3) (8.8), megestrol acetate (m/z 385.2–267.3–325.3) (8.4), melengestrol acetate (m/z 397.2–279.3–279.3) (8.8)

## KEY WORDS
fat; kidney; SPE

## REFERENCE
Hooijerink, H.; van Bennekom, E.O.; Nielen, M.W.F. Screening for gestagens in kidney fat using accelerated solvent extraction and liquid chromatography electrospray tandem mass spectrometry, *Anal.Chim.Acta*, **2003**, *483*, 51–59.

## ANNOTATED BIBLIOGRAPHY
Kabadi, M.B.; Chien, Y.W. Intravaginal controlled administration of flurogestone acetate II: development of an in vitro system for studying the intravaginal release and permeation of flurogestone acetate, *J.Pharm.Sci.*, **1984**, *73*, 1464–1468.

# Fluticasone propionate

**Molecular formula:** $C_{25}H_{31}F_3O_5S$
**Molecular weight:** 500.58
**CAS Registry No:** 80474-14-2
**Merck Index:** 13, 4237

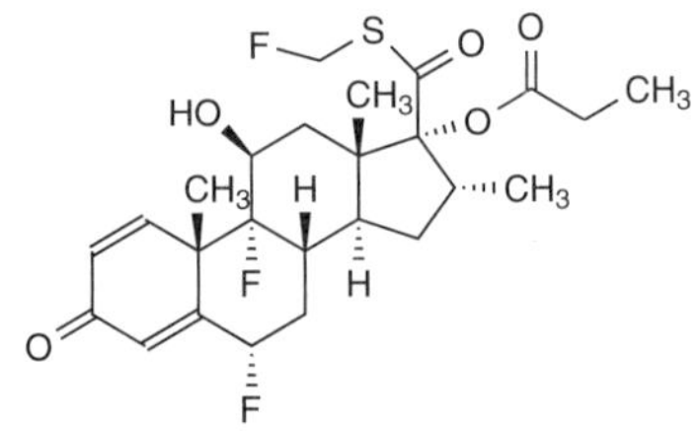

## SAMPLE

**Matrix:** blood

**Sample preparation:** The SPE cartridges were contained in 96 well MicroLute II SPE blocks (Porvair Sciences) with each cell packed with 50 mg Varian C18. Condition each cell with 400 μL MeOH and 400 μL water. Centrifuge plasma at 1000 g for 10 min. Mix 500 μL plasma with 500 μL 1 ng/mL IS in 100 mM pH 7.4 ammonium formate buffer, wash with 400 μL water, wash with 400 μL MeOH:water 40:60 (40% aqueous methanol), elute with 200 μL MeOH. Evaporate the eluate to dryness under a stream of heated nitrogen, reconstitute the residue with 100 μL MeOH:25 mM pH 5 ammonium formate 50:50, inject an 80 μL aliquot.

## HPLC VARIABLES

**Column:** 150 × 4.6 5 μm ResElut C8 BD (Varian)
**Column temperature:** 40
**Mobile phase:** MeOH:25 mM pH 5 ammonium formate 80:20
**Flow rate:** 1
**Injection volume:** 80
**Detector:** MS, PE Sciex API-III Plus triple quadrupole, column effluent split 1:30 so that 336 μL/min enters MS (?), TurboIonspray 500°, positive ion electrospray 5000 V, nebulizer gas nitrogen at 400 kPa, auxiliary gas nitrogen at 2.0 L/min, collision gas argon, collision energy 25 eV, m/z 501–313

## CHROMATOGRAM

**Retention time:** 3.0
**Internal standard:** $^{13}C_3$-fluticasone propionate (m/z 504–313)
**Limit of quantitation:** 20 pg/mL

## KEY WORDS

pharmacokinetics; plasma; SPE

## REFERENCE

Callejas, S.L.; Biddlecombe, R.A.; Jones, A.E.; Joyce, K.B.; Pereira, A.I.; Pleasance, S. Determination of the glucocorticoid fluticasone propionate in plasma by automated solid-phase extraction and liquid chromatography-tandem mass spectrometry, *J.Chromatogr.B*, **1998**, *718*, 243–250.

## SAMPLE

**Matrix:** bulk

**Sample preparation:** Inject a 200 μL aliquot of a 6.7 mg/mL solution in mobile phase.

## HPLC VARIABLES

**Column:** 250 × 4.6 5 μm Inertsil ODS-2
**Column temperature:** 40

**Mobile phase:** Gradient. A:B from 45:55 to 60:40 over 25 min, to 75:25 over 25 min. A was MeCN. B was deuterated water containing 0.05% trifluoroacetic acid.
**Flow rate:** 1
**Injection volume:** 200
**Detector:** UV 239 connected with NMR, Bruker AMX-600, stop-flow mode, 600.14 MHz; MS, Trio 1000 single quadrupole, thermospray, source 200°, orifice 200°, source ion repeller 100 V

---

## CHROMATOGRAM
**Retention time:** 14

---

## OTHER SUBSTANCES
**Simultaneous:** impurities

---

## REFERENCE
Mistry, N.; Ismail, I.M.; Smith, M.S.; Nicholson, J.K.; Lindon, J.C. Characterisation of impurities in bulk drug batches of fluticasone propionate using directly coupled HPLC-NMR spectroscopy and HPLC-MS, *J.Pharm.Biomed.Anal.*, **1997**, *16*, 697–705.

---

## SAMPLE
**Matrix:** formulations
**Sample preparation:** Mix 1 g of a topical product, 5 mL 100 mM pH 4 citrate buffer saturated with NaCl, and 10 mL ethyl acetate, shake vigorously by hand to ensure that no large clumps stick to the tube, mix with the tube on its side on an oscillating shaker for 15 min. Remove the upper ethyl acetate layer and wash with citrate buffer as before. Dry the ethyl acetate over anhydrous sodium sulfate and evaporate to dryness. Dissolve the residue in a mixture of 10 mL heptane and 5 mL MeCN:water 90:10. Remove the upper heptane layer and extract it with 2 mL MeCN:water 90:10. Combine the MeCN/water layers and evaporate them to dryness, dissolve the residue in 300 μL MeOH, filter (0.45 μm nylon), inject a 5 μL aliquot.

---

## HPLC VARIABLES
**Guard column:** 15 × 3.2 7 μm Brownlee NewGuard C18
**Column:** 75 × 4.6 3.5 μm Symmetry C18 (Waters)
**Mobile phase:** Gradient. MeCN:water 18:82 for 2 min, to 82:18 over 12 min, maintain at 82:18 for 3 min, re-equilibrate at initial conditions for 12 min.
**Flow rate:** 1
**Injection volume:** 5
**Detector:** UV 240

---

## CHROMATOGRAM
**Retention time:** 11.19 (17-propionate)
**Limit of detection:** 0.001%

---

## OTHER SUBSTANCES
**Simultaneous:** alclometasone 17,21-dipropionate (10.93), amcinonide (10.90), beclomethasone 17,21-dipropionate (11.90), betamethasone (6.52), betamethasone 21-acetate (8.47), betamethasone 17-benzoate (10.28), betamethasone 17,21-dipropionate (11.42), betamethasone 17-valerate (10.25), budesonide (8.55, 8.69 (epimers)), clobetasol 17-propionate (11.06), cortisone (5.62), cortisone 21-acetate (8.07), dexamethasone (6.57), dexamethasone 21-acetate (8.68), desonide (6.99), desoximetasone (7.60), desoxycorticosterone acetate (10.90), desoxycorticosterone pivalate (14.45), diflorasone 17,21-diacetate (9.81), fluocinolone acetonide (7.38), fluocinonide (9.79), flurandrenolide (7.36), fludrocortisone 21-acetate (7.77), fluorometholone (7.67), flumethasone 21-pivalate (11.20), flunisolide (7.14), fluprednisolone (5.46), halcinonide (10.72), halobetasol propionate (10.98), hydrocortisone (5.50), hydrocortisone 21-acetate (7.65), hydrocortisone 17-butyrate (8.66), hydrocortisone 21-cypionate (12.54),

hydrocortisone 17-valerate (9.53), mometasone 17-furoate (11.24), methylprednisolone (6.31), methylprednisolone 21-acetate (8.34), meprednisone (6.55), prednisolone (5.37), prednisolone 21-acetate (7.47), prednisolone 21-tebuate (11.01), paramethasone 21-acetate (8.64), prednicarbate (10.72), prednisone (5.46), triamcinolone (4.15), triamcinolone acetonide (7.04), triamcinolone 16,21-diacetate (7.49), triamcinolone hexacetonide (13.12)

## KEY WORDS

body wash; cream; gel; lotion; shampoo; spray

## REFERENCE

Reepmeyer, J.C. Screening for corticosteroids in topical pharmaceuticals by HPLC with a scanning ultraviolet detector, *J.Liq.Chromatogr.Rel.Technol.*, **2001**, *24*, 693–709.

## ANNOTATED BIBLIOGRAPHY

Bernal, J.L.; Del Nozal, M.J.; Martín, M.T.; Diez-Masa, J.C.; Cifuentes, A. Quantitation of active ingredients and excipients in nasal sprays by high-performance liquid chromatography, capillary electrophoresis and UV spectroscopy, *J.Chromatogr.A*, **1998**, *823*, 423–431. [beclomethasone dipropionate; fluticasone dipropionate; beclomethasone; fluticasone; benzalkonium; benzalkonium chloride; phenylethyl alcohol]

Borchard, G.; Cassará, M.L.; Roemelé, P.E.H.; Florea, B.I.; Junginger, H.E. Transport and local metabolism of budesonide and fluticasone propionate in a human bronchial epithelial cell line (Calu-3), *J.Pharm. Sci.*, **2002**, *91*, 1561–1567.

Krishnaswami, S.; Möllmann, H.; Derendorf, H.; Hochhaus, G. A sensitive LC-MS/MS method for the quantification of fluticasone propionate in human plasma, *J.Pharm.Biomed.Anal.*, **2000**, *22*, 123–129. [SPE]

Laugher, L.; Noctor, T.G.; Barrow, A.; Oxford, J.M.; Phillips, T. An improved method for the determination of fluticasone propionate in human plasma, *J.Pharm.Biomed.Anal.*, **1999**, *21*, 749–758. [SPE]

Li, Y.N.; Tattam, B.N.; Brown, K.F.; Seale, J.P. A sensitive method for the quantification of fluticasone propionate in human plasma by high-performance liquid chromatography/atmospheric pressure chemical ionisation mass spectrometry, *J.Pharm.Biomed.Anal.*, **1997**, *16*, 447–452.

Li, Y.N.; Tattam, B.; Brown, K.F.; Seale, J.P. Quantification of epimeric budesonide and fluticasone propionate in human plasma by liquid chromatography-atmospheric pressure chemical ionization tandem mass spectrometry, *J.Chromatogr.B*, **2001**, *761*, 177–185. [SPE]

# Flutrimazole

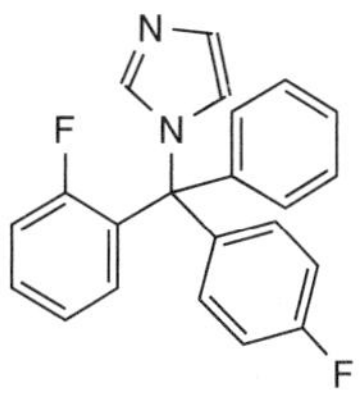


**Molecular formula:** $C_{22}H_{16}F_2N_2$
**Molecular weight:** 346.37
**CAS Registry No:** 119006-77-8
**Merck Index:** 13, 4241

## SAMPLE

**Matrix:** blood, feces, urine
**Sample preparation:** Mix 500 μL plasma or urine with 500 μL MeOH, shake, centrifuge, inject a 50 μL aliquot. Homogenize feces with an equal volume of MeOH, centrifuge at 5000 rpm for 20 min, inject a 100 μL aliquot.

## HPLC VARIABLES

**Guard column:** μBondapak C18
**Column:** 300 × 3.9 10 μm μBondapak C18
**Mobile phase:** MeOH:5 mM pH 7 phosphate buffer 75:25
**Flow rate:** 1
**Injection volume:** 50–100
**Detector:** UV 225

## CHROMATOGRAM

**Retention time:** 10

## OTHER SUBSTANCES

**Extracted:** clotrimazole (11.2), imidazole (3.3)

## KEY WORDS

dog; plasma

## REFERENCE

Conte, L.; Ramis, J.; Mis, R.; Vilageliu, J.; Basi, N.; Forn, J. Pharmacokinetic study of [$^{14}$C]flutrimazole after oral and intravenous administration in dogs. Comparison with clotrimazole, *Arzneimittelforschung*, **1992**, *42*, 854–858.

# Fomepizole

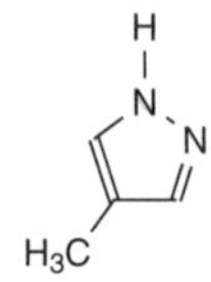


**Molecular formula:** $C_4H_6N_2$
**Molecular weight:** 82.10
**CAS Registry No:** 7554-65-6
**Merck Index:** 13, 4252

## SAMPLE
**Matrix:** blood
**Sample preparation:** Condition a 100 mg Bond Elut SCX SPE cartridge with two 1 mL portions of MeOH and two 1 mL portions of 5 mM HCl. Mix 200 μL plasma with 800 μL 6.25 μM IS in 5 mM HCl, add to the SPE cartridge, wash with 1 mL MeOH:5 mM HCl 5:95, elute with 1 mL MeOH:250 mM pH 7.4 potassium phosphate buffer, inject a 100 μL aliquot.

## HPLC VARIABLES
**Guard column:** 30 × 4 Nucleosil 120-5C18
**Column:** 100 × 4 Nucleosil 120-5C18
**Mobile phase:** MeOH:5 mM pH 6.0 potassium phosphate buffer 20:80
**Flow rate:** 0.8
**Injection volume:** 100
**Detector:** UV 220

## CHROMATOGRAM
**Retention time:** 7
**Internal standard:** 3-methylpyrazole (6)
**Limit of quantitation:** 2.5 μM

## KEY WORDS
plasma; SPE

## REFERENCE
Diczfalusy, U.; Eklöf, R. Determination of 4-methylpyrazole in plasma using solid phase extraction and HPLC, *Biomed.Chromatogr.*, **1987**, *2*, 226–227.

## SAMPLE
**Matrix:** blood, dialysate
**Sample preparation:** Mix 200 μL plasma with 10 μL 300 μg/mL IS in water, add 100 μL 15% trichloroacetic acid, mix, centrifuge at 10 000 g for 4 min. Mix a 150 μL aliquot of the supernatant with 100 μL 500 mM disodium hydrogen phosphate solution (pH ca. 7), inject a 50 μL aliquot. Inject a 200 μL aliquot of dialysate directly.

## HPLC VARIABLES
**Guard column:** 4 × 4 5 μm LiChrospher 60 RP-select B
**Column:** 125 × 4 5 μm LiChrospher 100 RP-18
**Column temperature:** 40
**Mobile phase:** MeCN:5 mM pH 6 potassium phosphate buffer 7.5:92.5
**Flow rate:** 1.5
**Injection volume:** 50–200
**Detector:** UV 220

## CHROMATOGRAM
**Retention time:** 4.9
**Internal standard:** 3-methylpyrazole (4.1)

**Limit of quantitation:** 300 ng/mL (plasma), 50 ng/mL (dialysate)

## KEY WORDS

plasma

## REFERENCE

Jobard, E.; Turcant, A.; Harry, P.; Le Bouil, A.; Allain, P. High-performance liquid chromatographic determination of 4-methylpyrazole in plasma and in dialysate, *J.Chromatogr.B*, **1997**, *695*, 444–447.

## ANNOTATED BIBLIOGRAPHY

McMartin, K.E.; Collins, T.D.; Hewlett, T.P. High pressure liquid chromatographic assay of 4-methylpyrazole. Measurements of plasma and urine levels, *J.Toxicol.Clin.Toxicol.*, **1984**, *22*, 133–148.

# Fomivirsen

**Molecular formula:** $C_{204}H_{263}N_{63}O_{114}P_{20}S_{20}$
**Molecular weight:** 6682.46
**CAS Registry No:** 144245-52-3, 160369-77-7 (Na salt)
**Merck Index:** 13, 4254

## SAMPLE

**Matrix:** tissue, vitreous humor
**Sample preparation:** Heat retina sample with 2 mg/mL proteinase K in buffer at 37° for 16–24 h, wash twice with phenol:chloroform 50:50 (Caution! Chloroform is a carcinogen!), wash with chloroform, evaporate to dryness under reduced pressure, suspend in water, inject an aliquot. Vitreous humor is similar, but buffer is not used. (Caution! Chloroform is a carcinogen! Buffer was 20 mM pH 8.0 Tris-HCl containing 0.5% Non-Idet P-40 (NP-40), 20 mM EDTA, and 100 mM NaCl.)

## HPLC VARIABLES

**Column:** 100 × 4.6 Gen Pak Fax strong anion exchange
**Mobile phase:** Gradient. A was MeOH:86 mM pH 8.0 Tris-HCl 20:80. B was 86 mM pH 8.0 Tris-HCl containing 1.5 M NaBr. A:B 100:0 for 5 min, to 40:60 over 45 min.
**Flow rate:** 0.5
**Detector:** Radioactivity ($^{14}C$)

## CHROMATOGRAM

**Retention time:** 50

## OTHER SUBSTANCES

**Extracted:** metabolites

## KEY WORDS

retina

## REFERENCE

Leeds, J.M.; Henry, S.P.; Truong, L.; Zutshi, A.; Levin, A.A.; Kornbrust, D. Pharmacokinetics of a potential human cytomegalovirus therapeutic, a phosphorothioate oligonucleotide, after intravitreal injection in the rabbit, *Drug Metab.Dispos.*, **1997**, *25*, 921–926.

# Fondaparinux

**Molecular formula:** $C_{31}H_{43}N_3Na_{10}O_{49}S_8$
**Molecular weight:** 1728.08
**CAS Registry No:** 114870-03-0
**Merck Index:** 13, 4257

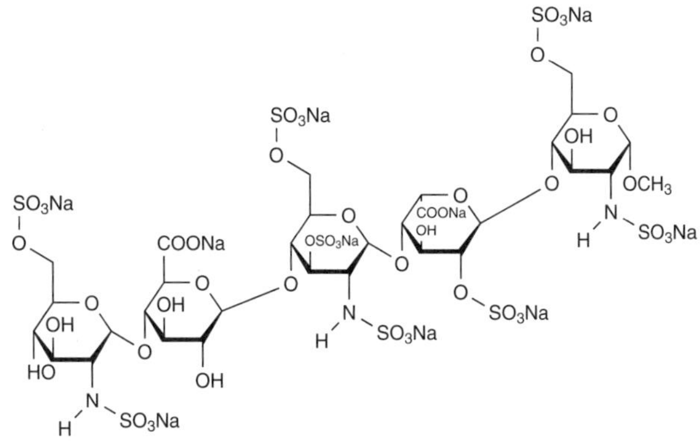


## SAMPLE
**Matrix:** solutions

## HPLC VARIABLES
**Column:** HR 5/5 mono-Q (Pharmacia)
**Mobile phase:** Gradient. A was 500 mM NaCl. B was 2 M NaCl. A:B from 100:0 to 0:100 over 25 min.
**Flow rate:** 1
**Detector:** Radioactivity ($^{35}S$)

## CHROMATOGRAM
**Retention time:** 18

## REFERENCE
Lieu, C.; Shi, J.; Donat, F.; Van Horn, R.; Brian, W.; Newton, J.; Delbressine, L.; Vos, R. Fondaparinux sodium is not metabolised in mammalian liver fractions and does not inhibit cytochrome P450-mediated metabolism of concomitant drugs, *Clin.Pharmacokinet.*, **2002**, *41*(Suppl. 2), 19–26.

# Formestane

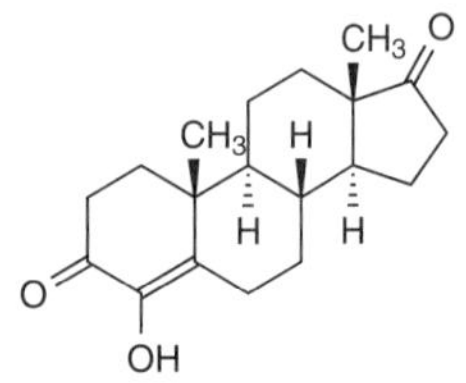

**Molecular formula:** $C_{19}H_{26}O_3$
**Molecular weight:** 302.41
**CAS Registry No:** 566-48-3
**Merck Index:** 13, 4264

## SAMPLE

**Matrix:** urine
**Sample preparation:** Add 500 μL β-glucuronidase (Boehringer Mannheim) to 5 mL urine, heat at 37° for 22 h, extract three times with 10 mL portions of ethyl acetate. Combine the extracts and evaporate to dryness, reconstitute the residue with 100 μL MeOH, inject a 20 μL aliquot.

## HPLC VARIABLES

**Column:** 150 × 4.6 Apex II 5ODS (Jones)
**Mobile phase:** MeOH:water:formic acid 60:40:0.01
**Flow rate:** 1.2
**Injection volume:** 20
**Detector:** MS, Finnigan MAT TSQ 700 triple quadrupole, thermospray, discharge-on mode (1000 V), positive ion mode, source block 200°, vaporizer 100°, repeller 50 V, m/z 303

## CHROMATOGRAM

**Retention time:** 8.1

## OTHER SUBSTANCES

**Extracted:** metabolites

## REFERENCE

Poon, G.K.; Jarman, M.; McCague, R.; Davies, J.H.; Heeremans, C.E.M.; van der Hoeven, R.A.M.; Niessen, W.M.A.; van der Greef, J. Identification of 4-hydroxyandrost-4-ene-3,17-dione metabolites in prostatic cancer patients by liquid chromatography-mass spectrometry, *J.Chromatogr.*, **1992**, *576*, 235–244.

# Formoterol

**Molecular formula:** $C_{19}H_{24}N_2O_4$
**Molecular weight:** 344.40
**CAS Registry No:** 73573-87-2
**Merck Index:** 13, 4269

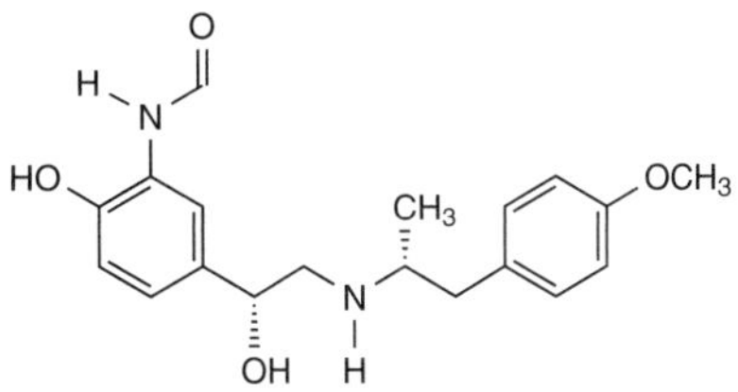

## SAMPLE

**Matrix:** blood

**Sample preparation:** Condition a 1 mL 100 mg polysulfonic sorbent strong cation-exchange SPE cartridge with 2 mL MeOH, 500 μL water, 3 mL 25 mM pH 6.6 phosphate buffer, and 1 mL water. Mix 1 mL plasma with 50 μL 5 ng/mL IS in water, 750 μL 50 mM pH 6.6 buffer, and 50 μL water, vortex for 5 s. Add to the SPE cartridge, wash with 2.5 mL 25 mM pH 6.6 phosphate buffer, wash with 500 μL water, wash with 250 μL MeOH:water 20:80, dry with 2 mL air, elute with two 150 μL portions of MeOH:pH 6.0 buffer 30:70, dilute the eluate with 200 μL water, mix, filter (0.45 μm nylon), inject a 300 μL aliquot. (50 mM pH 6.6 Buffer was 50 mM potassium dihydrogen phosphate:50 mM disodium hydrogen phosphate 62.7:37.3. Prepare 25 mM pH 6.6 buffer by dilution. The pH 6.0 buffer was 30 mM potassium dihydrogen phosphate:30 mM disodium hydrogen phosphate 87.7:12.3 containing 15 g/L KCl.)

## HPLC VARIABLES

**Guard column:** 10 × 2 R2 Chromsep
**Column:** 200 × 3 5 μm Hypersil ODS
**Column temperature:** 33
**Mobile phase:** MeCN:MeOH:buffer 5:25:70 (The buffer was 35 mM potassium dihydrogen phosphate:35 mM disodium hydrogen phosphate 87.7:12.3 containing 20 mg/L EDTA, pH 6.)
**Flow rate:** 0.4
**Injection volume:** 300
**Detector:** E, Model Decade (Antec, Leiden, Netherlands), VT-03 flowcell, working electrode with a 50 μm thickness spacer, potential +0.63 V in DC mode

## CHROMATOGRAM

**Retention time:** 15.6
**Internal standard:** 3-acetylamino-4-hydroxy-α-[*N*-[1-methyl-2-(*p*-methoxyphenyl) ethyl]aminomethyl]benzyl alcohol (CGP 47086A) (19)
**Limit of detection:** 4 pM
**Limit of quantitation:** 3 pg/mL

## KEY WORDS

pharmacokinetics; plasma; SPE

## REFERENCE

Campestrini, J.; Lecaillon, J.B.; Godbillon, J. Automated and sensitive method for the determination of formoterol in human plasma by high-performance liquid chromatography and electrochemical detection, *J.Chromatogr.B*, **1997**, *704*, 221–229.

## SAMPLE

**Matrix:** blood

**Sample preparation:** Condition an Isolute weak cation-exchange SPE cartridge with MeOH, water, and 100 mM pH 5.0 ammonium acetate buffer. Add plasma to the SPE cartridge, wash with 100 mM pH 5.0 ammonium acetate buffer, wash with MeOH, elute with MeOH:water:formic acid 75:23.75:1.25. Evaporate the eluate to dryness, reconstitute the residue with MeOH:water:acetic acid 10:89.5:0.5, inject an 80 μL aliquot onto column A, elute to waste with mobile phase A, divert the mobile phase containing formoterol onto column B, elute the contents of column B onto column C with mobile phase B, monitor the effluent from column C.

## HPLC VARIABLES

**Column:** A 10 × 2.1 Hypersil NH2 + 33 × 2.1 4 μm Genesis CN; B 10 × 2.1 4 μm Genesis C18; C Hichrom ACE C18
**Mobile phase:** A MeOH:water:acetic acid 15:84.5:0.5; B MeOH:water:acetic acid 35:64.5:0.5
**Flow rate:** 0.22
**Injection volume:** 80
**Detector:** MS, Finnigan TSQ 7000, atmospheric pressure ionization, positive ion electrospray

## CHROMATOGRAM

**Limit of quantitation:** 5 pM

## KEY WORDS

column-switching; pharmacokinetics; plasma; SPE

## REFERENCE

Tronde, A.; Nordén, B.; Marchner, H.; Wendel, A.-K.; Lennernäs, H.; Bengtsson, U.H. Pulmonary absorption rate and bioavailability of drugs in vivo in rats: structure-absorption relationships and physicochemical profiling of inhaled drugs, *J.Pharm.Sci.*, **2003**, *92*, 1216–1233.

## ANNOTATED BIBLIOGRAPHY

Rosenborg, J.; Larsson, P.; Tegnér, K.; Hallström, G. Mass balance and metabolism of [$^3$H]formoterol in healthy men after combined i.v. and oral administration-mimicking inhalation, *Drug Metab.Dispos.*, **1999**, *27*, 1104–1116.

# Fosamprenavir calcium

**Molecular formula:** $C_{25}H_{36}CaN_3O_9PS$
**Molecular weight:** 625.68
**CAS Registry No:** 226700-81-8

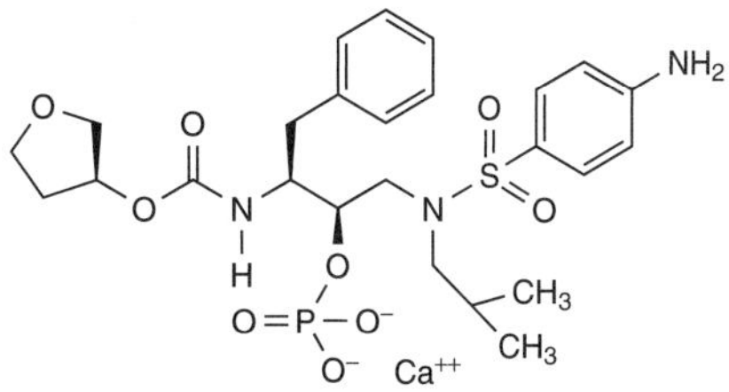

## SAMPLE

**Matrix:** blood
**Sample preparation:** Vortex 100 μL plasma with 200 μL 1 μg/mL IS in MeCN, centrifuge. Mix a 100 μL aliquot of the supernatant with 100 μL 0.1% formic acid and 400 μL MeCN:water 50:50, inject a 10–20 μL aliquot.

## HPLC VARIABLES

**Column:** 50 × 2 3 μm Phenomenex Aqua C18 or 30 × 2 3 μm Phenomenex Aqua C18
**Mobile phase:** Gradient. A was MeOH:10 mM pH 3.5 ammonium formate 1.5:98.5. B was MeCN. A:B 99:1 for 3 min, to 5:95 over 1.5 min, maintain at 5:95 for 1.4 min, return to initial conditions over 0.1 min, re-equilibrate for 3 min.
**Flow rate:** 0.325
**Injection volume:** 10–20
**Detector:** MS, PE Sciex API2000 turbo ion spray, electrospray ionization, positive ion mode

## CHROMATOGRAM

**Internal standard:** $^{13}C_6$-amprenavir

## OTHER SUBSTANCES

**Extracted:** amprenavir, nelfinavir

## KEY WORDS

plasma

## REFERENCE

Huang, L.; Wring, S.A.; Woolley, J.L.; Brouwer, K.R.; Serabjit-Singh, C.; Polli, J.W. Induction of P-glycoprotein and cytochrome P450 3A by HIV protease inhibitors, *Drug Metab.Dispos.*, **2001**, *29*, 754–760.

## SAMPLE

**Matrix:** blood
**Sample preparation:** Mix 100 μL plasma with 500 μL EtOH:MeCN 50:50, centrifuge at 15 800 g at 4° for 5 min. Evaporate the supernatant to dryness, reconstitute the residue with 100 μL initial mobile phase, inject a 10 μL aliquot.

## HPLC VARIABLES

**Column:** 100 × 2 5 μm Luna C18 (Phenomenex)
**Mobile phase:** Gradient. MeCN:buffer from 20:80 to 95:5 over 5 min, maintain at 95:5 for 4 min. (The buffer was 0.1% aqueous acetic acid adjusted to pH 5.4 with ammonium hydroxide.)
**Flow rate:** 0.2
**Injection volume:** 10

**Detector:** MS, PE Sciex API-365, triple quadrupole, collision energy 30 eV, m/z 586.3–418.2

## CHROMATOGRAM

**Retention time:** 5.5
**Limit of quantitation:** 300 pg/mL

## OTHER SUBSTANCES

**Extracted:** amprenavir (506.3–245.2) (7.0)

## KEY WORDS

dog; plasma; rat

## REFERENCE

Furfine, E.S.; Baker, C.T.; Hale, M.R.; Reynolds, D.J.; Salisbury, J.A.; Searle, A.D.; Studenberg, S.D.; Todd, D.; Tung, R.D.; Spaltenstein, A. Preclinical pharmacology and pharmacokinetics of GW433908, a water-soluble prodrug of the human immunodeficiency virus protease inhibitor amprenavir, *Antimicrob.Agents Chemother.*, **2004**, *48*, 791–798.

# Fosinopril

**Molecular formula:** $C_{30}H_{46}NO_7P$
**Molecular weight:** 563.66
**CAS Registry No:** 98048-97-6
**Merck Index:** 13, 4279

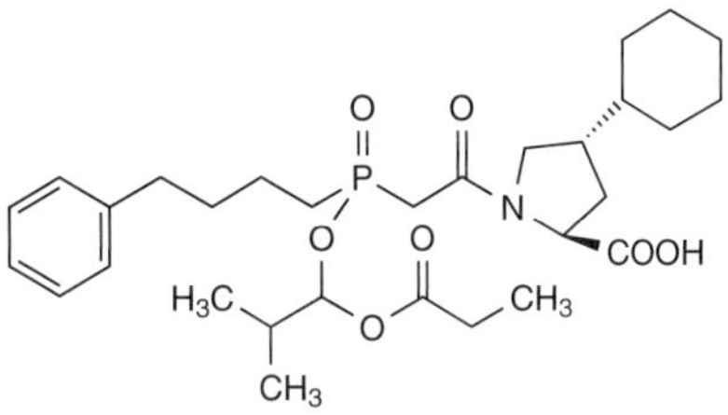


## SAMPLE

**Matrix:** blood

**Sample preparation:** Condition a 3 mL 200 mg Bond Elut cyclohexyl SPE cartridge with 3 mL MeOH and 3 mL 0.1 N phosphoric acid. Vortex 1 mL serum with 2 mL 0.2 N phosphoric acid and an aliquot of EtOH containing 1.5 μg/mL SQ-33055 and 1.5 μg/mL SQ-27133, add to the SPE cartridge, wash with 3 mL 0.1 N phosphoric acid, wash with 3 mL 10 mM pH 4.6 ammonium acetate buffer, dry under vacuum, elute with 1.5 mL MeOH containing 10 mM ammonium acetate. Evaporate the eluate to dryness under a stream of nitrogen at 40°, reconstitute the residue with 75 μL initial mobile phase, inject a 5 μL aliquot. (Use silanized glass or polypropylene containers. Silanize glass vials by immersing in Prosil-28 working solution (10 mL concentrate diluted to 1 L with water) for 10 min, rinsing with water several times, drying in air.)

## HPLC VARIABLES

**Column:** 50 × 2 5 μm Asahipak ODP PVA-C18

**Mobile phase:** Gradient. A was 770 mg ammonium acetate in 750 mL water and 250 mL MeOH, adjusted to pH 5.5 with glacial acetic acid. B was 770 mg ammonium acetate in MeOH. A:B from 70:30 to 5:95 over 1 min, maintain at 5:95 for 2 min, return to initial conditions over 2 min, re-equilibrate for 5 min.

**Flow rate:** 0.2

**Injection volume:** 5

**Detector:** MS, PE Sciex API-III Plus triple quadrupole, turbo ion spray, positive ion, sprayer 4200 V, orifice 35 V, declustering 4 V, nebulizer gas nitrogen at 65 psi, curtain gas nitrogen at 1.2 L/min, turbo ion spray nitrogen 4 L/min, collision gas argon 23 eV, m/z 581.3–436.2

## CHROMATOGRAM

**Retention time:** 4.2

**Internal standard:** SQ-33055 ((4*S*)-4-phenyl-1-[[(*R*)-[(1*S*)-2-methyl-1-(1-oxopropoxy) propoxy](4-phenylbutyl)phosphinyl]acetyl]-L-proline) (m/z 575.3–430.2) (4), SQ-27133 ((4*S*)-4-phenylthio-1-[[(*R*)-(4-phenylbutyl)phosphinyl]acetyl]-L-proline) (m/z 479.2–416.2) (3.3)

**Limit of quantitation:** 1.17 ng/mL

## OTHER SUBSTANCES

**Extracted:** fosinoprilat (m/z 453.2–390.2) (3.5)

## KEY WORDS

serum; SPE

## REFERENCE


Jemal, M.; Mulvana, D.E. Liquid chromatographic-electrospray tandem mass spectrometric method for the simultaneous quantitation of the prodrug fosinopril and the active drug fosinoprilat in human serum, *J.Chromatogr.B*, **2000**, *739*, 255–271.

# Fosphenytoin

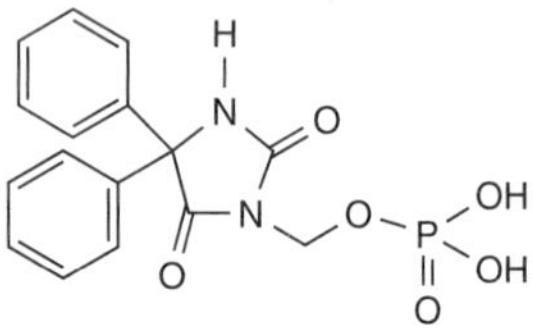

**Molecular formula:** $C_{16}H_{15}N_2O_6P$
**Molecular weight:** 362.27
**CAS Registry No:** 93390-81-9
**Merck Index:** 13, 4280

## SAMPLE
**Matrix:** blood
**Sample preparation:** Vortex 1 mL plasma, 3 mL MeCN, and 5 μg IS, centrifuge for 10 min, inject an aliquot of the supernatant.

## HPLC VARIABLES
**Column:** 150 × 3.9 μBondapak C18
**Mobile phase:** MeOH:10 mM pH 4.0 tetrabutylammonium dihydrogen phosphate 47:53
**Flow rate:** 1.7
**Detector:** UV 214

## CHROMATOGRAM
**Internal standard:** diphenylphosphate
**Limit of quantitation:** 100 ng/mL

## KEY WORDS
pharmacokinetics; plasma

## REFERENCE
Boucher, B.A.; Bombassaro, A.M.; Rasmussen, S.N.; Watridge, C.B.; Achari, R.; Turlapaty, P. Phenytoin prodrug 3-phosphoryloxymethyl phenytoin (ACC-9653): pharmacokinetics in patients following intravenous and intramuscular administration, *J.Pharm.Sci.*, **1989**, *78*, 929–932.

## SAMPLE
**Matrix:** blood
**Sample preparation:** Vortex 100 μL plasma, 100 μL 30 μg/mL IS in MeCN:water 3:97, and 100 μL 85% phosphoric acid for 10 s, add 2 mL diethyl ether, shake horizontally at 120 cycles/min for 20 min, centrifuge at 1300 g for 10 min, evaporate the supernatant to dryness under a stream of nitrogen at 40°, reconstitute the residue with 200 μL mobile phase, inject a 50 μL aliquot. Alternatively, filter (Amicon Centrifree with YMT membrane) 1 mL plasma at 1000 g for 20 min, collect ca. 150 μL, inject a 50 μL aliquot of the ultrafiltrate.

## HPLC VARIABLES
**Guard column:** Guard-Pak Resolve C18 (Waters)
**Column:** 150 × 3.9 5 μm Resolve C18 (Waters)
**Column temperature:** 30
**Mobile phase:** MeCN:water 20:80 containing 5 mM tetrabutylammonium hydrogen sulfate, adjusted to pH 2.2–2.5 with 800 μL/L 85% phosphoric acid
**Flow rate:** 2
**Injection volume:** 50
**Detector:** UV 210

## CHROMATOGRAM
**Retention time:** 8.8
**Internal standard:** 5-(*p*-methylphenyl)-5-phenylhydantion (11.3)
**Limit of detection:** 100 ng/mL (extraction), 15 ng/mL (ultrafiltrate)

**Limit of quantitation:** 400 ng/mL (extraction), 30 ng/mL (ultrafiltrate)

## OTHER SUBSTANCES

**Extracted:** phenytoin (5.8)

**Noninterfering:** *N*-acetylprocainamide, carbamazepine, diazepam, digoxin, ethosuximide, gentamicin, lithium, lorazepam, phenobarbital, primidone, procainamide, quinidine, theophylline, valproic acid

## KEY WORDS

pharmacokinetics; plasma; ultrafiltrate

## REFERENCE

Cwik, M.J.; Liang, M.; Deyo, K.; Andrews, C.; Fischer, J. Simultaneous rapid high-performance liquid chromatographic determination of phenytoin and its prodrug, fosphenytoin in human plasma and ultrafiltrate, *J.Chromatogr.B*, **1997**, *693*, 407–414.

## ANNOTATED BIBLIOGRAPHY

Kugler, A.R.; Annesley, T.M.; Nordblom, G.D.; Koup, J.R.; Olson, S.C. Cross-reactivity of fosphenytoin in two human plasma phenytoin assays, *Clin.Chem.*, **1998**, *44*, 1474–1480.

# Frovatriptan

**Molecular formula:** $C_{14}H_{17}N_3O$
**Molecular weight:** 243.30
**CAS Registry No:** 158747-02-5, 158930-17-7 (succinate)

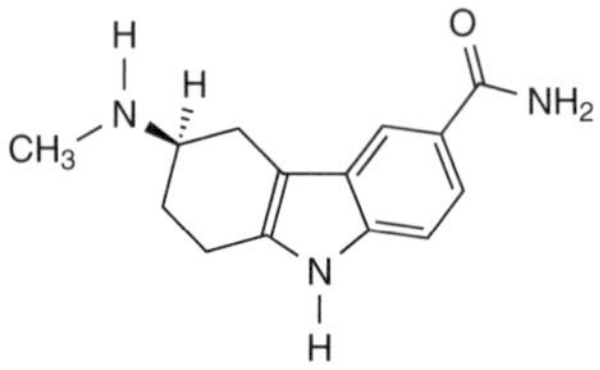

## SAMPLE
**Matrix:** blood
**Sample preparation:** Mix 200 μL blood with 40 μL 10 μg/mL IS and 100 μL 100 mM sodium carbonate, extract with 3 mL chloroform:isopropanol 50:50 (Caution! Chloroform is a carcinogen!). Evaporate the organic layer to dryness under a stream of nitrogen at 60°, reconstitute the residue with 150 μL mobile phase, inject an aliquot.

## HPLC VARIABLES
**Guard column:** 10 × 4.6 4 μm μBondapak C18
**Column:** 150 × 4.6 5 μm Hypersil BDS C18
**Mobile phase:** MeCN:10 mM ammonium acetate 6:94, adjusted to pH 4 with glacial acetic acid
**Flow rate:** 1
**Detector:** UV 244

## CHROMATOGRAM
**Retention time:** 5.5
**Internal standard:** 5,6,7,8-Tetrahydro-6-(ethylamino)carbazole-3-carboxamide (8)
**Limit of quantitation:** 10 ng/mL

## KEY WORDS
dog; mouse; rat; whole blood

## REFERENCE
Laugher, L.; Briggs, R.; Doughty, J. Development of an analytical methodology from toxicokinetic to clinical studies for the anti-migraine drug frovatriptan, *Chromatographia*, **2000**, *52*, S113–S119.

## SAMPLE
**Matrix:** blood
**Sample preparation:** Human. Mix 1 mL blood with 100 μL 200 ng/mL IS and 2 mL MeCN, centrifuge. Evaporate the supernatant to dryness under a stream of nitrogen at 40°, reconstitute the residue with 200 μL mobile phase, inject an aliquot. Rabbit. Mix 250 μL blood with 100 μL 500 ng/mL IS and 2 mL MeCN, centrifuge. Evaporate the supernatant to dryness under a stream of nitrogen at 40°, reconstitute the residue with 250 μL mobile phase, inject an aliquot.

## HPLC VARIABLES
**Guard column:** 10 × 3.2 5 μm Kromasil C8
**Column:** 50 × 4.6 3 μm Hypersil BDS C8
**Mobile phase:** MeCN:10 mM pH 4 ammonium acetate buffer 13:87
**Flow rate:** 1
**Injection volume:**
**Detector:** MS, PE Sciex API-III, ionspray, 40 μL/min entered the detector, m/z 244.3–213.0

## CHROMATOGRAM
**Retention time:** 1

**Internal standard:** 5,6,7,8-Tetrahydro-6-(ethylamino)carbazole-3-carboxamide (m/z 258.3–213.0) (1)
**Limit of quantitation:** 200 pg/mL (human), 10 ng/mL (rabbit)

---

## KEY WORDS

human; rabbit; whole blood

---

## REFERENCE

Laugher, L.; Briggs, R.; Doughty, J. Development of an analytical methodology from toxicokinetic to clinical studies for the anti-migraine drug frovatriptan, *Chromatographia*, **2000**, *52*, S113–S119.

# Fumagillin

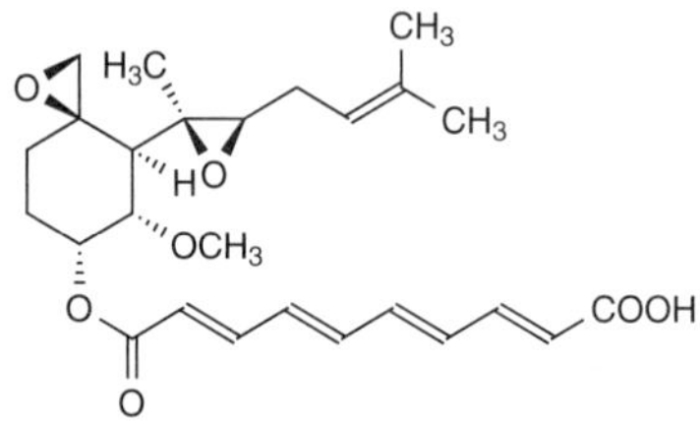

**Molecular formula:** $C_{26}H_{34}O_7$
**Molecular weight:** 458.54
**CAS Registry No:** 23110-15-8
**Merck Index:** 13, 4307

## SAMPLE

**Matrix:** formulations
**Sample preparation:** Mix 100 μL of a 70 μg/mL solution in 0.9% sodium chloride with 900 μL MeOH, vortex for 15 s, centrifuge for 10 min. Mix 100 μL of the supernatant with 900 μL MeOH, vortex for 15 s, centrifuge for 10 min, inject a 20 μL aliquot.

## HPLC VARIABLES

**Column:** Zorbax C18
**Mobile phase:** MeCN:MeOH:water:phosphoric acid 50:20:30:0.1
**Flow rate:** 0.5
**Injection volume:** 20
**Detector:** UV 350

## CHROMATOGRAM

**Retention time:** 3.7
**Limit of quantitation:** 200 ng/mL

## KEY WORDS

ophthalmic solutions; stability-indicating

## REFERENCE

Abdel-Rahman, S.M.; Nahata, M.C. Stability of fumagillin in an extemporaneously prepared ophthalmic solution, *Am.J.Health-Syst.Pharm.*, **1999**, *56*, 547–550.

## SAMPLE

**Matrix:** tissue
**Sample preparation:** Mix 2 g cut up tissue with dichloromethane:dioxane:isopropanol 75:16:9 (Caution! Dioxane is a carcinogen!), sonicate, shake, try to divide solid into smaller pieces with a glass rod, sonicate for 15 min, centrifuge at 5000 rpm for 10 min, filter. Evaporate the filtrate to 1–1.5 mL under a stream of nitrogen at 55°, make up to 2 mL with mobile phase, inject an aliquot.

## HPLC VARIABLES

**Column:** 250 × 4.6 5 μm silica (Perkin Elmer) (Before use, wash column with dichloromethane:acetic acid 95:5.)
**Mobile phase:** *n*-Hexane:dichloromethane:dioxane:isopropanol:acetic acid 43:43:9:5:0.1
**Flow rate:** 2
**Detector:** UV 340

## CHROMATOGRAM

**Retention time:** 10
**Limit of detection:** 5 ng/g
**Limit of quantitation:** 15 ng/g

## KEY WORDS

fat; fish; kidney; liver; muscle; normal phase; protect from light

## REFERENCE

Fekete, J.; Romvári, Z.; Gebefügi, I.; Kettrup, A. Comparative study on determination of fumagillin in fish by normal and reversed phase chromatography, *Chromatographia*, **1998**, *48*, 48–52.

## SAMPLE

**Matrix:** tissue

**Sample preparation:** Condition a Sep-Pak C18 SPE cartridge with 5 mL dichloromethane and 10 mL MeOH. Mix 2 g cut up tissue with dichloromethane:dioxane:isopropanol 75:16:9 (Caution! Dioxane is a carcinogen!), sonicate, shake, try to divide solid into smaller pieces with a glass rod, sonicate for 15 min, centrifuge at 5000 rpm for 10 min, filter. Evaporate the filtrate to 0.1–0.2 mL under a stream of nitrogen at 55°, make up to 2 mL with mobile phase, add to the SPE cartridge, discard the first 0.5 mL (?), collect the second 0.5 mL, inject a 100 μL aliquot.

## HPLC VARIABLES

**Column:** 250 × 4.6 5 μm Spherisorb ODS

**Mobile phase:** MeCN:water:bicyclohexylamine 70:30:0.05, adjusted to pH 5.0 with 1 M phosphoric acid

**Flow rate:** 2

**Injection volume:** 100

**Detector:** UV 340

## CHROMATOGRAM

**Retention time:** 5

**Limit of detection:** 5 ng/g

**Limit of quantitation:** 15 ng/g

## KEY WORDS

fat; fish; kidney; liver; muscle; protect from light; SPE

## REFERENCE

Fekete, J.; Romvári, Z.; Gebefügi, I.; Kettrup, A. Comparative study on determination of fumagillin in fish by normal and reversed phase chromatography, *Chromatographia*, **1998**, *48*, 48–52.

## ANNOTATED BIBLIOGRAPHY

Fekete, J.; Romvári, Z.; Szepesi, I.; Morovján, G. Liquid chromatographic determination of the antibiotic fumagillin in fish meat samples, *J.Chromatogr.A*, **1995**, *712*, 378–381. [trout; LOQ 5 ng/g]

Guyonnet, J.; Richard, M.; Hellings, P. Determination of fumagillin in muscle tissue of rainbow trout using automated ion-pairing liquid chromatography, *J.Chromatogr.B*, **1995**, *666*, 354–359. [SPE; LOQ 20 ng/g]

# Galantamine

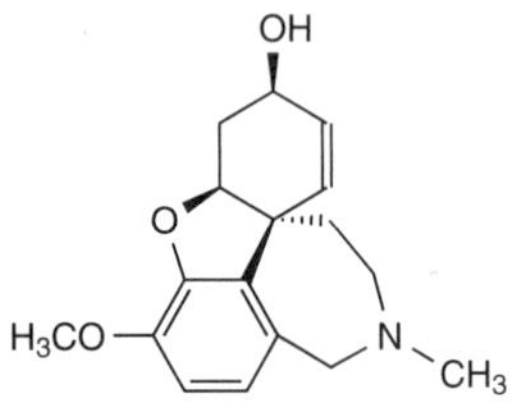


**Molecular formula:** $C_{17}H_{21}NO_3$
**Molecular weight:** 287.35
**CAS Registry No:** 357-70-0
**Merck Index:** 13, 4361

## SAMPLE

**Matrix:** blood
**Sample preparation:** Mix 1 mL plasma with 100 μL 10 μg/mL IS in MeOH, 1 mL saturated KCl solution, and 100 μL 1 M NaOH, vortex, extract twice with 2.5 mL portions of toluene. Combine the organic layers and evaporate to dryness under a stream of nitrogen at 65°, reconstitute the residue with 100 μL MeOH:10 mM ammonium acetate:diethylamine 90:10:1, inject an aliquot.

## HPLC VARIABLES

**Column:** 100 × 4.6 3 μm Hypersil C18 BDS
**Mobile phase:** MeCN:10 mM pH 7 ammonium acetate 15:85
**Flow rate:** 0.8
**Detector:** F ex 280 em 310

## CHROMATOGRAM

**Internal standard:** codeine phosphate
**Limit of quantitation:** 2–10 ng/mL

## OTHER SUBSTANCES

**Extracted:** norgalantamine

## KEY WORDS

dog; mouse; pharmacokinetics; plasma; rabbit; rat

## REFERENCE

Monbaliu, J.; Verhaeghe, T.; Willems, B.; Bode, W.; Lavrijsen, K.; Meuldermans, W. Pharmacokinetics of galantamine, a cholinestrase inhibitor, in several animal species, *Arzneimittelforschung*, **2003**, *53*, 486–495.

## SAMPLE

**Matrix:** blood
**Sample preparation:** Mix 200 μL plasma with 100 μL 50 ng/mL IS in 2% bovine serum albumin in 50 mM pH 7.5 phosphate buffer, add 1 mL 100 mM NaOH, add 500 μL saturated KCl, add 5 mL toluene, extract by over-the-top mixing at 15 rpm for 10 min, centrifuge at 2000 g for 10 min. Evaporate the organic layer to dryness under a stream of nitrogen at 65°, reconstitute the residue with 200 μL mobile phase, inject a 20 μL aliquot.

## HPLC VARIABLES

**Column:** 50 × 4.6 3.5 μm Symmetry Shield C18 (Waters)
**Mobile phase:** MeCN:10 mM ammonium acetate 15:85
**Flow rate:** 1.5
**Injection volume:** 20
**Detector:** MS, PE Sciex API 3000 triple quadrupole, TurboIonspray, positive ion mode, ionspray 4500 V, turbo gas nitrogen 350° 6.5 L/min, nebulizing gas air, collision gas nitrogen, orifice 31 V, ring potential 230 V, m/z 288.1–213.0

## CHROMATOGRAM

**Retention time:** 1.12
**Internal standard:** $^{13}C^{2}H_3$-galantamine (292.1–217.0) (1.10)
**Limit of quantitation:** 1 ng/mL

## OTHER SUBSTANCES

**Extracted:** epigalantamine

## KEY WORDS

pharmacokinetics; plasma

## REFERENCE

Verhaeghe, T.; Diels, L.; De Vries, R.; De Meulder, M.; de Jong, J. Development and validation of a liquid chromatographic-tandem mass spectrometric method for the determination of galantamine in human heparinized plasma, *J.Chromatogr.B*, **2003**, *789*, 337–346.

## ANNOTATED BIBLIOGRAPHY

Huang, F.; Lasseter, K.C.; Janssens, L.; Verhaeghe, T.; Lau, H.; Zhao, Q. Pharmacokinetic and safety assessments of galantamine and risperidone after the two drugs are administered alone and together, *J.Clin.Pharmacol.*, **2002**, *42*, 1341–1351.

Mannens, G.S.J.; Snel, C.A.W.; Hendrickx, J.; Verhaeghe, T.; Le Jeune, L.; Bode, W.; van Beijsterveldt, L.; Lavrijsen, K.; Leempoels, J.; Van Osselaer, N.; Van Peer, A.; Meuldermans, W. The metabolism and excretion of galantamine in rats, dogs, and humans, *Drug Metab.Dispos.*, **2002**, *30*, 553–563.

Zhao, Q.; Brett, M.; Van Osselaer, N.; Huang, F.; Raoult, A.; Van Peer, A.; Verhaeghe, T.; Hust, R. Galantamine pharmacokinetics, safety, and tolerability profiles are similar in healthy Caucasian and Japanese subjects, *J.Clin.Pharmacol.*, **2002**, *42*, 1002–1010.

Zhao, Q.; Iyer, G.R.; Verhaeghe, T.; Truyen, L. Pharmacokinetics and safety of galantamine in subjects with hepatic impairment and healthy volunteers, *J.Clin.Pharmacol.*, **2002**, *42*, 428–436.

# Ganirelix

**Molecular formula:** $C_{80}H_{113}ClN_{18}O_{13}$
**Molecular weight:** 1570.35
**CAS Registry No:** 124904-93-4
**Merck Index:** 13, 4380

### SAMPLE
**Matrix:** solutions

### HPLC VARIABLES
**Column:** 250 × 4.6 5 μm Zorbax Rx octyl
**Mobile phase:** MeCN:50 mM pH 2 ammonium phosphate buffer 27.5:72.5
**Flow rate:** 1
**Detector:** UV 225

### REFERENCE
Strickley, R.G.; Brandl, M.; Chan, K.W.; Straub, K.; Gu, L. High-performance liquid chromatographic (HPLC) and HPLC-mass spectrometric (MS) analysis of the degradation of the luteinizing hormone-releasing hormone (LH-RH) antagonist RS-26306 in aqueous solution, *Pharm.Res.*, **1990**, *7*, 530–536.

### SAMPLE
**Matrix:** solutions

### HPLC VARIABLES
**Column:** 250 × 4.6 5 μm Zorbax Rx octyl
**Mobile phase:** MeCN:100 mM pH 6 ammonium acetate buffer 42:58
**Flow rate:** 1.2
**Injection volume:** 100
**Detector:** UV 230; MS, Finnigan-MAT TSQ-70, thermospray, vaporizer 65–94°, jet 240°, collector 75–80 V, m/z 1570

### CHROMATOGRAM
**Retention time:** 23

### OTHER SUBSTANCES
**Simultaneous:** degradants

### REFERENCE
Strickley, R.G.; Brandl, M.; Chan, K.W.; Straub, K.; Gu, L. High-performance liquid chromatographic (HPLC) and HPLC-mass spectrometric (MS) analysis of the degradation of the luteinizing hormone-releasing hormone (LH-RH) antagonist RS-26306 in aqueous solution, *Pharm.Res.*, **1990**, *7*, 530–536.

# Gatifloxacin

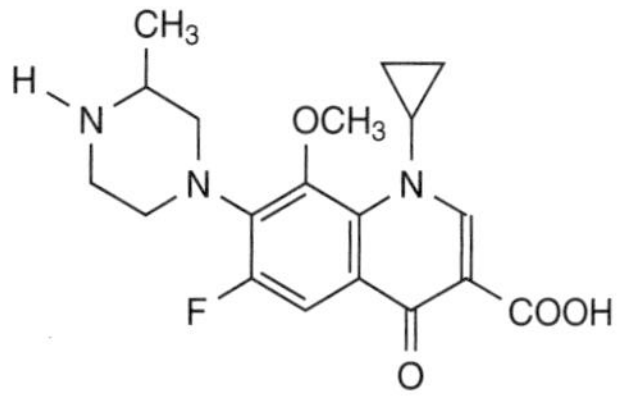

**Molecular formula:** $C_{19}H_{22}FN_3O_4$
**Molecular weight:** 375.39
**CAS Registry No:** 112811-59-3
**Merck Index:** 13, 4388

## SAMPLE

**Matrix:** blood, urine
**Sample preparation:** Serum. Mix 200 μL serum with 200 μL 500 μg/mL IS in MeCN: 100 mM phosphoric acid 90:10 and 400 μL MeCN, centrifuge. Mix 200 μL of the supernatant with 800 μL PIC A solution, inject a 20 μL aliquot. Urine. Dilute 10 μL urine with 990 μL PIC A solution, inject a 5 μL aliquot. (PIC A solution was 10 mM tetrabutylammonium phosphate (Waters PIC A) adjusted to pH 3.48.)

## HPLC VARIABLES

**Guard column:** 30 × 4 30–40 μm Perisorb RP-18
**Column:** 125 × 4 5 μm Nucleosil-100 5C18
**Mobile phase:** MeCN:10 mM tetrabutylammonium phosphate (Waters PIC A) 17:83, adjusted to pH 3.48 with ca. 1.3 mL concentrated phosphoric acid.
**Flow rate:** 1
**Injection volume:** 5–20
**Detector:** F ex 295 em 480

## CHROMATOGRAM

**Retention time:** 4.5
**Internal standard:** trovafloxacin (12.6)
**Limit of detection:** 30 ng/mL (serum), 1.7 μg/mL (urine)
**Limit of quantitation:** 60 ng/mL (serum), 3.5 μg/mL (urine)

## OTHER SUBSTANCES

**Simultaneous:** ciprofloxacin (3.6), grepafloxacin (11.1), moxifloxacin (7.7), ofloxacin (2.7), salicylic acid (22.2), tryptophan (1.9)
**Noninterfering:** furosemide, sparfloxacin

## KEY WORDS

pharmacokinetics; serum

## REFERENCE

Borner, K.; Hartwig, H.; Lode, H. Determination of gatifloxacin in human serum and urine by HPLC, *Chromatographia*, **2000**, *52*, S105–S107.

## SAMPLE

**Matrix:** blood, urine
**Sample preparation:** Serum. Add 50 μL water to 450 μL serum, vortex for 30 s, add 450 μL MeCN:75 mM pH 7.5 phosphate buffer containing 500 μg/mL sodium dodecyl sulfate 20:80, add 50 μL 30 μg/mL IS solution, vortex for 30 s, filter (Amicon Centrifree) while centrifuging at 1500 g for 30 min, inject a 20 μL aliquot of the ultrafiltrate. Urine. Dilute 50 μL urine with 1 mL mobile phase, add 50 μL 30 μg/mL IS solution, vortex for 30 s, inject a 20 μL aliquot.

## HPLC VARIABLES

**Guard column:** 7.5 × 4.6 5 μm Adsorbosphere HS C18

**Column:** 250 × 4.6 5 μm Adsorbosphere HS C18
**Mobile phase:** MeCN:buffer 50:50 (The buffer was 25 mM citric acid containing 10 mM sodium dodecyl sulfate and 10 mM tetrabutylammonium acetate.)
**Flow rate:** 1
**Injection volume:** 20
**Detector:** UV 293

## CHROMATOGRAM

**Retention time:** 6.3
**Internal standard:** ciprofloxacin (5.5)
**Limit of quantitation:** 100 ng/mL

## KEY WORDS

pharmacokinetics; serum; ultrafiltrate

## REFERENCE

Overholser, B.R.; Kays, M.B.; Sowinski, K.M. Determination of gatifloxacin in human serum and urine by high-performance liquid chromatography with ultraviolet detection, *J.Chromatogr.B*, **2003**, *798*, 167–173.

## ANNOTATED BIBLIOGRAPHY

Grant, E.M.; Nicolau, D.P.; Nightingale, C.; Quintiliani, R. Minimal interaction between gatifloxacin and oxycodone, *J.Clin.Pharmacol.*, **2002**, *42*, 928–932. [no HPLC of oxycodone]

Liang, H.; Kays, M.B.; Sowinski, K.M. Separation of levofloxacin, ciprofloxacin, gatifloxacin, moxifloxacin, trovafloxacin and cinoxacin by high-performance liquid chromatography: application to levofloxacin determination in human plasma, *J.Chromatogr.B*, **2002**, *772*, 53–63.

Wallace, A.W.; Victory, J.M.; Amsden, G.W. Lack of bioequivalence of gatifloxacin when coadministered with calcium-fortified orange juice in healthy volunteers, *J.Clin.Pharmacol.*, **2003**, *43*, 92–96. [moxifloxacin is internal standard; fluorescence detection]

# Gefitinib

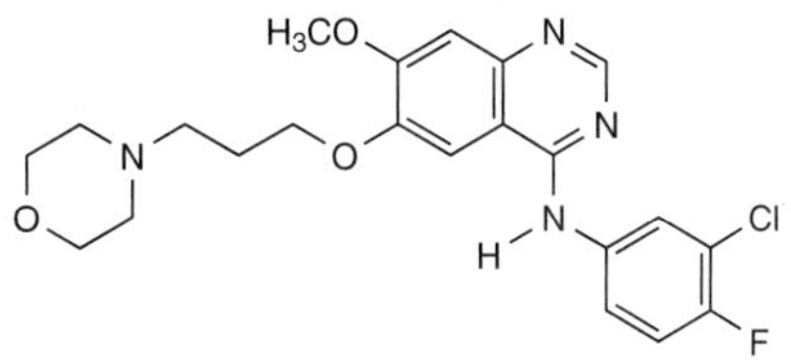


**Molecular formula:** $C_{22}H_{24}ClFN_4O_3$
**Molecular weight:** 446.90
**CAS Registry No:** 184475-35-2

## SAMPLE

**Matrix:** blood
**Sample preparation:** Vortex 500 μL plasma, 25 μL of 1 μg/mL IS in MeOH, 500 μL 1 M NaOH, and 6 mL MTBE for 2 min, centrifuge for 30 s. Remove 5.5 mL of the organic layer and evaporate it to dryness under a stream of nitrogen at 20°, reconstitute the residue with 250 μL mobile phase, inject a 15–100 μL aliquot.

## HPLC VARIABLES

**Guard column:** Inertsil ODS3
**Column:** 150 × 4.6 Inertsil ODS3
**Mobile phase:** MeCN:1% ammonium acetate 80:20
**Flow rate:** 1
**Injection volume:** 15–100
**Detector:** MS, PE Sciex API-III tandem, APCI, m/z 447.2–128

## CHROMATOGRAM

**Retention time:** 2.5
**Internal standard:** $d_8$-gefitinib (m/z 455.4–136) (2.5)
**Limit of quantitation:** 0.5 ng/mL

## KEY WORDS

pharmacokinetics; plasma

## REFERENCE

Jones, H.K.; Stafford, L.E.; Swaisland, H.C.; Payne, R. A sensitive assay for ZD1839 (Iressa) in human plasma by liquid-liquid extraction and high performance liquid chromatography with mass spectrometric detection: validation and use in Phase I clinical trials, *J.Pharm.Biomed.Anal.*, **2002**, *29*, 221–228.

# Gemcitabine

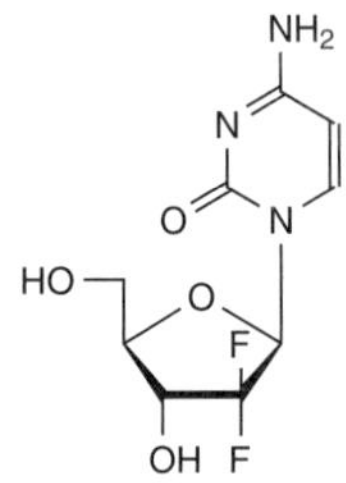


**Molecular formula:** $C_9H_{11}F_2N_3O_4$
**Molecular weight:** 263.20
**CAS Registry No:** 95058-81-4
**Merck Index:** 13, 4397

## SAMPLE

**Matrix:** blood

**Sample preparation:** Mix 6 mL blood with 100 μL 10 mg/mL tetrahydrouridine, a cytidine deaminase inhibitor, centrifuge at 1200 g for 15 min to obtain plasma. Vortex 100 μL plasma and 10 μL 40% trichloroacetic acid for 1 min, centrifuge at 10 000 g at 4° for 5 min, inject a 25 μL aliquot of the supernatant.

## HPLC VARIABLES

**Column:** 250 × 4.6 5 μm Thermo Hypersil BDS C18
**Mobile phase:** MeOH:buffer 17:83 (The buffer was 20 mM pH 3.1 ammonium dihydrogen phosphate buffer containing 10 mM sodium 1-heptanesulfonate.)
**Flow rate:** 0.8
**Injection volume:** 25
**Detector:** UV 272

## CHROMATOGRAM

**Retention time:** 24
**Limit of detection:** 50 ng/mL
**Limit of quantitation:** 80 ng/mL

## OTHER SUBSTANCES

**Extracted:** 2′,2′-difluorodeoxyuridine (dFdU) (6)

## KEY WORDS

pharmacokinetics; plasma

## REFERENCE

Wang, L.-Z.; Goh, B.-C.; Lee, H.-S.; Noordhuis, P.; Peters, G.J. An expedient assay for determination of gemcitabine and its metabolite in human plasma using isocratic ion-pair reversed-phase high-performance liquid chromatography, *Ther.Drug Monit.*, **2003**, *25*, 552–557.

## SAMPLE

**Matrix:** solutions

## HPLC VARIABLES

**Column:** 250 × 4.6 5 μm Zorbax Rx-C8
**Mobile phase:** 13.8 g/L sodium dihydrogen phosphate containing 2.5 mL/L phosphoric acid
**Flow rate:** 1.2
**Injection volume:** 5
**Detector:** UV 275

## REFERENCE

Jansen, P.J.; Akers, M.J.; Amos, R.M.; Baertschi, S.W.; Cooke, G.G.; Dorman, D.E.; Kemp, C.A.J.; Maple, S.R.; McCune, K.A. The degradation of the antitumor agent gemcitabine hydrochloride in an acidic aqueous solution at pH 3.2 and identification of degradation products, *J.Pharm.Sci.*, **2000**, *89*, 885–891.

## SAMPLE

**Matrix:** solutions

## HPLC VARIABLES

**Column:** 250 × 4.6 5 μm Zorbax Rx-C8

**Mobile phase:** Gradient. A was MeOH:buffer 3:97. B was MeOH:buffer 50:50. A:B 100:0 for 8 min, to 0:100 over 5 min, maintain at 0:100 for 7 min. (The buffer was 13.8 g/L sodium dihydrogen phosphate containing 2.5 mL/L phosphoric acid.)

**Flow rate:** 1.2

**Injection volume:** 5

**Detector:** UV 205

## CHROMATOGRAM

**Retention time:** 7.5

## OTHER SUBSTANCES

**Simultaneous:** degradants

**Interfering:** 6-hydroxy-5,6-dihydro-2′-deoxy-2′,2′-difluorouridine

## REFERENCE

Jansen, P.J.; Akers, M.J.; Amos, R.M.; Baertschi, S.W.; Cooke, G.G.; Dorman, D.E.; Kemp, C.A.J.; Maple, S.R.; McCune, K.A. The degradation of the antitumor agent gemcitabine hydrochloride in an acidic aqueous solution at pH 3.2 and identification of degradation products, *J.Pharm.Sci.*, **2000**, *89*, 885–891.

## ANNOTATED BIBLIOGRAPHY

Floridia, L.; Pietropaolo, A.M.; Tavazzani, M.; Rubino, F.M.; Colombi, A. Measurement of surface contamination from nucleoside analogue antineoplastic drugs by high-performance liquid chromatography in occupational hygiene studies of oncologic hospital departments, *J.Chromatogr.B*, **1999**, *724*, 325–334. [fluorouracil; bromouracil; iodouracil; cytarabine; gemcitabine; methotrexate; aminopterin; doxorubicin; daunorubicin; epirubicin; idarubicin]

Freeman, K.B.; Anliker, S.; Hamilton, M.; Osborne, D.; Dhahir, P.H.; Nelson, R.; Allerheiligen, S.R.B. Validated assays for the determination of gemcitabine in human plasma and urine using high-performance liquid chromatography with ultraviolet detection, *J.Chromatogr.B*, **1995**, *665*, 171–181. [normal phase; LOQ 50 ng/mL; pharmacokinetics]

Keith, B.; Xu, Y.; Grem, J.L. Measurement of the anti-cancer agent gemcitabine in human plasma by high-performance liquid chromatography, *J.Chromatogr.B*, **2003**, *785*, 65–72.

Sparidans, R.W.; Crul, M.; Schellens, J.H.M.; Beijnen, J.H. Isocratic ion-exchange chromatographic assay for the nucleotide gemcitabine triphosphate in human white blood cells, *J.Chromatogr.B*, **2002**, *780*, 423–430.

Yilmaz, B.; Kadioglu, Y.; Aksoy, Y. Simultaneous determination of gemcitabine and its metabolite in human plasma by high-performance liquid chromatography, *J.Chromatogr.B*, **2003**, *791*, 103–109.

# Gemifloxacin

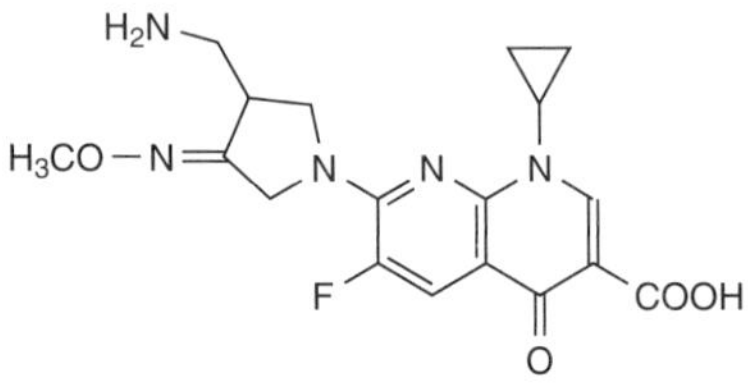


**Molecular formula:** $C_{18}H_{20}FN_5O_4$
**Molecular weight:** 389.38
**CAS Registry No:** 175463-14-6
**Merck Index:** 13, 4400

## SAMPLE
**Matrix:** blood
**Sample preparation:** Vortex 50 μL plasma and 250 μL 250 ng/mL IS in MeCN for 10 s, shake for 30 min, let stand at room temperature for 10 min, centrifuge at 14 000 g for 15 min, add the supernatant to 200 μL buffer in a silanized tube, inject a 10 μL aliquot. (The buffer was 10 mM ammonium acetate adjusted to pH 2.5 with trifluoroacetic acid.)

## HPLC VARIABLES
**Column:** 500 × 4.6 5 μm 100 Å PLRP-S
**Column temperature:** 40
**Mobile phase:** MeCN:buffer 30:70 (The buffer was 10 mM ammonium acetate adjusted to pH 2.5 with trifluoroacetic acid.)
**Flow rate:** 1
**Injection volume:** 10
**Detector:** MS, PE Sciex API 300 tandem, heat-assisted nebulization, electrospray, positive ion mode, nebulizer gas at 60 psi, curtain gas at 40 psi, collision gas thickness 4, auxiliary gas 7 L/min, dwell 400 ms, pause 5 ms, m/z 390–313

## CHROMATOGRAM
**Retention time:** 1
**Internal standard:** $^{13}C^{2}H_3$-gemifloxacin (m/z 394–313)
**Limit of quantitation:** 10 ng/mL

## KEY WORDS
pharmacokinetics; plasma

## REFERENCE
Doyle, E.; Fowles, S.E.; McDonnell, D.F.; McCarthy, R.; White, S.A. Rapid determination of gemifloxacin in human plasma by high-performance liquid chromatography-tandem mass spectrometry, *J.Chromatogr.B*, **2000**, *746*, 191–198.

## SAMPLE
**Matrix:** solutions
**Sample preparation:** Inject a 1–3 μL aliquot of a 2 mg/mL solution.

## HPLC VARIABLES
**Column:** 150 × 4 5 μm Crownpak CR(+) (Daicel)
**Mobile phase:** MeOH:water 15:85 containing 10 mM sulfuric acid
**Flow rate:** 1.2
**Injection volume:** 1–3
**Detector:** UV 272

## CHROMATOGRAM
**Retention time:** 38.88, 58.86 (enantiomers)

## KEY WORDS

chiral

## REFERENCE

Lee, W.; Hong, C.Y. Direct liquid chromatographic enantiomer separation of new fluoroquinolones including gemifloxacin, *J.Chromatogr.A*, **2000**, *879*, 113–120.

## ANNOTATED BIBLIOGRAPHY

Edelstein, P.H.; Shinzato, T.; Doyle, E.; Edelstein, M.A.C. In vitro activity of gemifloxacin (SB-265805, LB20304a) against Legionella pneumophila and its pharmacokinetics in guinea pigs with L-pneumophila pneumonia, *Antimicrob.Agents Chemother.*, **2001**, *45*, 2204–2209. [LC-MS]

Hyun, M.H.; Han, S.C. Liquid chromatographic separation of the enantiomers of fluoroquinolone antibacterials on a chiral stationary phase based on a chiral crown ether, *J.Biochem.Biophys.Methods*, **2002**, *54*, 235–243.

Hyun, M.H.; Han, S.C.; Cho, Y.J.; Jin, J.S.; Lee, W. Liquid chromatographic resolution of gemifloxacin mesylate on a chiral stationary phase derived from crown ether, *Biomed.Chromatogr.*, **2002**, *16*, 356–360.

Naber, C.K.; Hammer, M.; Kinzig-Schippers, M.; Sauber, C. Söurgel, F.; Bygate, E.A.; Fairless, A.J.; Machka, K.; Naber, K.G. Urinary excretion and bactericidal activities of gemifloxacin and ofloxacin after a single oral dose in healthy volunteers, *Antimicrob.Agents Chemother.*, **2001**, *45*, 3524–3530. [LC-MS; no HPLC of ofloxacin]

Ramji, J.V.; Austin, N.E.; Boyle, G.W.; Chalker, M.H.; Duncan, G.; Fairless, A.J.; Hollis, F.J.; McDonnell, D.F.; Musick, T.J.; Shardlow, P.C. The disposition of gemifloxacin, a new fluoroquinolone antibiotic, in rats and dogs, *Drug Metab.Dispos.*, **2001**, *29*, 435–442.

# Gestodene

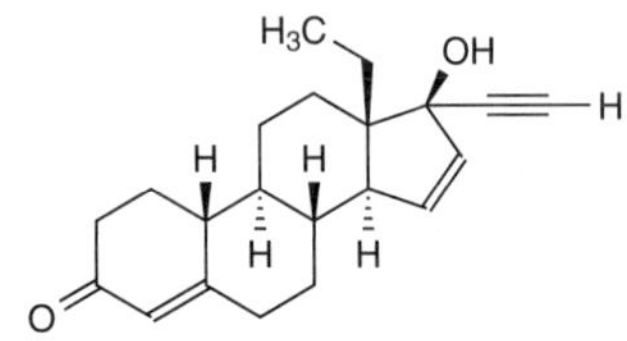

**Molecular formula:** $C_{21}H_{26}O_2$
**Molecular weight:** 310.43
**CAS Registry No:** 60282-87-3
**Merck Index:** 13, 4425

## SAMPLE
**Matrix:** formulations
**Sample preparation:** Powder 2 tablets, sonicate with 10 mL EtOH for 15 min, shake mechanically for 20 min, centrifuge. Dilute a 2.5 mL aliquot of the supernatant to 10 mL with water, inject a 20 μL aliquot.

## HPLC VARIABLES
**Column:** 150 × 3.9 4 μm Nova-Pak C18
**Mobile phase:** MeCN:MeOH:water 35:15:45
**Flow rate:** 1
**Injection volume:** 20
**Detector:** UV 215

## OTHER SUBSTANCES
**Extracted:** ethinyl estradiol

## KEY WORDS
tablets

## REFERENCE
Berzas, J.J.; Rodríguez, J.; Gastañeda, G. Determination of ethinylestradiol and gestodene in pharmaceuticals by a partial least-squares and principal component regression multivariate calibration, *Anal.Sci.*, **1997**, *13*, 1029–1032.

## SAMPLE
**Matrix:** solutions

## HPLC VARIABLES
**Column:** 150 × 4.6 3 μm Hypersil ODS
**Mobile phase:** Gradient. MeCN:water from 35:65 to 65:35 (time not given).
**Injection volume:** 50
**Detector:** UV 242, UV 227

## OTHER SUBSTANCES
**Simultaneous:** gestodene esters

## REFERENCE
Lipp, R.; Laurent, H.; Günther, C.; Riedl, J.; Esperling, P.; Täuber, U. Prodrugs of gestodene for matrix-type transdermal drug delivery systems, *Pharm.Res.*, **1998**, *15*, 1419–1424.

# Gestrinone

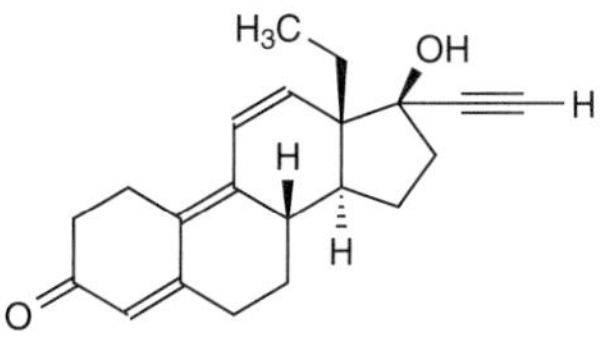


**Molecular formula:** $C_{21}H_{24}O_2$
**Molecular weight:** 308.41
**CAS Registry No:** 16320-04-0
**Merck Index:** 13, 4427

## SAMPLE

**Matrix:** blood
**Sample preparation:** Mix 200 μL serum with 50 μL 100 ng/mL IS in MeOH, add 2 mL diethyl ether, vortex for 2 min, freeze in dry ice/acetone, repeat the extraction. Combine the organic layers and evaporate to dryness, reconstitute the residue with 200 μL mobile phase, inject a 20 μL aliquot.

## HPLC VARIABLES

**Column:** 250 × 4.6 5 μm Kromasil C18
**Mobile phase:** 0.2% Formic acid in MeOH
**Flow rate:** 1
**Injection volume:** 20
**Detector:** MS, PE Sciex API 3000, 100 μL/min entered the MS, electrospray, positive ion mode, source 200°, ionspray 5000 V, nebulizer gas nitrogen at 1 L/min, curtain gas at 1.25 L/min, orifice 45 V, collision gas nitrogen 30 μtorr, collision energy 45 eV, m/z 309–241

## CHROMATOGRAM

**Retention time:** 3
**Internal standard:** mifepristone (m/z 430–372) (2.7)
**Limit of detection:** 0.8 ng/mL (S/N = 3)
**Limit of quantitation:** 3.5 ng/mL

## KEY WORDS

pharmacokinetics; serum; UV 340 can be used with MeOH:water 70:30

## REFERENCE

Wang, Q.; Wu, Z.; Wang, Y.; Luo, G.; Wu, E.; Gao, X. Determination of gestrinone in human serum by liquid chromatography-electrospray tandem mass spectrometry, *J.Chromatogr.B*, **2000**, *746*, 151–159.

# Glycerin

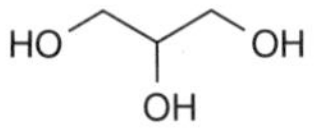

**Molecular formula:** $C_3H_8O_3$
**Molecular weight:** 92.09
**CAS Registry No:** 56-81-5
**Merck Index:** 13, 4497

## SAMPLE

**Matrix:** blood
**Sample preparation:** Mix 1 mL plasma with 2 mL 10% trichloroacetic acid, centrifuge at 2500 rpm for 20 min. Wash the clear supernatant several times with ether until the aqueous layer becomes neutral. Dry the aqueous layer under a stream of air, reconstitute with 50 μL water containing 5 μg ethylene glycol, inject a 25 μL aliquot.

## HPLC VARIABLES

**Column:** 20 × 7 5 μm IEX 215 cation-exchange (Toyo Soda)
**Mobile phase:** Gradient. MeCN:water 76:24 for 30 min, to 0:100 (time not specified)
**Flow rate:** 0.6
**Injection volume:** 25
**Detector:** F ex 412 em 475 following post-column reaction. The column effluent mixed with 25 mM periodic acid pumped at 0.1 mL/min and the resulting mixture flowed through an 0.4 mm ID reaction coil at 145° for 3 min. The effluent from this coil mixed with 8% acetylacetone in 2 M ammonium acetate (allowed to stand overnight before use) pumped at 0.2 mL/min and the mixture flowed through a similar coil at 64° to the detector.

## CHROMATOGRAM

**Retention time:** 26

## OTHER SUBSTANCES

**Extracted:** anhydroglucitol (29), dexteros (43), erythritol (32), ethylene glycol (20)

## KEY WORDS

plasma; post-column reaction

## REFERENCE

Akanuma, H.; Ogawa, K.; Lee, Y.; Akanuma, Y. Reduced levels of plasma 1,5-anhydroglucitol in diabetic patients, *J.Biochem.(Tokyo)*, **1981**, *90*, 157–162.

## SAMPLE

**Matrix:** blood
**Sample preparation:** Add 2 mL of a slurry of an anion-exchange resin (AG1-X8, 200–400 mesh, hydroxide form, Bio-Rad) in an equal volume of water to an 80 × 10 column. Add 2 mL of a slurry of a cation-exchange resin (AG50-WX8, 200–400 mesh, hydrogen form, Bio-Rad) in an equal volume of water to an 80 × 10 column. Place the anion column on top of the cation column. Vortex 1 mL plasma briefly, add 1 mL 0.3 N barium hydroxide, add 1.1 mL 0.3 N zinc sulfate, vortex, centrifuge at 2000 rpm for 10 min. Add the supernatant to the columns, vortex the pellet with 1 mL water, centrifuge at 2000 rpm, add the supernatant to the columns, repeat the extraction of the pellet, add two 1 mL portions of water to the columns. Collect all the eluate from the columns, add 100 μL 100 mM sucrose, vortex, lyophilize at − 20° for 15 h. Add 5 drops 2,2-dimethoxypropane, mix, dry under air (it is critical that the sample be completely dry), add 1 mL 30 mg/mL 4-dimethylaminopyridine in pyridine, vortex, sonicate at 45° for 5 min (no longer than 5 min), add 100 μL 25% benzoyl peroxide in

butyl acetate, vortex until a white precipitate appears, heat at 45° for 15 min, cool to room temperature, add 50 μL MeOH, vortex at 45° for 5 min. Evaporate to dryness under a stream of air at 45° for 1 h, reconstitute the residue with 4 mL chloroform with vortexing and sonicating (Caution! Chloroform is a carcinogen!), add 2 mL 1 M HCl, vortex for 15 s, centrifuge at 2000 rpm for 5 min, discard the aqueous layer. Repeat the HCl wash. Evaporate the chloroform layer to dryness under a stream of air, reconstitute the residue with 172.8 μL MeCN and 67.2 μL water, vortex, centrifuge for 2 min, inject a 200 μL aliquot.

## HPLC VARIABLES

**Column:** 250 mm long 5 μm Apex C18 (Jones)
**Mobile phase:** Gradient. MeCN:water 56:44 for 30 min, to 95:5 (step gradient), maintain at 95:5 for 10 min, re-equilibria at initial conditions for 20 min.
**Flow rate:** 1.3
**Injection volume:** 200
**Detector:** UV 254

## CHROMATOGRAM

**Retention time:** 26

## OTHER SUBSTANCES

**Extracted:** dextrose (28), sucrose (30)

## KEY WORDS

derivatization; plasma; SPE

## REFERENCE

Akanuma, H.; Ogawa, K.; Lee, Y.; Akanuma, Y. Reduced levels of plasma 1,5-anhydroglucitol in diabetic patients, *J.Biochem.*, **1981**, *90*, 157–162.

# Guanabenz

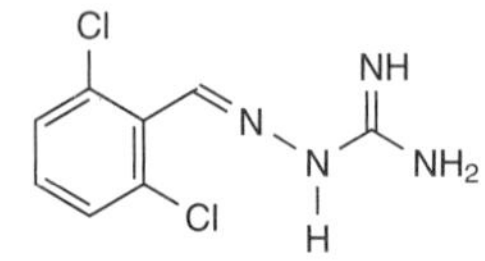

**Molecular formula:** $C_8H_8Cl_2N_4$
**Molecular weight:** 231.09
**CAS Registry No:** 5051-62-7
**Merck Index:** 13, 4574

## SAMPLE

**Matrix:** blood
**Sample preparation:** Mix 2 mL serum with 10 μL 10 μg/mL IS in MeOH, add 200 μL concentrated ammonium hydroxide, add 5 mL dichloromethane, shake on a reciprocating shaker for 30 min, centrifuge at 730 g for 20 min. (If excessive emulsion is present, mix by inversion and centrifuge again.). Evaporate the organic layer to dryness in a silanized glass tube under a stream of nitrogen below 40°, reconstitute the residue with 80 μL mobile phase, vortex, inject a 25 μL aliquot.

## HPLC VARIABLES

**Column:** 250 × 2 5 μm Luna phenyl hexyl (Phenomenex)
**Column temperature:** 30
**Mobile phase:** Gradient. MeCN:water:formic acid 5:95:0.05 for 1 min, to 90:10:0.05 over 4 min, maintain at 90:10:0.05 for 5 min, return to initial conditions over 0.5 min, re-equilibrate for 4.5 min.
**Flow rate:** 0.45
**Injection volume:** 25
**Detector:** MS, Micromass Quattro II, positive ion mode, collision gas argon 3 μbar, photomultiplier 750–800 V, source cone 26 V, collision energy – 23 ± 1 V, capillary + 3.1 kV, HV lens 520 V, source 120°, m/z 231–172

## CHROMATOGRAM

**Retention time:** 6.19
**Internal standard:** clenbuterol (277–203) (6.12)
**Limit of detection:** 0.3 ng/mL
**Limit of quantitation:** 3 ng/mL

## KEY WORDS

horse; pharmacokinetics; serum

## REFERENCE

Harkins, J.D.; Dirikolu, L.; Lehner, A.F.; Hughes, C.; Schroedter, D.; Mayer, B.; Bratton, C.; Fisher, M.V.; Tobin, T. The detection and biotransformation of guanabenz in horses: a preliminary report, *Vet.Ther.*, **2003**, *4*, 197–209.

# Guanadrel

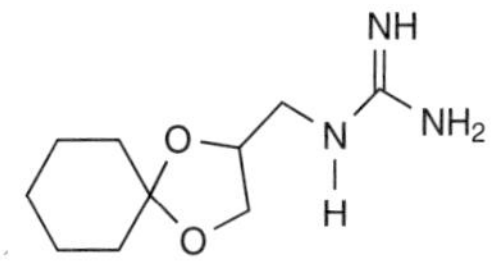


**Molecular formula:** $C_{10}H_{19}N_3O_2$
**Molecular weight:** 213.28
**CAS Registry No:** 40580-59-4
**Merck Index:** 13, 4575

## SAMPLE

**Matrix:** feed
**Sample preparation:** Vortex 1 g pulverized feed with 5 mL 50 mM ammonium dihydrogen phosphate, let stand for 15 min, add 1 mL IS solution, add 3 mL MeCN, shake on a reciprocating shaker at 2000 rpm for 20 min. Add 3 mL supernatant to 1 mL 1 M sodium bicarbonate, add 2.5 mL MeOH, add 2 mL acetylacetone, mix, heat at $120 \pm 5°$ for 100 min, cool, mix, centrifuge at 2000 rpm for 10 min, inject an aliquot of the supernatant.

## HPLC VARIABLES

**Column:** 250 × 4.6 10 μm LiChrosorb RP-8
**Mobile phase:** MeCN:THF:50 mM ammonium dihydrogen phosphate 40:10:50
**Flow rate:** 0.8
**Injection volume:** 40
**Detector:** F ex 238 em 360

## CHROMATOGRAM

**Retention time:** 10.5
**Internal standard:** cyclohexymethylguanidine (18)
**Limit of detection:** 1 ppm

## KEY WORDS

derivatization

## REFERENCE

Bombardt, P.A.; Adams, W.J. Liquid chromatographic determination of guanadrel in laboratory animal diet as the fluorescent acetylacetone derivative, *Anal.Chem.*, **1982**, *54*, 1087–1090.

# Halobetasol propionate

**Molecular formula:** $C_{25}H_{31}ClF_2O_5$
**Molecular weight:** 484.97
**CAS Registry No:** 66852-54-8
**Merck Index:** 13, 4608

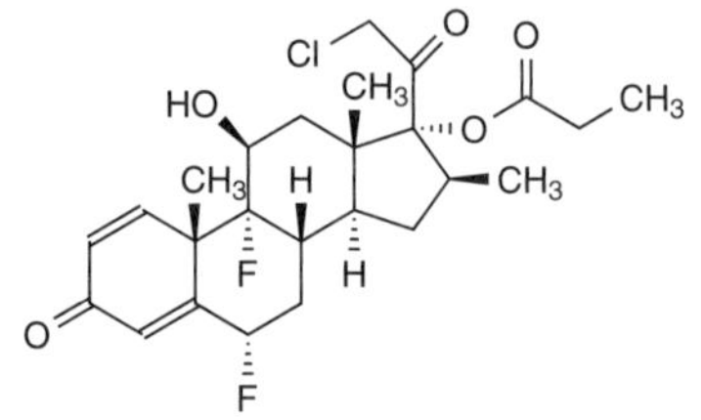


## SAMPLE

**Matrix:** formulations

**Sample preparation:** Mix 1 g of a topical product, 5 mL 100 mM pH 4 citrate buffer saturated with NaCl, and 10 mL ethyl acetate; shake vigorously by hand to ensure that no large clumps stick to the tube, mix with the tube on its side on an oscillating shaker for 15 min. Remove the upper ethyl acetate layer and wash with citrate buffer as before. Dry the ethyl acetate over anhydrous sodium sulfate and evaporate to dryness. Dissolve the residue in a mixture of 10 mL heptane and 5 mL MeCN:water 90:10. Remove the upper heptane layer and extract it with 2 mL MeCN:water 90:10. Combine the MeCN/water layers and evaporate them to dryness, dissolve the residue in 300 μL MeOH, filter (0.45 μm nylon), inject a 5 μL aliquot.

## HPLC VARIABLES

**Guard column:** 15 × 3.2 7 μm Brownlee NewGuard C18
**Column:** 75 × 4.6 3.5 μm Symmetry C18 (Waters)
**Mobile phase:** Gradient. MeCN:water 18:82 for 2 min, to 82:18 over 12 min, maintain at 82:18 for 3 min, re-equilibrate at initial conditions for 12 min.
**Flow rate:** 1
**Injection volume:** 5
**Detector:** UV 240

## CHROMATOGRAM

**Retention time:** 10.98
**Limit of detection:** 0.001%

## OTHER SUBSTANCES

**Simultaneous:** alclometasone 17,21-dipropionate (10.93), amcinonide (10.90), beclomethasone 17,21-dipropionate (11.90), betamethasone (6.52), betamethasone 21-acetate (8.47), betamethasone 17-benzoate (10.28), betamethasone 17,21-dipropionate (11.42), betamethasone 17-valerate (10.25), budesonide (8.55, 8.69 (epimers)), clobetasol 17-propionate (11.06), cortisone (5.62), cortisone 21-acetate (8.07), dexamethasone (6.57), dexamethasone 21-acetate (8.68), desonide (6.99), desoximetasone (7.60), desoxycorticosterone acetate (10.90), desoxycorticosterone pivalate (14.45), diflorasone 17,21-diacetate (9.81), fluocinolone acetonide (7.38), fluocinonide (9.79), flurandrenolide (7.36), fludrocortisone 21-acetate (7.77), fluorometholone (7.67), flumethasone 21-pivalate (11.20), flunisolide (7.14), fluprednisolone (5.46), fluticasone 17-propionate (11.19), halcinonide (10.72), hydrocortisone (5.50), hydrocortisone 21-acetate (7.65), hydrocortisone 17-butyrate (8.66), hydrocortisone 21-cypionate (12.54), hydrocortisone 17-valerate (9.53), mometasone 17-furoate (11.24), methylprednisolone (6.31), methylprednisolone 21-acetate (8.34), meprednisone (6.55), prednisolone (5.37), prednisolone 21-acetate (7.47), prednisolone 21-tebuate (11.01), paramethasone 21-acetate (8.64), prednicarbate (10.72), prednisone (5.46), triamcinolone (4.15), triamcinolone acetonide (7.04), triamcinolone 16,21-diacetate (7.49), triamcinolone hexacetonide (13.12)

## KEY WORDS

body wash, cream, gel, lotion, shampoo, spray

## REFERENCE

Reepmeyer, J.C. Screening for corticosteroids in topical pharmaceuticals by HPLC with a scanning ultraviolet detector, *J.Liq.Chromatogr.Rel.Technol.*, **2001**, *24*, 693–709.

# Halofuginone

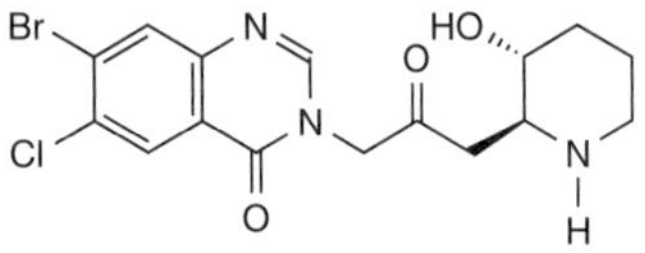


**Molecular formula:** $C_{16}H_{17}BrClN_3O_3$
**Molecular weight:** 414.69
**CAS Registry No:** 55837-20-2
**Merck Index:** 13, 4611

## SAMPLE
**Matrix:** blood
**Sample preparation:** Condition a 1 mL Bond Elut C8 SPE cartridge with 2 mL MeOH and 8 mL water adjusted to pH 4.3 with acetic acid. Mix 4 mL serum with 8 mL 10% acetic acid, add to the SPE cartridge, wash with 5 mL water adjusted to pH 4.3 with acetic acid, wash with 1 mL MeOH:water 35:65 adjusted to pH 4.3 with acetic acid, wash with 1 mL water adjusted to pH 4.3 with acetic acid, elute with 1 mL MeCN:water:acetic acid 20:79.9:0.1 containing 2.104 μL/mL decylamine (pH ca. 4.3), inject a 50 μL aliquot of the eluate. (Use silanized glassware.)

## HPLC VARIABLES
**Guard column:** 30–38 μm CO:PELL
**Column:** 250 × 4.6 5 μm Supelcosil LC-DB18
**Mobile phase:** MeCN:buffer:water 22:15:63 containing 210.4 μL/L decylamine, pH ca. 4.75 (The buffer was 250 mM pH 4.3 ammonium acetate.)
**Injection volume:** 50
**Detector:** UV 243

## CHROMATOGRAM
**Retention time:** 8
**Limit of detection:** 1.5 ng/mL

## KEY WORDS
chicken; serum; SPE

## REFERENCE
Beier, R.C.; Rowe, L.D.; Abd El-Aziz Nasr, M.I.; Elissalde, M.H.; Stanker, L.H. Extraction and HPLC analysis of halofuginone in chicken serum, *J.Liq.Chromatogr.*, **1994**, *17*, 2961–2970.

## SAMPLE
**Matrix:** eggs, tissue
**Sample preparation:** Condition a 3 mL 60 mg Oasis HLB SPE cartridge with 3 mL MeOH, 3 mL water, and 3 mL 125 mM pH 4.9 ammonium acetate buffer. Mix 2 g minced liver or homogenized egg with 2 mL 25 mg/mL trypsin in water, vortex, adjust pH to 7–8 with 10% sodium carbonate, shake on an orbital shaker at 40° overnight, cool, add 1 mL 10% sodium carbonate, add 10 (liver) or 15 (eggs) mL ethyl acetate, shake on a mechanical shaker for 3 min, centrifuge at 600 g at 4° for 2 min, extract the aqueous layer again with 10 mL ethyl acetate. Combine the ethyl acetate layers and add to 5 mL 125 mM pH 4.9 ammonium acetate buffer, shake mechanically for 1 min. Shake the aqueous layer with 5 mL hexane for 20 s, discard the hexane layer. Add the aqueous layer to the SPE cartridge, wash with 2 mL toluene, push air through the SPE cartridge at 19 mL/min, elute with 2 mL MeOH. Evaporate the eluate to dryness under a stream of nitrogen at 40°, reconstitute the residue with 500 μL 125 mM pH 4.9 ammonium acetate buffer, vortex for 30 s, inject a 25 μL aliquot.

## HPLC VARIABLES
**Column:** 250 × 4.6 5 μm Prodigy C18 (Phenomenex)

**Mobile phase:** MeOH:water:acetic acid 40:59.5:0.5
**Flow rate:** 1
**Injection volume:** 25
**Detector:** MS, Micromass Quattro LC, 0.2 mL/min entered the detector, source 150°, drying gas nitrogen 500 L/h, nebulizing gas nitrogen 80 L/h, positive mode, collision cell entrance 0 eV, collision cell exit 2 eV, cone 30 V, collision gas argon 2.3310 mbar, m/z 416–398–138–120–100

## CHROMATOGRAM

**Retention time:** 6.3
**Limit of quantitation:** 15 ng/g (liver), 5 ng/g (eggs)

## KEY WORDS

chicken; liver; SPE

## REFERENCE

Yakkundi, S.; Cannavan, A.; Elliott, C.T.; Lövgren, T.; Kennedy, D.G. Development and validation of a method for the confirmation of halofuginone in chicken liver and eggs using electrospray tandem mass spectrometry, *J.Chromatogr.B*, **2003**, *788*, 29–36.

## ANNOTATED BIBLIOGRAPHY

Anderson, A.; Christopher, D.H.; Woodhouse, R.N. Analysis of the anti-coccidial drug, halofuginone, in chicken feed using gas-liquid chromatography and high-performance liquid chromatography, *J.Chromatogr.*, **1979**, *168*, 471–480.

Anderson, A.; Goodall, E.; Bliss, G.W.; Woodhouse, R.N. Analysis of the anti-coccidial drug, halofuginone, in chicken tissue and chicken feed using high-performance liquid chromatography, *J.Chromatogr.*, **1981**, *212*, 347–355.

Holland, D.C.; Manns, R.K.; Roybal, J.E.; Hurlbut, J.A.; Long, A.R. Liquid chromatographic determination of the anticoccidial drug halofuginone hydrobromide in eggs, *J.AOAC Int.*, **1995**, *78*, 37–40.

Kinabo, L.D.B.; McKellar, Q.A.; Murray, M. Determination of halofuginone in bovine plasma by competing-ion high performance liquid chromatography after solid phase extraction, *Biomed.Chromatogr.*, **1989**, *3*, 136–138.

Mortier, L.; Daeseleire, E.; Delahaut, P. Simultaneous detection of five coccidiostats in eggs by liquid chromatography-tandem mass spectrometry, *Anal.Chim.Acta*, **2003**, *483*, 27–37. [diclazuril; dimetridazole; halofuginone; nicarbazin; robenidine]

Tillier, C.; Cagniant, E.; Devaux, P. Determination of halofuginone in poultry feeds by high-performance liquid chromatography, *J.Chromatogr.*, **1988**, *441*, 406–416.

Yamamoto, Y.; Kondo, F. Determination of halofuginone and amprolium in chicken muscle and egg by liquid chromatography, *J.AOAC Int.*, **2001**, *84*, 43–46. [SPE]

# Hetastarch

**CAS Registry No:** 9004-62-0
**Merck Index:** 13, 4692

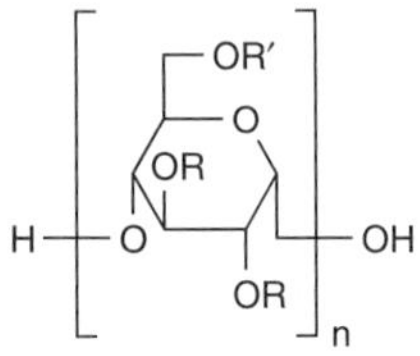


R or R′ = H or $CH_2CH_2OH$

## SAMPLE

**Matrix:** blood, lymph
**Sample preparation:** Mix 2 mL plasma or lymph with 150 μL 85% trichloroacetic acid, centrifuge. Add 100 μL Tris buffer and 25μL phenol red to 1.5 mL of the supernatant, neutralize excess acid with 5–25 μL 5 M KOH, inject an aliquot of the supernatant

## HPLC VARIABLES

**Column:** SEC-60 (TosoHaas) + SEC-50 (TosoHaas) + SEC-10 (Bio-Rad)
**Mobile phase:** pH 4.0 acetate buffer
**Detector:** Refractive Index

## KEY WORDS

plasma; sheep

## REFERENCE

Korent, V.A.; Conhaim, R.L.; McGrath, A.M.; DeAngeles, D.A.; Harms, B.A. Molecular distribution of hetastarch in plasma and lung lymph of unanesthetized sheep, *Am.J.Respir.Crit.Care Med.*, **1997**, *155*, 1302–1308.

# Hydroquinone

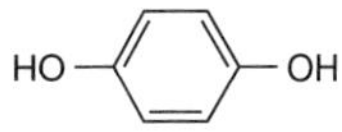

**Molecular formula:** $C_6H_6O_2$
**Molecular weight:** 110.11
**CAS Registry No:** 123-31-9
**Merck Index:** 13, 4833

## SAMPLE
**Matrix:** blood
**Sample preparation:** Filter (Amicon Centrifree) whole blood while centrifuging at 3° at 3000 g for 1 h, inject a 100 μL aliquot of the ultrafiltrate.

## HPLC VARIABLES
**Column:** 170 × 4.6 5 μm Supelcosil LC-18
**Mobile phase:** 50 mM pH 4.5 formate buffer
**Flow rate:** 1
**Injection volume:** 100
**Detector:** Radioactivity ($^{14}C$); UV (wavelength not specified, other papers have recommended 220–280)

## CHROMATOGRAM
**Retention time:** 7

## OTHER SUBSTANCES
**Extracted:** hydroquinone glucuronide (3), hydroquinone sulfate (5)

## KEY WORDS
ultrafiltrate; whole blood

## REFERENCE
Deisinger, P.J.; English, J.C. Bioavailability and metabolism of hydroquinone after intratracheal instillation in male rats, *Drug Metab.Dispos.*, **1999**, *27*, 442–448.

# Hygromycin B

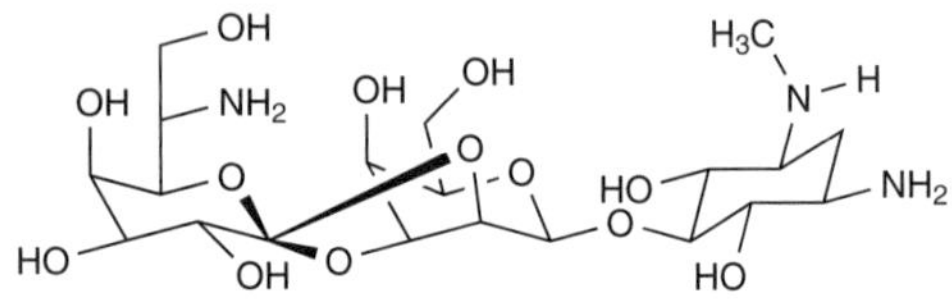


**Molecular formula:** $C_{20}H_{37}N_3O_{13}$
**Molecular weight:** 527.52
**CAS Registry No:** 31282-04-9
**Merck Index:** 13, 4878

## SAMPLE

**Matrix:** tissue

**Sample preparation:** Mix 500 mg homogenized kidney with 2 g 40 μm Bondesil preparative-grade end-capped cyanopropyl bulk packing material (Analytichem), blend with mortar and pestle for 2 min. (For a higher degree of purification, use 3 g Bondesil CN and blend for 3 min.) Place in a chromatography column, wash with 3 mL hexane, wash with 5 mL ethyl acetate, wash with 5 mL MeOH, wash with 5 mL MeOH:water 50:50, elute with 1 mL water and then with 8 mL 100 mM formic acid, filter (0.45 μm) a 4.5 mL aliquot of the eluate. Concentrate 4 mL of the filtrate to 75 μL under reduced pressure, add 7.5 μL 10% pentafluoropropionic acid in water, vortex, filter (0.20 μm), inject a 22 μL aliquot of the filtrate.

## HPLC VARIABLES

**Guard column:** 10 × 2 3 μm YMCbasic

**Column:** 100 × 2 3 μm YMCbasic

**Mobile phase:** Gradient. A was MeCN:water 60:40 containing 20 mM pentafluoropropionic acid. B was MeCN:water 5:95 containing 20 mM pentafluoropropionic acid. A:B 0:100 for 5.1 min, to 32:68 over 0.1 min, maintain at 32:68 for 13.3 min, to 100:0 over 0.1 min, maintain at 100:0 for 5.5 min, return to initial conditions over 0.1 min, re-equilibrate for 21.8 min.

**Flow rate:** 0.2 for 18.5 min, 0.22 for 0.1 min, 0.25 for 22.3 min, then 0.2

**Injection volume:** 22

**Detector:** MS, PE Sciex API III triple quadrupole, atmospheric pressure ion source, positive ion, laboratory-constructed ionspray interface (details provided), nebulizer gas nitrogen at 67 psi, curtain gas nitrogen, collision gas argon, 40 μL/min entered the detector, m/z 265–177 or 528–352–177

## CHROMATOGRAM

**Retention time:** 5.5

**Limit of detection:** 0.02 ppm

## OTHER SUBSTANCES

**Extracted:** dihydrostreptomycin (7), gentamicin (13.5–15), neomycin (16), spectinomycin (5), streptomycin (6.5), tobramycin (12)

## KEY WORDS

cow; kidney; MSPD

## REFERENCE

McLaughlin, L.G.; Henion, J.D.; Kijak, P.J. Multi-residue confirmation of aminoglycoside antibiotics and bovine kidney by ion spray high-performance liquid chromatography/tandem mass spectrometry, *Biol.Mass Spectrom.*, **1994**, *23*, 417–429.

# Iloprost

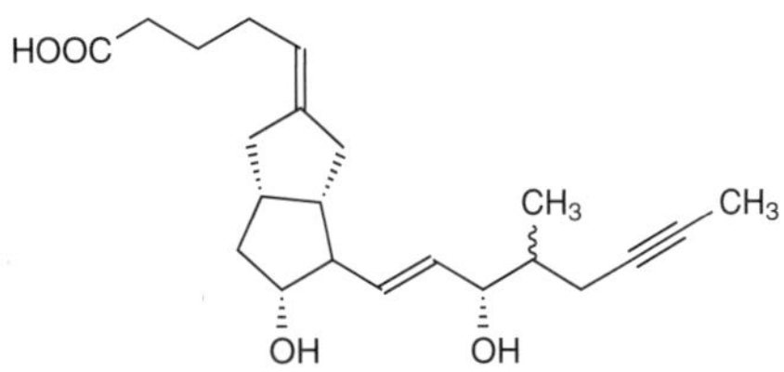

**Molecular formula:** $C_{22}H_{32}O_4$
**Molecular weight:** 360.49
**CAS Registry No:** 78919-13-8
**Merck Index:** 13, 4925

## SAMPLE

**Matrix:** blood
**Sample preparation:** Mix 200 μL plasma with 200 μL 100 ppm 2-naphthoic acid and 300 μL MeCN, shake well, centrifuge for 10 min, inject a 20 μL aliquot of the supernatant.

## HPLC VARIABLES

**Column:** 250 × 4.6 5 μm Zorbax C8
**Mobile phase:** MeCN:MeOH:20 mM pH 3.0 potassium phosphate buffer 30:24:46
**Flow rate:** 1.7
**Injection volume:** 20
**Detector:** UV 210; radioactivity ($^3H$)

## CHROMATOGRAM

**Retention time:** 14.7, 15.9 (diastereomers)
**Internal standard:** 2-naphthoic acid (6)
**Limit of quantitation:** 500 ng (UV), 42 pg (radioactivity detector)

## OTHER SUBSTANCES

**Extracted:** misoprostol (LOQ 1 μg (UV), 12 pg (radioactivity detector)) (18.5, diastereomers not resolved)

## KEY WORDS

plasma; rat

## REFERENCE


Womack, I.M.; Lee, A.S.; Kamath, B.; Agrawal, K.C.; Kishore, V. A high performance liquid radiochromatographic assay for the simultaneous analysis of iloprost and misoprostol, *Prostaglandins*, **1996**, *52*, 249–259.

# Imatinib

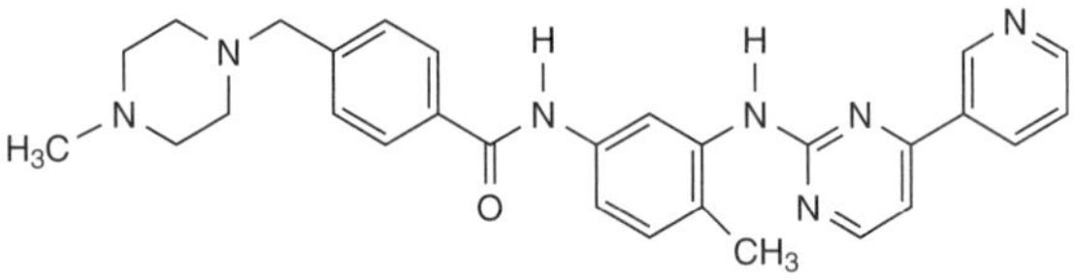

**Molecular formula:** $C_{29}H_{31}N_7O$
**Molecular weight:** 493.60
**CAS Registry No:** 152459-95-5, 220127-57-1 (mesylate)
**Merck Index:** 13, 4926

## SAMPLE
**Matrix:** blood
**Sample preparation:** Mix 200 μL plasma with 10 μL 10 μg/mL IS in MeOH, add 1 mL MeCN, vortex briefly, centrifuge at 12000 g for 5 min. Evaporate a 1 mL aliquot of the supernatant to dryness under a stream of nitrogen at 27°, reconstitute the residue with 100 μL MeOH:water 20:80, vortex briefly, inject a 3 μL aliquot.

## HPLC VARIABLES
**Column:** 50 × 4.6 5 μm Phenomenex Luna C18(2)
**Mobile phase:** Gradient. MeOH:water:formic acid from 20:80:0.1 to 60:40:0.1 over 6 min, to 100:0:0.1 over 1 min, maintain at 100:0:0.1 for 2 min, return to initial conditions over 1 min, re-equilibrate for 4 min.
**Flow rate:** 1 for 6 min, to 2 over 1 min, maintain at 2 for 6 min, to 1 over 0.1 min, maintain at 1 for 0.9 min
**Injection volume:** 3
**Detector:** MS, ThermoFinnigan aQa, electrospray, positive single-ion, 10% of column effluent entered detector, insert probe 250°, ionspray 5000 V, orifice 10 V, nitrogen 75 psi, m/z 493.7

## CHROMATOGRAM
**Retention time:** 4.0
**Internal standard:** $d_8$-imatinib (m/z 501.7) (3.9)
**Limit of quantitation:** 30 ng/mL

## OTHER SUBSTANCES
**Extracted:** CGP 74588 (metabolite) (m/z 479.7) (3.7)

## KEY WORDS
pharmacokinetics; plasma

## REFERENCE
Parise, R.A.; Ramanathan, R.K.; Hayes, M.J.; Egorin, M.J. Liquid chromatographic-mass spectrometric assay for quantitation of imatinib and its main metabolite (CGP 74588) in plasma, *J.Chromatogr.B*, **2003**, *791*, 39–44.

## SAMPLE
**Matrix:** formulations
**Sample preparation:** Dissolve capsules in MeOH:water 25:75 containing 40 μg/mL acetaminophen so as to achieve an imatinib concentration of 2 mg/mL, inject an aliquot.

## HPLC VARIABLES
**Column:** 150 × 4.6 5 μm XTerra C18
**Column temperature:** 25
**Mobile phase:** MeOH:water:triethylamine 25:74:1 (Mix 740 mL water with 10 mL triethylamine, adjust pH to 2.4 with 85% phosphoric acid, add 250 mL MeOH, adjust pH to 2.6 with 85% phosphoric acid, if necessary.)

**Flow rate:** 1
**Injection volume:** 20
**Detector:** UV 267

## CHROMATOGRAM

**Retention time:** 7.7
**Internal standard:** acetaminophen (2.9)
**Limit of detection:** 10 ng/mL
**Limit of quantitation:** 100 ng/mL

## OTHER SUBSTANCES

**Simultaneous:** impurity STI 509-00

## KEY WORDS

capsules; robust

## REFERENCE

Ivanovic, D.; Medenica, M.; Jancic, B.; Malenovic, A. Reversed-phase liquid chromatography analysis of imatinib mesylate and impurity product in Glivec capsules, *J.Chromatogr.B*, **2004**, *800*, 253–258.

## ANNOTATED BIBLIOGRAPHY

Bakhtiar, R.; Khemani, L.; Hayes, M.; Bedman, T.; Tse, F. Quantification of the anti-leukemia drug STI571 (Gleevec) and its metabolite (CGP 74588) in monkey plasma using a semi-automated solid phase extraction procedure and liquid chromatography-tandem mass spectrometry, *J.Pharm.Biomed.Anal.*, **2002**, *28*, 1183–1194. [SPE; LOQ 1 ng/mL]

Bakhtiar, R.; Lohne, J.; Ramos, L.; Khemani, L.; Hayes, M.; Tse, F. High-throughput quantification of the anti-leukemia drug STI571 (Gleevec™) and its main metabolite (CGP 74588) in human plasma using liquid chromatography-tandem mass spectrometry, *J.Chromatogr.B*, **2002**, *768*, 325–340. [LOQ 4 ng/mL]

Hamada, A.; Miyano, H.; Watanabe, H.; Saito, H. Interaction of imatinib mesilate with human P-glycoprotein, *J.Pharmacol.Exp.Ther.*, **2003**, *307*, 824–828. [cell cultures; UV detection]

Vivekanand, V.V.; Rao, D.S.; Vaidyanathan, G.; Sekhar, N.M.; Kelkar, S.A.; Puranik, P.R. A validated LC method for imatinib mesylate, *J.Pharm.Biomed.Anal.*, **2003**, *33*, 879–889. [bulk; UV detection]

# Imidocarb

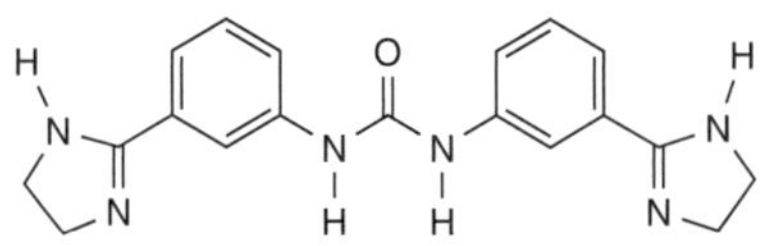

**Molecular formula:** $C_{19}H_{20}N_6O$
**Molecular weight:** 348.40
**CAS Registry No:** 27885-92-3
**Merck Index:** 13, 4938

## SAMPLE

**Matrix:** blood

**Sample preparation:** Condition a 100 mg Supelclean LC-18 SPE cartridge with 3 mL MeOH and 2 mL water. Vortex 1 mL plasma with 3 mL 50 mM disodium EDTA solution for 30 s, add to the SPE cartridge, wash with 2 mL MeOH:water 10:90, dry under vacuum for 5 min, elute with 3 mL MeCN:2% acetic acid in water containing 25 mM sodium octanesulfonate 90:10. Evaporate the eluate to dryness under reduced pressure at 40°, reconstitute the residue with 350 μL mobile phase, vortex, filter (0.45 μm), inject a 50 μL aliquot.

## HPLC VARIABLES

**Column:** 100 × 8 4 μm Nova Pak C18 radial compression

**Mobile phase:** MeCN:buffer 30:70 (The buffer was 5 mM sodium octanesulfonate containing 0.1% triethylamine adjusted to pH 3.2 with glacial acetic acid.)

**Flow rate:** 1.5

**Detector:** UV 254

## CHROMATOGRAM

**Retention time:** 5.4

## OTHER SUBSTANCES

**Extracted:** diminazene (UV 370) (3.7)

## KEY WORDS

cow; plasma; SPE; imidocarb is IS in original paper

## REFERENCE

Gummow, B.; Du Preez, J.L.; Swan, G.E. Paired-ion extraction and high-performance liquid chromatographic determination of diminazene in cattle plasma: a modified method, *Onderstepoort J.Vet.Res.*, **1995**, *62*, 1–4.

## SAMPLE

**Matrix:** tissue

**Sample preparation:** Homogenize (Ultra-Turrax) 10 g thinly sliced kidney with 2 mL 1 M sodium carbonate and 25 mL acetone for 2 min, sonicate for 3 min, centrifuge at 4 200 g for 5 min, extract the residue again with 2 mL 1 M sodium carbonate, 25 mL acetone, and 8 mL water. Combine the supernatants, add 50 mL chloroform (Caution! Chloroform is a carcinogen!), add 20 mL saturated aqueous NaCl, add 2 mL 40% NaOH, shake for 1 min. Filter the lower organic layer through anhydrous sodium sulfate and phase-separating filter paper (Whatman PS-1), evaporate to dryness under reduced pressure at 50°, reconstitute with three 3 mL portions of MeOH:10 mM pH 7 sodium acetate containing 10 mM sodium trifluoroacetate 80:20, centrifuge at 1860 g for 5 min, add the supernatant to a pre-washed (not otherwise described) Bond Elut CBA (carboxylic acid, weak cation exchange) SPE cartridge. Wash the residue with 1 mL MeOH:10 mM pH 7 sodium acetate containing 10 mM sodium trifluoroacetate 80:20 and add the supernatant to the SPE cartridge. Wash the SPE cartridge with 4 mL MeOH, elute with 5 mL MeOH:trifluoroacetic acid 95:5. Evaporate the eluate to dryness under reduced pressure at 50°, reconstitute the residue with 500 μL mobile phase, vortex for 15 s, sonicate for 3 min, centrifuge at 1860 g for 5 min, filter (0.45 μm), inject a 50 μL aliquot. (Use silanized glassware.)

## HPLC VARIABLES

**Column:** 100 × 4.6 3 μm Spherisorb S3W-C18
**Mobile phase:** Gradient. A was MeCN:10 mM pH 7 sodium acetate containing 10 mM sodium trifluoroacetate 15:85. B was MeCN:10 mM pH 2 sodium acetate containing 10 mM trifluoroacetic acid and 10 mM tetramethylammonium chloride 10:90. A:B 100:0 for 5 min, to 0:100 (step gradient), maintain at 0:100 for 15 min, return to initial conditions (step gradient), re-equilibrate for 10 min.
**Flow rate:** 1
**Injection volume:** 50
**Detector:** UV 260

## CHROMATOGRAM

**Retention time:** 14
**Limit of detection:** 1–2 ng/g

## KEY WORDS

cow; kidney; SPE

## REFERENCE

Tarbin, J.A.; Shearer, G. High-performance liquid chromatographic determination of imidocarb in cattle kidney with cation-exchange clean-up, *J.Chromatogr.*, **1992**, *577*, 376–381.

## ANNOTATED BIBLIOGRAPHY

Coldham, N.G.; Moore, A.S.; Sivapathasundaram, S.; Sauer, M.J. Imidocarb depletion from cattle liver and mechanism of retention in isolated bovine hepatocytes, *Analyst*, **1994**, *119*, 2549–2552. [LOD 74 ng/g]

Coldham, N.G.; Moore, A.S.; Dave, M.; Graham, P.J.; Sivapathasundaram, S.; Lake, B.G.; Sauer, M.J. Imidocarb residues in edible bovine tissues and in vitro assessment of imidocarb metabolism and cytotoxicity, *Drug Metab.Dispos.*, **1995**, *23*, 501–505.

# Iobenguane

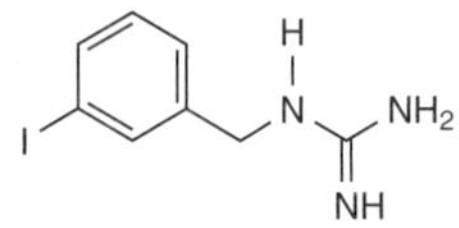


**Molecular formula:** $C_8H_{10}IN_3$
**Molecular weight:** 275.09
**CAS Registry No:** 80663-95-2
**Merck Index:** 13, 5028

## SAMPLE

**Matrix:** blood, urine
**Sample preparation:** Inject a 200 μL aliquot of serum or urine onto column A and elute to waste with mobile phase A; after 5 min, backflush the contents of column A onto column B with mobile phase B, monitor the effluent from column B. After 15 min, re-equilibrate column A with mobile phase A for 5 min. After 5 determinations, flush column A with MeOH.

## HPLC VARIABLES

**Column:** A 30 × 4 5 μm LiChrosorb C8; B 250 × 4 5 μm LiChrosorb C8
**Mobile phase:** A 60 mM pH 6.0 phosphate buffer; B MeCN:60 mM pH 6.0 phosphate buffer containing 1% sodium heptanesulfonate 35:65
**Flow rate:** 1
**Injection volume:** 200
**Detector:** UV 232 (urine), UV 254 (serum)

## CHROMATOGRAM

**Retention time:** 14
**Limit of detection:** 50 ng/mL (S/N = 3)

## OTHER SUBSTANCES

**Extracted:** metoclopramide (13), sulfamethoxazole/trimethoprim (?) (11)
**Noninterfering:** acetaminophen, diazepam, dimenhydrinate, etoposide, prolonium iodide

## KEY WORDS

column-switching; serum

## REFERENCE

Schwabe, D.; Rohrbach, E.; Köhl, U. Determination of *m*-iodobenzylguanidine in serum and urine by high-performance liquid chromatography, *J.Chromatogr.*, **1989**, *487*, 177–182.

## SAMPLE

**Matrix:** blood
**Sample preparation:** Plasma. Condition a 1 mL 100 mg Bakerbond cyano SPE cartridge with 1 mL MeOH and 2 mL water. Add 500 μL plasma to the SPE cartridge, wash with 2 mL water, wash with 2 mL MeOH, elute with 1 mL 100 mM HCl in MeOH. Evaporate the eluate to dryness under a stream of nitrogen at 60°, reconstitute the residue with 200 μL mobile phase, vortex for 15 s, centrifuge at 9 500 g for 2 min, inject a 100 μL aliquot. Whole blood. Condition a 1 mL 100 mg Bakerbond cyano SPE cartridge with 1 mL MeOH and 2 mL water, retaining 0.5 mL water above the sorbent. Place a 3 mL Bakerbond SPE filtration column above the SPE cartridge. Mix 500 μL whole blood with 1.5 mL water, add to the filtration column, wash with 1 mL water, remove the filtration column. Wash the SPE cartridge with 1 mL water, wash with 2 mL MeOH, elute with 1 mL 100 mM HCl in MeOH. Evaporate the eluate to dryness under a stream of nitrogen at 60°, reconstitute the residue with 200 μL mobile phase, vortex for 15 s, centrifuge at 9 500 g for 2 min, inject a 100 μL aliquot.

## HPLC VARIABLES

**Column:** 300 × 3.9 10 μm μBondapak C18
**Mobile phase:** MeCN:25 mM pH 4.0 ammonium phosphate buffer 20:80
**Flow rate:** 1
**Injection volume:** 100
**Detector:** UV 254

## CHROMATOGRAM

**Limit of detection:** 75 ng/mL
**Limit of quantitation:** 100 ng/mL

## KEY WORDS

plasma; SPE; whole blood

## REFERENCE

Wafelman, A.R.; Konings, M.C.P.; Rosing, H.; Hoefnagel, C.A.; Taal, B.G.; Maes, R.A.A.; Beijnen, J.H. High-performance liquid chromatographic determination of metaiodobenzylguanidine in whole blood and plasma of cancer patients, *J.Pharm.Biomed.Anal.*, **1994**, *12*, 1173–1179.

## ANNOTATED BIBLIOGRAPHY

Sparidans, R.W.; Taal, B.G.; Beijnen, J.H. Bioanalysis of *m*-iodobenzylguanidine in plasma by high-performance liquid chromatography after derivatization with benzoin, *J.Chromatogr.B*, **1999**, *730*, 193–199. [SPE; fluorescence detection; LOQ 2 ng/mL]

Wafelman, A.R.; Nortier, Y.L.M.; Rosing, H.; Underberg, W.J.M.; Maes, R.A.A.; Beijnen, J.H. High-performance liquid chromatographic determination of *m*-iodobenzylguanidine in urine of cancer patients, *J.Chromatogr.*, **1993**, *622*, 71–77. [SPE; LOQ 200 ng/mL]

# Iodixanol

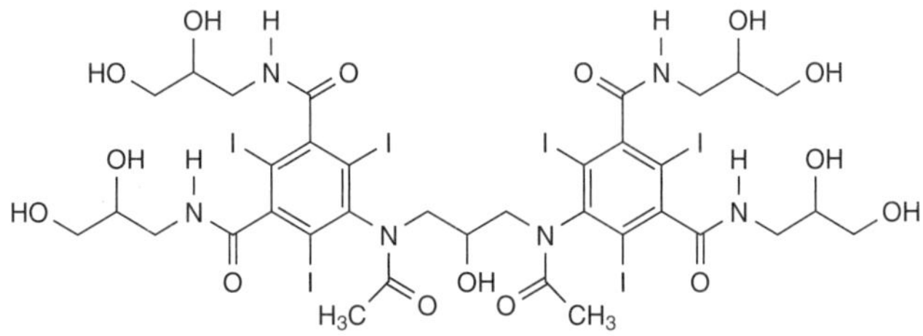


**Molecular formula:** $C_{35}H_{44}I_6N_6O_{15}$
**Molecular weight:** 1550.18
**CAS Registry No:** 92339-11-2
**Merck Index:** 13, 5045

## SAMPLE

**Matrix:** blood
**Sample preparation:** Dialyze 110 μL plasma against 175 μL water in the recipient channel using a Cuprophane cellulose dialysis membrane (molecular weight cut-off 15 000), pump 4 mL water in a pulsed fashion through the recipient channel, pass the water through column A, elute the contents of column A onto column B using the mobile phase, monitor the effluent from column B. (Purge the system with 0.01% Triton X-100 in water.) (ASTED XL system)

## HPLC VARIABLES

**Column:** A 5 × 1.6 10 μm Hypersil ODS; B 15 × 3.2 7 μm Brownlee RP-18 Newguard + 250 × 4.6 5 μm Brownlee OD-5A, Spheri-5, RP-18
**Mobile phase:** MeCN:water 9:91
**Flow rate:** 1
**Detector:** UV 244

## CHROMATOGRAM

**Retention time:** 7 (exo isomer), 9 (endo isomer)
**Limit of detection:** 19–130 ng/mL
**Limit of quantitation:** 52–340 ng/mL

## KEY WORDS

dialysis; monkey; human; plasma; rat

## REFERENCE

Jacobsen, P.B. On-line dialysis and quantitative high-performance liquid chromatography analysis of iodixanol in human, rat and monkey plasma, *J.Chromatogr.B*, **2000**, *749*, 135–142.

## SAMPLE

**Matrix:** urine
**Sample preparation:** Dilute urine 1:20 with water, inject an aliquot.

## HPLC VARIABLES

**Column:** 250 × 4.6 5 μm C18 (Becton Dickinson)
**Mobile phase:** Gradient. MeCN:water 8:92 for 5 min, to 16:84 (step gradient), maintain at 16:84 for 10 min, return to initial conditions (step gradient), re-equilibrate for 5 min.
**Detector:** UV 250

## CHROMATOGRAM

**Retention time:** 11.6 (exo), 13.8 (endo)
**Limit of detection:** 640 ng/mL

## REFERENCE

Kerr, S.W.; Wolyniec, W.W.; Filipovic, Z.; Nodop, S.G.; Braza, F.; Winquist, R.J.; Noonan, T.C. Repeated measurement of intestinal permeability as an assessment of colitis severity in HLA-B27 transgenic rats, *J.Pharmacol.Exp.Ther.*, **1999**, *291*, 903–910.

## ANNOTATED BIBLIOGRAPHY

Molander, P.; Theodorsen, M.; Lundanes, E.; Soerenssen, D.M.; Greibrokk, T. Temperature effects on packed-capillary liquid chromatography of the X-ray contrast agent iodixanol, *J.Chromatogr.Sci.*, **2000**, *38*, 157–161.

Nomura, H.; Teshima, E.; Hakusui, H. Simple isocratic high-performance liquid chromatographic method for measurement of iodixanol in human plasma, *J.Chromatogr.*, **1991**, *572*, 333–338. [size exclusion chromatography]

# Iopanoic acid

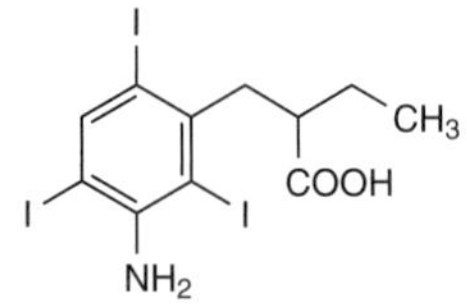

**Molecular formula:** $C_{11}H_{12}I_3NO_2$
**Molecular weight:** 570.93
**CAS Registry No:** 96-83-3
**Merck Index:** 13, 5074

## SAMPLE

**Matrix:** blood
**Sample preparation:** Mix 50 μL plasma with 50 μL 4 mM IS in MeOH, centrifuge at 14 000 rpm for 4 min, inject a 1–5 μL aliquot of the supernatant.

## HPLC VARIABLES

**Guard column:** Present but not specified
**Column:** 150 × 4.6 Hisep shielded hydrophobic phase
**Mobile phase:** Gradient. MeOH:50 mM pH 3.4 phosphate buffer from 12:88 to 70:30 over 2 min, maintain at 70:30 for 30 min, return to initial conditions over 5 min.
**Flow rate:** 1.5
**Injection volume:** 1–5
**Detector:** UV 231, UV 254

## CHROMATOGRAM

**Retention time:** 7.1
**Internal standard:** 2,4,6-triiodobenzoic acid (23.5)
**Limit of detection:** 6.25 ng

## KEY WORDS

dog; plasma

## REFERENCE

Andeejani, A.M.; Hughes, H.; Feuchuk, D.M.; Aboul-Enein, H.Y. Rapid assay of iopanoic acid in dog plasma using a Hisep column, *Biomed.Chromatogr.*, **1994**, *8*, 26–28.

## SAMPLE

**Matrix:** blood, tissue
**Sample preparation:** Blood. Shake 500 μL whole blood with 5 mL MeOH for 30 s, centrifuge at 2000 rpm for 2 min, inject a 20 μL aliquot of the supernatant. Liver. Shake 1 g liver homogenate with 10 mL MeOH for 30 s, centrifuge at 2000 rpm for 2 min, inject a 20 μL aliquot of the supernatant.

## HPLC VARIABLES

**Column:** 300 × 4 10 μm LiChrosorb RP-18
**Mobile phase:** MeOH:water 60:40 containing 5 mM tetrabutylammonium phosphate
**Flow rate:** 2
**Injection volume:** 20
**Detector:** UV 232

## CHROMATOGRAM

**Retention time:** 7.15
**Limit of detection:** 1 μg/mL

## OTHER SUBSTANCES

**Extracted:** diatrizoic acid (UV 242) (1.30), iothalamic acid (UV 242) (1.30), ioxaglic acid (UV 242) (1.33)

## KEY WORDS

whole blood; liver

## REFERENCE

Crowley, R.; Kacprzak, J. The determination of commonly used iodinated contrast media in postmortem samples using HPLC and TLC, *J.Anal.Toxicol.*, **1986**, *10*, 53–55.

# Iopromide

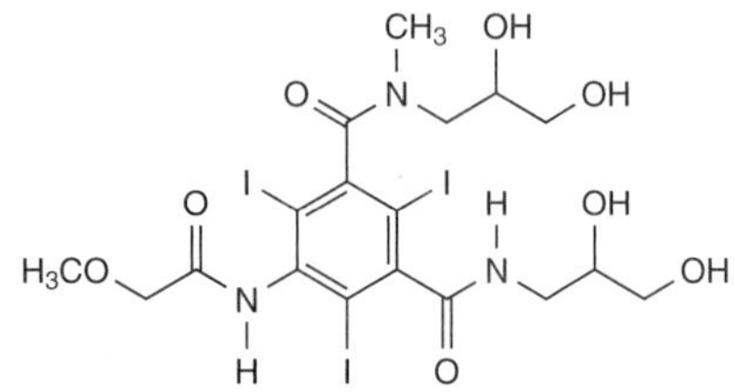


**Molecular formula:** $C_{18}H_{24}I_3N_3O_8$
**Molecular weight:** 791.11
**CAS Registry No:** 73334-07-3
**Merck Index:** 13, 5078

### SAMPLE
**Matrix:** blood
**Sample preparation:** Mix plasma with 4 vol 5% perchloric acid, centrifuge, inject a 20 μL aliquot of the supernatant.

### HPLC VARIABLES
**Column:** 250 × 4 LiChrosorb C18
**Mobile phase:** MeCN:water 4:96 adjusted to pH 2.5 with phosphoric acid
**Flow rate:** 1.5
**Injection volume:** 20
**Detector:** UV 254

### CHROMATOGRAM
**Retention time:** 11, 13 (isomers)
**Limit of detection:** 500 ng/mL

### OTHER SUBSTANCES
**Extracted:** iohexol (4.5, 5.5 (isomers))

### KEY WORDS
plasma

### REFERENCE
Gaspari, F.; Perico, N.; Ruggenenti, P.; Mosconi, L.; Amuchastegui, C.S.; Guerini, E.; Daina, E.; Remuzzi, G. Plasma clearance of nonradioactive iohexol as a measure of glomerular filtration rate, *J.Am.Soc. Nephrol.*, **1995**, *6*, 257–263.

### SAMPLE
**Matrix:** water
**Sample preparation:** Condition a 3 mL 200 mg LiChrolut EN SPE cartridge with water, 9 mL MeOH, and 9 mL water adjusted to pH 3.5 with nitric acid. Condition a 3 mL 250 mg LiChrolut Envi-Carb SPE cartridge (Supelco) with water, 9 mL MeOH, and 9 mL water adjusted to pH 2 with nitric acid. Adjust pH of 500–1000 mL water to 3.5 with nitric acid, pass through the EN cartridge at 200 mL/h. Adjust the pH of the eluate to 2 with nitric acid and pass through the Envi-Carb SPE cartridge at 300 mL/h. Dry cartridges under vacuum for 1 min. Elute the EN cartridge with 6 mL MeOH. Elute the Envi-Carb cartridge with 8 mL MeCN:water 50:50 containing a trace of ammonium acetate in the reverse direction. Combine the eluates and evaporate to dryness under a stream of nitrogen, reconstitute the residue with 500 μL mobile phase A, inject a 10 μL aliquot.

### HPLC VARIABLES
**Column:** 150 × 2 3 μm Luna C18(2)
**Column temperature:** 45

**Mobile phase:** Gradient. A was 0.05% trifluoroacetic acid in water. B was 0.05% trifluoroacetic acid in MeOH. A:B from 100:0 to 95:5 over 10 min, to 75:25 over 15 min, return to initial conditions over 2 min, re-equilibrate for 5 min.
**Flow rate:** 0.25
**Injection volume:** 10
**Detector:** MS, Micromass Quattro-LC, electrospray, positive ion mode, drying gas nitrogen, nebulizing gas nitrogen, collision gas argon, m/z 792–573; UV 242

## CHROMATOGRAM
**Retention time:** 21.5, 22.5 (isomers)
**Limit of detection:** ca. 50 pg/mL

## OTHER SUBSTANCES
**Extracted:** diatrizoate (m/z 615–361) (11), iohexol (m/z 822–803) (16), iotrolan (m/z 814) (18)

## KEY WORDS
SPE

## REFERENCE
Putschew, A.; Schittko, S.; Jekel, M. Quantification of triiodinated benzene derivatives and X-ray contrast media in water samples by liquid chromatography-electrospray tandem mass spectrometry, *J.Chromatogr.A*, **2001**, *930*, 127–134.

## ANNOTATED BIBLIOGRAPHY
Putschew, A.; Wischnack, S.; Jekel, M. Occurrence of triiodinated X-ray contrast agents in the aquatic environment, *Sci.Total Environ.*, **2000**, *255*, 129–134. [iopromide; diatrizoate; iotrolan; iotroxin acid; iotroxin]

Sacher, F.; Lange, F.T.; Brauch, H.-J.; Blankenhorn, I. Pharmaceuticals in groundwaters. Analytical methods and results of a monitoring program in Baden-Württemberg, Germany, *J.Chromatogr.A*, **2001**, *938*, 199–210. [amidotrizoic acid; amoxicillin; anhydro-erythromycin; atenolol; betaxolol; bezafibrate; bisoprolol; carbamazepine; chloramphenicol; clarithromycin; clenbuterol; clofibric acid; cloxacillin; cyclophosphamide; dapsone; diazepam; diclofenac; dicloxacillin; dimethylaminophenazone; erythromycin; etofibrate; fenofibrate; fenoprofen; furazolidone; gemfibrozil; ibuprofen; ifosfamide; indomethacin; iomeprol; iopamidol; iopromide; ketoprofen; metoprolol; metronidazole; nafcillin; naproxen; oleandomycin; oxacillin; penicillin G; penicillin V; pentoxifylline; phenazone; pindolol; propranolol; propyphenazone; ronidazole; roxithromycin; albuterol; simvastatin; sotalol; spiramycin; sulfadiazine; sulfadimidine; sulfamerazine; sulfamethoxazole; terbutaline; trimethoprim; tylosin; virginiamycin; sulfamethazine]

Vanderford, B.J.; Pearson, R.A.; Rexing, D.J.; Snyder, S.A. Analysis of endocrine disruptors, pharmaceuticals, and personal care products in water using liquid chromatography/tandem mass spectrometry, *Anal.Chem.*, **2003**, *75*, 6265–6274. [hydrocodone; trimethoprim; acetaminophen; caffeine; erythromycin; sulfamethoxazole; fluoxetine; pentoxifylline; meprobamate; phenytoin; carbamazepine; DEET; diazepam; oxybenzone; progesterone; iopromide; naproxen; ibuprofen; diclofenac; triclosan; gemfibrozil; ethinyl estradiol; estradiol; testosterone; SPE]

# Ioversol

**Molecular formula:** $C_{18}H_{24}I_3N_3O_9$
**Molecular weight:** 807.11
**CAS Registry No:** 87771-40-2
**Merck Index:** 13, 5085

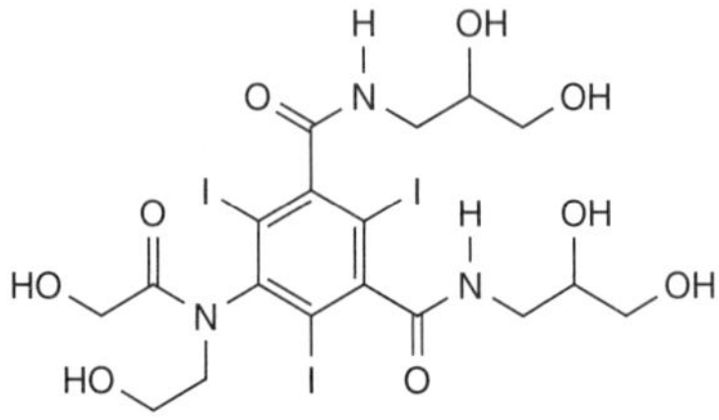


## SAMPLE

**Matrix:** blood, CSF, tissue, urine
**Sample preparation:** Blood. Mix plasma or serum with 1 vol of MeCN:EtOH:water 60:38.4:1.6 or 20% trifluoroacetic acid at 4°, let stand overnight at 4°, centrifuge at 13 000 g at 4° for 10 min, inject an aliquot of the supernatant. Brain. Homogenize (Janke & Kunkel Ikawerk homogenizer) with an equal volume of saline at 20 000 rpm at room temperature for 3 min, centrifuge at 12 000 g at 4° for 30 min, filter (Centrisart 1) while centrifuging at 4°, inject an aliquot of the ultrafiltrate. CSF. Dilute 50–200-fold with water, inject an aliquot. Urine. Dilute with 4 vol of water. Mix with 1 vol of MeCN:EtOH:water 60:38.4:1.6 at 4°, let stand overnight at 4°, centrifuge at 13 000 g at 4° for 10 min, inject an aliquot of the supernatant. Alternatively, filter (Centrisart 1 with 5000 Da cut-off) while centrifuging at 4°, inject an aliquot of the ultrafiltrate.

## HPLC VARIABLES

**Column:** 250 × 4.6 Spheri-5 RP-18
**Mobile phase:** Gradient. MeCN:water from 1:99 to 30:70 over 20 min
**Flow rate:** 1
**Injection volume:** 10
**Detector:** UV 220–280

## CHROMATOGRAM

**Retention time:** 9.9

## OTHER SUBSTANCES

**Extracted:** iohexol (11.1 (endo), 11.3 (exo))
**Noninterfering:** diatrizoate, iothalamate

## KEY WORDS

brain; plasma; serum; ultrafiltrate

## REFERENCE

Jacobsen, P.B. High performance liquid chromatography with multiwavelength detection: a technique for identification of iodinated x-ray contrast agents in human body fluids and brain tissue, *Am.J.Neuroradiol.*, **1992**, *13*, 1521–1525.

# Ipratropium bromide

**Molecular formula:** $C_{20}H_{30}BrNO_3$
**Molecular weight:** 412.37
**CAS Registry No:** 22254-24-6
**Merck Index:** 13, 5092

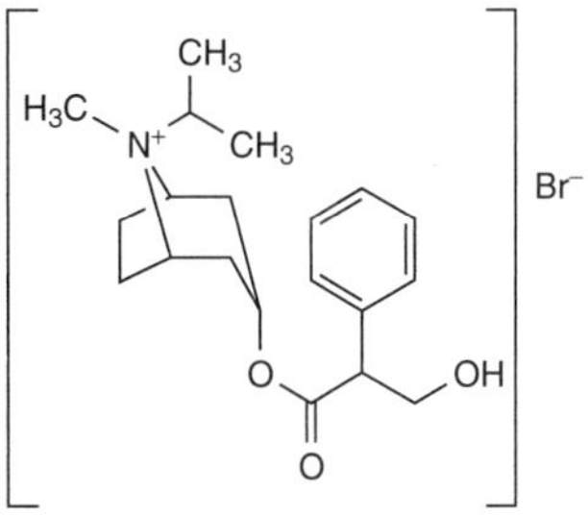


## SAMPLE

**Matrix:** blood
**Sample preparation:** Inject 100 μL plasma onto column A and elute to waste with mobile phase A; after 10 min, backflush the contents of column A onto column B with mobile phase B; after 2 min, remove column A from the circuit, elute column B with mobile phase B, monitor the effluent from column B, re-equilibrate column A with mobile phase A for 5 min.

## HPLC VARIABLES

**Column:** A 25 × 4 25 μm LiChrospher 60 XDS ($SO_3$/diol); B 250 × 4 5 μm LiChrospher 60 RP-Select B
**Column temperature:** 25 ± 0.1
**Mobile phase:** A MeOH:2 mM lithium perchlorate adjusted to pH 3.0 with 1 M perchloric acid 3:97; B MeCN:50 mM pH 3.0 phosphate buffer containing 0.5 mM sodium butanesulfonate 20:80
**Flow rate:** 1
**Injection volume:** 100
**Detector:** UV 220

## OTHER SUBSTANCES

**Extracted:** atropine (16)

## KEY WORDS

column-switching; plasma

## REFERENCE

Chiap, P.; Rbeida, O.; Christiaens, B.; Hubert, P.; Lubda, D.; Boos, K.-S.; Crommen, J. Use of a novel cation-exchange restricted-access material for automated sample clean-up prior to the determination of basic drugs in plasma by liquid chromatography, *J.Chromatogr.A*, **2002**, *975*, 145–155.

# Ipriflavone

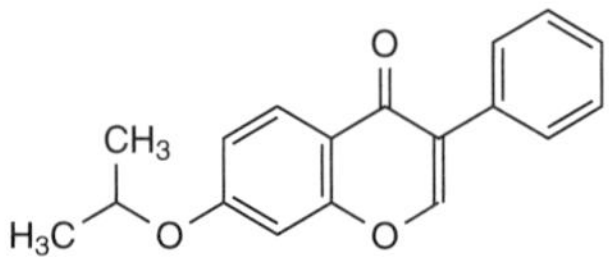

**Molecular formula:** $C_{18}H_{16}O_3$
**Molecular weight:** 280.32
**CAS Registry No:** 35212-22-7
**Merck Index:** 13, 5093

## SAMPLE

**Matrix:** blood, tissue, urine
**Sample preparation:** Homogenize (Ultra-Turrax T25) tissue with 4 vol of water. Vortex 50 μL plasma, urine, or tissue homogenate with 125 μL 1 μg/mL IS in MeCN, centrifuge at 14 000 g for 1 min, inject a 50 μL aliquot of the supernatant.

## HPLC VARIABLES

**Column:** 250 × 4.6 10 μm LiChrosorb RP-18
**Mobile phase:** MeCN:MeOH:50 mM pH 3 acetate buffer 35:25:40
**Flow rate:** 1.5
**Injection volume:** 50
**Detector:** UV 254

## CHROMATOGRAM

**Retention time:** 12
**Internal standard:** testosterone (6)
**Limit of detection:** 20 ng/mL (plasma), 50 ng/mL (tissue), 100 ng/mL (urine)

## KEY WORDS

brain; fat; heart; intestine; kidney; liver; lung; muscle; plasma; rat; spleen; stomach

## REFERENCE

Kim, S.H.; Lee, J.S.; Lee, M.G. Determination of a isoflavone derivative, ipriflavone, and its metabolites, M1 and M5, in rat plasma, urine, and tissue homogenate by high-performance liquid chromatography, *Res.Commun.Mol.Pathol.Pharmacol.*, **1997**, *98*, 313–324.

# Isoflupredone

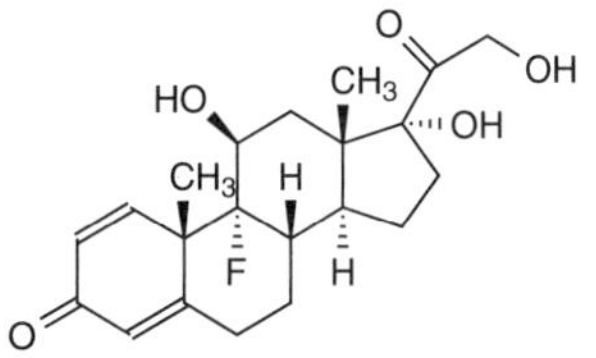

**Molecular formula:** $C_{21}H_{27}FO_5$
**Molecular weight:** 378.43
**CAS Registry No:** 338-95-4, 338-98-7 (21-acetate)
**Merck Index:** 13, 5192

## SAMPLE

**Matrix:** blood, synovial fluid
**Sample preparation:** Mix 500 μL plasma or synovial fluid with 4 mL pH 4 potassium phosphate buffer and 5 mL dichloromethane, centrifuge at 3000 g. Evaporate the organic layer to dryness, reconstitute the residue with 1 mL mobile phase, inject a 70 μL aliquot.

## HPLC VARIABLES

**Column:** 250 × 4.6 CR-J01 C18 (Shimadzu)
**Mobile phase:** MeCN:50 mM pH 3.5 potassium phosphate buffer (ratio not given)
**Flow rate:** 0.5
**Injection volume:** 70
**Detector:** UV 240

## CHROMATOGRAM

**Limit of quantitation:** 10 ng/mL

## KEY WORDS

horse; plasma; for isoflupredone and isoflupredone acetate

## REFERENCE

Lillich, J.D.; Bertone, A.L.; Schmall, L.M.; Ruggles, A.J.; Sams, R.A. Plasma, urine, and synovial fluid disposition of methylprednisolone acetate and isoflupredone acetate after intra-articular administration in horses, *Am.J.Vet.Res.*, **1996**, *57*, 187–192.

# Isopropamide iodide

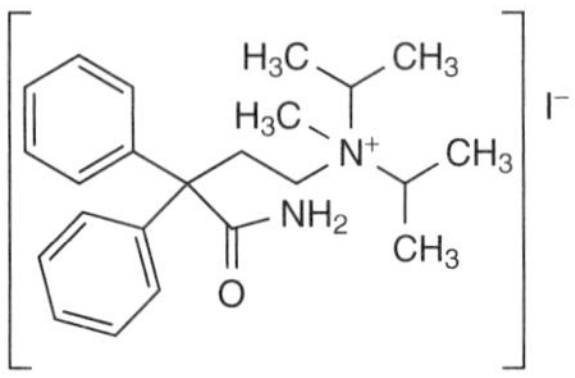

**Molecular formula:** $C_{23}H_{33}IN_2O$
**Molecular weight:** 480.42
**CAS Registry No:** 71-81-8
**Merck Index:** 13, 5222

## SAMPLE
**Matrix:** blood
**Sample preparation:** Condition a 3 mL diol SPE cartridge (Baker) with 2 mL MeOH and 2 mL water. Add 1 mL plasma to the SPE cartridge, wash with 2 mL water, wash with 2 mL MeOH, elute with 3 mL 60 mM KBr in MeOH. Evaporate the eluate to dryness under a stream of nitrogen at 60°, reconstitute the residue with 200 μL mobile phase, inject an aliquot. (Prepare 60 mM KBr in MeOH by sonicating for 45 min.)

## HPLC VARIABLES
**Column:** 250 × 4 LiChrosorb RP18
**Mobile phase:** MeOH:water 55:45 containing 4.325 g/L sodium octanesulfonate and 2 mL/L *N*,*N*-dimethyloctylamine, adjusted to pH 3.0 with phosphoric acid
**Flow rate:** 1
**Detector:** UV 220

## CHROMATOGRAM
**Retention time:** 5

## OTHER SUBSTANCES
**Extracted:** clidinium bromide (9.5), mepenzolate bromide (8)

## KEY WORDS
plasma; SPE

## REFERENCE
Russ-Kirschenbaum, R.; Koziol, T.; Woolf, E. Solid phase extraction of quaternary ammonium compounds on diol columns: Application to the HPLC determination of CK-1649 in plasma, *J.Liq.Chromatogr.*, **1989**, *12*, 3051–3059.

## SAMPLE
**Matrix:** formulations
**Sample preparation:** Sonicate a tablet in 35 mL MeOH:water 45:55 and 10 mL 40 μg/mL IS in MeOH:water 45:55, make up to 50 mL with MeOH:water 45:55, centrifuge at 3020 g for 10 min, inject an aliquot of the supernatant. Mix 1 mL oral solution with 10 mL 40 μg/mL IS in MeOH:water 45:55, make up to 50 mL with MeOH:water 45:55, inject an aliquot.

## HPLC VARIABLES
**Column:** 150 × 4.1 5 μm Rsil C18 (RSL, Belgium)
**Column temperature:** 25
**Mobile phase:** MeOH:water 55:45 containing 20 mM sodium 1-octanesulfonate and 10 mM *N*,*N*-dimethyloctylamine, adjusted to pH 3.0 with orthophosphoric acid
**Flow rate:** 1
**Injection volume:** 10
**Detector:** UV 220

## CHROMATOGRAM
**Retention time:** 4
**Internal standard:** fenpiverinium bromide (6)

## OTHER SUBSTANCES
**Simultaneous:** chlorpheniramine (10), glycopyrrolate (10), mepenzolate bromide (6), methyl paraben (2.5), pentienate bromide (7), phenylpropanolamine (3), tiemonium iodide (5)
**Noninterfering:** cinnarizine, haloperidol

## KEY WORDS
tablets; oral solutions; stability-indicating

## REFERENCE
De Schutter, J.A.; Van den Bossche, W.; De Moerloose, P. Separation and determination of isopropamide iodide in pharmaceutical formulations by reversed-phase ion-pair high-performance liquid chromatography, *J.Chromatogr.*, **1986**, *366*, 321–328.

---

## SAMPLE
**Matrix:** solutions

## HPLC VARIABLES
**Column:** 150 × 4.1 5 μm PRP-1 (Hamilton)
**Mobile phase:** Gradient. MeCN:50 mM ammonium acetate from 10:90 to 50:50 over 30 min.
**Flow rate:** 0.5
**Injection volume:** 20
**Detector:** MS, Finnigan MAT TSQ-70, thermospray, discharge-off mode, vaporizer 90°, ion-source 250°, repeller 50–100 V, make-up flow MeCN:50 mM ammonium acetate 10:90 at 1 mL/min, collision gas pressure 0.5 Pa, collision energy 20 eV, m/z 353–238

## CHROMATOGRAM
**Limit of detection:** 50 pg

## OTHER SUBSTANCES
**Simultaneous:** antrenyl, clidinium, mepenzolate (14.5), methylbenacyzine, neostigmine, pipenzolate (15.5), propantheline (22), valethamate

## REFERENCE
van der Hoeven, R.A.M.; Reeuwijk, H.J.E.M.; Tjaden, U.R.; van der Greef, J. Analysis of quaternary ammonium drugs by thermospray liquid chromatography-mass spectrometry using a resin-based stationary phase, *J.Chromatogr.A*, **1996**, *741*, 75–84.

# Itopride

**Molecular formula:** $C_{20}H_{26}N_2O_4$
**Molecular weight:** 358.43
**CAS Registry No:** 122898-67-3

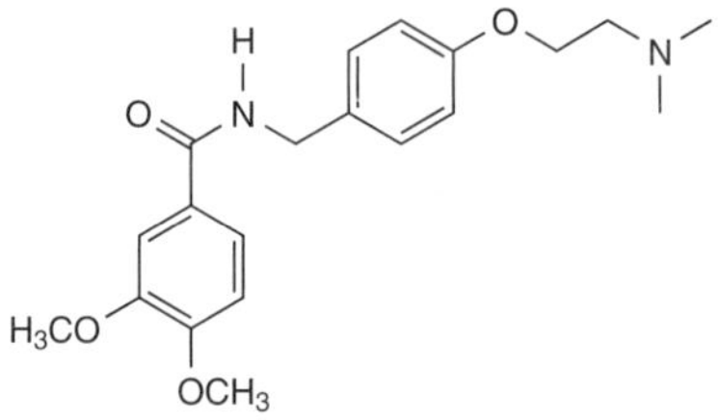

## SAMPLE

**Matrix:** blood, urine
**Sample preparation:** Inject 200 μL serum or urine onto column A and elute to waste with mobile phase A. After 4 min, backflush the contents of column A onto column B with mobile phase B; after 1 min, remove column A from the circuit, elute column B with mobile phase B, monitor the effluent from column B. Re-equilibrate column A with mobile phase A.

## HPLC VARIABLES

**Column:** A 10 × 4 25–40 μm Nucleosil CN (Macherey-Nagel); B 10 × 4.6 5 μm TSKgel ODS-80TM + 150 × 4.6 5 μm TSKgel ODS-80TM (Tosoh)
**Mobile phase:** A 100 mM pH 7.0 phosphate buffer; B MeCN:50 mM pH 5.5 phosphate buffer 20:80
**Flow rate:** 1
**Injection volume:** 200
**Detector:** F ex 308 em 344

## CHROMATOGRAM

**Retention time:** 13
**Limit of quantitation:** 5 ng/mL (serum), 20 ng/mL (urine)

## OTHER SUBSTANCES

**Extracted:** metabolites

## KEY WORDS

column-switching; pharmacokinetics; serum

## REFERENCE

Takahara, E.; Fukuoka, H.; Takagi, T.; Nagata, O.; Kato, H. Simultaneous determination of a new gastrointestinal prokinetic agent (HSR-803) and its metabolites in human serum and urine by high-performance liquid chromatography using automated column-switching, *J.Chromatogr.*, **1992**, *576*, 174–178.

# Kinetin

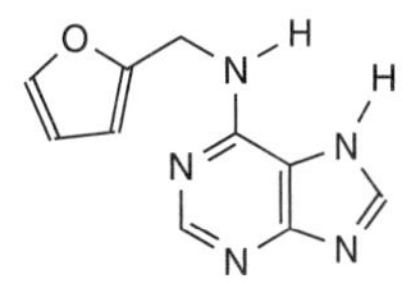

**Molecular formula:** $C_{10}H_9N_5O$
**Molecular weight:** 215.21
**CAS Registry No:** 525-79-1
**Merck Index:** 13, 5329

## SAMPLE
**Matrix:** solutions
**Sample preparation:** Inject an aliquot of a solution in 200 mM HCl.

## HPLC VARIABLES
**Guard column:** present but not specified
**Column:** 150 × 3.3 Vydac
**Mobile phase:** MeOH:buffer 4:96 (The buffer was 12.5 mM citric acid containing 25 mM sodium acetate, 30 mM NaOH, and 10 mM acetic acid.)
**Detector:** E, ESA Coulochem II, 650 mV; UV 260

## CHROMATOGRAM
**Retention time:** 8

## REFERENCE
Barciszewski, J.; Siboska, G.E.; Pedersen, B.O.; Clark, B.F.C.; Rattan, S.I.S. Evidence for the presence of kinetin in DNA and cell extracts, *FEBS Lett.*, **1996**, *393*, 197–200.

# Lafutidine

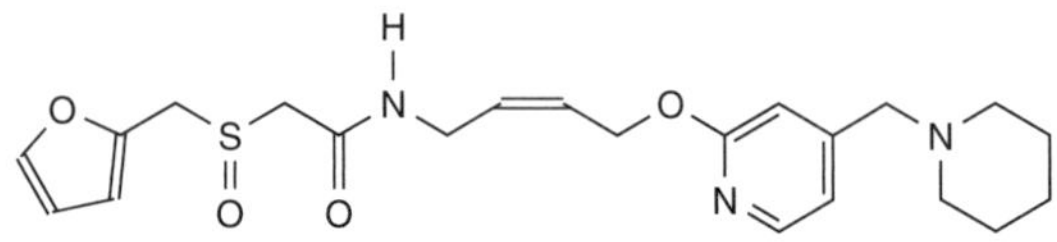

**Molecular formula:** $C_{22}H_{29}N_3O_4S$
**Molecular weight:** 431.56
**CAS Registry No:** 118288-08-7
**Merck Index:** 13, 5362

## SAMPLE

**Matrix:** blood
**Sample preparation:** Mix 1 mL plasma with 500 μL 1 M NaOH, extract with 3 mL ethyl acetate. Evaporate the extract to dryness under reduced pressure, reconstitute the residue with 100 μL 100 mM HCl, wash with 1 mL ethyl acetate. Add 750 μL 100 mM NaOH to the aqueous layer, extract with 3 mL ethyl acetate containing 20 ng IS. Evaporate the organic layer to dryness, reconstitute the residue with 200 μL mobile phase, inject an aliquot.

## HPLC VARIABLES

**Guard column:** Guard-Pak
**Column:** Cosmosil 5C18-AR
**Column temperature:** 40
**Mobile phase:** MeCN:10 mM pH 5.9 phosphate buffer 20:80
**Flow rate:** 1
**Detector:** UV 230

## CHROMATOGRAM

**Internal standard:** 4-amino-3-nitroanisole

## KEY WORDS

plasma

## REFERENCE

Itoh, H.; Naito, T.; Takeyama, M. Lafutidine changes levels of somatostatin, calcitonin gene-related peptide, and secretin in human plasma, *Biol.Pharm.Bull.*, **2002**, *25*, 379–382.

# Lamivudine

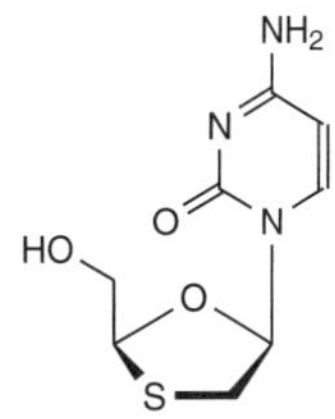

**Molecular formula:** $C_8H_{11}N_3O_3S$
**Molecular weight:** 229.26
**CAS Registry No:** 134678-17-4
**Merck Index:** 13, 5367

## SAMPLE

**Matrix:** blood
**Sample preparation:** Condition a 1 mL 100 mg Bond Elut-C SPE cartridge with 1 mL MeOH and 1 mL 100 mM pH 7.0 ammonium acetate buffer. Heat plasma at 58° for 1 h to inactivate HIV. Vortex 800 μL plasma with 300 μL 2 μg/mL hexobarbital in 25 mM pH 7.0 ammonium acetate buffer for 30 s and centrifuge at 18 000 g for 5 min. Add 1 mL of the supernatant to the SPE cartridge, wash with 1 mL 100 mM pH 7.0 ammonium acetate buffer, suck dry for 1 min, elute with 800 μL MeOH. Evaporate the eluate to dryness under a stream of nitrogen at 40° and reconstitute the residue with 100 μL mobile phase. Vortex for 30 s, centrifuge at 18 000 g for 3 min, and inject an 80 μL aliquot.

## HPLC VARIABLES

**Guard column:** 20 × 3.9 5 μm Polarity dC18 (Waters)
**Column:** 150 × 3.9 5 μm Polarity dC18 (Waters)
**Column temperature:** 40
**Mobile phase:** Gradient. A was 10 mM pH 6.5 ammonium acetate buffer. B was 10 mM pH 6.5 ammonium acetate buffer:MeCN:MeOH 20:50:30. A:B 96:4 for 15 min, to 36:64 over 15 min, maintain at 36:64 for 3 min, re-equilibrate at initial conditions for 7 min.
**Flow rate:** 1.1
**Injection volume:** 80
**Detector:** UV 269 for 11 min, UV 250 for 3 min, UV 271 for 10 min, UV 230 for 9 min

## CHROMATOGRAM

**Retention time:** 8.6
**Internal standard:** hexobarbital (30.6)
**Limit of quantitation:** 10.0 ng/mL

## OTHER SUBSTANCES

**Extracted:** abacavir (25.1), didanosine (13.6), nevirapine (27.3), stavudine (15.7), zalcitabine (5.9), zidovudine (23.8)
**Noninterfering:** tenofovir

## KEY WORDS

plasma; SPE

## REFERENCE

Rezk, N.L.; Tidwell, R.R.; Kashuba, A.D.M. Simultaneous determination of six HIV nucleoside analogue reverse transcriptase inhibitors and nevirapine by liquid chromatography with ultraviolet absorbance detection, *J.Chromatogr.B*, **2003**, *791*, 137–147.

## SAMPLE

**Matrix:** blood
**Sample preparation:** Condition a 100 mg Dual Zone C18 SPE cartridge (Diazem) with 2 mL MeOH and 2 mL water. Dilute 500 μL serum with 1 mL water, add to the SPE cartridge, wash with 500 μL water, elute with 1 mL MeOH. Evaporate the eluate to

dryness with vortexing under reduced pressure at 40° and reconstitute the residue with 300 μL MeOH, inject a 10 μL aliquot.

---

## HPLC VARIABLES

**Column:** two 150 × 4.6 3 μm Luna C18 columns in series
**Column temperature:** 60
**Mobile phase:** Gradient. MeCN:water from 5:95 to 45:55 over 20 min.
**Flow rate:** 0.85
**Injection volume:** 10
**Detector:** UV 250

---

## CHROMATOGRAM

**Retention time:** 9.5
**Limit of detection:** 260 ng/mL

---

## OTHER SUBSTANCES

**Extracted:** abacavir (17, LOD 75 ng/mL), didanosine (10.5, LOD 120 ng/mL), stavudine (11.5, LOD 40 ng/mL), zalcitabine (7.5, LOD 440 ng/mL), zidovudine (16, LOD 30 ng/mL)

---

## KEY WORDS

serum; SPE

---

## REFERENCE

Simon, V.A.; Thiam, M.D.; Lipford, L.C. Determination of serum levels of thirteen human immunodeficiency virus-suppressing drugs by high-performance liquid chromatography, *J.Chromatogr.A*, **2001**, *913*, 447–453.

---

## SAMPLE

**Matrix:** blood
**Sample preparation:** Vortex 100 μL serum with 20 μL 20% trichloroacetic acid for 10 s, centrifuge at 16 000 g for 5 min, inject an aliquot of the supernatant.

---

## HPLC VARIABLES

**Column:** 250 × 4.6 5 μm Hypersil C18
**Mobile phase:** MeOH: buffer 11.7:88.3 (The buffer was 43 mM orthophosphoric acid containing 10 mM triethylammonium acetate, adjusted to pH 7.0 with 5 M KOH.)
**Flow rate:** 1
**Injection volume:** 100
**Detector:** UV 280

---

## CHROMATOGRAM

**Retention time:** 9.5
**Limit of detection:** 5 ng/mL
**Limit of quantitation:** 20 ng/mL

---

## KEY WORDS

pharmacokinetics; serum

---

## REFERENCE

Zhou, X.-J.; Sommadossi, J.-P. Rapid quantitation of (−)-2′-deoxy-3′-thiacytidine in human serum by high-performance liquid chromatography with ultraviolet detection, *J.Chromatogr.B*, **1997**, *691*, 417–424.

---

## ANNOTATED BIBLIOGRAPHY

Aboul-Enein, H.Y.; Hefnawy, M.M. High throughput analysis of lamivudine in pharmaceutical preparations using monolithic silica HPLC column, *Anal.Lett.*, **2003**, *36*, 2527–2538. [LOD 12.5 ng/mL]

Aymard, G.; Legrand, M.; Trichereau, N.; Diquet, B. Determination of twelve antiretroviral agents in human plasma sample using reversed-phase high-performance liquid chromatography, *J.Chromatogr.B*, **2000**, *744*, 227–240. [column-switching; LOQ 20 ng/mL for lamivudine; amprenavir; efavirenz; indinavir; nelfinavir; ritonavir; saquinavir; abacavir; didanosine; lamivudine; stavudine; nevirapine; zidovudine]

Fan, B.; Bartlett, M.G.; Stewart, J.T. Determination of lamivudine/stavudine/efavirenz in human serum using liquid chromatography/electrospray tandem mass spectrometry with ionization polarity switch, *Biomed.Chromatogr.*, **2002**, *16*, 383–389. [SPE; LOQ 1.1 ng/mL for lamivudine]

Fan, B.; Stewart, J.T. Determination of zidovudine/lamivudine/nevirapine in human plasma using ion-pair HPLC, *J.Pharm.Biomed.Anal.*, **2002**, *28*, 903–908. [SPE; LOQ 59 ng/mL for lamivudine]

Gibbs, J.E.; Rashid, T.; Thomas, S.A. Effect of transport inhibitors and additional anti-HIV drugs on the movement of lamivudine (3TC) across the guinea pig brain barriers, *J.Pharmacol.Exp.Ther.*, **2003**, *306*, 1035–1041.

Harker, A.J.; Evans, G.L.; Hawley, A.E.; Morris, D.M. High-performance liquid chromatographic assay for 2′-deoxy-3′-thiacytidine in human serum, *J.Chromatogr.B*, **1994**, *657*, 227–232. [SPE; carbovir is internal standard]

Hoetelmans, R.M.; Profijt, M.; Mennhorst, P.L.; Mulder, J.W.; Beijnen, J.H. Quantitative determination of (−)-2′-deoxy-3′-thiacytidine (lamivudine) in human plasma, saliva and cerebrospinal fluid by high-performance liquid chromatography with ultraviolet detection, *J.Chromatogr.B*, **1998**, *713*, 387–394. [SPE; LOQ 10 ng/mL]

Hsyu, P.-H.; Lloyd, T.L. Automated high-performance liquid chromatographic analysis of (−)- 2′-deoxy-3′-thiacytidine in biological fluids using the automated sequential trace enrichment of dialysate systems, *J.Chromatogr.B*, **1994**, *655*, 253–259. [dialysate; serum; urine; column-switching]

Kakuda, T.N.; Page, L.M.; Anderson, P.L.; Henry, K.; Schacker, T.W.; Rhame, F.S.; Acosta, E.P.; Brundage, R.C.; Fletcher, C.V. Pharmacological basis for concentration-controlled therapy with zidovudine, lamivudine, and indinavir, *Antimicrob.Agents Chemother.*, **2001**, *45*, 236–242. [SPE; no HPLC of indinavir]

Kenney, K.B.; Wring, S.A.; Carr, R.M.; Wells, G.N.; Dunn, J.A. Simultaneous determination of zidovudine and lamivudine in human serum using HPLC with tandem mass spectrometry, *J.Pharm.Biomed.Anal.*, **2000**, *22*, 967–983. [ultrafiltration; LOQ 2.5 ng/mL]

Morris, D.; Selinger, K. Determination of 2′-deoxy-3′-thiacytidine (3TC) in human urine by liquid chromatography: direct injection with column switching, *J.Pharm.Biomed.Anal.*, **1994**, *12*, 255–264. [LOQ 500 ng/mL]

Moyer, T.P.; Temesgen, Z.; Enger, R.; Estes, L.; Charlson, J.; Oliver, L.; Wight, A. Drug monitoring of antiretroviral therapy for HIV-1 infection: method validation and results of a pilot study, *Clin.Chem.*, **1999**, *45*, 1465–1476. [SPE; LOQ 100 ng/mL for lamivudine; didanosine; lamivudine; stavudine; zalcitabine; zidovudine; delavirdine; nevirapine; indinavir; nelfinavir; ritonavir; saquinavir]

Ozkan, S.A.; Uslu, B. Rapid HPLC assay for lamivudine in pharmaceuticals and human serum, *J.Liq.Chromatogr.Rel.Technol.*, **2002**, *25*, 1447–1456. [LOQ 14 ng/mL; deflazacort is internal standard]

Pereira, A.S.; Kenney, K.B.; Cohen, M.S.; Hall, J.E.; Eron, J.J.; Tidwell, R.R.; Dunn, J.A. Simultaneous determination of lamivudine and zidovudine concentrations in human seminal plasma using high-performance liquid chromatography and tandem mass spectrometry, *J.Chromatogr.B*, **2000**, *742*, 173–183. [ultrafiltration; LOQ 5 ng/mL]

Rajagopalan, P.; Boudinot, F.D.; Chu, C.K.; Tennant, B.C.; Baldwin, B.H.; Schinazi, R.F. Pharmacokinetics of (−)-2′-3′-dideoxy-3′-thiacytidine in woodchucks, *Antimicrob.Agents Chemother.*, **1996**, *40*, 642–645. [LOQ 250 ng/mL]

Rezk, N.L.; Tidwell, R.R.; Kashuba, A.D.M. Simultaneous determination of six HIV nucleoside analogue reverse transcriptase inhibitors and nevirapine by liquid chromatography with ultraviolet absorbance detection, *J.Chromatogr.B*, **2003**, *791*, 137–147. [plasma; LOQ 10 ng/mL; zalcitabine; lamivudine; didanosine; stavudine; zidovudine; abacavir; hexobarbital is internal standard; SPE]

Rodriguez, J.F.; Rodriguez, J.L.; Santana, J.; García, H.; Rosario, O. Simultaneous quantitation of intracellular zidovudine and lamivudine triphosphates in human immunodeficiency virus-infected individuals, *Antimicrob.Agents Chemother.*, **2000**, *44*, 3097–3100. [LC-MS]

Santos-Magalhaes, N.S.; Pontes, A.; Cavalcante, R.M.; Costa, R.M.R.; Rangel, F.A.; Guimaraes, M.I.V.; de Carvalho, J.N.L.; de Souza, S.D.F.; de Oliveira, H.M.; Esteves, I.L.C.; Ramalho, M.S.; Vieira, S.L.A.; Alves, A.J. Bioequivalence of two lamivudine tablet formulations, *Arzneimittelforschung*, **2001**, *51*, 310–314.

Simon, V.A.; Thiam, M.D.; Lipford, L.C. Determination of serum levels of thirteen human immunodeficiency virus-suppressing drugs by high-performance liquid chromatography, *J.Chromatogr.A*, **2001**,

*913*, 447–453. [SPE; LOD 260 ng/mL for lamivudine; zalcitabine; lamivudine; stavudine; didanosine; zidovudine; nevirapine; abacavir; indinavir; delavirdine; nelfinavir; saquinavir; ritonavir; efavirenz]

Solas, C.; Li, Y.-F.; Xie, M.-Y.; Sommadossi, J.-P.; Zhou, X.-J. Intracellular nucleotides of (−)-2′,3′-deoxy-3′-thiacytidine in peripheral blood mononuclear cells of a patient infected with human immunodeficiency virus, *Antimicrob.Agents Chemother.*, **1998**, *42*, 2989–2995.

Uslu, B.; Ozkan, S.A. Determination of lamivudine and zidovudine in binary mixtures using first derivative spectrophotometric, first derivative of the ratio-spectra and high-performance liquid chromatography-UV methods, *Anal.Chim.Acta*, **2002**, *466*, 175–185. [LOQ 25 ng/mL for lamivudine; finasteride is internal standard]

Wang, L.H.; Chittick, G.E.; McDowell, J.A. Single-dose pharmacokinetics and safety of abacavir (1592U89), zidovudine, and lamivudine administered alone and in combination in adults with human immunodeficiency virus infection, *Antimicrob.Agents Chemother.*, **1999**, *43*, 1708–1715.

Williams, L.D.; Von Tungeln, L.S.; Beland, F.A.; Doerge, D.R. Liquid chromatographic-mass spectrometric determination of the metabolism and disposition of the anti-retroviral nucleoside analogs zidovudine and lamivudine in C57BL/6N and B6C3F1 mice, *J.Chromatogr.B*, **2003**, *798*, 55–62. [SPE; LC-MS]

Zheng, J.J.; Wu, S.T.; Emm, T.A. High-performance liquid chromatographic assay for the determination of 2′-deoxy-3′-thiacytidine (lamivudine) in human plasma, *J.Chromatogr.B*, **2001**, *761*, 195–201. [SPE; LOQ 10 ng/mL]

# Latanoprost

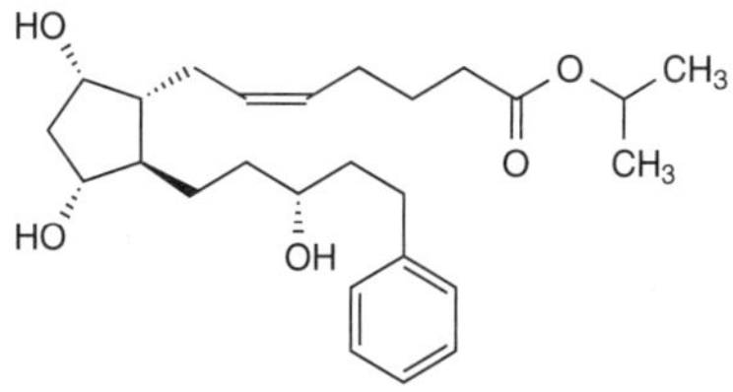


**Molecular formula:** $C_{26}H_{40}O_5$
**Molecular weight:** 432.59
**CAS Registry No:** 130209-82-4
**Merck Index:** 13, 5391

## SAMPLE

**Matrix:** aqueous humor, tissue

**Sample preparation:** Homogenize (Polytron) rabbit eye tissue with 3 mL EtOH, centrifuge, evaporate the supernatant under a stream of nitrogen, reconstitute with EtOH, inject an aliquot. Acidify aqueous humor to pH 3.4 with 1 M formic acid, extract with 3 mL ethyl acetate. Evaporate the ethyl acetate to dryness under a stream of nitrogen, reconstitute with EtOH, inject an aliquot.

## HPLC VARIABLES

**Column:** 5 μm Nucleosil C18

**Mobile phase:** Gradient. MeCN:water:acetic acid 35:65:0.1 for 11 min, to 46:54:0.1 over 1 min, maintain at 46:54:0.1 for 6 min, return to initial conditions over 3 min, re-equilibrate for about 10 min. Alternatively, MeCN:water:acetic acid 25:75:0.1 for 15 min, to 40:60:0.1 over 10 min, return to initial conditions over 5 min, re-equilibrate for 5 min.

**Flow rate:** 1

**Detector:** Radioactivity ($^3H$)

## CHROMATOGRAM

**Retention time:** 22

## OTHER SUBSTANCES

**Extracted:** metabolites

## KEY WORDS

ciliary body; cornea; eye; pharmacokinetics; rabbit

## REFERENCE

Sjöquist, B.; Basu, S.; Byding, P.; Bergh, K.; Stjernschantz, J. The pharmacokinetics of a new antiglaucoma drug, latanoprost, in the rabbit, *Drug Metab.Dispos.*, **1998**, *26*, 745–754.

## SAMPLE

**Matrix:** blood, feces, urine

**Sample preparation:** Blood, urine. Condition a Sep-Pak C18 with 10 mL EtOH and 10 mL water. Centrifuge plasma or urine at 800 g for 10 min, acidify to pH 3.5–4.0 with 1 M HCl, add to the SPE cartridge, wash with 10 mL water, wash with 10 mL petroleum ether, elute with 10 mL methyl formate. Evaporate the eluate to dryness under a stream of nitrogen, reconstitute the residue with EtOH, inject an aliquot. Homogenize feces with 3 volumes of EtOH, centrifuge at 800 g for 15 min, evaporate 2 mL of the supernatant to dryness, reconstitute with 200 μL EtOH, add 1.8 mL water, acidify to pH 3.5–4.0 with 1 M HCl, extract twice with 4 mL portions of ethyl acetate. Evaporate the organic layer to dryness under a stream of nitrogen, reconstitute the residue with EtOH, inject an aliquot.

## HPLC VARIABLES

**Guard column:** 30 × 4 30–40 μm Presorb RP-18

**Column:** 125 × 4.6 5 μm Nucleosil C18
**Mobile phase:** Gradient. MeCN:water:acetic acid 20:80:0.1 for 20 min, to 40:60:0.1 over 20 min, return to initial conditions over 5 min, re-equilibrate for 10 min.
**Flow rate:** 1
**Detector:** Radioactivity ($^3H$)

---

## CHROMATOGRAM
**Retention time:** 20

---

## OTHER SUBSTANCES
**Extracted:** metabolites

---

## KEY WORDS
monkey; pharmacokinetics; plasma

---

## REFERENCE
Sjöquist, B.; Tajallaei, S.; Stjernschantz, J. Pharmacokinetics of latanoprost in the cynomolgus monkey. 1st communication: single intravenous, oral or topical administration on the eye, *Arzneimittelforschung*, **1999**, *49*, 225–233.

# Leflunomide

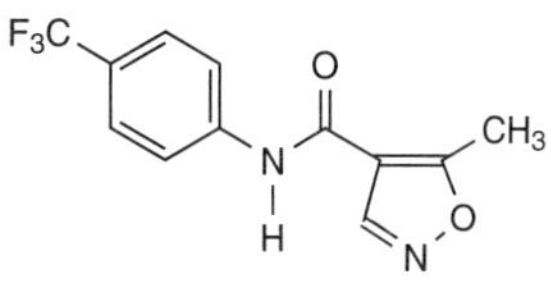


**Molecular formula:** $C_{12}H_9F_3N_2O_2$
**Molecular weight:** 270.21
**CAS Registry No:** 75706-12-6
**Merck Index:** 13, 5451

## SAMPLE

**Matrix:** blood
**Sample preparation:** Mix 75 μL plasma with 200 μL 50 ng/mL IS in MeCN, centrifuge, inject.

## HPLC VARIABLES

**Column:** 150 × 4.6 Zorbax Eclipse XDB-C8
**Mobile phase:** Gradient. MeCN:10 mM ammonium formate containing 0.1% formic acid 0:100 for 3 min, to 90:10 over 17 min, return to initial conditions over 5 min.
**Flow rate:** 1
**Detector:** MS, PE Sciex API 2000 triple quadrupole, 50 μL/min entered the detector, negative ionization, ionspray interface 150°, ionspray 4.5 kV, orifice 30 eV, nebulizer gas nitrogen, m/z 269–82; UV 254

## CHROMATOGRAM

**Retention time:** 16.5
**Internal standard:** indomethacin

## OTHER SUBSTANCES

**Extracted:** metabolite A771726 (82) (12.7), 3-methylleflunomide (16.5)

## KEY WORDS

human; plasma; rat

## REFERENCE

Kalgutkar, A.S.; Nguyen, H.T.; Vaz, A.D.N.; Doan, A.; Dalvie, D.K.; McLeod, D.G.; Murray, J.C. In vitro metabolism studies on the isoxazole ring scission in the anti-inflammatory agent leflunomide to its active α-cyanoenol metabolite A771726: Mechanistic similarities with the cytochrome P450-catalyzed dehydration of aldoximes, *Drug Metab.Dispos.*, **2003**, *31*, 1240–1250.

## SAMPLE

**Matrix:** blood
**Sample preparation:** Mix 250 μL plasma, 500 μL 100 mM pH 5 sodium acetate buffer, and 100 μL 25 μg/mL IS in water, add 10 mL ethyl acetate, shake for 15 min, centrifuge at 4000 rpm for 5 min. Evaporate the organic layer to dryness under a stream of nitrogen at room temperature, reconstitute the residue with 200 μL mobile phase, vortex for 1 min, inject a 50 μL aliquot.

## HPLC VARIABLES

**Guard column:** C18
**Column:** 125 × 3 Nucleosil 100-5C18
**Column temperature:** 21
**Mobile phase:** MeCN:water:formic acid 40:59.8:0.2
**Flow rate:** 0.5
**Injection volume:** 50
**Detector:** UV 261

## CHROMATOGRAM

**Retention time:** 16.2
**Internal standard:** warfarin (12.2)
**Limit of quantitation:** 50 ng/mL

## OTHER SUBSTANCES

**Extracted:** metabolite A771726 (8.2)

## KEY WORDS

plasma

## REFERENCE

Schmidt, A.; Schwind, B.; Gillich, M.; Brune, K.; Hinz, B. Simultaneous determination of leflunomide and its active metabolite, A77 1726, in human plasma by high-performance liquid chromatography, *Biomed.Chromatogr.*, **2003**, *17*, 276–281.

# Lercanidipine

**Molecular formula:** $C_{36}H_{41}N_3O_6$
**Molecular weight:** 611.73
**CAS Registry No:** 100427-26-7
**Merck Index:** 13, 5465

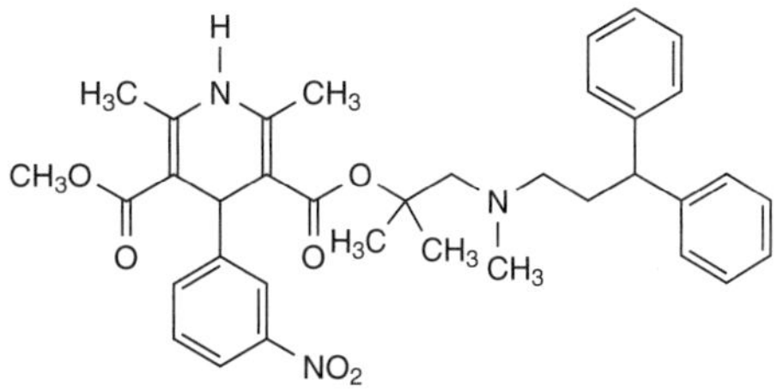

## SAMPLE

**Matrix:** blood
**Sample preparation:** Mix 1 mL plasma with 25 μL 660 ng/mL IS in MeOH and 50 μL 100 mM NaOH, add 4.5 mL hexane:isopropanol 99:1, vortex for 2 min, centrifuge at 2000 g for 10 min. Evaporate the organic layer to dryness under reduced pressure at 25°, reconstitute the residue with 50 μL mobile phase, vortex for 20 s, inject a 20 μL aliquot. (Carry out all manipulations under yellow light.)

## HPLC VARIABLES

**Guard column:** 4 × 4 5 μm 100 RP-18
**Column:** 250 × 4.6 10 μm Chiralpak AD
**Column temperature:** 30 ± 1
**Mobile phase:** Hexane:EtOH:diethylamine 95:5:0.1
**Flow rate:** 1.3
**Injection volume:** 20
**Detector:** MS, Micromass Quatro Micro LC triple quadrupole, electrospray interface, positive ion mode, capillary voltage 3.5 kV, source 100°, desolvation 150°, nebulizing gas nitrogen at 365 L/h, collision gas argon at 3.5 μbar, cone 40 V, collision energy 38.0 eV, m/z 612.40–100.10 (The make-up liquid was EtOH:10 mM ammonium acetate 95:5 pumped at 0.25 mL/min, which mixed with the column effluent. A splitter was used so that 0.2 mL/min entered the detector.)

## CHROMATOGRAM

**Retention time:** 5.98 (S), 6.50 (R)
**Internal standard:** amiodarone (collision energy 30.0 eV) (m/z 646.30–100.30) (3.47)
**Limit of quantitation:** 25 pg/mL

## KEY WORDS

chiral; pharmacokinetics; plasma

## REFERENCE

Jabor, V.A.P.; Coelho, E.B.; Ifa, D.R.; Bonato, P.S.; dos Santos, N.A.G.; Lanchote, V.L. Enantioselective determination of lercanidipine in human plasma for pharmacokinetic studies by normal-phase liquid chromatography-tandem mass spectrometry, *J.Chromatogr.B*, **2003**, *796*, 429–437.

## SAMPLE

**Matrix:** formulations
**Sample preparation:** Sonicate tablet in 5 mL EtOH, make up to 10 mL with mobile phase, centrifuge at 2700 g for 10 min. Dilute a 500 μL aliquot of the supernatant to 10 mL with mobile phase, inject an aliquot.

## HPLC VARIABLES

**Guard column:** 30 × 4.6 Bondapak C18
**Column:** 150 × 3.9 5 μm Symmetry C18
**Column temperature:** 25 ± 1
**Mobile phase:** MeCN:10 mM pH 4.0 phosphate buffer 45:55

**Flow rate:** 1
**Injection volume:** 20
**Detector:** UV 356; E, glassy carbon working electrode 1000 mV, Ag/AgCl reference electrode, platinum rod auxiliary electrode

---

## CHROMATOGRAM
**Retention time:** 5
**Limit of detection:** 930 nM (UV), 750 nM (E)
**Limit of quantitation:** 1.2 μM (UV), 3.2 μM (E)

---

## KEY WORDS
tablets

---

## REFERENCE
Alvarez-Lueje, A.; Pujol, S.; Squella, J.A.; Núñz-Vergara, L.J. A selective HPLC method for determination of lercanidipine in tablets, *J.Pharm.Biomed.Anal.*, **2003**, *31*, 1–9.

---

## ANNOTATED BIBLIOGRAPHY
Boatto, G.; Nieddu, M.; Faedda, M.V.; De Caprariis, P. Enantiomeric separation by HPLC of 1,4-dihydropyridines with vancomycin as chiral selector, *Chirality*, **2003**, *15*, 494–497. [isradipine; nimodipine; nisoldipine; felodipine; lercanidipine; amlodipine]

Calleri, E.; De Lorenzi, E.; Siluk, D.; Markuszewski, M.; Kaliszan, R.; Massolini, G. Column liquid chromatography riboflavin binding protein-chiral stationary phase: Investigation of retention mechanism, *Chromatographia*, **2002**, *55*, 651–658. [fenfluramine; dichloroisoproterenol; propranolol; oxprenolol; alprenolol; acebutolol; ketoprofen; indoprofen; warfarin; lorazepam; oxazepam; isradipine; nimodipine; amlodipine; nicardipine; lercanidipine; gallopamil; verapamil; bepridil]

Garzotti, M.; Hamdan, M. Supercritical fluid chromatography coupled to electrospray mass spectrometry: a powerful tool for the analysis of chiral mixtures, *J.Chromatogr.B*, **2002**, *770*, 53–61. [alprenolol; lercanidipine; disopyramide; propafenone; tropicamide; atenolol; ofloxacin; albuterol; econazole; miconazole; homatropine; nicardipine; sulpiride; verapamil; pindolol; flecainide; atropine; clenbuterol; sulconazole; ketoconazole]

# Letrozole

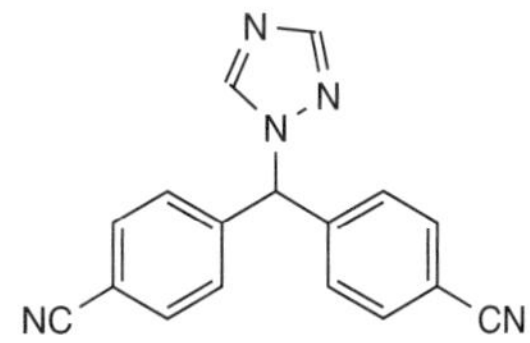


**Molecular formula:** $C_{17}H_{11}N_5$
**Molecular weight:** 285.30
**CAS Registry No:** 112809-51-5
**Merck Index:** 13, 5469

## SAMPLE
**Matrix:** blood, urine
**Sample preparation:** Condition a 1 mL 100 mg Bond Elut C8 SPE cartridge with 1 mL MeOH and 1 mL 100 mM HCl. Dilute urine with an equal volume of water. Mix 1 mL plasma or diluted urine with 100 μL IS in water and 1 mL 100 mM HCl, add to the SPE cartridge, wash with 2 mL 10 mM pH 7 phosphate buffer, wash with 500 μL MeCN:10 mM pH 7 phosphate buffer 20:80, elute with 2 mL MeCN:10 mM pH 7 phosphate buffer 40:60, inject a 100 μL aliquot of the eluate.

## HPLC VARIABLES
**Guard column:** 20 × 4.6 5 μm Supelguard LC
**Column:** 250 × 4.6 5 μm ODS Hypersil C18
**Mobile phase:** MeCN:10 mM pH 7 phosphate buffer 30:70 (The buffer was 1.36 g potassium dihydrogen phosphate and 1.42 g disodium hydrogen phosphate in 1 L water.)
**Flow rate:** 1.5
**Injection volume:** 100
**Detector:** F ex 230 em 295

## CHROMATOGRAM
**Retention time:** 11.3
**Internal standard:** CGP 47 645 (1-[bis(4-cyanophenyl)fluoromethyl]-1,2,4-triazole) (16.7)
**Limit of quantitation:** 1.40 nM (plasma), 2.80 nM (urine)

## OTHER SUBSTANCES
**Extracted:** metabolite CGP 44 645 (12.3)

## KEY WORDS
plasma; SPE

## REFERENCE
Marfil, F.; Pineau, V.; Sioufi, A.; Godbillon, J. High-performance liquid chromatography of the aromatase inhibitor, letrozole, and its metabolite in biological fluids with automated liquid-solid extraction and fluorescence detection, *J.Chromatogr.B*, **1996**, *683*, 251–258.

# Levetiracetam

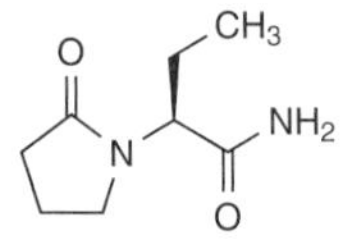

**Molecular formula:** $C_8H_{14}N_2O_2$
**Molecular weight:** 170.21
**CAS Registry No:** 102767-28-2
**Merck Index:** 13, 5482

### SAMPLE
**Matrix:** blood
**Sample preparation:** Condition a 100 mg C18 SPE cartridge (Baker) with 2 mL MeOH and 2 mL water. Add 200 μL serum and 100 μL 20 μg/mL IS in MeOH:water 2:98 to the SPE cartridge, wash with 750 μL water, dry under vacuum for 3 min, elute with 1 mL MeOH:water 1:5, inject a 50 μL aliquot of the eluate.

### HPLC VARIABLES
**Column:** 150 × 4.6 Spherisorb 3ODS2
**Mobile phase:** Gradient. MeCN:water from 6:94 to 20:80 over 5 min, to 40:60 over 1 min, maintain at 40:60 for 4 min, return to initial conditions over 1 min, re-equilibrate for 4 min.
**Flow rate:** 1
**Injection volume:** 150
**Detector:** UV 205

### CHROMATOGRAM
**Retention time:** 6.4
**Internal standard:** G025 (α-methyl-5,5-dimethyl-2-oxo-1-pyrrolidine acetamide) (7.8)
**Limit of quantitation:** 360 ng/mL

### OTHER SUBSTANCES
**Noninterfering:** carbamazepine, carbamazepine dihydrodiol, carbamazepine epoxide, *N*-desmethylsuximide, ethosuximide, lamotrigine, loreclezole, monohydroxycarbamazepine, phenobarbital, phenytoin, primidone, valproic acid, vigabatrin

### KEY WORDS
comparison with GC; pharmacokinetics; serum; SPE

### REFERENCE
Vermeij, T.A.C.; Edelbroek, P.M. High-performance liquid chromatographic and megabore gas-liquid chromatographic determination of levetiracetam (ucb L059) in human serum after solid-phase extraction, *J.Chromatogr.B*, **1994**, *662*, 134–139.

### SAMPLE
**Matrix:** blood, saliva
**Sample preparation:** Vortex 100 μL serum or saliva with 100 μL IS solution and 25 μL 5 M NaOH for 10 s, add 1 mL dichloromethane, vortex for 1 min, centrifuge at 3000 rpm for 5 min. Evaporate the organic layer to dryness, reconstitute the residue with 200 μL mobile phase, inject a 50 μL aliquot.

### HPLC VARIABLES
**Guard column:** PRP-1 (Burdick & Jackson)
**Column:** 100 × 8 4 μm Nova-pak C18 Radial-Pak
**Mobile phase:** Gradient. A was MeCN:buffer 12:88. B was MeCN:buffer 45:55. A:B 100:0 for 7 min, to A:B 0:100 (step gradient), maintain at 0:100 for 5 min, return to initial conditions (step gradient), re-equilibrate for 5 min. (Prepare A by dissolving

2.28 g dipotassium hydrogen phosphate trihydrate and 5.44 g potassium dihydrogen phosphate in 880 mL water and then adding 120 mL MeCN. Prepare B by dissolving 2.28 g dipotassium hydrogen phosphate trihydrate and 5.44 g potassium dihydrogen phosphate in 550 mL water and then adding 450 mL MeCN.)
**Flow rate:** 1
**Injection volume:** 50
**Detector:** UV 208

## CHROMATOGRAM
**Internal standard:** ucb 17025
**Limit of quantitation:** $\leq$1.3 μg/mL

## KEY WORDS
serum

## REFERENCE
Grim, S.A.; Ryan, M.; Miles, M.V.; Tang, P.H.; Strawsburg, R.H.; deGrauw, T.J.; Fakhoury, T.A.; Baumann, R.J. Correlation of levetiracetam concentrations between serum and saliva, *Ther.Drug Monit.*, **2003**, *25*, 61–66.

## ANNOTATED BIBLIOGRAPHY
Ratnaraj, N.; Doheny, H.C.; Patsalos, P.N. A micromethod for the determination of the new antiepileptic drug levetiracetam (ucb L059) in serum or plasma by high performance liquid chromatography, *Ther.Drug Monit.*, **1996**, *18*, 154–157. [LOQ 5 μM]

# Levonordefrin

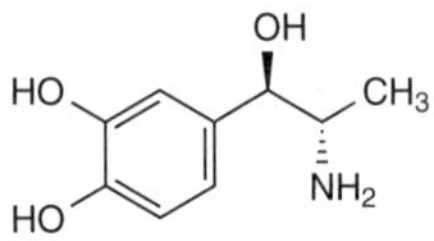


**Molecular formula:** $C_9H_{13}NO_3$
**Molecular weight:** 183.20
**CAS Registry No:** 829-74-3
**Merck Index:** 13, 6725

## SAMPLE

**Matrix:** blood
**Sample preparation:** Prepare an SPE column by packing 50–100 mesh Bio-Rex 70 ($Na^+$) cation exchange resin (Bio-Rad) into a 40 × 10 polypropylene column and washing three times with 1 M HCl, three times with 1 M NaOH, once with 1 M pH 6.5 sodium acetate, and once with 0.01% disodium EDTA. Mix 2 mL plasma with 5 mL 0.1% disodium EDTA, 500 µL 1 M pH 6.5 sodium acetate, and 4 ng IS, add to the SPE column, wash with 10 mL water, elute with 1 mL 700 mM sulfuric acid and 3.5 mL 2 M ammonium sulfate containing 0.1% disodium EDTA. Adjust the pH of the eluate to 8.6 with 3 mL 1 M pH 8.6 Tris-HCl buffer containing 2% disodium EDTA, add 100 mg activated alumina, shake for 3 min. Centrifuge, discard the liquid, wash the alumina three times with 10 mL distilled water. Centrifuge to remove excess water, elute with 200 µL 300 mM perchloric acid, inject a 50–100 µL aliquot of the eluate.

## HPLC VARIABLES

**Column:** 150 × 4.6 5 µm Spherisorb ODS2
**Mobile phase:** 100 mM pH 5.0 sodium dihydrogen phosphate containing 2 mM sodium heptanesulfonate and 0.001% disodium EDTA
**Flow rate:** 1.5
**Injection volume:** 50–100
**Detector:** E, Bioanalytical Systems LC-4A, glassy carbon electrode +0.50 V, Ag/AgCl reference electrode

## CHROMATOGRAM

**Retention time:** 12
**Internal standard:** 3,4-dihydroxybenzylamine (20)
**Limit of detection:** 50 pg/mL

## OTHER SUBSTANCES

**Extracted:** epinephrine (14), norepinephrine (6)

## KEY WORDS

human; plasma; rabbit; SPE

## REFERENCE

Jackman, G.P.; Oddie, C.J.; Skews, H.; Bobik, A. High-performance liquid chromatographic determination of plasma catecholamines during α-methyldopa therapy, *J.Chromatogr.*, **1984**, *308*, 301–305.

# Levosimendan

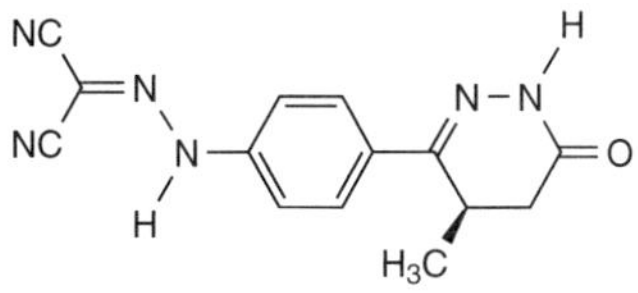


**Molecular formula:** $C_{14}H_{12}N_6O$
**Molecular weight:** 280.28
**CAS Registry No:** 141505-33-1
**Merck Index:** 13, 5491

## SAMPLE

**Matrix:** blood
**Sample preparation:** Mix 500 μL plasma with 25 μL 1.5 μg/mL IS in 50 mM pH 7.2 phosphate buffer. Using a Cuprophan cellulose membrane (cut-off 15 kDa) dialyze a 370 μL aliquot against 18 mL 40 mM pH 4.0 ammonium acetate buffer. Pump the buffer through a 650 μL recipient channel over 18 min and allow it to flow through column A. At the end of this time, backflush the contents of column A onto column B with the mobile phase, monitor the effluent from column B. (ASTED system. After use, flush the donor side with 10 mL 0.05% Triton X-100 in 0.86% NaCl solution. Flush the recipient side with 4 mL 40 mM pH 4.0 ammonium acetate buffer. Regenerate column A with 1 mL 40 mM pH 4.0 ammonium acetate buffer.)

## HPLC VARIABLES

**Column:** A 5.8 × 4.6 10 μm Hypersil ODS; B 250 × 4 10 μm LiChrosorb RP-18
**Mobile phase:** MeOH:THF:32 mM sodium dihydrogen phosphate 65:1:45, adjusted to apparent pH 3.5 with phosphoric acid
**Flow rate:** 1
**Detector:** UV 380

## CHROMATOGRAM

**Retention time:** 6.4
**Internal standard:** OR-1097 ([[4-(1,4,5,6-tetrahydro-6-oxo-3-pyridazinyl)phenyl]hydrazono]propanedinitrile) (5.4)
**Limit of detection:** 5 ng/mL (S/N = 3)

## OTHER SUBSTANCES

**Interfering:** carbamazepine

## KEY WORDS

ASTED; dialysate; plasma

## REFERENCE

Karlsson, M.; Korkolainen, T.; Wikberg, T. Automated analysis of levosimendan in human plasma by on-line dialysis and liquid chromatography, *Biomed.Chromatogr.*, **1997**, *11*, 54–58.

## SAMPLE

**Matrix:** blood
**Sample preparation:** Mix 200 (rat) or 500 (dog) μL plasma with 200 μL 50 mM pH 9 phosphate buffer, add 200 μL 1 M HCl, add 4 mL ethyl acetate:*n*-hexane 50:50, vortex for 2 min, centrifuge at 3500 g for 5 min. Evaporate 3 mL of the organic layer to dryness under a stream of nitrogen at 40°, reconstitute the residue with 200 μL 0.5% pH 9 triethylammonium acetate buffer, inject a 60 μL aliquot. Prepare human plasma samples in the same fashion, but use 1 mL plasma and twice the volumes of reagents. Alternatively filter (Ultrafree-MC 10000) 400 μL plasma while centrifuging at 1200 g for 30 min, inject a 40 μL aliquot of the ultrafiltrate.

## HPLC VARIABLES

**Guard column:** Guard-Pak CN
**Column:** 250 × 4.6 5 μm Cyclobond I
**Column temperature:** 30
**Mobile phase:** MeOH:buffer 30:70-33:67 (Prepare the buffer by dissolving 10 mL triethylamine in 2 L and adjusting pH to 6.0 with acetic acid.)
**Flow rate:** 1
**Injection volume:** 40–60
**Detector:** UV 380

## CHROMATOGRAM

**Retention time:** 19 (+), 21 (−)
**Limit of quantitation:** 10 ng/mL

## KEY WORDS

chiral; dog; human; plasma; rat; ultrafiltrate

## REFERENCE

Wikberg, T.; Korkolainen, T.; Karlsson, M. Enantiomeric bioanalysis of simendan and levosimendan by chiral high-performance liquid chromatography, *Chirality*, **1996**, *8*, 511–517.

# Lidamidine

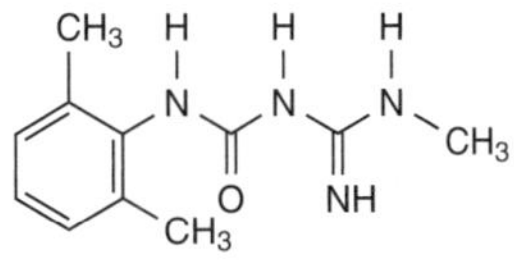


**Molecular formula:** $C_{11}H_{16}N_4O$
**Molecular weight:** 220.27
**CAS Registry No:** 66871-56-5, 65009-35-0 (HCl)
**Merck Index:** 13, 5502

## SAMPLE

**Matrix:** solutions

## HPLC VARIABLES

**Column:** 250 × 4 ODS
**Mobile phase:** MeOH:water:ammonium carbonate 60:40:0.02
**Detector:** UV 254

## REFERENCE

Zalipsky, J.J.; Won, C.M.; Patel, D.M. Analytical-physical profile of lidamidine hydrochloride (WHR-1142A), a novel antidiarrheal agent, *Arzneimittelforschung*, **1978**, *28*, 1441–1447.

## SAMPLE

**Matrix:** solutions

## HPLC VARIABLES

**Column:** 250 × 2.1 Zorbax Sil
**Mobile phase:** MeOH:dichloromethane:ammonium hydroxide 50:50:0.1
**Detector:** UV 254

## REFERENCE

Zalipsky, J.J.; Won, C.M.; Patel, D.M. Analytical-physical profile of lidamidine hydrochloride (WHR-1142A), a novel antidiarrheal agent, *Arzneimittelforschung*, **1978**, *28*, 1441–1447.

# Lincomycin

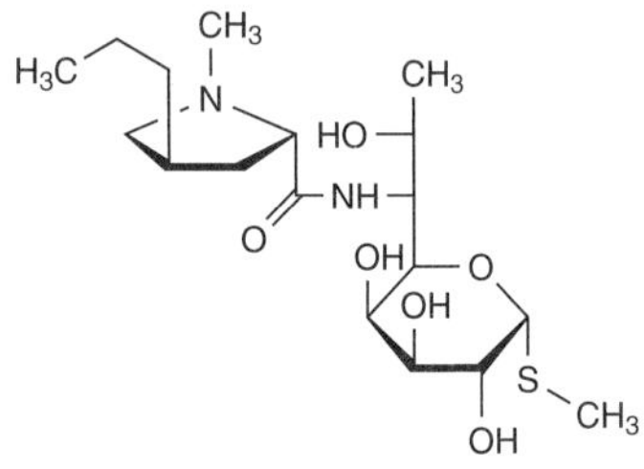


**Molecular formula:** $C_{18}H_{34}N_2O_6S$
**Molecular weight:** 406.54
**CAS Registry No:** 154-21-2, 7179-49-9 (HCl monohydrate)
**Merck Index:** 13, 5522

## SAMPLE
**Matrix:** blood
**Sample preparation:** Mix 500 μL serum with 500 μL MeCN, let stand for 20 min, centrifuge at 10 000 g for 5 min, inject a 10 μL aliquot of the supernatant.

## HPLC VARIABLES
**Guard column:** LichroCart 4-4 RP-18
**Column:** 125 × 4 5 μm Supersphere RP-18
**Column temperature:** 35
**Mobile phase:** MeCN:0.02% trifluoroacetic acid 40:60
**Flow rate:** 0.75
**Injection volume:** 10
**Detector:** MS, ThermoQuest Finnigan LCQ ion trap, APCI positive mode, vaporizer 450°, capillary 200°, capillary 9 V, discharge current 5 μA, sheath gas 0.6 L/min, auxiliary gas 3 L/min, m/z 405

## CHROMATOGRAM
**Retention time:** 2.2

## OTHER SUBSTANCES
**Extracted:** clindamycin (m/z 425) (3.7)

## KEY WORDS
lincomycin is IS in original paper; serum

## REFERENCE
Martens-Lobenhoffer, J.; Banditt, P. Sensitive and specific determination of clindamycin in human serum and bone tissue applying liquid chromatography-atmospheric pressure chemical ionization-mass spectrometry, *J.Chromatogr.B*, **2001**, *755*, 143–149.

## SAMPLE
**Matrix:** blood
**Sample preparation:** Vortex 250 μL plasma with 50 μL water for 15 s, add 50 μL 20% trichloroacetic acid in water, vortex for 15 s, centrifuge at 6800 g for 10 min, inject a 20 μL aliquot of the supernatant.

## HPLC VARIABLES
**Guard column:** 25 × 3 5 μm Hypersil RP-18
**Column:** 100 × 3 5 μm Hypersil RP-18
**Mobile phase:** Gradient. MeCN:10 mM ammonium acetate from 30:70 to 70:30 over 0.5 min, maintain at 70:30 for 7.5 min, return to initial conditions over 0.5 min, re-equilibrate for 9.5 min.
**Injection volume:** 20
**Detector:** MS, ThermoQuest Finnigan MAT, ESI, first 3 min and last 8 min of run diverted to waste, spray 3.5 kV, capillary 5 V, octapole 1 offset – 2 V, lens – 16 V,

octapole 2 offset – 5 V, RF amplitude 400 V, collision energy 1.2 V, m/z 407.3–389.1–359.2–126.3

---

## CHROMATOGRAM

**Retention time:** 4.0
**Limit of detection:** 1.3 ng/mL (for clindamycin)
**Limit of quantitation:** 50 ng/mL (for clindamycin)

---

## OTHER SUBSTANCES

**Extracted:** clindamycin (m/z 425.3–389.2–377.2–126.3) (7.4)

---

## KEY WORDS

dog; lincomycin is IS in original; plasma

---

## REFERENCE

Cherlet, M.; Croubels, S.; De Backer, P. Determination of clindamycin in animal plasma by high-performance liquid chromatography combined with electrospray ionization mass spectrometry, *J.Mass Spectrom.*, **2002**, *37*, 848–853.

# Lindane

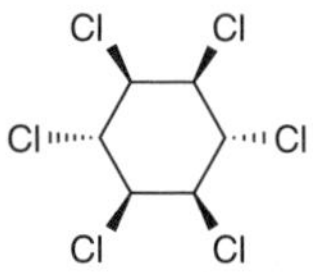


**Molecular formula:** $C_6H_6Cl_6$
**Molecular weight:** 290.83
**CAS Registry No:** 58-89-9
**Merck Index:** 13, 5523

## SAMPLE
**Matrix:** solutions

## HPLC VARIABLES
**Column:** 300 × 3.9 10 μm μBondapak C18
**Mobile phase:** MeCN:water 60:40
**Flow rate:** 2.5
**Detector:** UV 254

## OTHER SUBSTANCES
**Simultaneous:** hexachlorobenzene

## REFERENCE
Gopalaswamy, U.V.; Aiyar, A.S. Biotransformation of lindane in the rat, *Bull.Environ.Contam.Toxicol.*, **1984**, *32*, 148–156.

## SAMPLE
**Matrix:** solutions

## HPLC VARIABLES
**Column:** 600 × 0.2 5 μm ODS (Phenomenex)
**Mobile phase:** Carbon dioxide
**Detector:** UV 280

## CHROMATOGRAM
**Retention time:** 13

## OTHER SUBSTANCES
**Simultaneous:** carbofuran (8), chlorpyrifos (9.9), naphthol (10.1)

## KEY WORDS
SFC; UV restrictor at 100°; pressure from 100 to 300 atm at 4 atm/min

## REFERENCE
Pyo, D.; Kim, H.; Li, W.; Lee, M.L. Supercritical fluid chromatographic detection by use of a parallel flow restrictor, *J.Liq.Chromatogr.Rel.Technol.*, **1997**, *20*, 3389–3399.

# Linezolid

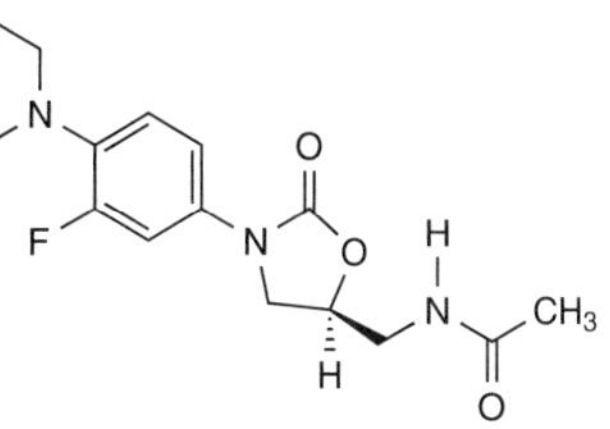

**Molecular formula:** $C_{16}H_{20}FN_3O_4$
**Molecular weight:** 337.35
**CAS Registry No:** 165800-03-3
**Merck Index:** 13, 5526

## SAMPLE

**Matrix:** blood
**Sample preparation:** Condition a 1 mL 100 mg C2 SPE cartridge (Varian) with two 1 mL portions of MeCN and two 1 mL portions of water. Mix 50 μL plasma with 25 μL water and 1 mL 100 ng/mL IS in water, add to the SPE cartridge, wash with 1 mL MeCN:water 5:95, elute with 500 μL MeOH. Evaporate the eluate to dryness under a stream of nitrogen at 40°, reconstitute the residue with 200 μL mobile phase, inject a 100 μL aliquot.

## HPLC VARIABLES

**Column:** 150 × 4.6 5 μm Zorbax RXC8
**Mobile phase:** MeCN:water 20:80
**Flow rate:** 1
**Injection volume:** 100
**Detector:** UV 251

## CHROMATOGRAM

**Retention time:** 11
**Internal standard:** PNU-101145 ((*S*)-*N*-[[3-[3-fluoro-4-[4-[(1-hydroxycyclopropyl)carbonyl]-1-piperazinyl]phenyl]- 2-oxo-5-oxazolidinyl]methyl]-acetamide) (9)
**Limit of detection:** 10 ng/mL

## KEY WORDS

dog; mouse; plasma; rabbit; rat; SPE

## REFERENCE

Peng, G.W.; Stryd, R.P.; Murata, S.; Igarashi, M.; Chiba, K.; Aoyama, H.; Aoyama, M.; Zenki, T.; Ozawa, N. Determination of linezolid in plasma by reversed-phase high-performance liquid chromatography, *J.Pharm.Biomed.Anal.*, **1999**, *20*, 65–73.

## SAMPLE

**Matrix:** blood
**Sample preparation:** Condition a Sep-Pak C18 with 1 mL MeCN and 1 mL water. Mix 500 μL plasma with 50 μL 10 μg/mL IS in MeCN, add to the SPE cartridge, wash with 1 mL water, elute with 300 μL MeCN. Evaporate the eluate to dryness under a stream of nitrogen at 50°, reconstitute the residue with 100 μL MeCN, inject a 10 μL aliquot.

## HPLC VARIABLES

**Column:** 150 × 4.6 5 μm Shim-Pack CLC-CN C18
**Mobile phase:** MeCN:20 mM ammonium acetate 80:20
**Flow rate:** 1
**Injection volume:** 10
**Detector:** MS, Finnigan LCQ, APCI, corona discharge 4.5 kV, vaporization 430°, capillary 150° 32 V, positive mode, m/z 296.2

## CHROMATOGRAM

**Retention time:** 3.44
**Internal standard:** *N*-carbobenzoxy-3-fluoro-4-morpholinylaniline (m/z 223.2) (2.91)
**Limit of detection:** 50 ng/mL
**Limit of quantitation:** 100 ng/mL

## KEY WORDS

plasma; SPE

## REFERENCE

Phillips, O.A.; Abdel-Hamid, M.E.; Al-Hassawi, N.A. Determination of linezolid in human plasma by LC-MS-MS, *Analyst*, **2001**, *126*, 609–614.

## ANNOTATED BIBLIOGRAPHY

Ehrlich, M.; Trittler, R.; Daschner, F.D.; Kümmerer, K. A new and rapid method for monitoring the new oxazolidinone antibiotic linezolid in serum and urine by high performance liquid chromatography-integrated sample preparation, *J.Chromatogr.B*, **2001**, *755*, 373–377. [column-switching]

Gentry-Nielsen, M.J.; Olsen, K.M.; Preheim, L.C. Pharmacodynamic activity and efficacy of linezolid in a rat model of pneumococcal pneumonia, *Antimicrob.Agents Chemother.*, **2002**, *46*, 1345–1351. [linezolid; rat; ceftriaxone; cephalexin is internal standard]

Gross, M.; Bürli, R.; Jones, P.; Garcia, M.; Batiste, B.; Kaizerman, J.; Moser, H.; Jiang, V.; Hoch, U.; Duan, J.-X.; Tanaka, R.; Johnson, K.W. Pharmacology of novel heteroaromatic polycycle antibacterials, *Antimicrob.Agents Chemother.*, **2003**, *47*, 3448–3457. [LC-MS; linezolid; vancomycin]

Johnson, R.A.; Haan, D.E.; James, C.A.; Hopkins, N.K. Determination of linezolid, PNU-100766, in human plasma and urine using high-performance liquid chromatography with ultraviolet detection (Abstract 2487), *Pharm.Res.*, **1997**, *14*, S374.

Sisson, T.L.; Jungbluth, G.L.; Hopkins, N.K. A pharmacokinetic evaluation of concomitant administration of linezolid and aztreonam, *J.Clin.Pharmacol.*, **1999**, *39*, 1277–1282. [SPE]

Welshman, I.R.; Sisson, T.A.; Jungbluth, G.L.; Stalker, D.J.; Hopkins, N.K. Linezolid absolute bioavailability and the effect of food on oral bioavailability, *Biopharm.Drug Dispos.*, **2001**, *22*, 91–97.

# Liothyronine

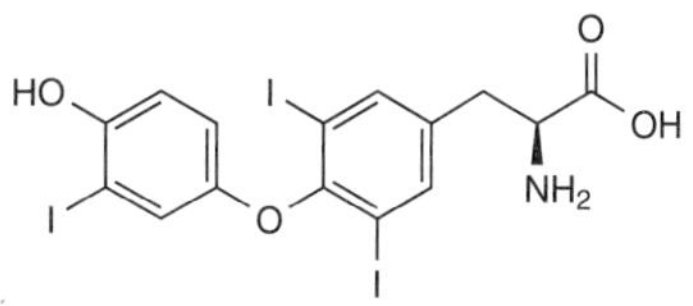


**Molecular formula:** $C_{15}H_{12}I_3NO_4$
**Molecular weight:** 650.97
**CAS Registry No:** 6893-02-3
**Merck Index:** 13, 5532

## SAMPLE

**Matrix:** blood, urine
**Sample preparation:** Condition a diol SPE cartridge with 3 mL MeOH and 3 mL water. Serum. Vortex 40 µL serum with 50 µL MeCN, dilute with 200 µL water, centrifuge at 3500 rpm for 15 min, remove organic solvents under a stream of nitrogen at 45°, add to the SPE cartridge, elute with 3 mL MeOH. Evaporate the eluate to dryness under a stream of nitrogen at 45°, reconstitute the residue with 200 µL 1 µg/mL IS in water, inject a 20 µL aliquot. Urine. Add 100 µL urine to the SPE cartridge, elute with 3 mL MeOH. Evaporate the eluate to dryness under a stream of nitrogen at 45°, reconstitute the residue with 200 µL 1 µg/mL IS in water, inject a 20 µL aliquot.

## HPLC VARIABLES

**Column:** 150 × 4.0 5 µm Inertsil ODS-3
**Mobile phase:** MeOH:2% acetic acid 65:35
**Flow rate:** 1
**Injection volume:** 20
**Detector:** UV 240

## CHROMATOGRAM

**Retention time:** 3
**Internal standard:** anthraquinone (10.4)
**Limit of detection:** 1 ng

## OTHER SUBSTANCES

**Extracted:** levothyroxine (LOD 2 ng) (4.7)

## KEY WORDS

serum; SPE

## REFERENCE

Samanidou, V.F.; Gika, H.G.; Papadoyannis, I.N. Rapid HPLC analysis of thyroid gland hormones tri-iodothyronine (T3) and thyroxine (T4) in human biological fluids after SPE, *J.Liq.Chromatogr.Rel.Technol.*, **2000**, *23*, 681–692.

# Lomerizine

**Molecular formula:** $C_{27}H_{30}F_2N_2O_3$
**Molecular weight:** 468.53
**CAS Registry No:** 101477-55-8, 101477-54-7 (di HCl)
**Merck Index:** 13, 5584

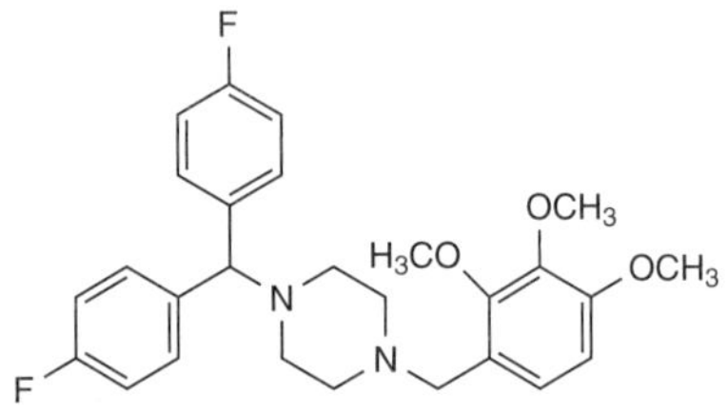

## SAMPLE

**Matrix:** blood, tissue

**Sample preparation:** Plasma. Mix 100 μL plasma with 10 μL 1 μg/mL IS in MeOH and 90 μL MeOH, add 200 μL 2 M NaOH, shake for 5 min, vortex with 750 μL *n*-hexane, centrifuge at 1500 g for 5 min, repeat the extraction of the aqueous layer. Evaporate the combined extracts to dryness under a stream of nitrogen, reconstitute the residue with 2 mL MeOH:1 M HCl 90:10, add to a 5 × 5 44–88 μm SP-Toyopearl SPE column, wash with 1 mL MeOH:water 90:10, elute with 1 mL MeOH:1 M ammonia solution 90:10. Evaporate the eluate to dryness under a stream of nitrogen, reconstitute the residue with 50 μL MeCN:20 mM pH 5.5 phosphate buffer 70:30, inject a 5–20 μL aliquot. Tissue. Homogenize brain tissue with 320 mM sucrose, centrifuge at 1300 g for 3 min. remove the supernatant and centrifuge it at 17 000 g for 10 min to give the P2 fraction. Mix 400 μL of the P2 fraction with 20 μL 1 μg/mL IS in MeOH, add 800 μL 2 M NaOH, shake for 5 min, vortex with 3 mL *n*-hexane, centrifuge at 1500 g for 5 min, repeat the extraction of the aqueous layer. Evaporate the combined extracts to dryness under a stream of nitrogen, reconstitute the residue with 2 mL MeOH:1 M HCl 90:10, add to a 5 × 5 44–88 μm SP-Toyopearl SPE column, wash with 1 mL MeOH:water 90:10, elute with 1 mL MeOH:1 M ammonia solution 90:10. Evaporate the eluate to dryness under a stream of nitrogen, reconstitute the residue with 50 μL MeCN:20 mM pH 5.5 phosphate buffer 70:30, inject a 5–20 μL aliquot.

## HPLC VARIABLES

**Column:** 200 × 2.1 5 μm Inertsil C8
**Mobile phase:** MeCN:20 mM pH 5.5 potassium phosphate buffer 69:31
**Flow rate:** 0.2
**Injection volume:** 5–20
**Detector:** UV 210

## CHROMATOGRAM

**Retention time:** 7.7
**Internal standard:** cinnarizine (8.6)
**Limit of detection:** 100 pg

## OTHER SUBSTANCES

**Extracted:** flunarizine (9.5)

## KEY WORDS

brain; plasma; rat; SPE

## REFERENCE

Waki, H.; Ando, S. Column liquid chromatography of calcium channel blockers, *J.Chromatogr.*, **1989**, *494*, 408–412.

# Lopinavir

**Molecular formula:** $C_{37}H_{48}N_4O_5$
**Molecular weight:** 628.80
**CAS Registry No:** 192725-17-0
**Merck Index:** 13, 5594

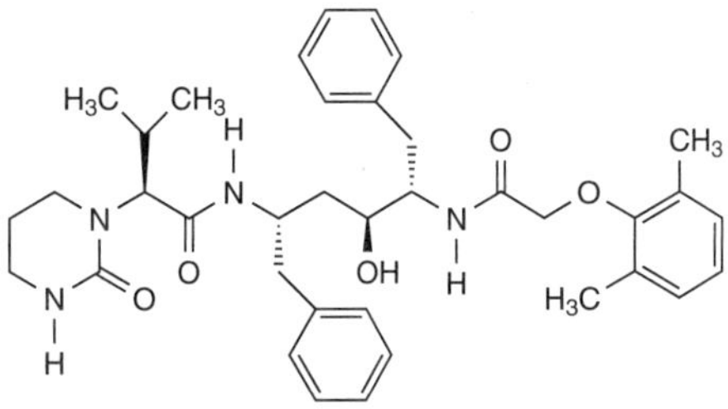

## SAMPLE

**Matrix:** blood
**Sample preparation:** Mix 250 μL plasma with 50 μL MeOH, add 100 μL 2 μg/mL IS in MeOH, add 250 μL 1 M NaOH, add 3 mL hexane:ethyl acetate 50:50, shake at high speed for 25 min, centrifuge at 3000 g for 15 min. Evaporate the organic layer to dryness under a stream of air, reconstitute the residue with 1 mL initial mobile phase, inject a 20 μL aliquot.

## HPLC VARIABLES

**Guard column:** 10 × 2.1 Symmetry Shield
**Column:** 30 × 2.1 3.5 μm Symmetry C18
**Mobile phase:** Gradient. MeCN:5 mM pH 3.25 acetate buffer from 25:75 to 80:20 over 4 min using a nonlinear gradient (not specified).
**Flow rate:** 0.35
**Injection volume:** 20
**Detector:** MS, PE Sciex API 3000, turbo ionspray source, column effluent split 1:1 before entering source

## CHROMATOGRAM

**Retention time:** 3.1
**Internal standard:** Abbott A-86093 (3.2)
**Limit of detection:** 750 pg/mL
**Limit of quantitation:** 16.3 ng/mL

## OTHER SUBSTANCES

**Extracted:** amprenavir (2.7, LOQ 16.3 ng/mL, LOD 380 pg/mL), indinavir (2.0, LOQ 16.3 ng/mL, LOD 1.5 ng/mL), nelfinavir (2.5, LOQ 16.3 ng/mL, LOD 330 pg/mL), ritonavir (2.9, LOQ 51.2 ng/mL, LOD 650 pg/mL), saquinavir (2.4, LOQ 16.3 ng/mL, LOD 780 pg/mL)

## KEY WORDS

plasma

## REFERENCE

Frerichs, V.A.; DiFrancesco, R.; Morse, G.D. Determination of protease inhibitors using liquid chromatography-tandem mass spectrometry, *J.Chromatogr.B*, **2003**, *787*, 393–403.

## SAMPLE

**Matrix:** blood
**Sample preparation:** Mix 1 mL plasma with 200 μL 10 μg/mL IS in water, add 200 μL 100 mM NaOH, mix, add 4 mL diethyl ether, shake for 5 min, centrifuge at 2500 rpm for 5 min. Evaporate the organic layer to dryness under a stream of nitrogen, reconstitute the residue with 200 μL initial mobile phase, inject a 100 μL aliquot.

## HPLC VARIABLES

**Column:** 250 × 4.6 Stability RP18 (CIL, France)

**Mobile phase:** Gradient. MeCN:50 mM pH 5.65 phosphate buffer from 36:64 to 64:36 over 25 min, to 80:20 (step gradient), maintain at 80:20 for 10 min, re-equilibrate at initial conditions for 5 min.

**Flow rate:** 1.5

**Injection volume:** 100

**Detector:** UV 240 for 5 min, UV 215 for 22 min, UV 260 for rest of run

## CHROMATOGRAM

**Retention time:** 18.9

**Internal standard:** JR051012 (Janssen Cilag) (28.2)

**Limit of quantitation:** 50 ng/mL

## OTHER SUBSTANCES

**Extracted:** amprenavir (11.2), efavirenz (19.9), indinavir (8.5), nelfinavir (24.1), nevirapine (3.3), ritonavir (17.6), saquinavir (16.7)

**Noninterfering:** acetaminophen, amineptine, amphotericin B, aspirin, bromazepam, buspirone, citalopram, clobazam, diazepam, didanosine, fluconazole, flunitrazepam, fluvoxamine, hydroxyitraconazole, isoniazid, itraconazole, lamivudine, loprazolam, lorazepam, metronidazole, minalcipram, nordiazepam, omeprazole, paroxetine, pyrimethamine, rifampin, sertraline, stavudine, sulfadiazine, trimethoprim, venlafaxine, zalcitabine, zidovudine, zolpidem, zopiclone

## KEY WORDS

plasma

## REFERENCE

Titier, K.; Lagrange, F.; Péhourcq, F.; Edno-Mcheik, L.; Moore, N.; Molimard, M. High-performance liquid chromatographic method for the simultaneous determination of the six HIV-protease inhibitors and two non-nucleoside reverse transcriptase inhibitors in human plasma, *Ther.Drug Monit.*, **2002**, *24*, 417–424.

## ANNOTATED BIBLIOGRAPHY

Droste, J.A.H.; Verweij-van Wissen, C.P.W.G.M.; Burger, D.M. Simultaneous determination of the HIV drugs indinavir, amprenavir, saquinavir, ritonavir, lopinavir, nelfinavir, the nelfinavir hydroxymetabolite M8, and nevirapine in human plasma by reversed-phase high-performance liquid chromatography, *Ther.Drug Monit.*, **2003**, *25*, 393–399. [indinavir; amprenavir; saquinavir; ritonavir; lopinavir; nelfinavir; nevirapine]

Faux, J.; Venisse, N.; Olivier, J.C.; Bouquet, S. Rapid high-performance liquid chromatography determination of lopinavir, a novel HIV-1 protease inhibitor, in human plasma, *Chromatographia*, **2001**, *54*, 469–473. [SPE; LOQ 187 ng/mL]

Faux, J.; Venisse, N.; Le Moal, G.; Dupuis, A.; Bouquet, S. Simultaneous determination of six HIV protease inhibitors, one metabolite, and two non-nucleoside reverse transcriptase inhibitors in human plasma by isocratic reversed-phase liquid chromatography after solid-phase extraction, *Chromatographia*, **2003**, *58*, 421–426. [SPE; amprenavir; indinavir; lopinavir; nelfinavir; ritonavir; saquinavir; efavirenz; nevirapine; prazepam is internal standard]

Justesen, U.S.; Pedersen, C.; Klitgaard, N.A. Simultaneous quantitative determination of the HIV protease inhibitors indinavir, amprenavir, ritonavir, lopinavir, saquinavir, nelfinavir and the nelfinavir active metabolite M8 in plasma by liquid chromatography, *J.Chromatogr.B*, **2003**, *783*, 491–500. [indinavir; amprenavir; ritonavir; lopinavir; saquinavir; nelfinavir; LOQ 25 ng/mL]

Keil, K.; Frerichs, V.A.; DiFrancesco, R.; Morse, G. Reverse phase high-performance liquid chromatography method for the analysis of amprenavir, efavirenz, indinavir, lopinavir, nelfinavir and its active metabolite (M8), ritonavir, and saquinavir in heparinized human plasma, *Ther.Drug Monit.*, **2003**, *25*, 340–346. [LOQ 100 ng/mL]

Leibenguth, P.; Le Guellec, C.; Besnier, J.-M.; Bastides, F.; Macé, M.; Gaudet, M.-L.; Autret-Leca, E.; Paintaud, G. Therapeutic drug monitoring of HIV protease inhibitors using high-performance liquid chromatography with ultraviolet or photodiode array detection, *Ther.Drug Monit.*, **2001**, *23*, 679–688. [indinavir; saquinavir; lopinavir; ritonavir; nelfinavir; amprenavir; carbamazepine is internal standard; LOQ 50-100 ng/mL]

Marzolini, C.; Béguin, A.; Telenti, A.; Schreyer, A.; Buclin, T.; Biollaz, J.; Decosterd, L.A. Determination of lopinavir and nevirapine by high-performance liquid chromatography after solid-phase extraction: application for the assessment of their transplacental passage at delivery, *J.Chromatogr.B*, **2002**, *774*, 127–140. [LOQ 100 ng/mL; clozapine is internal standard]

Poirier, J.-M.; Robidou, P.; Jaillon, P. Simultaneous determination of the six HIV protease inhibitors (amprenavir, indinavir, lopinavir, nelfinavir, ritonavir, and saquinavir) plus M8 nelfinavir metabolite and the nonnucleoside reverse transcription inhibitor efavirenz in human plasma by solid-phase extraction and column liquid chromatography, *Ther.Drug Monit.*, **2002**, *24*, 302–309. [LOQ 25 ng/mL]

Ray, J.; Pang, E.; Carey, D. Simultaneous determination of indinavir, ritonavir and lopinavir (ABT 378) in human plasma by high-performance liquid chromatography, *J.Chromatogr.B*, **2002**, *775*, 225–230. [100 ng/mL]

Rentsch, K.M. Sensitive and specific determination of eight antiretroviral agents in plasma by high-performance liquid chromatography-mass spectrometry, *J.Chromatogr.B*, **2003**, *788*, 339–350. [SPE; amprenavir; efavirenz; indinavir; lopinavir; nelfinavir; nevirapine; ritonavir; saquinavir; LOQ 1–250 ng/mL]

Tribut, O.; Arvieux, C.; Michelet, C.; Chapplain, J.-M.; Allain, H.; Bentué-Ferrer, D. Simultaneous quantitative assay of six HIV protease inhibitors, one metabolite, and two non-nucleoside reverse transcriptase inhibitors in human plasma by isocratic reversed-phase liquid chromatography, *Ther.Drug Monit.*, **2002**, *24*, 554–562. [nevirapine; efavirenz; indinavir; amprenavir; nelfinavir; ritonavir; lopinavir; saquinavir; LOQ 25 ng/mL]

Turner, M.L.; Reed-Walker, K.; King, J.R.; Acosta, E.P. Simultaneous determination of nine antiretroviral compounds in human plasma using liquid chromatography, *J.Chromatogr.B*, **2003**, *784*, 331–341. [indinavir; nelfinavir; saquinavir; ritonavir; amprenavir; delavirdine; efavirenz; lopinavir; metabolites; LOQ 50 ng/mL]

Usami, Y.; Oki, T.; Nakai, M.; Sagisaka, M.; Kaneda, T. A simple HPLC method for simultaneous determination of lopinavir, ritonavir and efavirenz, *Chem.Pharm.Bull.*, **2003**, *26*, 715–718. [LOQ 21–60 ng/mL]

# Loteprednol etabonate

**Molecular formula:** $C_{24}H_{31}ClO_7$
**Molecular weight:** 466.96
**CAS Registry No:** 82034-46-6
**Merck Index:** 13, 5605

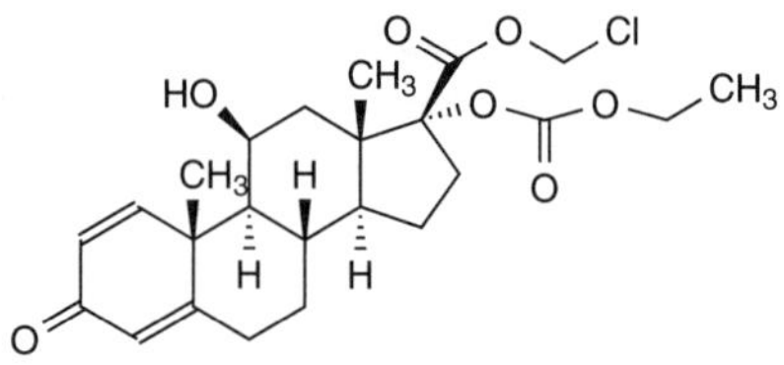

## SAMPLE
**Matrix:** bile, blood, urine
**Sample preparation:** Vortex 100 μL blood with 200 μL MeCN:DMSO 95:5, cool at 0° for several min, centrifuge at 3000 rpm for 10 min, inject an aliquot of the supernatant. (A similar procedure can also be used for bile and urine, see Bodor,N.; Wu,W.-M.; Murakami,T.; Engel,S. Soft drugs 19. Pharmacokinetics, metabolism and excretion of a novel soft corticosteroid, loteprednol etabonate, in rats. *Pharm.Res.* **1995**, 12, 875–879.)

## HPLC VARIABLES
**Column:** 75 × 3.9 4 μm Nova-Pak phenyl
**Mobile phase:** MeCN:water:acetic acid 45:54:1
**Flow rate:** 1
**Injection volume:** 20
**Detector:** UV 254

## CHROMATOGRAM
**Retention time:** 5.12
**Limit of detection:** <100 ng/mL

## KEY WORDS
rat; whole blood

## REFERENCE
Bodor, N.; Murakami, T.; Wu, W.-M. Soft drugs 18. Oral and rectal delivery of loteprednol etabonate, a novel soft corticosteroid, in rats – for safer treatment of gastrointestinal inflammation, *Pharm.Res.*, **1995**, *12*, 869–874.

## SAMPLE
**Matrix:** solutions

## HPLC VARIABLES
**Column:** 75 × 3.9 4 μm Nova-Pak phenyl
**Mobile phase:** MeCN:water:acetic acid 50:50:1
**Flow rate:** 1.5
**Detector:** UV 254

## CHROMATOGRAM
**Retention time:** 2.2

## OTHER SUBSTANCES

**Simultaneous:** metabolites

## REFERENCE

Bodor, N.; Drustrup, J.; Wu, W. Effect of cyclodextrins on the solubility and stability of a novel soft corticosteroid, loteprednol etabonate, *Pharmazie*, **2000**, *55*, 206–209.

# Marbofloxacin

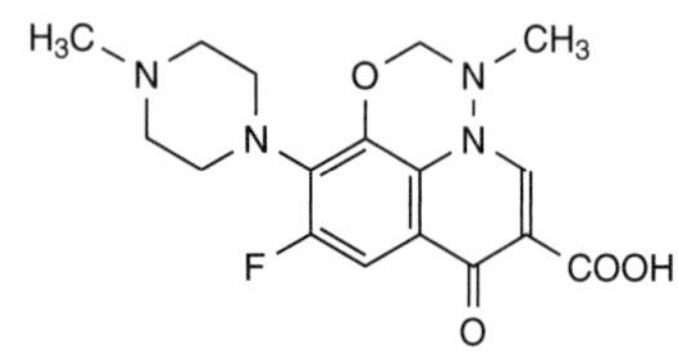

**Molecular formula:** $C_{17}H_{19}FN_4O_4$
**Molecular weight:** 362.35
**CAS Registry No:** 115550-35-1
**Merck Index:** 13, 5774

## SAMPLE

**Matrix:** blood
**Sample preparation:** Mix 200 μL plasma with 800 μL 1.5 μg/mL IS in 100 mM pH 7.4 phosphate buffer, add 6 mL chloroform (Caution! Chloroform is a carcinogen!), shake at 200 oscillations/min for 30 min, centrifuge at 13 000 g for 6 min. Evaporate the organic layer to dryness under a stream of nitrogen at 40°, reconstitute the residue with 200 μL phosphate-buffered saline, inject a 10–80 μL aliquot.

## HPLC VARIABLES

**Guard column:** Novapak C18 Guard-Pak
**Column:** 150 × 3.9 5 μm Novapak C18
**Mobile phase:** MeCN:buffer 20:80 (The buffer was 20 mM potassium dihydrogen phosphate, 6 mM phosphoric acid, and 12 mM tetraethylammonium bromide, adjusted to pH 3.0 with 2 M NaOH.)
**Flow rate:** 1
**Injection volume:** 10–80
**Detector:** F ex 338 em 425

## CHROMATOGRAM

**Retention time:** 2.20
**Internal standard:** enrofloxacin (3.30)
**Limit of detection:** 15 ng/mL

## OTHER SUBSTANCES

**Simultaneous:** danofloxacin (2.80), difloxacin (4.52), orbifloxacin (3.09), sarafloxacin (4.40)
**Interfering:** ciprofloxacin (2.28), norfloxacin (2.16)

## KEY WORDS

plasma

## REFERENCE

Garcia, M.A.; Solans, C.; Aramayona, J.J.; Rueda, S.; Bregante, M.A. Determination of marbofloxacin in plasma samples by high-performance liquid chromatography using fluorescence detection, *J.Chromatogr. B*, **1999**, *729*, 157–161.

## SAMPLE

**Matrix:** tissue
**Sample preparation:** Condition a 6 mL 10 mm SDB-PRS or MPC-SD (mixed C8 and cation-exchange) SPE disc (3M) with two 1 mL portions of MeCN:96% acetic acid 95:5. Mix 1 g minced kidney with 150 ng IS, 10 mL MeCN, and 2.5 g sodium sulfate, homogenize (turrax) for 2 min, centrifuge at 3000 g for 5 min, filter the supernatant through 2.5 g sodium sulfate and Whatman paper. Acidify the filtrate with 2.5 mL 96% acetic acid and add to the SPE disc, dry under vacuum for 5 min, elute with four 1 mL portions of MeOH:1 M ammonia 75:25. Evaporate the eluate to dryness under a stream

of nitrogen at 35°, reconstitute the residue with 300 μL dilute formic acid (pH 2.5), filter (0.45 μm), inject a 50 μL aliquot.

## HPLC VARIABLES

**Column:** 70 × 4 5 μm Nucleosil 100-5 C18
**Column temperature:** 25
**Mobile phase:** Gradient. MeCN:dilute formic acid (pH 2.5) from 2:98 to 70:30 over 5 min, maintain at 70:30 for 1 min.
**Flow rate:** 1
**Injection volume:** 50
**Detector:** MS, Micromass Quattro triple-stage quadrupole, ESI, 100 μL/min of column effluent entered the detector, capillary 3.2 kV, source 130°, desolvation 400°, desolvation gas nitrogen 650 L/h, nebulizer gas nitrogen 75 L/h, collision gas argon, cone 30 V, collision energy 15 eV m/z 363–320

## CHROMATOGRAM

**Retention time:** 2.97
**Internal standard:** quinine (m/z 325–160) (2.74)
**Limit of detection:** <20 ng/g

## OTHER SUBSTANCES

**Extracted:** cinoxacin (m/z 263–217) (3.91), ciprofloxacin (m/z 332–245) (3.06), danofloxacin (m/z 358–96) (3.10), enoxacin (m/z 321–206) (2.97), enrofloxacin (m/z 360–245) (3.19), flumequine (m/z 262–202) (4.88), nalidixic acid (m/z 233–215) (4.78), norfloxacin (m/z 320–233) (3.01), ofloxacin (m/z 362–261) (3.01), oxolinic acid (m/z 262–216) (4.18)

## KEY WORDS

kidney; pig; SPE

## REFERENCE

Van Vyncht, G.; Jànosi, A.; Bordin, G.; Toussaint, B.; Maghuin-Rogister, G.; De Pauw, E.; Rodriguez, A.R. Multiresidue determination of (fluoro)quinolone antibiotics in swine kidney using liquid chromatography-tandem mass spectrometry, *J.Chromatogr.A*, **2002**, *952*, 121–129.

## ANNOTATED BIBLIOGRAPHY

Ballesteros, O.; Toro, I.; Sanz-Nebot, V.; Navalóan, A.; Vílchez, J.L.; Barbosa, J. Optimization of the composition and pH of the mobile phase used for separation and determination of a series of quinolone antibacterials regulated by the European Union, *Chromatographia*, **2002**, *56*, 413–421. [ciprofloxacin; enrofloxacin; norfloxacin; marbofloxacin; danofloxacin; pipemidic acid; difloxacin; sarafloxacin; flumequine; oxolinic acid; LOD 10 ng/mL; fluorescence detection]

Garcia, M.A.; Solans, C.; Calvo, A.; Royo, M.; Hernandez, E.; Rey, R.; Bregante, M.A. HPLC separation and quantification of ofloxacin enantiomers in rabbit plasma. Application to pharmacokinetic studies, *Chromatographia*, **2002**, *56*, 39–42. [marbofloxacin is internal standard; fluorescence detection]

Gigosos, P.G.; Revesado, P.R.; Cadahía, O.; Fente, C.A.; Vazquez, B.I.; Franco, C.M.; Cepeda, A. Determination of quinolones in animal tissues and eggs by high-performance liquid chromatography with photodiode-array detection, *J.Chromatogr.A*, **2000**, *871*, 31–36. [enrofloxacin; ciprofloxacin; difloxacin; norfloxacin; marbofloxacin; SPE; LOD 4 ng; UV 280]

Hernández-Arteseros, J.A.; Boronat, I.; Compaño, R.; Prat, M.D. Liquid chromatographic separation of fluoroquinolone antibacterials used as veterinary drugs, *Chromatographia*, **2000**, *52*, 295–300. [danofloxacin; sarafloxacin; difloxacin; marbofloxacin; enrofloxacin; ciprofloxacin; fluorescence detection]

Johnson, G.; Westwood, S.; Houghton, R. A generic approach for the HPLC separation of quinolone antibiotics, *Chromatographia*, **2002**, *55*, S127–S131. [nalidixic acid; oxolinic acid; sarafloxacin; difloxacin; pipemidic acid; flumequine; ofloxacin; marbofloxacin; enrofloxacin; ciprofloxacin; danofloxacin]

Schneider, M.; Thomas, V.; Boisrame, B.; Deleforge, J. Pharmacokinetics of marbofloxacin in dogs after oral and parenteral administration, *J.Vet.Pharmacol.Ther.*, **1996**, *19*, 56–61. [ofloxacin is internal standard]

Toussaint, B.; Bordin, G.; Janosi, A.; Rodriguez, A.R. Validation of a liquid chromatography-tandem mass spectrometry method for the simultaneous quantification of 11 (fluoro)quinolone antibiotics in swine kidney, *J.Chromatogr.A*, **2002**, *976*, 195–206. [danofloxacin; cinoxacin; ciprofloxacin; enoxacin; enrofloxacin; flumequine; marbofloxacin; nalidixic acid; norfloxacin; ofloxacin; oxolinic acid; SPE]

Yorke, J.C.; Froc, P. Quantitation of nine quinolones in chicken tissues by high-performance liquid chromatography with fluorescence detection, *J.Chromatogr.A*, **2000**, *882*, 63–77. [ciprofloxacin; danofloxacin; enrofloxacin; nalidixic acid; oxolinic acid; sarafloxacin; difloxacin; flumequine; marbofloxacin; LOQ 7.5–150 ng/g]

# Masoprocol

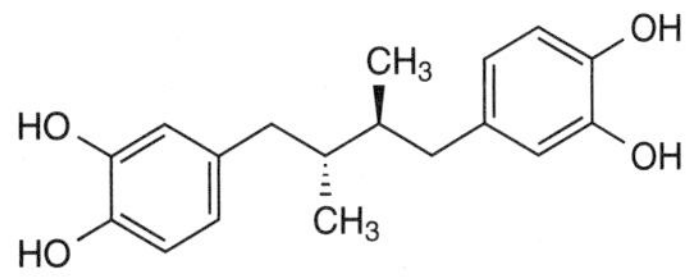

**Molecular formula:** $C_{18}H_{22}O_4$
**Molecular weight:** 302.36
**CAS Registry No:** 500-38-9, 27686-84-6 (meso)
**Merck Index:** 13, 6726

## SAMPLE

**Matrix:** blood
**Sample preparation:** Condition a Bond Elut C18 SPE cartridge with three 3 mL portions of MeOH and three 3 mL portions of water. Mix 500 μL serum with 2.5 μg IS, add to the SPE cartridge, wash with 3 mL water, elute with 300 μL 100 mM HCl in MeOH, inject a 100 μL aliquot.

## HPLC VARIABLES

**Column:** 250 × 4.6 5 μm Adsorbosphere C18
**Mobile phase:** Gradient MeCN:water:trifluoroacetic acid from 30:70:0.1 to 70:30:0.1 over 15 min, maintain at 70:30:0.1 for 15 min.
**Flow rate:** 1.5
**Injection volume:** 100
**Detector:** UV (wavelength not specified)

## CHROMATOGRAM

**Retention time:** 8
**Internal standard:** tetra-*O*-methylnordihydroguaiaretic acid (Prepare IS by dissolving 1 equiv masoprocol and 6 equiv potassium carbonate in acetone, add 4 equiv dimethyl sulfate (Caution! Dimethyl sulfate is highly toxic, particularly upon inhalation, and a carcinogen! Perform the reaction in a chemical fume hood!), reflux for 8 h, add 6 equiv 1 M HCl, evaporate the acetone under reduced pressure, extract three times with 100 mL portions of ethyl acetate. Combine the organic layers, wash three times with 100 mL portions of 100 mM HCl, wash three times with 100 mL portions of 1 M NaCl, evaporate the organic layer to dryness, recrystallize the product from dichloromethane.) (20)
**Limit of detection:** 500 ng/mL

## KEY WORDS

mouse; pharmacokinetics; plasma; SPE

## REFERENCE

Lambert, J.D.; Meyers, R.O.; Timmermann, B.N.; Dorr, R.T. Pharmacokinetic analysis by high-performance liquid chromatography of intravenous nordihydroguaiaretic acid in the mouse, *J.Chromatogr.B*, **2001**, *754*, 85–90.

# Maxacalcitol

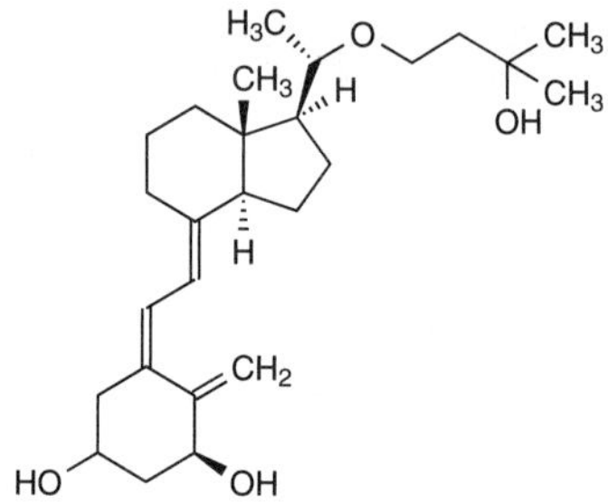


**Molecular formula:** $C_{29}H_{42}O_4$
**Molecular weight:** 418.61
**CAS Registry No:** 103909-75-7
**Merck Index:** 13, 5783

## SAMPLE

**Matrix:** blood
**Sample preparation:** Condition a 3 mL 200 mg Bond Elut C18 SPE cartridge with 3 mL MeOH:water 80:20, 2 mL water, and 2 mL 1 M HCl. Condition a 1 mL 100 mg Bond Elut NH2 SPE cartridge with 1 mL hexane:isopropanol 70:30 and 1 mL hexane:isopropanol 90:10. Mix 20 μL 16 ng/mL IS in EtOH with 1 mL serum, add 200 μL 1 M HCl, mix, add to the C18 SPE cartridge, wash with 2 mL water, wash with 3 mL MeOH:water 50:50, elute with 4 mL MeOH:water 80:20. Evaporate the eluate to dryness under a stream of nitrogen at 60°, reconstitute the residue with 100 μL hexane:isopropanol 90:10, add to the NH2 SPE cartridge, wash with 1 mL hexane:isopropanol 90:10, elute with 2 mL hexane:isopropanol 70:30. Evaporate the eluate to dryness, reconstitute the residue with 50 μL mobile phase, inject a 20 μL aliquot.

## HPLC VARIABLES

**Column:** 250 × 2 5 μm Capcell Pak C18 UG120
**Mobile phase:** MeOH:10 mM ammonium acetate 90:10
**Flow rate:** 0.2
**Injection volume:** 20
**Detector:** MS, Finnigan TSQ-700 triple quadrupole, electrospray 4.5 kV, positive ion, tube lens 120 V, collision gas argon 2.0 mtorr, collision energy -20 eV, (m/z 436–297)

## CHROMATOGRAM

**Retention time:** 4.5
**Internal standard:** ED-94 ((5Z,7E,20S)-20α-(2-hydroxy-2-methylpropoxy)-9,10-seco-pregna-5,7,10(19)-trien-1α,3β-diol) (m/z 422–297) (4.5)
**Limit of quantitation:** 20 pg/mL

## KEY WORDS1

serum; SPE

## REFERENCE

Ishigai, M.; Asoh, Y.; Kumaki, K. Determination of 22-oxacalcitriol, a new analog of 1α,25-dihydroxyvitamin $D_3$, in human serum by liquid chromatography-mass spectrometry, *J.Chromatogr.B*, **1998**, *706*, 261–267.

# Medetomidine

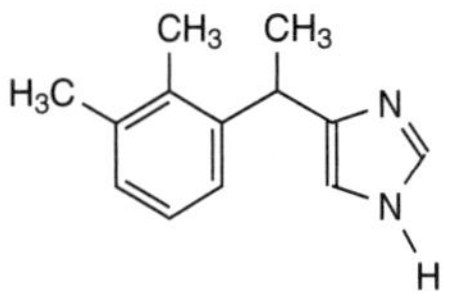


**Molecular formula:** $C_{13}H_{16}N_2$
**Molecular weight:** 200.28
**CAS Registry No:** 86347-14-0
**Merck Index:** 13, 5811

## SAMPLE
**Matrix:** blood
**Sample preparation:** Condition a Sep-Pak C18 SPE cartridge with water, MeOH, and 100 mM ammonium acetate. Add 5 mL plasma to the SPE cartridge, wash with 100 mM ammonium acetate, elute with MeOH:100 mM ammonium acetate 75:25. Evaporate the eluate to dryness under reduced pressure, reconstitute the residue in 200 μL mobile phase, inject a 50 μL aliquot.

## HPLC VARIABLES
**Column:** 150 × 4.6 5 μm Hitachi gel #3056
**Mobile phase:** MeOH:100 mM ammonium acetate 65:35
**Flow rate:** 1
**Injection volume:** 50
**Detector:** MS, Hitachi M-1000, APCI interface, drift voltage 21 V, nebulizer 260°, vaporizer 399°, multiplier voltage 1500 VF, m/z 201

## CHROMATOGRAM
**Retention time:** 7.5
**Internal standard:** detomidine (m/z 187) (6.5)
**Limit of quantitation:** 1–2 ng/mL

## OTHER SUBSTANCES
**Extracted:** atipamazole (m/z 213) (8.5), midazolam (m/z 326) (10.5)

## KEY WORDS
pharmacokinetics; pig; plasma; SPE

## REFERENCE
Kanazawa, H.; Nishimura, R.; Sasaki, N.; Takeuchi, A.; Takai, N.; Nagata, Y.; Matsushima, Y. Determination of medetomidine, atipamazole and midazolam by liquid chromatography-mass spectrometry, *Biomed.Chromatogr.*, **1995**, *9*, 188–191.

## SAMPLE
**Matrix:** blood
**Sample preparation:** Mix 200 μL 250 mM NaOH with 500 μL plasma, add 6 mL dichloromethane, mix gently for 10 min, centrifuge at 1700 g for 10 min. Evaporate 4 mL of the lower organic layer to dryness under a stream of nitrogen, reconstitute the residue with 100 μL 50 mM pH 3.2 phosphate buffer, vortex for 1.5 min, centrifuge at 1700 g for 10 min, inject a 50 μL aliquot.

## HPLC VARIABLES
**Guard column:** 20 × 4.6 5 μm Supelguard LC-DP
**Column:** 250 × 4.6 5 μm Supelcosil LC-DP
**Mobile phase:** MeCN:50 mM phosphate buffer:triethylamine 27:73:0.05, adjusted to pH 3.2
**Flow rate:** 1
**Injection volume:** 50

**Detector:** UV 215

---

## CHROMATOGRAM

**Retention time:** 14.6
**Limit of detection:** 5 ng/mL

---

## OTHER SUBSTANCES

**Extracted:** atipamezole (16.2)

---

## KEY WORDS

pharmacokinetics; plasma; reindeer

---

## REFERENCE

Ranheim, B.; Horsberg, T.E.; Nymoen, U.; Soli, N.E.; Tyler, N.J.; Arnemo, J.M. Reversal of medetomidine-induced sedation in reindeer (*Rangifer tarandus tarandus*) with atipamezole increases the medetomidine concentration in plasma, *J.Vet.Pharmacol.Ther.*, **1997**, *20*, 350–354.

---

## ANNOTATED BIBLIOGRAPHY

Kanazawa, H.; Nagata, Y.; Matsushima, Y.; Takai, N.; Uchiyama, H.; Nishimura, R.; Takeuchi, A. Liquid chromatography-mass spectrometry for the determination of medetomidine and other anaesthetics in plasma, *J.Chromatogr.*, **1993**, *631*, 215–220. [SPE; flumazenil; butorphanol; atropine; ketamine; xylazine; medetomidine; atipamezole; midazolam; dog; LOD 0.5–2.5 ng/mL]

# Meglutol

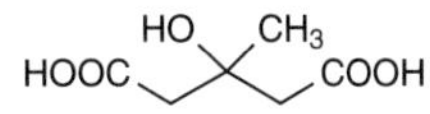


**Molecular formula:** $C_6H_{10}O_5$
**Molecular weight:** 162.14
**CAS Registry No:** 503-49-1
**Merck Index:** 13, 5831

## SAMPLE

**Matrix:** blood
**Sample preparation:** Condition a 30 × 10 column of DEAE-cellulose (Serva) with 10 vol of 100 mM pH 7.0 sodium perchlorate and 15 vol of water. Mix 6 mL plasma with 6 mL MeCN, centrifuge at 4000 g for 10 min, suspend the pellet in 6 mL MeCN:water 50:50, centrifuge at 4000 g for 10 min. Combine the supernatants, adjust the pH of a 10 mL aliquot to 7.0 with 100 mM NaOH, add to the DEAE-cellulose column, wash with 10 mL water, elute with 10 mL 100 mM perchloric acid. Freeze-dry the eluate overnight at 0°, reconstitute the residue with 500 μL mobile phase, centrifuge at 12 000 g for 2 min, inject a 100 μL aliquot.

## HPLC VARIABLES

**Guard column:** 50 × 4 micro-Guard-NH2
**Column:** 300 × 7.8 Aminex HPX-87 cation-exchange (Bio-Rad)
**Mobile phase:** 6.5 mM sulfuric acid
**Flow rate:** 0.5
**Injection volume:** 100
**Detector:** UV 210

## CHROMATOGRAM

**Retention time:** 8

## OTHER SUBSTANCES

**Extracted:** adipic acid (12), citric acid (5), 2-oxoglutaric acid (6.4)

## KEY WORDS

plasma; SPE

## REFERENCE

Lippe, G.; Trevisan, R.; Nosadini, R.; Fabris, R.; Deana, R. 3-Hydroxy-3-methylglutaric, adipic, and 2-oxoglutaric acids measured by HPLC in the plasma from diabetic patients, *Clin.Biochem.*, **1987**, *20*, 275–279.

# Melatonin

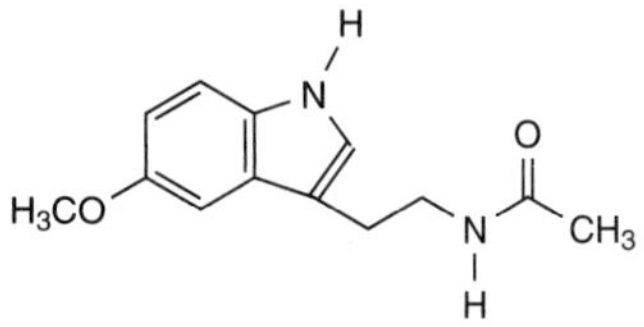


**Molecular formula:** $C_{13}H_{16}N_2O_2$
**Molecular weight:** 232.28
**CAS Registry No:** 73-31-4
**Merck Index:** 13, 5838

## SAMPLE

**Matrix:** blood, CSF
**Sample preparation:** For total melatonin, mix 300 μL 60 mM trichloroacetic acid with 1 mL plasma or CSF, let stand at 0° for 10 min, centrifuge at 5000 g for 10 min, adjust the pH of the supernatant to 7.4 with 20 μL 1 M NaOH, add to a 1 mL Chem-Elut 1001 SPE cartridge (without preconditioning), let stand for 3–5 min, elute with two 4 mL portions of dichloromethane. Evaporate the eluate to dryness at 37°, reconstitute the residue with 60 μL mobile phase, inject a 40 μL aliquot. For free melatonin add 1 mL plasma or CSF directly to the SPE cartridge.

## HPLC VARIABLES

**Column:** 150 × 4.6 3 μm Extrasil ODS-2 (EsseCi, Italy)
**Mobile phase:** MeCN:75 mM sodium acetate 28:72 (pH 5.0)
**Flow rate:** 1
**Injection volume:** 40
**Detector:** F ex 275 em 345; E, Chromsystem EC 41000, 0.9 V

## CHROMATOGRAM

**Retention time:** 5
**Limit of quantitation:** 0.5 pg/mL (F)

## KEY WORDS

plasma; SPE

## REFERENCE

Rizzo, V.; Porta, C.; Moroni, M.; Scoglio, E.; Moratti, R. Determination of free and total (free plus protein-bound) melatonin in plasma and cerebrospinal fluid by high-performance liquid chromatography with fluorescence detection, *J.Chromatogr.B*, **2002**, *774*, 17–24.

## SAMPLE

**Matrix:** formulations
**Sample preparation:** Powder tablets and weigh out an amount equivalent to 100 mg melatonin, add 100 mL 50 mM pH 3.0 phosphate buffer, sonicate for 10 min, centrifuge at 3000 rpm for 10 min, filter (0.20 μm), inject an aliquot.

## HPLC VARIABLES

**Guard column:** 50 × 4.6 5 μm Hypersil ODS
**Column:** 250 × 4.6 5 μm Hypersil ODS
**Mobile phase:** MeCN:MeOH:15 mM pH 2.43 phosphate buffer 14:19:67 containing 0.25% triethylamine (pH 3.0)
**Flow rate:** 1
**Injection volume:** 20
**Detector:** UV 223

## CHROMATOGRAM

**Retention time:** 6.0
**Internal standard:** adenosine (2.2)

**Limit of detection:** 65 ng/mL
**Limit of quantitation:** 125 ng/mL

## KEY WORDS

tablets

## REFERENCE

Raggi, M.A.; Bugamelli, F.; Pucci, V. Determination of melatonin in galenic preparations by LC and voltammetry, *J.Pharm.Biomed.Anal.*, **2002**, *29*, 283–289.

## ANNOTATED BIBLIOGRAPHY

Andrisano, V.; Bertucci, C.; Battaglia, A.; Cavrini, V. Photostability of drugs: photodegradation of melatonin and its determination in commercial formulations, *J.Pharm.Biomed.Anal.*, **2000**, *23*, 15–23. [post-column photochemical derivatization]

Azekawa, T.; Sano, A.; Aoi, K.; Sei, H.; Morita, Y. Concurrent on-line sampling of melatonin in pineal microdialysates from conscious rat and its analysis by high-performance liquid chromatography with electrochemical detection, *J.Chromatogr.*, **1990**, *530*, 47–55.

Bechgaard, E.; Lindhardt, K.; Martinsen, L. High-performance liquid chromatographic analysis of melatonin in human plasma and rabbit serum with on-line column enrichment, *J.Chromatogr.B*, **1998**, *712*, 177–181. [column-switching; LOD <1 ng/mL]

Chanut, E.; Nguyen-Legros, J.; Versaux-Botteri, C.; Trouvin, J.H.; Launay, J.M. Determination of melatonin in rat pineal, plasma and retina by high- performance liquid chromatography with electrochemical detection, *J.Chromatogr.B*, **1998**, *709*, 11–18. [LOD 15 pg/mL]

Chin, J.R. Determination of six indolic compounds, including melatonin, in rat pineal using high-performance liquid chromatography with serial fluorimetric-electrochemical detection, *J.Chromatogr.*, **1990**, *528*, 111–121.

Dubbels, R.; Reiter, R.J.; Klenke, E.; Goebel, A.; Schnakenberg, E.; Ehlers, C.; Schiwara, H.W.; Schloot, W. Melatonin in edible plants identified by radioimmunoassay and by high performance liquid chromatography-mass spectrometry, *J.Pineal Res.*, **1995**, *18*, 28–31.

Eriksson, K.; Ostin, A.; Levin, J.-O. Quantification of melatonin in human saliva by liquid chromatography-tandem mass spectrometry using stable isotope dilution, *J.Chromatogr.B*, **2003**, *794*, 115–123. [SPE; LOD 1 pg/mL]

Goldman, M.E.; Hamm, H.; Erickson, C.K. Determination of melatonin by high-performance liquid chromatography with electrochemical detection, *J.Chromatogr.*, **1980**, *190*, 217–220.

Hall, F.; Tengerdy, C.; Morita, M.; Pautler, E. Determination of bovine retinal melatonin with HPLC-EC, *Curr.Eye Res.*, **1985**, *4*, 847–850.

Hamase, K.; Tomita, T.; Kiyomizu, A.; Zaitsu, K. Determination of pineal melatonin by precolumn derivatization reversed-phase high-performance liquid chromatography and its application to the study of circadian rhythm in rats and mice, *Anal.Biochem.*, **2000**, *279*, 106–110.

Harumi, T.; Akutsu, H.; Matsushima, S. Simultaneous determination of serotonin, *N*-acetylserotonin and melatonin in the pineal gland of the juvenile golden hamster by high-performance liquid chromatography with electrochemical detection, *J.Chromatogr.B*, **1996**, *675*, 152–156.

Härtter, S.; Morita, S.; Bodin, K.; Ursing, C.; Tybring, G.; Bertilsson, L. Determination of exogenous melatonin and its 6-hydroxy metabolite in human plasma by liquid chromatography-mass spectrometry, *Ther.Drug Monit.*, **2001**, *23*, 282–286. [LC-MS; LOD <2 ng/mL]

Hernandez, G.; Abreu, P.; Alonso, R.; Calzadilla, C.H. Determination of pineal melatonin by high-performance liquid chromatography with electrochemical detection: application for rhythm studies and tissue explants, *J.Pineal Res.*, **1990**, *8*, 11–19.

Iinuma, F.; Hamase, K.; Matsubayashi, S.; Takahashi, M.; Watanabe, M.; Zaitsu, K. Sensitive determination of melatonin by precolumn derivatization and reversed-phase high-performance liquid chromatography, *J.Chromatogr.A*, **1999**, *835*, 67–72. [derivatization; LOD 500 amole; rat pineal gland]

Itoh, M.T.; Hattori, A.; Sumi, Y. Hydroxyindole-*O*-methyltransferase activity assay using high-performance liquid chromatography with fluorometric detection: determination of melatonin enzymatically formed from *N*-acetylserotonin and *S*-adenosyl-L-methionine, *J.Chromatogr.B*, **1997**, *692*, 217–221.

Kulczykowska, E.; Iuvone, P.M. Highly sensitive and specific assay of plasma melatonin using high-performance liquid chromatography with fluorescence detection preceded by solid-phase extraction, *J.Chromatogr.Sci.*, **1998**, *36*, 175–178. [fish; LOD 3 pg/mL]

Mills, M.H.; King, M.G.; Keats, N.G.; McDonald, R.A. Melatonin determination in human urine by high-performance liquid chromatography with fluorescence detection, *J.Chromatogr.*, **1986**, *377*, 350–355.

Naylor, S.; Johnson, K.L.; Williamson, B.L.; Klarskov, K.; Gleich, G.J. Structural characterization of contaminants in commercial preparations of melatonin by on-line HPLC-electrospray ionization-tandem mass spectrometry, *Adv.Exp.Med.Biol.*, **1999**, *467*, 769–777.

Peniston-Bird, J.F.; Di, W.L.; Street, C.A.; Kadva, A.; Stalteri, M.A.; Silman, R.E. HPLC assay of melatonin in plasma with fluorescence detection, *Clin.Chem.*, **1993**, *39*, 2242–2247.

Presits, P.; Molnár-Perl, I. HPLC of tryptophan and its metabolites using simultaneously UV, native fluorescence and pre-column fluorescence derivatization, *Chromatographia*, **2003**, *57*, S87–S92. [UV detection; fluorescence detection; niacin; niacinamide; melatonin]

Sagara, Y.; Okatani, Y.; Yamanaka, S.; Kiriyama, T. Determination of plasma 5-hydroxytryptophan, 5-hydroxytryptamine, 5-hydroxyindoleacetic acid, tryptophan and melatonin by high-performance liquid chromatography with electrochemical detection, *J.Chromatogr.*, **1988**, *431*, 170–176.

Torano, J.S.; van Rijn-Bikker, P.; Merkus, P.; Guchelaar, H.-J. Quantitative determination of melatonin in human plasma and cerebrospinal fluid with high-performance liquid chromatography and fluorescence detection, *Biomed.Chromatogr.*, **2000**, *14*, 306–310. [fluorescence detection; 5-fluorotryptamine is internal standard; LOD 8 pg/mL]

Vitale, A.A.; Ferrari, C.C.; Aldana, H.; Affanni, J.M. Highly sensitive method for the determination of melatonin by normal-phase high-performance liquid chromatography with fluorometric detection, *J.Chromatogr.B*, **1996**, *681*, 381–384.

Wakabayashi, H.; Shimada, K.; Aizawa, Y. Determination of serotonin and melatonin in rat pineal gland by high-performance liquid chromatography with ultraviolet and fluorometric dual detection, *Chem.Pharm.Bull.*, **1985**, *33*, 3875–3880.

Wakabayashi, H.; Shimada, K.; Aizawa, Y. Variation of melatonin and serotonin content in rat pineal gland with sex and oestrous phase difference determined by high-performance liquid chromatography with fluorimetric detection, *J.Chromatogr.*, **1986**, *381*, 21–28.

Yamada, K.; Sakamoto, Y.; Satoh, T. A simple method for the measurement of pineal melatonin with a high performance liquid chromatography-UV detection, *Res.Commun.Chem.Pathol.Pharmacol.*, **1984**, *46*, 283–287.

# Melengestrol acetate

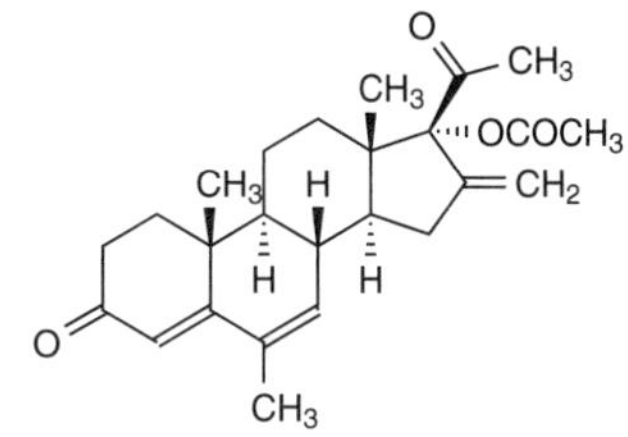

**Molecular formula:** $C_{25}H_{32}O_4$
**Molecular weight:** 396.52
**CAS Registry No:** 2919-66-6, 5633-18-1 (melengestrol only)
**Merck Index:** 13, 5840

## SAMPLE

**Matrix:** tissue

**Sample preparation:** Condition a 3 mL 500 mg CN SPE cartridge with 3 mL ethyl acetate and 5 mL hexane. Cut about 10 g partially thawed fat into small pieces, place in a glass funnel with the outlet covered by glass wool, heat the fat in a microwave oven at high power for 2 or 3 min, collect the rendered fat in a beaker. Add 5 mL MeCN to 2 g liquid fat, heat at 60°, shake for 5 min on a reciprocating shaker at high speed, cool in an ice bath for 5 min, centrifuge at 3000 g for 5 min, decant the acetonitrile extract, repeat the extraction procedure, combine the extracts. Add 3 mL hexane to the combined extracts, vortex for 30 s, centrifuge at 1800 g for 5 min, discard the hexane, repeat the hexane wash. Dry the MeCN layer under a stream of nitrogen at 60°, dissolve the residue in 4 mL hexane, add 1 mL 100 mM NaOH, add 500 μL 1 M magnesium chloride, shake for 30 s, heat at 60° for 15 min, centrifuge at 1800 g for 5 min. Remove the hexane, add 4 mL hexane to the residue, shake for 30 s, centrifuge at 1800 g for 5 min. Combine the hexane extracts and evaporate to dryness under a stream of nitrogen at 60° (to remove any water), reconstitute with 1 mL hexane. Add the sample to the SPE cartridge, wash with 5 mL hexane, wash with 6 mL ethyl acetate:hexane 5:95, elute with three 1 mL and one 500 μL portions of ethyl acetate:hexane 20:80. Evaporate the eluate to dryness under a stream of nitrogen at 60°, reconstitute the residue with 1 mL MeCN:water 70:30, add 10 μL 10% HCl, shake, filter (0.22 μm), inject a 40 μL aliquot.

## HPLC VARIABLES

**Guard column:** 20 × 4.6 5 μm C18 end-capped with 20% carbon load
**Column:** 250 × 4.6 5 μm C18 end-capped with 20% carbon load
**Mobile phase:** MeCN:water 70:30
**Flow rate:** 1
**Injection volume:** 40
**Detector:** UV 291

## CHROMATOGRAM

**Retention time:** 9.32
**Limit of detection:** 3 ppb
**Limit of quantitation:** 10 ppb

## OTHER SUBSTANCES

**Extracted:** chlormadinone (8.57), megestrol (8.92)

## KEY WORDS

cow; fat; pig; SPE

## REFERENCE

Andresen, M.T.; Fesser, A.C.E. Liquid chromatographic determination of progestogens in animal fat, *J.AOAC Int.*, **1996**, *79*, 1037–1042.

## SAMPLE

**Matrix:** tissue

**Sample preparation:** Condition a 100 mg C18 SPE cartridge (Varian) with MeOH and 10 mM pH 8.5 Tris buffer. Fill a 22 mL vessel (from bottom to top) with 5 g alumina (dried at 120° for 48 h before use), 6 g anhydrous sodium sulfate, and 2 g melted (microwave) kidney fat. There is filter paper between the alumina and sodium sulfate. Pass hexane down through the layers at 60° at 1500 psi followed by MeCN at 50° at 1500 psi. Store the MeCN at −20° for 30 min to precipitate fat, filter through a plug of glass wool, evaporate to dryness under a stream of nitrogen at 40°, reconstitute the residue with 2 mL MeOH, mix with 5 mL water, add to the SPE cartridge, wash with 2 mL 20 mM pH 8.5 Tris buffer, wash with 2 mL MeOH:water 40:60, elute with 2 mL MeOH. Evaporate the eluate to dryness under a stream of nitrogen at 40°, reconstitute the residue with 75 μL MeCN, add 75 μL 0.5% formic acid, mix, inject a 75 μL aliquot. (The extraction is done in a Dionex ASE system (advanced solvent extraction).)

---

## HPLC VARIABLES

**Column:** 150 × 3 5 μm Symmetry (Waters)
**Column temperature:** 40
**Mobile phase:** Gradient. A:B from 45:55 to 100:0 over 12 min. A was MeCN:water:formic acid 10:90:0.5. B was MeCN:water:formic acid 90:10:0.5.
**Flow rate:** 0.4
**Injection volume:** 75
**Detector:** MS, Micromass Quattro Ultima, positive electrospray, capillary 2.5 kV, cone 40 V, source 120°, cone gas 189 L/h, desolvation gas 652 L/h, m/z 397.2–279.3–279.3

---

## CHROMATOGRAM

**Retention time:** 8.8
**Limit of detection:** <2 ng/g

---

## OTHER SUBSTANCES

**Extracted:** chloromadinone acetate (m/z 405.2–309.2–345.3) (8.6), chlorotestosterone acetate (m/z 365.2–305.2–323.3) (12.2), delmadinone acetate (m/z 403.2–205.1–181.1) (7.8), flurogestone acetate (m/z 407.2–267.4–225.4) (4.5), medroxyprogesterone acetate (m/z 387.2–327.3–285.3) (8.8), megestrol acetate (m/z 385.2–267.3–325.3) (8.4)

---

## KEY WORDS

fat; kidney; SPE

---

## REFERENCE

Hooijerink, H.; van Bennekom, E.O.; Nielen, M.W.F. Screening for gestagens in kidney fat using accelerated solvent extraction and liquid chromatography electrospray tandem mass spectrometry, *Anal.Chim.Acta*, **2003**, *483*, 51–59.

---

## ANNOTATED BIBLIOGRAPHY

Campbell, H.M.; Sauvé, F. Liquid chromatographic determination of melengestrol acetate in feeds, *J.AOAC Int.*, **1993**, *76*, 1163–1167.

Chichila, T.M.; Edlund, P.O.; Henion, J.D.; Epstein, R.L. Determination of melengestrol acetate in bovine tissues by automated coupled-column normal-phase high-performance liquid chromatography, *J.Chromatogr.*, **1989**, *488*, 389–406. [column-switching]

Huopalahti, R.P.; Henion, J.D. Application of supercritical fluid extraction and high performance liquid chromatography/mass spectrometry for the determination of some anabolic agents directly from bovine tissue samples, *J.Liq.Chromatogr.Rel.Technol.*, **1996**, *19*, 69–87. [UV detection; LC-MS; cow; muscle; liver; LOD 100 ppb; SFE; dexamethasone; trenbolone; triamcinolone acetonide; zeranol; diethylstilbestrol; medroxyprogesterone; melengestrol acetate]

Joos, P.E.; Van Ryckeghem, M. Liquid chromatography-tandem mass spectrometry of some anabolic steroids, *Anal.Chem.*, **1999**, *71*, 4701–4710. [taleranol; estriol; trenbolone; boldenone; fluoxymesterone; nortestosterone; nandrolone; methylboldenone; zeranol; estrone; ethinyl estradiol; diethylstilbestrol;

hexestrol; estradiol; dienestrol; testosterone; estranediol; acetoxyprogesterone; delmadinone; norgestrel; methyltestosterone; methandriol; medroxyprogesterone; progesterone; megestrol; chlormadinone; melengestrol; norethandrolone; chlortestosterone; stanozolol; hydroxyprogesterone; caproxyprogesterone]

Parks, O.W.; Shadwell, R.J.; Lightfield, A.R.; Maxwell, R.J. Determination of melengestrol acetate in supercritical fluid-solid phase extracts of bovine fat tissue by HPLC-UV and GC-MS, *J.Chromatogr.Sci.*, **1996**, *34*, 353–357.

Roybal, J.E. High pressure liquid chromatographic determination of melengestrol acetate in dry feed supplements, *J.Assoc.Off.Anal.Chem.*, **1981**, *64*, 661–664.

Weigand, J.L.; Dille, D.S. Determination of melengestrol acetate in feedstuffs with liquid chromatographic preparatory column cleanup and quantitative analysis, *J.Assoc.Off.Anal.Chem.*, **1988**, *71*, 707–709.

# Memantine

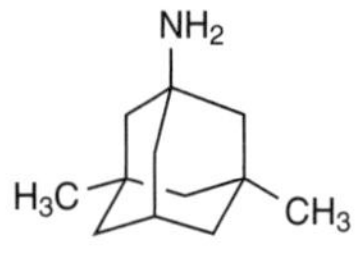

**Molecular formula:** $C_{12}H_{21}N$
**Molecular weight:** 179.30
**CAS Registry No:** 19982-08-2
**Merck Index:** 13, 5851

## SAMPLE

**Matrix:** blood

**Sample preparation:** Mix 1 mL plasma with 50 μL 1 μg/mL IS in 10 mM HCl, 1 mL 1 M NaOH, and 4 mL *n*-heptane:isoamyl alcohol 98.5:1.5 on a rocking platform at low speed for 10 min, centrifuge at 1500 g for 10 min. Add the organic supernatant to 150 μL 100 mM HCl, mix at moderate speed for 10 min, centrifuge at 1500 g for 10 min. Evaporate the aqueous layer to dryness under reduced pressure at 45° for 1 h, reconstitute the residue with 50 μL 1 M pH 10.3 carbonate buffer, add 25 μL 1% dansyl chloride in MeCN, vortex, let stand at room temperature for 45 min. Evaporate to dryness under reduced pressure at 45° for 30 min, reconstitute the residue with 125 μL MeCN:water 75:25, vortex briefly, centrifuge for 3–5 min, inject a 40 μL aliquot of the supernatant.

## HPLC VARIABLES

**Column:** 250 × 4.6 5 μm Supelcosil LC-18

**Mobile phase:** MeCN:buffer 73:27 (Prepare mobile phase by mixing 730 mL MeCN, 270 mL 25 mM potassium dihydrogen phosphate, 500 μL orthophosphoric acid, and 600 μL *n*-butylamine.)

**Flow rate:** 1.8

**Injection volume:** 40

**Detector:** F ex 235 em 470

## CHROMATOGRAM

**Retention time:** 12.1

**Internal standard:** amantadine (7.6)

**Limit of quantitation:** 3 ng/mL (S/N 20)

## OTHER SUBSTANCES

**Extracted:** amoxapine (13.72), *m*-chlorophenylpiperazine (8.21), clovoxamine (8.45), desipramine (14.54), desmethylcitalopram (6.68), desmethylclomipramine (20.05), desmethyldoxepin (10.55), desmethylmaprotiline (9.56), desmethylmianserin (12.87), desmethylsertraline (15.41), desmethyltrimipramine (19.06), fenfluramine (10.02), fluoxetine (12.94), fluvoxamine (8.36), *trans*-10-hydroxynortriptyline (5.48), *cis*-10-hydroxynortriptyline (6.18), maprotiline (15.67), norclozapine (8.35), norfenfluramine (5.32), norfluoxetine (8.50), nortriptyline (17.45), paroxetine (10.15), propranolol (7.11), protriptyline (14.59), rimantadine (13.48), sertraline (31.70)

**Noninterfering:** amitriptyline, bupropion, citalopram, clomipramine, clozapine, doxepin, haloperidol, 2-hydroxydesipramine, 2-hydroxyimipramine, imipramine, loxapine, moclobemide, olanzapine, risperidone

## KEY WORDS

derivatization; plasma

## REFERENCE

Suckow, R.F.; Zhang, M.F.; Collins, E.D.; Fischman, M.W.; Cooper, T.B. Sensitive and selective liquid chromatographic assay of memantine in plasma with fluorescence detection after pre-column derivatization, *J.Chromatogr.B*, **1999**, *729*, 217–224.

## SAMPLE

**Matrix:** solutions

## HPLC VARIABLES

**Column:** 100 × 4.6 5 μm Prodigy ODS(3) (Phenomenex)
**Mobile phase:** Gradient. MeOH:water:formic acid from 50:50:0.1 to 70:30:0.1 over 3 min, maintain at 70:30:0.1 for 2 min, return to initial conditions over 0.5 min, re-equilibrate for 3.5 min.
**Flow rate:** 0.8
**Injection volume:** 100
**Detector:** MS, Agilent HP-1100 MSD, drying gas 30 L/min at 300°, nebulizer pressure 30 psi, capillary entrance 3500 V, capillary exit 70 V, m/z 180

## CHROMATOGRAM

**Retention time:** 3.6
**Limit of quantitation:** 100 pM (water), 500 pM (PBS)

## KEY WORDS

use silanized glassware

## REFERENCE

Koeberle, M.J.; Hughes, P.M.; Wilson, C.G.; Skellern, G.G. Development of a liquid chromatography-mass spectrometric method for measuring the binding of memantine to different melanins, *J.Chromatogr.B*, **2003**, *787*, 313–322.

## ANNOTATED BIBLIOGRAPHY

Duh, T.-H.; Wu, H.-L.; Kou, H.-S.; Lu, C.-Y. (2-Naphthoxy)acetyl chloride, a simple fluorescent reagent, *J.Chromatogr.A*, **2003**, *987*, 205–209. [derivatization; amantadine; memantine]

# Menthol

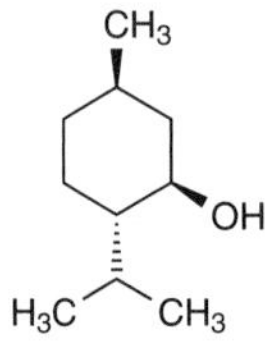


**Molecular formula:** $C_{10}H_{20}O$
**Molecular weight:** 156.26
**CAS Registry No:** 89-78-1
**Merck Index:** 13, 5861

## SAMPLE

**Matrix:** blood
**Sample preparation:** Heat 5 μL serum, 45 μL water, and 200 mM NaOH in EtOH:water 95:5 at 100° for 15 min, cool, add 1 mL water, add 3 mL hexane, vortex for 2 min, centrifuge at 3000 rpm for 5 min. Remove the organic layer and evaporate it to dryness under a stream of nitrogen, reconstitute the residue in 20 μL acetone, add 50 μL 5 mM 4-dimethylaminopyridine in acetone, mix well, add 100 μL 20 mM 3-(2-phthalimidyl)benzoyl azide in acetone, mix well, heat at 125° in a stoppered vial for 30 min, cool in a water bath, inject a 10 μL aliquot. (Omit 4-dimethylaminopyridine for cholesterol determinations. Synthesis of 3-(2-phthalimidyl)benzoyl azide is as follows. Mix 680 mg *m*-aminobenzoic acid in 50 mL diethyl ether with 670 mg *o*-phthalaldehyde in 100 mL diethyl ether, mix, stir at room temperature for 2 days, filter to obtain 3-(2-phthalimidyl)benzoic acid as a white solid. Dissolve 500 mg 3-(2-phthalimidyl)benzoic acid in 15 mL DMF, add 560 mg diphenylphosphoryl azide in 4 mL DMF, add 200 mg triethylamine, stir at 0° for 2 h, add 60 mL 5% sodium bicarbonate in water, add 100 mL diethyl ether, shake. Wash the organic layer twice with 100 mL portions of cold water. Collect the solid at the interface and combine it with the residue obtained when the organic layer is evaporated under reduced pressure. Take up the crude product in acetone, add water to precipitate pure product, repeat this procedure twice to obtain 3-(2-phthalimidyl)benzoyl azide as white needles (mp 123–126°).)

## HPLC VARIABLES

**Column:** 100 × 6 ERC-ODS-1161 (Erma, Tokyo)
**Mobile phase:** MeCN:water 70:30
**Flow rate:** 1
**Injection volume:** 10
**Detector:** F ex 302 em 440

## CHROMATOGRAM

**Retention time:** 12
**Limit of detection:** 100–400 fmol

## OTHER SUBSTANCES

**Extracted:** decanol (21), nonanol (14), octanol (10),

## KEY WORDS

derivatization; serum; serum extraction only validated for cholesterol

## REFERENCE

Tsuruta, Y.; Date, Y.; Kohashi, K. (2-Phthalimidyl)benzoylazides as fluorescence labeling reagents for alcohols in high-performance liquid chromatography, *Anal.Sci.*, **1991**, 7, 411–414.

# Mepenzolate bromide

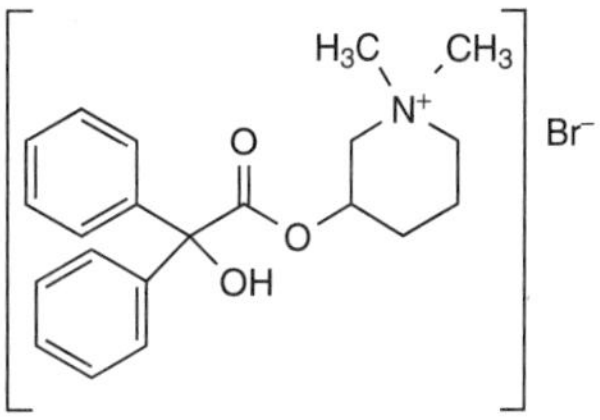


**Molecular formula:** $C_{21}H_{26}BrNO_3$
**Molecular weight:** 420.35
**CAS Registry No:** 76-90-4
**Merck Index:** 13, 5873

## SAMPLE
**Matrix:** blood
**Sample preparation:** Condition a 3 mL diol SPE cartridge (Baker) with 2 mL MeOH and 2 mL water. Add 1 mL plasma to the SPE cartridge, wash with 2 mL water, wash with 2 mL MeOH, elute with 3 mL 60 mM KBr in MeOH. Evaporate the eluate to dryness under a stream of nitrogen at 60°, reconstitute the residue with 200 μL mobile phase, inject an aliquot. (Prepare 60 mM KBr in MeOH by sonicating for 45 min.)

## HPLC VARIABLES
**Column:** 250 × 4 LiChrosorb RP18
**Mobile phase:** MeOH:water 55:45 containing 4.325 g/L sodium octanesulfonate and 2 mL/L *N,N*-dimethyloctylamine, adjusted to pH 3.0 with phosphoric acid
**Flow rate:** 1
**Detector:** UV 220

## CHROMATOGRAM
**Retention time:** 8

## OTHER SUBSTANCES
**Extracted:** clidinium bromide (9.5), isopropamide iodide (5)

## KEY WORDS
plasma; SPE

## REFERENCE
Russ-Kirschenbaum, R.; Koziol, T.; Woolf, E. Solid phase extraction of quaternary ammonium compounds on diol columns: Application to the HPLC determination of CK-1649 in plasma, *J.Liq.Chromatogr.*, **1989**, *12*, 3051–3059.

## SAMPLE
**Matrix:** solutions

## HPLC VARIABLES
**Column:** 150 × 4.1 5 μm PRP-1 (Hamilton)
**Mobile phase:** Gradient. MeCN:50 mM ammonium acetate from 10:90 to 50:50 over 30 min.
**Flow rate:** 0.5
**Injection volume:** 20
**Detector:** MS, Finnigan MAT TSQ-70, thermospray, vaporizer 90°, ion-source 250°, repeller 50–100 V, make up flow 1 mL/min of MeCN:50 mM ammonium acetate 10:90, collision gas pressure 0.5 Pa, 35 eV, m/z 340

## CHROMATOGRAM
**Retention time:** 14
**Limit of detection:** 50 pg

## OTHER SUBSTANCES

**Simultaneous:** antrenyl (m/z 348), clidinium (m/z 352) (15), isopropamide (m/z 353), pipenzolate (m/z 354) (15), propantheline (m/z 368) (22), valethanate (m/z 306)

## REFERENCE

van der Hoeven, R.A.M.; Reeuwijk, H.J.E.M.; Tjaden, U.R.; van der Greef, J. Analysis of quaternary ammonium drugs by thermospray liquid chromatography-mass spectrometry using a resin-based stationary phase, *J.Chromatogr.A*, **1996**, *741*, 75–84.

# Mepixanox

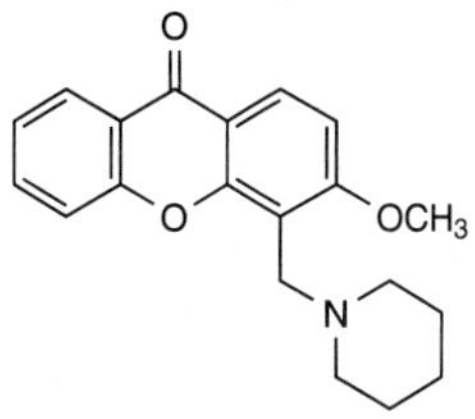

**Molecular formula:** $C_{20}H_{21}NO_3$
**Molecular weight:** 323.38
**CAS Registry No:** 17854-59-0
**Merck Index:** 13, 5885

## SAMPLE

**Matrix:** blood
**Sample preparation:** Add 1 mL plasma or serum and 4 mL 900 ng/mL IS in water to a previously mixed Toxi Tube A (Analytical Systems), rotate for 5 min, centrifuge at 3000 g for 5 min. Evaporate the organic layer to dryness under a stream of nitrogen, reconstitute the residue with 200 μL MeCN:diethylamine 100:0.004, inject an aliquot. (Toxi Tube A contains 3.5% sodium carbonate, 3.5% sodium bicarbonate, 18% dichloromethane, and 17% dichloroethane, buffered at pH 9.0.)

## HPLC VARIABLES

**Column:** 250 × 4 5 μm Si 60 (Merck)
**Mobile phase:** MeCN:isopropanol:diethylamine 50:50:0.004
**Flow rate:** 1
**Injection volume:** 50
**Detector:** UV 237

## CHROMATOGRAM

**Retention time:** 6.1
**Internal standard:** chlorpromazine (8.6)
**Limit of detection:** 1 ng/mL
**Limit of quantitation:** 5 ng/mL

## OTHER SUBSTANCES

**Simultaneous:** amitriptyline (12.2), chlordiazepoxide (5.4), dimefline (6.4), dipyrone (9.1), doxapram (4.0), fluphenazine (2.7), imipramine (15.4), levomepromazine (4.0), nikethamide (5.5), promazine (12.7), promethazine (8.1), thioridazine (12.0)

## KEY WORDS

normal phase; pharmacokinetics; plasma; serum

## REFERENCE

Grossi, G.; Lippi, A.; Battistoni, R.; Martelli, E.; Sturani, C. High-performance liquid chromatographic determination of mepixanthone in serum, *J.Chromatogr.*, **1984**, *309*, 214–218.

# Mequinol

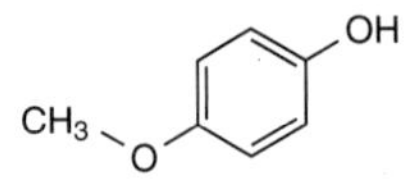


**Molecular formula:** $C_7H_8O_2$

**Molecular weight:** 124.14

**CAS Registry No:** 150-76-5

## SAMPLE

**Matrix:** blood

**Sample preparation:** Mix 100 μL serum with 1 mL 20 μg/mL IS in mobile phase, inject a 20 μL aliquot.

## HPLC VARIABLES

**Guard column:** present but not specified

**Column:** 150 × 4.6 5 μm Pinkerton ISRP (Regis)

**Mobile phase:** MeOH:100 mM pH 6.8 potassium phosphate buffer 2.5:97.5

**Flow rate:** 1.2

**Injection volume:** 20

**Detector:** UV 280

## CHROMATOGRAM

**Retention time:** 5

**Internal standard:** 4-ethoxyphenol (8)

**Limit of detection:** 2 μg/mL

## OTHER SUBSTANCES

**Extracted:** acetaminophen (2.9), caffeine (3.8), metoclopramide (3.8), salicylic acid (3.2), theophylline (4)

## KEY WORDS

serum

## REFERENCE

Dawson, C.M.; Belcher, H.J.C.R.; Rainbow, S.J.; Tickner, T.R. Measurement of 4-hydroxyanisole in serum by direct injection high-performance liquid chromatography, *J.Chromatogr.*, **1990**, *534*, 267–270.

# Methenamine

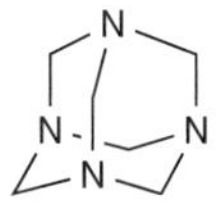

**Molecular formula:** $C_6H_{12}N_4$
**Molecular weight:** 140.19
**CAS Registry No:** 100-97-0
**Merck Index:** 13, 5994

## SAMPLE

**Matrix:** air

**Sample preparation:** Pull air through a coated filter at 150 mL/min for 20 min, shake filter with 3 mL MeCN for 1 min, filter (Millex SR) the solution, inject a 10 μL aliquot. (Prepare filter as follows. Mix 300 mg 2,4-dinitrophenylhydrazine hydrochloride, 500 μL 85% phosphoric acid, 1.5 mL glycerol:EtOH 20:80, and 9 mL MeCN. Dip glass fiber filter in this solution for a few seconds, let dry on glass surface at room temperature for 2 h, store in a desiccator over saturated NaCl. Before use, recrystallize 2,4-dinitrophenylhydrazine hydrochloride twice from 4 M HCl. Methenamine is converted to formaldehyde on the filter and the formaldehyde 2,4-dinitrophenylhydrazone is detected by HPLC.)

## HPLC VARIABLES

**Column:** 100 × 5 10 μm Waters Radial Pak A C18
**Mobile phase:** MeOH:water 40:60
**Flow rate:** 0.8
**Injection volume:** 10
**Detector:** UV 365

## CHROMATOGRAM

**Retention time:** k′ 2.2
**Limit of quantitation:** 500 ng/mL

## KEY WORDS

derivatization

## REFERENCE

Levin, J.-O.; Fängmark, I. High-performance liquid chromatographic determination of hexamethylenetetramine in air, *Analyst*, **1988**, *113*, 511–513.

# Methoprene

**Molecular formula:** $C_{19}H_{34}O_3$
**Molecular weight:** 310.47
**CAS Registry No:** 40596-69-8
**Merck Index:** 13, 6013

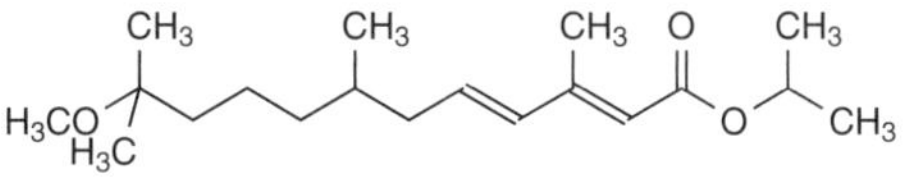


## SAMPLE

**Matrix:** blood, urine
**Sample preparation:** Condition a 3 mL 500 mg Sep-Pak C18 SPE cartridge with 3 mL MeCN and 3 mL water. Vortex 500 μL plasma or urine with 100 μL 1 M pH 5.0 acetic acid for 30 s, centrifuge at 1000 g for 5 min, add the supernatant to the SPE cartridge, wash with 3 mL water, elute with 2 mL MeOH, elute with 2 mL MeCN. Evaporate the combined eluates to 500 μL under a stream of nitrogen, inject a 10 μL aliquot.

## HPLC VARIABLES

**Guard column:** 20 × 4 5 μm Supelco
**Column:** 3000 × 3.9 10 μm μBondapak C18
**Mobile phase:** Gradient. MeCN:water adjusted to pH 4.0 with 1 M acetic acid 55:45 to 60:40 over 10 min, to 80:20 over 3 min, return to initial conditions over 5 min, re-equilibrate for 2 min.
**Flow rate:** 0.6 for 9 min, to 1 over 1 min, maintain at 1 for 8 min
**Injection volume:** 10
**Detector:** UV 254

## CHROMATOGRAM

**Retention time:** 18.4
**Limit of detection:** 100 ng/mL (S/N 3)
**Limit of quantitation:** 150 ng/mL

## OTHER SUBSTANCES

**Extracted:** methoprene acid (12.3), permethrin (UV 210) (17.6), *m*-phenoxybenzoic acid (UV 210) (7.9), *m*-phenoxybenzyl alcohol (UV 210) (7.3)

## KEY WORDS

plasma; rat; SPE

## REFERENCE

Abu-Qare, A.W.; Abou-Donia, M.B. A solid phase extraction reversed-phase HPLC method for the simultaneous determination of methoprene, permethrin and selected metabolites in rat plasma and urine, *Biomed.Chromatogr.*, **2001**, *15*, 464–470.

# Methoxychlor

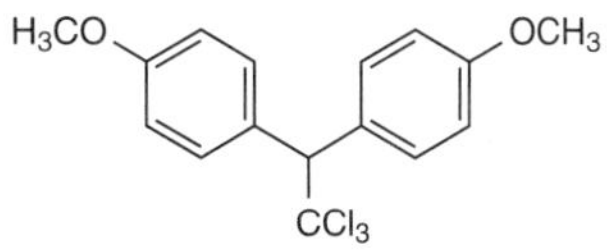


**Molecular formula:** $C_{16}H_{15}Cl_3O_2$
**Molecular weight:** 345.65
**CAS Registry No:** 72-43-5
**Merck Index:** 13, 6020

## SAMPLE

**Matrix:** blood, urine
**Sample preparation:** Add 1 mL urine or whole blood and 3 mL water to a Toxi-Tube A, mix by gentle inversion for 5 min, centrifuge at 1500 g for 5 min. Evaporate the organic layer to dryness under a stream of nitrogen at 40°, reconstitute the residue with 50 μL MeCN:water 50:50, vortex, inject a 10 (urine) or 30 (blood) μL aliquot.

## HPLC VARIABLES

**Guard column:** 20 mm long Symmetry C18 (Waters)
**Column:** 250 × 4.6 5 μm Symmetry C8 (Waters)
**Column temperature:** 30
**Mobile phase:** Gradient. MeCN:buffer 15:85 for 6.5 min, 35:65 for 18.5 min (step gradient), 80:20 for 3 min (step gradient), re-equilibrate at initial conditions for 7 min. (The buffer was 6 g sodium dihydrogen phosphate in 1 L water, adjusted to pH 3.8 with 10% phosphoric acid.)
**Flow rate:** 1 for 6.5 min, to 1.5 over 18.5 min, maintain at 1.5 until re-equilibration is complete
**Injection volume:** 10–30
**Detector:** UV 200

## CHROMATOGRAM

**Retention time:** 27.368

## OTHER SUBSTANCES

**Extracted:** acamprosate (abs 200.5) (2.97), acebutolol (abs 233.4) (10.233), acefylline heptaminol (abs 207.5) (3.628), acenocoumarol (abs 204) (20.052), acepromazine (abs 249.9) (10.763), aceprometazine (abs 240.5) (14.225), acetaminophen (abs 200.5) (5.592), acetazolamide (abs 265.3) (6.927), acetiamin (abs 200.5) (11.78), acetorphan (abs 200.5) (22.495), acetylleucine (abs 200.5) (9.825), acetylmethionine (abs 200.5) (5.03), 6-acetylmorphine (abs 208.7) (7.32), actinoquinol (abs 202.8) (4.637), acyclovir (abs 200.5) (3.073), adenosine (abs 206.4) (2.697), adrafinil (abs 200.5) (13.833), albendazole (abs 218.1) (17.777), aldicarb (abs 200.5) (15.143), alfuzosin (abs 244) (10.37), alimemazine (abs 253.5) (15.257), alminoprofen (abs 201.7) (18.695), almitrine (abs 200.5) (25.905), alprazolam (abs 220.5) (16.972), altazide (abs 226.3) (16.368), altretamine (abs 229.9) (17.833), ambemonium (abs 200.5) (10.543), amfepramone (abs 200.5) (8.688), amiloride (abs 214.6) (3.608), amineptine (abs 200.5) (14.032), amiodarone (abs 204) (21.915), amisulpride (abs 225.2) (8.923), amitriptyline (abs 206.4) (15.878), amlodipine (abs 200.5) (15.093), amobarbital (abs 200.5) (16.617), amoxapine (abs 211.l) (14.187), amoxicillin (abs 200.5) (3.067), amphetamine (abs 200.5) (3.71), amphotericin B (abs 346.2) (15.718), ampicillin (abs 200.5) (3.827), anethole (abs 205.2) (23.902), apomorphine (abs 206.4) (8.962), aprindine (abs 200.5) (16.967), arecoline (abs 207.5) (3.1), aspartame (abs 200.5) (9.8), astemizole (abs 200.5) (13.16), atenolol (abs 200.5) (3.637), atrazine (abs 221.6) (18.177), atropine (abs 200.5) (10.388), bacampicillin (abs 200.5) (14.687), bamethan (abs 200.5) (5.91), bamifylline (abs 207.5) (10.288), barbital (abs 200.5) (10.445), benazepril (abs 205.2) (17.003), bendroflumethiazide (abs 208.7) (18.632), benfluorex (abs 200.5) (16.378), bentazone (abs 224) (13.615), benzbromarone (abs 204) (26.075), benzene (abs 209.9) (19.465), benzophenone (abs 200.5) (16.967), benzoylecgonine (abs 200.5) (9.678), benzydamine (abs 215.8) (14.955), bepridil (abs 202.8) (19.503), betahistine (abs 200.5) (3.155), betamethasone (abs 200.5) (13.277), betaxolol (abs 200.5) (13.405), bezafibrate (abs 200.5) (18.268), BHT (abs 200.5) (22.015), bidesethylchloroquine (abs 219.3) (3.867), biperiden (abs 200.5) (14.847), biscoumacetate (abs 208.7) (19.172), bisoprolol (abs 200.5) (12.283), boldine (abs 218.1) (8.068), bromadiolone (abs 202.8) (25.363), bromazepam (abs 232.2) (14.732), bromocriptine (abs 200.5) (16.652), brompheniramine (abs 200.5) (13.935), buclizide (abs 200.5) (22.752), buflomedil (abs 201.7) (11.3), bumetanide (abs 200.5) (19.055), buprenorphine (abs 212.2) (14.035), buspirone (abs 236.9) (12.523), butobarbital (abs 200.5) (14.858), *N*-butylscopolammonium bromide (abs 200.5) (12.232), butylparaben (abs 200.5) (20.332), cafedrine (abs 206.4) (10.235), caffeic acid (abs 321.2) (8.963), caffeine (abs 205.2) (6.647), canrenoate (abs 287.8) (16.52), captodiamine (abs 200.5) (20.15), captopril (abs 200.5) (9.652), carbamazepine (abs 213.4) (15.763), carbaryl (abs 220.5) (17.968), carbimazole (abs 200.5) (11.138), carbinoxamine (abs 200.5) (12.81), carbutamide (abs 200.5) (14.547), carbutamine (abs 200.5) (14.475), carnitine (abs 287.8) (16.058), carpipramine (abs 200.5) (16.18), carteolol (abs 214.6) (5,935), cefadroxil (abs 200.5) (3.188), cefatrizine (abs 200.5) (3.808), cefazolin (abs 200.5) (8.423), cefixime (abs 287.8) (4.823), cefpodoxime (abs 200.5) (18.945), ceftriaxone (abs 245.2) (5.34), cefuroxine (abs 278.3) (16.293, 16.552), celiprolol (abs 232.2) (11.493), cephalexin (abs 200.5) (3.875, 4.792), cetirizine (abs 200.5) (15.683), chlorambucil (abs 201.7) (22.387), chloramphenicol (abs 200.5) (14.105), chlordiazepoxide (abs 244) (15.223), chlorhexidine (abs 200.5) (13.532), chloridazone (abs 227.5) (12.293), chlormadinone (abs 283.1) (24.11), chlormezanone (abs 200.5) (15.493), chlorobenzoic acid (abs 200.5) (16.143), chlorophacinone (abs 200.5) (20.978), 4-chlorophenylbiguanide (abs 200.5) (10.068), chloropicrin (abs 202.8) (19.433), chloroquine (abs 221.6) (5.442), chlorpheniramine (abs 200.5) (12.925), chlorpromazine (abs 254.7) (16.035), chlorpropamide (abs 200.5) (17.657), chlortoluron (abs 209.9) (17.573), cibenzoline (abs 200.5) (13.143), cicletanine (abs 200.5) (13.808), cilazapril (abs 200.5) (14.367), cimetidine (abs 200.5) (3.602), cinchonine (abs 202.8) (10.198), cinnarizine (abs 200.5) (19.258), ciprofibrate (abs 200.5) (21.22), ciprofloxacin (abs 278.3) (9.102), cisapride (abs 214.6) (14.627), clenbuterol (abs 211.1) (10.802), clidinium (abs 200.5) (13.27), clindamycin (abs 200.5) (11.962), clobazam (abs 229.9) (19.19), clobenzorex (abs 200.5) (13.912), clocinizine (abs 200.5) (20.432), clofibrate (abs 200.5) (18.267), clofibride (abs 200.5) (21.067), clomethiazole (abs 249.9) (15.958), clomipramine (abs 200.5) (16.442), clonazepam (abs 200.5) (17.417), clonidine (abs 200.5) (6.128), clorazepate (abs 227.5) (18.44), clotiazepam (abs 211.1) (21.652), cloxacillin (abs 200.5) (15.702), cocaine (abs 200.5) (11.92), codeine (abs 212.2) (4.975), codethyline (abs 211.1) (8.735), colchicine (abs 244) (13.118), colchicoside (abs 244) (5.32), colchicosine (abs 200.5) (5.078), cortivazol (abs 207.5) (12.002), cotinine (abs 200.5) (4.7), coumachlor (abs 204) (22.153), coumafen (abs 205.2) (20.355), coumatetralyl (abs 264.1) (4.993), crimidine (abs 251.1) (3.703), cyamemazine (abs 270) (14.993), cyclandelate (abs 200.5) (26.383), cycloguanil (abs 200.5) (10,793), cyproheptadine (abs 224) (15.015), cyromazine (abs 214.6) (3.283), dacarbazine (abs 323.5) (3.6), dapsone (abs 200.5) (12.583), desethylatrazine (abs 213.4) (12.583), desipramine (abs 200.5) (14.87), desoxy-2-phenobarbital (abs 200.5) (11.047), dexamethasone (abs 241.7) (13.127), dextromethorphan (abs 200.5) (13.312), dextromoramide (abs 200.5) (15.835), dextropropoxyphene (abs 200.5) (15.82), di-syston (abs 200.5) (26.795), diamorphine (abs 206.4) (11.152), diaveridine (abs 200.5) (7.022), diazepam (abs 200.5) (20.327), diazinon (abs 200.5) (25.763), dibekacin (abs 200.5) (14.053), dichlorprop (abs 200.5) (16.947), diclofenac (abs 200.5) (22.115), diethylstilbestrol (abs 200.5) (20.882), difemerine (abs 200.5) (13.222), digitoxin (abs 219.3) (18.725), digoxin (abs 220.5) (13.852), dihydralazine (abs 219.3) (2.81), dihydrocodeine (abs 208.7) (4.7), diltiazem (abs 200.5) (13.992), dimerazole (abs 200.5) (8.007), dimethyl phthalate (abs 200.5) (17.04), dimetridazole (abs 320) (9.955, 10.118), dinoseb (abs 213.4) (23.652), diphacinone (abs 200.5) (19.63), dipheniramine (abs 200.5) (14.092), diprophylline (abs 206.4) (3.625), dipropyline (abs 207.5) (16.305), dipyridamole (abs 285.5) (13.197), disopyramide (abs 200.5) (11.445), disulfiram (abs 216.9) (25.13), diuron (abs 211.1) (18.503), domperidone (abs 207.5) (12.355), dosulepin (abs 200.5) (14.943), doxepin (abs 206.4) (14.095), doxorubicin (abs 232.2) (12.057), doxylamine (abs 200.5) (11.147), droperidol (abs 202.8) (21.18), dropropizine (abs 200.5) (7.243), econazole (abs 200.5) (20.137), EDDP (abs 200.5) (14.655), embutramide (abs 200.5) (17.365), emetine (abs 202.8) (9.385), enalapril (abs 211.1) (3.432), enoxacin (abs 268.9) (7.672), ephedrine (abs 206.4) (5.655), epinephrine (abs 200.5) (2.87), esculin (abs 202.8) (5.277), eserine (abs 204) (8.253), estazolam (abs 221.6) (16.495), β-estradiol (abs 200.5) (18.202), estriol (abs 200.5) (13.142), ethosuximide (abs 200.5) (10.485), ethyl paraben (abs 200.5) (16.19), etodolac (abs 225.2) (21.503), famotidine (abs 202.8) (3.487), felodipine (abs 200.5) (24.423), fenbufen (abs 200.5) (19.292), fenfluramine (abs 207.5) (13.055), fenofibrate (abs 200.5) (18.262), fenoprofen (abs 200.5) (21.16), fenoverine (abs 200.5) (15.57), fenozolone (abs 220.5) (12.913), fenproporex (abs 200.5) (19.263), fentanyl (abs 255.8) (14.202), flavone (abs 201.7) (20.768), flecaine (abs 204) (14.417), floctafenine (abs 209.9) (17.177), flubendazole (abs 211.1) (16.745), fluconazole (abs 200.5) (11.398), flucytosine (abs 200.5) (3.052), flunarizine (abs 200.5) (19.317), flunindione (abs 222.8) (17.892), flunitrazepam (abs 200.5) (18.558), fluorouracil (abs 204) (3.433), fluoxetine (abs 200.5) (16.185), flupenthixol (abs 228.7) (17.358), fluphenazine (abs 259.4) (17.357), flurbiprofen (abs 200.5) (21.337), flutamide (abs 200.5) (22.248), fluvoxamine (abs 200.5) (15.347), folic acid (abs 200.5) (3.583), fumaric acid (abs 206.4) (2.985), furaltadone (abs 347.4) (8,927), furazolidone (abs 347.4) (12.24), furosemide (abs 234.6) (15.17), fusidic acid (abs 200.5) (24.86), gentamicin (abs 200.5) (14.037), glibenclamide (abs 200.5) (21.953), gliclazide (abs 200.5) (20.5), glipizide (abs 200.5) (17.603), glutathione (abs 200.5) (2.835), griseofulvin (abs 292.6) (18.392), guaiacol (abs 200.5) (13.787), guaifenesin (abs 200.5) (11.435), guanfacine (abs 200.5) (11.387), halofantrine (abs 258.2) (22.993), haloperidol (abs 200.5) (14.415), harpagoside (abs 279.5) (13.48), hematoporphyrin (abs 347.4) (18.66), histidine (abs 211.1) (2.613), hydrochlorothiazide (abs 226.3) (9.397), hydrocortisone (abs 242.9) (17.735), 4-hydroxybutyrate (abs 207.5) (2.772), 3-hydroxytyramine (abs 200.5) (3.002), hydroxyzine (abs 200.5) (15.267), hyoscyamine (abs 200.5) (9.657), ibuprofen (abs 200.5) (23.815), imidazole (abs 205.2) (2.74), imipramine (abs 200.5) (15.113), inchonidine (abs 202.8) (10.480), indomethacin (abs 201.7) (21.748), indoramine (abs 200.5) (12.533), iodoform (abs 200.5) (22.05), isoquercitrin (abs 205.2) (10.582), isosorbide (abs 200.5) (18.48), isothipendyl (abs 248.8) (13.467), isradipine (abs 200.5) (22.352), josamycine (abs 231.1) (16.763), ketamine (abs 202.8) (9.637), ketoconazole (abs 202.8) (15.738), ketoprofen (abs 200.5) (19.628), lacidipine (abs 239.3) (27.235), lactic acid (abs 200.5) (3.227), lanzoprazole (abs 200.5) (16.613), levamisole (abs 213.4) (6.97), levodopa (abs 200.5) (3.575), levomepromazine (abs 251.1) (15.842), levopenbutolol (abs 200.5) (15.928), lidocaine (abs 200.5) (9.922), linuron (abs 209.9) (21.253), lisuride (abs 209.9) (4.53), lobeline (abs 200.5) (14.625), loflazepate (abs 228.7) (21.027), lomustine (abs 229.9) (22.982), loprazolam (abs 200.5) (13.387), loratadine (abs 200.5) (22.943), lorazepam (abs 228.7) (17.175), loxapine (abs 209.9) (14.553), LSD (abs 200.5) (12.003), lufenuron (abs 209.9) (27.658), maprotiline (abs 200.5) (15.508), MCPA (abs 200.5) (15.877), MDA (abs 200.5) (8.058), MDMA (abs 200.5) (9.058), MDPA (abs 200.5) (11.185), mebendazole (abs 209.9) (16.077), mebezonium (abs 225.2) (3.4), meclozine (abs 200.5) (19.955), mecoprop (abs 200.5) (17.705), medazepam (abs 200.5) (15.83), medifoxamine (abs 200.5) (13.833), medroxyprogesterone (abs 241.7) (24.203), mefenorex (abs 207.5) (11.897), mefloquine (abs 222.8) (16.597), megestrol (abs 290.2) (23.815), melatonin (abs 200.5) (12.923), melphalan (abs 201.7) (12.93), meperidine (abs 200.5) (11.77), mesalamine (abs 205.2) (4.737), metamitron (abs 200.5) (11.853), metaproterenol (abs 200.5) (4.15), metformin (abs 233.4) (2.803), methacycline (abs 242.9) (11.493), methadone (abs 200.5) (15.753), methamphetamine (abs 206.4) (8.433), methionine (abs 200.5) (2.988), methyclothiazide (abs 226.3) (15.36), methyldopa (abs 200.5) (2.96), 3-methyldopamine (abs 200.5) (3.23), 4-methyldopamine (abs 200.5) (4.185), methylhydroxyprogesterone (abs 242.9) (21.852), methylparaben (abs 200.5) (14.04), methylprednisolone (abs 245.2) (18.887), metoclopramide (abs 213.4) (9.915), metoprolol (abs 200.5) (10.722), metronidazole (abs 320) (6.778), mexiletine (abs 200.5) (11.468), mianserin (abs 200.5) (13.787), miconazole (abs 202.8) (21.53), midazolam (abs 200.5) (14.873), minaprin (abs 204) (11.225), minocycline (abs 200.5) (22.637), minoxidil (abs 231.1) (9.76), moclobemide (abs 200.5) (10.218), molsidomine (abs 312.8) (10.007), monodesbutylhalofantrine (abs 258.2) (20.45), monodesethylchloroquine (abs 219.3) (4.675), moroxydine (abs 236.9) (2.873), morphine (abs 211.1) (3.315), *N*-acetyl-L-tyrosine ethyl ester (abs 200.5) (11.997), nadolol (abs 200.5) (6.77), nadoxolol (abs 211.1) (12.445), naftidrofuryl (abs 225.2) (15.832), nalidixic acid (abs 258.2) (16.008), nalorphine (abs 211.1) (4.76), naloxone (abs 200.5) (14.028), naltrexone (abs 205.2) (6.087), nefopam (abs 200.5) (12.653), nemonapride (abs 212.2) (14.985), neopynamine (abs 209.9) (3.233), neostigmine (abs 200.5) (4.782), netilmicin (abs 206.4) (12.082), niacin (abs 209.9) (3.163), niacinamide (abs 214.6) (3.61), niaprazine (abs 200.5) (11.092), nicardipine (abs 205.2) (15.528), nicergoline (abs 202.8) (15.452), niclosamide (abs 205.2) (23.92), nifedipine (abs 236.9) (19.485), niflumic acid (abs 200.5) (21.968), nifuroxazide (abs 200.5) (13.43), nikethamide (abs 200.5) (10.623), nitrazepam (abs 200.5) (16.927), nitrendipine (abs 236.9) (22.087), nitrofural (abs 260.6) (10.323), nizatidine (abs 316.4) (3.302), noramidopyrine (abs 200.5) (9.157), norclomipramine (abs 200.5) (16.617), norephedrine (abs 205.2) (5.015), norepinephrine (abs 200.5) (2.8), norethisterone (abs 240.5) (24.038), norgestrel (abs 241.7) (21.565), norpropoxyphene (abs 200.5) (15.478), nortriptyline (abs 206.4) (15.603), noscapine (abs 213.4) (12.827), ofloxacin (abs 295) (8.648), omeprazole (abs 200.5) (14.065), opiclone (abs 200.5) (10.372), opipramol (abs 255.8) (14.163), oxacillin (abs 200.5) (14.76), oxatomide (abs 205.2) (15.797), oxazepam (abs 228.7) (16.745), oxprenolol (abs 200.5) (12.017), pancuronium (abs 200.5) (3.027), papaverine (abs 251.1) (12.12), paraminosalicylic acid (abs 207.5) (5.875), parbendazole (abs 207.5) (18.15), parconazole (abs 202.8) (15.5), paroxetine (abs 200.5) (15.275), pefloxacin (abs 278.3) (8.942), penfluridol (abs 200.5) (20.183), pentaerythritol (abs 200.5) (23.082), pentazocine (abs 200.5) (12.522), pentobarbital (abs 200.5) (16.437), pentoxifylline (abs 206.4) (11.477), pentoxyverine (abs 200.5) (15.582), perindopril (abs 206.4) (13.698), perphenazine (abs 255.8) (15.96), phenindione (abs 226.3) (18.062), phenobarbital (abs 200.5) (13.993), phenol (abs 200.5) (13.422), phenoxyethanol (abs 200.5) (13.375), phenylbutazone (abs 200.5) (24.098), phenytoin (abs 200.5) (16.288), phloroglucinol (abs 202.8) (4.172), pholcodine (abs 211.1) (2.687), phorate (abs 200.5) (26.477), phosdrin (abs 215.8) (13.615), phthalic acid (abs 201.7) (6.59), pilocarpine (abs 214.6) (4.622), pimaricin (abs 303.3) (13.637), pimozide (abs 205.2) (17.192), pinaverium (abs 213.4) (21.337), pindolol (abs 215.8) (8.568), pipamperone (abs 200.5) (10.918), piperacillin (abs 200.5) (12.758), piperine (abs 341.5) (20.373), piperonyl butoxide (abs 202.8) (27.627), pipothiazine (abs 262.9) (14.695), piracetam (abs 200.5) (3.295), piretanide (abs 200.5) (17.8), pirisudanol (abs 209.9) (3.462), piroxicam (abs 200.5) (16.57), pivampicillin (abs 200.5) (15.813), pizotifene (abs 200.5) (15.197), pralidoxime (abs 295) (2.863), prazepam (abs 200.5) (23.36), prazosin (abs 246.4) (10.608), prednisolone (abs 246.4) (14.113), prednisone (abs 241.7) (14.178), prifinium (abs 200.5) (15.622), primidone (abs 200.5) (11.13), pristinamycin (abs 227.5) (17.235), procaine (abs 292.6) (5.218), progesterone (abs 242.9) (23.835), proguanil (abs 200.5) (13.61), promethazine (abs 251.1) (14.482), prometrine (abs 221.6) (21.89), propafenone (abs 211.1) (15.128), propazine (abs 221.6) (20.433), propericyazine (abs 270) (14.168), propoxur (abs 200.5) (17.293), propranolol (abs 213.4) (13.06), propylparaben (abs 200.5) (18.34), proscillaridine a (abs 200.5) (15.585), protopine (abs 205.2) (11.707), psilocybin (abs 220.5) (3.343), pyrazinamide (abs 208.7) (3.823), pyridostigmine (abs 200.5) (3.228), pyrimethamine (abs 208.7) (12.497), pyroglutamic acid (abs 200.5) (3.008), quercetin (abs 202.8) (14.168), quinapril (abs 200.5) (16.782), quinidine (abs 208.7) (11.025), quinine (abs 208) (11.253), quinupramine (abs 200.5) (15.168), ramipril (abs 206.4) (15.678), ranitidine (abs 228.7) (3.74), reserpine (abs 218.1) (16.433), resorcinol (abs 200) (8.027), rifabutin (abs 208.7) (17.583), rifampin (abs 236.9) (16.167), rifamycin (abs 225.2) (20.85), rilmenidine (abs 200.5) (9.8), ronidazole (abs 308) (8.265), rosmarinic acid (abs 200.5) (10.637), rosoxacin (abs 271.2) (13.467), roxithromycin (abs 200.5) (15.833), rutine (abs 204) (10.1), saccharin (abs 200.5) (5.907), salicylic acid (abs 202.8) (12.12), scopolamine (abs 200.5) (7.39), secnidazole (abs 318.8) (9.668), secobarbital (abs 200.5) (17.42), selegiline (abs 207.5) (10.712), simazine (abs 220.5) (15.752), sisomicin (abs 200.5) (14.032), sotalol (abs 200.5) (3.842), spironolactone (abs 239.3) (20.68), strychnine (abs 207.5) (9.202), sulbutiamine (abs 200.5) (12.97), sulfachloropyrazine (abs 271.2) (14.763), sulfadiazine (abs 200.5) (8.375), sulfadimethoxine (abs 200.5) (14.735), sulfadoxine (abs 200.5) (13.345), sulfaguanidine (abs 200.5) (3.795), sulfamethoxazole (abs 200.5) (13.445), sulfamethoxypyridazine (abs 200.5) (11.217), sulfanilamide (abs 200.5) (5.047), sulfathiazole (abs 200.5) (9.022), sulindac (abs 200.5) (16.627), sulpiride (abs 212.2) (3.858), sultopride (abs 212.2) (13.012), synephrine (abs 200.5) (3.02), tamoxifen (abs 200.5) (20.652), temazepam (abs 200.5) (18.562), tenoxicam (abs 200.5) (12.733), terbutaline (abs 200.5) (3.683), terfenadine (abs 200.5) (19.118), terpine (abs 312.8) (13.868), tetracaine (abs 312.8) (13.69), tetracycline (abs 274.8) (9.888), tetramisole (abs 214.6) (6.97), tetrazepam (abs 226.3) (22.378), theobromine (abs 204) (3.79), theodrenaline (abs 201.7) (3.682), theophylline (abs 202.8) (4.877), thiamphenicol (abs 200.5) (7), thiopental (abs 285.5) (19.202), thioproperazine (abs 265.3) (15.212), thioridazine (abs 262.9) (17.168), tianeptine (abs 206.4) (14.877), tiapride (abs 213.4) (5.468), tiaprofenic acid (abs 200.5) (17.653), ticlopidine (abs 200.5) (13.813), tiemonium (abs 200.5) (11.795), timolol (abs 296.1) (10.295), tinidazole (abs 317.6) (10.563), tioclomarol (abs 200.5) (22.525), tobramycin (abs 200.5) (13.393), tofisopam (abs 204) (18.508), toloxatone (abs 204) (14.107), toluene (abs 215.8) (21.88), trandolapril (abs 206.4) (16.993), trazodone (abs 211.1) (12.683), triamterene (abs 215.8) (8.705), triazolam (abs 220.5) (17.353), trichlorocarbanilide (abs 264.1) (25.573), trichloromethiazide (abs 225.2) (14.907), trifluoperazine (abs 258.2) (17.747), trifluperidol (abs 200.5) (21.638), trihexyphenidyl (abs 200.5) (15.298), trimebutine (abs 213.4) (14.588), trimetazidine (abs 206.4) (6.06), trimethoprim (abs 205.2) (8.282), trimipramine (abs 200.5) (15.943), triprolidine (abs 200.5) (13.133), tritoqualine (abs 214.6) (20.968), tropatepine (abs 200.5) (15.425), vanillin (abs 231.1) (11.687), verapamil (abs 201.7) (15.365), veratrine (abs 220.5) (13.647), veratrole (abs 201.7) (16.377), viloxazine (abs 200.5) (10.993), vinblastine (abs 200.5) (8.37), vincamine (abs 221.6) (12.08), vincristine (abs 220.5) (13.765), vinylbital (abs 200.5) (16.583), virginiamycin (abs 227.5) (17.21), vitamin B1 (abs 200.5) (2.597), vitamin B2 (abs 267.7) (7.182), vitamin B5 (abs 200.5) (3.772), vitamin B6 (abs 200.5) (2.895), vitamin B12 (abs 207.5) (3.777), vitamin C (abs 249.9) (2.928), vitamin H (abs 200.5) (8.89), warfarin (abs 205.2) (20.358), xanthydrol (abs 209.9) (18.857), xipamide (abs 218.1) (18.823), xylene (abs 218.1) (23.985), yohimbine (abs 220.5) (11.51), zipeprol (abs 205.2) (13.45), zolpidem (abs 208.7) (11.882), zuclopenthixol (abs 206.4) (16.325) [abs = wavelength of maximum absorbance]

## KEY WORDS

whole blood

## REFERENCE

Gaillard, Y.; Pépin, G. Use of high-performance liquid chromatography with photodiode-array UV detection for the creation of a 600-compound library. Application to forensic toxicology, *J.Chromatogr.A*, **1997**, *763*, 149–163.

## SAMPLE

**Matrix:** cell cultures
**Sample preparation:** Add 20% MeCN to cell culture, let stand at room temperature for 15 min, centrifuge at 600 g for 5 min, inject a 400 μL aliquot of the supernatant.

## HPLC VARIABLES

**Column:** 150 × 4.6 5 μm Zorbax SB-C8
**Mobile phase:** Gradient. MeCN:water:acetic acid 40:60:1 for 2 min, to 90:10:1 over 8 min, maintain at 90:10:1 for 15 min
**Injection volume:** 400
**Detector:** UV 254

## CHROMATOGRAM

**Retention time:** 12

## OTHER SUBSTANCES

**Extracted:** metabolites

## REFERENCE

Charles, G.D.; Bartels, M.J.; Gennings, C.; Zacharewski, T.R.; Freshour, N.L.; Gollapudi, B.B.; Carney, E.W. Incorporation of S-9 activation into an ER-$\alpha$ transactivation assay, *Repro.Toxicol.*, **2000**, *14*, 207–216.

# Methyltestosterone

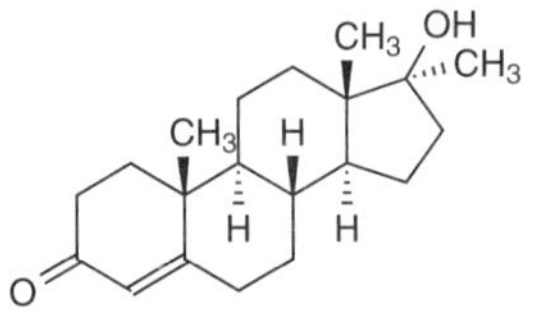


**Molecular formula:** $C_{20}H_{30}O_2$
**Molecular weight:** 302.45
**CAS Registry No:** 58-18-4
**Merck Index:** 13, 6148

## SAMPLE

**Matrix:** blood
**Sample preparation:** Condition a 1 mL 100 mg Bond Elut C1 SPE cartridge with 2 column vol of MeCN and 2 column vol of water. Mix 500 μL serum with 25 μL 1 μg/mL IS in MeCN, add 500 μL water, vortex for 5 s, add to the SPE cartridge, wash with 3 column vol of water, wash with 1 column vol of MeCN:water 20:80, air dry at high vacuum for 10 min. Elute with 1 mL hexane:chloroform 50:50 (Caution! Chloroform is a carcinogen!), evaporate the eluate to dryness under a stream of nitrogen at 45°, reconstitute with 50 μL MeCN, vortex while adding 200 μL water, inject a 50 μL aliquot.

## HPLC VARIABLES

**Guard column:** 15 × 4.6 7 μm silica (Brownlee)
**Column:** 220 × 4.6 5 μm silica (Brownlee)
**Column temperature:** 50
**Mobile phase:** MeCN:50 mM pH 2.5 sodium phosphate buffer 21:79
**Flow rate:** 1
**Injection volume:** 50
**Detector:** UV 247

## CHROMATOGRAM

**Retention time:** 5.8
**Internal standard:** testosterone propionate (9.1)
**Limit of detection:** 2 ng/mL

## KEY WORDS

serum: SPE

## REFERENCE

Lampert, B.L.; Stewart, J.T. Determination of anabolic steroids and zeranol in human serum by isocratic reverse phase HPLC on silica, *J.Liq.Chromatogr.*, **1989**, *12*, 3231–3249.

# Metrizamide

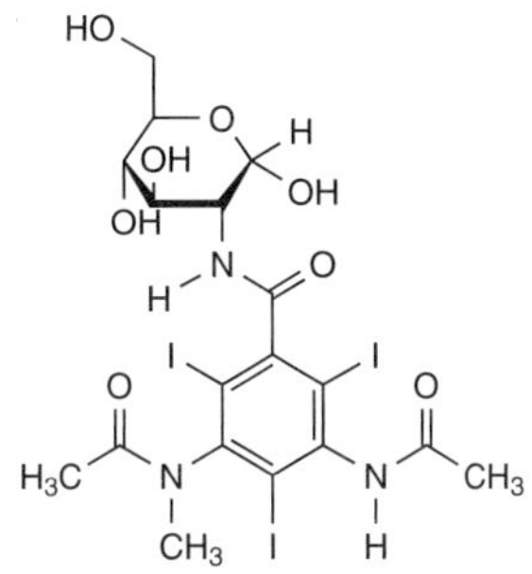

**Molecular formula:** $C_{18}H_{22}I_3N_3O_8$
**Molecular weight:** 789.10
**CAS Registry No:** 31112-62-6
**Merck Index:** 13, 6176

## SAMPLE

**Matrix:** blood
**Sample preparation:** Mix 1 mL plasma with 50 μL 1.5 mg/mL IS and 200 μL 20% zinc sulfate in water, add 200 μL saturated aqueous barium hydroxide, mix, heat on a steam bath for 2 min, cool, centrifuge, inject a 25 μL aliquot.

## HPLC VARIABLES

**Guard column:** 60 × 4 Corasil C18
**Column:** 250 × 4.6 Partisil 10 ODS
**Mobile phase:** MeOH:water:acetic acid 5:94.5:0.5
**Flow rate:** 2
**Injection volume:** 25
**Detector:** UV 244

## CHROMATOGRAM

**Retention time:** 3.9
**Internal standard:** 3-dimethylaminomethyleneimino-2,4,6-triiodobenzoic acid (11.2)
**Limit of detection:** 720 ng/mL
**Limit of quantitation:** 2.5 μg/mL

## KEY WORDS

pharmacokinetics; plasma

## REFERENCE

Kullberg, M.P.; Biddlecome, C.E.; Ross, R.W.; Edelson, J. High-performance liquid chromatographic determination of metrizamide in plasma, *J.Chromatogr.*, **1979**, *168*, 533–537.

# Metyrosine

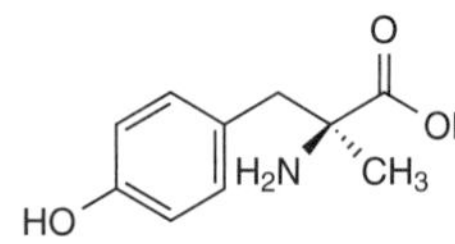


**Molecular formula:** $C_{10}H_{13}NO_3$
**Molecular weight:** 195.21
**CAS Registry No:** 672-87-7
**Merck Index:** 13, 6183

## SAMPLE

**Matrix:** blood
**Sample preparation:** Condition a 1 mL Bond Elut SCX SPE cartridge with 2 mL MeOH and 2 mL 100 mM HCl. Vortex 1 mL serum with 20 μL 100 μg/mL IS in MeOH and 1 mL water for 2 min, add to the SPE cartridge, wash with 2 mL water, air dry for 3 min, elute with four 250 μL portions of buffer, inject a 100 μL aliquot. (The buffer was 1 M dipotassium hydrogen phosphate adjusted to pH 5.0 with 1 M phosphoric acid.)

## HPLC VARIABLES

**Column:** 250 × 4.6 5 μm Spherisorb ODS2
**Mobile phase:** MeCN:MeOH:buffer 5:2.5:92.5 (Prepare mobile phase by dissolving 30 mg sodium 1-heptanesulfonate hydrate, 1.379 g sodium dihydrogen phosphate monohydrate, 2.5 mL MeOH, and 5 mL MeCN in 100 mL water. Adjust pH to 3.0 with 100 mM phosphoric acid.)
**Flow rate:** 1
**Injection volume:** 100
**Detector:** F ex 282 em 370

## CHROMATOGRAM

**Retention time:** 10.4
**Internal standard:** dopamine hydrochloride (8.9)
**Limit of detection:** 1 μg/mL

## OTHER SUBSTANCES

**Extracted:** methyldopa (LOD 100 ng/mL) (6.2)

## KEY WORDS

cow; serum; SPE

## REFERENCE

Hefnawy, M.M.; Stewart, J.T. Determination of metyrosine and its metabolite in serum by reversed phase high performance liquid chromatography using solid phase extraction and fluorescence detection, *J.Liq.Chromatogr.Rel.Technol.*, **1997**, *20*, 3009–3016.

## SAMPLE

**Matrix:** solutions
**Sample preparation:** Prepare metyrosine methyl esters by treatment of racemic metyrosine with absolute MeOH and concentrated sulfuric acid (*J.Am.Chem.Soc.* **1978**, 100, 6536). Dissolve 5 mg of the racemic metyrosine methyl esters in 5 mL MeCN:water 50:50 containing 0.5% v/v triethylamine. Heat a 50 μL aliquot of this solution at 45° for 15 min, add 10 μL 0.5% w/v 2,3,4,6-tetra-*O*-acetyl-*β*-D-glucopyranosyl isothiocyanate in MeCN, heat at 45° for 2 h, add 10 μL 0.5% v/v hydrazine in MeCN (Caution! Hydrazine is carcinogenic and toxic!), heat at 45° for 15 min. Evaporate the mixture under a stream of nitrogen, reconstitute the residue with 250 μL mobile phase, inject a 100 μL aliquot. (Prepare 2,3,4,6-tetra-*O*-acetyl-*β*-D-glucopyranosyl isothiocyanate solution and hydrazine solution fresh each day.)

## HPLC VARIABLES

**Column:** 250 × 4.6 5 μm Spherisorb ODS
**Mobile phase:** MeCN:water 35:65 containing 0.1% triethylamine, adjusted to pH 4.0 with trifluoroacetic acid
**Flow rate:** 1
**Injection volume:** 100
**Detector:** UV 250

## CHROMATOGRAM

**Retention time:** 7.32 (L), 8.73 (D)

## KEY WORDS

chiral; derivatization

## REFERENCE

Hefnawy, M.M.; Stewart, J.T. HPLC separation of metyrosine enantiomers as methyl esters derivatized with 2,3,4,6-tetra-*O*-acetyl-β-D-glucopyranosyl isothiocyanate, *J.Liq.Chromatogr.Rel.Technol.*, **1998**, *21*, 381–389.

# Micafungin

**Molecular formula:** $C_{56}H_{71}N_9O_{23}S$
**Molecular weight:** 1270.29
**CAS Registry No:** 235114-32-6, 208538-73-2 (Na salt)
**Merck Index:** 13, 6199

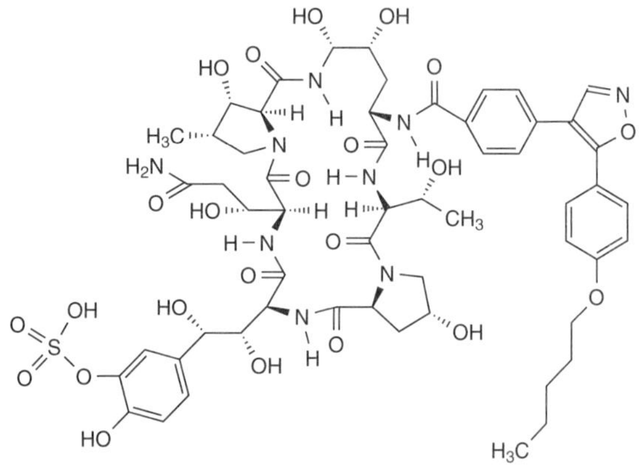

## SAMPLE
**Matrix:** blood
**Sample preparation:** Acidify plasma, add MeCN, centrifuge, dilute the supernatant with phosphate buffer, inject a 75 μL aliquot.

## HPLC VARIABLES
**Column:** 150 × 4.6 5 μm TSK-GEL ODS80TM
**Column temperature:** 50
**Mobile phase:** MeCN:20 mM potassium dihydrogen phosphate 41:59
**Flow rate:** 1
**Injection volume:** 75
**Detector:** F ex 273 em 464

## CHROMATOGRAM
**Retention time:** 12
**Limit of detection:** 10 ng/mL
**Limit of quantitation:** 100 ng/mL

## KEY WORDS
pharmacokinetics; plasma; rabbit

## REFERENCE
Groll, A.H.; Mickiene, D.; Petraitis, V.; Petraitiene, R.; Ibrahim, K.H.; Piscitelli, S.C.; Bekersky, I.; Walsh, T.J. Compartmental pharmacokinetics and tissue distribution of the antifungal echinocandin lipopeptide micafungin (FK463) in rabbits, *Antimicrob.Agents Chemother.*, **2001**, *45*, 3322–3327.

## SAMPLE
**Matrix:** tissue
**Sample preparation:** Homogenize (Ultra-Turrax 10N head) tissue with 4 vol of ice-cold pH 7.4 phosphate-buffered saline at 0° twice for 30 s each time, let stand at 4°C for 30 min, centrifuge at 2000 g for 10 min, add to a C8 SPE cartridge (Varian), wash and elute using MeCN/50 mM pH 4.0 ammonium acetate mixtures (unspecified). Evaporate the eluate to dryness under a stream of nitrogen at 40°, reconstitute the residue with MeOH:50 mM pH 4.0 ammonium acetate 50:50, inject a 75 μL aliquot.

## HPLC VARIABLES
**Column:** 150 × 4.6 5 μm Alltech Inertsil C8
**Mobile phase:** MeCN:20 mM pH 4.0 ammonium acetate 45:55
**Flow rate:** 0.75

**Injection volume:** 75
**Detector:** UV 273

## CHROMATOGRAM

**Retention time:** 8
**Limit of detection:** 10 ng/mL
**Limit of quantitation:** 50 ng/mL

## KEY WORDS

brain; kidney; liver; lung; plasma; SPE; spleen

## REFERENCE

Groll, A.H.; Mickiene, D.; Petraitis, V.; Petraitiene, R.; Ibrahim, K.H.; Piscitelli, S.C.; Bekersky, I.; Walsh, T.J. Compartmental pharmacokinetics and tissue distribution of the antifungal echinocandin lipopeptide micafungin (FK463) in rabbits, *Antimicrob.Agents Chemother.*, **2001**, *45*, 3322–3327.

# Milnacipran

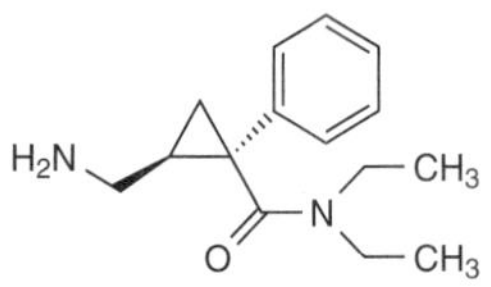

**Molecular formula:** $C_{15}H_{22}N_2O$
**Molecular weight:** 246.35
**CAS Registry No:** 92623-85-3
**Merck Index:** 13, 6216

## SAMPLE

**Matrix:** blood
**Sample preparation:** Vortex 1 mL whole blood with 3 mL water and 20 μL 100 μg/mL IS in MeOH for 1 min, add to a Toxi Tube A, shake gently for 5 min, centrifuge at 2800 rpm for 10 min. Evaporate the organic layer to dryness, reconstitute the residue with 100 μL MeOH, inject a 20 μL aliquot.

## HPLC VARIABLES

**Column:** 250 × 4.6 5 μm Symmetry (Waters)
**Column temperature:** 30
**Mobile phase:** MeCN:MeOH:100 mM ammonium acetate 30:30:40
**Flow rate:** 1
**Injection volume:** 20
**Detector:** UV 220

## CHROMATOGRAM

**Retention time:** 4.64
**Internal standard:** tetrazepam (21.7)
**Limit of quantitation:** 10 ng/mL

## OTHER SUBSTANCES

**Extracted:** bromazepam (5.7), nordiazepam (10.4)

## KEY WORDS

whole blood

## REFERENCE

Rop, P.P.; Sournac, M.H.; Burle, J.; Fornaris, M.; Coiffait, P.E. Blood concentration of milnacipran in a case of a fatal automobile accident, *J.Anal.Toxicol.*, **2002**, *26*, 123–126.

## SAMPLE

**Matrix:** solutions

## HPLC VARIABLES

**Column:** 250 × 4.6 5 μm Satisfaction RP 18 AB (CIL, France)
**Column temperature:** 45
**Mobile phase:** MeCN:buffer 35:65 (The buffer was 25 mM potassium dihydrogen phosphate and 10 mM triethylamine adjusted to pH 4.80 with phosphoric acid.)
**Flow rate:** 1
**Injection volume:** 20
**Detector:** UV 225

## CHROMATOGRAM

**Retention time:** 4.26

## OTHER SUBSTANCES

**Simultaneous:** citalopram (7.66), desmethylcitalopram (6.98), desmethylmirtazapine (3.63), desmethylvenlafaxine (3.16), fluoxetine (17.94), fluvoxamine (12.44), mirtazapine (5.14), norfluoxetine (15.22), paroxetine (10.24), sertraline (20.00), venlafaxine (4.74)

## REFERENCE

Dallet, P.; Labat, L.; Richard, M.; Langlois, M.H.; Dubost, J.P. A reversed-phase HPLC method development for the separation of new antidepressants, *J.Liq.Chromatogr.Rel.Technol.*, **2002**, *25*, 101–111.

## ANNOTATED BIBLIOGRAPHY

Duverneuil, C.; de la Grandmaison, G.L.; De Mazancourt, P.; Alvarez, J.-C. A high-performance liquid chromatography method with photodiode-array UV detection for therapeutic drug monitoring of the nontricyclic antidepressant drugs, *Ther.Drug Monit.*, **2003**, *25*, 565–573. [LOD 2.5–10 ng/mL; plasma; fluoxetine; norfluoxetine; sertraline; paroxetine; citalopram; fluvoxamine; moclobemide; mirtazapine; milnacipran; toloxatone; venlafaxine; viloxazine]

Titier, K.; Castaing, N.; Scotto-Gomez, E.; Pehourcq, F.; Moore, N.; Molimard, M. High-performance liquid chromatographic method with diode array detection for identification and quantification of the eight new antidepressants and five of their active metabolites in plasma after overdose, *Ther.Drug Monit.*, **2003**, *25*, 581–587. [LOQ 25–100 ng/mL; fluvoxamine; paroxetine; sertraline; fluoxetine; citalopram; mirtazapine; milnacipran; venlafaxine; norfluoxetine]

# Mirtazapine

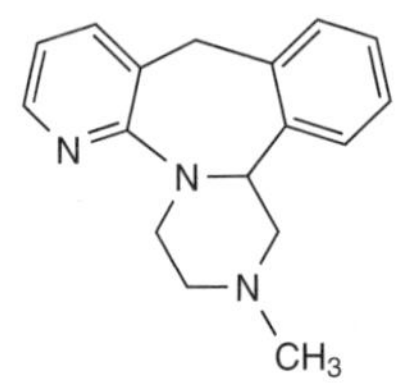

**Molecular formula:** $C_{17}H_{19}N_3$
**Molecular weight:** 265.35
**CAS Registry No:** 61337-67-5
**Merck Index:** 13, 6230

## SAMPLE

**Matrix:** blood
**Sample preparation:** Mix 1 mL whole blood with 20 μL 1 μg/mL dibenzepin in MeOH:water 50:50, add 300 μL pH 11 Tris buffer, mix, add 500 μL butyl acetate, vortex for 2 min, centrifuge. Preserve the aqueous layer (A). Remove the organic layer and add it to 75 μL 10 mM ammonium acetate buffer containing 0.1% formic acid (pH 3.2), evaporate (?). Add 75 μL MeCN, sonicate for 5 min, centrifuge at 5000 rpm for 5 min, keep the extract as B. Add 20 μL 1 μg/mL enalapril in MeOH:water 50:50 and 120 mg NaCl to the aqueous layer (A), mix, add 500 μL pH 3 phosphate buffer, add 600 μL 8.5% phosphoric acid, add 5 mL dichloromethane:isopropanol 95:5, shake at 250 cycles/min in a bench-top shaker for 30 min, centrifuge at 5000 rpm for 5 min. Remove the lower organic layer and evaporate it to dryness under a stream of air at 45°. Reconstitute the residue with 150 μL initial mobile phase, sonicate, centrifuge, combine with extract B, inject a 30 μL aliquot. (Sample preparation from Gergov,M.; Robson,J.N.; Ojanperä,I.; Heinonen,O.P.; Vuori,E. Simultaneous screening and quantitation of 18 antihistamine drugs in blood by liquid chromatography ionspray tandem mass spectrometry. *Forensic Sci.Inter.* **2001**, *121*, 108–115.)

## HPLC VARIABLES

**Guard column:** 40 mm long 4 μm Purospher RP-18 LiChro Cart 4-4
**Column:** 100 × 2.1 4 μm Genesis C18 (Jones Chromatography)
**Column temperature:** 35
**Mobile phase:** Gradient. MeCN:buffer from 20:80 to 100:0 over 10 min, maintain at 0:100 for 3 min, re-equilibrate at initial conditions for 5 min. (The buffer was 10 mM ammonium acetate containing 0.1% formic acid (pH 3.2).)
**Flow rate:** 0.2
**Injection volume:** 30
**Detector:** MS, PE Sciex API 365 triple-stage quadrupole LC-MS-MS, PE Sciex Turbo Ion Spray interface, positive ion mode, needle voltage 5.2 kV, nebulizer gas air at 60 psi, curtain gas nitrogen at 40 psi, collision cell gas nitrogen at 40 psi, turbo ionspray heater 375°, heater gas flow 7 L/min

## CHROMATOGRAM

**Retention time:** 4.4
**Internal standard:** dibenzepin, enalapril
**Limit of detection:** <20 ng/mL

## OTHER SUBSTANCES

**Extracted:** acebutolol (3.8, LOD 0.1 μg/mL), acrivastine (5.7, LOD <0.02 μg/mL), alprazolam (6.1, LOD <0.02 μg/mL), alprenolol (5.4, LOD 0.01 μg/mL), amantadine (3.4, LOD 0.1 μg/mL), amiloride (2.0, LOD 0.1 μg/mL), aminophenazone (2.8, LOD <5 μg/mL), amiodarone (10.2, LOD 0.05 μg/mL), amitriptyline (6.6, LOD <0.02 μg/mL), astemizole (5.8, LOD <0.02 μg/mL), atenolol (1.7, LOD 0.30 μg/mL), azacyclonol (5.1, LOD 0.02 μg/mL), benzhexol (6.6, LOD <0.02 μg/mL), benzoylecgonine (3.3, LOD 0.01 μg/mL), betaxolol (5.5, LOD 0.01 μg/mL), biperidine (6.2, LOD <0.02 μg/mL), bisoprolol (5.0, LOD <0.02 μg/mL), brompheniramine (5.3, LOD 0.002 μg/mL), bupivacaine (5.1, LOD <0.02 μg/mL), buprenorphine (5.9, LOD 0.01 μg/mL), buspirone (5.1, LOD 0.002 μg/mL), caffeine (2.8, LOD 1 μg/mL), carbamazepine

(6.1, LOD <0.02 μg/mL), carbinoxamine (5.1, LOD 0.002 μg/mL), carisoprodol (6.7, LOD <5 μg/mL), carvedilol (6.2, LOD <0.02 μg/mL), celiprolol (4.3, LOD 0.05 μg/mL), cetirizine (6.3, LOD 0.05 μg/mL), chlorcyclizine (6.6, LOD <0.02 μg/mL), chlordiazepoxide (5.7, LOD <0.02 μg/mL), chlormezanone (5.8, LOD <5 μg/mL), chloroquine (2.7, LOD 0.02 μg/mL), chlorpheniramine (5.1, LOD 0.002 μg/mL), chlorpromazine (7.0, LOD 0.02 μg/mL), chlorpropamide (6.7, LOD <5 μg/mL), chlorprothixene (7.0, LOD <0.02 μg/mL), cinnarizine (7.9, LOD <0.02 μg/mL), citalopram (5.7, LOD <0.02 μg/mL), clemastine (7.7, LOD 0.02 μg/mL), clobazam (7.3, LOD <0.02 μg/mL), clobutinol (5.3, LOD 0.02 μg/mL), clomethiazole (6.2, LOD 0.5 μg/mL), clomipramine (7.1, LOD <0.02 μg/mL), clonazepam (6.6, LOD <0.02 μg/mL), clonidine (2.8, LOD 0.1 μg/mL), clozapine (5.6, LOD <0.02 μg/mL), cocaine (4.6, LOD <0.02 μg/mL), codeine (2.5, LOD 0.1 μg/mL), coumatetralyl (8.4, LOD 0.05 μg/mL), cyclizine (5.8, LOD <0.02 μg/mL), dextropropoxyphene (6.6, LOD 0.05 μg/mL), demoxepam (5.8, LOD 0.02 μg/mL), dextromethorphan (5.5, LOD <0.02 μg/mL), diazepam (8.1, LOD 0.02 μg/mL), diltiazem (5.8, LOD <0.02 μg/mL), diphenhydramine (5.7, LOD <0.02 μg/mL), dipyridamole (5.4, LOD 0.005 μg/mL), disopyramine (4.4, LOD <0.02 μg/mL), dixyrazine (6.8, LOD 0.005 μg/mL), doxapram (4.8, LOD <0.02 μg/mL), doxepin (5.9, LOD <0.02 μg/mL), ebastine (9.6, LOD 0.005 μg/mL), embutramide (6.7, LOD 0.005 μg/mL), ergotamine (5.5, LOD 0.005 μg/mL), ethenzamide (5.0, LOD 0.05 μg/mL), ethylmorphine (3.2, LOD 0.05 μg/mL), ethylparathion (9.7, LOD <5 μg/mL), etodroxizine (6.4, LOD <0.02 μg/mL), felodipine (9.6, LOD 0.02 μg/mL), fenazepam (7.5, LOD <0.02 μg/mL), fenfluramine (5.3, LOD <0.02 μg/mL), fenkamfamine (5.1, LOD <0.02 μg/mL), fentanyl (5.5, LOD <0.02 μg/mL), fexofenadine (6.3, LOD <0.02 μg/mL), flecainide (5.9, LOD <0.02 μg/mL), fluconazole (4.0, LOD 0.1 μg/mL), flumazenil (5.2, LOD <0.02 μg/mL), flunitrazepam (7.1, LOD 0.002 μg/mL), fluoxetine (6.8, LOD 0.1 μg/mL), flupentixol (7.5, LOD 0.18 μg/mL), fluvoxamine (6.3, LOD 0.02 μg/mL), glibenclamide (8.5, LOD <0.02 μg/mL), glipizide (6.8, LOD <0.05 μg/mL), haloperidol (6.1, LOD <0.02 μg/mL), histapyrrodine (6.3, LOD 0.02 μg/mL), hydrocodone (3.0, LOD 0.05 μg/mL), hydroxychloroquine (2.4, LOD <0.3 μg/mL), hydroxyzine (6.3, LOD <0.02 μg/mL), imipramine (6.4, LOD 0.05 μg/mL), indomethacin (8.6, LOD 0.05 μg/mL), isoniazid (2.2, LOD 3 μg/mL), isradipine (8.6, LOD 0.05 μg/mL), ketamine (3.6, LOD <0.05 μg/mL), ketobemidone (3.3, LOD <0.05 μg/mL), ketoprofen (7.3, LOD 0.1 μg/mL), ketorolac (6.2, LOD 0.05 μg/mL), labetalol (4.9, LOD 0.05 μg/mL), lamotrigine (4.0, LOD 0.1 μg/mL), levocabastine (5.8, LOD 0.01 μg/mL), levomepromazine (6.5, LOD 0.02 μg/mL), lidocaine (3.7, LOD <0.05 μg/mL), loratadine (9.3, LOD 0.002 μg/mL), lorazepam (6.6, LOD 0.02 μg/mL), lormetazepam (7.4, LOD <0.02 μg/mL), LSD (4.7, LOD <0.02 μg/mL), malathion (8.9, LOD 10 μg/mL), maprotiline (6.4, LOD <0.02 μg/mL), MDMA (3.3, LOD 0.02 μg/mL), meclozine (8.5, LOD <0.02 μg/mL), medazepam (6.3, LOD <0.02 μg/mL), meloxicam (7.1, LOD 0.01 μg/mL), melperone (5.0, LOD <0.02 μg/mL), meperidine (4.7, LOD <0.02 μg/mL), mepivacaine (3.7, LOD <0.02 μg/mL), meprobamate (4.9, LOD 0.1 μg/mL), mesoridazine (5.4, LOD <0.02 μg/mL), methamphetamine (3.3, LOD 0.05 μg/mL), methadone (6.7, LOD <0.02 μg/mL), methylparathion (8.6, LOD 10 μg/mL), methylphenidate (4.2, LOD <0.02 μg/mL), metoclopramide (3.8, LOD <0.02 μg/mL), metoprolol (4.1, LOD 0.02 μg/mL), metronidazole (2.6, LOD 1 μg/mL), mexiletine (4.4, LOD 0.05 μg/mL), mianserin (5.7, LOD <0.02 μg/mL), midazolam (5.9, LOD <0.02 μg/mL), mizolastine (5.5, LOD 0.01 μg/mL), moclobemide (3.7, LOD 0.05 μg/mL), molindone (4.0, LOD <0.02 μg/mL), monoacetylmorphine (2.7, LOD 0.1 μg/mL), morphine (2.0, LOD 0.1 μg/mL), nicotine (2.2, LOD 0.05 μg/mL), nifedipine (7.5, LOD 0.02 μg/mL), nikethamide (3.6, LOD <0.02 μg/mL), nitrazepam (6.5, LOD <0.02 μg/mL), nizatidine (1.7, LOD 1 μg/mL), nomifensine (4.6, LOD <0.02 μg/mL), nortriptyline (6.4, LOD <0.02 μg/mL), norverapamil (6.2, LOD 1 μg/mL), noscapine (5.0, LOD <0.02 μg/mL), olanzapine (3.0, LOD 0.05 μg/mL), ondansetron (4.6, LOD <0.02 μg/mL), orphenadrine (6.1, LOD <0.02 μg/mL), oxazepam (6.3, LOD <0.02 μg/mL), oxcarbazepine (5.3, LOD 0.02 μg/mL), oxprenolol (4.7, LOD 0.02 μg/mL), oxycodone (2.8, LOD 0.05 μg/mL), papaverine (4.8, LOD <0.02 μg/mL), acetaminophen (2.5, LOD <5 μg/mL), paroxetine (6.2, LOD 0.02 μg/mL), pemoline (3.3, LOD 0.05 μg/mL), pentazocine (5.0, LOD <0.02 μg/mL), pentifylline (7.3, LOD <5 μg/mL), pentoxyverine (6.6, LOD <0.02 μg/mL), perphenazine (6.9, LOD 0.002 μg/mL), phenazone (3.9, LOD 0.05 μg/mL), phencyclidine (5.3, LOD 0.05 μg/mL), pheniramine

(4.1, LOD 0.02 µg/mL), phenylbutazone (9.0, LOD <5 µg/mL), phenylpropanolamine (2.5, LOD 0.3 µg/mL), phenytoin (6.1, LOD 0.05 µg/mL), pindolol (3.3, LOD 0.05 µg/mL), piroxicam (6.6, LOD 0.02 µg/mL), pitofenone (5.4, LOD <0.02 µg/mL), pizotifen (6.5, LOD <0.02 µg/mL), practolol (1.8, LOD 0.1 µg/mL), prazosin (4.1, LOD 0.05 µg/mL), prilocaine (3.8, LOD <0.02 µg/mL), primidone (4.0, LOD <5 µg/mL), procainamide (2.2, LOD 0.05 µg/mL), prochlorperazine (7.5, LOD 0.02 µg/mL), promazine (6.2, LOD <0.02 µg/mL), promethazine (6.0, LOD 0.05 µg/mL), propafenone (6.3, LOD <0.02 µg/mL), propranolol (5.4, LOD 0.02 µg/mL), propyphenazone (6.6, LOD 0.50 µg/mL), pseudoephedrine (2.6, LOD 1 µg/mL), quinine (4.2, LOD 0.02 µg/mL), ranitidine (1.8, LOD 0.1 µg/mL), risperidone (4.9, LOD <0.02 µg/mL), rocurone (3.8, LOD 0.1 µg/mL), ropivacaine (4.6, LOD <0.02 µg/mL), salicylamide (4.2, LOD <5 µg/mL), selegiline (4.1, LOD 0.05 µg/mL), sertindole (7.2, LOD <0.02 µg/mL), sertraline (6.8, LOD 0.02 µg/mL), sulindac (6.5, LOD 0.02 µg/mL), simazine (6.0, LOD 0.1 µg/mL), sincocaine (6.5, LOD <0.02 µg/mL), sisapride (5.9, LOD <0.02 µg/mL), sotalol (2.1, LOD 0.1 µg/mL), strychnine (5.3, LOD 0.05 µg/mL), sulpiride (1.9, LOD 0.1 µg/mL), sulthiame (4.1, LOD 0.05 µg/mL), temazepam (7.2, LOD <0.02 µg/mL), terbutaline (2.3, LOD 0.1 µg/mL), terfenadine (8.1, LOD 0.002 µg/mL), terodiline (6.7, LOD <0.02 µg/mL), tetracaine (5.7, LOD <0.02 µg/mL), dronabinol (12.3, LOD 0.05 µg/mL), tetrahydrozoline (3.6, LOD 0.1 µg/mL), theobromine (2.3, LOD <5 µg/mL), theophylline (2.4, LOD <5 µg/mL), thioridazine (7.5, LOD 0.02 µg/mL), timolol (3.8, LOD 0.05 µg/mL), thiothixene (6.7, LOD 0.02 µg/mL), tolbutamide (7.1, LOD <5 µg/mL), toremifene (8.7, LOD 0.02 µg/mL), tramadol (4.2, LOD 0.02 µg/mL), trazodone (5.2, LOD <0.02 µg/mL), triamterene (3.2, LOD 0.1 µg/mL), triazolam (6.7, LOD 0.002 µg/mL), trimeprazine (6.4, LOD <0.02 µg/mL), trimethoprim (3.1, LOD 0.05 µg/mL), trimipramine (6.7, LOD <0.02 µg/mL), venlafaxine (4.9, LOD 0.02 µg/mL), verapamil (6.5, LOD <0.02 µg/mL), warfarin (7.9, LOD <0.02 µg/mL), yohimbine (4.5, LOD <0.02 µg/mL), zolpidem (4.7, LOD <0.02 µg/mL), zopiclone (4.0, LOD 0.1 µg/mL)

## KEY WORDS

whole blood

## REFERENCE

Gergov, M.; Ojanperä, I.; Vuori, E. Simultaneous screening for 238 drugs in blood by liquid chromatography-ionspray tandem mass spectrometry with multiple-reaction monitoring, *J.Chromatogr.B*, **2003**, *795*, 41–53.

## SAMPLE

**Matrix:** blood

**Sample preparation:** Condition a 3 mL 130 mg Bond Elut Certify SPE cartridge with 3 mL MeOH, 3 mL water, and 1 mL 100 mM pH 6.0 potassium phosphate buffer. Vortex 1 mL plasma, 2 mL 100 mM pH 6.0 potassium phosphate buffer, and 100 µL 1 µg/mL IS in EtOH, add to the SPE cartridge, wash with 3 mL water, wash with 1 mL 1 M acetic acid, wash with 3 mL MeOH, draw air through the column for 5 min, elute with 3 mL dichloromethane:isopropanol:ammonium hydroxide 78:20:2. Evaporate the eluate to dryness under a stream of air at 55°, reconstitute the residue with 120 µL mobile phase, inject a 80 µL aliquot.

## HPLC VARIABLES

**Guard column:** 25 × 4.6 10 µm Chiralpak AD
**Column:** 50 × 4.6 10 µm Chiralpak AD
**Mobile phase:** Hexane:EtOH:isopropanol 98:1:1
**Flow rate:** 1.5
**Injection volume:** 80
**Detector:** UV 290

## CHROMATOGRAM

**Retention time:** 12.5 (+), 15.0 (−)

**Internal standard:** imipramine (3.8)
**Limit of quantitation:** 10 ng/mL

### KEY WORDS
chiral; plasma; SPE

### REFERENCE
Dodd, S.; Burrows, G.D.; Norman, T.R. Chiral determination of mirtazapine in human blood plasma by high-performance liquid chromatography, *J.Chromatogr.B*, **2000**, *748*, 439–443.

### SAMPLE
**Matrix:** formulations
**Sample preparation:** Extract a 300 mg amount powdered tablets with 100 mL MeCN:water 50:50, inject a 10 μL aliquot.

### HPLC VARIABLES
**Column:** 250 × 4.6 5 μm Hypersil ODS
**Column temperature:** 40
**Mobile phase:** MeCN:MeOH:THF:buffer 14.875:12.67:7.455:65 (The buffer was 18 g/L tetramethylammonium hydroxide pentahydrate in water adjusted to pH 7.4 with 85% phosphoric acid.)
**Flow rate:** 1.5
**Injection volume:** 10
**Detector:** UV 210

### CHROMATOGRAM
**Limit of detection:** 0.01–0.04%
**Limit of quantitation:** 0.02–0.12%

### OTHER SUBSTANCES
**Simultaneous:** impurities

### KEY WORDS
comparison with capillary electrophoresis; tablets

### REFERENCE
Wynia, G.S.; Windhorst, G.; Post, P.C.; Maris, F.A. Development and validation of a capillary electrophoresis method within a pharmaceutical quality control environment and comparison with high-performance liquid chromatography, *J.Chromatogr.A*, **1997**, *773*, 339–350.

## ANNOTATED BIBLIOGRAPHY

Dallet, P.; Labat, L.; Richard, M.; Langlois, M.H.; Dubost, J.P. A reversed-phase HPLC method development for the separation of new antidepressants, *J.Liq.Chromatogr.Rel.Technol.*, **2002**, *25*, 101–111. [fluvoxamine; fluoxetine; sertraline; paroxetine; citalopram; venlafaxine; milnacipran; mirtazapine]

Duverneuil, C.; de la Grandmaison, G.L.; De Mazancourt, P.; Alvarez, J.-C. A high-performance liquid chromatography method with photodiode-array UV detection for therapeutic drug monitoring of the nontricyclic antidepressant drugs, *Ther.Drug Monit.*, **2003**, *25*, 565–573. [LOD 2.5–10 ng/mL; plasma; fluoxetine; norfluoxetine; sertraline; paroxetine; citalopram; fluvoxamine; moclobemide; mirtazapine; milnacipran; toloxatone; venlafaxine; viloxazine]

Frahnert, C.; Rao, M.L.; Grasmäder, K. Analysis of eighteen antidepressants, four atypical antipsychotics and active metabolites in serum by liquid chromatography: a simple tool for therapeutic drug monitoring, *J.Chromatogr.B*, **2003**, *794*, 35–47. [SPE; sulpiride; moclobemide; amisulpride; venlafaxine; normirtazapine; melperone; reboxetine; zolpidem; nordoxepin; diazepam; risperidone; benperidol; normaprotiline; dibenzepine; opipramol; fluvoxamine; quetiapine; desipramine; citalopram; norfluoxetine; norclozapine; nortriptyline; haloperidol; paroxetine; maprotiline; mirtazapine; fluoxetine; doxepin; norclomipramine; imipramine; trifluperidol; olanzapine; trimipramine; amitriptyline;

ziprasidone; promethazine; mianserin; clomipramine; clozapine; fluphenazine; nefazodone; sertraline; chlorprothixene; thioridazine; pimozide; LOQ 5–10 ng/mL]

Labat, L.; Dallet, P.; Kummer, E.; Dubost, J.P. Spectrophotometric, spectrofluorimetric, HPLC and CZE determination of mirtazapine in pharmaceutical tablets, *J.Pharm.Biomed.Anal.*, **2002**, *28*, 365–371. [pirenzepine is internal standard]

Maris, F.A.; Dingler, E.; Niehues, S. High-performance liquid chromatographic assay with fluorescence detection for the routine monitoring of the antidepressant mirtazapine and its demethyl metabolite in human plasma, *J.Chromatogr.B*, **1999**, *721*, 309–316. [LOQ 0.5 ng/mL]

Moody, J.D.; Freeman, J.P.; Fu, P.P.; Cerniglia, C.E. Biotransformation of mirtazapine by Cunninghamella elegans, *Drug Metab.Dispos.*, **2002**, *30*, 1274–1279.

Morgan, P.E.; Tapper, J.; Spencer, E.P. Measurement of total mirtazapine and normirtazapine in plasma/serum by liquid chromatography with fluorescence detection, *J.Chromatogr.B*, **2003**, *798*, 211–215. [LOQ 1 ng/mL; imipramine is internal standard]

Ptácek, P.; Klíma, J.; Macek, J. Determination of mirtazapine in human plasma by liquid chromatography, *J.Chromatogr.B*, **2003**, *794*, 323–328. [fluorescence detection; LOQ 1.5 ng/mL; zolpidem is internal standard]

Romiguieres, T.; Pehourcq, F.; Matoga, M.; Begaud, B.; Jarry, C. Determination of mirtazapine and its demethyl metabolite in plasma by high-performance liquid chromatography with ultraviolet detection; Application to management of acute intoxication, *J.Chromatogr.B*, **2002**, *775*, 163–168. [LOQ 20 ng/mL; opipramol is internal standard]

Thieme, D.; Sachs, H. Improved screening capabilities in forensic toxicology by application of liquid chromatography-tandem mass spectrometry, *Anal.Chim.Acta*, **2003**, *492*, 171–186. [alprazolam; dothiepin; piritramide; cocaine; lorazepam; lormetazepam; clonazepam; flunitrazepam; bromazepam; midazolam; flurazepam; nitrazepam; temazepam; medazepam; nordazepam; diazepam; methylclonazepam; triazolam; oxazepam; haloperidol; benperidol; sulpiride; amisulpride; mirtazapine; citalopram; olanzapine; paroxetine; fluoxetine; sertraline; zopiclone; zolpidem; risperidone; quetiapine; fentanyl; pipamperone; meperidine; buprenorphine; propoxyphene; pentazocine; phenazocine; EDDP; tilidine; methadone; morphine; codeine; dihydrocodeine; acetylmorphine; amphetamine; ephedrine; norephedrine; pseudoephedrine; methylephedrine; amphetaminil; benzphetamine; methylphenidate; nikethamide; amfeparone; clobenzorex; atropine; scopolamine; ajmaline; aconitine; colchicine; strychnine; metoprolol; acebutolol; propranolol; sotalol; atenolol; bisoprolol; amiloride; triamterene; warfarin; brodifacoum; coumatetralyl; phenprocoumon; methaqualone; clomethiazole; acetaminophen; methoxamine; vecuronium; neostigmine; LSD]

Titier, K.; Castaing, N.; Scotto-Gomez, E.; Pehourcq, F.; Moore, N.; Molimard, M. High-performance liquid chromatographic method with diode array detection for identification and quantification of the eight new antidepressants and five of their active metabolites in plasma after overdose, *Ther.Drug Monit.*, **2003**, *25*, 581–587. [LOQ 25–100 ng/mL; fluvoxamine; paroxetine; sertraline; fluoxetine; citalopram; mirtazapine; milnacipran; venlafaxine; norfluoxetine]

# Misoprostol

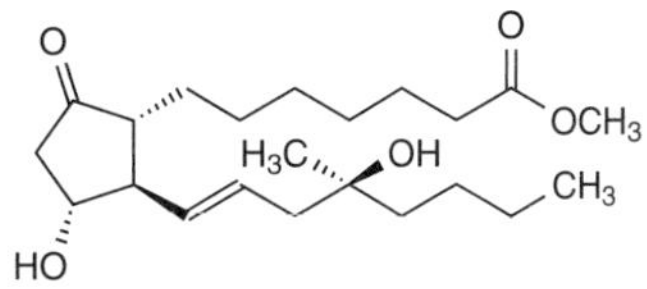


**Molecular formula:** $C_{22}H_{38}O_5$
**Molecular weight:** 382.53
**CAS Registry No:** 59122-46-2
**Merck Index:** 13, 6232

## SAMPLE

**Matrix:** blood
**Sample preparation:** Mix 200 μL plasma with 200 μL 100 ppm 2-naphthoic acid and 300 μL MeCN, shake well, centrifuge for 10 min, inject a 20 μL aliquot of the supernatant.

## HPLC VARIABLES

**Column:** 250 × 4.6 5 μm Zorbax C8
**Mobile phase:** MeCN:MeOH:20 mM pH 3.0 potassium phosphate buffer 30:24:46
**Flow rate:** 1.7
**Injection volume:** 20
**Detector:** UV 210; Radioactivity ($^3H$)

## CHROMATOGRAM

**Retention time:** 18.5 (diastereomers not resolved)
**Internal standard:** 2-naphthoic acid (6)
**Limit of quantitation:** LOQ 1 μg (UV), 12 pg (radioactivity detector)

## OTHER SUBSTANCES

**Extracted:** iloprost (LOQ 500 ng (UV), 42 pg (radioactivity detector)) (14.7, 15.9 (diastereomers))

## KEY WORDS

plasma; rat

## REFERENCE

Womack, I.M.; Lee, A.S.; Kamath, B.; Agrawal, K.C.; Kishore, V. A high performance liquid radiochromatographic assay for the simultaneous analysis of iloprost and misoprostol, *Prostaglandins*, **1996**, *52*, 249–259.

# Mizolastine

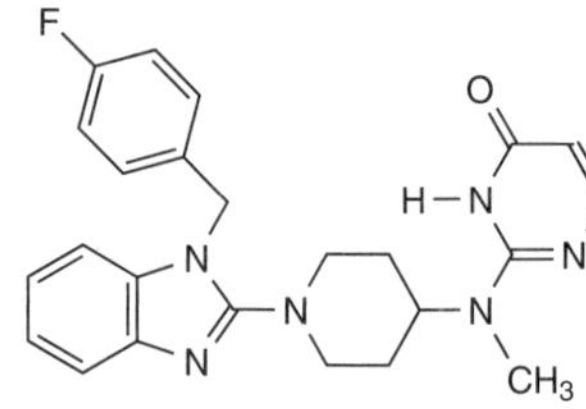


**Molecular formula:** $C_{24}H_{25}FN_6O$
**Molecular weight:** 432.49
**CAS Registry No:** 108612-45-9
**Merck Index:** 13, 6242

## SAMPLE

**Matrix:** blood

**Sample preparation:** Mix 1 mL whole blood with 20 μL 1 μg/mL dibenzepin in MeOH:water 50:50, add 300 μL pH 11 Tris buffer, mix, add 500 μL butyl acetate, vortex for 2 min, centrifuge. Preserve the aqueous layer (A). Remove the organic layer and add it to 75 μL 10 mM ammonium acetate buffer containing 0.1% formic acid (pH 3.2), evaporate (?). Add 75 μL MeCN, sonicate for 5 min, centrifuge at 5000 rpm for 5 min, keep the extract as B. Add 20 μL 1 μg/mL enalapril in MeOH:water 50:50 and 120 mg NaCl to the aqueous layer (A), mix, add 500 μL pH 3 phosphate buffer, add 600 μL 8.5% phosphoric acid, add 5 mL dichloromethane:isopropanol 95:5, shake at 250 cycles/min in a bench-top shaker for 30 min, centrifuge at 5000 rpm for 5 min. Remove the lower organic layer and evaporate it to dryness under a stream of air at 45°. Reconstitute the residue with 150 μL initial mobile phase, sonicate, centrifuge, combine with extract B, inject a 30 μL aliquot. (Sample preparation from Gergov,M.; Robson,J.N.; Ojanperä,I.; Heinonen,O.P.; Vuori,E. Simultaneous screening and quantitation of 18 antihistamine drugs in blood by liquid chromatography ionspray tandem mass spectrometry. *Forensic Sci.Inter.* **2001**, *121*, 108–115.)

## HPLC VARIABLES

**Guard column:** 40 mm long 4 μm Purospher RP-18 LiChro Cart 4-4
**Column:** 100 × 2.1 4 μm Genesis C18 (Jones Chromatography)
**Column temperature:** 35
**Mobile phase:** Gradient. MeCN:buffer from 20:80 to 100:0 over 10 min, maintain at 0:100 for 3 min, re-equilibrate at initial conditions for 5 min. (The buffer was 10 mM ammonium acetate containing 0.1% formic acid (pH 3.2).)
**Flow rate:** 0.2
**Injection volume:** 30
**Detector:** MS, PE Sciex API 365 triple-stage quadrupole LC-MS-MS, PE Sciex Turbo Ion Spray interface, positive ion mode, needle voltage 5.2 kV, nebulizer gas air at 60 psi, curtain gas nitrogen at 40 psi, collision cell gas nitrogen at 40 psi, turbo ionspray heater 375°, heater gas flow 7 L/min

## CHROMATOGRAM

**Retention time:** 5.5
**Internal standard:** dibenzepin, enalapril
**Limit of detection:** 10 ng/mL

## OTHER SUBSTANCES

**Extracted:** acebutolol (3.8, LOD 0.1 μg/mL), acrivastine (5.7, LOD <0.02 μg/mL), alprazolam (6.1, LOD <0.02 μg/mL), alprenolol (5.4, LOD 0.01 μg/mL), amantadine (3.4, LOD 0.1 μg/mL), amiloride (2.0, LOD 0.1 μg/mL), aminophenazone (2.8, LOD <5 μg/mL), amiodarone (10.2, LOD 0.05 μg/mL), amitriptyline (6.6, LOD <0.02 μg/mL), astemizole (5.8, LOD <0.02 μg/mL), atenolol (1.7, LOD 0.30 μg/mL), azacyclonol (5.1, LOD 0.02 μg/mL), benzhexol (6.6, LOD <0.02 μg/mL), benzoylecgonine (3.3, LOD 0.01 μg/mL), betaxolol (5.5, LOD 0.01 μg/mL), biperidine (6.2, LOD <0.02 μg/mL), bisoprolol (5.0, LOD <0.02 μg/mL), brompheniramine (5.3, LOD 0.002 μg/mL), bupivacaine (5.1, LOD <0.02 μg/mL), buprenorphine (5.9, LOD 0.01 μg/mL), buspirone (5.1, LOD 0.002 μg/mL), caffeine (2.8, LOD 1 μg/mL), carbamazepine

(6.1, LOD <0.02 μg/mL), carbinoxamine (5.1, LOD 0.002 μg/mL), carisoprodol (6.7, LOD <5 μg/mL), carvedilol (6.2, LOD <0.02 μg/mL), celiprolol (4.3, LOD 0.05 μg/mL), cetirizine (6.3, LOD 0.05 μg/mL), chlorcyclizine (6.6, LOD <0.02 μg/mL), chlordiazepoxide (5.7, LOD <0.02 μg/mL), chlormezanone (5.8, LOD <5 μg/mL), chloroquine (2.7, LOD 0.02 μg/mL), chlorpheniramine (5.1, LOD 0.002 μg/mL), chlorpromazine (7.0, LOD 0.02 μg/mL), chlorpropamide (6.7, LOD <5 μg/mL), chlorprothixene (7.0, LOD <0.02 μg/mL), cinnarizine (7.9, LOD <0.02 μg/mL), citalopram (5.7, LOD <0.02 μg/mL), clemastine (7.7, LOD 0.02 μg/mL), clobazam (7.3, LOD <0.02 μg/mL), clobutinol (5.3, LOD 0.02 μg/mL), clomethiazole (6.2, LOD 0.5 μg/mL), clomipramine (7.1, LOD <0.02 μg/mL), clonazepam (6.6, LOD <0.02 μg/mL), clonidine (2.8, LOD 0.1 μg/mL), clozapine (5.6, LOD <0.02 μg/mL), cocaine (4.6, LOD <0.02 μg/mL), codeine (2.5, LOD 0.1 μg/mL), coumatetralyl (8.4, LOD 0.05 μg/mL), cyclizine (5.8, LOD <0.02 μg/mL), dextropropoxyphene (6.6, LOD 0.05 μg/mL), demoxepam (5.8, LOD 0.02 μg/mL), dextromethorphan (5.5, LOD <0.02 μg/mL), diazepam (8.1, LOD 0.02 μg/mL), diltiazem (5.8, LOD <0.02 μg/mL), diphenhydramine (5.7, LOD <0.02 μg/mL), dipyridamole (5.4, LOD 0.005 μg/mL), disopyramine (4.4, LOD <0.02 μg/mL), dixyrazine (6.8, LOD 0.005 μg/mL), doxapram (4.8, LOD <0.02 μg/mL), doxepin (5.9, LOD <0.02 μg/mL), ebastine (9.6, LOD 0.005 μg/mL), embutramide (6.7, LOD 0.005 μg/mL), ergotamine (5.5, LOD 0.005 μg/mL), ethenzamide (5.0, LOD 0.05 μg/mL), ethylmorphine (3.2, LOD 0.05 μg/mL), ethylparathion (9.7, LOD <5 μg/mL), etodroxizine (6.4, LOD <0.02 μg/mL), felodipine (9.6, LOD 0.02 μg/mL), fenazepam (7.5, LOD <0.02 μg/mL), fenfluramine (5.3, LOD <0.02 μg/mL), fenkamfamine (5.1, LOD <0.02 μg/mL), fentanyl (5.5, LOD <0.02 μg/mL), fexofenadine (6.3, LOD <0.02 μg/mL), flecainide (5.9, LOD <0.02 μg/mL), fluconazole (4.0, LOD 0.1 μg/mL), flumazenil (5.2, LOD <0.02 μg/mL), flunitrazepam (7.1, LOD 0.002 μg/mL), fluoxetine (6.8, LOD 0.1 μg/mL), flupentixol (7.5, LOD 0.18 μg/mL), fluvoxamine (6.3, LOD 0.02 μg/mL), glibenclamide (8.5, LOD <0.02 μg/mL), glipizide (6.8, LOD <0.05 μg/mL), haloperidol (6.1, LOD <0.02 μg/mL), histapyrrodine (6.3, LOD 0.02 μg/mL), hydrocodone (3.0, LOD 0.05 μg/mL), hydroxychloroquine (2.4, LOD <0.3 μg/mL), hydroxyzine (6.3, LOD <0.02 μg/mL), imipramine (6.4, LOD 0.05 μg/mL), indomethacin (8.6, LOD 0.05 μg/mL), isoniazid (2.2, LOD 3 μg/mL), isradipine (8.6, LOD 0.05 μg/mL), ketamine (3.6, LOD <0.05 μg/mL), ketobemidone (3.3, LOD <0.05 μg/mL), ketoprofen (7.3, LOD 0.1 μg/mL), ketorolac (6.2, LOD 0.05 μg/mL), labetalol (4.9, LOD 0.05 μg/mL), lamotrigine (4.0, LOD 0.1 μg/mL), levocabastine (5.8, LOD 0.01 μg/mL), levomepromazine (6.5, LOD 0.02 μg/mL), lidocaine (3.7, LOD <0.05 μg/mL), loratadine (9.3, LOD 0.002 μg/mL), lorazepam (6.6, LOD 0.02 μg/mL), lormetazepam (7.4, LOD <0.02 μg/mL), LSD (4.7, LOD <0.02 μg/mL), malathion (8.9, LOD 10 μg/mL), maprotiline (6.4, LOD <0.02 μg/mL), MDMA (3.3, LOD 0.02 μg/mL), meclozine (8.5, LOD <0.02 μg/mL), medazepam (6.3, LOD <0.02 μg/mL), meloxicam (7.1, LOD 0.01 μg/mL), melperone (5.0, LOD <0.02 μg/mL), meperidine (4.7, LOD <0.02 μg/mL), mepivacaine (3.7, LOD <0.02 μg/mL), meprobamate (4.9, LOD 0.1 μg/mL), mesoridazine (5.4, LOD <0.02 μg/mL), methamphetamine (3.3, LOD 0.05 μg/mL), methadone (6.7, LOD <0.02 μg/mL), methylparathion (8.6, LOD 10 μg/mL), methylphenidate (4.2, LOD <0.02 μg/mL), metoclopramide (3.8, LOD <0.02 μg/mL), metoprolol (4.1, LOD 0.02 μg/mL), metronidazole (2.6, LOD 1 μg/mL), mexiletine (4.4, LOD 0.05 μg/mL), mianserin (5.7, LOD <0.02 μg/mL), midazolam (5.9, LOD <0.02 μg/mL), mirtazapine (4.4, LOD <0.02 μg/mL), moclobemide (3.7, LOD 0.05 μg/mL), molindone (4.0, LOD <0.02 μg/mL), monoacetylmorphine (2.7, LOD 0.1 μg/mL), morphine (2.0, LOD 0.1 μg/mL), nicotine (2.2, LOD 0.05 μg/mL), nifedipine (7.5, LOD 0.02 μg/mL), nikethamide (3.6, LOD <0.02 μg/mL), nitrazepam (6.5, LOD <0.02 μg/mL), nizatidine (1.7, LOD 1 μg/mL), nomifensine (4.6, LOD <0.02 μg/mL), nortriptyline (6.4, LOD <0.02 μg/mL), norverapamil (6.2, LOD 1 μg/mL), noscapine (5.0, LOD <0.02 μg/mL), olanzapine (3.0, LOD 0.05 μg/mL), ondansetron (4.6, LOD <0.02 μg/mL), orphenadrine (6.1, LOD <0.02 μg/mL), oxazepam (6.3, LOD <0.02 μg/mL), oxcarbazepine (5.3, LOD 0.02 μg/mL), oxprenolol (4.7, LOD 0.02 μg/mL), oxycodone (2.8, LOD 0.05 μg/mL), papaverine (4.8, LOD <0.02 μg/mL), acetaminophen (2.5, LOD <5 μg/mL), paroxetine (6.2, LOD 0.02 μg/mL), pemoline (3.3, LOD 0.05 μg/mL), pentazocine (5.0, LOD <0.02 μg/mL), pentifylline (7.3, LOD <5 μg/mL), pentoxyverine (6.6, LOD <0.02 μg/mL), perphenazine (6.9, LOD 0.002 μg/mL), phenazone (3.9, LOD 0.05 μg/mL), phencyclidine (5.3, LOD 0.05 μg/mL), pheniramine

(4.1, LOD 0.02 μg/mL), phenylbutazone (9.0, LOD <5 μg/mL), phenylpropanolamine (2.5, LOD 0.3 μg/mL), phenytoin (6.1, LOD 0.05 μg/mL), pindolol (3.3, LOD 0.05 μg/mL), piroxicam (6.6, LOD 0.02 μg/mL), pitofenone (5.4, LOD <0.02 μg/mL), pizotifen (6.5, LOD <0.02 μg/mL), practolol (1.8, LOD 0.1 μg/mL), prazosin (4.1, LOD 0.05 μg/mL), prilocaine (3.8, LOD <0.02 μg/mL), primidone (4.0, LOD <5 μg/mL), procainamide (2.2, LOD 0.05 μg/mL), prochlorperazine (7.5, LOD 0.02 μg/mL), promazine (6.2, LOD <0.02 μg/mL), promethazine (6.0, LOD 0.05 μg/mL), propafenone (6.3, LOD <0.02 μg/mL), propranolol (5.4, LOD 0.02 μg/mL), propyphenazone (6.6, LOD 0.50 μg/mL), pseudoephedrine (2.6, LOD 1 μg/mL), quinine (4.2, LOD 0.02 μg/mL), ranitidine (1.8, LOD 0.1 μg/mL), risperidone (4.9, LOD <0.02 μg/mL), rocurone (3.8, LOD 0.1 μg/mL), ropivacaine (4.6, LOD <0.02 μg/mL), salicylamide (4.2, LOD <5 μg/mL), selegiline (4.1, LOD 0.05 μg/mL), sertindole (7.2, LOD <0.02 μg/mL), sertraline (6.8, LOD 0.02 μg/mL), sulindac (6.5, LOD 0.02 μg/mL), simazine (6.0, LOD 0.1 μg/mL), sincocaine (6.5, LOD <0.02 μg/mL), sisapride (5.9, LOD <0.02 μg/mL), sotalol (2.1, LOD 0.1 μg/mL), strychnine (5.3, LOD 0.05 μg/mL), sulpiride (1.9, LOD 0.1 μg/mL), sulthiame (4.1, LOD 0.05 μg/mL), temazepam (7.2, LOD <0.02 μg/mL), terbutaline (2.3, LOD 0.1 μg/mL), terfenadine (8.1, LOD 0.002 μg/mL), terodiline (6.7, LOD <0.02 μg/mL), tetracaine (5.7, LOD <0.02 μg/mL), dronabinol (12.3, LOD 0.05 μg/mL), tetrahydrozoline (3.6, LOD 0.1 μg/mL), theobromine (2.3, LOD <5 μg/mL), theophylline (2.4, LOD <5 μg/mL), thioridazine (7.5, LOD 0.02 μg/mL), timolol (3.8, LOD 0.05 μg/mL), thiothixene (6.7, LOD 0.02 μg/mL), tolbutamide (7.1, LOD <5 μg/mL), toremifene (8.7, LOD 0.02 μg/mL), tramadol (4.2, LOD 0.02 μg/mL), trazodone (5.2, LOD <0.02 μg/mL), triamterene (3.2, LOD 0.1 μg/mL), triazolam (6.7, LOD 0.002 μg/mL), trimeprazine (6.4, LOD <0.02 μg/mL), trimethoprim (3.1, LOD 0.05 μg/mL), trimipramine (6.7, LOD <0.02 μg/mL), venlafaxine (4.9, LOD 0.02 μg/mL), verapamil (6.5, LOD <0.02 μg/mL), warfarin (7.9, LOD <0.02 μg/mL), yohimbine (4.5, LOD <0.02 μg/mL), zolpidem (4.7, LOD <0.02 μg/mL), zopiclone (4.0, LOD 0.1 μg/mL)

## KEY WORDS
whole blood

## REFERENCE
Gergov, M.; Ojanperä, I.; Vuori, E. Simultaneous screening for 238 drugs in blood by liquid chromatography-ionspray tandem mass spectrometry with multiple-reaction monitoring, *J.Chromatogr.B*, **2003**, *795*, 41–53.

## SAMPLE
**Matrix:** blood
**Sample preparation:** Mix 1 g plasma with 20 μL 15 μg/mL IS in MeOH, add 200 μL MeCN, vortex, centrifuge at 11 000 g for 4 min, inject a 200 μL aliquot onto column A and elute to waste with mobile phase A; after 2 min, backflush the contents of column A onto column B with mobile phase B; after 1.5 min, remove column A from the circuit, elute column B with mobile phase B, monitor the effluent from column B. Backflush column A with MeCN:water 50:50, MeCN, MeOH:water 50:50, re-equilibrate with mobile phase A.

## HPLC VARIABLES
**Column:** A 75 × 2.1 30–40 μm Perisorb C18; B 20 × 4.6 40 μm Pelliguard C18 (Supelchem) + 150 × 4.6 5 μm Hypersil BDS C8
**Mobile phase:** A MeCN:water 10:90; B MeCN:25 mM pH 4.5 phosphate buffer 40:60
**Flow rate:** 1
**Injection volume:** 200
**Detector:** UV 285

## CHROMATOGRAM
**Retention time:** 6.6
**Internal standard:** SL 86.0116 (chloro analogue of mizolastine) (8.2)
**Limit of quantitation:** 2.5 ng/mL

## KEY WORDS

column-switching; comparison with liquid–liquid extraction and SPE; plasma

## REFERENCE

Ascalone, V.; Guinebault, P.; Rouchouse, A. Determination of mizolastine, a new antihistaminic drug, in human plasma by liquid-liquid extraction and column-switching techniques in combination with high-performance liquid chromatography, *J.Chromatogr.*, **1993**, *619*, 275–284.

## SAMPLE

**Matrix:** blood

**Sample preparation:** Mix 1 g plasma with 20 μL 15 μg/mL IS in MeOH, add 500 μL 700 mM pH 9 borate buffer, add 7 mL diethyl ether, shake on a tumble extractor at 40 rpm for 10 min, centrifuge at 2000 g at 5° for 5 min. Remove the organic layer and add it to 25 mM pH 2.6 phosphate buffer, shake on a tumble extractor at 20 rpm for 10 min. Discard the organic layer, remove traces of ether from the aqueous layer with a stream of nitrogen at 40°, inject a 150 μL aliquot. (Prepare 700 mM pH 9 borate buffer by dissolving 6.18 g boric acid and 7.46 g KCl in 100 mL water and adjusting to pH 9 with ca. 50 mL 1 M NaOH.)

## HPLC VARIABLES

**Guard column:** 20 × 4.6 40 μm Pelliguard C18 (Supelchem)
**Column:** 150 × 4.6 5 μm Hypersil BDS C8
**Mobile phase:** MeCN:25 mM pH 2.5 phosphate buffer:triethylamine 35:65:0.1
**Flow rate:** 1
**Injection volume:** 150
**Detector:** UV 285

## CHROMATOGRAM

**Retention time:** 3.6
**Internal standard:** SL 86.0116 (chloro analogue of mizolastine) (4.6)
**Limit of quantitation:** 0.5 ng/mL

## KEY WORDS

comparison with column-switching and SPE; liquid–liquid extraction; plasma

## REFERENCE

Ascalone, V.; Guinebault, P.; Rouchouse, A. Determination of mizolastine, a new antihistaminic drug, in human plasma by liquid-liquid extraction and column-switching techniques in combination with high-performance liquid chromatography, *J.Chromatogr.*, **1993**, *619*, 275–284.

## SAMPLE

**Matrix:** blood

**Sample preparation:** Condition a 100 mg 40 μm Bakerbond CN SPE cartridge with 1 mL MeOH saturated with ammonium dihydrogen phosphate and 2 mL water. Mix 1 mL plasma with 500 μL 200 ng/mL IS in MeCN:water, mix, add a 1.4 mL aliquot to the SPE cartridge, wash twice with 2 mL portions of water, wash twice with 1 mL portions of MeOH:water 30:70, dry, elute with 600 μL MeOH saturated with ammonium dihydrogen phosphate, mix the eluate (ca. 300 μL) with 400 μL 50 mM pH 4.6 phosphate buffer, inject a 650 μL aliquot.

## HPLC VARIABLES

**Column:** 250 × 4.6 5 μm Ultrabase C8
**Mobile phase:** MeCN:50 mM pH 4.6 phosphate buffer:triethylamine 45:55:0.1
**Flow rate:** 1
**Injection volume:** 650
**Detector:** UV 285

## CHROMATOGRAM
**Retention time:** 8
**Internal standard:** SL 86.0116 (chloro analogue of mizolastine) (10)
**Limit of quantitation:** 1 ng/mL

## KEY WORDS
comparison with column-switching and liquid–liquid extraction; plasma; SPE

## REFERENCE
Ascalone, V.; Guinebault, P.; Rouchouse, A. Determination of mizolastine, a new antihistaminic drug, in human plasma by liquid-liquid extraction and column-switching techniques in combination with high-performance liquid chromatography, *J.Chromatogr.*, **1993**, *619*, 275–284.

# Moexipril

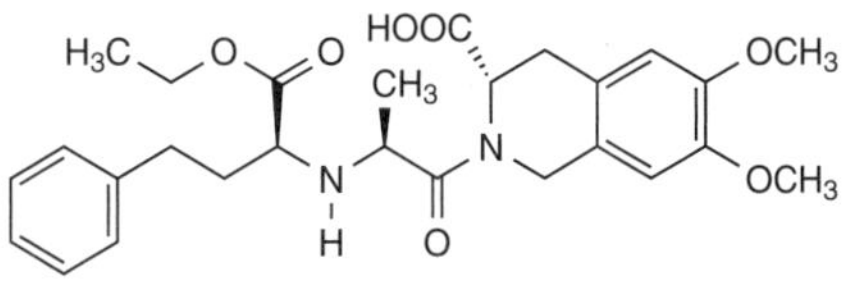


**Molecular formula:** $C_{27}H_{34}N_2O_7$
**Molecular weight:** 498.57
**CAS Registry No:** 103775-10-6, 82586-52-5 (HCl)
**Merck Index:** 13, 6250

## SAMPLE
**Matrix:** formulations
**Sample preparation:** Inject an aliquot of a 50 μg/mL solution in water.

## HPLC VARIABLES
**Column:** 5 μm Ultrasphere-ODS
**Mobile phase:** MeCN:THF:50 mM pH 2 ammonium phosphate 35:10:55
**Flow rate:** 1
**Detector:** UV 220

## OTHER SUBSTANCES
**Simultaneous:** degradants

## KEY WORDS
lyophilized powder

## REFERENCE
Strickley, R.G.; Visor, G.C.; Lin, L.H.; Gu, L. An unexpected pH effect on the stability of moexipril lyophilized powder, *Pharm.Res.*, **1989**, *6*, 971–975.

# Mofezolac

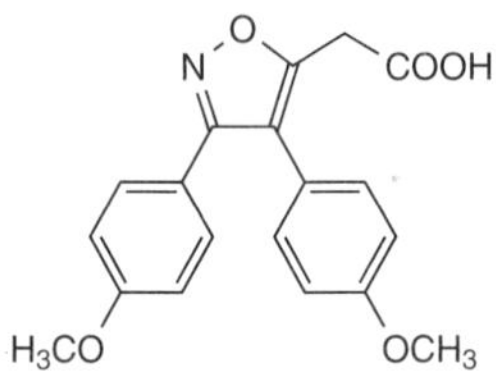


**Molecular formula:** $C_{19}H_{17}NO_5$
**Molecular weight:** 339.34
**CAS Registry No:** 78967-07-4
**Merck Index:** 13, 6254

## SAMPLE

**Matrix:** blood, urine
**Sample preparation:** Condition a Sep-Pak C18 SPE cartridge with 5 mL MeOH and 5 mL water. Plasma. Add 4 mL ice-cold MeCN to 500 μL plasma, vortex for 30 s, place on ice for 30 min, centrifuge refrigerated at 2000 g for 15 min. Evaporate the supernatant to dryness under a stream of nitrogen, reconstitute the residue with 2 mL water, add to the SPE cartridge, wash with 3 mL water, elute with 4 mL MeOH. Evaporate the eluate to dryness under a stream of nitrogen, reconstitute the residue with 100 μL MeOH, inject a 20 μL aliquot. Urine. Mix 1 mL urine with 1 mL 10 mM pH 7.0 phosphate buffer, add to the SPE cartridge, wash with 3 mL water, elute with 4 mL MeOH. Evaporate the eluate to dryness under a stream of nitrogen, reconstitute the residue with 200 μL MeOH:water 40:60, inject a 20 μL aliquot. (10 mM pH 7.0 Phosphate buffer was 35.81 g disodium hydrogen phosphate dodecahydrate and 13.61 g potassium dihydrogen phosphate in 1 L water, pH adjusted to 7.0 with NaOH or phosphoric acid.)

## HPLC VARIABLES

**Column:** 150 × 6 5 μm Shimadzu Shim-pack CLC ODS
**Column temperature:** 35
**Mobile phase:** Gradient. MeOH:20 mM pH 6.4 potassium dihydrogen phosphate buffer from 0:100 to 100:0 over 33 min.
**Flow rate:** 1.5
**Injection volume:** 20
**Detector:** UV 254

## CHROMATOGRAM

**Retention time:** 23.9
**Limit of detection:** 10 ng/mL (plasma), 100 ng/mL (urine)

## OTHER SUBSTANCES

**Extracted:** metabolites

## KEY WORDS

pharmacokinetics; plasma; SPE

## REFERENCE

Marunaka, T.; Maniwa, M. High-performance of liquid chromatographic determination of [3,4-di-(4-methoxyphenyl)-5-isoxazolyl]acetic acid and its metabolites in human plasma and urine, *J.Chromatogr.*, **1987**, *422*, 227–233.

# Mometasone furoate

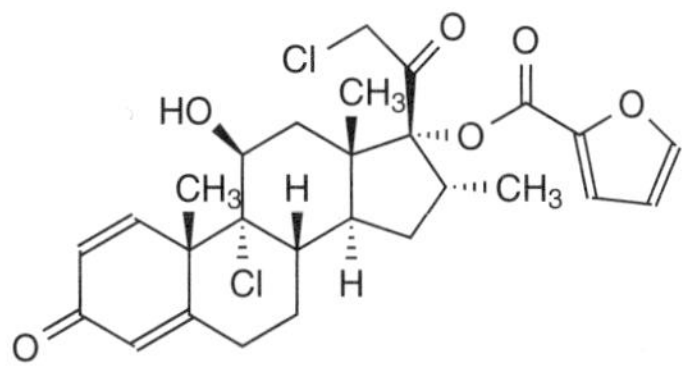

**Molecular formula:** $C_{27}H_{30}Cl_2O_6$
**Molecular weight:** 521.44
**CAS Registry No:** 83919-23-7
**Merck Index:** 13, 6264

## SAMPLE
**Matrix:** blood
**Sample preparation:** Mix 400 μL plasma with IS, extract with 8 mL *n*-chlorobutane:ethyl acetate 95:5. Evaporate the organic layer to dryness, reconstitute the residue with 35 μL MeOH, inject an 8 μL aliquot.

## HPLC VARIABLES
**Column:** 33 × 4.6 3 μm LC-18-DB
**Mobile phase:** MeOH:25 mM ammonium acetate 80:20
**Flow rate:** 1
**Injection volume:** 8
**Detector:** MS, PE Sciex API-III, positive ion daughter, m/z 355.0

## CHROMATOGRAM
**Retention time:** 24
**Internal standard:** betamethasone 17,21-dipropionate (m/z 279.3)
**Limit of quantitation:** 49.7 pg/mL

## OTHER SUBSTANCES
**Extracted:** mometasone (15), metabolites

## KEY WORDS
pharmacokinetics; plasma

## REFERENCE
Affrime, M.B.; Cuss, F.; Padhi, D.; Wirth, M.; Pai, S.; Clement, R.P.; Lim, J.; Kantesaria, B.; Alton, K.; Cayen, M.N. Bioavailability and metabolism of mometasone furoate following administration by metered-dose and dry-powder inhalers in healthy human volunteers, *J.Clin.Pharmacol.*, **2000**, *40*, 1227–1236.

## SAMPLE
**Matrix:** blood, simulated gastric fluid, simulated intestinal fluid, simulated lung fluid
**Sample preparation:** Mix 500 μL plasma, simulated gastric fluid, simulated intestinal fluid, or simulated lung fluid with 500 μL 10 μg/mL IS in EtOH, add 4 mL dichloromethane, mix on a roller mixer for 30 min, centrifuge at 2500 rpm for 15 min. Evaporate the organic layer to dryness under a stream of nitrogen at 37°, reconstitute the residue with 250 μL mobile phase, centrifuge at 15 000 rpm for 3 min, inject a 50 μL aliquot of the supernatant.

## HPLC VARIABLES
**Column:** 150 × 4.6 5 μm Ultrasphere C8
**Mobile phase:** MeOH:water 59:41
**Flow rate:** 1.5
**Injection volume:** 50
**Detector:** UV 248

## CHROMATOGRAM

**Retention time:** 17
**Internal standard:** dexamethasone 21-acetate (7)
**Limit of detection:** 50 ng/mL
**Limit of quantitation:** 200 ng/mL

## OTHER SUBSTANCES

**Extracted:** degradants
**Simultaneous:** beclomethasone dipropionate (33.9), budesonide (12.3), fluticasone propionate (16.3), hydrocortisone (3.2), prednisone (2.7)

## KEY WORDS

plasma

## REFERENCE

Teng, X.W.; Foe, K.; Brown, K.F.; Cutler, D.J.; Davies, N.M. High-performance liquid chromatographic analysis of mometasone furoate and its degradation products. Application to in vitro degradation studies, *J.Pharm.Biomed.Anal.*, **2001**, *26*, 313–319.

## ANNOTATED BIBLIOGRAPHY

Teng, X.W.; Cutler, D.J.; Davies, N.M. Mometasone furoate degradation and metabolism in human biological fluids and tissues, *Biopharm.Drug Dispos.*, **2003**, *24*, 321–333. [plasma; urine; microsomal incubations; lung tissue]

# Monensin

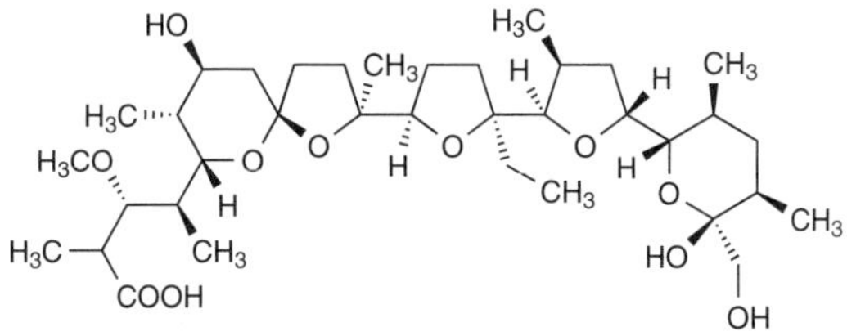

**Molecular formula:** $C_{36}H_{62}O_{11}$
**Molecular weight:** 670.87
**CAS Registry No:** 17090-79-8
**Merck Index:** 13, 6270

## SAMPLE

**Matrix:** blood
**Sample preparation:** Vortex 100 μL MeOH:water 80:20 and 400 μL trichloroacetic acid solution with 500 μL plasma, centrifuge at 4000 rpm for 4 min, filter (Spin-X) the supernatant, inject a 30 μL aliquot of the filtrate. (Prepare trichloroacetic acid solution as follows. Dissolve 85 g trichloroacetic acid in 15 mL water. Store this solution in the refrigerator. Dilute 150 μL of this solution with 100 mL acetone.)

## HPLC VARIABLES

**Guard column:** C18
**Column:** 250 × 4.6 5 μm Supelco Discovery C18
**Mobile phase:** MeOH:10 mM ammonium acetate 85:15
**Flow rate:** 0.8 for 15 min, 1 for 10 min
**Injection volume:** 30
**Detector:** MS, PE Sciex API 100 single quadrupole, turbo ionspray, turbo probe 150°, air 6 L/min, 50 μL/min flowed into detector, m/z 693.7

## CHROMATOGRAM

**Retention time:** 15
**Limit of detection:** 4 ng/mL
**Limit of quantitation:** 8 ng/mL

## OTHER SUBSTANCES

**Extracted:** lasalocid (10), narasin (23), salinomycin (20)

## KEY WORDS

chicken; plasma

## REFERENCE

Hormazábal, V.; Yndestad, M. Determination of amprolium, ethopabate, lasalocid, monensin, narasin, and salinomycin in chicken tissues, plasma, and egg using liquid chromatography-mass spectrometry, *J.Liq.Chromatogr.Rel.Technol.*, **2000**, *23*, 1585–1598.

# Morantel

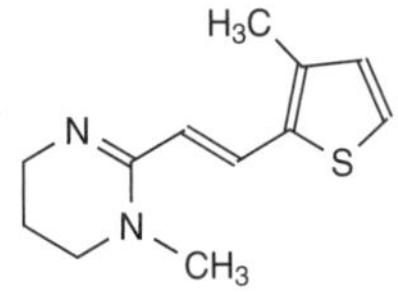


**Molecular formula:** $C_{12}H_{16}N_2S$
**Molecular weight:** 220.34
**CAS Registry No:** 20574-50-9, 26155-31-7 (tartrate)
**Merck Index:** 13, 6289

## SAMPLE

**Matrix:** abomasal fluid, blood, feces, intestinal fluid, ruminal fluid
**Sample preparation:** Add 100 μL 50 μg/mL IS to 5 mL plasma, abomasal fluid, intestinal fluid, ruminal fluid or 5 g feces, mix, add 10 mL 200 mM ammonium hydroxide, add 30 mL chloroform (Caution! Chloroform is a carcinogen!), shake mechanically for 30 min, centrifuge at 2000 g for 15 min, filter (Whatman 1 PS) the chloroform layer. Add 3 mL 1 M sulfuric acid to the filtrate, shake for 15 min, centrifuge at 2000 g for 5 min. Remove a 1 mL aliquot, neutralize with 2 drops concentrated ammonium hydroxide, inject a 50 μL aliquot.

## HPLC VARIABLES

**Column:** 10 μm Bondex C18 (Phenomenex)
**Mobile phase:** MeCN:pH 3.5 ammonium acetate buffer 28:72
**Flow rate:** 1.3
**Injection volume:** 50
**Detector:** UV 318

## CHROMATOGRAM

**Retention time:** 5.64
**Internal standard:** pyrantel tartrate (3.97)
**Limit of detection:** 25 ng/mL (plasma, ruminal fluid), 20 ng/mL (abomasal fluid, intestinal fluid), 50 ng/g (feces)

## KEY WORDS

cow; plasma

## REFERENCE

Lanusse, C.E.; Gascon, L.H.; Ranjan, S.; Prichard, R.K. Morantel tartrate release from a long-acting intraruminal device in cattle: pharmacokinetics and gastrointestinal distribution, *J.Vet.Pharmacol.Ther.*, **1992**, *15*, 117–123.

# Mosapride

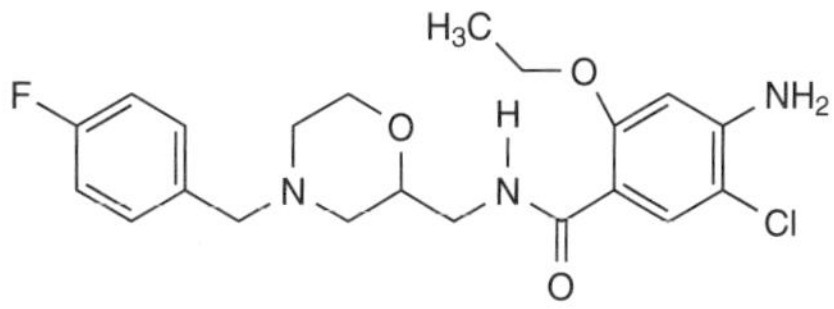


**Molecular formula:** $C_{21}H_{25}ClFN_3O_3$
**Molecular weight:** 421.90
**CAS Registry No:** 112885-41-3, 112885-42-4 (citrate)
**Merck Index:** 13, 6306

## SAMPLE

**Matrix:** blood
**Sample preparation:** Condition a Sep-Pak C18 SPE cartridge with two 3 mL portions of MeOH, two 3 mL portions of water, and two 2 mL portions of 20 mM pH 7.0 phosphate buffer. Mix 50 μL 100 μg/mL IS in water, 2 mL pH 7.0 sodium phosphate buffer, and 1 mL plasma, add to the SPE cartridge, wash with two 3 mL portions of water, elute with 2 mL MeOH, evaporate the eluate to dryness under reduced pressure at 50°, dissolve the residue in 400 μL MeOH:0.05% acetic acid 45:55, centrifuge at 5000 rpm for 10 min, inject a 180 μL aliquot.

## HPLC VARIABLES

**Guard column:** 20 × 9 μBondapak C18
**Column:** 150 × 4 5 μm Shim-pack CLC-ODS (Shimadzu)
**Mobile phase:** MeOH:0.05% acetic acid from 20:80 to 38.7:61.3 over 7 min, to 90:10 in 2 min, maintain at 90:10 for 4 min, from 90:10 to 20:80 in 0.5 min, maintain at 20:80 for 4.5 min.
**Flow rate:** 1.3
**Injection volume:** 180
**Detector:** F ex 314 em 354

## CHROMATOGRAM

**Retention time:** 13
**Internal standard:** AD-9675 ((+)-4-amino-5-chloro-2-ethoxy-*N*-[[4-propyl-2-morpholinyl]methyl]benzamide citrate) (10.5)

## OTHER SUBSTANCES

**Extracted:** metabolites

## KEY WORDS

plasma; rat; SPE

## REFERENCE

Yokoyama, I.; Mizuki, Y.; Yamaguchi, T.; Fujii, T. Simultaneous enantiomeric determination of a gastroprokinetic agent mosapride citrate and its metabolite in plasma using $\alpha_1$-acid glycoprotein HPLC column, *J.Pharm.Biomed.Anal.*, **1997**, *15*, 1527–1535.

## SAMPLE

**Matrix:** formulations
**Sample preparation:** Let powdered 50 mg tablet stand in 25 mL MeOH for 6 h with occasional sonication, filter (0.45 μm). Dilute the filtrate to 10 μg/mL with mobile phase.

## HPLC VARIABLES

**Guard column:** 20 mm long Pelliguard LC-18 (Supelco)
**Column:** 150 × 4.6 5 μm RP C-18

**Column temperature:** 40
**Mobile phase:** MeOH:20 mM pH 4.0 potassium phosphate buffer 70:30
**Flow rate:** 1.1
**Injection volume:** 20
**Detector:** UV 274

## CHROMATOGRAM
**Retention time:** 6.1
**Internal standard:** risperidone (4.1)

## KEY WORDS
tablets

## REFERENCE
Krishnaiah, Y.S.R.; Murthy, T.K.; Sankar, D.G.; Satyanarayana, V. The determination of mosapride citrate in bulk drug samples and pharmaceutical dosage forms using HPLC, *Anal.Sci.*, **2002**, *18*, 1269–1271.

## SAMPLE
**Matrix:** solutions
**Sample preparation:** Inject a 100 μL aliquot of a solution in 200 μL 20 mM disodium hydrogen phosphate:20 mM citric acid 65:35.

## HPLC VARIABLES
**Guard column:** 10 × 3 Chiral AGP
**Column:** 100 × 4 10 μm Chiral AGP (Chrom Tech)
**Mobile phase:** Gradient. A was MeOH. B was 20 mM disodium hydrogen phosphate. C was 20 mM citric acid. A:B:C for 0:65:35 for 4 min, to 5:60:35 over 6 min, maintain at 5:60:35 for 3 min, to 15:48:37 over 1 min, to 25:40:35 over 11 min, to 0:65:35 over 1 min, maintain at 0:65:35 for 4 min. (The pH of the mobile phase was 4.4 between 4 and 6 min and 5.0 between 20 and 22 min.)
**Flow rate:** 1
**Injection volume:** 100
**Detector:** F ex 314 em 354

## CHROMATOGRAM
**Retention time:** 20.8 (R), 21.8 (S)
**Internal standard:** AD-9675 ((+)-4-amino-5-chloro-2-ethoxy-*N*-[[4-propyl-2-morpholinyl]methyl]benzamide citrate) (9.5)

## KEY WORDS
chiral

## REFERENCE
Yokoyama, I.; Mizuki, Y.; Yamaguchi, T.; Fujii, T. Simultaneous enantiomeric determination of a gastroprokinetic agent mosapride citrate and its metabolite in plasma using $\alpha_1$-acid glycoprotein HPLC column, *J.Pharm.Biomed.Anal.*, **1997**, *15*, 1527–1535.

## ANNOTATED BIBLIOGRAPHY
Itoh, H.; Nagano, T.; Takeyama, M. Effects of mosapride citrate on human plasma levels of motilin, gastrin, somatostatin, and secretin, *Biol.Pharm.Bull.*, **2001**, *24*, 1072–1075. [LOQ 10 ng/mL]

Karlsson, A.; Aspegren, A. The use of mobile phase pH and column temperature to reverse the retention order of enantiomers on Chiral-AGP, *Chromatographia*, **1998**, *47*, 189–196.

Krishnaiah, Y.S.R.; Murthy, T.K.; Sankar, D.C.; Satyanarayana, V. A validated RP-HPLC method for the determination of mosapride citrate in bulk drug samples and pharmaceutical formulations, *Pharmazie*, **2002**, *57*, 814–816.

Kumar, Y.R.; Babu, J.M.; Sarma, M.S.P.; Seshidhar, B.; Reddy, S.S.; Reddy, G.S.; Vyas, K. Application of LC-MS/MS for the identification of a polar impurity in mosapride, a gastroprokinetic drug, *J.Pharm.Biomed.Anal.*, **2003**, *32*, 361–368.

# Moxifloxacin

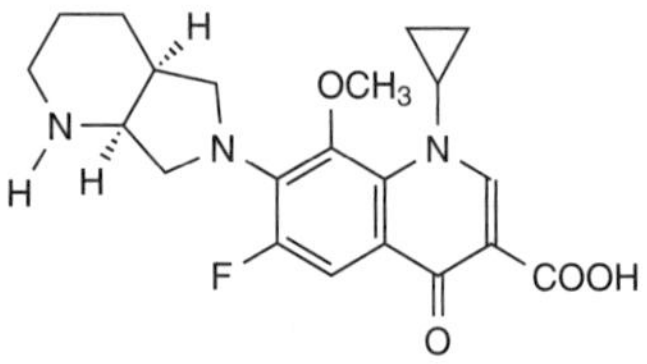

**Molecular formula:** $C_{21}H_{24}FN_3O_4$
**Molecular weight:** 401.43
**CAS Registry No:** 151096-09-2
**Merck Index:** 13, 6316

## SAMPLE

**Matrix:** blood

**Sample preparation:** Condition a 1 mL 30 mg OASIS HLB SPE cartridge with 1 mL MeOH and 1 mL water. Mix 1 mL plasma with 100 μL 500 ng/mL IS in water, add to the SPE cartridge, wash with two 1 mL portions of water, elute with 1 mL MeOH:trifluoroacetic acid 99.9:0.1. Evaporate the eluate to dryness under reduced pressure, reconstitute the residue with 100 μL mobile phase, inject a 10 μL aliquot.

## HPLC VARIABLES

**Column:** 100 × 4.6 5 μm Hypersil C18
**Mobile phase:** MeCN:0.1% formic acid 40:60
**Flow rate:** 1
**Injection volume:** 10
**Detector:** MS, PE Sciex API 3000, TurboIonSpray, positive ion mode, source 400°, ionspray 3500 V, declustering 65 V, focusing 300 V, collision energy 20 eV, collision gas nitrogen, column effluent split 1:5 before entering the detector, m/z 402–384

## CHROMATOGRAM

**Retention time:** 2.75
**Internal standard:** lomefloxacin (m/z 352–308) (2.32)
**Limit of detection:** 50 pg/mL
**Limit of quantitation:** 1 ng/mL

## KEY WORDS

plasma; SPE

## REFERENCE

Vishwanathan, K.; Bartlett, M.G.; Stewart, J.T. Determination of moxifloxacin in human plasma by liquid chromatography electrospray ionization tandem mass spectrometry, *J.Pharm.Biomed.Anal.*, **2002**, *30*, 961–968.

## SAMPLE

**Matrix:** blood, tissue

**Sample preparation:** Condition a 1 mL 30 mg OASIS HLB SPE cartridge with 1 mL MeOH and 1 mL water. Homogenize (Ultra Turrax) 250 mg lung tissue with 2.5 mL PBS, centrifuge at 3080 g for 15 min. Mix 1.2 mL plasma or lung homogenate with 300 μL 10 μg/mL IS in water, add a 1.5 mL aliquot to the SPE cartridge, wash with 1 mL water, dry with air, elute with 1 mL MeOH:trifluoroacetic acid 99.9:0.1. Evaporate the eluate to dryness under a stream of nitrogen at 50°, reconstitute the residue with 500 μL mobile phase, inject a 50 μL aliquot.

## HPLC VARIABLES

**Column:** 150 × 4.6 5 μm Supelcosil ABZ+
**Mobile phase:** MeCN:buffer 18:82 (The buffer was 10 mM potassium dihydrogen phosphate adjusted to pH 4 with orthophosphoric acid.)
**Flow rate:** 1.25

**Injection volume:** 50
**Detector:** UV 296

---

## CHROMATOGRAM

**Retention time:** 6.2
**Internal standard:** enrofloxacin (3.5)
**Limit of detection:** 6.5 ng/mL (plasma), 50 ng/g (lung)
**Limit of quantitation:** 30 ng/mL (plasma), 400 ng/g (lung)

---

## KEY WORDS

lung; plasma; SPE

---

## REFERENCE

Lemoine, T.; Breilh, D.; Ducint, D.; Dubrez, J.; Jougon, J.; Velly, J.F.; Saux, M.C. Determination of moxifloxacin (BAY 12–8039) in plasma and lung tissue by high-performance liquid chromatography with ultraviolet detection using a fully automated extraction method with a new polymeric cartridge, *J.Chromatogr.B*, **2000**, *742*, 247–254.

---

## SAMPLE

**Matrix:** blood, vitreous humor
**Sample preparation:** Serum. Rotate 400 μL serum with 3.2 mL dichloromethane at 20 rpm for 10 min, shake for 1 min. Shake the organic layer with 200 μL 20 mM pH 2.0 orthophosphoric acid for 1 min, centrifuge at 2000 g at 10° for 10 min, inject a 20 μL aliquot of the aqueous layer. Vitreous humor. Centrifuge vitreous humor briefly, inject a 20 μL aliquot.

---

## HPLC VARIABLES

**Column:** 75 × 4.6 3 μm Ultrasphere C18
**Mobile phase:** MeCN:buffer 12:88 (serum) or 10:90 (vitreous humor) (The buffer was 5 mM tetrabutylammonium bromide adjusted to pH 1.8 with phosphoric acid.)
**Flow rate:** 1.5
**Injection volume:** 20
**Detector:** UV 296

---

## CHROMATOGRAM

**Limit of detection:** 10 ng/mL

---

## KEY WORDS

rabbit; serum

---

## REFERENCE

Bronner, S.; Jehl, F.; Peter, J.-D.; Ploy, M.-C.; Renault, C.; Arvis, P.; Monteil, H.; Prevost, G. Moxifloxacin efficacy and vitreous penetration in a rabbit model of Staphylococcus aureus endophthalmitis and effect on gene expression of leucotoxins and virulence regulator factors, *Antimicrob.Agents Chemother.*, **2003**, *47*, 1621–1629.

---

## ANNOTATED BIBLIOGRAPHY

Ba, B.B.; Etienne, R.; Ducint, D.; Quentin, C.; Saux, M.-C. Determination of moxifloxacin in growth media by high-performance liquid chromatography, *J.Chromatogr.B*, **2001**, *754*, 107–112. [fluorescence detection; LOQ 50 ng/mL]

Djabarouti, S.; Boselli, E.; Allaouchiche, B.; Ba, B.; Nguyen, A.T.; Gordien, J.B.; Bernadou, J.M.; Saux, M.C.; Breilh, D. Determination of levofloxacin in plasma, bronchoalveolar lavage and bone tissues by high-performance liquid chromatography with ultraviolet detection using a fully automated extraction method, *J.Chromatogr.B*, **2004**, *799*, 165–172. [SPE; moxifloxacin is internal standard]

Liang, H.; Kays, M.B.; Sowinski, K.M. Separation of levofloxacin, ciprofloxacin, gatifloxacin, moxifloxacin, trovafloxacin and cinoxacin by high-performance liquid chromatography: application to levofloxacin determination in human plasma, *J.Chromatogr.B*, **2002**, *772*, 53–63.

Paillard, D.; Grellet, J.; Dubois, V.; Saux, M.-C.; Quentin, C. Discrepancy between uptake and intracellular activity of moxifloxacin in a Staphylococcus aureus-human THP-1 monocytic cell model, *Antimicrob.Agents Chemother.*, **2002**, *46*, 288–293. [column-switching; fluorescence detection; LOD 1 ng/mL]

Stass, H.; Dalhoff, A. Determination of BAY 12–8039, a new 8-methoxyquinolone, in human body fluids by high-performance liquid chromatography with fluorescence detection using on-column focusing, *J.Chromatogr.B*, **1997**, *702*, 163–174.

Tobin, C.M.; Sunderland, J.; White, L.O.; MacGowan, A.P.; Reeves, D.S. An isocratic high performance liquid chromatography (HPLC) assay for moxifloxacin, a new 8-methoxyquinolone, *J.Antimicrob.Chemother.*, **1998**, *42*, 278–279.

Wallace, A.W.; Victory, J.M.; Amsden, G.W. Lack of bioequivalence of gatifloxacin when coadministered with calcium-fortified orange juice in healthy volunteers, *J.Clin.Pharmacol.*, **2003**, *43*, 92–96. [moxifloxacin is internal standard]

Xuan, D.; Zhong, M.; Mattoes, H.; Bui, K.Q.; McNabb, J.; Nicolau, D.P.; Quintiliani, R.; Nightingale, C.H. *Streptococcus pneumoniae* response to repeated moxifloxacin or levofloxacin exposure in a rabbit tissue cage model, *Antimicrob.Agents Chemother.*, **2001**, *45*, 794–799. [serum; fluorescence detection; LOQ 100 ng/mL; ciprofloxacin is IS]

# Moxonidine

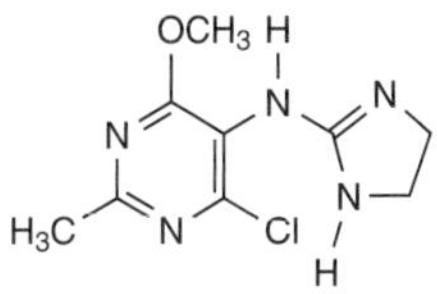


**Molecular formula:** $C_9H_{12}ClN_5O$
**Molecular weight:** 241.68
**CAS Registry No:** 75438-57-2
**Merck Index:** 13, 6318

## SAMPLE

**Matrix:** blood, urine

**Sample preparation:** Plasma. Condition a 10 mm 6 mL SPE disk (3M Empore) with MeOH and water. Mix 1 mL plasma with 1 mL water at 0°, add 10 ng IS, add to the SPE disk, wash with MeOH:water 15:85, elute with MeOH:water:trifluoroacetic acid 95:5:3. Evaporate the eluate to dryness under reduced pressure at 45°, reconstitute the residue with 150 μL 100 mM ammonium acetate, inject a 125 μL aliquot. Urine. Condition a 1 mL Bakerbond carboxylic acid SPE cartridge with MeOH and water. Mix 1 mL urine with 1 mL water, add 100 ng IS, vortex, add to the SPE cartridge, wash with 1 mL water, wash with 1 mL MeOH:water 50:50, elute with MeOH:water:trifluoroacetic acid 95:5:3. Evaporate the eluate to dryness under reduced pressure at 45°, reconstitute the residue with 150 μL 100 mM ammonium acetate, inject a 60 μL aliquot.

## HPLC VARIABLES

**Column:** 250 × 4.6 5 μm Supelcosil C18 LC-ABZ

**Mobile phase:** Gradient. MeCN:25 mM pH 5 ammonium acetate 0:100 for 2 min, to 4:96 over 20 min, maintain at 4:96 for 4 min, return to initial conditions over 1 min.

**Flow rate:** 1

**Injection volume:** 60–125

**Detector:** MS, PE Sciex API-III Plus, APCI, heated nebulizer, 25% of column effluent entered the detector, m/z 242–44

## CHROMATOGRAM

**Retention time:** 27

**Internal standard:** clonidine (m/z 230–213)

**Limit of quantitation:** 50 pg/mL

## OTHER SUBSTANCES

**Extracted:** metabolites

## KEY WORDS

plasma; SPE

## REFERENCE

He, M.M.; Abraham, T.L.; Lindsay, T.J.; Schaefer, H.C.; Pouliquen, I.J.; Payne, C.; Czeskis, B.; Shipley, L.A.; Oliver, S.D.; Mitchell, M.I. Metabolism and disposition of the antihypertensive agent moxonidine in humans, *Drug Metab.Dispos.*, **2003**, *31*, 334–342.

# Nadifloxacin

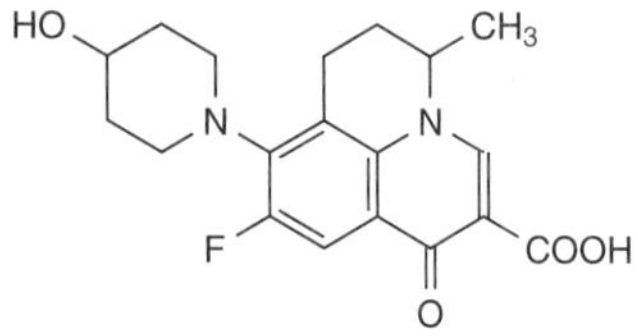


**Molecular formula:** $C_{19}H_{21}FN_2O_4$
**Molecular weight:** 360.38
**CAS Registry No:** 124858-35-1
**Merck Index:** 13, 6371

## SAMPLE

**Matrix:** blood
**Sample preparation:** Shake 1 mL plasma, 10 μL 100 ng/mL IS in MeOH, 1 mL 25 mM pH 6.86 phosphate buffer, and 5 mL chloroform for 10 min (Caution! Chloroform is a carcinogen!), centrifuge at 1800 g for 10 min. Remove a 3.5 mL aliquot of the lower chloroform layer and evaporate it to dryness under a stream of air at 40°, reconstitute the residue with 120 μL MeOH, inject a 40 μL aliquot.

## HPLC VARIABLES

**Column:** 150 × 4.6 5 μm TSK-gel ODS-80TM
**Mobile phase:** MeCN:THF:25 mM pH 5.5 ammonium phosphate buffer 35:3:65
**Flow rate:** 1
**Injection volume:** 40
**Detector:** UV 295

## CHROMATOGRAM

**Retention time:** 6.8
**Internal standard:** OPC-7258 (9-fluoro-6,7-dihydro-5-methyl-8-morpholinyl-1-oxo-1H, 5H-benzo[*i,j*]quinolizine-2-carboxylic acid) (11.3)
**Limit of detection:** 1 ng/mL
**Limit of quantitation:** 5 ng/mL

## KEY WORDS

human; pharmacokinetics; plasma; rat

## REFERENCE

Koike, M.; Akiyama, H.; Shimizu, T. High-performance liquid chromatographic procedure for the determination of rat plasma concentrations of a new antibacterial agent, (±)-9-fluoro-6,7-dihydro-8-(4-hydroxy-1-piperidyl)-5-methyl-1-oxo-1*H*,5*H*- benzo[*i,j*]quinolizine-2-carboxylic acid, for topical use, *J.Chromatogr.*, **1990**, *526*, 235–239.

# Naftopidil

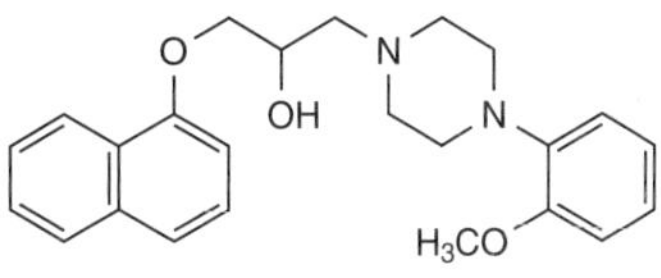

**Molecular formula:** $C_{24}H_{28}N_2O_3$
**Molecular weight:** 392.49
**CAS Registry No:** 57149-07-2
**Merck Index:** 13, 6382

## SAMPLE

**Matrix:** blood
**Sample preparation:** Mix 1 mL plasma, 10 μL 5 μg/mL IS in MeOH, 500 μL 1 M pH 9.2 dipotassium hydrogen phosphate buffer, and 8 mL diethyl ether, rotate for 10 min. Add 5 mL of the organic layer to 300 μL 50 mM sulfuric acid, rotate for 10 min, centrifuge at 1000 g, discard the organic layer, remove traces of ether from the aqueous layer with a stream of nitrogen for 30 s. Remove a 250 μL aliquot of the aqueous layer and add it to 50 μL 1 M pH 9.2 dipotassium hydrogen phosphate buffer, mix, inject a 100 μL aliquot.

## HPLC VARIABLES

**Column:** 250 × 4.6 5 μm RP-LiChrosorb Select B (Hibar)
**Mobile phase:** MeCN:MeOH:buffer 22.5:22.5:55 (The buffer was 20 mM potassium dihydrogen phosphate adjusted to pH 1.8 with phosphoric acid.)
**Flow rate:** 0.8
**Injection volume:** 100
**Detector:** F ex 215 em 320

## CHROMATOGRAM

**Retention time:** 9.3
**Internal standard:** carvedilol (7.7)
**Limit of detection:** 1 ng/mL
**Limit of quantitation:** 2 ng/mL

## OTHER SUBSTANCES

**Extracted:** metabolites

## KEY WORDS

pharmacokinetics; plasma

## REFERENCE

Niebch, G.; Borbe, H.O.; Besenfelder, E. High-performance liquid chromatography of naftopidil, a novel antihypertensive drug, and two metabolites in human plasma, *J.Chromatogr.*, **1990**, *534*, 247–252.

## SAMPLE

**Matrix:** tissue
**Sample preparation:** Mix 200 mg frozen (−80°) tissue with 100 μL 1.16 μg/mL IS in MeOH:water 50:50 and 200 μL water in a PTFE tube precooled in liquid nitrogen, homogenize (Braun microdismembrator), pour the contents into a centrifuge tube, wash in with 500 μL water, wash in with 500 μL acetone, vortex for 2 min, centrifuge at 4000 rpm at 4° for 20 min. Evaporate the supernatant to dryness under a stream of nitrogen at 40°, reconstitute the residue with 1 mL 500 mM pH 3.5 potassium acetate buffer, centrifuge, add the supernatant dropwise to a 1 mL 100 mg 40 μm Bond-Elut cyano SPE cartridge, wash twice with 1 mL portions of water, elute with two 250 μL aliquots of MeCN:100 mM pH 5 potassium acetate buffer 80:20, inject a 100 μL aliquot.

## HPLC VARIABLES

**Guard column:** 17 × 4 5 μm Spherisorb C6
**Column:** 150 × 4.6 5 μm Spherisorb C6
**Mobile phase:** MeCN:250 mM pH 4 potassium acetate buffer 65:35
**Flow rate:** 1
**Injection volume:** 100
**Detector:** F ex 285 em 360

## CHROMATOGRAM

**Retention time:** 5.4
**Internal standard:** naftopidil (5.4)
**Limit of quantitation:** 10 pg/mg (S/N 10) (for carvedilol)

## OTHER SUBSTANCES

**Extracted:** carvedilol (4.7)

## KEY WORDS

heart; naftopidil is IS in original; SPE

## REFERENCE

Behn, F.; Läer, S.; Scholz, H. Determination of carvedilol in human cardiac tissue by high-performance liquid chromatography, *J.Chromatogr.Sci.*, **2001**, *39*, 121–124.

# Nandrolone

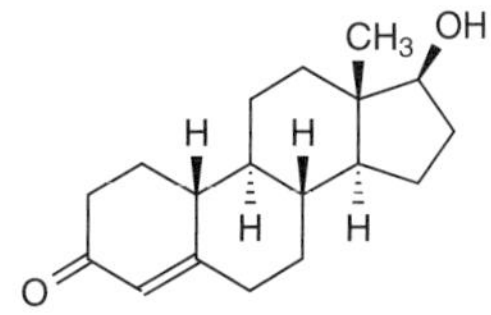

**Molecular formula:** $C_{18}H_{26}O_2$
**Molecular weight:** 274.40
**CAS Registry No:** 434-22-0
**Merck Index:** 13, 6391

## SAMPLE
**Matrix:** blood
**Sample preparation:** Condition a 1 mL 100 mg Bond Elut C1 SPE cartridge with 2 column vol of MeCN and 2 column vol of water. Mix 500 μL serum with 25 μL 5 μg/mL IS in MeCN, add 500 μL water, vortex for 5 s, add to the SPE cartridge, wash with 2 column vol of water, wash with 1 column vol of MeCN:water 10:90, elute with 1 mL MeCN:water 45:55, concentrate the eluate to 250 μL under a stream of nitrogen at 45°, inject a 50 μL aliquot.

## HPLC VARIABLES
**Guard column:** 15 × 4.6 7 μm silica (Brownlee)
**Column:** 220 × 4.6 5 μm silica (Brownlee)
**Column temperature:** 60
**Mobile phase:** MeCN:100 mM pH 2.5 sodium phosphate buffer 15:85
**Flow rate:** 1
**Injection volume:** 50
**Detector:** UV 247

## CHROMATOGRAM
**Retention time:** 8.2
**Internal standard:** spironolactone (7.8)
**Limit of detection:** 3 ng/mL

## OTHER SUBSTANCES
**Extracted:** fluoxymesterone (5.4), methandrostenolone (9.3), methyltestosterone (9.6), stanozolol (UV 230; LOD 7 ng/mL) (12.7), testosterone (8.6), zeranol (UV 263) (3.6)

## KEY WORDS
serum: SPE

## REFERENCE
Lampert, B.L.; Stewart, J.T. Determination of anabolic steroids and zeranol in human serum by isocratic reverse phase HPLC on silica, *J.Liq.Chromatogr.*, **1989**, *12*, 3231–3249.

## SAMPLE
**Matrix:** blood, urine
**Sample preparation:** Condition a 3 mL 500 mg Baker C18 SPE cartridge with 2.5 mL MeOH and 5 mL water. Serum. Mix 2 mL serum with 4 ng IS and 15 mL 150 mM pH 5 acetate buffer, sonicate for 5 min, add to the SPE cartridge, wash with 5 mL 150 mM pH 5 acetate buffer, wash with 10 mL water, wash with 2.5 mL MeOH:water 40:60, elute with 4 mL MeOH. Evaporate the eluate to dryness under a stream of nitrogen, reconstitute the residue with 100 μL MeOH, inject a 5 μL aliquot. Urine. Heat 2 mL urine, 10 ng IS, and 20 μL β-glucuronidase/aryl sulfatase (Helix pomatia) (Boehringer Mannheim) at 37° for 12 h, cool, add 15 mL 150 mM pH 5 acetate buffer, add to the SPE cartridge, wash with 5 mL 150 mM pH 5 acetate buffer, wash with 10 mL water, wash with 2.5 mL MeOH:water 40:60, elute with 4 mL MeOH. Evaporate the eluate to

dryness under a stream of nitrogen, reconstitute the residue with 100 μL MeOH, inject a 5 μL aliquot.

## HPLC VARIABLES

**Column:** 250 × 2 5 μm Kingsorb C18 (Phenomenex)
**Mobile phase:** MeCN:water 70:30 containing 2 mM ammonium acetate
**Flow rate:** 0.15
**Injection volume:** 5
**Detector:** MS, PE Sciex API-III Plus triple quadrupole, APCI, positive ion mode, nebulizer gas air at 400 kPa, curtain gas nitrogen at 0.6 L/min, auxiliary gas air 1.5 L/min, collision gas argon, nebulizer 350°, discharge current 4 μA, orifice 90 V, m/z 275–83, 275–109

## CHROMATOGRAM

**Retention time:** 5.6
**Internal standard:** $d_2$-17β-testosterone (m/z 291–99) (6.1)
**Limit of quantitation:** 100 pg/mL

## OTHER SUBSTANCES

**Extracted:** 17α-nortestosterone (m/z 275–83, 275–109) (6.4), progesterone (m/z 315–97) (11.4), 17α-testosterone (m/z 289–107, 289–109) (7.2), 17β-testosterone (m/z 289–107, 289–109) (6.1)

## KEY WORDS

cow; serum; SPE

## REFERENCE

Draisci, R.; Palleschi, L.; Ferretti, E.; Lucentini, L.; Cammarata, P. Quantitation of anabolic hormones and their metabolites in bovine serum and urine by liquid chromatography-tandem mass spectrometry, *J.Chromatogr.A*, **2000**, *870*, 511–522.

# Narasin

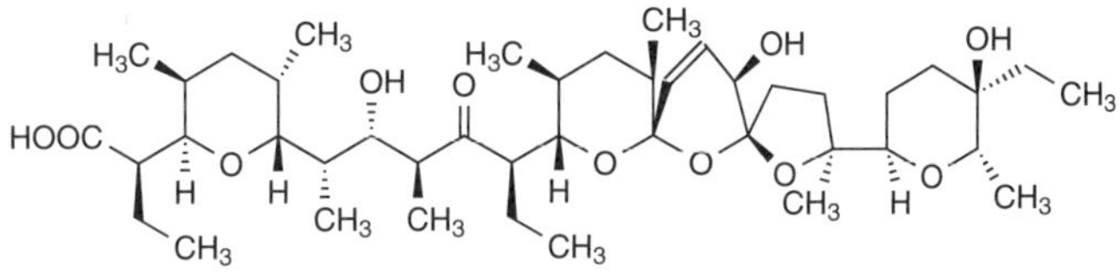


**Molecular formula:** $C_{43}H_{72}O_{11}$
**Molecular weight:** 765.02
**CAS Registry No:** 55134-13-9
**Merck Index:** 13, 6445

## SAMPLE

**Matrix:** blood
**Sample preparation:** Vortex 100 μL MeOH:water 80:20 and 400 μL trichloroacetic acid solution with 500 μL plasma, centrifuge at 4000 rpm for 4 min, filter (Spin-X) the supernatant, inject a 30 μL aliquot of the filtrate. (Prepare trichloroacetic acid solution as follows. Dissolve 85 g trichloroacetic acid in 15 mL water. Store this solution in the refrigerator. Dilute 150 μL of this solution with 100 mL acetone.)

## HPLC VARIABLES

**Guard column:** C18
**Column:** 250 × 4.6 5 μm Supelco Discovery C18
**Mobile phase:** MeOH:10 mM ammonium acetate 85:15
**Flow rate:** 0.8 for 15 min, 1 for 10 min
**Injection volume:** 30
**Detector:** MS, PE Sciex API 100 single quadrupole, turbo ionspray, turbo probe 150°, air 6 L/min, 50 μL/min flowed into detector, m/z 693.7

## CHROMATOGRAM

**Retention time:** 23
**Limit of detection:** 5 ng/mL
**Limit of quantitation:** 10 ng/mL

## OTHER SUBSTANCES

**Extracted:** lasalocid (10), monensin (15), salinomycin (20)

## KEY WORDS

chicken; plasma

## REFERENCE

Hormazábal, V.; Yndestad, M. Determination of amprolium, ethopabate, lasalocid, monensin, narasin, and salinomycin in chicken tissues, plasma, and egg using liquid chromatography-mass spectrometry, *J.Liq.Chromatogr.Rel.Technol.*, **2000**, *23*, 1585–1598.

# Nartograstim

**Molecular formula:** $C_{850}H_{1344}N_{226}O_{245}S_8$
**Molecular weight:** 18905.67
**CAS Registry No:** 134088-74-7
**Merck Index:** 13, 4552

## SAMPLE

**Matrix:** solutions

## HPLC VARIABLES

**Column:** TSK gel ODS-120T (Tosoh)
**Mobile phase:** Gradient. MeCN:0.1% trifluoroacetic acid from 40:60 to 90:10 over 80 (?) min.
**Flow rate:** 1

## CHROMATOGRAM

**Retention time:** 39 (oxidized form), 42 (reduced form)

## REFERENCE

Yamasaki, M.; Konishi, N.; Yamaguchi, K.; Itoh, S.; Yokoo, Y. Purification and characterization of recombinant human granulocyte colony-stimulating factor (rhG-CSF) derivatives: KW-2228 and other derivatives, *Biosci.Biotechnol.Biochem.*, **1998**, *62*, 1528–1534.

# Nateglinide

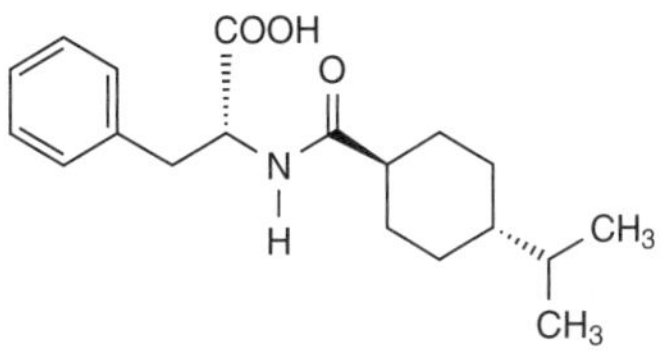


**Molecular formula:** $C_{19}H_{27}NO_3$
**Molecular weight:** 317.42
**CAS Registry No:** 105816-04-4
**Merck Index:** 13, 6454

## SAMPLE

**Matrix:** blood
**Sample preparation:** Mix 50 μL plasma with 10 μL 5 μg/mL IS in MeOH and 100 μL MeCN for 10 min, centrifuge at 3000 g for 10 min. Evaporate the supernatant to dryness under a stream of nitrogen at 50°, reconstitute the residue with 30 μL mobile phase, inject a 20 μL aliquot.

## HPLC VARIABLES

**Guard column:** present but not specified
**Column:** 250 × 3 5 μm ProntoSIL 120-5-C18 AQ (Bischoff)
**Column temperature:** 50
**Mobile phase:** MeCN:MeOH:buffer 30:8:70 (The buffer was 100 mM potassium hydrogen phosphate (sic) adjusted to pH 4.0 with 30% KOH.)
**Flow rate:** 1
**Injection volume:** 20
**Detector:** UV 210

## CHROMATOGRAM

**Retention time:** 14.1
**Internal standard:** carbamazepine (6.7)
**Limit of quantitation:** 100 pg/mL

## KEY WORDS

pharmacokinetics; plasma

## REFERENCE

Bauer, S.; Störmer, E.; Kirchheiner, J.; Michael, C.; Brockmöller, J.; Roots, I. Rapid and simple method for the analysis of nateglinide in human plasma using HPLC analysis with UV detection, *J.Pharm.Biomed.Anal.*, **2003**, *31*, 551–555.

## SAMPLE

**Matrix:** blood
**Sample preparation:** Mix 500 μL plasma with IS and 2 mL 50 mM pH 6.6 phosphate buffer, add to a conditioned (unspecified) Sep-Pak SPE cartridge, wash with two 3 mL portions of water, elute with 5 mL MeOH. Evaporate the eluate to dryness, reconstitute the residue with 250 μL mobile phase, mix, centrifuge, inject a 150 μL aliquot.

## HPLC VARIABLES

**Column:** 250 × 4.6 5 μm Supelcosil LC-ABZ
**Mobile phase:** MeCN:50 mM pH 6.27 phosphate buffer 35:65
**Injection volume:** 150
**Detector:** UV 210

## CHROMATOGRAM

**Internal standard:** *N*-(*trans*-4-*t*-butylcyclohexylcarbonyl)-D-phenylalanine
**Limit of quantitation:** 50 ng/mL

## KEY WORDS

pharmacokinetics; plasma

## REFERENCE

Weaver, M.L.; Orwig, B.A.; Rodriguez, L.C.; Graham, E.D.; Chin, J.A.; Shapiro, M.J.; McLeod, J.F.; Mangold, J.B. Pharmacokinetics and metabolism of nateglinide in humans, *Drug Metab.Dispos.*, **2001**, *29*, 415–421.

## ANNOTATED BIBLIOGRAPHY

Anderson, D.M.; Shelley, S.; Crick, N.; Buraglio, M. No effect of the novel antidiabetic agent nateglinide on the pharmacokinetics and anticoagulant properties of warfarin in healthy volunteers, *J.Clin.Pharmacol.*, **2002**, *42*, 1358–1365. [LOQ 50 ng/mL; SPE; column-switching]

Choudhury, S.; Hirschberg, Y.; Filipek, R.; Lasseter, K.; McLeod, J.F. Single-dose pharmacokinetics of nateglinide in subjects with hepatic cirrhosis, *J.Clin.Pharmacol.*, **2000**, *40*, 634–640. [SPE; LOQ 25.8 ng/mL]

Ono, I.; Matsuda, K.; Kanno, S. Determination of *N*-(*trans*-4-isopropylcyclohexylcarbonyl)-D-phenylalanine in human plasma by solid-phase extraction and column-switching high-performance liquid chromatography with ultraviolet detection, *J.Chromatogr.B*, **1996**, *678*, 384–387. [LOQ 50 ng/mL]

Ono, I.; Matsuda, K.; Kanno, S. Determination of *N*-(*trans*-4-isopropylcyclohexanecarbonyl)-D-phenylalanine and its metabolites in human plasma and urine by column-switching high-performance liquid chromatography with ultraviolet detection, *J.Chromatogr.B*, **1997**, *692*, 397–404.

Qi, M.; Wang, P.; Sun, Y.; Li, Y. Determination of the L-enantiomer of nateglinide in a bulk drug substance by chiral reversed-phase liquid chromatography, *J.Liq.Chromatogr.Rel.Technol.*, **2003**, *26*, 1839–1845. [LOD (L) 300 ng/mL (D) 400 ng/mL; LOQ (L) 800 ng/mL (D) 1000 ng/mL]

# Nebivolol

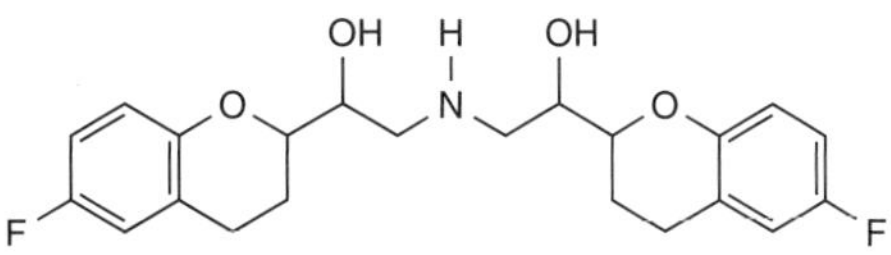

**Molecular formula:** $C_{22}H_{25}F_2NO_4$
**Molecular weight:** 405.43
**CAS Registry No:** 99200-09-6
**Merck Index:** 13, 6459

## SAMPLE
**Matrix:** solutions
**Sample preparation:** Inject a 20 μL aliquot of a 100 μg/mL solution in EtOH.

## HPLC VARIABLES
**Column:** 150 × 4.6 5 μm Chiralpak AD-RH
**Column temperature:** 23 ± 1
**Mobile phase:** Isopropanol
**Flow rate:** 0.5
**Injection volume:** 20
**Detector:** UV 220

## CHROMATOGRAM
**Retention time:** 13.17 (+), 17.45 (−)

## KEY WORDS
chiral

## REFERENCE
Aboul-Enein, H.Y.; Ali, I. HPLC enantiomeric resolution of nebivolol on normal and reversed amylose based chiral phases, *Pharmazie*, **2001**, *56*, 214–216.

## SAMPLE
**Matrix:** urine
**Sample preparation:** Mix 5 mL urine with 50 μL 10 μg/mL IS, add 500 μL 1 M pH 5.2 sodium acetate buffer, add 50 μL β-glucuronidase/aryl sulfatase (Helix pomatia) (Boehringer Mannheim), heat at 50° for 1 h, cool, add 500 mg sodium hydrogen carbonate:potassium carbonate 2:1, add 1 mL *t*-butanol, add 5 mL MTBE, add 1 g sodium sulfate, shake for 15 min, centrifuge for 10 min. Evaporate the organic layer to dryness under reduced pressure at 50°, reconstitute the residue with 50 μL MeCN, inject a 5 μL aliquot.

## HPLC VARIABLES
**Column:** 55 × 4 3 μm Purospher STAR RP-18e (Merck)
**Column temperature:** 25
**Mobile phase:** Gradient. A MeCN. B was pH 3.5 ammonium acetate buffer (4 mM ammonium acetate and 1% acetic acid). A:B from 0:100 to 80:20 over 4 min, re-equilibrate at initial conditions for 3 min.
**Flow rate:** 1
**Injection volume:** 5
**Detector:** MS, PE Sciex API 2000 triple quadrupole, APCI, 10% of column effluent entered the detector, positive ionization, interface 400°, ionization 5.9 kV, m/z 151.0

## CHROMATOGRAM
**Retention time:** 3.82
**Internal standard:** bupranolol (216.1)
**Limit of detection:** 10 ng/mL

## OTHER SUBSTANCES

**Extracted:** acebutolol (m/z 116.2), alprenolol (m/z 116.3), atenolol (m/z 145.1), betaxolol (m/z 116.1), befunolol(m/z 56.0), bisoprolol (m/z 116.3), bunitrolol (m/z 193.2) (2.97), butofilolol (m/z 256.4), carazolol (m/z 116.3), carteolol (m/z 237.3), carvedilol (m/z 100.2) (3.69), celiprolol (m/z 251.1), cloranolol (m/z 236.1), esmolol (m/z 145.1), indenolol (m/z 171.1), labetalol (m/z 91.1), levobunolol (m/z 236.3), mepindolol (m/z 116.2), metipranolol (m/z 116.2), metoprolol (m/z 116.2), moprolol (m/z 116.2), nadolol (m/z 254.4), nifenalol (m/z 165.2), oxprenolol (m/z 72.0), penbutolol (m/z 236.4), pindolol (m/z 116.2), propranolol (m/z 116.1), sotalol (m/z 133.2), talinolol (m/z 308.3) (3.83), timolol (m/z 261.2), toliprolol (m/z 147.2)

## REFERENCE

Thevis, M.; Opfermann, G.; Schänzer, W. High speed determination of beta-receptor blocking agents in human urine by liquid chromatography/tandem mass spectrometry, *Biomed.Chromatogr.*, **2001**, *15*, 393–402.

## ANNOTATED BIBLIOGRAPHY

Ali, I.; Aboul-Enein, H.Y. Enantioseparation of some clinically used drugs by HPLC using cellulose Tris (3,5-dichlorophenylcarbamate) chiral stationary phase, *Biomed.Chromatogr.*, **2003**, *17*, 113–117. [metoprolol; teratolol; tolamolol; nebivolol; econazole; miconazole; cromakalim; etodolac]

Woestenborghs, R.; Embrechts, L.; Heykants, J. HPLC-fluorescence method for the determination of the new $\beta_1$-adrenoreceptor blocking agent nebivolol in human plasma, *Methodol.Surv.Biochem.Anal.*, **1988**, *18*, 215–216.

# Nelfinavir

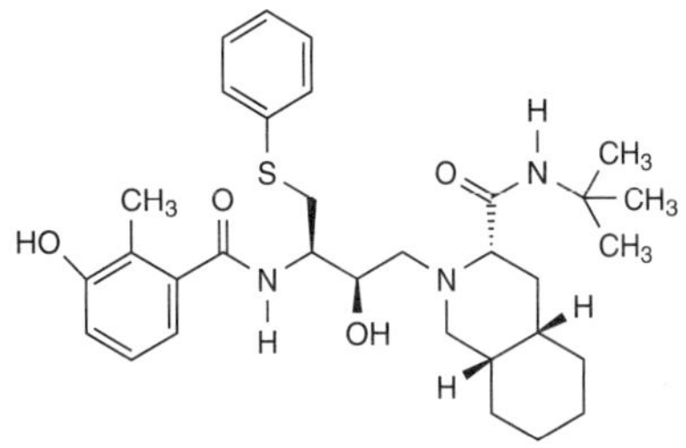

**Molecular formula:** $C_{32}H_{45}N_3O_4S$
**Molecular weight:** 567.79
**CAS Registry No:** 159989-64-7, 159989-65-8 (mesylate)
**Merck Index:** 13, 6471

## SAMPLE

**Matrix:** blood
**Sample preparation:** Condition an Extrasep C18 SPE cartridge (Lida) with 2 mL MeOH and 2 mL water. Dilute 500 μL serum with 500 μL water, add to the SPE cartridge, wash with 500 μL water, elute with 1 mL MeOH. Evaporate the eluate to dryness with vortexing under reduced pressure at 40° and reconstitute the residue with 300 μL MeOH, inject a 10 μL aliquot.

## HPLC VARIABLES

**Column:** two 150 × 4.6 3 μm Luna C18 columns in series
**Column temperature:** 60
**Mobile phase:** Gradient. MeCN:4 mM sulfuric acid from 8:92 to 63:37 over 45 min, maintain at 63:37 for 5 min.
**Flow rate:** 0.85
**Injection volume:** 10
**Detector:** UV 265 for 31 min then UV 240

## CHROMATOGRAM

**Retention time:** 33.5
**Limit of detection:** 400 ng/mL

## OTHER SUBSTANCES

**Extracted:** delavirdine (25.5, LOD 110 ng/mL), efavirenz (51, LOD 62 ng/mL), indinavir (24.5, LOD 210 ng/mL), nelfinavir (33.5, LOD 400 ng/mL), nevirapine (23.5, LOD 84 ng/mL), ritonavir (50.5, LOD 510 ng/mL), saquinavir (35, LOD 100 ng/mL)

## KEY WORDS

SPE; serum

## REFERENCE

Simon, V.A.; Thiam, M.D.; Lipford, L.C. Determination of serum levels of thirteen human immunodeficiency virus-suppressing drugs by high-performance liquid chromatography, *J.Chromatogr.A*, **2001**, *913*, 447–453.

## SAMPLE

**Matrix:** blood
**Sample preparation:** Mix 250 μL plasma with 50 μL MeOH, add 100 μL 2 μg/mL IS in MeOH, add 250 μL 1 M NaOH, add 3 mL hexane:ethyl acetate 50:50, shake at high speed for 25 min, centrifuge at 3000 g for 15 min. Evaporate the organic layer to dryness under a stream of air, reconstitute the residue with 1 mL initial mobile phase, inject a 20 μL aliquot.

## HPLC VARIABLES

**Guard column:** 10 × 2.1 Symmetry Shield

**Column:** 30 × 2.1 3.5 μm Symmetry C18
**Mobile phase:** Gradient. MeCN:5 mM pH 3.25 acetate buffer from 25:75 to 80:20 over 4 min using a nonlinear gradient (not specified).
**Flow rate:** 0.35
**Injection volume:** 20
**Detector:** MS, PE Sciex API 3000, turbo ionspray source, column effluent split 1:1 before entering source

---

## CHROMATOGRAM
**Retention time:** 2.5
**Internal standard:** Abbott A-86093 (3.2)
**Limit of detection:** 330 pg/mL
**Limit of quantitation:** 16.3 ng/mL

---

## OTHER SUBSTANCES
**Extracted:** amprenavir (2.7, LOQ 16.3 ng/mL, LOD 380 pg/mL), indinavir (2.0, LOQ 16.3 ng/mL, LOD 1.5 ng/mL), lopinavir (3.1, LOQ 16.3 ng/mL, LOD 750 pg/mL), ritonavir (2.9, LOQ 51.2 ng/mL, LOD 650 pg/mL), saquinavir (2.4, LOQ 16.3 ng/mL, LOD 780 pg/mL)

---

## KEY WORDS
plasma

---

## REFERENCE
Frerichs, V.A.; DiFrancesco, R.; Morse, G.D. Determination of protease inhibitors using liquid chromatography-tandem mass spectrometry, *J.Chromatogr.B*, **2003**, *787*, 393–403.

---

## SAMPLE
**Matrix:** blood
**Sample preparation:** Mix 1 mL plasma with 200 μL 10 μg/mL IS in water, add 200 μL 100 mM NaOH, mix, add 4 mL diethyl ether, shake for 5 min, centrifuge at 2500 rpm for 5 min. Evaporate the organic layer to dryness under a stream of nitrogen, reconstitute the residue with 200 μL initial mobile phase, inject a 100 μL aliquot.

---

## HPLC VARIABLES
**Column:** 250 × 4.6 Stability RP18 (CIL, France)
**Mobile phase:** Gradient. MeCN:50 mM pH 5.65 phosphate buffer from 36:64 to 64:36 over 25 min, to 80:20 (step gradient), maintain at 80:20 for 10 min, re-equilibrate at initial conditions for 5 min.
**Flow rate:** 1.5
**Injection volume:** 100
**Detector:** UV 240 for 5 min, UV 215 for 22 min, UV 260 for rest of run

---

## CHROMATOGRAM
**Retention time:** 24.1
**Internal standard:** JR051012 (Janssen Cilag) (28.2)
**Limit of quantitation:** 50 ng/mL

---

## OTHER SUBSTANCES
**Extracted:** amprenavir (11.2), efavirenz (19.9), indinavir (8.5), lopinavir (18.9), nevirapine (3.3), ritonavir (17.6), saquinavir (16.7)
**Noninterfering:** acetaminophen, amineptine, amphotericin B, aspirin, bromazepam, buspirone, citalopram, clobazam, diazepam, didanosine, fluconazole, flunitrazepam, fluvoxamine, hydroxyitraconazole, isoniazid, itraconazole, lamivudine, loprazolam, lorazepam, metronidazole, minalcipram, nordiazepam, omeprazole, paroxetine, pyrimethamine, rifampin, sertraline, stavudine, sulfadiazine, trimethoprim, venlafaxine, zalcitabine, zidovudine, zolpidem, zopiclone

## KEY WORDS
plasma

## REFERENCE
Titier, K.; Lagrange, F.; Péhourcq, F.; Edno-Mcheik, L.; Moore, N.; Molimard, M. High-performance liquid chromatographic method for the simultaneous determination of the six HIV-protease inhibitors and two non-nucleoside reverse transcriptase inhibitors in human plasma, *Ther.Drug Monit.*, **2002**, *24*, 417–424.

## SAMPLE
**Matrix:** blood
**Sample preparation:** Briefly vortex 250 μL 1 μg/mL IS in MeOH with 250 μL plasma, add 500 μL 100 mM pH 10.5 ammonium hydroxide, vortex, add 2 mL MeCN:ethyl acetate 10:90, vortex for 4 min, centrifuge at 2060 g, evaporate the organic layer to dryness under a stream of nitrogen at 50°, reconstitute with 150 μL mobile phase, vortex for 2 min, inject a 100 μL aliquot.

## HPLC VARIABLES
**Guard column:** Nova-Pak C18
**Column:** 250 × 4.6 5 μm Symmetry C18 (Waters)
**Mobile phase:** MeCN:25 mM sodium dihydrogen phosphate adjusted to pH 3.4 with phosphoric acid 42:58
**Flow rate:** 1.3
**Injection volume:** 100
**Detector:** UV 220

## CHROMATOGRAM
**Retention time:** 8.2
**Internal standard:** 6,7-dimethyl-2,3-di(2-pyridyl)quinoxaline (9.9)
**Limit of quantitation:** 50 ng/mL

## KEY WORDS
plasma

## REFERENCE
Wu, E.Y.; Wilkinson, J.M.I.I.; Naret, D.G.; Daniels, V.L.; Williams, L.J.; Khalil, D.A.; Shetty, B.V. High-performance liquid chromatographic method for the determination of nelfinavir, a novel HIV-1 protease inhibitor, in human plasma, *J.Chromatogr.B*, **1997**, *695*, 373–380.

## ANNOTATED BIBLIOGRAPHY

Aymard, G.; Legrand, M.; Trichereau, N.; Diquet, B. Determination of twelve antiretroviral agents in human plasma sample using reversed-phase high-performance liquid chromatography, *J.Chromatogr.B*, **2000**, *744*, 227–240. [amprenavir; efavirenz; indinavir; nelfinavir; ritonavir; saquinavir; abacavir; didanosine; lamivudine; stavudine; nevirapine; zidovudine; column-switching; SPE; LOQ 5–100 ng/mL]

Bouley, M.; Briere, C.; Padoin, C.; Petitjean, O.; Tod, M. Sensitive and rapid method for the simultaneous quantification of the HIV-protease inhibitors indinavir, nelfinavir, ritonavir, and saquinavir in human plasma by reversed-phase liquid chromatography, *Ther.Drug Monit.*, **2001**, *23*, 56–60. [LOQ 25–250 ng/mL]

Chi, J.; Jayewardene, A.L.; Stone, J.A.; Motoya, T.; Aweeka, F.T. Simultaneous determination of five HIV protease inhibitors nelfinavir, indinavir, ritonavir, saquinavir and amprenavir in human plasma by LC/MS/MS, *J.Pharm.Biomed.Anal.*, **2002**, *30*, 675–684. [LOQ 5 ng/mL]

Cociglio, M.; Hillaire-Buys, D.; Peyrière, H.; Alric, R. Performance analysis of a rapid HPLC determination with the solvent demixing extraction of HIV antiproteases and efavirenz in plasma, *J.Chromatogr.Sci.*, **2003**, *41*, 80–86. [LOD 1–175 ng/mL]

Dailly, E.; Thomas, L.; Kergueris, M.F.; Jolliet, P.; Bourin, M. High-performance liquid chromatographic assay to determine the plasma levels of HIV-protease inhibitors (amprenavir, indinavir, nelfinavir, ritonavir and saquinavir) and the non-nucleoside reverse transcriptase inhibitor (nevirapine) after liquid-liquid extraction, *J.Chromatogr.B*, **2001**, *758*, 129–135. [LOQ 50–400 ng/mL]

Droste, J.A.H.; Verweij-van Wissen, C.P.W.G.M.; Burger, D.M. Simultaneous determination of the HIV drugs indinavir, amprenavir, saquinavir, ritonavir, lopinavir, nelfinavir, the nelfinavir hydroxymetabolite M8, and nevirapine in human plasma by reversed-phase high-performance liquid chromatography, *Ther.Drug Monit.*, **2003**, *25*, 393–399. [LOQ 50–70 ng/mL]

Faux, J.; Venisse, N.; Le Moal, G.; Dupuis, A.; Bouquet, S. Simultaneous determination of six HIV protease inhibitors, one metabolite, and two non-nucleoside reverse transcriptase inhibitors in human plasma by isocratic reversed-phase liquid chromatography after solid-phase extraction, *Chromatographia*, **2003**, *58*, 421–426. [LOQ 50 – 120 ng/mL; amprenavir; indinavir; lopinavir; nelfinavir; ritonavir; saquinavir; efavirenz; nevirapine; prazepam is internal standard]

Hugen, P.W.H.; Verweij-van Wissen, C.P.W.G.M.; Burger, D.M.; Wuis, E.W.; Koopmans, P.P.; Hekster, Y.A. Simultaneous determination of the HIV-protease inhibitors indinavir, nelfinavir, saquinavir and ritonavir in human plasma by reversed-phase high-performance liquid chromatography, *J. Chromatogr.B*, **1999**, *727*, 139–149. [LOQ 45 ng/mL]

Janoly, A.; Bleyzac, N.; Favetta, P.; Gagneu, M.C.; Bourhis, Y.; Coudray, S.; Oger, I.; Aulagner, G. Simple and rapid high-performance liquid chromatographic method for nelfinavir, M8 nelfinavir metabolite, ritonavir and saquinavir assay in plasma, *J.Chromatogr.B*, **2002**, *780*, 155–160. [LOD 200 ng/mL]

Justesen, U.S.; Pedersen, C.; Klitgaard, N.A. Simultaneous quantitative determination of the HIV protease inhibitors indinavir, amprenavir, ritonavir, lopinavir, saquinavir, nelfinavir and the nelfinavir active metabolite M8 in plasma by liquid chromatography, *J.Chromatogr.B*, **2003**, *783*, 491–500. [LOQ 25 ng/mL]

Keil, K.; Frerichs, V.A.; DiFrancesco, R.; Morse, G. Reverse phase high-performance liquid chromatography method for the analysis of amprenavir, efavirenz, indinavir, lopinavir, nelfinavir and its active metabolite (M8), ritonavir, and saquinavir in heparinized human plasma, *Ther.Drug Monit.*, **2003**, *25*, 340–346. [LOQ 100–200 ng/mL]

Lamotte, C.; Peytavin, G.; Farinotti, R. Determination of nelfinavir, a potent HIV protease inhibitor, and its active metabolite M8 in human plasma by high-performance liquid chromatography with photodiode-array detection, *J.Chromatogr.B*, **1999**, *735*, 159–170. [LOQ 25 ng/mL]

Poirier, J.-M.; Radembino, N.; Robidou, P.; Jaillon, P. Simultaneous determination of the five HIV-protease inhibitors: Amprenavir, indinavir, nelfinavir, ritonavir, and saquinavir in human plasma by solid-phase extraction and column liquid chromatography, *Ther.Drug Monit.*, **2000**, *22*, 465–473. [LOQ 25 ng/mL]

Poirier, J.-M.; Robidou, P.; Jaillon, P. Simultaneous determination of the six HIV protease inhibitors (amprenavir, indinavir, lopinavir, nelfinavir, ritonavir, and saquinavir) plus M8 nelfinavir metabolite and the nonnucleoside reverse transcription inhibitor efavirenz in human plasma by solid-phase extraction and column liquid chromatography, *Ther.Drug Monit.*, **2002**, *24*, 302–309. [LOQ 25 ng/mL]

Proust, V.; Toth, K.; Hulin, A.; Taburet, A.-M.; Gimenez, F.; Singlas, E. Simultaneous high-performance liquid chromatographic determination of the antiretroviral agents amprenavir, nelfinavir, ritonavir, saquinavir, delavirdine and efavirenz in human plasma, *J.Chromatogr.B*, **2000**, *742*, 453–458. [LOQ 50–150 ng/mL]

Remmel, R.P.; Kawle, S.P.; Weller, D.; Fletcher, C.V. Simultaneous HPLC assay for quantification of indinavir, nelfinavir, ritonavir, and saquinavir in human plasma, *Clin.Chem.*, **2000**, *46*, 73–81. [LOQ 20–50 ng/mL]

Rentsch, K.M. Sensitive and specific determination of eight antiretroviral agents in plasma by high-performance liquid chromatography-mass spectrometry, *J.Chromatogr.B*, **2003**, *788*, 339–350. [SPE; LOQ 1–250 ng/mL; amprenavir; efavirenz; indinavir; lopinavir; nelfinavir; nevirapine; ritonavir; saquinavir]

Sarasa-Nacenta, M.; López-Púa, Y.; Mallolas, J.; Blanco, J.L.; Gatell, J.M.; Carné, X. Simultaneous determination of the HIV-protease inhibitors indinavir, amprenavir, ritonavir, saquinavir and nelfinavir in human plasma by reversed-phase high-performance liquid chromatography, *J.Chromatogr.B*, **2001**, *757*, 325–332. [SPE; LOQ 10–85 ng/mL]

Simon, V.A.; Thiam, M.D.; Lipford, L.C. Determination of serum levels of thirteen human immunodeficiency virus-suppressing drugs by high-performance liquid chromatography, *J.Chromatogr.A*, **2001**, *913*, 447–453. [SPE; LOD 30–440 ng/mL; zalcitabine; lamivudine; stavudine; didanosine; zidovudine; nevirapine; abacavir; indinavir; delavirdine; nelfinavir; saquinavir; ritonavir; efavirenz]

Tribut, O.; Arvieux, C.; Michelet, C.; Chapplain, J.-M.; Allain, H.; Bentué-Ferrer, D. Simultaneous quantitative assay of six HIV protease inhibitors, one metabolite, and two non-nucleoside reverse transcriptase inhibitors in human plasma by isocratic reversed-phase liquid chromatography, *Ther.Drug*

*Monit.*, **2002**, *24*, 554–562. [LOQ 25 ng/mL; nevirapine; efavirenz; indinavir; amprenavir; nelfinavir; ritonavir; lopinavir; saquinavir]

Turner, M.L.; Reed-Walker, K.; King, J.R.; Acosta, E.P. Simultaneous determination of nine antiretroviral compounds in human plasma using liquid chromatography, *J.Chromatogr.B*, **2003**, *784*, 331–341. [LOQ 50 ng/mL; indinavir; nelfinavir; saquinavir; ritonavir; amprenavir; delavirdine; efavirenz; lopinavir]

van Heeswijk, R.P.G.; Hoetelmans, R.M.W.; Harms, R.; Meenhorst, P.L.; Mulder, J.W.; Lange, J.M.A.; Beijnen, J.H. Simultaneous quantitative determination of the HIV protease inhibitors amprenavir, indinavir, nelfinavir, ritonavir and saquinavir in human plasma by ion-pair high-performance liquid chromatography with ultraviolet detection, *J.Chromatogr.B*, **1998**, *719*, 159–168. [SPE; LOQ 25–50 ng/mL]

Villani, P.; Feroggio, M.; Gianelli, L.; Bartoli, A.; Montagna, M.; Maserati, R.; Regazzi, M.B. Antiretrovirals: simultaneous determination of five protease inhibitors and three nonnucleoside transcriptase inhibitors in human plasma by a rapid high-performance liquid chromatography-mass spectrometry assay, *Ther.Drug Monit.*, **2001**, *23*, 380–388. [LOD 10–25 ng/mL; LOQ 20–40 ng/mL; saquinavir; indinavir; ritonavir; nelfinavir; amprenavir; nevirapine; delavirdine; efavirenz]

Yamada, H.; Kotaki, H.; Nakamura, T.; Iwamoto, A. Simultaneous determination of the HIV protease inhibitors indinavir, amprenavir, saquinavir, ritonavir and nelfinavir in human plasma by high-performance liquid chromatography, *J.Chromatogr.B*, **2001**, *755*, 85–89. [LOQ 50 ng/mL]

Zhang, K.E.; Wu, E.; Patick, A.K.; Kerr, B.; Zorbas, M.; Lankford, A.; Kobayashi, T.; Maeda, Y.; Shetty, B.; Webber, S. Circulating metabolites of the human immunodeficiency virus protease inhibitor nelfinavir in humans: Structural identification, levels in plasma, and antiviral activities, *Antimicrob.Agents Chemother.*, **2001**, *45*, 1086–1093. [LC-MS; reserpine is internal standard; LOQ 20 ng/mL]

# Nequinate

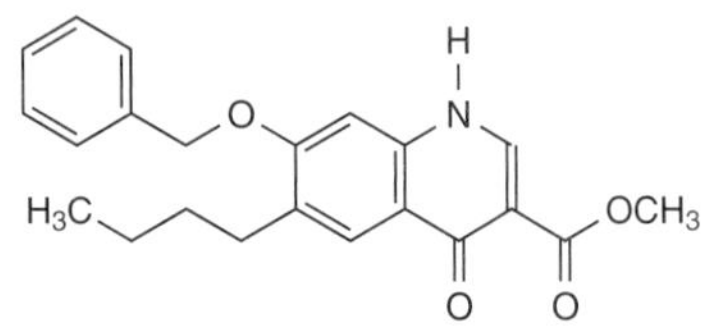


**Molecular formula:** $C_{22}H_{23}NO_4$
**Molecular weight:** 365.42
**CAS Registry No:** 13997-19-8
**Merck Index:** 13, 6498

## SAMPLE

**Matrix:** feed
**Sample preparation:** Slurry 100 g 100–200 mesh Dowex 50 W-X8 ($H^+$) cation-exchange resin with 500 mL 10% HCl, heat to boiling with continuous stirring, cool, discard liquid, wash resin twice with 500 mL portions of water, wash with 250 mL MeOH, filter (Whatman No. 541 paper), rinse with 200 mL MeOH, air dry. Slurry 5 g resin with 50 mL 10% HCl, pour into a chromatography column, wash with water until the effluent is neutral to litmus, wash with 50 mL MeOH. Do not allow to go dry at any stage. Shake 20 g finely ground feed with 100 mL 2% methanesulfonic acid in MeOH for 30 min, filter (Whatman No. 541 paper). Add 25 mL filtrate to 100 mL 10% HCl and 100 mL dichloromethane, shake for 1 min, extract the aqueous phase twice with 40 mL portions of dichloromethane. Combine the organic layers and evaporate to dryness under reduced pressure at 40°, reconstitute the residue with 20–25 mL MeOH, add to the ion-exchange column, rinse flask with 10 mL MeOH, add the rinse to the column, wash the column with 50 mL MeOH, elute the column with 150 mL 2% methanesulfonic acid in MeOH. Mix the eluate with 300 mL 10% HCl and 130 mL dichloromethane, shake for 1 min, extract the aqueous layer twice with 70 mL portions of dichloromethane. Combine the organic layers and evaporate to dryness under reduced pressure at 40°, reconstitute the residue with 10 mL MeOH, inject a 20 μL aliquot.

## HPLC VARIABLES

**Column:** 300 × 3.9 μBondapak C18
**Mobile phase:** MeOH:water 75:25
**Flow rate:** 1.5
**Injection volume:** 20
**Detector:** UV 265

## CHROMATOGRAM

**Retention time:** 8
**Limit of quantitation:** 1 μg/g

## OTHER SUBSTANCES

**Noninterfering:** clopidol

## KEY WORDS

SPE

## REFERENCE

Merson, G.H.J.; Hill, L.A.; Johnson, S.F. Determination of methyl benzoquate in poultry feeding stuffs using high-performance liquid chromatography, *Analyst*, **1985**, *110*, 761–764.

# Neridronic acid

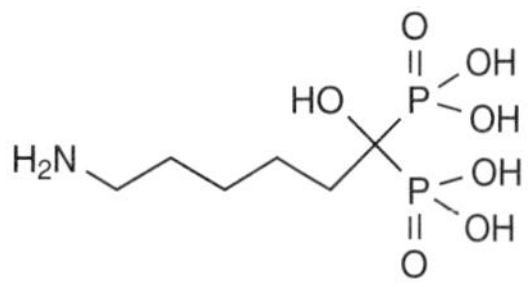

**Molecular formula:** $C_6H_{17}NO_7P_2$
**Molecular weight:** 277.15
**CAS Registry No:** 79778-41-9

## SAMPLE

**Matrix:** formulations
**Sample preparation:** Dilute injections 100-fold, inject a 20 μL aliquot. Disintegrate a 5 mg tablet in 100 mL water, sonicate for 5 min, centrifuge an aliquot at 3600 g for 4 min, inject a 20 μL aliquot of the supernatant.

## HPLC VARIABLES

**Column:** 150 × 4.6 10 μm IC-PAK Anion HC (Waters)
**Column temperature:** 30
**Mobile phase:** 1.5 mM nitric acid containing 0.5 mM copper(II) nitrate (Prepare column by pumping ILC Regenerant A (Waters) and 100 mM nitric acid for 30 min.)
**Flow rate:** 1
**Injection volume:** 20
**Detector:** UV 245

## CHROMATOGRAM

**Retention time:** 1.5

## OTHER SUBSTANCES

**Simultaneous:** alendronate (k′ 0.48), clodronate (k′ 26.4), etidronate (5.8 min), olpadronate (k′ 0.82), pamidronate (2 min)

## KEY WORDS

complexation; injections; tablets

## REFERENCE

Sparidans, R.W.; Den Hartigh, J.; Vermeij, P. High-performance ion-exchange chromatography with in-line complexation of bisphosphonates and their quality control in pharmaceutical preparations, *J.Pharm.Biomed.Anal.*, **1995**, *13*, 1545–1550.

## SAMPLE

**Matrix:** solutions
**Sample preparation:** Vortex 100 μL of an 80–500 μg/mL solution in water with 50 μL EtOH, 40 μL pyridine, 10 μL triethylamine, and 2 μL phenylisothiocyanate, heat at 80° for 5 min, evaporate under nitrogen at 80°, reconstitute with 1 mL water, wash twice with 2 mL portions of chloroform:1-pentanol 90:10 (Caution! Chloroform is a carcinogen!), inject a 20 μL aliquot of the aqueous layer.

## HPLC VARIABLES

**Column:** 100 × 3 5 μm Chromspher C18
**Mobile phase:** MeCN:20 mM pH 7–8 phosphate buffer containing 5 mM tetraethylammonium hydroxide and 0.5 mM etidronate (adsorption suppressor) 3:97
**Flow rate:** 0.4
**Injection volume:** 20
**Detector:** UV 240

## OTHER SUBSTANCES

**Simultaneous:** alendronate, pamidronate

## KEY WORDS

derivatization

## REFERENCE

Sparidans, R.W.; Den Hartigh, J.; Beijnen, J.H.; Vermeij, P. Derivatization of pamidronate and other amino(bis)phosphonates with different isothiocyanates prior to ion-pair liquid chromatography, *J.Chromatogr.A*, **1997**, *782*, 211–217.

# Nevirapine

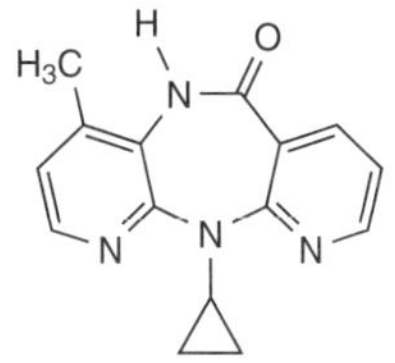

**Molecular formula:** $C_{15}H_{14}N_4O$
**Molecular weight:** 266.30
**CAS Registry No:** 129618-40-2
**Merck Index:** 13, 6514

## SAMPLE
**Matrix:** blood
**Sample preparation:** Condition an Extrasep C18 SPE cartridge (Lida) with 2 mL MeOH and 2 mL water. Dilute 500 μL serum with 500 μL water, add to the SPE cartridge, wash with 500 μL water, elute with 1 mL MeOH. Evaporate the eluate to dryness with vortexing under reduced pressure at 40° and reconstitute the residue with 300 μL MeOH, inject a 10 μL aliquot.

## HPLC VARIABLES
**Column:** two 150 × 4.6 3 μm Luna C18 columns in series
**Column temperature:** 60
**Mobile phase:** Gradient. MeCN:4 mM sulfuric acid from 8:92 to 63:37 over 45 min, maintain at 63:37 for 5 min.
**Flow rate:** 0.85
**Injection volume:** 10
**Detector:** UV 265 for 31 min then UV 240

## CHROMATOGRAM
**Retention time:** 23.5
**Limit of detection:** 84 ng/mL

## OTHER SUBSTANCES
**Extracted:** delavirdine (25.5, LOD 110 ng/mL), efavirenz (51, LOD 62 ng/mL), indinavir (24.5, LOD 210 ng/mL), nelfinavir (33.5, LOD 400 ng/mL), ritonavir (50.5, LOD 510 ng/mL), saquinavir (35, LOD 100 ng/mL)

## KEY WORDS
SPE; serum

## REFERENCE
Simon, V.A.; Thiam, M.D.; Lipford, L.C. Determination of serum levels of thirteen human immunodeficiency virus-suppressing drugs by high-performance liquid chromatography, *J.Chromatogr.A*, **2001**, *913*, 447–453.

## SAMPLE
**Matrix:** blood
**Sample preparation:** Mix 1 mL plasma with 200 μL 10 μg/mL IS in water, add 200 μL 100 mM NaOH, mix, add 4 mL diethyl ether, shake for 5 min, centrifuge at 2500 rpm for 5 min. Evaporate the organic layer to dryness under a stream of nitrogen, reconstitute the residue with 200 μL initial mobile phase, inject a 100 μL aliquot.

## HPLC VARIABLES
**Column:** 250 × 4.6 Stability RP18 (CIL, France)
**Mobile phase:** Gradient. MeCN:50 mM pH 5.65 phosphate buffer from 36:64 to 64:36 over 25 min, to 80:20 (step gradient), maintain at 80:20 for 10 min, re-equilibrate at initial conditions for 5 min.
**Flow rate:** 1.5

**Injection volume:** 100
**Detector:** UV 240 for 5 min, UV 215 for 22 min, UV 260 for rest of run

---

## CHROMATOGRAM
**Retention time:** 3.3
**Internal standard:** JR051012 (Janssen Cilag) (28.2)
**Limit of quantitation:** 100 ng/mL

---

## OTHER SUBSTANCES
**Extracted:** amprenavir (11.2), efavirenz (19.9), indinavir (8.5), lopinavir (18.9), nelfinavir (24.1), ritonavir (17.6), saquinavir (16.7)
**Noninterfering:** acetaminophen, amineptine, amphotericin B, aspirin, bromazepam, buspirone, citalopram, clobazam, diazepam, didanosine, fluconazole, flunitrazepam, fluvoxamine, hydroxyitraconazole, isoniazid, itraconazole, lamivudine, loprazolam, lorazepam, metronidazole, minalcipram, nordiazepam, omeprazole, paroxetine, pyrimethamine, rifampin, sertraline, stavudine, sulfadiazine, trimethoprim, venlafaxine, zalcitabine, zidovudine, zolpidem, zopiclone

---

## KEY WORDS
plasma

---

## REFERENCE
Titier, K.; Lagrange, F.; Péhourcq, F.; Edno-Mcheik, L.; Moore, N.; Molimard, M. High-performance liquid chromatographic method for the simultaneous determination of the six HIV-protease inhibitors and two non-nucleoside reverse transcriptase inhibitors in human plasma, *Ther.Drug Monit.*, **2002**, *24*, 417–424.

---

## SAMPLE
**Matrix:** bulk
**Sample preparation:** Mix 24 mg drug substance with 4 mL MeCN and 80 mL mobile phase, sonicate until all solid dissolves, cool, make up to 100 mL with mobile phase, inject a 50 μL aliquot.

---

## HPLC VARIABLES
**Column:** 150 × 4.6 5 μm Supelcosil LC-ABZ
**Column temperature:** 35
**Mobile phase:** MeCN:25 mM pH 5.0 ammonium dihydrogen phosphate buffer 20:80
**Flow rate:** 1
**Injection volume:** 50
**Detector:** UV 220

---

## CHROMATOGRAM
**Retention time:** 7.44
**Limit of detection:** 0.001% (S/N 3)
**Limit of quantitation:** 0.003% (S/N 10)

---

## OTHER SUBSTANCES
**Simultaneous:** impurities

---

## KEY WORDS
robust; stability-indicating

---

## REFERENCE
Li, Q.C.; Tougas, T.; Cohen, K.; Lee, R.; Meagan, P.; Corson, M.; Muchnick, T. Validation of a high-performance liquid chromatography method for the assay of and determination of related organic impurities in nevirapine drug substance, *J.Chromatogr.Sci.*, **2000**, *38*, 246–254.

## ANNOTATED BIBLIOGRAPHY

Aymard, G.; Legrand, M.; Trichereau, N.; Diquet, B. Determination of twelve antiretroviral agents in human plasma sample using reversed-phase high-performance liquid chromatography, *J.Chromatogr.B*, **2000**, *744*, 227–240. [amprenavir; efavirenz; indinavir; nelfinavir; ritonavir; saquinavir; abacavir; didanosine; lamivudine; stavudine; nevirapine; zidovudine; column-switching; SPE; LOQ 5–100 ng/mL]

Chi, J.; Jayewardene, A.L.; Stone, J.A.; Aweeka, F.T. An LC-MS-MS method for the determination of nevirapine, a non-nucleoside reverse transcriptase inhibitor, in human plasma, *J.Pharm.Biomed.Anal.*, **2003**, *31*, 953–959. [LOQ 25 ng/mL]

Dailly, E.; Thomas, L.; Kergueris, M.F.; Jolliet, P.; Bourin, M. High-performance liquid chromatographic assay to determine the plasma levels of HIV-protease inhibitors (amprenavir, indinavir, nelfinavir, ritonavir and saquinavir) and the non-nucleoside reverse transcriptase inhibitor (nevirapine) after liquid-liquid extraction, *J.Chromatogr.B*, **2001**, *758*, 129–135. [LOQ 50–400 ng/mL]

Droste, J.A.H.; Verweij-van Wissen, C.P.W.G.M.; Burger, D.M. Simultaneous determination of the HIV drugs indinavir, amprenavir, saquinavir, ritonavir, lopinavir, nelfinavir, the nelfinavir hydroxymetabolite M8, and nevirapine in human plasma by reversed-phase high-performance liquid chromatography, *Ther.Drug Monit.*, **2003**, *25*, 393–399. [LOQ 50–70 ng/mL]

Erickson, D.A.; Mather, G.; Trager, W.F.; Levy, R.H.; Keirns, J.J. Characterization of the in vitro biotransformation of the HIV-1 reverse transcriptase inhibitor nevirapine by human hepatic cytochromes P-450, *Drug Metab.Dispos.*, **1999**, *27*, 1488–1495.

Fan, B.; Stewart, J.T. Determination of zidovudine/zalcitabine/nevirapine in human plasma by ion-pair HPLC, *J.Liq.Chromatogr.Rel.Technol.*, **2001**, *24*, 3017–3026. [SPE; LOQ 20–58 ng/mL; aprobarbital is internal standard]

Fan, B.; Stewart, J.T. Determination of zidovudine/lamivudine/nevirapine in human plasma using ion-pair HPLC, *J.Pharm.Biomed.Anal.*, **2002**, *28*, 903–908. [SPE; LOQ 53–59 ng/mL]

Faux, J.; Venisse, N.; Le Moal, G.; Dupuis, A.; Bouquet, S. Simultaneous determination of six HIV protease inhibitors, one metabolite, and two non-nucleoside reverse transcriptase inhibitors in human plasma by isocratic reversed-phase liquid chromatography after solid-phase extraction, *Chromatographia*, **2003**, *58*, 421–426. [LOQ 50 – 120 ng/mL; amprenavir; indinavir; lopinavir; nelfinavir; ritonavir; saquinavir; efavirenz; nevirapine; prazepam is internal standard]

Hollanders, R.M.F.; van Ewijk-Beneken Kolmer, E.W.J.; Burger, D.M.; Wuis, E.W.; Koopmans, P.P.; Hekster, Y.A. Determination of nevirapine, an HIV-1 non-nucleoside reverse transcriptase inhibitor, in human plasma by reversed-phase high-performance liquid chromatography, *J.Chromatogr.B*, **2000**, *744*, 65–71. [LOD 50 ng/mL; LOQ 100 ng/mL]

Kishimoto, W.; Takano, J.; Senda, C.; Ishiguro, N.; Sakai, K.; Igarashi, T. Quantitative prediction of in vivo drug interactions between nevirapine and antifungal agents from in vitro data in rats, *Biol.Pharm.Bull.*, **2000**, *23*, 1027–1032.

Lips, A.G.A.M.; Lameijer, W.; Fokkens, R.H.; Nibbering, N.M.M. Methodology for the development of a drug library based upon collision-induced fragmentation for the identification of toxicologically relevant drugs in plasma samples, *J.Chromatogr.B*, **2001**, *759*, 191–207. [plasma; LC-MS; morphine; sotalol; acetaminophen; codeine; olanzapine; caffeine; amphetamine; lamotrigine; phenytoin; nevirapine; phenobarbital; clozapine; midazolam; flurazepam; paroxetine; promethazine; flecainide; haloperidol; oxazepam; phenprocoumon; lorazepam; naproxen; amitriptyline; zopiclone; sertraline; zuclopenthixol; triazolam; flunitrazepam; saquinavir; pimozide; clomipramine; clobazam; valproic acid; chlordiazepoxide; diazepam; ritonavir; ibuprofen; promazine is internal standard]

Lopez, R.M.; Pou, L.; Gomez, M.R.; Ruiz, I.; Monterde, J. Simple and rapid determination of nevirapine in human serum by reversed-phase high-performance liquid chromatography, *J.Chromatogr.B*, **2001**, *751*, 371–376. [LOQ 100 ng/mL]

Marchei, E.; Valvo, L.; Pacifici, R.; Pellegrini, M.; Tossini, G.; Zuccaro, P. Simultaneous determination of zidovudine and nevirapine in human plasma by RP-LC, *J.Pharm.Biomed.Anal.*, **2002**, *29*, 1081–1088. [SPE; LOD 25–50 ng/mL; LOQ 50–150 ng/mL]

Marzolini, C.; Béguin, A.; Telenti, A.; Schreyer, A.; Buclin, T.; Biollaz, J.; Decosterd, L.A. Determination of lopinavir and nevirapine by high-performance liquid chromatography after solid-phase extraction: application for the assessment of their transplacental passage at delivery, *J.Chromatogr.B*, **2002**, *774*, 127–140. [SPE; LOD 15–50 ng/mL; LOQ 60–100 ng/mL; clozapine is internal standard]

Moyer, T.P.; Temesgen, Z.; Enger, R.; Estes, L.; Charlson, J.; Oliver, L.; Wight, A. Drug monitoring of antiretroviral therapy for HIV-1 infection: Method validation and results of a pilot study, *Clin.Chem.*, **1999**, *45*, 1465–1476. [serum; LOD 10–100 ng/mL; didanosine; lamivudine; stavudine; zalcitabine; zidovudine; delavirdine; nevirapine; indinavir; nelfinavir; ritonavir; saquinavir]

Palladino, D.E.; Hopkins, J.L.; Ingraham, R.H.; Warren, T.C.; Kapadia, S.R.; Van Moffaert, G.J.; Grob, P.M.; Stevenson, J.M.; Cohen, K.A. High-performance liquid chromatography and photoaffinity crosslinking to explore the binding environment of nevirapine to reverse transcriptase of human immunodeficiency virus, *J.Chromatogr.A*, **1994**, *676*, 99–112.

Pav, J.W.; Rowland, L.S.; Korpalski, D.J.; Elvin, A.T. HPLC-UV method for the quantitation of nevirapine (Viramune$^{TM}$) in human plasma, serum, and milk following solid phase extraction (Abstract 4214), *Pharm.Res.*, **1997**, *14*, S697.

Pav, J.W.; Rowland, L.S.; Korpalski, D.J. HPLC-UV method for the quantitation of nevirapine in biological matrices following solid phase extraction, *J.Pharm.Biomed.Anal.*, **1999**, *20*, 91–98. [SPE; LOQ 25 ng/mL; plasma; serum; milk; CSF]

Rabin, L.; Hincenbergs, M.; Moreno, M.B.; Warren, S.; Linquist, V.; Datema, R.; Charpiot, B.; Seifert, J.; Kaneshima, H.; McCune, J.M. Use of standardized SCID-hu Thy/Liv mouse model for preclinical efficacy testing if anti-human immunodeficiency virus type I compounds, *Antimicrob.Agents Chemother.*, **1996**, *40*, 755–762. [plasma; LOD 8 ng/mL]

Rentsch, K.M. Sensitive and specific determination of eight antiretroviral agents in plasma by high-performance liquid chromatography-mass spectrometry, *J.Chromatogr.B*, **2003**, *788*, 339–350. [SPE; LOQ 1–250 ng/mL; amprenavir; efavirenz; indinavir; lopinavir; nelfinavir; nevirapine; ritonavir; saquinavir]

Rezk, N.L.; Tidwell, R.R.; Kashuba, A.D.M. Simple and rapid quantification of the non-nucleoside reverse transcriptase inhibitors nevirapine, delavirdine, and efavirenz in human blood plasma using high-performance liquid chromatography with ultraviolet absorbance detection, *J.Chromatogr.B*, **2002**, *774*, 79–88. [LOQ 10–25 ng/mL; SPE; hexobarbital is internal standard]

Rezk, N.L.; Tidwell, R.R.; Kashuba, A.D.M. Simultaneous determination of six HIV nucleoside analogue reverse transcriptase inhibitors and nevirapine by liquid chromatography with ultraviolet absorbance detection, *J.Chromatogr.B*, **2003**, *791*, 137–147. [SPE; LOQ 10 ng/mL; zalcitabine; lamivudine; didanosine; stavudine; zidovudine; abacavir; hexobarbital is internal standard]

Riska, P.; Lamson, M.; Macgregor, T.; Sabo, J.; Hattox, S.; Pav, J.; Keirns, J. Disposition and biotransformation of the antiretroviral drug nevirapine in humans, *Drug Metab.Dispos.*, **1999**, *27*, 895–901.

Riska, P.S.; Joseph, D.P.; Dinallo, R.M.; Davidson, W.C.; Keirns, J.J.; Hattox, S.E. Biotransformation of nevirapine, a non-nucleoside HIV-1 reverse transcriptase inhibitor, in mice, rats, rabbits, dogs, monkeys, and chimpanzees, *Drug Metab.Dispos.*, **1999**, *27*, 1434–1447. [SPE]

Simon, V.A.; Thiam, M.D.; Lipford, L.C. Determination of serum levels of thirteen human immunodeficiency virus-suppressing drugs by high-performance liquid chromatography, *J.Chromatogr.A*, **2001**, *913*, 447–453. [SPE; LOD 30–440 ng/mL; zalcitabine; lamivudine; stavudine; didanosine; zidovudine; nevirapine; abacavir; indinavir; delavirdine; nelfinavir; saquinavir; ritonavir; efavirenz]

Störmer, E.; von Moltke, L.L.; Perloff, M.D.; Greenblatt, D.J. Differential modulation of P-glycoprotein expression and activity by non-nucleoside HIV-1 reverse transcriptase inhibitors in cell culture, *Pharm.Res.*, **2002**, *19*, 1038–1045. [verapamil; efavirenz; nevirapine; delavirdine]

Tribut, O.; Arvieux, C.; Michelet, C.; Chapplain, J.-M.; Allain, H.; Bentué-Ferrer, D. Simultaneous quantitative assay of six HIV protease inhibitors, one metabolite, and two non-nucleoside reverse transcriptase inhibitors in human plasma by isocratic reversed-phase liquid chromatography, *Ther.Drug Monit.*, **2002**, *24*, 554–562. [LOQ 25 ng/mL; nevirapine; efavirenz; indinavir; amprenavir; nelfinavir; ritonavir; lopinavir; saquinavir]

van Heeswijk, R.P.; Hoetelmans, R.M.; Meenhorst, P.L.; Mulder, J.W.; Beijnen, J.H. Rapid determination of nevirapine in human plasma by ion-pair reversed- phase high-performance liquid chromatography with ultraviolet detection, *J.Chromatogr.B*, **1998**, *713*, 395–399. [LOQ 52 ng/mL]

Villani, P.; Feroggio, M.; Gianelli, L.; Bartoli, A.; Montagna, M.; Maserati, R.; Regazzi, M.B. Antiretrovirals: simultaneous determination of five protease inhibitors and three nonnucleoside transcriptase inhibitors in human plasma by a rapid high-performance liquid chromatography-mass spectrometry assay, *Ther.Drug Monit.*, **2001**, *23*, 380–388. [LOD 10–25 ng/mL; LOQ 20–40 ng/mL; saquinavir; indinavir; ritonavir; nelfinavir; amprenavir; nevirapine; delavirdine; efavirenz]

# Nicarbazin

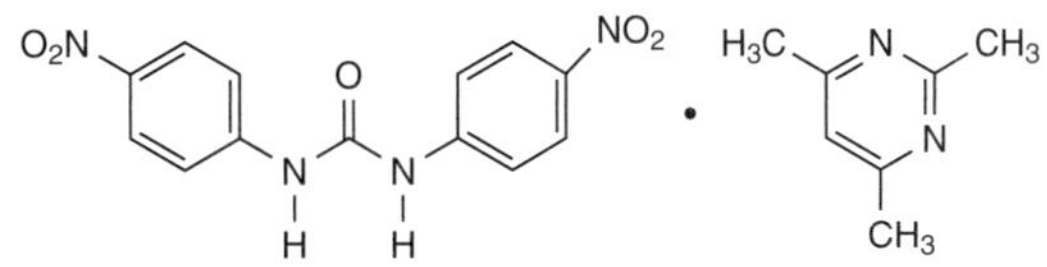


**Molecular formula:** $C_{19}H_{18}N_6O_6$
**Molecular weight:** 426.38
**CAS Registry No:** 330-95-0
**Merck Index:** 13, 6519

## SAMPLE

**Matrix:** blood
**Sample preparation:** Vortex 100 μL plasma with 200 μL MeCN, centrifuge for 5 min, inject a 60 μL aliquot of the supernatant.

## HPLC VARIABLES

**Guard column:** 15 × 4.6 C18
**Column:** 250 × 4.6 5 μm Keystone ODS/H
**Column temperature:** 35
**Mobile phase:** MeCN:water 60:40
**Flow rate:** 1
**Injection volume:** 60
**Detector:** UV 347

## CHROMATOGRAM

**Retention time:** 8.5 (dinitrocarbanilide)
**Limit of detection:** 27–35 ng/mL

## KEY WORDS

chicken; duck; goose; plasma

## REFERENCE

Primus, T.M.; Kohler, D.J.; Goodall, M.A.; Yoder, C.; Griffin, D.; Miller, L.; Johnston, J.J. Determination of 4,4′-dinitrocarbanilide (DNC), the active component of the antifertility agent nicarbazin, in chicken, duck, and goose plasma, *J.Agric.Food Chem.*, **2001**, *49*, 3589–3593.

# Nilutamide

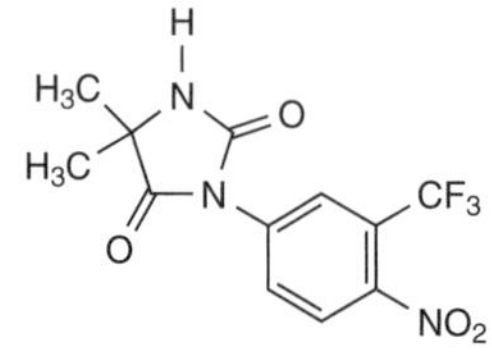

**Molecular formula:** $C_{12}H_{10}F_3N_3O_4$
**Molecular weight:** 317.22
**CAS Registry No:** 63612-50-0
**Merck Index:** 13, 6572

## SAMPLE

**Matrix:** blood, urine
**Sample preparation:** Extract 1 mL plasma or urine with 2 mL chloroform (Caution! Chloroform is a carcinogen!). Evaporate an aliquot of the organic layer to dryness, reconstitute the residue with 1 mL mobile phase, inject a 10–20 μL aliquot.

## HPLC VARIABLES

**Column:** 250 mm long 5 μm C18
**Mobile phase:** MeOH
**Flow rate:** 1.2
**Injection volume:** 10–20
**Detector:** UV 254

## CHROMATOGRAM

**Retention time:** 4.5
**Limit of detection:** 50 ng/mL

## KEY WORDS

plasma

## REFERENCE

Pendyala, L.; Creaven, P.J.; Huben, R.; Tremblay, D.; Bertagna, C. Pharmacokinetics of Anandron in patients with advanced carcinoma of the prostate, *Cancer Chemother.Pharmacol.*, **1988**, *22*, 69–76.

# Nipradilol

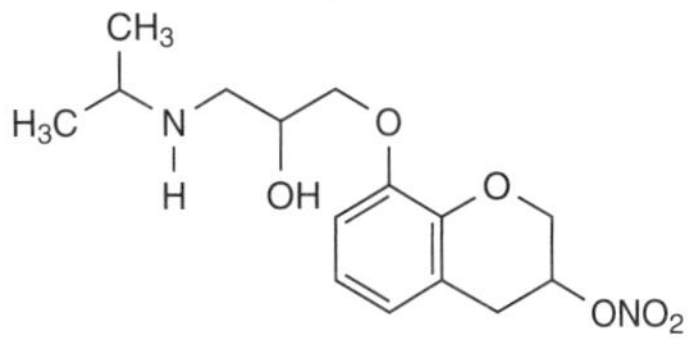


**Molecular formula:** $C_{15}H_{22}N_2O_6$
**Molecular weight:** 326.34
**CAS Registry No:** 81486-22-8
**Merck Index:** 13, 6590

## SAMPLE

**Matrix:** solutions
**Sample preparation:** Add 50 μL L-menthoxyacetyl chloride to a solution of 5–10 mg nipradilol in 500 μL dry pyridine, let stand at room temperature for 30 min, add 50 μL water, evaporate to dryness under reduced pressure, reconstitute with 5 mL chloroform (Caution! Chloroform is a carcinogen!), wash twice with 3 mL portions of 1 M HCl, wash with 3 mL saturated sodium bicarbonate solution, dry over anhydrous sodium sulfate, inject an aliquot.

## HPLC VARIABLES

**Column:** 200 × 4 10 μm Partisil-10
**Mobile phase:** Hexane:ethyl acetate 100:20
**Flow rate:** 1.5
**Detector:** UV 275

## CHROMATOGRAM

**Retention time:** 14 (2′R,3S), 15 (2′S,3R), 17 (2′S,3S), 20 (2′R,3R) [ 2′ hydroxy, 3 nitroxy]

## KEY WORDS

chiral; derivatization; normal phase

## REFERENCE

Yoneda, M.; Shiratsuchi, M.; Yoshimura, M.; Ohkawa, Y.; Muramatsu, T. Optical resolution and determination of absolute configuration of nipradilol, *Chem.Pharm.Bull.*, **1985**, *33*, 2735–2742.

# Nitazoxanide

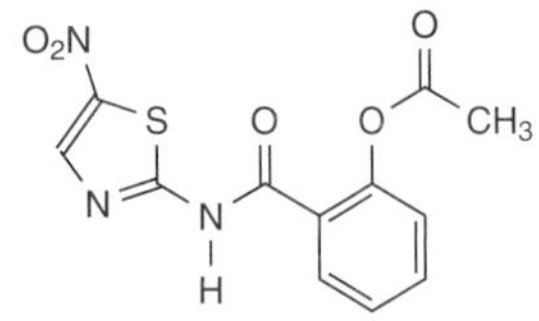

**Molecular formula:** $C_{12}H_9N_3O_5S$
**Molecular weight:** 307.29
**CAS Registry No:** 55981-09-4
**Merck Index:** 13, 6595

## SAMPLE
**Matrix:** blood, urine
**Sample preparation:** Plasma. Vortex 500 µL plasma and IS with 1 mL MeCN, centrifuge, inject a 20 µL aliquot of the supernatant. Urine. Mix 500 µL urine with 500 µL 1% Helix Pomatia juice (type H2, Sigma) in 100 mM pH 5 acetate buffer, heat at 37° overnight, inject an aliquot.

## HPLC VARIABLES
**Column:** 100 × 4.6 5 µm Nucleosil C18
**Mobile phase:** Gradient. MeCN:20 mM pH 2.5 phosphate buffer from10:90 to 35:65 over 8 min.
**Injection volume:** 20
**Detector:** UV 360

## CHROMATOGRAM
**Internal standard:** nifuroxazide
**Limit of quantitation:** 20 ng/mL (plasma), 200 ng/mL (urine)

## KEY WORDS
only desacetylnitazoxanide is detected in plasma and urine; plasma

## REFERENCE
Stockis, A.; Deroubaix, X.; Lins, R.; Jeanbaptiste, B.; Calderon, P.; Rossignol, J.F. Pharmacokinetics of nitazoxanide after single oral dose administration in 6 healthy volunteers, *Int.J.Clin.Pharmacol.Ther.*, **1996**, *34*, 349–351.

## SAMPLE
**Matrix:** microsomal incubations, urine
**Sample preparation:** Urine. Mix 2 mL urine with 10 µL 100 U/mL $\beta$-glucuronidase, 5 µL arylsulfatase (Helix pomatia) (Boehringer Mannheim), and 200 µL 1 M pH 5.5 acetate buffer, heat at 37° for 3 h, add 200 µL 1 M formic acid, centrifuge, inject a 100 µL aliquot. Microsomal incubations. Mix 1 mL microsomal incubation with 100 µL 2 M formic acid, add 550 µL MeCN, mix, centrifuge, inject an aliquot. (This method is also said to work for plasma and feces, but there are no details.) (Only the deacylated compound tizanoxide is detected in biological fluids.)

## HPLC VARIABLES
**Column:** Nucleosil 5C18
**Mobile phase:** Gradient. MeCN:10 mM formic acid from 10:90 to 90:10 over (?)
**Injection volume:** 100
**Detector:** MS, VG Quattro II

## KEY WORDS

dog; human; liver; monkey; rat

## REFERENCE

Broekhuysen, J.; Stockis, A.; Lins, R.L.; De Graeve, J.; Rossignol, J.F. Nitazoxanide: pharmacokinetics and metabolism in man, *Int.J.Clin.Pharmacol.Ther.*, **2000**, *38*, 387–394.

# Nitenpyram

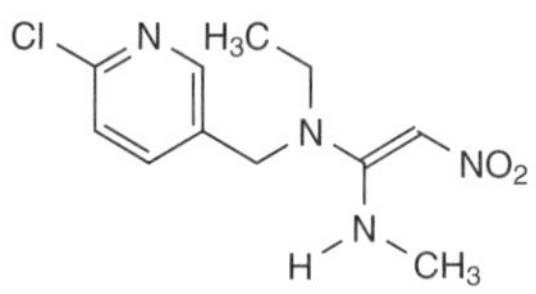


**Molecular formula:** $C_{11}H_{15}ClN_4O_2$
**Molecular weight:** 270.72
**CAS Registry No:** 150824-47-8
**Merck Index:** 13, 6596

## SAMPLE

**Matrix:** fruit, vegetables
**Sample preparation:** Condition a 3 mL 500 mg Bond Elut PSA (weak anion-exchange) SPE cartridge with 10 mL acetone and 10 mL acetone:hexane 50:50. Condition a 6 mL 1 g Mega Bond Elut silica SPE cartridge with 10 mL acetone and 10 mL acetone:hexane 30:70. Blend (Polytron) 20 g sample with 100 mL MeCN for 2 min, filter (paper). Add 5 g NaCl to the filtrate and shake mechanically for 1 min. Evaporate 50 mL of the MeCN layer to near dryness, take up in 2 mL acetone, wash out flask with 2 mL acetone. Combine the acetone solutions with 4 mL hexane, add to the PSA SPE cartridge, elute with 5 mL acetone:hexane 50:50. Evaporate the eluates and dissolve in 2 mL acetone:hexane 30:70, add to the silica SPE cartridge, wash with 10 mL acetone:hexane 30:70. Discard all effluent from the silica SPE cartridge. Elute with 10 mL acetone:hexane 40:60 (for acetamiprid and imidacloprid) and 20 mL acetone (for nitenpyram). Evaporate each eluate to dryness, reconstitute the residue with 2 mL MeOH, inject a 10 μL aliquot.

## HPLC VARIABLES

**Column:** 75 × 4.6 3 μm Cadenza CD-C18 (Imtakt, Kyoto)
**Column temperature:** 50
**Mobile phase:** Gradient. MeCN:50 mM potassium dihydrogen phosphate 3:97 for 3 min, to 40:60 over 7 min, maintain at 40:60 for 5 min, to 100:0 over 5 min, maintain at 100:0 for 5 min, to 5:95 over 5 min, maintain at 5:95 for 5 min.
**Flow rate:** 0.8
**Injection volume:** 10
**Detector:** UV 270

## CHROMATOGRAM

**Retention time:** 11
**Limit of detection:** 20 ng/g

## OTHER SUBSTANCES

**Extracted:** acetamiprid (UV 245; LOD 10 ng/g) (15), imidacloprid (LOD 10 ng/g) (14)

## KEY WORDS

cucumber; eggplant; grape; Japanese radish; potato; SPE; tomato

## REFERENCE

Obana, H.; Okihashi, M.; Akutsu, K.; Kitagawa, Y.; Hori, S. Determination of acetamiprid, imidacloprid, and nitenpyram residues in vegetables and fruits by high-performance liquid chromatography with diode-array detection, *J.Agric.Food Chem.*, **2002**, *50*, 4464–4467.

# Nomegestrol

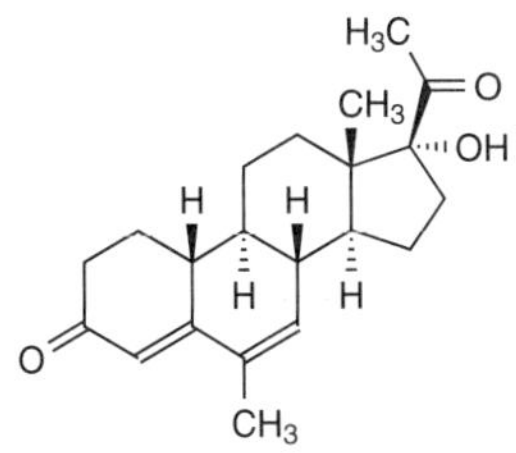

**Molecular formula:** $C_{21}H_{28}O_3$
**Molecular weight:** 328.45
**CAS Registry No:** 58691-88-6

## SAMPLE

**Matrix:** blood
**Sample preparation:** Condition a Sep Pak C18 SPE cartridge with MeOH and water. Add 1 mL plasma to the SPE cartridge, wash with 5 mL water, wash with 2 mL MeOH:water 70:30, elute with 2 mL MeOH. Evaporate the eluate to dryness, reconstitute the residue with 150 μL EtOH, inject a 20 μL aliquot.

## HPLC VARIABLES

**Column:** 250 × 4.6 5 μm Ultrasphere
**Mobile phase:** MeOH:water 70:30
**Flow rate:** 1.2
**Injection volume:** 20
**Detector:** UV 280

## CHROMATOGRAM

**Retention time:** 13

## KEY WORDS

plasma; SPE; for nomegestrol acetate

## REFERENCE

Ezan, E.; Benech, H.; Bucourt, R.; Ardouin, T.; Tchernatinsky, C.; Thomas, J.L.; Paris, J.; Grognet, J.M. Enzyme immunoassay for nomegestrol acetate in human plasma, *J.Steroid Biochem.Mol.Biol.*, **1993**, *46*, 507–514.

# Nonoxynol-9

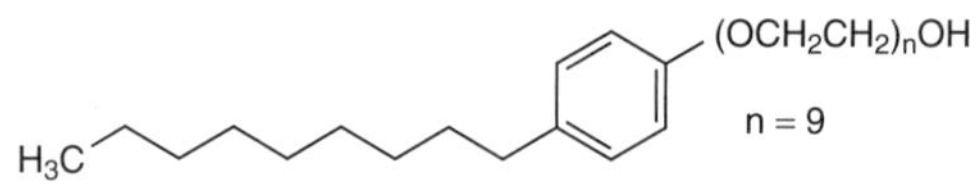


**CAS Registry No:** 26027-38-3

**Merck Index:** 13, 6711

## SAMPLE

**Matrix:** blood

**Sample preparation:** Mix 1 mL whole blood with 1 g solid NaCl, grind (sic) for 5 min, extract twice with 4 mL portions of benzene (Caution! Benzene is a carcinogen!), centrifuge at 8000 rpm for 30 min. Combine the extracts and evaporate them to dryness under a stream of nitrogen, dry under vacuum at 50° for 1 h, reconstitute the residue with 100 μL MeOH, centrifuge, inject a 20 μL aliquot of the supernatant.

## HPLC VARIABLES

**Column:** 250 × 4.6 TSK-Gel C18

**Mobile phase:** MeOH:water (ratio not given)

**Injection volume:** 20

**Detector:** UV 221

## CHROMATOGRAM

**Retention time:** 7.3

**Limit of quantitation:** 50 ng/mL

## KEY WORDS

whole blood; also, using normal phase with a silica column and *n*-hexane at UV 276 gives numerous peaks for oligomers.

## REFERENCE

Yang, J.; Zhao, Z. Quantitative analysis of nonoxynol-9 in the blood, *Contraception*, **1991**, *43*, 161–166.

# Nystatin

**Molecular formula:** $C_{47}H_{75}NO_{17}$ ($A_1$)
**Molecular weight:** 926.09 ($A_1$)
**CAS Registry No:** 1400-61-9, 34786-70-4 ($A_1$)
**Merck Index:** 13, 6770

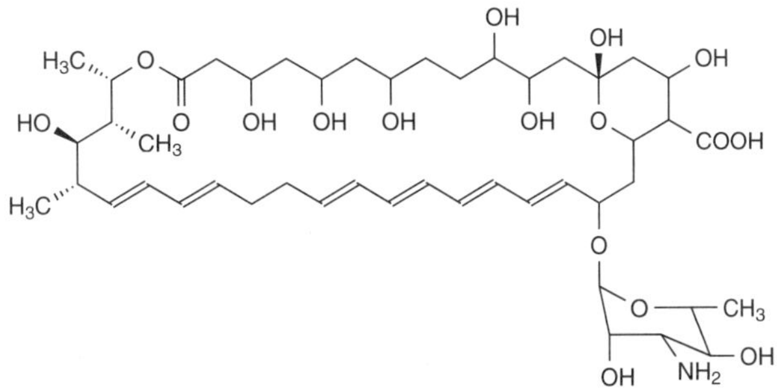

## SAMPLE

**Matrix:** blood, tissue

**Sample preparation:** Plasma. Mix 800 μL MeOH with 400 μL plasma, let stand at 4° for 30 min, centrifuge at 2000 g for 10 min, centrifuge the supernatant in a separate tube at 10 000 g for 4 min. Filter (Durapore 0.22 μm) 400 μL of the supernatant while centrifuging at 4000 g for 4 min, inject a 200 μL aliquot of the filtrate. Tissue. Homogenize (Tissuemizer 10 N head) tissue with 2 vol of ice-cold MeOH at 0°, let stand at 4° for 30 min, centrifuge at 2000 g for 10 min, centrifuge the supernatant in a separate tube at 10 000 g for 4 min. Filter (Durapore 0.22 μm) 400 μL of the supernatant while centrifuging at 4000 g for 4 min, inject a 200 μL aliquot of the filtrate.

## HPLC VARIABLES

**Guard column:** 15 × 3.2 5 μm NewGuard RP-18
**Column:** 300 × 3.9 10 μm μBondapak C18
**Column temperature:** 30
**Mobile phase:** MeCN:MeOH:10 mM sodium phosphate buffer containing 1 mM EDTA 30:30:40, adjusted to pH 6.0 with 85% phosphoric acid
**Flow rate:** 2
**Injection volume:** 200
**Detector:** UV 305

## CHROMATOGRAM

**Retention time:** 8.1, 10.0
**Limit of quantitation:** 50 ng/mL

## OTHER SUBSTANCES

**Noninterfering:** acetaminophen, amikacin, amitriptyline, carbamazepine, ceftazidime, cyclosporine, digoxin, disopyramide, ethosuximide, gentamicin, lidocaine, lithium, methotrexate, phenobarbital, phenytoin, primidone, procainamide, propranolol, quinidine, salicylic acid, theophylline, tobramycin, valproic acid, vancomycin

## KEY WORDS

human; pharmacokinetics; plasma; rabbit

## REFERENCE

Groll, A.H.; Mickiene, D.; Werner, K.; Piscitelli, S.C.; Walsh, T.J. High-performance liquid chromatographic determination of liposomal nystatin in plasma and tissues for pharmacokinetic and tissue distribution studies, *J.Chromatogr.B*, **1999**, *735*, 51–62.

# Octocrylene

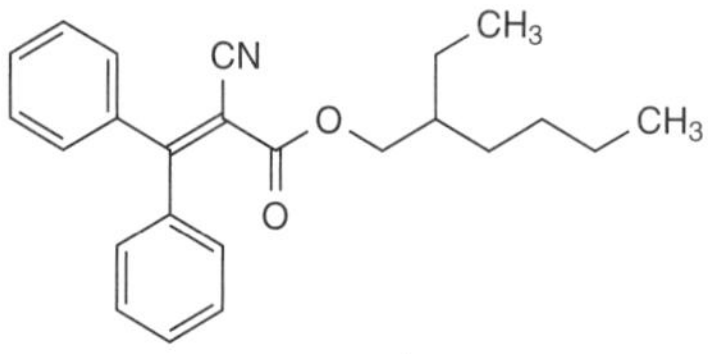


**Molecular formula:** $C_{24}H_{27}NO_2$
**Molecular weight:** 361.48
**CAS Registry No:** 6197-30-4

## SAMPLE

**Matrix:** sunscreen
**Sample preparation:** Heat 2 g sunscreen in 40 mL MeOH and 250 μL 2 M sulfuric acid at 60° for 5 min until a homogeneous solution develops, cool, make up to 50 mL with MeOH, dilute a 1 mL aliquot to 5 mL with initial mobile phase, inject a 5–50 μL aliquot.

## HPLC VARIABLES

**Column:** 150 × 4.6 5 μm PLRP-S (Polymer Labs)
**Column temperature:** 25
**Mobile phase:** Gradient. MeCN:THF:buffer 10:10:80 for 2.5 min, to 45:45:10 over 22.5 min, maintain at 45:45:10 for 10 min, return to initial conditions over 5 min, re-equilibrate for 5 min. (The buffer was 1.4 g/L citric acid monohydrate containing 6.8 g/L tetrabutylammonium hydroxide, adjusted to pH 9.0 with concentrated ammonia.)
**Flow rate:** 0.8 for 2.5 min, to 0.6 over 2.5 min, maintain at 0.6 for 10 min, return to 0.8 over 5 min, maintain at 0.8 for 5 min
**Injection volume:** 5–50
**Detector:** UV 302

## CHROMATOGRAM

**Retention time:** 29.13
**Limit of detection:** 50–500 ppm

## OTHER SUBSTANCES

**Simultaneous:** benzophenone-3 (25.93 min, UV 288), benzophenone-4 (12.47 min, UV 241), 3-benzylidene-*d,l*-camphor (26.47 min, UV 293), benzylidene camphor sulfonic acid (15.77 min, UV 297), butyl methoxydibenzoylmethane (29.71 min, UV 359), camphor benzalkonium methosulfate (6.178 min, UV 288), 3-(4′-ethylbenzylidene)-*d,l*-camphor (27.30 min, UV 297), homosalate (30.32 min, UV 241), isoamyl *p*-methoxycinnamate (26.93 min, UV 307), isopropyl dibenzoylmethane (29.86 min, UV 350), megasol (29.17 min, UV 241), octyl dimethyl PABA (28.83 min, UV 312), octyl methoxycinnamate (29.43 min, UV 307), octyl salicylate (30.03 min, UV 241), octyl triazone (33.72 min, UV 312), PEG–25 *p*-aminobenzoic acid (10.79 min, UV 307), phenylbenzimidazole sulfonic acid (8.898 min, UV 302), terephthalydiene dicamphor sulfonic acid (13.18 min, UV 340), urocanic acid (2.468 min, UV 278)

## REFERENCE

Rastogi, S.C.; Jensen, G.H. Identification of UV filters in sunscreen products by high-performance liquid chromatography-diode-array detection, *J.Chromatogr.A*, **1998**, *828*, 311–316.

# Oleic acid

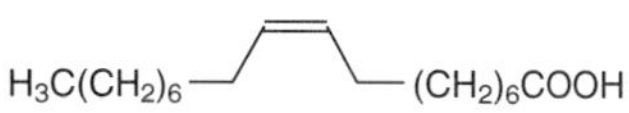


**Molecular formula:** $C_{18}H_{34}O_2$
**Molecular weight:** 282.46
**CAS Registry No:** 112-80-1
**Merck Index:** 13, 6898

## SAMPLE
**Matrix:** blood
**Sample preparation:** Mix 10 μL plasma with 200 μL 40 μg/mL IS in EtOH and 90 μL EtOH, add 100 μL 2 M KOH in EtOH:water 50:50, heat at 80° for 20 min, cool to room temperature. Add 200 μL 20 mM 2-nitrophenylhydrazine in 300 mM HCl:EtOH 50:50, add 200 μL 250 mM 1-ethyl-3-(3-dimethylaminopropyl)carbodiimide hydrochloride in EtOH:pyridine 97:3, heat at 80° for 5 min. Cool the solution to room temperature, add 10% KOH in MeOH:water 50:50, heat again at 80° for 5 min, cool to room temperature. Add 4 mL 33 mM pH 4.6 potassium phosphate buffer:500 mM HCl 70:10, extract with 4 mL hexane by shaking vigorously for 2 min. Centrifuge at 20 000 rpm for 10 min, evaporate the upper layer under a stream of nitrogen, resuspend the residue in 400 μL MeOH, inject a 50 μL aliquot.

## HPLC VARIABLES
**Guard column:** 20 × 2 20–40 μm LiChroprep C18
**Column:** 250 × 6 YMC FA C8 (Hichrom) + Apex 3 μm ODS (Jones Chromatography) in series
**Column temperature:** 45
**Mobile phase:** Gradient. A was MeCN. B was MeCN:MeOH:water 75:11:14. A:B 0:100 for 20 min, from 0:100 to 100:0 over 5 min, maintain at 100:0 for 10 min, return to initial conditions over for 2 min, re-equilibrate.
**Flow rate:** 1.5
**Injection volume:** 50
**Detector:** UV 400

## CHROMATOGRAM
**Retention time:** 20
**Internal standard:** margaric acid

## OTHER SUBSTANCES
**Extracted:** arachidonic acid (14.5), docosanoic acid (38), eicosanoic acid (34), eicosatrienoic acid (17), eicosapentanoic acid (12), lauric acid (10), linoleic acid (15.5), linolenic acid (12.5), gamma-linolenic acid (13), myristic acid (13), palmitic acid (18.5), palmitoleic acid (15), stearic acid (27), tetracosanoic acid (42)

## KEY WORDS
derivatization; plasma

## REFERENCE
Bailey, A.L.; Southon, S. Determination of total long-chain fatty acids in human plasma and lipoproteins, before and during copper-stimulated oxidation, by high-performance liquid chromatography, *Anal.Chem.*, **1998**, *70*, 415–419.

## SAMPLE
**Matrix:** blood
**Sample preparation:** Vortex 5 μL serum, 5 μL EtOH, and 50 μL 4% pyridine in EtOH containing 20 mM HCl for 10 s, add 25 μL 50 mM reagent in DMF, add 15 μL

2 M 1-ethyl-3-(3-dimethylaminopropyl)carbodiimide in water, heat at 37° for 10 min, centrifuge at 1000 g for 5 min, inject a 10 μL aliquot of the supernatant. (The reagent was 6,7-dimethoxy-1-methyl-2(1H)-quinoxalinone-3-propionylcarboxylic acid hydrazide. Synthesis is as follows. Stir 483 g veratrole in 1.45 L acetic acid at 15°, add 683 g concentrated nitric acid (s.g. 1.05) over 1 h keeping the temperature below 40° (cool if necessary), add 2.127 L fuming nitric acid (s.g. 1.50) over 1 h keeping the temperature below 30°, allow to stand for 2 h, pour into a large volume of cold water, filter, wash the solid until it is free from acid, recrystallize from EtOH to give 4,5-dinitroveratrole (mp 129.5–130.5°) (*J.Am.Chem.Soc.* **1946**, *68*, 1536). Reflux 5 g 4,5-dinitroveratrole in 200 mL benzene (Caution! Benzene is a carcinogen!), add 100 g 60 mesh iron powder and 20 mL concentrated HCl in small portions over 1 h, reflux for 4 h, add 10 mL water, reflux for 2 h, cool, make alkaline with 2.5 M NaOH, extract several times with 200 mL portions of benzene. Combine the extracts and evaporate to dryness, add 10 mL concentrated HCl, recrystallize from EtOH to give 1,2-diamino-4,5-dimethoxybenzene hydrochloride as slightly pink needles (mp 240° d) (*Anal.Chim.Acta* **1982**, *134*, 39). Dissolve 2.5 mmol 1,2-diamino-4,5-dimethoxybenzene monohydrochloride and 2.4 mmol $\alpha$-ketoglutaric acid in 30 mL 500 mM HCl, heat in a boiling water bath for 2 h, cool in ice, filter, wash the precipitate with water, dry under vacuum, recrystallize from MeOH:water 90:10 to give 6,7-dimethoxy-2(1H)-quinoxalinone-3-propionylcarboxylic acid as yellow needles (mp 240°) (*Chem.Pharm.Bull.* **1985**, *33*, 3493). Treat 1.5 g 6,7-dimethoxy-2(1H)-quinoxalinone-3-propionylcarboxylic acid in 100 mL MeOH with ethereal diazomethane, evaporate to dryness under reduced pressure, dissolve the residue in 30 mL chloroform (Caution! Chloroform is a carcinogen!), chromatograph on a 250 × 35 column of 70–230 mesh silica gel 60 (Merck) with hexane:ethyl acetate 50:50 to give methyl 6,7-dimethoxy-1-methyl-2(1H)-quinoxalinone-3-propionylcarboxylate as colorless needles (mp 178–179°). Dissolve 900 mg methyl 6,7-dimethoxy-1-methyl-2(1H)-quinoxalinone-3-propionylcarboxylate in 100 mL 45% hydrazine hydrate in water, heat at 100° for 1 h, recrystallize the precipitate from EtOH to give 6,7-dimethoxy-1-methyl-2(1H)-quinoxalinone-3-propionylcarboxylic acid hydrazide as colorless needles (mp 205–206°) (*Analyst* **1990**, *115*, 1363).)

## HPLC VARIABLES

**Column:** 250 × 4.6 10 μm YMC-Pack C8 (Yamamura, Kyoto)
**Column temperature:** 30 ± 0.2
**Mobile phase:** Gradient. MeCN:water 55:45 for 56 min, to 95:5 over 16 min, maintain at 95:5 for 4 min, re-equilibrate at initial conditions for 4 min.
**Flow rate:** 1
**Injection volume:** 100
**Detector:** F ex 360 em 435

## CHROMATOGRAM

**Retention time:** 62.5
**Limit of detection:** 2–7 fmol

## OTHER SUBSTANCES

**Extracted:** arachidonic acid (41), dihomo-gamma-linolenic acid (51), docosahexaenoic acid (41), eicosapentaenoic acid (20.5), lauric acid (12), linoleic acid (39), linolenic acid (27), margaric acid (66), myristic acid (24), myristoleic acid (16), palmitic acid (49), palmitoleic acid (30), stearic acid (72)
**Noninterfering:** adipic acid, alcohols, aldehydes, amines, amino acids, benzoic acid, cinnamic acid, $\alpha$-keto acids, lactic acid, malic acid, malonic acid, oxalic acid, phenols, salicylic acid, succinic acid, sugars

## KEY WORDS

derivatization; serum

## REFERENCE

Iwata, T.; Inoue, K.; Nakamura, M.; Yamaguchi, M. Simple and highly sensitive determination of free fatty acids in human serum by high performance liquid chromatography with fluorescence detection, *Biomed.Chromatogr.*, **1992**, *6*, 120–123.

## SAMPLE

**Matrix:** blood

**Sample preparation:** Mix 100 μL serum with 20 μL 5 mM IS in isopropanol, add 500 μL isopropanol:*n*-heptane:2 M phosphoric acid 40:10:1, mix, let stand at room temperature for 5–10 min, add 200 μL *n*-heptane, add 300 μL water, vortex thoroughly, centrifuge at 1000 g for 5 min. Remove a 200 μL aliquot of the upper organic layer and evaporate it to dryness under a stream of nitrogen, reconstitute the residue in 6 μL reagent, 500 μL MeCN, and ca. 1 mg potassium bicarbonate, flush the tube with nitrogen, close the PTFE-lined cap tightly, heat at 85° with vigorous stirring for 45 min (weigh vial before and after heating to check for leakage), cool, remove stir bar, centrifuge, inject a 10–25 μL aliquot of the supernatant. (The reagent was 50 mM *p*-bromophenacyl bromide in MeCN containing 5 mM 18-crown-6, store protected from light.)

## HPLC VARIABLES

**Guard column:** 4 × 4 5 μm CN
**Column:** 25 × 4 3 μm Spherisorb C6
**Column temperature:** 30
**Mobile phase:** MeCN:water 77:23
**Flow rate:** 1.3
**Injection volume:** 10–25
**Detector:** UV 254

## CHROMATOGRAM

**Retention time:** 13
**Internal standard:** heptadecanoic acid (margaric acid) (15)
**Limit of detection:** 800 nM

## OTHER SUBSTANCES

**Extracted:** arachidonic acid (9.8), docosahexaenoic acid (9), eicosapentaenoic acid (8), elaidic acid (14), lauric acid (6), linoleic acid (10.5), linolenic acid (8.5), myristic acid (8.8), myristoleic acid (7), palmitic acid (12), palmitoleic acid (9.5), stearic acid (18)

## KEY WORDS

derivatization; serum

## REFERENCE

Puttmann, M.; Krug, H.; von Ochsenstein, E.; Kattermann, R. Fast HPLC determination of serum free fatty acids in the picomole range, *Clin.Chem.*, **1993**, *39*, 825–832.

## SAMPLE

**Matrix:** blood

**Sample preparation:** Sonicate 10 μL serum, 44 μL MeOH, and 1 μL pyridine for 5 min, add 25 μL 100 mM reagent in DMF, add 20 μL 400 mM 1-ethyl-3-(3-dimethylaminopropyl)carbodiimide in MeOH, let stand at 25° for 2 h, centrifuge, inject an aliquot. (The reagent was 2-(5-hydrazinocarbonyl-2-furyl)-5,6-dimethoxybenzothiazole, which is synthesized as follows. Pass dry hydrogen chloride into a mixture of 12.6 g methyl 2-furoate, 4.5 g paraformaldehyde, and 3.4 g anhydrous zinc chloride in 50 mL dry chloroform for 3 h while holding the reaction temperature at 30° (Caution! Chloroform is a carcinogen!). After cooling, pour the contents of the flask into 100 mL cold water, remove the chloroform layer, extract the aqueous layer with chloroform (cf *Coll.Czech.Chem.Commun.* **1960**, *25*, 1058). Combine the chloroform layers, neutralize, dry over anhydrous calcium chloride, evaporate, distil to give 5-chloromethyl furyl-2-carboxylic acid methyl ester (bp 108°/4 mm Hg). Reflux 10 g 5-chloromethyl furyl-2-carboxylic acid methyl ester and 25 g silver carbonate in 100 mL THF:water 70:30 for 5 h, filter through Celite, concentrate the filtrate under reduced pressure, chromatograph the product on silica gel with chloroform to give 5-hydroxymethyl furyl-2-carboxylic acid methyl ester as a light yellow oil. Add a solution of 2.9 g 5-hydroxymethyl furyl-2-carboxylic acid methyl ester in 30 mL dichloromethane to 12 g pyridinium chlorochromate in 100 mL dichloromethane, stir at room temperature for 4 h, evaporate to dryness under reduced pressure, chromatograph on silica with dichloromethane to give 5-formyl furyl-2-carboxylic acid methyl ester as a light yellow powder. Add 10 mL concentrated nitric acid dropwise to 20 g 4-bromoveratrole in 60 mL acetic acid while keeping the temperature at 10–30° with occasional cooling; when the addition is complete, pour the reaction mixture into ice water. Collect the precipitate and dissolve it in 500 mL hot EtOH, add activated charcoal, filter, add 40 mL water to the filtrate to give 4,5-dimethoxy-2-nitrobromobenzene as a light yellow crystalline solid (mp 121 122°). Prepare sodium sulfide by melting together 5 g sodium sulfide nonahydrate and 700 mg sulfur, add this mixture to 5 g 4,5-dimethoxy-2-nitrobromobenzene in 50 mL EtOH:water 95:5, reflux for 30 min, pour into ice water, collect the solid, recrystallize from dichloromethane to give di(4,5-dimethoxy-2-nitrophenyl)sulfide as yellow needles (mp 231–232°). Add 15 mL concentrated HCl dropwise to 1.5 g di(4,5-dimethoxy-2-nitrophenyl)sulfide and 4.5 g tin powder stirred at 40–50° in 150 mL EtOH, reflux for 1 h, cool to room temperature, filter, add 1.17 g 5-formyl furyl-2-carboxylic acid methyl ester to the filtrate, reflux for 1 h, cool, filter, chromatograph the solid on silica gel with dichloromethane, recrystallize from EtOH to give 5-(5′,6′-dimethoxybenzothiazolyl)-*N*-furan-2-carboxylic acid methyl ester as a yellow powder (mp 192–202°). Add 2 mL hydrazine hydrate (Caution! Hydrazine hydrate is a carcinogen!) to 800 mg 5-(5′,6′-dimethoxybenzothiazolyl)-*N*-furan-2-carboxylic acid methyl ester in 20 mL EtOH, reflux for 30 min, collect the solid, wash with MeOH, dry under vacuum over phosphorus pentoxide to give 2-(5-hydrazinocarbonyl-2-furyl)-5,6-dimethoxybenzothiazole as a light yellow solid (mp 226–228°).)

## HPLC VARIABLES

**Column:** 250 × 4.6 5 μm Wakosil-II 5C18 HG
**Column temperature:** 40
**Mobile phase:** Gradient. MeCN:water from 70:30 to 75:25 over 25 min, to 100:0 over 15 min, maintain at 100:0.
**Flow rate:** 1
**Injection volume:** 10
**Detector:** F ex 363 em 452

## CHROMATOGRAM

**Retention time:** 35
**Limit of detection:** 50 fmol

## OTHER SUBSTANCES

**Extracted:** alprostadil (51), arachidonic acid (24), dinoprost (41), dinoprostone (50), lauric acid (10), linoleic acid (25), linolenic acid (18), margaric acid (38), myristic acid (19), myristoleic acid (12), palmitic acid (32), palmitoleic acid (21), stearic acid (42)

## KEY WORDS

derivatization; serum

## REFERENCE

Saito, M.; Ushijima, T.; Sasamoto, K.; Ohkura, Y.; Ueno, K. 2-(5-Hydrazinocarbonyl-2-furyl)-5,6-dimethoxybenzothiazole as a precolumn fluorescence derivatization reagent for carboxylic acids in high-performance liquid chromatography and its application to the assay of fatty acids in human serum, *Anal.Sci.*, **1995**, *11*, 103–107.

## SAMPLE

**Matrix:** blood

**Sample preparation:** Vortex 10 μL plasma, 200 μL 500 mM pH 6.5 phosphate buffer, 50 μL 20 μM IS in MeOH, and 2 mL *n*-heptane:chloroform 50:50 for 2 min (Caution! Chloroform is a carcinogen!), centrifuge at 1000 g for 10 min. Remove the lower organic layer and evaporate it to dryness, reconstitute the residue in two 100 μL aliquots of acetone. Evaporate the acetone solution to dryness, add 2–3 mg finely powdered potassium bicarbonate:sodium sulfate 50:50, add 50 μL 800 μM dibenzo-18-crown-6 in acetone, add 50 μL 2 mM 7-acetoxy-4-bromomethylcoumarin in acetone, heat at 50° in the dark for 30 min, inject a 50 μL aliquot. (7-Acetoxy-4-bromomethylcoumarin is available from TCI America or Tokyo Kasei, or it may be prepared as follows. Reflux 50 g 7-hydroxy-4-methylcoumarin ($\beta$-methylumbelliferone) and 100 mL acetic anhydride for 1 h, cool, pour into 500 mL cold water, filter, dry the solid, recrystallize from EtOH to give 4-methyl-7-acetoxycoumarin. Reflux 10 g 4-methyl-7-acetoxycoumarin, 9 g *N*-bromosuccinimide, a little 2,2′-(azobis(2-methylpropionitrile) ($\alpha$,$\alpha'$-azobisisobutyronitrile, Eastman), and 100 mL carbon tetrachloride for 20 h, cool, evaporate under reduced pressure to remove the solvent, wash the residue with water, filter, dry, recrystallize from ethyl acetate/cyclohexane to give 7-acetoxy-4-bromomethyl-coumarin (mp 184–185°) (*J.Chromatogr.* **1982**, *234*, 121).)

## HPLC VARIABLES

**Column:** 250 × 4 5 μm LiChrosorb RP-18

**Column temperature:** 40

**Mobile phase:** Gradient. MeCN:MeOH:water from 35:35:30 to 0:90:10 over 70 min (convex gradient).

**Flow rate:** 1.2

**Injection volume:** 50

**Detector:** F ex 365 em 460 following post-column reaction. The column effluent mixed with 200 mM NaOH in MeOH:water 80:20 pumped at 0.4 mL/min and the mixture flowed through a 3.5 m ×0.5 mm ID stainless steel coil at 50° to the detector.

## CHROMATOGRAM

**Retention time:** 51

**Internal standard:** margaric acid (57)

**Limit of detection:** 5 pmol

## OTHER SUBSTANCES

**Extracted:** arachidonic acid (37), capric acid (17), caproic acid (6), caprylic acid (11), heptanoic acid (9), lauric acid (26), linoleic acid (42), linolenic acid (35), myristic acid (36), myristoleic acid (28), nonanoic acid (15), palmitic acid (50), palmitoleic acid (39), stearic acid (66), tridecanoic acid (30), undecanoic acid (21)

## KEY WORDS

derivatization; plasma; post-column reaction

## REFERENCE

Tsuchiya, H.; Hayashi, T.; Sato, M.; Tatsumi, M.; Takagi, N. Simultaneous separation and sensitive determination of free fatty acids in blood plasma by high-performance liquid chromatography, *J.Chromatogr.*, **1984**, *309*, 43–52.

## SAMPLE

**Matrix:** blood

**Sample preparation:** Vortex 50 μL plasma, 10 μL IS solution, and 450 μL micelle solution for 10 s, add 25 μL 28 mg/mL 9-bromomethylacridine in acetone, mix. Remove a 50 μL aliquot and heat it to 60° for 6 min, inject the whole amount through a 2 μm stainless steel filter of 8 sq mm area onto column A, wash to waste with 400 μL mobile phase A, backflush the contents of column A onto column B with the mobile phase B, monitor the effluent from column B. After each injection, backflush the stainless steel filter with 1 mL buffer. (The micelle solution was 25 mM Arkopal N-130 (a polyoxyethylene(13)nonylphenol, Hoechst Holland, Amsterdam) in 10 mM pH 7.0 phosphate buffer containing 6 mM tetrakis(decyl)ammonium bromide. 9-Bromomethylacridine is available from TCI America or Tokyo Kasei, or it may be prepared as follows. Heat 10 g diphenylamine, 10 mL glacial acetic acid, and 40 g anhydrous zinc chloride to 220°, evaporate excess acetic acid with stirring, heat at 220–230° for 6 h, digest with hot 10% sulfuric acid, make strongly alkaline with 25% ammonia to dissolve the zinc chloride. Extract the insoluble residue with toluene. Extract the organic layer with 10% sulfuric acid, make the aqueous layer alkaline with aqueous ammonia. Collect the yellow precipitate that separates and recrystallize it twice from petroleum ether to give 9-methyl acridine as pale yellow needles (*Chromatographia* **1989**, *28*, 267). Reflux 560 mg 9-methylacridine, 445 mg *N*-bromosuccinimide, and 10 mg benzoyl peroxide in 30 mL carbon tetrachloride for more than 2 h, cool, chromatograph on silica gel with benzene:ethyl acetate 30:1 (Caution! Benzene is a carcinogen!) to obtain 9-bromomethylacridine as yellow crystals (mp 147–151°) (*Anal.Lett.* **1987**, *20*, 1581).)

## HPLC VARIABLES
**Column:** A 10 × 2.1 40 μm Chromsep C18 (Chrompack); B 100 × 3 5 μm Chromspher C18 (Chrompack)
**Mobile phase:** A 10 mM pH 7.0 phosphate buffer; B Gradient. MeOH:water 75:25 for 3 min, to 100:0 over 12 min (concave gradient).
**Injection volume:** 50
**Detector:** UV 254; F ex 362 em 418

## CHROMATOGRAM
**Retention time:** 10
**Internal standard:** heptadecanoic acid (13)
**Limit of detection:** 300 nM

## OTHER SUBSTANCES
**Extracted:** arachidonic acid (7), linoleic acid (9), linolenic acid (6), myristic acid (8), palmitic acid (10.5), palmitoleic acid (7.5), stearic acid (13.5)

## KEY WORDS
column-switching; derivatization; plasma

## REFERENCE
van der Horst, F.A.L.; Post, M.H.; Holthuis, J.J.M.; Brinkman, U.A.T. Automated high-performance liquid chromatographic determination of plasma free fatty acids using on-line derivatization with 9-bromomethylacridine based on micellar phase-transfer catalysis, *J.Chromatogr.*, **1990**, *500*, 443–452.

---

## SAMPLE
**Matrix:** solutions
**Sample preparation:** Mix 100 μL of a solution in EtOH, EtOH/water, or water with 400 μL reagent solution and 200 μL 20 mM 2-nitrophenylhydrazine hydrochloride in water, heat at 60° for 20 min, add 100 μL 15% KOH in MeOH:water 80:20, heat at 60° for 15 min, cool, inject a 1–2 μL aliquot. (Prepare the reagent by mixing equal volumes of 3% pyridine in EtOH and 250 mM 1-(3-dimethylaminopropyl)-3-ethylcarbodiimide hydrochloride in EtOH.)

## HPLC VARIABLES
**Column:** 250 × 4.6 5 μm YMC-C8 (Yamamuta Chemical Research, Kyoto)
**Column temperature:** 50

**Mobile phase:** MeOH:water 86:14 adjusted to pH 4.5 with 100 mM HCl
**Flow rate:** 1.2
**Injection volume:** 1–2
**Detector:** UV 230; UV 400

## CHROMATOGRAM
**Retention time:** 10.7
**Limit of detection:** 2.5–5 pmol (UV 230), 10–15 pmol (UV 400)

## OTHER SUBSTANCES
**Simultaneous:** capric acid (3), lauric acid (5), linoleic acid (8.5), linolenic acid (7), myristic acid (6.5), palmitic acid (10), palmitoleic acid (7.5), stearic acid (14)

## KEY WORDS
derivatization

## REFERENCE
Miwa, H.; Hiyama, C.; Yamamoto, M. High-performance liquid chromatography of short- and long-chain fatty acids as 2-nitrophenylhydrazides, *J.Chromatogr.*, **1985**, *321*, 165–174.

## ANNOTATED BIBLIOGRAPHY
Abushufa, R.; Reed, P.; Weinkove, C. Fatty acids in erythrocytes measured by isocratic HPLC, *Clin.Chem.*, **1994**, *40*, 1707–1712. [oleic acid; arachidonic acid; palmitoleic acid; linoleic acid; eicosatrienoic acid; palmitic acid; stearic acid]

Akasaka, K.; Suzuki, T.; Ohrui, H.; Meguro, H.; Shindo, Y.; Takahashi, H. 9-Bromomethylacridine a novel fluorescent labeling reagent of carboxylic group for HPLC, *Anal.Lett.*, **1987**, *20*, 1581–1594. [caprylic acid; capric acid; lauric acid; linolenic acid; arachidonic acid; palmitoleic acid; myristic acid; linoleic acid; oleic acid; palmitic acid; stearic acid]

Akasaka, K.; Ohrui, H.; Meguro, H. Determination of carboxylic acids by high-performance liquid chromatography with 2-(2,3-anthracenedicarboximido)ethyl trifluoromethanesulfonate as highly sensitive fluorescent labelling reagent, *Analyst*, **1993**, *118*, 765–768. [LOD 1.4–3.8 pmol; caprylic acid; caproic acid; lauric acid; myristoleic acid; eicosapentaenoic acid; linolenic acid; myristic acid; docosahexaenoic acid; palmitoleic acid; arachidonic acid; linoleic acid; eicosatrienoic acid; palmitic acid; oleic acid; eicosadienoic acid; margaric acid; stearic acid; gondoic acid]

Baty, J.D.; Pazouki, S.; Dolphin, J. Analysis of fatty acids as their anthrylmethyl esters by high-performance liquid chromatography with fluorescence detection, *J.Chromatogr.*, **1987**, *395*, 403–411. [plasma; LOD 50 ng; lauric acid; linolenic acid; arachidonic acid; myristic acid; palmitoleic acid; linoleic acid; palmitic acid; oleic acid; stearic acid; arachidic acid; behenic acid; lignoceric acid; margaric acid]

Borch, R.F. Separation of long chain fatty acids as phenacyl esters by high pressure liquid chromatography, *Anal.Chem.*, **1975**, *47*, 2437–2439. [LOD 100 ng; lauric acid; myristoleic acid; linolenic acid; myristic acid; palmitoleic acid; arachidonic acid; linoleic acid; pentadecanoic acid; linolelaidic acid; eicosatrienoic acid; palmitic acid; oleic acid; vaccenic acid; petroselinic acid; elaidic acid; eicosadienoic acid; heptadecanoic acid; stearic acid; eicosaenoic acid; nonadecanoic acid; arachidic acid; erucic acid; heneicosanoic acid; behenic acid; nervonic acid; lignoceric acid]

Carrier, A.; Parent, J. Liquid chromatography-mass spectrometry determination of free fatty acids in phospholipid-based formulations, *J.Liq.Chromatogr.Rel.Technol.*, **2001**, *24*, 97–107. [stearic acid; linoleic acid; linolenic acid; palmitic acid; oleic acid]

Cooper, M.J.; Anders, M.W. Determination of long chain fatty acids as 2-naphthacyl esters by high pressure liquid chromatography and mass spectrometry, *Anal.Chem.*, **1974**, *46*, 1849–1852. [oleic acid; linolenic acid; linoleic acid; dihomolinolenic acid; arachidonic acid]

Czauderna, M.; Kowalczyk, J. Separation of some mono-, di- and tri-unsaturated fatty acids containing 18 carbon atoms by high-performance liquid chromatography and photodiode array detection, *J.Chromatogr.B*, **2001**, *760*, 165–178. [derivatization; linoleic acid; linolenic acid; oleic acid; ricinoleic acid; myristic acid; palmitic acid; vaccenic acid; stearic acid]

Eguchi, Y. Analysis of lipoprotein lipase activity using high-performance liquid chromatography, *Biomed.Chromatogr.*, **2002**, *16*, 500–503. [oleic acid; derivatization]

Ergan, F.; Andre, G. Simple high performance liquid chromatography methods for monitoring lipase reactions, *Lipids*, **1989**, *24*, 76–78. [oleic; acid; linolenic acid; linoleic acid; palmitic acid; stearic acid]

Fuse, T.; Kusu, F.; Takamura, K. Determination of higher fatty acids in oils by high-performance liquid chromatography with electrochemical detection, *J.Chromatogr.A*, **1997**, *764*, 177–182. [linoleic acid; oleic acid; palmitic acid; stearic acid; post-column reaction; derivatization; camellia oil; olive oil; rapeseed oil; corn oil; soybean oil; LOD 20 pmol]

Ghiggeri, G.M.; Candiano, G.; Delfino, G.; Queirolo, C.; Ginevri, F.; Perfamo, F.; Gusmano, R. Separation of the 9-anthryldiazomethane derivates of fatty acids by high-performance liquid chromatography on a fatty acid analysis column. Application to albumin-bound fatty acid analysis, *J.Chromatogr.*, **1986**, *381*, 411–418. [derivatization; caproic acid; caprylic acid; capric acid; lauric acid; myristoleic acid; myristic acid; linolenic acid; linoleic acid; palmitic acid; arachidonic acid; oleic acid; stearic acid; eicosanoic acid; arachidic acid; erucic acid; behenic acid; neuronic acid]

Golden, K.D.; Williams, O.J. Amino acid, fatty acid, and carbohydrate content of Artocarpus altilis (breadfruit), *J.Chromatogr.Sci.*, **2001**, *39*, 243–250. [derivatization; oleic acid; linoleic acid]

Hamilton, J.G.; Comai, K. Separation of neutral lipid free fatty acid and phospholipid classes by normal phase HPLC, *Lipids*, **1988**, *23*, 1150–1153. [oleic acid; phosphatidylcholine; phosphatidylethanolamine; phosphatidylinositol]

Hanis, T.; Smrz, M.; Klir, P.; Macek, K.; Klima, J.; Base, J.; Deyl, Z. Determination of fatty acids as phenacyl esters in rat adipose tissue and blood vessel walls by high-performance liquid chromatography, *J.Chromatogr.*, **1988**, *452*, 443–457. [derivatization; LOD 0.8–1.2 ng; column temp 40; caproic acid; caprylic acid; capric acid; lauric acid; myristic acid; myristoleic acid; pentadecanoic acid; palmitic acid; palmitoleic acid; heptadecanoic acid; stearic acid; oleic acid; elaidic acid; linoleic acid; linoelaidic acid; eicosenoic acid; eicosadienoic acid; eicosatrienoic acid; arachidonic acid; erucic acid; docosahexaenoic acid]

Hoffman, N.E.; Liao, J.C. High pressure liquid chromatography of p-methoxyanilides of fatty acids, *Anal.Chem.*, **1976**, *48*, 1104–1106. [gradient; stearic acid; palmitoleic acid; oleic acid; linoleic acid; linolenic acid; arachidonic acid; erucic acid; docoshexaenoic acid; nervonic acid; tetradecanoic acid; pentadecanoic acid; hexadecanoic acid; octadecanoic acid; eicosanoic acid; docosanoic acid; tetracosanoic acid; palmitic acid; myristic acid]

Holcapek, M.; Jandera, P.; Fischer, J.; Prokes, B. Analytical monitoring of the production of biodiesel by high-performance liquid chromatography with various detection methods, *J.Chromatogr.A*, **1999**, *858*, 13–31.

Hu, M.; Niculescu, M.; Zhang, X.M.; Hui, A. High-performance liquid chromatographic determination of polysorbate 80 in pharmaceutical suspensions, *J.Chromatogr.A*, **2003**, *984*, 233–236. [derivatization; oleic acid]

Ikeda, M.; Shimada, K.; Sakaguchi, T. High-performance liquid chromatographic determination of free fatty acids with 1-naphthylamine, *J.Chromatogr.*, **1983**, *272*, 251–259. [derivatization; serum; column temp 40; LOD 4 ng; margaric acid; myristic acid; palmitoleic acid; linoleic acid; palmitic acid; oleic acid; stearic acid]

Ikeda, M.; Shimada, K.; Sakaguchi, T.; Matsumoto, U. Fluorometric high-performance liquid chromatography of 9-aminophenanthrene-derivatized free fatty acids, *J.Chromatogr.*, **1984**, *305*, 261–270. [derivatization; column temp 40; LOD 10–15 pmol; myristic acid; palmitic acid; palmitoleic acid; stearic acid; linoleic acid; arachidonic acid; oleic acid; margaric acid]

Ikenoya, S.; Hiroshima, O.; Ohmae, M.; Kawabe, K. Electrochemical detector for high-performance liquid chromatography. IV. Analysis of fatty acids, bile acids and prostaglandins by derivatization to an electrochemically active form, *Chem.Pharm.Bull.*, **1980**, *28*, 2941–2947. [derivatization; dinoprost; dinoprostone; guinea pig; plasma; column temp 25; LOD 0.5–2 ng/mL; palmitic acid; palmitoleic acid; stearic acid; oleic acid; linoleic acid; linolenic acid; cholic acid; chenodiol; deoxycholic acid; lithocholic acid]

Iwata, T.; Hirose, T.; Nakamura, M.; Yamaguchi, M. 4-(1-Methylphenanthro[9,10-*d*]imidazol-2-yl) benzohydrazide as derivatization reagent for carboxylic acids in high-performance liquid chromatography with conventional and laser-induced fluorescence detection, *Analyst*, **1994**, *119*, 1747–1751. [LOD 0.4–2.3 fmol; arachidic acid; margaric acid; palmitic acid; stearic acid; arachidonic acid; docosahexaenoic acid; linoleic acid; oleic acid; prostaglandin $D_2$; dinoprostone; prostaglandin $E_{2\alpha}$; thromboxane $A_2$; androsterone glucuronide; dehydroisoandrosterone glucuronide; etiocholanolone glucuronide; testosterone glucuronide]

Jones, P.J.; Pencharz, P.B.; Clandinin, M.T. Absorption of $^{13}C$-labeled stearic, oleic, and linoleic acids in humans: application to breath tests, *J.Lab.Clin.Med.*, **1985**, *105*, 647–652. [oleic acid; linoleic acid]

Jordi, H.C. Separation of long and short chain fatty acids as naphthacyl and substituted phenacyl esters by high performance liquid chromatography, *J.Liq.Chromatogr.*, **1978**, *1*, 215–230. [acetic acid;

propionic acid; butyric acid; caproic acid; caprylic acid; capric acid; lauric acid; myristic acid; palmitic acid; stearic acid; arachidic acid; behenic acid; lignoceric acid; palmitoleic acid; oleic acid; elaidic acid; vaccenic acid; linoleic acid; linolenic acid; nervonic acid; arachidonic acid; lactic acid; valeric acid; heptanoic acid; nonanoic acid; undecanoic acid; tridecanoic acid; pentadecanoic acid; heptadecanoic acid; nonadecanoic acid; heneicosanoic acid; tricosanoic acid; myristoleic acid; palmitelaidic acid; linolelaidic acid; eicosenoic acid; erucic acid; eicosatrienoic acid; brassidic acid; docosahexenoic acid; eicosapentaenoic acid]

Juanéda, P. Utilisation of reversed-phase high-performance liquid chromatography as an alternative to silver-ion chromatography for the separation of cis- and trans-C18:1 fatty acid isomers, *J.Chromatogr.A*, **2002**, *954*, 285–289. [derivatization; oleic acid]

Kaneda, H.; Kano, Y.; Kamimura, M.; Osawa, T.; Kawakishi, S. Analysis of long-chain fatty acids in beer by HPLC-fluorescence detection method, *J.Agric.Food Chem.*, **1990**, *38*, 1363–1367. [column temp 40; derivatization; gradient; LOD 0.2 ng/mL; LOQ 0.4 ng/mL; lauric acid; linolenic acid; myristic acid; palmitoleic acid; linoleic acid; palmitic acid; oleic acid; heptadecanoic acid; stearic acid]

Kelly, R.A.; O'Hara, D.S.; Kelley, V. High-performance liquid chromatographic separation of femtomolar quantities of endogenous carboxylic acids, including arachidonic acid metabolites, as 4-bromomethyl-7-acetoxycoumarin derivatives, *J.Chromatogr.*, **1987**, *416*, 247–254. [arachidonic acid; fluorescence detection; gradient; post-column reaction; rat; plasma; SPE; LOD 10 fmol; lauric acid; myristoleic acid; linoleic acid; myristic acid; palmitoleic acid; eicosatrienoic acid; palmitic acid; oleic acid; margaric acid; stearic acid]

Korte, W.D. 9-(Chloromethyl)anthracene: a useful derivatizing reagent for enhanced ultraviolet and fluorescence detection of carboxylic acids with liquid chromatography, *J.Chromatogr.*, **1982**, *243*, 153–157. [chenodiol; cholic acid; deoxycholic acid; lauric acid; myristic acid; linoleic acid; palmitic acid; oleic acid; stearic acid; glycocholic acid; glycochenodeoxycholic acid; glycodeoxycholic acid]

Kotani, A.; Fuse, T.; Kusu, F. Determination of plasma free fatty acids by high-performance liquid chromatography with electrochemical detection, *Anal.Biochem.*, **2000**, *284*, 65–69. [linoleic acid; oleic; acid]

Lawrence, J.F.; Charbonneau, C.F. Direct, sensitive and selective detection of free fatty acids by high-performance liquid chromatography with post-column ion-pair extraction and absorbance detection, *J.Chromatogr.*, **1988**, *445*, 189–197. [gradient; post-column extraction; butter; margarine; orange juice; LOD 26–83 ng; caprylic acid; capric acid; lauric acid; linolenic acid; myristic acid; palmitoleic acid; linoleic acid; linolelaidic acid; oleic acid; palmitic acid; elaidic acid; stearic acid]

Lee, Y.M.; Nakamura, H.; Nakajima, T. Rapid determination by high-performance liquid chromatography of free fatty acids released from rat platelets after derivatization with monodansylcadaverine, *J.Chromatogr.*, **1990**, *515*, 467–473. [gradient; platelets; fluorescence; fluorescence detection; column temp 40; LOD 0.1 pmol; lauric acid; myristoleic acid; linolenic acid; myristic acid; palmitoleic acid; linoleic acid; palmitic acid; oleic acid; stearic acid]

Li, Z.; Gu, T.; Kelder, B.; Kopchick, J.J. Analysis of fatty acids in mouse cells using reversed-phase high-performance liquid chromatography, *Chromatographia*, **2001**, *54*, 463–467. [oleic acid; linoleic acid; alprostadil; dinoprostone; prostaglandin $E_3$; prostaglandin $F_{1\alpha}$]

Lu, C.Y.; Wu, H.L.; Chen, S.H.; Kou, H.S. Highly sensitive determination of long-chain free fatty acids by fluorogenic derivatization and liquid chromatography, *Chromatographia*, **2001**, *53*, S250–S253. [lauric acid; myristic acid; palmitic acid; stearic acid; palmitoleic acid; oleic acid; linoleic acid; margaric acid]

Mehta, A.; Oeser, A.M.; Carlson, M.G. Rapid quantitation of free fatty acids in human plasma by high-performance liquid chromatography, *J.Chromatogr.B*, **1998**, *719*, 9–23. [linoleic acid; derivatization; oleic acid]

Miwa, H.; Yamamoto, M. Improved method of determination of biologically important $C_{10:0}$-$C_{22:6}$ fatty acids as their 2-nitrophenylhydrazides by reversed-phase high-performance liquid chromatography, *J.Chromatogr.*, **1986**, *351*, 275–282. [column temp 30; LOQ 2.5 pmol; capric acid; lauric acid; myristoleic acid; eicosapentenoic acid; linolenic acid; myristic acid; docosahexenoic acid; palmitoleic acid; arachidonic acid; linoleic acid; eicosatrienoic acid; palmitic acid; oleic acid; margaric acid is internal standard; stearic acid]

Miwa, H.; Yamamoto, M.; Nishida, T.; Nunoi, K.; Kikuchi, M. High-performance liquid chromatographic analysis of serum long-chain fatty acids by direct derivatization method, *J.Chromatogr.*, **1987**, *416*, 237–245. [LOD 100–200 fmol; column temp 30; capric acid; lauric acid; myristoleic acid; eicosapentaenoic acid; linolenic acid; myristic acid; docosahexaenoic acid; palmitoleic acid; arachidonic acid; linoleic acid; dihomolinolenic acid; palmitic acid; oleic acid; margaric acid is internal standard; stearic acid]

Miwa, H.; Yamamoto, M. Liquid chromatographic determination of free and total fatty acids in milk and milk products as their 2-nitrophenylhydrazides, *J.Chromatogr.*, **1990**, *523*, 235–246. [milk; butter; cheese; condensed milk; ice cream; gradient; column temp 35; LOD 0.5–2 pmol; lactic acid; acetic acid; propionic acid; isobutyric acid; butyric acid; isovaleric acid; valeric acid; 2-ethylbutyric acid; isocaproic acid; caproic acid; caprylic acid; capric acid; lauric acid; myristoleic acid; eicosapentaenoic acid; linolenic acid; myristic acid; docosahexaenoic acid; palmitoleic acid; arachidonic acid; linoleic acid; linoelaidic acid; eicosatrienoic acid; palmitic acid; docosatetraenoic acid; oleic acid; elaidic acid; eicosadienoic acid; margaric acid; docosatrienoic acid; stearic acid; eicosaenoic acid; docosadienoic acid; arachidic acid; erucic acid]

Miwa, H.; Yamamoto, M. Rapid liquid chromatographic determination of fatty acids as 2-nitrophenylhydrazine derivatives, *J.AOAC Int.*, **1996**, *79*, 493–497. [column temp 50; derivatization; caprylic acid; capric acid; lauric acid; myristoleic acid; octadecatetraenoic acid; eicosapentaenoic acid; linolenic acid; myristic acid; docosahexaenoic acid; palmitoleic acid; arachidonic acid; linoleic acid; linoelaidic acid; eicosatrienoic acid; dihomolinolenic acid; palmitic acid; docosatetraenoic acid; oleic acid; elaidic acid; eicosadienoic acid; margaric acid is internal standard; docosatrienoic acid; stearic acid; eicosenoic acid; docosadienoic acid; arachidic acid; erucic acid]

Morita, H.; Konishi, M. Electrogenerated chemiluminescence derivatization reagents for carboxylic acids and amines in high-performance liquid chromatography using tris(2,2′-bipyridine)ruthenium(II), *Anal.Chem.*, **2002**, *74*, 1584–1589. [histamine; post-column reaction; linoleic acid; oleic acid]

Morita, H.; Konishi, M. Electrogenerated chemiluminescence derivatization reagent, 3-isobutyl-9,10-dimethoxy-1,3,4,6,7,11b-hexahydro-2H-pyrido[2,1-a]isoquinolin-2-ylamine, for carboxylic acid in high-performance liquid chromatography using tris(2,2′-bipyridine)ruthenium(II), *Anal.Chem.*, **2003**, *74*, 940–946. [lauric acid; myristoleic acid; myristic acid; linolenic acid; palmitoleic acid; arachidonic acid; linoleic acid; oleic acid; stearic acid]

Nakashima, K.; Taguchi, Y.; Kuroda, N.; Akiyama, S.; Duan, G. 2-(4-Hydrazinocarbonylphenyl)-4,5-diphenylimidazole as a versatile fluorescent derivatization reagent for the high-performance liquid chromatographic analysis of free fatty acids, *J.Chromatogr.*, **1993**, *619*, 1–8. [lauric acid; myristic acid; linolenic acid; palmitoleic acid; linoleic acid; arachidonic acid; palmitic acid; oleic acid; stearic acid; margaric acid; LOD 7–57 fmol; serum; gradient; fluorescence detection]

Netting, A.G.; Duffield, A.M. Pentafluorobenzyl esters as derivatives for the semi-preparative high-performance liquid chromatography of fatty acids, *J.Chromatogr.*, **1983**, *257*, 174–179. [derivatization; SPE; normal phase; stearic acid; oleic acid; linoleic acid; linolenic acid]

Nimura, N.; Kinoshita, T. Fluorescent labeling of fatty acids with 9-anthryldiazomethane (ADAM) for high performance liquid chromatography, *Anal.Lett.*, **1980**, *13*, 191–202. [derivatization; propionic acid; butyric acid; valeric acid; caproic acid; heptanoic acid; caprylic acid; capric acid; lauric acid; myristic acid; linolenic acid; linoleic acid; palmitic acid; stearic acid; oleic acid]

Nimura, N.; Kinoshita, T.; Yoshida, T.; Uetake, A.; Nakai, C. 1-Pyrenyldiazomethane as a fluorescent labeling reagent for liquid chromatographic determination of carboxylic acids, *Anal.Chem.*, **1988**, *60*, 2067–2070. [derivatization; gradient; isocratic; fluorescence detection; LOD 20–30 fmol; lactic acid; formic acid; propionic acid; dinoprostone; dinoprost; alprostadil; linolenic acid; palmitoleic acid; myristic acid; linoleic acid; palmitic acid; oleic acid; stearic acid; arachidic acid]

Pei, P.T.-S.; Kossa, W.C.; Ramachandran, S.; Henly, R.S. High pressure reverse phase liquid chromatography of fatty acid *p*-bromophenacyl esters, *Lipids*, **1976**, *11*, 814–816. [linoleic acid; derivatization; linolenic acid; linoleaidic acid; oleic acid; elaidic acid; stearic acid; palmitoleic acid; palmitelaidic acid; petroselinic acid]

Ryan, P.J.; Honeyman, T.W. Determination of fatty acids by high-performance liquid chromatography of Dns-ethanolamine derivatives, *J.Chromatogr.*, **1984**, *312*, 461–466. [derivatization; fluorescence detection; bulk; linolenic acid; palmitoleic acid; linoleic acid; eicosadienoic acid; palmitic acid; oleic acid; elaidic acid; margaric acid; arachidonic acid]

Saito, M.; Ushijima, T.; Sasamoto, K.; Ohkura, Y.; Ueno, K. 2-(5-Hydrazinocarbonyl-2-oxazolyl)-5,6-dimethoxybenzothiazole as a precolumn fluorescence derivatization reagent for carboxylic acids in high-performance liquid chromatography and its application to the assay of fatty acids in human serum, *J.Chromatogr.B*, **1995**, *674*, 167–175. [column temp 40; fluorescence detection; lauric acid; myristic acid; gradient; palmitic acid; stearic acid; margaric acid; nonadecanoic acid; myristoleic acid; palmitoleic acid; linoleic acid; oleic acid; arachidonic acid; dinoprost; dinoprostone; alprostadil; prostaglandin $F_{1\alpha}$]

Sasamoto, K.; Ushijima, T.; Saito, M.; Ohkura, Y. Precolumn fluorescence derivatization of carboxylic acids using 4-aminomethyl-6,7-dimethoxycoumarin in a two-phase medium, *Anal.Sci.*, **1996**, *12*, 189–193. [serum; column temp 40; LOD 20–50 fmol; linoleic acid; palmitic acid; oleic acid; lauric acid;

myristoleic acid; linolenic acid; myristic acid; palmitoleic acid; arachidonic acid; margaric acid; stearic acid; nonadecanoic acid]

Tojo, H.; Ono, T.; Okamoto, M. Reverse-phase high-performance liquid chromatographic assay of phospholipases: application of spectrophotometric detection to rat phospholipase $A_2$ isozymes, *J.Lipid Res.*, **1993**, *34*, 837–844. [derivatization; enzyme incubations; column temp 20; palmitic acid; margaric acid; stearic acid; oleic acid; linoleic acid; linolenic acid; arachidonic acid]

Vioque, E.; Maza, M.P.; Millán, F. High-performance liquid chromatography of fatty acids as their *p*-phenylazophenacyl esters, *J.Chromatogr.*, **1985**, *331*, 187–192. [derivatization; bulk; formic acid; acetic acid; propionic acid; butyric acid; valeric acid; caproic acid; heptanoic acid; caprylic acid; nonanoic acid; capric acid; undecanoic acid; lauric acid; tridecanoic acid; myristic acid; pentadecanoic acid; palmitic acid; margaric acid; stearic acid; arachidic acid; linolenic acid; linoleic acid; oleic acid]

Wolf, J.H.; Korf, J. Improved automated precolumn derivatization reaction of fatty acids with bromomethylmethoxycoumarin as label, *J.Chromatogr.*, **1990**, *502*, 423–430. [derivatization; whole blood; gradient; fluorescence detection; rat; palmitoleic acid; myristic acid; oleic acid; palmitic acid; stearic acid; arachidonic acid]

Wood, R.; Lee, T. High-performance liquid chromatography of fatty acids: quantitative analysis of saturated, monoenoic, polyenoic and geometrical isomers, *J.Chromatogr.*, **1983**, *254*, 237–246. [bulk; derivatization; gradient; LOQ 10 ng; lauric acid; myristoleic acid; tridecanoic acid; linolenic acid; myristic acid; arachidonic acid; linoleic acid; pentadecanoic acid; palmitic acid; oleic acid; elaidic acid; stearic acid; heptadecanoic acid]

Yamaguchi, M.; Hara, S.; Matsunaga, R.; Nakamura, M.; Ohkura, Y. 3-Bromomethyl-6,7-dimethoxy-1-methyl-2(1H)-quinoxalinone as a new fluorescence derivatization reagent for carboxylic acids in high-performance liquid chromatography, *J.Chromatogr.*, **1985**, *346*, 227–236. [gradient; fluorescence detection; LOD 0.3–1 fmol; propionic acid; butyric acid; valeric acid; caproic acid; caprylic acid; capric acid; lauric acid; myristic acid; palmitic acid; margaric acid; stearic acid; arachidic acid; myristoleic acid; palmitoleic acid; linolenic acid; arachidonic acid; linoleic acid; oleic acid; benzoic acid; salicylic acid; p-aminobenzoic acid; aminobenzoic acid; uridine; deoxyuridine; thymidine; imidazoleacetic acid; methylimidazoleacetic acid; glucuronic acid]

Yanagisawa, I.; Yamane, M.; Urayama, T. Simultaneous separation and sensitive determination of free fatty acids in blood plasma by high-performance liquid chromatography, *J.Chromatogr.*, **1985**, *345*, 229–240. [fluorescence detection; serum; SPE; gradient; acetic acid; propionic acid; pyruvic acid; lactic acid; butyric acid; isobutyric acid; hydroxybutyric acid; valeric acid; isovaleric acid; pentenoic acid; caproic acid; isocaproic acid; hexenoic acid; heptanoic acid; heptenoic acid; caprylic acid; octenoic acid; nonanoic acid; capric acid; undecanoic acid; lauric acid; tridecanoic acid; myristic acid; myristoleic acid; pentadecanoic acid; palmitic acid; palmitoleic acid; palmitelaidic acid; heptadecanoic acid; stearic acid; petroselinic acid; oleic acid; elaidic acid; vaccenic acid; linoleic acid; linoelaidic acid; linolenic acid; nonadecanoic acid; arachidic acid; eicosenoic acid; eicosadienoic acid; homolinolenic acid; eicosatrienoic acid; arachidonic acid; eicospentaenoic acid; heneicosanoic acid; behenic acid; erucic acid; brassidic acid; docosahexaenoic acid; tricosanoic acid; lignoceric acid; nervonic acid]

Yasaka, Y.; Tanaka, M.; Matsumoto, T.; Katakawa, J.; Tetsumi, T.; Shono, T. 2-(Phthalimino)ethyl trifluoromethanesulfonate as a highly reactive ultraviolet-labeling agent for carboxylic acids in high-performance liquid chromatography, *Anal.Sci.*, **1990**, *6*, 49–52. [mouse; brain; LOD 200 fmol; palmitic acid; oleic acid; margaric acid; stearic acid]

Yasaka, Y.; Tanaka, M.; Shono, T.; Tetsumi, T.; Katakawa, J. 2-(2,3-Naphthalimino)ethyl trifluoromethanesulfonate as a highly reactive ultraviolet and fluorescent labelling agent for the liquid chromatographic determination of carboxylic acids, *J.Chromatogr.*, **1990**, *508*, 133–140. [fluorescence detection; UV detection; mouse; brain; LOD 4 fmol; docosahexaenoic acid; arachidonic acid; palmitic acid; oleic acid; margaric acid; stearic acid]

Yoshida, T.; Uetake, A.; Yamaguchi, H.; Nimura, N.; Kinoshita, T. New preparation method for 9-anthryldiazomethane (ADAM) as a fluorescent labeling reagent for fatty acids and derivatives, *Anal.Biochem.*, **1988**, *173*, 70–74. [derivatization; LOQ 125 pmol; gradient; isocratic; fluorescence; fluorescence detection; column temp 50; linolenic acid; palmitoleic acid; arachidic acid; stearic acid; myristic acid; linoleic acid; palmitic acid; oleic acid; lactic acid; acetic acid; propionic acid; pyruvic acid; levulinic acid; ketobutyric acid; ketovaleric acid; ketocaproic acid; glycolic acid; hydroxyisobutyric acid; hydroxyisocaproic acid; hydroxycaprylic acid; hydroxynaphthoic acid; hydroxymyristic acid; hydroxystearic acid]

You, J.; Zhang, W.; Jia, X.; Zhang, Y. An improved derivatization method for sensitive determination of fatty acids by high-performance liquid chromatography using 9-(2-hydroxylethyl)-carbazole as derivatization reagent with fluorescence detection, *Chromatographia*, **2001**, *54*, 316–322. [derivatization; fluorescence detection; linoleic acid; oleic acid; stearic acid]

You, J.; Zhang, W.; Zhang, Y. Simple derivatization method for sensitive determination of fatty acids with fluorescence detection by high-performance liquid chromatography using 9-(2-hydroxyethyl)-carbazole as derivatization reagent, *Anal.Chim.Acta*, **2001**, *436*, 163–172. [derivatization; linoleic acid; oleic acid]

Zaitseva, I.; Ajmal, M.; Cersosimo, E. Application of high-performance liquid chromatography of plasma fatty acids as their phenacyl esters to evaluate splanchnic and renal fatty acid balance in vivo, *J.Chromatogr.B*, **1999**, *727*, 15–22. [derivatization; linoleic acid; oleic acid]

# Olmesartan

**Molecular formula:** $C_{29}H_{30}N_6O_6$
**Molecular weight:** 558.58
**CAS Registry No:** 144689-63-4
**Merck Index:** 13, 6909

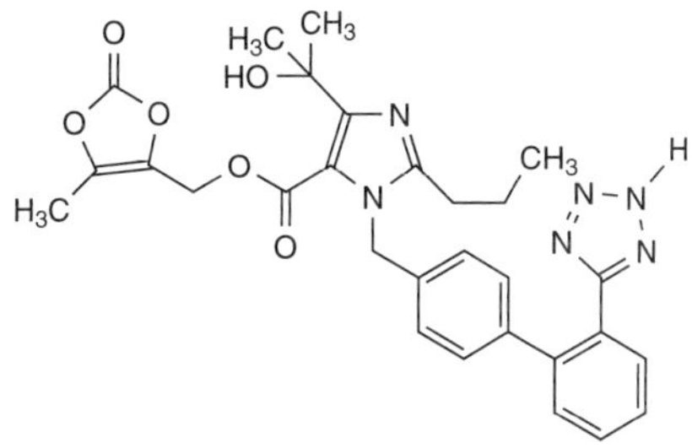

## SAMPLE

**Matrix:** microsomal incubations
**Sample preparation:** Mix 100 μL microsomal incubation with 100 μL MeCN, centrifuge at 10 000 g for 3 min, inject a 10 μL aliquot.

## HPLC VARIABLES

**Column:** 150 × 6.9 YMC-Pack ODS-A-312 C18
**Mobile phase:** MeCN:water:PIC A 45:55:2.2
**Flow rate:** 1
**Injection volume:** 10
**Detector:** UV 254

## CHROMATOGRAM

**Retention time:** 6

## KEY WORDS

human; intestine; liver; rat

## REFERENCE

Kobayashi, N.; Fujimori, I.; Watanabe, M.; Ikeda, T. Real-time monitoring of metabolic reactions by microdialysis in combination with tandem mass spectrometry: hydrolysis of CS-866 in vitro in human and rat plasma, livers, and small intestines, *Anal.Biochem.*, **2000**, *287*, 272–278.

# Olopatadine

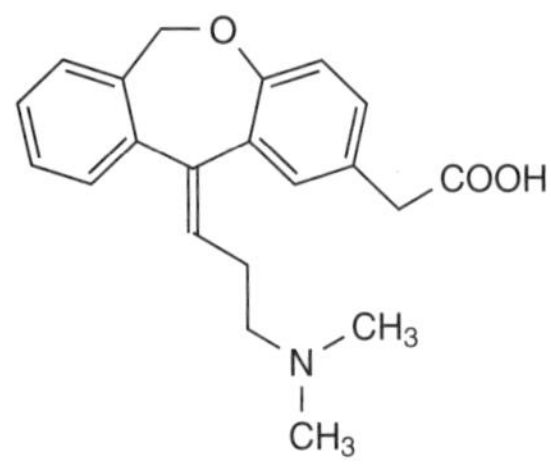


**Molecular formula:** $C_{21}H_{23}NO_3$
**Molecular weight:** 337.41
**CAS Registry No:** 113806-05-6, 140462-76-6 (HCl)
**Merck Index:** 13, 6910

## SAMPLE

**Matrix:** blood
**Sample preparation:** Condition a Bond Elut C18 SPE cartridge with 1 mL MeOH, 2 mL water, 2 mL 1% bovine serum albumin in water, and 5 mL water. Vortex 250 μL plasma, 50 μL 200 ng/mL IS in water, and 250 μL water, add to the SPE cartridge, wash with 2 mL water, elute with 1 mL MeOH. Evaporate the eluate to dryness under a stream of nitrogen, reconstitute the residue with 50 μL mobile phase, filter (0.2 μm PTFE), inject a 10 μL aliquot.

## HPLC VARIABLES

**Column:** 50 × 2 5 μm Develosil ODS HG-5
**Mobile phase:** MeOH:10 mM acetic acid 45:55
**Flow rate:** 0.1
**Injection volume:** 10
**Detector:** MS, Micromass Quattro, electrospray, ion source 120°, capillary 4.0 kV, counter current electrode 1.2 kV, cone 20 V, positive mode, collision gas argon 0.2 Pa, collision energy 20 eV, m/z 338–165

## CHROMATOGRAM

**Retention time:** 2.5
**Internal standard:** KF11796 ((11*Z*)-11-[3-(dimethylamino)propylidene]-6,11-dihydrodibenz[*b*,*e*]oxepin-2-propionic acid) (m/z 352–247–179) (4)
**Limit of quantitation:** 1 ng/mL

## OTHER SUBSTANCES

**Extracted:** metabolites

## KEY WORDS

pharmacokinetics; plasma; SPE

## REFERENCE

Fujita, K.; Magara, H.; Kobayashi, H. Determination of olopatadine, a new antiallergic agent, and its metabolites in human plasma by high-performance liquid chromatography with electrospray ionization tandem mass spectrometry, *J.Chromatogr.B*, **1999**, *731*, 345–352.

## SAMPLE

**Matrix:** microsomal incubations
**Sample preparation:** Mix 100 μL microsomal incubation with 100 μL ice-cold MeCN, centrifuge at 14 020 g for 10 min, filter the supernatant, inject an aliquot of the filtrate.

## HPLC VARIABLES

**Column:** 150 × 6 5 μm YMC-Pack AM312
**Mobile phase:** MeCN:0.1% trifluoroacetic acid 20:80
**Flow rate:** 1
**Detector:** Radioactivity ($^{14}C$)

## OTHER SUBSTANCES

**Extracted:** metabolites

## KEY WORDS

human; liver

## REFERENCE

Kajita, J.; Inano, K.; Fuse, E.; Kuwabara, T.; Kobayashi, H. Effects of olopatadine, a new antiallergic agent, on human liver microsomal cytochrome P450 activities, *Drug Metab.Dispos.*, **2002**, *30*, 1504–1511.

# Orbifloxacin

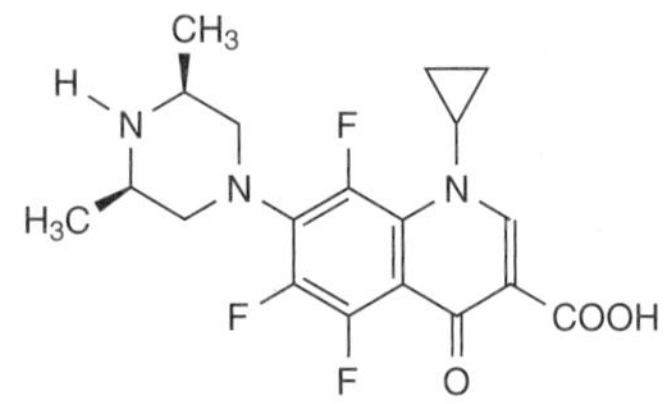


**Molecular formula:** $C_{19}H_{20}F_3N_3O_3$
**Molecular weight:** 395.38
**CAS Registry No:** 113617-63-3

## SAMPLE

**Matrix:** blood
**Sample preparation:** Mix 200 μL plasma with 800 μL 1.5 μg/mL IS in 100 mM pH 7.4 phosphate buffer, add 6 mL chloroform (Caution! Chloroform is a carcinogen!), shake at 200 oscillations/min for 30 min, centrifuge at 13 000 g for 6 min. Evaporate the organic layer to dryness under a stream of nitrogen at 40°, reconstitute the residue with 200 μL phosphate-buffered saline, inject an aliquot.

## HPLC VARIABLES

**Guard column:** Novapack C18 Guard-Pak
**Column:** 150 × 3.9 5 μm Novapack C18
**Mobile phase:** MeCN:buffer 20:80 (The buffer was 20 mM potassium dihydrogen phosphate containing 6 mM phosphoric acid and 12 mM tetraethylammonium bromide, pH adjusted to 3.0 with 2 M NaOH.)
**Flow rate:** 1
**Detector:** F ex 338 em 425

## CHROMATOGRAM

**Retention time:** 3.09
**Internal standard:** norfloxacin (2.16)
**Limit of detection:** 9 ng/mL (sic)
**Limit of quantitation:** 4 ng/mL

## OTHER SUBSTANCES

**Simultaneous:** ciprofloxacin (2.28), danofloxacin (2.80), difloxacin (4.52), enrofloxacin (3.30), marbofloxacin (2.20), sarafloxacin (4.40)

## KEY WORDS

plasma; rabbit

## REFERENCE

García, M.A.; Solans, C.; Aramayona, J.J.; Rueda, S.; Bregante, M.A. Determination of orbifloxacin in rabbit plasma by high-performance liquid chromatography with fluorescence detection, *J.Chromatogr. Sci.*, **1999**, *37*, 199–202.

## SAMPLE

**Matrix:** solutions

## HPLC VARIABLES

**Column:** 250 × 4 Develosil ODS-7
**Mobile phase:** MeOH:dioxane:100 mM pH 3.5 citrate buffer 12:5:84 (Caution! Dioxane is a carcinogen!)
**Flow rate:** 1.2
**Detector:** UV 290

## CHROMATOGRAM
**Retention time:** 6

## OTHER SUBSTANCES
**Simultaneous:** degradation products

## KEY WORDS
photodegradation

## REFERENCE
Morimura, T.; Kohno, K.; Nobuhara, Y.; Matsukura, H. Photoreaction and active oxygen generation by photosensitization of a new antibacterial fluoroquinolone derivative, orbifloxacin, in the presence of chloride ion, *Chem.Pharm.Bull.*, **1997**, *45*, 1828–1832.

---

## SAMPLE
**Matrix:** tissue
**Sample preparation:** Condition a 3 mL 250 mg Supelclean ENVI C18 SPE cartridge with 2 mL MeOH and 2 mL MeCN. Pack 250 mg 200–400 mesh AG MP-1 resin (Bio-Rad) between frits in an empty 3 mL tube, condition with MeOH, water, and 5 mL 8 mM NaOH. Add 5 mL MeCN to 1.5 g pureed fish, sonicate (Sonifier 450, 30% duty cycle, 40% power) for 3 min, rotate for 10 min, centrifuge at 3000 rpm for 5 min, repeat extraction with 5 mL MeCN. Combine the extracts and evaporate them to ca. 4 mL under a stream of nitrogen at 45°, add to the C18 SPE cartridge, elute with another 1 mL MeCN. Dilute all the eluates to 40 mL with 8 mM NaOH. Add this solution to the AG MP-1 cartridge, wash with 2 mL water, wash with 2 mL MeOH, dry with a stream of nitrogen, elute with 3 mL MeCN:2% formic acid 20:80, add 90 μL IS solution to the eluate, inject a 10 μL aliquot. (IS solution contained 500 ng/mL clenbuterol and 50 ng/mL penbutolol in MeCN:water 20:80.)

## HPLC VARIABLES
**Column:** 150 × 2.1 5 μm Zorbax Extend C18
**Column temperature:** 30
**Mobile phase:** Gradient. MeCN:2% formic acid:water 20:10:70 for 3 min, to 55:10:35 over 0.1 min, maintain at 55:10:35 for 6.9 min, return to initial conditions over 0.1 min, re-equilibrate for 6.9 min.
**Flow rate:** 0.2
**Injection volume:** 10
**Detector:** MS, Micromass Quattro LC triple quadrupole, positive ion mode, nebulizer gas nitrogen 100 L/h, drying gas nitrogen 700 L/h, collision gas argon at 1 μbar, source block 110°, desolvation 350°, cone 32 V, collision energy 16 eV, m/z 396.36–352.11

## CHROMATOGRAM
**Retention time:** 2.57
**Internal standard:** clenbuterol (m/z 277.00–203.00, cone 20 V, collision 15 eV) (3.13), penbutolol (m/z 292.27–236.13, cone 28 V, collision 16 eV) (8.53)
**Limit of detection:** 5 ng/g

## OTHER SUBSTANCES
**Extracted:** ciprofloxacin (m/z 332.15–288.07, cone 40 V, collision 16 eV; LOD 10 ng/g) (2.17), danofloxacin (m/z 358.08–95.91, cone 38 V, collision 22 eV) (2.23), enrofloxacin (m/z 360.24–316.14, cone 40 V, collision 16 eV) (2.41), flumequine (m/z 262.14–202.03, cone 28 V, collision 34 eV) (9.31), nalidixic acid (m/z 233.09–187.05, cone 28 V, collision 28 eV) (9.14), oxolinic acid (m/z 262.12–160.02, cone 28 V, collision 40 eV) (8.11), piromidic acid (m/z 289.17–243.06, cone 34 V, collision 28 eV) (10.05), sarafloxacin (m/z 386.20–299.06, cone 40 V, collision 28 eV) (3.04)

## KEY WORDS

abalone; fish; SPE; trout

## REFERENCE

Johnston, L.; Mackay, L.; Croft, M. Determination of quinolones and fluoroquinolones in fish tissue and seafood by high-performance liquid chromatography with electrospray ionisation tandem mass spectrometric detection, *J.Chromatogr.A*, **2002**, *982*, 97–109.

## ANNOTATED BIBLIOGRAPHY

Morimura, T.; Ohno, T.; Matsukura, H.; Nobuhara, Y. Photodegradation kinetics of the new antibacterial fluoroquinolone derivative, orbifloxacin, in aqueous solution, *Chem.Pharm.Bull.*, **1995**, *43*, 1000–1004.

Morimura, T.; Ohno, T.; Matsukura, H.; Nobuhara, Y. Degradation kinetics of the new antibacterial fluoroquinolone derivative, orbifloxacin, in aqueous solution, *Chem.Pharm.Bull.*, **1995**, *43*, 1052–1054.

Morimura, T.; Nobuhara, Y.; Matsukura, H. Photodegradation products of a new antibacterial fluoroquinolone derivative, orbifloxacin, in aqueous solution, *Biol.Pharm.Bull.*, **1997**, *45*, 373–377.

Schneider, M.J.; Donoghue, D.J. Multiresidue analysis of fluoroquinolone antibiotics in chicken tissue using liquid chromatography-fluorescence-multiple mass spectrometry, *J.Chromatogr.B*, **2002**, *780*, 83–92. [ciprofloxacin; norfloxacin; danofloxacin; enrofloxacin; orbifloxacin; sarafloxacin; difloxacin; LOQ 10 ng/g]

Schneider, M.J.; Donoghue, D.J. Multiresidue determination of fluoroquinolone antibiotics in eggs using liquid chromatography-fluorescence-mass spectrometry, *Anal.Chim.Acta*, **2003**, *483*, 39–49. [norfloxacin; ciprofloxacin; danofloxacin; enrofloxacin; orbifloxacin; sarafloxacin; difloxacin; LOQ 10 ng/g]

# Orlistat

**Molecular formula:** $C_{29}H_{53}NO_5$
**Molecular weight:** 495.73
**CAS Registry No:** 96829-58-2
**Merck Index:** 13, 6935

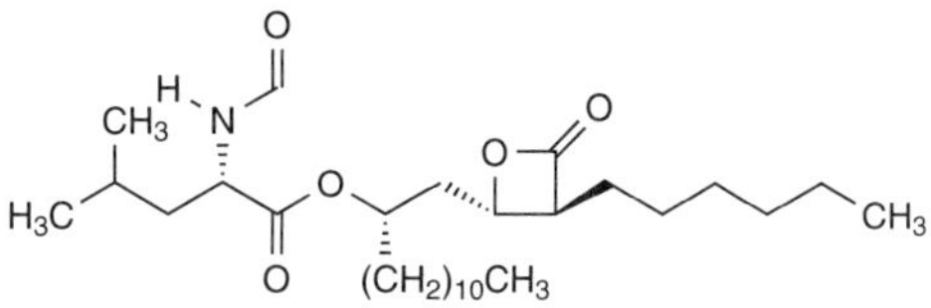

## SAMPLE

**Matrix:** blood
**Sample preparation:** Mix 1 mL plasma with 50 μL 100 ng/mL IS in MeOH, add 1 mL MeCN, vortex, centrifuge at RCF 834 for 5 min. Remove the upper layer and add it to 5 mL hexane, rotate for 20 min, centrifuge at RCF 834 for 5 min. Evaporate the organic layer to dryness under a stream of nitrogen at 25°, reconstitute the residue with 30 μL MeCN:2 mM ammonium acetate 70:30, inject a 10 μL aliquot.

## HPLC VARIABLES

**Column:** 50 × 2 5 μm Deltabond phenyl (Keystone)
**Mobile phase:** MeCN:2 mM ammonium acetate 90:10
**Flow rate:** 0.2
**Injection volume:** 10
**Detector:** MS, PE Sciex API-III, APCI, tandem triple quadrupole, positive ion mode, declustering potential 38 V, sprayer 4600 V, multiplier −4200 V, nebulizer gas nitrogen at 60 psi, curtain gas nitrogen at 1.2 L/min, collision gas argon, m/z 496−160

## CHROMATOGRAM

**Retention time:** 1.2
**Internal standard:** $d_5$-orlistat (m/z 501−160)
**Limit of detection:** 0.1 ng/mL
**Limit of quantitation:** 0.2 ng/mL

## KEY WORDS

plasma

## REFERENCE

Bennett, P.K.; Li, Y.T.; Edom, R.; Henion, J. Quantitative determination of Orlistat (tetrahydrolipostatin, Ro 18−0647) in human plasma by high-performance liquid chromatography coupled with ion spray tandem mass spectrometry, *J.Mass Spectrom.*, **1997**, *32*, 739−749.

## SAMPLE

**Matrix:** blood
**Sample preparation:** Mix 1 mL plasma, IS, and 1 mL MeCN, centrifuge. Mix the supernatant with 5 mL hexane, shake, centrifuge. Remove the upper hexane layer and evaporate it to dryness, reconstitute the residue with 50 μL MeCN:10 mM ammonium acetate 70:30, inject a 20 μL aliquot.

## HPLC VARIABLES

**Guard column:** 2 mm long
**Column:** 100 × 2 Spherisorb C6
**Mobile phase:** A MeCN:0.1% formic acid 95:5; or B MeOH:water 85:15 for 2.4 min, to 100:0 (step gradient), maintain at 100:0 for 3 min, re-equilibrate at 85:15 for 3 min.
**Flow rate:** A 0.15; B 0.15 for 2.4 min then 0.3
**Injection volume:** 20

**Detector:** MS Finnigan LCQ, quadrupolar ion trap, capillary electrospray, needle voltage 4 kV; nebulizer gas flow at 60% of maximum; capillary temperature 250°; capillary potential 20 V; lens potential 10 V, positive ion MS-MS mode, m/z 140–350

## CHROMATOGRAM

**Retention time:** 1.1 (A), 2.4 (B)
**Internal standard:** $d_5$-orlistat
**Limit of quantitation:** 0.3 ng/mL

## KEY WORDS

plasma

## REFERENCE

Wieboldt, R.; Campbell, D.A.; Henion, J. Quantitative liquid chromatographic-tandem mass spectrometric determination of orlistat in plasma with a quadrupole ion trap, *J.Chromatogr.B*, **1998**, *708*, 121–129.

# Oseltamivir

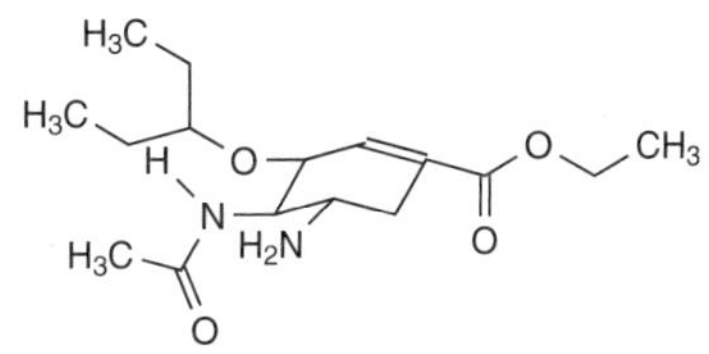


**Molecular formula:** $C_{16}H_{28}N_2O_4$
**Molecular weight:** 312.40
**CAS Registry No:** 196618-13-0, 204255-11-8 (phosphate)
**Merck Index:** 13, 6958

## SAMPLE

**Matrix:** blood
**Sample preparation:** Condition a 100 mg C18 SPE cartridge (Varian) with 1 mL MeCN:water 75:25 and 1 mL 10 mM HCl in MeCN:water 5:95. Mix 100 μL plasma with 25 μL 1 M citric acid and 100 μL 1 μg/mL IS in 50 mM sodium dihydrogen phosphate, add to the SPE cartridge, wash with 1 mL 10 mM HCl in MeCN:water 5:95, elute with 400 μL MeCN:water 75:25, add 50 μL 20 mM KCN in 200 mM pH 6.5 phosphate buffer to the eluate, add 50 μL 20 mM naphthalene-2,3-dialdehyde in MeCN, vortex, heat at 40° for 45 min. Evaporate to dryness under reduced pressure at room temperature, reconstitute the residue with 100 μL 50 mM sodium dihydrogen phosphate, centrifuge, inject a 40 μL aliquot.

## HPLC VARIABLES

**Column:** 150 × 4.6 5 μm Prodigy (ODS-2) (Phenomenex)
**Column temperature:** 40
**Mobile phase:** MeCN:water 27:73 containing 50 mM sodium acetate
**Flow rate:** 2
**Injection volume:** 40
**Detector:** F ex 420 em 472

## CHROMATOGRAM

**Retention time:** 5.2 (for GS4071 the free acid active metabolite)
**Internal standard:** GS4057 ((3*R*, 4*R*, 5*S*)-4-acetamido- 5 -amino-3-(1-cyclopentanoxy)-1-cyclohexene-1-carboxylic acid) (4.3)
**Limit of detection:** 20 ng/mL (S/N 3)
**Limit of quantitation:** 50 ng/mL

## KEY WORDS

derivatization; plasma; rat; SPE

## REFERENCE

Eisenberg, E.J.; Cundy, K.C. High-performance liquid chromatographic determination of GS4071, a potent inhibitor of influenza neuraminidase, in plasma by precolumn fluorescence derivatization with naphthalenedialdehyde, *J.Chromatogr.B*, **1998**, *716*, 267–273.

## SAMPLE

**Matrix:** blood, tissue, urine
**Sample preparation:** Homogenize (Ultraturrax) 1 g lung or liver, centrifuge at 1300 g at 4° for 10 min, filter (0.2 μm Gelman Z-spin) while centrifuging at 15 000 g for 15 min, inject an aliquot of the filtrate. Filter (Millipore 10 000 MW cut-off) plasma or urine while centrifuging at 15 000 g for 15 min, inject an aliquot of the ultrafiltrate.

## HPLC VARIABLES

**Column:** 125 × 4 5 μm Inertsil C-18
**Mobile phase:** Gradient. A was 50 mM ammonium acetate. B was MeCN:50 mM ammonium acetate 60:40. A:B from 90:10 to 0:100 over 20 min.

**Flow rate:** 1
**Detector:** UV 220; Radioactivity ($^{14}C$)

---

## CHROMATOGRAM
**Retention time:** 14.5

---

## OTHER SUBSTANCES
**Extracted:** GS4071 (free acid active metabolite) (7), other metabolites

---

## KEY WORDS
liver; lung; plasma; rat; ultrafiltrate

---

## REFERENCE
Sweeny, D.J.; Lynch, G.; Bidgood, A.M.; Lew, W.; Wang, K.-Y.; Cundy, K.C. Metabolism of the influenza neuraminidase inhibitor prodrug oseltamivir in the rat, *Drug Metab.Dispos.*, **2000**, *28*, 737–741.

---

## SAMPLE
**Matrix:** blood, urine
**Sample preparation:** Condition a 3 mL 7 mm Empore Mixed Phase Cation-MPC SPE disc with 1 mL MeOH, three 3 mL portions of MeOH:50 mM ammonium acetate 90:10, and 1 mL 5 mM pH 3.5 ammonium acetate. Mix 100 μL plasma or urine with 50 μL IS solution and 1 mL 5 mM pH 3.5 ammonium acetate buffer, add to the SPE disc, wash with 1 mL water, wash with 1 mL MeOH, wash with 1 mL MeOH:water 90:10, dry under vacuum, elute with 1 mL MeOH:50 mM ammonium acetate 90:10. Evaporate the eluate to dryness under a stream of nitrogen at ca. 50°, reconstitute the residue with 150 μL water, centrifuge at >500 rpm, inject a 100 μL aliquot. (IS solution contained 50 ng/ml of $d_3$-oseltamivir and 2 μg/ml of $d_3$-GS4071 in water).

---

## HPLC VARIABLES
**Column:** 100 × 5 4 μm Nova-Pak CN HP radial compression
**Mobile phase:** MeOH:80 mM pH 3 formic acid 50:50
**Flow rate:** 0.5
**Injection volume:** 100
**Detector:** MS, Finnigan TSQ 7000 tandem quadrupole, electrospray, collision gas argon, m/z 313–224

---

## CHROMATOGRAM
**Retention time:** 5
**Internal standard:** $d_3$-oseltamivir (m/z 316–227); $d_3$-GS4071 (m/z 288–201)
**Limit of quantitation:** 1 ng/mL (plasma), 5 ng/mL (urine)

---

## OTHER SUBSTANCES
**Extracted:** GS4071 (de-ethylated active metabolite) (LOQ 10 ng/mL (plasma), 30 ng/mL (urine); m/z 285–198) (3.5)

---

## KEY WORDS
ferret; human; marmoset; mouse; plasma; rabbit; rat; SPE

---

## REFERENCE
Wiltshire, H.; Wiltshire, B.; Citron, A.; Clarke, T.; Serpe, C.; Gray, D.; Herron, W. Development of a high-performance liquid chromatographic-mass spectrometric assay for the specific and sensitive quantification of Ro 64–0802, an anti-influenza drug, and its pro-drug, oseltamivir, in human and animal plasma and urine, *J.Chromatogr.B*, **2000**, *745*, 373–388.

# Oxaliplatin

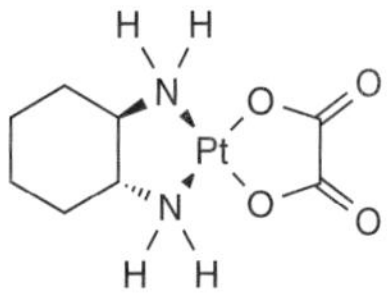

**Molecular formula:** $C_8H_{14}N_2O_4Pt$
**Molecular weight:** 397.29
**CAS Registry No:** 61825-94-3
**Merck Index:** 13, 6981

## SAMPLE
**Matrix:** blood
**Sample preparation:** Filter (Sartorius Centrisart I 10 000 MW cut-off) while centrifuging at 4000 g at 4° for 1 h, inject an aliquot of the ultrafiltrate.

## HPLC VARIABLES
**Column:** 250 × 4 5 μm Hypercarb
**Mobile phase:** MeOH:250 mM pH 7.0 succinic acid buffer 90:10
**Flow rate:** 0.5
**Injection volume:** 25
**Detector:** UV 344 following post-column reaction. The column effluent mixed with 2.7 mM sodium *N*, *N*-diethyldithiocarbamate in MeOH pumped at 0.17 mL/min and the mixture flowed through a 2.3 m long 0.51 mm ID length of PTFE tubing to the detector. A 2 m length of the tubing was heated in a microwave field (ca. 130 W, Smithcreator, Personal Chemistry, Sweden) at about 110°.

## CHROMATOGRAM
**Retention time:** 12
**Limit of quantitation:** 40 ng/mL (S/N 10)

## KEY WORDS
pharmacokinetics; post-column reaction; ultrafiltrate; whole blood

## REFERENCE
Ehrsson, H.; Wallin, I. Liquid chromatographic determination of oxaliplatin in blood using post-column derivatization in a microwave field followed by photometric detection, *J.Chromatogr.B*, **2003**, *795*, 291–294.

## SAMPLE
**Matrix:** blood, urine
**Sample preparation:** Plasma. Filter (Amicon MPS-1 with a YMT membrane) while centrifuging at 4° at 3000 g for 15 min, inject an aliquot of the ultrafiltrate. Urine. Directly inject an aliquot of urine. (Protect from light.)

## HPLC VARIABLES
**Column:** 250 × 4.6 Inertsil ODS-2
**Column temperature:** 40
**Mobile phase:** MeCN:10 mM pH 5.5 acetate buffer 5:95
**Flow rate:** 1
**Injection volume:** 100
**Detector:** UV 290 following post-column reaction. The column effluent mixed with the reagent (maintained at 0°) pumped at 0.3 mL/min and the mixture flowed through a 10 m ×0.5 mm ID PTFE coil at 60° to the detector. (The reagent was 10 mM pH 5.5 acetate buffer containing 40 mM sodium bisulfite.)

## CHROMATOGRAM
**Retention time:** 11

**Limit of detection:** 60 nM

## OTHER SUBSTANCES
**Extracted:** carboplatin (6.5), tetraplatin (8)

## KEY WORDS
plasma; post-column reaction; rabbit

## REFERENCE
Kizu, R.; Yamamoto, T.; Yokoyama, T.; Tanaka, M.; Miyazaki, M. A sensitive postcolumn derivatization/UV detection system for HPLC determination of antitumor divalent and quadrivalent platinum complexes, *Chem.Pharm.Bull.*, **1995**, *43*, 108–114.

## SAMPLE
**Matrix:** solutions
**Sample preparation:** Inject an aliquot of a 700 μg/mL solution of oxaliplatin in MeOH containing 800 μg/mL IS.

## HPLC VARIABLES
**Column:** 250 × 4.6 10 μm Vydac C18
**Column temperature:** 24
**Mobile phase:** MeCN:water 80:20
**Flow rate:** 0.8
**Detector:** UV 255

## CHROMATOGRAM
**Retention time:** 3.4
**Internal standard:** flavone (4.6)

## KEY WORDS
robust; validated

## REFERENCE
Ficarra, R.; Calabrò, M.L.; Cutroneo, P.; Tommasini, S.; Melardi, S.; Semreen, M.; Furlanetto, S.; Ficarra, P.; Altavilla, G. Validation of a LC method for the analysis of oxaliplatin in a pharmaceutical formulation using an experimental design, *J.Pharm.Biomed.Anal.*, **2002**, *29*, 1097–1103.

## ANNOTATED BIBLIOGRAPHY
Heudi, O.; Mercier-Jobard, S.; Cailleux, A.; Allain, P. Mechanisms of reaction of L-methionine with carboplatin and oxaliplatin in different media: a comparison with cisplatin, *Biopharm.Drug Dispos.*, **1999**, *20*, 107–116.

Luo, F.R.; Yen, T.-Y.; Wyrick, S.D.; Chaney, S.G. High-performance liquid chromatographic separation of the biotransformation products of oxaliplatin, *J.Chromatogr.B*, **1999**, *724*, 345–356.

# Oxiconazole

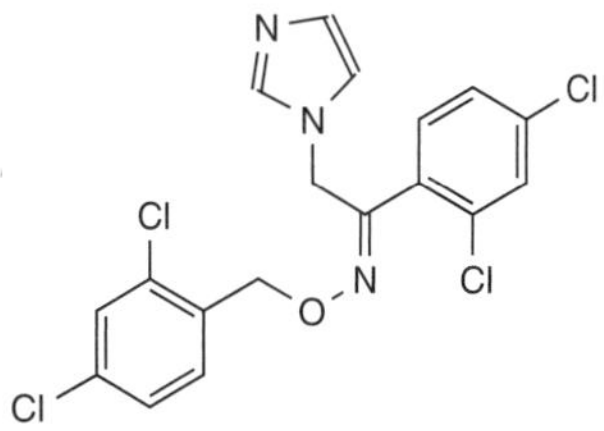


**Molecular formula:** $C_{18}H_{13}Cl_4N_3O$
**Molecular weight:** 492.14
**CAS Registry No:** 64211-45-6, 64211-46-7 (nitrate)
**Merck Index:** 13, 7006

### SAMPLE

**Matrix:** formulations
**Sample preparation:** Dissolve lotion containing 50 mg oxiconazole in 50 mL MeOH, dilute a 5 mL aliquot to 50 mL with mobile phase, inject an aliquot. Shake cream containing 50 mg oxiconazole with 50 mL chloroform:MeOH 50:50 for 30 min (Caution! Chloroform is a carcinogen!), make up to 100 mL with MeOH, centrifuge at 4000 rpm at 4° for 15 min, filter, dilute fivefold with MeOH.

### HPLC VARIABLES

**Guard column:** LiChrocart 4-4
**Column:** 125 × 4 5 μm LiChrocart C8
**Mobile phase:** MeOH:20 mM ammonium acetate buffer 85:15
**Flow rate:** 1
**Injection volume:** 20
**Detector:** UV 254

### CHROMATOGRAM

**Retention time:** 2.7
**Limit of detection:** 1.2 μg/mL
**Limit of quantitation:** 3.75 μg/mL

### OTHER SUBSTANCES

**Noninterfering:** benzyl alcohol

### KEY WORDS

cream; lotion; stability-indicating

### REFERENCE

Milano, J.; Morsch, L.M.; Gonalves Cardoso, S. LC method for the analysis of Oxiconazole in pharmaceutical formulations, *J.Pharm.Biomed.Anal.*, **2002**, *30*, 175–180.

### SAMPLE

**Matrix:** solutions
**Sample preparation:** Inject an aliquot of a 100 μg/mL solution in MeOH.

### HPLC VARIABLES

**Column:** 125 × 4 5 μm LiChrospher 100 RP8
**Mobile phase:** MeCN:20 mM ammonium acetate buffer 75:25
**Flow rate:** 1
**Detector:** UV 254

### CHROMATOGRAM

**Retention time:** 4.93

## OTHER SUBSTANCES

**Simultaneous:** bifonazole, photodegradation products

## REFERENCE

Thoma, K.; Kübler, N. Untersuchung der Photostabilität von Antimykotika. 1. Mitt.: Photostabilität von Azolantimykotika [Photodegradation of antimycotic drugs. Part 1: Photodegradation of azole antimycotics], *Pharmazie*, **1996**, *51*, 885–892.

# Panipenem

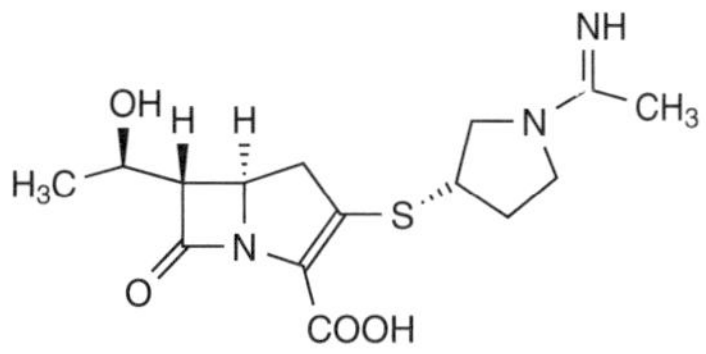

**Molecular formula:** $C_{15}H_{21}N_3O_4S$
**Molecular weight:** 339.42
**CAS Registry No:** 87726-17-8
**Merck Index:** 13, 7079

## SAMPLE

**Matrix:** blood
**Sample preparation:** Vortex 10 μL plasma with 10 μL of 200 mM MOPS (3-(*N*-morpholino)propanesulfonic acid) solution, add 100 μL MeOH, mix, centrifuge at 12 000 rpm for 5 min, inject a 25 μL aliquot of the supernatant.

## HPLC VARIABLES

**Column:** 150 × 4.6 5 μm Inertsil ODS-2
**Mobile phase:** MeOH:buffer 35:65 (The buffer was 5 mM pH 5.8 sodium dihydrogen phosphate containing 5 mM sodium *n*-dodecylsulfate.)
**Flow rate:** 0.8
**Injection volume:** 25
**Detector:** UV 300

## CHROMATOGRAM

**Retention time:** 6
**Limit of quantitation:** 500 ng/mL

## KEY WORDS

plasma

## REFERENCE

Kokubun, H.; Kimura, T.; Murase, S.; Shimada, S.; Kubo, H.; Matsumoto, M.; Nowatari, M.; Matsuura, N. Determination of panipenem in neonatal plasma by HPLC, *Anal.Sci.*, **2000**, *16*, 1077–1078.

# Parecoxib

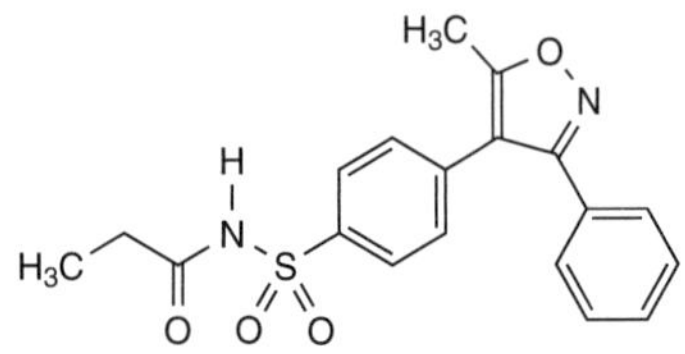


**Molecular formula:** $C_{19}H_{18}N_2O_4S$
**Molecular weight:** 370.42
**CAS Registry No:** 198470-84-7, 197502-82-2 (Na salt)

## SAMPLE
**Matrix:** formulations
**Sample preparation:** Inject a 10 μL aliquot.

## HPLC VARIABLES
**Column:** 150 × 4.6 5 μm YMC ODS-AQ
**Column temperature:** 40
**Mobile phase:** MeCN:10 mM pH 3.0 phosphate buffer 40:60
**Flow rate:** 2
**Injection volume:** 10
**Detector:** UV 215

## CHROMATOGRAM
**Retention time:** 9.5

## OTHER SUBSTANCES
**Simultaneous:** valdecoxib (5)

## KEY WORDS
injections; stability-indicating

## REFERENCE
Crane, I.M.; Mulhern, M.G.; Nema, S. Stability of reconstituted parecoxib for injection with commonly used diluents, *J.Clin.Pharm.Ther.*, **2003**, *28*, 363–369.

# Paricalcitol

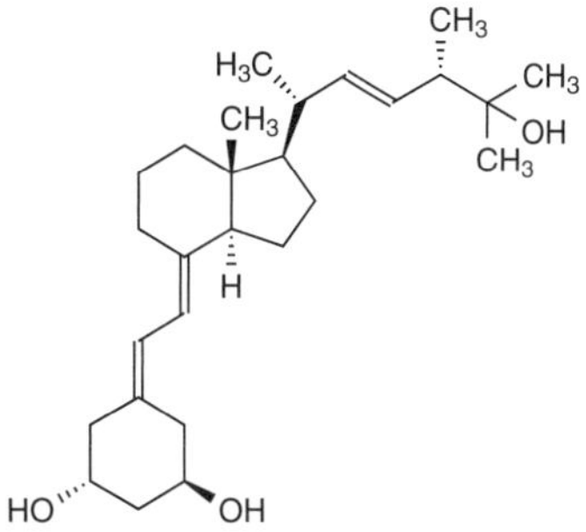


**Molecular formula:** $C_{27}H_{44}O_3$
**Molecular weight:** 416.63
**CAS Registry No:** 131918-61-1
**Merck Index:** 13, 7111

## SAMPLE
**Matrix:** blood
**Sample preparation:** Extract plasma with ether, inject an aliquot onto column A and column B in series and elute with mobile phase. After 1.5 min, remove column A from the circuit. Continue to elute column B with mobile phase and collect fractions. Regenerate column A by backflushing with a stronger mobile phase

## HPLC VARIABLES
**Column:** A 23 × 4 PVA-Sil; B 250 × 4.6 PVA-Sil
**Mobile phase:** MTBE:isopropanol 97.5:2.5
**Detector:** Radio receptor assay (after evaporation of fractions)

## KEY WORDS
column-switching; normal phase; plasma

## REFERENCE
Chang, M.; Qasawa, B.; Chu, S. Determination of 19-nor-1$\alpha$,25 dihydroxy-vitamin D2 (ABT-358) in plasma using a combination of HPLC and radio-receptor assay techniques (Abstract 2648), *Pharm.Res.*, **1997**, *14*, S427.

## SAMPLE
**Matrix:** solutions
**Sample preparation:** Inject an aliquot of a solution in mobile phase.

## HPLC VARIABLES
**Column:** 80 × 6.2 3 μm Zorbax SIL
**Mobile phase:** Hexane:isopropanol:MeOH 91:7:2
**Flow rate:** 1
**Detector:** UV 251

## CHROMATOGRAM
**Retention time:** 11.9

## OTHER SUBSTANCES
**Simultaneous:** metabolites

## KEY WORDS
normal phase

## REFERENCE

Shankar, V.N.; Propp, A.E.; Schroeder, N.; Surber, B.W.; Makin, H.L.J.; Jones, G. In vitro metabolism of 19-nor-1$\alpha$,25-$(OH)_2D_2$ in cultured cell lines: inducible synthesis of lipid- and water-soluble metabolites, *Arch.Biochem.Biophys.*, **2001**, *387*, 297–306.

## SAMPLE

**Matrix:** solutions
**Sample preparation:** Inject an aliquot of a solution in mobile phase.

## HPLC VARIABLES

**Column:** 250 × 4.6 6 μm Zorbax CN
**Mobile phase:** Hexane:isopropanol:MeOH 88:10:2
**Flow rate:** 1
**Detector:** UV 251

## CHROMATOGRAM

**Retention time:** 8.57

## OTHER SUBSTANCES

**Simultaneous:** metabolites

## REFERENCE

Shankar, V.N.; Propp, A.E.; Schroeder, N.; Surber, B.W.; Makin, H.L.J.; Jones, G. In vitro metabolism of 19-nor-1$\alpha$,25-$(OH)_2D_2$ in cultured cell lines: inducible synthesis of lipid- and water-soluble metabolites, *Arch.Biochem.Biophys.*, **2001**, *387*, 297–306.

# Pazufloxacin

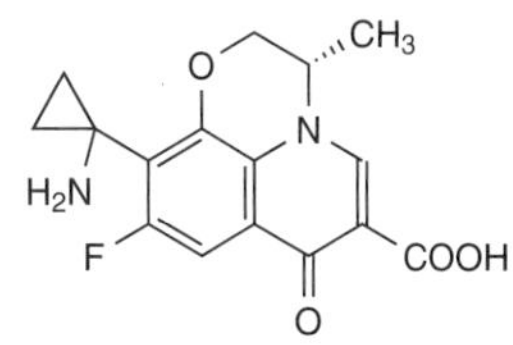

**Molecular formula:** $C_{16}H_{15}FN_2O_4$
**Molecular weight:** 318.30
**CAS Registry No:** 127045-41-4
**Merck Index:** 13, 7128

## SAMPLE
**Matrix:** blood
**Sample preparation:** Vortex 50 μL plasma with 350 μL 1 μg/mL IS in MeOH, centrifuge. Evaporate the supernatant to dryness under a stream of nitrogen at 40°, reconstitute the residue with mobile phase, inject an aliquot.

## HPLC VARIABLES
**Column:** 150 × 4.6 Cosmosil 5C18
**Column temperature:** 40
**Mobile phase:** MeCN:1 M ammonium acetate:50 mM citric acid 15:1:84
**Flow rate:** 1
**Detector:** F ex 335 em 450

## CHROMATOGRAM
**Internal standard:** lomefloxacin
**Limit of quantitation:** 100 ng/mL

## KEY WORDS
plasma; rat

## REFERENCE
Hasegawa, T.; Nadai, M.; Haghgoo, S.; Yamaki, K.; Takagi, K.; Nabeshima, T. Influence of a newly developed quinolone, T-3761, on pharmacokinetics of theophylline in rats, *Antimicrob.Agents Chemother.*, **1995**, *39*, 2138–2140.

## SAMPLE
**Matrix:** blood, urine
**Sample preparation:** Dilute plasma with an equal volume of MeOH, centrifuge, inject an aliquot. Dilute urine with mobile phase, inject an aliquot.

## HPLC VARIABLES
**Column:** 150 × 4 Develosil ODS-HG-5
**Mobile phase:** MeCN:200 mM pH 7.0 phosphate buffer:water 9:5:86 containing 0.68% tetra-*n*-butylammonium hydrogen sulfate
**Detector:** UV 330

## KEY WORDS
plasma; rat

## REFERENCE
Hayakawa, H.; Takagi, K.; Takano, Y.F.; Kawamura, Y.; Tsuji, A. Determinant of the distribution volume at steady state for novel quinolone pazufloxacin in rats, *J.Pharm.Pharmacol.*, **2002**, *54*, 1229–1236.

# Penciclovir

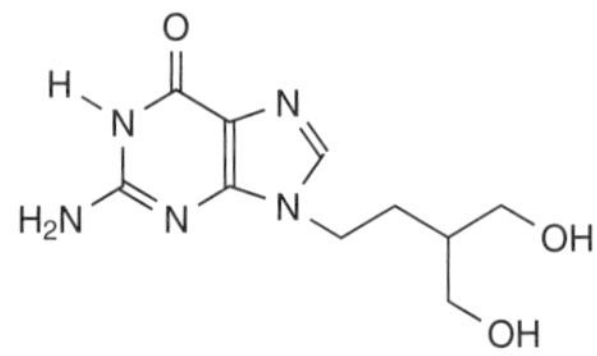

**Molecular formula:** $C_{10}H_{15}N_5O_3$
**Molecular weight:** 253.26
**CAS Registry No:** 39809-25-1
**Merck Index:** 13, 7154

## SAMPLE
**Matrix:** blood
**Sample preparation:** Mix 200 μL blood with 600 μL 16% trichloroacetic acid, centrifuge, inject an aliquot of the supernatant.

## HPLC VARIABLES
**Guard column:** C18 Guard-Pak (Waters)
**Column:** 100 × 8 Nova-Pak C18 (in a Z module)
**Mobile phase:** Gradient. A was 50 mM sodium hydrogen phosphate. B was MeOH:water 80:20 containing 5 mM sodium dihydrogen phosphate. A:B 99:1 for 1.5 min, to 5:95 over 18.5 min, re-equilibrate at initial conditions for 10 min.
**Flow rate:** 1.6
**Detector:** UV 254

## CHROMATOGRAM
**Retention time:** 13.5
**Limit of detection:** 200 ng/mL

## OTHER SUBSTANCES
**Extracted:** acyclovir (LOD 250 ng/mL) (11.6)

## KEY WORDS
mouse; pharmacokinetics; whole blood

## REFERENCE
Boyd, M.R.; Bacon, T.H.; Sutton, D. Antiherpesvirus activity of 9-(4-hydroxy-3-hydroxymethylbut-1-yl) guanine (BRL 39123) in animals, *Antimicrob.Agents Chemother.*, **1988**, *32*, 358–363.

## SAMPLE
**Matrix:** blood
**Sample preparation:** Condition a 1 mL Bond Elut SCX SPE cartridge with 1 mL MeOH, 1 mL water, and 1 mL 1 mM pH 7 sodium dihydrogen phosphate buffer. Mix 500 μL plasma with 50 μL 100 μg/mL IS in water, add 500 μL 16% trichloroacetic acid, mix, centrifuge (12 cm rotor arm) at 3000 rpm for 5 min. Add the supernatant to 500 μL 1 mM pH 7 sodium dihydrogen phosphate buffer, add to the SPE cartridge, wash with 1 mL 1 mM pH 7 sodium dihydrogen phosphate buffer, elute with 500 μL MeOH:250 mM pH 11 potassium dihydrogen phosphate 20:80, inject a 25–50 μL aliquot of the cluate.

## HPLC VARIABLES
**Guard column:** 15 × 3.2 7 μm Newguard RP-18
**Column:** 250 × 4.6 5 μm Spherisorb ODS2
**Mobile phase:** MeOH:10 mM pH 7 sodium dihydrogen phosphate buffer 10:90
**Flow rate:** 1
**Injection volume:** 25–50
**Detector:** UV 254

## CHROMATOGRAM
**Retention time:** 9
**Internal standard:** BRL 42377 (9-(4-hydroxy-2-hydroxymethylbut-1-yl)guanine) (10.5)
**Limit of quantitation:** 100 ng/mL

## KEY WORDS
pharmacokinetics; plasma; SPE

## REFERENCE
Fowles, S.E.; Pierce, D.M. High-performance liquid chromatographic method for the determination of 9-(4-hydroxy-3-hydroxymethylbut-1-yl)guanine (BRL-39123) in human plasma and urine, *Analyst*, **1989**, *114*, 1373–1375.

---

## SAMPLE
**Matrix:** blood, urine
**Sample preparation:** Condition a Bond Elut SCX strong cation exchange SPE cartridge with 1 mL MeOH, 1 mL water, and 1 mL 1 mM pH 7.0 disodium hydrogen phosphate buffer. Mix 500 μL plasma or 100 μL urine with 100 μL 20–50 μg/mL IS in water and 100–500 μL 16% trichloroacetic acid, add the supernatant to the SPE cartridge, wash with 1 mL MeOH:1 mM pH 7.0 disodium hydrogen phosphate buffer 20:80, elute with 1 mL MeOH:100 mM pH 11.0 dipotassium hydrogen phosphate buffer 25:75, inject an aliquot.

## HPLC VARIABLES
**Column:** 3 μm Apex 1 ODS
**Mobile phase:** Gradient. A was MeOH:10 mM pH 7.0 disodium hydrogen phosphate buffer 7:93. B was MeOH:10 mM pH 7.0 disodium hydrogen phosphate buffer 35:65. A:B from 100:0 to 0:100 over 4 min, maintain at 0:100 for 1.5 min, return to 0:100 over 1 min.
**Flow rate:** 2
**Detector:** UV 254

## CHROMATOGRAM
**Retention time:** 1.5
**Internal standard:** 6-deoxy-9-(4-hydroxy-2-hydroxymethylbut-1-yl)guanine (BRL 44056) (UV 305) (2.8)
**Limit of detection:** 500 ng/mL (plasma); 50 μg/mL (urine)

## OTHER SUBSTANCES
**Extracted:** famciclovir (UV 305) (6); metabolites

## KEY WORDS
dog; human; plasma; rat; SPE

## REFERENCE
Winton, C.F.; Fowles, S.E.; Pierce, D.M.; Hodge, A.V. Gradient high-performance liquid chromatographic method for the analysis of the pro-drug famciclovir and its metabolites, including the active anti-viral agent penciclovir, in plasma and urine, *Anal.Proc.*, **1990**, *27*, 181–182.

# Pentaerythritol tetranitrate

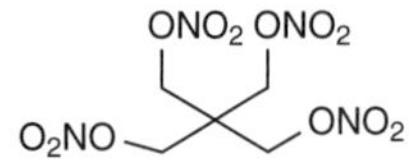


**Molecular formula:** $C_5H_8N_4O_{12}$
**Molecular weight:** 316.14
**CAS Registry No:** 78-11-5
**Merck Index:** 13, 7186

## SAMPLE

**Matrix:** blood
**Sample preparation:** Vigorously shake 5 mL plasma with 8 mL ethyl acetate for 1 min, sonicate for 5 min, centrifuge at 3000 rpm for 3 min, repeat extraction with 8 mL ethyl acetate, repeat extraction with 6 mL ethyl acetate. Pass the extracts through a C18 SPE cartridge, filter (0.5 μm) the eluate. Evaporate the eluate to 200 μL under a stream of nitrogen at 35°, inject a 25 μL aliquot.

## HPLC VARIABLES

**Column:** 300 × 3.9 10 μm μBondapak CN
**Mobile phase:** Iso-octane:dichloromethane:MeOH 75:20:5
**Flow rate:** 1.5
**Injection volume:** 25
**Detector:** TEA (Thermal Energy Analyzer; nitrosyl specific), Thermo Electron Corporation Model 502, pyrolyzer 500°, carrier gas nitrogen at 5 mL/min, reaction chamber 0.6 torr, cryogenic trap -78°

## CHROMATOGRAM

**Retention time:** 7
**Limit of quantitation:** 1 ng/mL

## OTHER SUBSTANCES

**Extracted:** isosorbide dinitrate (4), nitroglycerin (5), metabolites

## KEY WORDS

plasma

## REFERENCE

Yu, W.C.; Goff, E.U. Measurement of plasma concentrations of vasodilators and metabolites by the TEA Analyzer, *Biopharm.Drug Dispos.*, **1983**, *4*, 311–319.

# Pentosan polysulfate

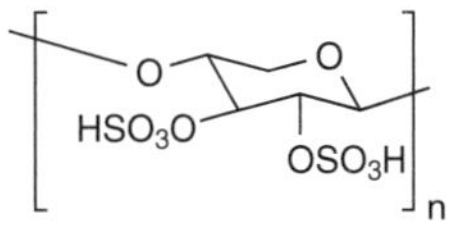

n = 6–12

**Molecular weight:** 1500–5000

**CAS Registry No:** 37300-21-3, 37319-17-8 (sodium salt)

**Merck Index:** 13, 7212

## SAMPLE

**Matrix:** solutions

## HPLC VARIABLES

**Column:** LiChrospher Si 100 diol
**Column temperature:** 25
**Mobile phase:** 200 mM NaCl
**Flow rate:** 1
**Detector:** Refractive Index

## CHROMATOGRAM

**Retention time:** 2

## REFERENCE

Muller, D.; Ndoume-Nze, M.; Jozefonvicz, J. High-pressure size-exclusion chromatography of anticoagulant materials, *J.Chromatogr.*, **1984**, *297*, 351–358.

# Perflubron

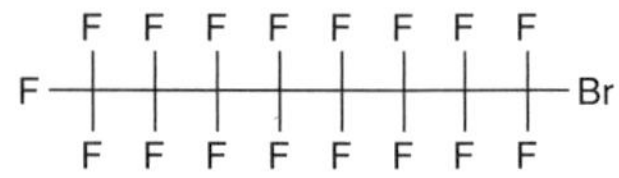


**Molecular formula:** $C_8BrF_{17}$
**Molecular weight:** 498.96
**CAS Registry No:** 423-55-2
**Merck Index:** 13, 7235

## SAMPLE
**Matrix:** solutions

## HPLC VARIABLES
**Column:** Dynamax Microsorb silica (Rainin)
**Mobile phase:** Hexane:isopropanol 50:1
**Flow rate:** 2
**Injection volume:** 20
**Detector:** UV 254

## CHROMATOGRAM
**Retention time:** 1.6

## OTHER SUBSTANCES
**Simultaneous:** phenol (3.8)
**Interfering:** 1-(4-perfluorobutylphenyl)-1-hexanone

## KEY WORDS
normal phase

## REFERENCE
Williams, T.D.; Jay, M.; Lehmler, H.-J.; Clark, M.E.; Stalker, D.J.; Bummer, P.M. Solubility enhancement of phenol and phenol derivatives in perfluorooctyl bromide, *J.Pharm.Sci.*, **1998**, *87*, 1585–1589.

# Perospirone

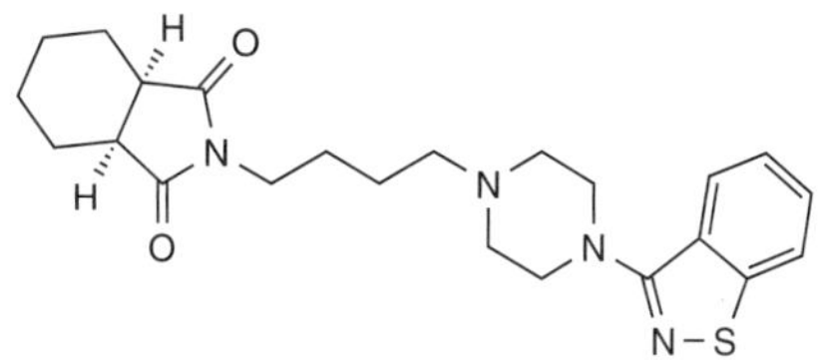

**Molecular formula:** $C_{23}H_{30}N_4O_2S$
**Molecular weight:** 426.58
**CAS Registry No:** 150915-41-6, 129273-38-7 (HCl)
**Merck Index:** 13, 7259

## SAMPLE

**Matrix:** blood
**Sample preparation:** Vortex 2 mL plasma, 500 μL water, 500 μL 1 mM NaOH, and 100 μL 250 ng/mL IS in water for 10 s, add 5 mL *n*-hexane:chloroform 70:30 (Caution! Chloroform is a carcinogen!), shake for 10 min, centrifuge at 1700 g at 4° for 10 min. Evaporate the organic layer to dryness under reduced pressure at 45°, reconstitute the residue with 750 μL mobile phase A, inject a 500 μL aliquot onto column A and elute to waste with mobile phase A; after 13.5 min, elute the contents of column A onto column B with mobile phase B; after another 3.5, min remove column A from the circuit, elute column B with mobile phase B, monitor the effluent from column B. (Re-equilibrate column A with mobile phase A for 18 min.)

## HPLC VARIABLES

**Column:** A 35 × 4.6 10 μm TSK gel PW (methacrylate polymer) (Tosoh); B 150 × 4.6 5 μm C18 STR ODS II (Shinwa, Kyoto)
**Column temperature:** 30 (column B only)
**Mobile phase:** A MeCN:50 mM pH 4.6 phosphate buffer:6 M perchloric acid 4.6:94.9:0.5; B MeCN:50 mM pH 4.6 phosphate buffer:6 M perchloric acid 41.5:58.0:0.5
**Flow rate:** A 1.2; B 0.6
**Injection volume:** 500
**Detector:** F ex 315 em 405

## CHROMATOGRAM

**Retention time:** 25
**Internal standard:** tiospirone (29.5)
**Limit of detection:** 60 pg/mL
**Limit of quantitation:** 100 pg/mL

## OTHER SUBSTANCES

**Extracted:** metabolite (ID-15036) (21.5)
**Noninterfering:** alprazolam, bromperidol, chlorpromazine, diazepam, haloperidol, levomepromazine, nitrazepam, risperidone

## KEY WORDS

column-switching; pharmacokinetics; plasma

## REFERENCE

Yasui-Furukori, N.; Inoue, Y.; Tateishi, T. Determination of a new atypical antipsychotic agent perospirone and its metabolite in human plasma by automated column-switching high-performance liquid chromatography, *J.Chromatogr.B*, **2003**, *789*, 239–245.

# Phenazopyridine hydrochloride

**Molecular formula:** $C_{11}H_{12}ClN_5$
**Molecular weight:** 249.71
**CAS Registry No:** 136-40-3
**Merck Index:** 13, 7296

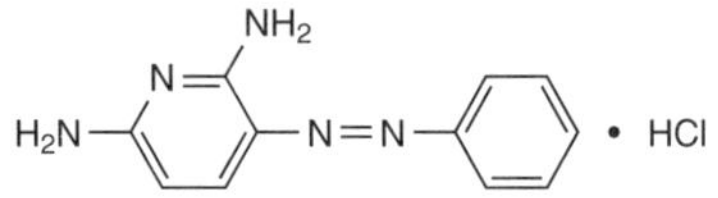

## SAMPLE

**Matrix:** blood
**Sample preparation:** Vigorously mix 1 mL plasma, 50 μL 40 μg/mL IS, 200 μL 100 mM NaOH, and 4 mL MTBE, centrifuge at 3000 rpm for 5 min. Evaporate the organic layer to dryness under a stream of nitrogen, reconstitute the residue with 1 mL 20 mM phosphoric acid, inject an 80 μL aliquot.

## HPLC VARIABLES

**Guard column:** BDS C8
**Column:** 100 × 4.6 5 μm LiChrospher 60 RP-Select B
**Mobile phase:** MeCN:20 mM pH 2.7 potassium dihydrogen phosphate buffer 30:70
**Flow rate:** 1.5
**Injection volume:** 80
**Detector:** UV 405

## CHROMATOGRAM

**Retention time:** 2.5
**Internal standard:** diltiazem (UV 238) (3.9)
**Limit of detection:** 10 ng/mL
**Limit of quantitation:** 50 ng/mL

## KEY WORDS

plasma

## REFERENCE

Farin, D.; Piva, G.; Kitzes-Cohen, R. Determination of phenazopyridine in human plasma by high performance liquid chromatography, *Chromatographia*, **2000**, *52*, 179–180.

## SAMPLE

**Matrix:** blood, urine
**Sample preparation:** Condition a 3 mL 130 mg Bond Elut Certify SPE cartridge with 3 mL MeOH and 3 mL 200 mM pH 6.0 phosphate buffer. Vortex 3 mL serum or urine with 3 mL 400 mM pH 6.0 phosphate buffer for 1 min, add to the SPE cartridge at 1 mL/min, wash with 3 mL MeOH:200 mM pH 6.0 phosphate buffer 5:95, pass air through the SPE cartridge for 10 s, elute with 750 μL MeOH:10% ammonia 100:20 at 0.5 mL/min. Add a few drops of 1 M HCl to the eluate and evaporate to dryness under a stream of nitrogen, reconstitute the residue with 150 μL mobile phase, inject an aliquot.

## HPLC VARIABLES

**Guard column:** 10 × 2.1 5 μm Hypersil C18
**Column:** 150 × 4.6 5 μm Hypersil C18
**Column temperature:** 45
**Mobile phase:** Gradient. A was MeCN:buffer 5:95. B was MeCN:buffer 50:50. A:B from 85:15 to 10:90 over 20 min, maintain at 10:90 for 5 min, return to initial conditions over 3 min, re-equilibrate for 6 min. (The buffer was 50 mM pH 3.0 phosphate buffer containing 375 mg/L sodium octyl sulfate and 3 mL/L triethylamine.)

**Flow rate:** 1
**Detector:** UV 240

---

**CHROMATOGRAM**
**Retention time:** 17.1

---

**OTHER SUBSTANCES**
**Extracted:** acebutolol (UV 326,235) (12.8), acetaminophen (UV 247) (2.7), *N*-acetylprocainamide (UV 268) (4.9), amineptine (UV 268) (19.3), amitriptyline (UV 240,211) (22.3), amobarbital (UV 211) (15.3), amoxapine (UV 296,211.5) (19.1), amphetamine (UV 254) (11.2), aprobarbital (14.7), atenolol (UV 275,230) (4.3), atropine (UV 260) (11.3), barbital (UV 212) (5.5), benzhexol (UV 260) (22.5), benztropine (UV 257,223) (22.9), benzoylecgonine (UV 278,235) (6.7), bromazepam (UV 235) (13.3), buprenorphine (UV 284,213) (19.4), buspirone (UV 302,239,211) (16.2), butabarbital (UV 211) (10.1), caffeine (UV 270) (3.5), carbamazepine (UV 283,215) (15.1), chloramphenicol (UV 277) (10.6), chlordiazepoxide (UV 303,245) (15.1), chlormethiazole (UV 252,211) (12.4), chlorpromazine (UV 310,255) (23.9), cimetidine (UV 219) (3.5), citalopram (UV 239) (19.0), clobazam (UV 292,231) (20.5), clomipramine (UV 250,211) (23.6), clonazepam (UV 310,240,211) (18.5), clonidine (UV 265) (6.9), cloperastine (UV 238) (26.2), cloxazolam (UV 340,265,239.5) (18.4), cocaine (UV 278,235) (14.7), codeine (UV 280,211) (5.0), desipramine (UV 250) (20.5), diazepam (UV 313,229) (22.6), dihydrocodeine (UV 282,211) (4.8), diltiazem (UV 242,211.5) (19.2), diphenhydramine (UV 258) (19.4), disopyramide (UV 260) (14.9), dothiepin (UV 303,263,231,211) (21.5), doxepin (UV 291) (19.1), ephedrine (UV 257) (6.2), erythromycin (UV 211) (23.3), fentanyl (UV 262) (18.4), flecainide (UV 299) (19.5), flumazenil (UV 250) (11.0), fluoxetine (UV 262,228) (22.6), fluphenazine (UV 310,257) (23.7), flupenthixol (UV 268,231) (27.2), flurazepam (UV 315,231) (20.6), fluvoxamine (UV 251) (20.8), furosemide (UV 344,275,235) (13.2), glutethimide (15.8), guaifenesin (UV 275,224) (6.7), haloperidol (UV 245,220) (19.6), hydralazine (UV 304,260,211) (3.5), hydroxyzine (UV 231) (21.2), ibuprofen (UV 263,219.5) (28.0), imipramine (UV 250,211) (21.6), indomethacin (UV 299,240) (27.9), ketamine (UV 270) (12.7), ketoconazole (UV 295,240) (20.8), labetalol (UV 303) (15.7), lidocaine (UV 266) (13.7), lorazepam (UV 318,231.5) (18.4), lormetazepam (UV 315,231) (20.5), loxapine (UV 295,250) (19.8), maprotiline (UV 272) (22.0), MDMA (UV 287,235) (13.0), methadone (23.0), methamphetamine (UV 254) (12.0), methaqualone (UV 308,267,227) (18.1), methylephedrine (UV 254) (7.7), methylphenidate (UV 260) (15.0), metoclopramide (UV 311,275,215) (13.0), metoprolol (UV 275,223) (13.8), mianserin (UV 280) (19.3), midazolam (UV 221) (19.3), moclobemide (UV 238) (14.1), molindone (UV 301,256,211) (13.6), morphine (UV 286,212) (2.7), nadolol (UV 272,219) (6.6), naloxone (UV 282,209) (5.0), naproxen (UV 270,231.5) (20.2), nifedipine (UV 335,235) (20.8), nitrazepam (UV 310,260,219) (16.3), nordiazepam (UV 310,231) (19.7), norfluoxetine (UV 262,228) (21.4), nortriptyline (UV 240) (21.0), ofloxacin (UV 327,295.5,230) (9.5), oxycodone (UV 280) (9.9), oxymorphone (UV 280) (6.7), paroxetine (UV 295,235) (19.8), pentobarbital (UV 211) (14.9), pericyazine (UV 269,232) (19.6), perphenazine (UV 320,255,210) (22.2), pethidine (15.5), phenazopyridine (UV 240) (17.1), phencyclidine (UV 265) (17.8), pheniramine (UV 260) (13.0), phenobarbital (9.9), phentermine (UV 254) (12.5), phenylephrine (UV 302,259,219) (9.2), phenylpropanolamine (UV 258) (4.9), phenyltoloxamine (UV 272) (20.6), phenytoin (15.3), pholcodine (UV 284,211.5) (4.0), pimozide (UV 280) (25.3), pinazepam (UV 310,227.5) (25.5), piroxicam (UV 248) (15.9), prazepam (UV 311,231.5) (25.1), primidone (6.2), procainamide (UV 280,210) (3.6), promethazine (UV 297,250) (21.7), propoxyphene (UV 259) (21.8), protriptyline (UV 290,211) (21.4), pseudoephedrine (UV 257) (6.2), quinidine (UV 330,240,212) (13.8), ranitidine (UV 315,227) (4.6), salicylic acid (UV 300,234) (6.5), secobarbital (UV 211) (16.0), sertraline (UV 273) (23.2), spironolactone (UV 240) (24.5), sulfamethoxazole (UV 269.5,211.5) (8.0), sulpiride (UV 293,213) (4.0), temazepam (UV 231.5,312) (19.5), terfenadine (UV 260,219) (27.4), theophylline (UV 270) (2.5), thiopentone (UV 288,238) (18.7), thioridazine (UV 314,265) (24.5), thiothixene (UV 308,259,231) (22.0), tocainide (UV 276) (8.8), tolbutamide (UV 228) (18.9), trazodone (UV 250) (14.9), triazolam (UV 223) (19.3), triflupromazine (UV

308,257) (25.8), trimeprazine (UV 250) (26.9), trimethoprim (UV 272) (10.9), trimipramine (UV 250,211) (23.9), verapamil (UV 280,230) (21.6), zopiclone (UV 305,218) (12.3), zuclopenthixol (UV 325,267,227) (22.0)

**Interfering:** alprazolam (UV 225) (17.5), brompheniramine (UV 227, UV 263) (17.2), chlorpheniramine (UV 263, UV 223) (17.3), clozapine (UV 211, UV 290) (17.0), fenfluramine (UV 263) (17.2), flunitrazepam (UV 220, UV 253, UV 313) (17.5), mesoridazine (UV 220, UV 273) (17.2), oxazepam (UV 229, UV 310) (17.5), oxazolam (UV 234) (17.0), pentazocine (UV 280) (16.7), phenolphthalein (UV 231, UV 278) (16.8), propranolol (UV215, UV 230, UV 290) (17.2), propyphenazone (UV 249, UV 274) (17.0), risperidone (UV 236, UV 276) (16.7)

## KEY WORDS

serum; SPE

## REFERENCE

Lai, C.-K.; Lee, T.; Au, K.-M.; Chan, A.Y.-W. Uniform solid-phase extraction procedure for toxicological drug screening in serum and urine by HPLC with photodiode-array detection, *Clin.Chem.*, **1997**, *43*, 312–325.

# Phentermine

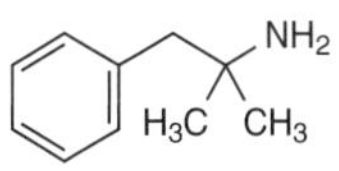


**Molecular formula:** $C_{10}H_{15}N$
**Molecular weight:** 149.23
**CAS Registry No:** 122-09-8, 1197-21-3 (HCl)
**Merck Index:** 13, 7346

## SAMPLE

**Matrix:** blood

**Sample preparation:** Condition a 200 mg Bond Elut C18 SPE cartridge with 1 mL MeOH, 1 mL water, and 2 mL 10 mM pH 9.3 ammonium carbonate buffer. Centrifuge serum, whole blood, or urine at 14 000 g for 5 min. Vortex 0.2–1 mL aliquots of the supernatant with 2 mL 10 mM pH 9.3 ammonium carbonate buffer and IS, centrifuge at 5000 g for 10 min, add 2 mL of the supernatant to the SPE cartridge, wash with 2 mL 10 mM pH 9.3 ammonium carbonate buffer, dry SPE cartridge under vacuum for 5 min, elute with 500 μL MeOH:500 mM acetic acid 90:10. Add 10 μL 1 mM HCl to the eluate, evaporate to dryness under a stream of nitrogen, reconstitute the residue with 100 μL mobile phase, centrifuge at 14 000 g for 4 min, inject a 5–20 μL aliquot.

## HPLC VARIABLES

**Column:** 125 × 3 Superspher RP 18
**Mobile phase:** MeCN:50 mM pH 3.0 ammonium formate buffer 25:75
**Flow rate:** 0.3
**Injection volume:** 5–20
**Detector:** MS, Finnigan MAT SSQ 7000, APCI, positive ion mode, sheath gas nitrogen 70 psi, auxiliary gas nitrogen 20 mL/min, capillary 190°, vaporizer 450°, corona current 5 μA

## CHROMATOGRAM

**Retention time:** 4.3
**Internal standard:** $d_{10}$-methamphetamine
**Limit of detection:** 1 ng/mL

## OTHER SUBSTANCES

**Extracted:** amphetamine (3.4), cathinone (2.6), ephedrine (3.1), fenfluramine (7.0), MDA (3.5), MDEA (5.2), MDMA (4.2), methamphetamine (4.1), norfenfluramine (4.2), phenylpropanolamine (2.5)

## KEY WORDS

serum; SPE; urine; whole blood

## REFERENCE

Bogusz, M.J. Liquid chromatography-mass spectrometry as a routine method in forensic sciences: a proof of maturity, *J.Chromatogr.B*, **2000**, *748*, 3–19.

## SAMPLE

**Matrix:** blood

**Sample preparation:** Mix 100 μL plasma with 10 μL 2 μM IS in water, add 200 μL 100 mM pH 10.6 borate buffer, add 750 μL ethyl acetate, vortex for 1 min, centrifuge at 1700 g at 4° for 10 min. Add 10 μL acetic acid to the organic layer, evaporate to dryness under reduced pressure at 45° over 15 min. Vortex the residue with 75 μL 4.7 mM dansyl chloride in MeCN and 25 μL 20 mM pH 9.0 carbonate buffer, heat at 45° for 30 min, add 3 μL MeCN:methylamine 80:20, let stand for 10 min, add 4 μL 3 M HCl, inject a 20 μL aliquot.

## HPLC VARIABLES

**Column:** 250 × 4.6 5 μm Diasopak SP-120-5-ODS (Daiso, Osaka)
**Mobile phase:** Gradient. A was MeOH:100 mM acetic acid 60:40. B was MeCN. A:B 80:20 for 20 min, to 45:55 over 2 min, maintain at 45:55 for 17 min, return to initial conditions.
**Flow rate:** 1
**Injection volume:** 20
**Detector:** F ex 325 em 530

## CHROMATOGRAM

**Retention time:** 31
**Internal standard:** fluoxetine (45)
**Limit of quantitation:** 5 nM

## OTHER SUBSTANCES

**Extracted:** 4-bromo-2,5-dimethoxyphenylethylamine (30), epinephrine (21), fenfluramine (39), norepinephrine (16), 2-phenylethylamine (25)

## KEY WORDS

derivatization; human; pharmacokinetics; plasma; rat

## REFERENCE

Kaddoumi, A.; Nakashima, M.N.; Wada, M.; Kuroda, N.; Nakahara, Y.; Nakashima, K. HPLC of (±)-fenfluramine and phentermine in plasma after derivatization with dansyl chloride, *J.Liq.Chromatogr. Rel.Technol.*, **2001**, *24*, 57–67.

---

## SAMPLE

**Matrix:** blood, urine
**Sample preparation:** Condition a 3 mL 130 mg Bond Elut Certify SPE cartridge with 3 mL MeOH and 3 mL 200 mM pH 6.0 phosphate buffer. Vortex 3 mL serum or urine with 3 mL 400 mM pH 6.0 phosphate buffer for 1 min, add to the SPE cartridge at 1 mL/min, wash with 3 mL MeOH:200 mM pH 6.0 phosphate buffer 5:95, pass air through the SPE cartridge for 10 s, elute with 750 μL MeOH:10% ammonia 100:20 at 0.5 mL/min. Add a few drops of 1 M HCl to the eluate and evaporate to dryness under a stream of nitrogen, reconstitute the residue with 150 μL mobile phase, inject an aliquot.

## HPLC VARIABLES

**Guard column:** 10 × 2.1 5 μm Hypersil C18
**Column:** 150 × 4.6 5 μm Hypersil C18
**Column temperature:** 45
**Mobile phase:** Gradient. A was MeCN:buffer 5:95. B was MeCN:buffer 50:50. A:B from 85:15 to 10:90 over 20 min, maintain at 10:90 for 5 min, return to initial conditions over 3 min, re-equilibrate for 6 min. (The buffer was 50 mM pH 3.0 phosphate buffer containing 375 mg/L sodium octyl sulfate and 3 mL/L triethylamine.)
**Flow rate:** 1
**Detector:** UV 254

## CHROMATOGRAM

**Retention time:** 12.5

## OTHER SUBSTANCES

**Extracted:** acetaminophen (UV 247) (2.7), *N*-acetylprocainamide (UV 268) (4.9), alprazolam (UV 225) (17.5), amineptine (UV 268) (19.3), amitriptyline (UV 240,211) (22.3), amobarbital (UV 211) (15.3), amoxapine (UV 296,211.5) (19.1), amphetamine (UV 254) (11.2), aprobarbital (14.7), atenolol (UV 275,230) (4.3), atropine (UV 260) (11.3), barbital (UV 212) (5.5), benzhexol (UV 260) (22.5), benztropine (UV 257,223) (22.9), benzoylecgonine (UV 278,235) (6.7), bromazepam (UV 235) (13.3), brompheniramine (UV 263,227) (17.2), buprenorphine (UV 284,213) (19.4), buspirone (UV 302,239,211) (16.2), butabarbital (UV 211) (10.1), caffeine (UV 270) (3.5), carbamazepine (UV 283,215) (15.1), chloramphenicol (UV 277) (10.6), chlordiazepoxide (UV 303,245) (15.1), chlorpheniramine (UV 263,223) (17.3), chlorpromazine (UV 310,255) (23.9), cimetidine (UV 219) (3.5), citalopram (UV 239) (19.0), clobazam (UV 292,231) (20.5), clomipramine (UV 250,211) (23.6), clonazepam (UV 310,240,211) (18.5), clonidine (UV 265) (6.9), cloperastine (UV 238) (26.2), cloxazolam (UV 340,265,239.5) (18.4), clozapine (UV 290,211) (17.0), cocaine (UV 278,235) (14.7), codeine (UV 280,211) (5.0), desipramine (UV 250) (20.5), diazepam (UV 313,229) (22.6), dihydrocodeine (UV 282,211) (4.8), diltiazem (UV 242,211.5) (19.2), diphenhydramine (UV 258) (19.4), disopyramide (UV 260) (14.9), dothiepin (UV 303,263,231,211) (21.5), doxepin (UV 291) (19.1), ephedrine (UV 257) (6.2), erythromycin (UV 211) (23.3), fenfluramine (UV 263) (17.2), fentanyl (UV 262) (18.4), flecainide (UV 299) (19.5), flumazenil (UV 250) (11.0), flunitrazepam (UV 313,253,220) (17.5), fluoxetine (UV 262,228) (22.6), fluphenazine (UV 310,257) (23.7), flupenthixol (UV 268,231) (27.2), flurazepam (UV 315,231) (20.6), fluvoxamine (UV 251) (20.8), furosemide (UV 344,275,235) (13.2), glutethimide (15.8), guaifenesin (UV 275,224) (6.7), haloperidol (UV 245,220) (19.6), hydralazine (UV 304,260,211) (3.5), hydroxyzine (UV 231) (21.2), ibuprofen (UV 263,219.5) (28.0), imipramine (UV 250,211) (21.6), indomethacin (UV 299,240) (27.9), ketoconazole (UV 295,240) (20.8), labetalol (UV 303) (15.7), lidocaine (UV 266) (13.7), lorazepam (UV 318,231.5) (18.4), lormetazepam (UV 315,231) (20.5), loxapine (UV 295,250) (19.8), maprotiline (UV 272) (22.0), mesoridazine (UV 273,220) (17.2), methadone (23.0), methaqualone (UV 308,267,227) (18.1), methylephedrine (UV 254) (7.7), methylphenidate (UV 260) (15.0), metoprolol (UV 275,223) (13.8), mianserin (UV 280) (19.3), midazolam (UV 221) (19.3), moclobemide (UV 238) (14.1), molindone (UV 301,256,211) (13.6), morphine (UV 286,212) (2.7), nadolol (UV 272,219) (6.6), naloxone (UV 282,209) (5.0), naproxen (UV 270,231.5) (20.2), nifedipine (UV 335,235) (20.8), nitrazepam (UV 310,260,219) (16.3), nordiazepam (UV 310,231) (19.7), norfluoxetine (UV 262,228) (21.4), nortriptyline (UV 240) (21.0), ofloxacin (UV 327,295.5,230) (9.5), oxazepam (UV 310,229) (17.6), oxazolam (UV 234) (17.0), oxycodone (UV 280) (9.9), oxymorphone (UV 280) (6.7), paroxetine (UV 295,235) (19.8), pentazocine (UV 280) (16.7), pentobarbital (UV 211) (14.9), pericyazine (UV 269,232) (19.6), perphenazine (UV 320,255,210) (22.2), pethidine (15.5), phenazopyridine (UV 240) (17.1), phencyclidine (UV 265) (17.8), phenobarbital (9.9), phenolphthalein (UV 278,231) (16.8), phentermine (UV 254) (12.5), phenylephrine (UV 302,259,219) (9.2), phenylpropanolamine (UV 258) (4.9), phenyltoloxamine (UV 272) (20.6), phenytoin (15.3), pholcodine (UV 284,211.5) (4.0), pimozide (UV 280) (25.3), pinazepam (UV 310,227.5) (25.5), piroxicam (UV 248) (15.9), prazepam (UV 311,231.5) (25.1), primidone (6.2), procainamide (UV 280,210) (3.6), promethazine (UV 297,250) (21.7), propoxyphene (UV 259) (21.8), propranolol (UV 290,230,215) (17.2), propyphenazone (UV 274,249) (17.0), protriptyline (UV 290,211) (21.4), pseudoephedrine (UV 257) (6.2), quinidine (UV 330,240,212) (13.8), ranitidine (UV 315,227) (4.6), risperidone (UV 276,236) (16.7), salicylic acid (UV 300,234) (6.5), secobarbital (UV 211) (16.0), sertraline (UV 273) (23.2), spironolactone (UV 240) (24.5), sulfamethoxazole (UV 269.5,211.5) (8.0), sulpiride (UV 293,213) (4.0), temazepam (UV 231.5,312) (19.5), terfenadine (UV 260,219) (27.4), theophylline (UV 270) (2.5), thiopentone (UV 288,238) (18.7), thioridazine (UV 314,265) (24.5), thiothixene (UV 308,259,231) (22.0), tocainide (UV 276) (8.8), tolbutamide (UV 228) (18.9), trazodone (UV 250) (14.9), triazolam (UV 223) (19.3), triflupromazine (UV 308,257) (25.8), trimeprazine (UV 250) (26.9), trimethoprim (UV 272) (10.9), trimipramine (UV 250,211) (23.9), verapamil (UV 280,230) (21.6), zuclopenthixol (UV 325,267,227) (22.0)

**Interfering:** acebutolol (UV 235, UV 326) (12.8), chlormethiazole (UV 211, UV 252) (12.4), ketamine (UV 270) (12.7), MDA (UV 235, UV 287) (12.4), MDMA (UV 235, UV 287) (13.0), methamphetamine (12.0), metoclopramide (UV 215, UV 275, UV 311) (13.0), pheniramine (UV 260) (13.0), zopiclone (UV 218, UV 305) (12.3),

---

**KEY WORDS**

serum; SPE

## REFERENCE

Lai, C.-K.; Lee, T.; Au, K.-M.; Chan, A.Y.-W. Uniform solid-phase extraction procedure for toxicological drug screening in serum and urine by HPLC with photodiode-array detection, *Clin.Chem.*, **1997**, *43*, 312–325.

## ANNOTATED BIBLIOGRAPHY

Bogusz, M.J.; Krüger, K.-D.; Maier, R.-D. Analysis of underivatized amphetamines and related phenethylamines with high-performance liquid chromatography-atmospheric pressure chemical ionization mass spectrometry, *J.Anal.Toxicol.*, **2000**, *24*, 77–84. [SPE; LOD 1–5 ng/mL; amphetamine; fenfluramine; methamphetamine; phentermine; cathinone; propylhexedrine; ephedrine; phenylpropanolamine; norfenfluramine; phenylethylamine]

Kaddoumi, A.; Kubota, A.; Nakashima, M.N.; Takahashi, M.; Nakashima, K. High performance liquid chromatography with UV detection for the simultaneous determination of sympathomimetic amines using 4-(4,5-diphenyl-1H-imidazole-2-yl)benzoyl chloride as a label, *Biomed.Chromatogr.*, **2001**, *15*, 379–388. [human; rat; plasma; derivatization; LOD 50–100 nM; ephedrine; norephedrine; 2-phenylethylamine; fenfluramine; phentermine; cyclohexylamine; fluoxetine is internal standard; pharmacokinetics]

Kaddoumi, A.; Nakashima, M.N.; Nakashima, K. Fluorometric determination of DL-fenfluramine, DL-norfenfluramine and phentermine in plasma by achiral and chiral high-performance liquid chromatography, *J.Chromatogr.B*, **2001**, *763*, 79–90. [fluorescence detection; derivatization; human; rat; LOD 10 nM; ephedrine; norephedrine]

Kaddoumi, A.; Nakashima, M.N.; Maki, T.; Matsumura, Y.; Nakamura, J.; Nakashima, K. Liquid chromatography studies on the pharmacokinetics of phentermine and fenfluramine in brain and blood microdialysates after intraperitoneal administration to rats, *J.Chromatogr.B*, **2003**, *791*, 291–303. [fluorescence detection; derivatization; LOQ 5 nM; fluoxetine is internal standard]

Unterhalt, B.; Wenning, C. Zur Trennung von Oxetacain und seinen Metaboliten [Separation of oxetacaine and its metabolites], *Pharmazie*, **2001**, *56*, 58–60. [oxetacaine; mephentermine; phentermine]

# Phosphatidylcholine

**CAS Registry No:** 8002-43-5
**Merck Index:** 13, 5447

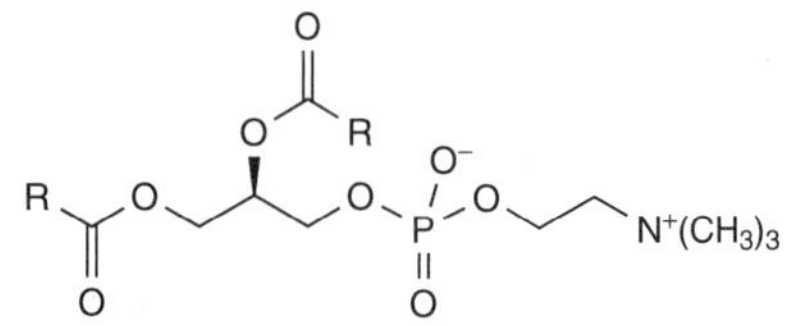


R = fatty acid

## SAMPLE

**Matrix:** blood
**Sample preparation:** Shake 2 mL blood, 4 mL chloroform (Caution! Chloroform is a carcinogen!), and 2 mL MeOH vigorously for 2 min, cool in ice, centrifuge at 800 g. Evaporate the lower chloroform layer to dryness under a stream of nitrogen, reconstitute the residue with chloroform:MeOH 2:1, inject an aliquot.

## HPLC VARIABLES

**Guard column:** 60 × 4.6 5 μm SI 60 (Merck)
**Column:** 250 × 4.6 5 μm Lichrosorb Diol
**Column temperature:** 55
**Mobile phase:** Gradient. A was MeCN. B was MeCN:water 80:20. A:B 88:12 for 5 min, to 77:23 over 3 min, to 30:70 over 4 min, to 25:75 over 1 min, maintain at 25:75 for 5 min, return to initial conditions over 1 min.
**Flow rate:** 2.5
**Detector:** UV 201

## CHROMATOGRAM

**Retention time:** 14

## OTHER SUBSTANCES

**Extracted:** phosphatidylethanolamine (13), sphingomyelin (16)

## KEY WORDS

whole blood

## REFERENCE

Heinze, T.; Kynast, G.; Dudenhausen, J.W.; Saling, E. Effect of blood and meconium on the determination of phospholipids in amniotic fluid using high pressure liquid chromatography, *J.Perinat.Med.*, **1988**, *16*, 53–60.

## SAMPLE

**Matrix:** blood
**Sample preparation:** Sonicate 300 μL whole blood, 600 μL water, and 5 mL MeOH containing 0.01% BHT for 1 min, add 10 mL chloroform (Caution! Chloroform is a carcinogen!), sonicate for 1 min, vortex for 30 s, let stand at room temperature for 1 h, add 5 mL water, mix for 5 s, centrifuge at 2600 g for 10 min. Evaporate the lower organic layer to dryness under a stream of nitrogen, reconstitute the residue with 300 μL chloroform:MeOH 95:5, inject a 10 μL aliquot.

## HPLC VARIABLES

**Guard column:** 10 × 2 5 μm LiChrospher 100 diol
**Column:** 250 × 2 5 μm LiChrospher 100 diol
**Column temperature:** 30
**Mobile phase:** Gradient. A was chloroform. B was MeOH:formic acid 99.9:0.1 with ammonia added to pH 5.3 (ca. 0.05% ammonia) and 0.05% triethylamine. A:B from 95:5

to 70:30 over 11 min, to 20:80 over 3 min, maintain at 20:80 for 4 min, return to initial conditions over 2 min, re-equilibrate for 10 min.
**Flow rate:** 0.3
**Injection volume:** 10
**Detector:** MS, Finnigan LCQ, electrospray, ionization needle 4.5 kV, capillary 230°, sheath gas flow 90 units

---

### CHROMATOGRAM
**Retention time:** 7.07 (m/z 878.5; C18:0, C22:6), 7.28 (m/z 826.5; C16:0, C20:4), 7.84 (m/z 778.5; C16:0, C16:0)
**Limit of detection:** 0.1–5 ng

---

### OTHER SUBSTANCES
**Extracted:** phosphatidylethanolamine (10), phosphatidylglycerol (6), phosphatidylinositol (13), phosphatidylserine (17)

---

### KEY WORDS
whole blood

---

### REFERENCE
Uran, S.; Larsen, A.; Jacobsen, P.B.; Skotland, T. Analysis of phospholipid species in human blood using normal-phase liquid chromatography coupled with electrospray ionization ion-trap tandem mass spectrometry, *J.Chromatogr.B*, **2001**, *758*, 265–275.

---

### SAMPLE
**Matrix:** lung lavage fluid
**Sample preparation:** Centrifuge lung lavage fluid at 4° at 450 g for 10 min. Shake 10 mL supernatant and 40 mL chloroform:MeOH 2:1 at 4° for 3 min (Caution! Chloroform is a carcinogen!). Remove the lower organic phase and wash it with 2 mL 50 mM NaCl, centrifuge, dry under a stream of nitrogen at 45°, reconstitute with 500 μL mobile phase, vortex at 4° for 1 min, inject a 100 μL aliquot.

---

### HPLC VARIABLES
**Guard column:** 20 × 2.1 5 μm Encapharm 100 spherical silica gel (Molnar, Berlin)
**Column:** 120 × 4.6 5 μm Encapharm 100 spherical silica gel (Molnar, Berlin)
**Column temperature:** 30
**Mobile phase:** Gradient. A was chloroform:MeOH:ammonium hydroxide 80:19.5:0.5. B was chloroform:MeOH:water:ammonium hydroxide 60:34:5.5:0.5. A:B from 100:0 to 0:100 over 14 min, return to initial conditions over 7 min, re-equilibrate for 10 min.
**Flow rate:** 1
**Injection volume:** 100
**Detector:** Evaporative Light-Scattering Detector, SEDERE Sedex-45, evaporation temperature 50°, nebulization gas nitrogen, pressure 200 kPa, flow 6 L/min, response is nonlinear but proportional to the power 1.7 of the mass and must be calibrated for each compound

---

### CHROMATOGRAM
**Retention time:** 13.11
**Limit of detection:** 100 ng

---

### OTHER SUBSTANCES
**Extracted:** diarachidoylphosphatidylcholine (12.75), dilinoleylphosphatidylcholine (12.69), diphosphatidylglycerol (6.30), lysophosphatidylcholine (18.46, 18.81), phosphatidic acid (13.78), phosphatidylethanolamine (9.00), phosphatidylglycerol (5.37), phosphatidylinositol (10.75), phosphatidylserine (11.81), sphingomyelin (15.24, 15.72)

## REFERENCE

Bünger, H.; Pison, U. Quantitative analysis of pulmonary surfactant phospholipids by high-performance liquid chromatography and light-scattering detection, *J.Chromatogr.B*, **1995**, *672*, 25–31.

## ANNOTATED BIBLIOGRAPHY

Wada, M.; Nakashima, K.; Kuroda, N.; Akiyama, S.; Imai, K. Sensitive flow-injection method with peroxylate chemiluminescence detection combined with preparative high-performance liquid chromatography for determination of choline-containing phospholipids in human serum, *J.Chromatogr.B*, **1996**, *678*, 129–136.

# Phosphatidylglycerol

**Merck Index:** 13, 8030

## SAMPLE

**Matrix:** blood

**Sample preparation:** Sonicate 300 μL whole blood, 600 μL water, and 5 mL MeOH containing 0.01% BHT for 1 min, add 10 mL chloroform (Caution! Chloroform is a carcinogen!), sonicate for 1 min, vortex for 30 s, let stand at room temperature for 1 h, add 5 mL water, mix for 5 s, centrifuge at 2600 g for 10 min. Evaporate the lower organic layer to dryness under a stream of nitrogen, reconstitute the residue with 300 μL chloroform:MeOH 95:5, inject a 10 μL aliquot.

## HPLC VARIABLES

**Guard column:** 10 × 2 5 μm LiChrospher 100 diol

**Column:** 250 × 2 5 μm LiChrospher 100 diol

**Column temperature:** 30

**Mobile phase:** Gradient. A was chloroform (Caution! Chloroform is a carcinogen!). B was MeOH:formic acid 99.9:0.1 with ammonia added to pH 5.3 (ca. 0.05% ammonia) and 0.05% triethylamine. A:B from 95:5 to 70:30 over 11 min, to 20:80 over 3 min, maintain at 20:80 for 4 min, return to initial conditions over 2 min, re-equilibrate for 10 min.

**Flow rate:** 0.3

**Injection volume:** 10

**Detector:** MS, Finnigan LCQ, electrospray, ionization needle 4.5 kV, capillary 230°, sheath gas flow 90 units

## CHROMATOGRAM

**Retention time:** 6 (m/z 721.5; C16:0, C16:0), 7.28 (m/z 749.5; C16:0, C18:0), 7.84 (m/z 777.5; C18:0, C18:0)

**Limit of detection:** 0.1–5 ng

## OTHER SUBSTANCES

**Extracted:** phosphatidylcholine (7.5), phosphatidylethanolamine (10), phosphatidylinositol (13), phosphatidylserine (17)

## KEY WORDS

whole blood

## REFERENCE

Uran, S.; Larsen, A.; Jacobsen, P.B.; Skotland, T. Analysis of phospholipid species in human blood using normal-phase liquid chromatography coupled with electrospray ionization ion-trap tandem mass spectrometry, *J.Chromatogr.B*, **2001**, *758*, 265–275.

## SAMPLE

**Matrix:** formulations

**Sample preparation:** Dilute liposome dispersions 10-fold with chloroform:MeOH 60:40, centrifuge at 2700 g for 15 min, inject an aliquot of the supernatant. (Caution! Chloroform is a carcinogen!)

## HPLC VARIABLES

**Column:** 250 × 4.6 5 μm Zorbax amino

**Mobile phase:** MeCN:MeOH:buffer 64:28:8 (The buffer was 10 mM phosphoric acid adjusted to pH 4.8 with dilute ammonium hydroxide solution. To prepare mobile phase, mix MeCN and MeOH and then add buffer.)

**Flow rate:** 1.5

**Injection volume:** 5–20
**Detector:** Refractive Index

## CHROMATOGRAM
**Retention time:** 6
**Limit of detection:** 29 μg/mL

## OTHER SUBSTANCES
**Simultaneous:** acyl lysophosphatidylcholine, acyl lysophosphatidylglycerol, phosphatidylcholine (5)

## KEY WORDS
liposome dispersions

## REFERENCE
Grit, M.; Crommelin, D.J.A.; Lang, J. Determination of phosphatidylcholine, phosphatidylglycerol and their lyso forms from liposome dispersions by high-performance liquid chromatography using high-sensitivity refractive index detection, *J.Chromatogr.*, **1991**, *585*, 239–246.

---

## SAMPLE
**Matrix:** lung lavage fluid
**Sample preparation:** Centrifuge lung lavage fluid at 4° at 450 g for 10 min. Shake 10 mL supernatant and 40 mL chloroform:MeOH 2:1 at 4° for 3 min (Caution! Chloroform is a carcinogen!). Remove the lower organic phase and wash it with 2 mL 50 mM NaCl, centrifuge, dry under a stream of nitrogen at 45°, reconstitute with 500 μL mobile phase, vortex at 4° for 1 min, inject a 100 μL aliquot.

## HPLC VARIABLES
**Guard column:** 20 × 2.1 5 μm Encapharm 100 spherical silica gel (Molnar, Berlin)
**Column:** 120 × 4.6 5 μm Encapharm 100 spherical silica gel (Molnar, Berlin)
**Column temperature:** 30
**Mobile phase:** Gradient. A was chloroform:MeOH:ammonium hydroxide 80:19.5:0.5. B was chloroform:MeOH:water:ammonium hydroxide 60:34:5.5:0.5. A:B from 100:0 to 0:100 over 14 min, return to initial conditions over 7 min, re-equilibrate for 10 min.
**Flow rate:** 1
**Injection volume:** 100
**Detector:** Evaporative Light-Scattering Detector, SEDERE Sedex-45, evaporation temperature 50°, nebulization gas nitrogen, pressure 200 kPa, flow 6 L/min, response is nonlinear but proportional to the power 1.7 of the mass and must be calibrated for each compound

## CHROMATOGRAM
**Retention time:** 5.37
**Limit of detection:** 100 ng

## OTHER SUBSTANCES
**Extracted:** diarachidoylphosphatidylcholine (12.75), dilinoleylphosphatidylcholine (12.69), diphosphatidylglycerol (6.30), lysophosphatidylcholine (18.46, 18.81), phosphatidic acid (13.78), phosphatidylcholine (13.11), phosphatidylethanolamine (9.00), phosphatidylinositol (10.75), phosphatidylserine (11.81), sphingomyelin (15.24, 15.72)

## REFERENCE
Bünger, H.; Pison, U. Quantitative analysis of pulmonary surfactant phospholipids by high-performance liquid chromatography and light-scattering detection, *J.Chromatogr.B*, **1995**, *672*, 25–31.

## SAMPLE

**Matrix:** tissue

**Sample preparation:** Homogenize (Potter-Elvehjem) liver or lung with 20 volumes chloroform:MeOH 2:1, filter (paper), wash with a volume of 50 mM NaCl equal to one-fifth the volume of extract, centrifuge (*J.Biol.Chem.* **1957**, *226*, 497). Remove the lower organic layer and evaporate it to dryness under a stream of nitrogen, reconstitute with chloroform:MeOH 25:75, inject an aliquot. (Caution! Chloroform is a carcinogen!)

## HPLC VARIABLES

**Guard column:** 10 × 4.6 5 μm Nucleosil 5 NH2

**Column:** 50 × 4.6 5 μm Nucleosil 5 NH2 + 175 × 4.6 5 μm Nucleosil 5 NH2

**Column temperature:** 50

**Mobile phase:** MeCN:MeOH:water:50% methylphosphonic acid in water 73:25:1.5:0.03, adjusted to pH 3 with 25% ammonium hydroxide in water

**Flow rate:** 1

**Detector:** F ex 340 em 460 following post-column derivatization. The column effluent mixed with the reagent pumped at 4.5 mL/min and the mixture flowed through a 2 m ×0.8 mm ID coil of PTFE tubing at 50° to the detector. (The reagent was 1 L water to which was added 250 μL Brij 35 (30% w/v), 50 mg sodium azide, and 150 μL 3 mM diphenylhexatriene in MeCN.)

## CHROMATOGRAM

**Retention time:** 22

## OTHER SUBSTANCES

**Extracted:** dipalmitolyphosphatidylcholine

## KEY WORDS

gastric mucosa; liver; lung; pig; post-column reaction

## REFERENCE

Bernhard, W.; Linck, M.; Creutzburg, H.; Postle, A.D.; Arning, A.; Martin-Carrera, I.; Sewing, K.-F. High-performance liquid chromatographic analysis of phospholipids from different sources with combined fluorescence and ultraviolet detection, *Anal.Biochem.*, **1994**, *220*, 172–180.

## ANNOTATED BIBLIOGRAPHY

Alvarez, J.G.; Touchstone, J.C. Separation of acidic and neutral lipids by aminopropyl-bonded silica gel column chromatography, *J.Chromatogr.*, **1992**, *577*, 142–145.

Bonanno, L.M.; Denizot, B.A.; Tchoreloff, P.C.; Puisieux, F.; Cardot, P.J. Determination of phospholipids from pulmonary surfactant using an on-line coupled silica/reversed-phase high-performance liquid chromatography system, *Anal.Chem.*, **1992**, *64*, 371–379.

Dethloff, L.A.; Gilmore, L.B.; Hook, G.E. Separation of pulmonary surfactant phospholipids by high-performance liquid chromatography, *J.Chromatogr.*, **1986**, *382*, 79–87.

Ishizuka, T.; Ishikawa, K.; Maseki, M.; Tomoda, Y.; Tsuda, T. Determination of phosphatidylglycerol in amniotic fluid for prediction of foetal lung maturity by microbore-column liquid chromatography, *J.Chromatogr.*, **1986**, *380*, 43–53.

Jaaskelainen, I.; Urtti, A. Liquid chromatography determination of liposome components using a light-scattering evaporative detector, *J.Pharm.Biomed.Anal.*, **1994**, *12*, 977–982.

Jungalwala, F.B.; Evans, J.E.; McCluer, R.H. Compositional and molecular species analysis of phospholipids by high performance liquid chromatography coupled with chemical ionization mass spectrometry, *J.Lipid Res.*, **1984**, *25*, 738–749.

Kynast, G.; Schmitz, C. Simplified quantitative determination of phospholipids in amniotic fluid, alveolar lavages and milk using high performance liquid chromatography (HPLC), *J.Perinat.Med.*, **1989**, *17*, 203–212.

Nissen, H.P.; Kreysel, H.W. Analysis of phospholipids in human semen by high-performance liquid chromatography, *J.Chromatogr.*, **1983**, *276*, 29–35.

Pison, U.; Gono, E.; Joka, T.; Obertacke, U.; Obladen, M. High-performance liquid chromatography of adult human bronchoalveolar lavage: assay for phospholipid lung profile, *J.Chromatogr.*, **1986**, *377*, 79–89.

Sas, B.; Peys, E.; Helsen, M. Efficient method for (lyso)phospholipid class separation by high-performance liquid chromatography using an evaporative light-scattering detector, *J.Chromatogr.A*, **1999**, *864*, 179–182.

Scarim, J.; Ghanbari, H.; Taylor, V.; Menon, G. Determination of phosphatidylcholine and disaturated phosphatidylcholine content in lung surfactant by high performance liquid chromatography, *J.Lipid Res.*, **1989**, *30*, 607–611.

Smith, L.A.; Norman, H.A.; Cho, S.H.; Thompson, G.A. Jr. Isolation and quantitative analysis of phosphatidylglycerol and glycolipid molecular species using reversed-phase high-performance liquid chromatography with flame ionization detection, *J.Chromatogr.*, **1985**, *346*, 291–299.

# Piketoprofen

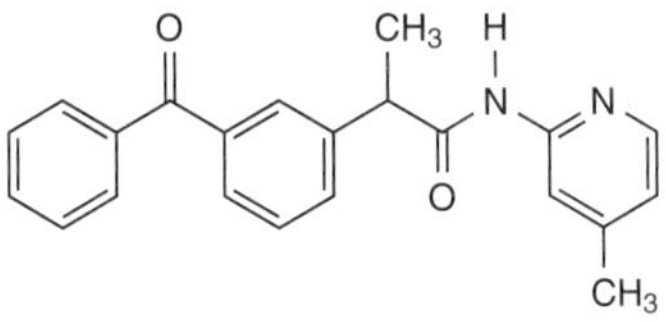

**Molecular formula:** $C_{22}H_{20}N_2O_2$
**Molecular weight:** 344.41
**CAS Registry No:** 60576-13-8
**Merck Index:** 13, 7503

## SAMPLE
**Matrix:** solutions

## HPLC VARIABLES
**Column:** 250 × 4.6 (S,S)-Whelk-O (Regis)
**Mobile phase:** Hexane:isopropanol:acetic acid 70:30:0.5
**Flow rate:** 1
**Injection volume:** 5
**Detector:** UV 270

## CHROMATOGRAM
**Retention time:** k′ 2.97 ($\alpha$ 1.17)

## KEY WORDS
chiral

## REFERENCE
Baeyens, W.R.G.; Van Der Weken, G.; Aboul-Enein, H.Y.; Reygaerts, S.; Smet, E. Comparison of the enantiomeric separation of some 2-arylpropionic acids on a novel Pirkle-concept stationary packing by narrow-bore and conventional liquid chromatography, *Biomed.Chromatogr.*, **2000**, *14*, 58–60.

## SAMPLE
**Matrix:** bulk
**Sample preparation:** Prepare a 1.0 mg/mL solution. Inject an aliquot.

## HPLC VARIABLES
**Column:** 150 × 4.6 5 μm Chiralcel OJ-R
**Mobile phase:** MeCN:buffer 35:65 (A) or MeCN:MeOH:buffer 35:35:30 (B) (Prepare the buffer by dissolving 70.25 g sodium perchlorate in water, adjust to pH 2.0 with concentrated perchloric acid, make up to 1 L with water.)
**Flow rate:** 0.5
**Detector:** UV 230; UV 254

## CHROMATOGRAM
**Retention time:** k′ 3.53 (mobile phase A, $\alpha = 1.29$), k′ 0.61 (mobile phase B, $\alpha = 1.17$)

## KEY WORDS
chiral

## REFERENCE
Van Overbeke, A.; Baeyens, W.; Oda, H.; Aboul-Enein, H.Y. Direct enantiomeric HPLC separation of several 2-arylpropionic acids, barbituric acids and benzodiazepines on Chiralcel OJ-R chiral stationary phase, *Chromatographia*, **1996**, *43*, 599–606.

# Pilsicainide

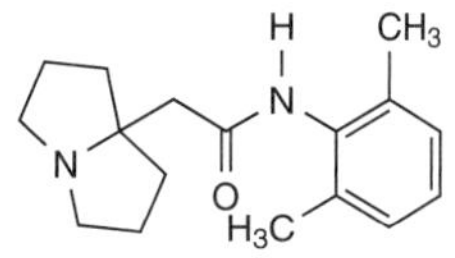

**Molecular formula:** $C_{17}H_{24}N_2O$
**Molecular weight:** 272.38
**CAS Registry No:** 88069-67-4, 88069-49-2 (HCl hemihydrate)
**Merck Index:** 13, 7508

## SAMPLE

**Matrix:** blood
**Sample preparation:** Mix 500 μL serum with 1 mL 100 mM NaOH, add 100 μL 10 μg/mL IS, add 5 mL chloroform (Caution! Chloroform is a carcinogen!), shake for 10 min, centrifuge at 2270 g for 5 min. Evaporate 4 mL of the organic layer to dryness under a stream of nitrogen at 40°, reconstitute the residue with 50 μL MeCN, inject a 40 μL aliquot.

## HPLC VARIABLES

**Column:** 100 × 5 10 μm Radial-pak 5CN
**Mobile phase:** MeCN:10 mM ammonium acetate 70:30
**Flow rate:** 2.5
**Injection volume:** 40
**Detector:** UV 210

## CHROMATOGRAM

**Internal standard:** mexiletine
**Limit of quantitation:** 250 ng/mL

## KEY WORDS

comparison with ELISA; rabbit; serum

## REFERENCE

Saita, T.; Fujito, H.; Mori, M. ELISA for the quantification of pilsicainide, *Biol.Pharm.Bull.*, **2001**, *24*, 1113–1116.

## SAMPLE

**Matrix:** blood, urine
**Sample preparation:** Plasma. Shake 500 μL plasma, 200 μL 500 mM NaOH, and 2 mL diethyl ether for 5 min, centrifuge at 3000 g for 5 min. Evaporate the organic layer to dryness under a stream of nitrogen at 40°, reconstitute the residue with 200 μL mobile phase, inject a 50 μL aliquot. Urine. Condition a Bond Elut CN SPE cartridge with two 1 mL portions of water, 1 mL MeCN:200 mM sodium perchlorate 1:2, and three 1 mL portions of water. Heat 10 mL urine with 40 μL β-glucuronidase (8 units) at 37° for 16 h, add 1 mL to the SPE cartridge, centrifuge the SPE cartridge at 1000 g for 5 min, wash with 1 mL MeCN:water 1:2, centrifuge at 2000 g for 5 min, elute with 1 mL MeCN:water 1:2, centrifuge at 2000 g for 5 min, inject a 100 μL aliquot of the eluate.

## HPLC VARIABLES

**Column:** 150 × 4.6 STR ODS-II (Shimadzu)
**Column temperature:** 40
**Mobile phase:** MeCN:MeOH:25 mM pH 4.7 phosphate buffer 10:3:87
**Flow rate:** 1.5
**Injection volume:** 50–100
**Detector:** UV 210

## CHROMATOGRAM

**Limit of detection:** 10 ng/mL
**Limit of quantitation:** 50 ng/mL (plasma), 1 μg/mL (urine)

## OTHER SUBSTANCES

**Noninterfering:** cimetidine, probenecid

## KEY WORDS

pharmacokinetics; plasma; SPE

## REFERENCE

Shiga, T.; Hashiguchi, M.; Urae, A.; Kasanuki, H.; Rikihisa, T. Effect of cimetidine and probenecid on pilsicainide renal clearance in humans, *Clin.Pharmacol.Ther.*, **2000**, *67*, 222–228.

# Pioglitazone

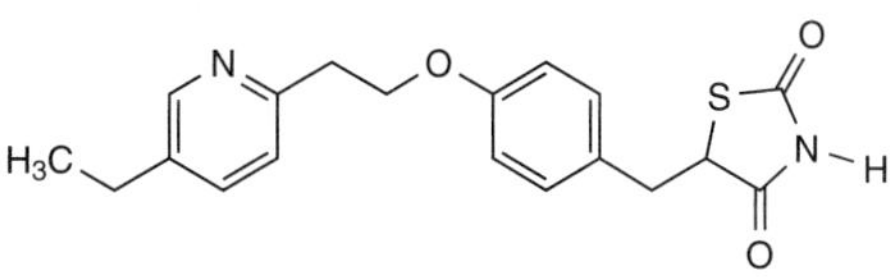


**Molecular formula:** $C_{19}H_{20}N_2O_3S$
**Molecular weight:** 356.45
**CAS Registry No:** 111025-46-8
**Merck Index:** 13, 7533

## SAMPLE
**Matrix:** blood
**Sample preparation:** Mix 200 μL plasma with 100 μL 50 ng/mL IS in MeCN and 1 mL 100 mM pH 4 ammonium acetate buffer, add 4 mL MTBE:1-chlorobutane 50:50, shake at high speed for 20 min, centrifuge at 4000 rpm for 20 min. Evaporate the organic layer to dryness under a stream of nitrogen at 45°, reconstitute the residue with 100 μL mobile phase, vortex for 1 min, centrifuge at 4000 rpm for 5 min, inject a 10 μL aliquot of the supernatant.

## HPLC VARIABLES
**Column:** 50 × 2 3 μm MetaChem Polaris C18-A
**Mobile phase:** MeCN:water 60:40 containing 10 mM ammonium acetate and 0.02% trifluoroacetic acid
**Flow rate:** 0.2
**Injection volume:** 10
**Detector:** MS, Micromass Quattro LC, triple quadrupole, electrospray atmospheric pressure ionization, positive ion mode, capillary 3.5 kV, cone 35 V, extractor 3 V, desolvation 400°, ion source 80°, collision energy 35 eV, argon 1.9 mtorr, m/z 257–134

## CHROMATOGRAM
**Retention time:** 1.37
**Internal standard:** (5-(4-(2-(5-(2-methyl-1,3-dioxolan-2-yl)-2-pyridyl)ethoxy)benzylidene)-2,4-thiazolidinedione (m/z 413–178) (1.26)
**Limit of quantitation:** 0.5 ng/mL (S/N 30)

## OTHER SUBSTANCES
**Extracted:** metabolites

## KEY WORDS
plasma

## REFERENCE
Lin, Z.J.; Ji, W.; Desai-Krieger, D.; Shum, L. Simultaneous determination of pioglitazone and its two active metabolites in human plasma by LC-MS/MS, *J.Pharm.Biomed.Anal.*, **2003**, *33*, 101–108.

## SAMPLE
**Matrix:** formulations
**Sample preparation:** Sonicate powdered tablet containing 500 mg metformin and 30 mg pioglitazone with 100 mL 100 mM HCl for 10 min with intermittent shaking, add 40 mL MeCN, sonicate for 25 min with intermittent shaking, cool to room temperature, add 10 mL MeCN, make up to 200 mL with 100 mM HCl, mix well, centrifuge at 10 000 rpm for 10 min. Mix a 2 mL aliquot with 1 mL 250 μg/mL IS in MeCN:100 mM HCl 25:75, make up to 50 mL with diluent, inject an aliquot. (The diluent was MeCN:5 mM disodium hydrogen phosphate 30:70, adjusted to pH 2.3 with HCl.)

## HPLC VARIABLES
**Column:** 150 × 4.6 5 μm Zorbax XDB C18

**Column temperature:** 40
**Mobile phase:** MeCN:10 mM disodium hydrogen phosphate containing 5 mM sodium dodecyl sulfate 34:66, adjusted to pH 7.1 with orthophosphoric acid
**Flow rate:** 1
**Injection volume:** 25
**Detector:** UV 226

## CHROMATOGRAM
**Retention time:** 7.82
**Internal standard:** methylparaben (3.30)
**Limit of detection:** 3 ng/mL
**Limit of quantitation:** 10 ng/mL

## OTHER SUBSTANCES
**Extracted:** metformin (LOD 22 ng/mL, LOQ 70 ng/mL) (5.06)

## KEY WORDS
tablets; validation details

## REFERENCE
Kolte, B.L.; Raut, B.B.; Deo, A.A.; Bagool, B.A.; Shinde, D.B. Simultaneous high-performance liquid chromatographic determination of pioglitazone and metformin in pharmaceutical-dosage form, *J.Chromatogr.Sci.*, **2004**, *42*, 27–31.

## SAMPLE
**Matrix:** solutions
**Sample preparation:** Inject a 1 μL aliquot of a solution in MeOH:water 10:90.

## HPLC VARIABLES
**Column:** 20 × 2 5 μm DASH BetaBasic C8 (ThermoHypersil Keystone)
**Mobile phase:** Gradient. A was MeCN:water:formic acid 5:95:0.1. B was MeCN:water: formic acid 95:5:0.1. A:B 100:0 for 0.2 min, to 0:100 over 1.5 min.
**Flow rate:** 1.5
**Injection volume:** 1
**Detector:** MS, PE Sciex API-3000, TurboIonspray, electrospray 4500 V, ring 290 V, orifice 60 V, drying gas 400°, 20% of column effluent entered the detector, m/z 357.2–134

## CHROMATOGRAM
**Retention time:** 0.87

## OTHER SUBSTANCES
**Simultaneous:** amitriptyline (m/z 278.3–233) (1.1), aprepitant (MK-869) (m/z 535.3–277) (1.4), diclofenac (m/z 296.1–215) (1.35), enoxacin (m/z 321.2–234) (0.7), fenofibrate (m/z 360.9–233) (1.6), finasteride (m/z 373.2–317) (1.2), indinavir (m/z 614.4–421) (0.93), raloxifene (m/z 474.1–112) (0.97)

## REFERENCE
Romanyshyn, L.A.; Tiller, P.R. Ultra-short columns and ballistic gradients: considerations for ultra-fast chromatographic liquid chromatographic-tandem mass spectrometric analysis, *J.Chromatogr.A*, **2001**, *928*, 41–51.

## ANNOTATED BIBLIOGRAPHY
Kiyota, Y.; Kondo, T.; Maeshiba, Y.; Hashimoto, A.; Yamashita, K.; Yoshimura, Y.; Motohashi, M.; Tanayama, S. Studies on the metabolism of the new antidiabetic agent pioglitazone. Identification of metabolites in rats and dogs, *Arzneimittelforschung*, **1997**, *47*, 22–28. [LC-MS]

Kolte, B.L.; Raut, B.B.; Deo, A.A.; Bagool, M.A.; Shinde, D.B. Liquid chromatographic method for the determination of rosiglitazone in human plasma, *J.Chromatogr.B*, **2003**, *788*, 37–44. [pioglitazone is internal standard]

Radhakrishna, T.; Rao, D.S.; Reddy, G.O. Determination of pioglitazone hydrochloride in bulk and pharmaceutical formulations by HPLC and MEKC methods, *J.Pharm.Biomed.Anal.*, **2002**, *29*, 593–607. [impurities]

Shen, Z.; Reed, J.R.; Creighton, M.; Liu, D.Q.; Tang, Y.S.; Hora, D.F.; Feeney, W.; Szewczyk, J.; Bakhtiar, R.; Franklin, R.B.; Vincent, S.H. Identification of novel metabolites of pioglitazone in rat and dog, *Xenobiotica*, **2003**, *33*, 499–509. [LC-MS; dog; rat; bile; plasma; urine]

Xue, Y.-J.; Turner, K.C.; Meeker, J.B.; Pursley, J.; Arnold, M.; Unger, S. Quantitative determination of pioglitazone in human serum by direct-injection high-performance liquid chromatography mass spectrometry and its application to a bioequivalence study, *J.Chromatogr.B*, **2003**, *795*, 215–226. [LC-MS; column-switching; LOQ 9 ng/mL]

Yamashita, K.; Murakami, H.; Okuda, T.; Motohashi, M. High-performance liquid chromatographic determination of pioglitazone and its metabolites in human serum and urine, *J.Chromatogr.B*, **1996**, *677*, 141–146. [SPE; LOD 10 ng/mL]

Zhong, W.Z.; Lakings, D.B. Determination of pioglitazone in dog serum using solid-phase extraction and high-performance liquid chromatography with ultraviolet (229 nm) detection, *J.Chromatogr.*, **1989**, *490*, 377–385.

Zhong, W.Z.; Williams, M.G. Simultaneous quantitation of pioglitazone and its metabolites in human serum by liquid chromatography and solid phase extraction, *J.Pharm.Biomed.Anal.*, **1996**, *14*, 465–473. [LOQ 20 ng/mL]

# Pipercuronium bromide

**Molecular formula:** $C_{35}H_{62}Br_2N_4O_4$
**Molecular weight:** 762.71
**CAS Registry No:** 52212-02-9
**Merck Index:** 13, 7540

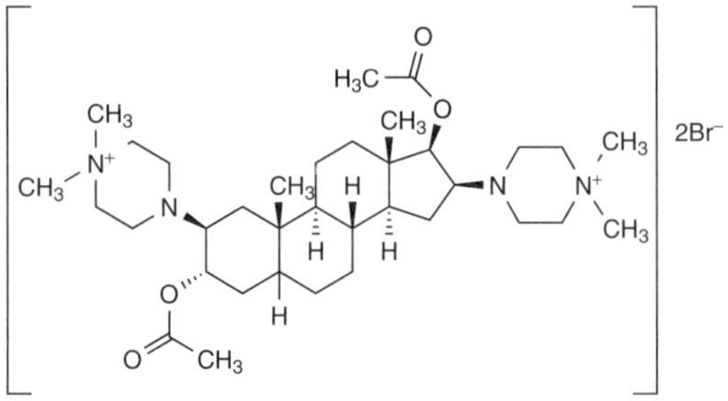

## SAMPLE

**Matrix:** blood
**Sample preparation:** Condition a 1 mL Bond-Elut C18 SPE cartridge with 1 mL MeOH:MeCN 2:1 and 1 mL water. Acidify 1 mL plasma with 150 μL 1 M sodium dihydrogen phosphate. Add 20–200 ng IS to 1 mL acidified plasma, add to the SPE cartridge, wash with 1 mL water, wash with 1 mL 100 mM pH 3 sodium dihydrogen phosphate, elute with 400 μL mobile phase, discard first 100 μL eluate, inject a 200 μL aliquot of the remaining eluate. (Although the original method was only validated for vecuronium, it can also be used to determine pipecuronium; see: D'Honneur,G.; Khalil,M.; Dominique,C.; Haberer,J.P.; Kleef,U.W.; Duvaldestin,P. Pharmacokinetics and pharmacodynamics of pipecuronium in patients with cirrhosis. *Anesth.Analg.* **1993**, *77*, 1203–1206.)

## HPLC VARIABLES

**Column:** 150 × 3.9 4 μm Nova-Pak C18
**Mobile phase:** Dioxane:water 20:80 containing 100 mM sodium dihydrogen phosphate and 0.44 mM 9,10-dimethoxyanthracene-2-sulfonate, pH adjusted to 3 with phosphoric acid. (Caution! Dioxane is a carcinogen!) (After each series of analyses, flush column with 200 mL MeOH and then re-equilibrate with 120 mL mobile phase.)
**Flow rate:** 1
**Injection volume:** 200
**Detector:** F ex 380 em 452 following post-column extraction. The column effluent mixed with dichloroethane pumped at 1.6 mL/min and the mixture flowed through a 1 m ×0.25 mm ID stainless steel coil to a phase separator (*Anal.Chim.Acta* **1987**, *192*, 267) and then the organic phase flowed through the detector.

## CHROMATOGRAM

**Internal standard:** 1-(3α,17β-diacetoxy-2β-piperidino-5α-androstan-16β,5α-yl)piperidine (16)
**Limit of detection:** 20 ng/mL

## OTHER SUBSTANCES

**Extracted:** vecuronium (LOD 5 ng/mL) (13)

## KEY WORDS

pharmacokinetics; SPE

## REFERENCE

Paanakker, J.E.; Thio, J.M.S.L.; Van den Wildenberg, H.M.; Kaspersen, F.M. Assay of vecuronium in plasma using solid-phase extraction, high-performance liquid chromatography and post-column ion-pair extraction with fluorimetric detection, *J.Chromatogr.*, **1987**, *421*, 327–335.

# Pirlimycin

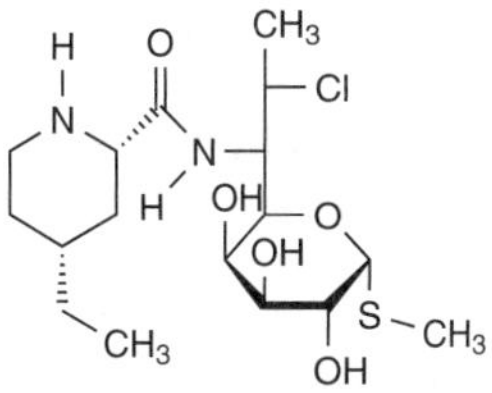

**Molecular formula:** $C_{17}H_{31}ClN_2O_5S$
**Molecular weight:** 410.96
**CAS Registry No:** 79548-73-5, 78822-40-9 (HCl)
**Merck Index:** 13, 7583

## SAMPLE

**Matrix:** milk
**Sample preparation:** Shake 3 mL extracting solution and 1 mL milk for 30 s, centrifuge at 1000 g for 4 min. Remove the supernatant and add 2 mL extracting solution to the pellet. Shake for 30 s and centrifuge at 1000 g for 4 min. Combine the supernatants, add 6 mL 1-chlorobutane, add 500 μL water, shake for 30 s, centrifuge at 1000 g for 4 min. Remove the lower aqueous phase and add 2 mL water to the organic layer. Shake for 30 s and centrifuge at 1000 g for 4 min. Combine the aqueous layers, evaporate to 2 mL under nitrogen at 55–60°, add 1 mL 15% ammonia, mix. Within 30 s, add 15 mL dichloromethane and shake for 30 s. (Caution! Wear gloves and eye protection, ammonia may escape under pressure!) Centrifuge at 1000 g for 4 min. Remove the water layer and evaporate the organic layer to dryness under nitrogen at 60°. Add 500 μL 250 μM NaOH, vortex for 10 s, add 500 μL 100 μg/mL 9-fluorenylmethyl chloroformate in MeCN, vortex for 10 s, let stand for 1 h, vortex for 10 s, let stand at room temperature for at least another 2 h. Inject a 50 μL aliquot. (The extracting solution was 125 μL concentrated HCl in 100 mL MeCN.)

## HPLC VARIABLES

**Column:** 100 × 2.1 5 μm Hypersil ODS
**Column temperature:** 35
**Mobile phase:** MeOH:MeCN:1% acetic acid 30:30:40
**Flow rate:** 0.6
**Injection volume:** 50
**Detector:** UV 264

## CHROMATOGRAM

**Retention time:** 8.8
**Limit of quantitation:** 100 ppb

## KEY WORDS

cow; derivatization

## REFERENCE

Heller, D.N. Determination of pirlimycin residue in milk by liquid chromatographic analysis of the 9-fluorenylmethyl chloroformate derivative, *J.AOAC Int.*, **1997**, *80*, 975–981.

## SAMPLE

**Matrix:** solutions
**Sample preparation:** Prepare a solution in mobile phase, inject a 25 μL aliquot.

## HPLC VARIABLES

**Column:** 250 × 4.6 Zorbax C8
**Mobile phase:** MeCN:water 12:88 containing 0.25 g/L tetrabutylammonium perchlorate and 2 mL/L 70% perchloric acid, apparent pH adjusted to 2.5 with 50% NaOH
**Flow rate:** 1.5

**Injection volume:** 25
**Detector:** UV 214

---

## CHROMATOGRAM
**Retention time:** 15

---

## OTHER SUBSTANCES
**Simultaneous:** benzyl alcohol, clindamycin (22.5), lincomycin (k′ 0.9)

---

## REFERENCE
Theis, D.L. Ion-pairing liquid chromatographic method for the determination of pirlimycin hydrochloride, *J.Chromatogr.*, **1987**, *402*, 335–343.

---

## SAMPLE
**Matrix:** tissue
**Sample preparation:** Mix 2 mL ground liver with 100 μL 10 μg/mL IS in water, add 10 mL MeCN:trifluoroacetic acid 99.75:0.25, homogenize (Polytron) at medium–high speed for 30 s, filter, wash tube and filter cake twice with 5 mL portions of MeCN:water:trifluoroacetic acid 84.75:15:0.25, filter the washes. Combine the filtrates and shake with 25 mL *n*-butyl chloride for 10 s, centrifuge at 500 g for 1 min, remove the aqueous layer, add 4 mL water to the organic layer, shake for 5–10 s, centrifuge at 500 g for 1 min. Combine the aqueous layers, evaporate to 1.5–2.0 mL under a stream of nitrogen at 80°, cool to room temperature, add 1 mL 15% ammonium hydroxide, extract with 15 mL dichloromethane. Evaporate the organic layer to dryness under a stream of nitrogen at 60°, reconstitute the residue with 2 mL MeCN:100 mM pH 6.8 ammonium acetate buffer 20:80, inject a 100 μL aliquot. (Sample preparation from Hornish,R.E.; Cazers,A.R.; Chester,S.T.,Jr.; Roof,R.D. Identification and determination of pirlimycin residue in bovine milk and liver by high-performance liquid chromatography-thermospray mass spectrometry. *J.Chromatogr.B* **1995**, *674*, 219–235.)

---

## HPLC VARIABLES
**Column:** 150 × 2.1 Zorbax C18
**Mobile phase:** MeCN:100 mM ammonium formate 30:70
**Flow rate:** 0.18
**Injection volume:** 100
**Detector:** MS, Finnigan TSQ 7000, APCI, single quad mode, Q3 as analyzer, corona 4 kV, corona current 4.5 kV, capillary 180–200°, m/z 411–413

---

## CHROMATOGRAM
**Retention time:** 14
**Internal standard:** iso-pirlimycin (10.5)
**Limit of quantitation:** 25 ng/g

---

## OTHER SUBSTANCES
**Extracted:** pirlimycin sulfone (m/z 443), pirlimycin sulfoxide (m/z 427–429)

---

## KEY WORDS
cow; liver

---

## REFERENCE
Hornish, R.E.; Roof, R.D.; Wiest, J.R. Pirlimycin residue in bovine liver-a case of reverse metabolism, *Analyst*, **1998**, *123*, 2463–2467.

## ANNOTATED BIBLIOGRAPHY

Heller, D.N. Determination and confirmation of pirlimycin residue in bovine milk and liver by liquid chromatography/thermospray mass spectrometry: Interlaboratory study, *J.AOAC Int.*, **1996**, *79*, 1054–1061.

Hornish, R.E.; Cazers, A.R.; Chester, S.T. Jr.; Roof, R.D. Identification and determination of pirlimycin residue in bovine milk and liver by high-performance liquid chromatography-thermospray mass spectrometry, *J.Chromatogr.B*, **1995**, *674*, 219–235. [SPE; LOQ 50 ng/mL; LOQ 25 ng/g]

Shah, J.A.; Weber, D.J. High-performance liquid chromatographic assay of pirlimycin in human serum and urine using 9-fluorenylmethylchloroformate, *J.Chromatogr.*, **1984**, *309*, 95–105. [derivatization]

# Poloxalene

**Molecular weight:** ca. 3000
**CAS Registry No:** 9003-11-6
**Merck Index:** 13, 7644

## SAMPLE
**Matrix:** solutions
**Sample preparation:** Inject a 20 μL aliquot of a 3% solution in mobile phase

## HPLC VARIABLES
**Column:** 300 × 7.5 10 μm HP Plgel 500A
**Mobile phase:** THF
**Flow rate:** 1
**Injection volume:** 20
**Detector:** Refractive Index

## CHROMATOGRAM
**Retention time:** 9

## REFERENCE
Wigman, L.S.; Abdel-Kader, H.; Menon, G.K. Size exclusion chromatography of poloxalene poloxamers: polyethylene glycol-polypropylene glycol co-polymers used to control cattle bloat, *J.Pharm.Biomed.Anal.*, **1994**, *12*, 719–722.

## SAMPLE
**Matrix:** solutions
**Sample preparation:** Inject a 20 μL aliquot of a 3% solution in mobile phase

## HPLC VARIABLES
**Column:** 300 × 7.5 10 μm Spherogel TSK2000SW 250A
**Mobile phase:** MeOH:water 30:70
**Flow rate:** 1
**Injection volume:** 20
**Detector:** Refractive Index

## CHROMATOGRAM
**Retention time:** 6

## REFERENCE
Wigman, L.S.; Abdel-Kader, H.; Menon, G.K. Size exclusion chromatography of poloxalene poloxamers: polyethylene glycol-polypropylene glycol co-polymers used to control cattle bloat, *J.Pharm.Biomed.Anal.*, **1994**, *12*, 719–722.

# Pramipexole

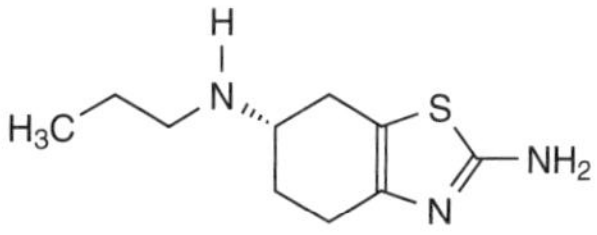

**Molecular formula:** $C_{10}H_{17}N_3S$
**Molecular weight:** 211.33
**CAS Registry No:** 104632-26-0
**Merck Index:** 13, 7790

## SAMPLE

**Matrix:** blood
**Sample preparation:** Vortex 1 mL plasma, 50 μL 180 ng/mL IS in 1% acetic acid, 100 μL 1 M NaOH, and 6 mL MTBE for 5 min, centrifuge at 630 g for 5 min, freeze in dry ice acetone. Evaporate the organic layer to dryness under a stream of nitrogen, reconstitute the residue with 100 μL MeOH:water 95:5, inject a 70 μL aliquot.

## HPLC VARIABLES

**Column:** 150 × 4.6 5 μm Zorbax SB CN
**Mobile phase:** MeOH:water:100 mM ammonium acetate 80:15:5
**Flow rate:** 1.2
**Injection volume:** 70
**Detector:** MS, PE Sciex API-III, triple quadrupole, orifice 45 V, nebulizer probe 470°, nebulizer gas at 550 kPa, auxiliary flow 1.5 L/min, corona discharge needle 3 μA, m/z 212–153

## CHROMATOGRAM

**Retention time:** 3.4
**Internal standard:** BHT-920 (m/z 210–141) (2.2)
**Limit of quantitation:** 50 pg/mL

## KEY WORDS

plasma; use polypropylene containers

## REFERENCE

Lau, Y.Y.; Selenka, J.M.; Hanson, G.D.; Talaat, R.; Ichhpurani, N. Determination of pramipexole (U-98,528) in human plasma by high-performance liquid chromatography with atmospheric pressure chemical ionization tandem mass spectrometry, *J.Chromatogr.B*, **1996**, *683*, 209–216.

## SAMPLE

**Matrix:** blood, urine
**Sample preparation:** Vortex 1 mL plasma or urine, 50 μL 30 ng/mL (plasma) or 15 μg/mL (urine) IS in 1% acetic acid, 100 μL 1 M NaOH, and 6 mL diethyl ether for 5 min, centrifuge at 2500 g for 5 min, freeze in dry ice/acetone. Extract the organic layer with 100 (plasma) or 200 (urine) μL buffer. Freeze in dry ice/acetone, discard organic layer, remove all traces of ether with a stream of nitrogen, thaw, inject a 50 μL aliquot. (The buffer was 1.7 g potassium dihydrogen phosphate, 1.7 g sodium acetate, and 0.5 g sodium heptanesulfonate in 500 mL water, adjusted to pH 3.5 with acetic acid.)

## HPLC VARIABLES

**Guard column:** 15 × 3.2 7 μm Brownlee RP-8
**Column:** 250 × 4.6 5 μm Zorbax Rx C8
**Mobile phase:** MeCN:buffer 45:55 (The buffer was 10.2 g potassium dihydrogen phosphate, 10.2 g sodium acetate, and 4.5 g sodium heptanesulfonate in 3 L water, adjusted to pH 3.5 with acetic acid.)
**Flow rate:** 1.2
**Injection volume:** 50

**Detector:** E, ESA 5100A, 5011 analytical cell, 5020 guard cell, electrode 2 0.6 V (detection), electrode 1 0.2 V (impurity screening), guard cell 0.65 V (mobile-phase conditioning) (plasma); UV 286 (urine)

---

## CHROMATOGRAM
**Retention time:** 14.4
**Internal standard:** BHT-920 (10.7)
**Limit of quantitation:** 50 pg/mL (plasma), 10 ng/mL (urine)

---

## KEY WORDS
plasma

---

## REFERENCE
Lau, Y.Y.; Hanson, G.D.; Ichhpurani, N. Determination of pramipexole (U-98,528) in human plasma and urine by high-performance liquid chromatography with electrochemical and ultraviolet detection, *J.Chromatogr.B*, **1996**, *683*, 217–223.

# Pranlukast

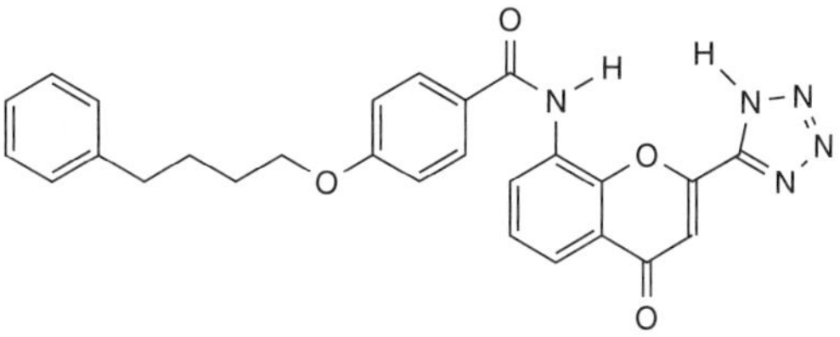

**Molecular formula:** $C_{27}H_{23}N_5O_4$
**Molecular weight:** 481.50
**CAS Registry No:** 103177-37-3
**Merck Index:** 13, 7795

## SAMPLE

**Matrix:** blood
**Sample preparation:** Add IS to 200 µL plasma, wash with hexane:isopropanol 95:5, add MeCN to the aqueous layer, centrifuge. Evaporate the supernatant to dryness, reconstitute the residue with mobile phase, inject an aliquot.

## HPLC VARIABLES

**Column:** 300 × 4.6 C18
**Mobile phase:** MeCN:70 mM ammonium formate buffer 45:55
**Flow rate:** 1
**Detector:** UV 262

## CHROMATOGRAM

**Internal standard:** ONO-RS-425 (4-oxo-6-methyl-8-(4-phenylbutoxy)benzoylamino)-2-(tetrazol-5-yl)-4H-1-benzopyran)
**Limit of quantitation:** 10 ng/mL

## KEY WORDS

pharmacokinetics; plasma

## REFERENCE

Brocks, D.R.; Upward, J.; Hust, R.; Köester, F.E.; Collie, H.; Qian, Y.; Dennis, M.J. The pharmacokinetics of pranlukast in healthy young and elderly subjects, *Int.J.Clin.Pharmacol.Ther.*, **1996**, *34*, 375–379.

## SAMPLE

**Matrix:** blood
**Sample preparation:** Mix 100 µL plasma with 100 µL 250 ng/mL IS in 1 M citric acid solution, inject a 30–50 µL aliquot onto column A and elute to waste with water at 1 mL/min; after 25 s, elute to waste with water at 7 mL/min; after another 15 s, elute to waste with water at 1 mL/min; after 70 s, backflush column A with water for 70 s, elute the contents of column A onto column B with the mobile phase.

## HPLC VARIABLES

**Column:** A 10 × 2 40 µm Ph-EC-IST SPE cartridge (Jones Chromatography); B 20 × 2 unspecified guard column + 30 × 2 3 µm Hypersil BDS C-18
**Mobile phase:** Gradient. MeOH:20 mM ammonium acetate 25:75 for 0.3 min, to 70:30 over 0.1 min, maintain at 70:30 for 2 min, to 90:10 over 0.1 min, maintain at 90:10 over 0.8 min, return to initial conditions over 0.1 min.
**Flow rate:** 0.3
**Injection volume:** 30–50
**Detector:** MS, PE Sciex API-III Plus triple quadrupole, turbo ionspray, negative ion mode, ionspray -3200 V, interface -650 V, orifice -68 V, collision energy 23 V, nebulizer gas air at 42 psi and 0.6 L/min, curtain gas nitrogen at 1.0 L/min, auxiliary gas air at 5.0 L/min, turbo-ionspray probe 350°, collision gas argon, ca. 60 µL/mL of column effluent entered the detector, m/z 480.0–291.0

## CHROMATOGRAM

**Internal standard:** SK&F 108566 (m/z 423.0–244.0) (1.9)
**Limit of quantitation:** 10 ng/mL

## KEY WORDS

column-switching; plasma

## REFERENCE

Marchese, A.; McHugh, C.; Kehler, J.; Bi, H. Determination of pranlukast and its metabolites in human plasma by LC/MS/MS with PROSPEKT$^{TM}$ on-line solid-phase extraction, *J.Mass.Spectrom.*, **1998**, *33*, 1071–1079.

## SAMPLE

**Matrix:** solutions

## HPLC VARIABLES

**Column:** 250 × 4.6 10 μm Cosmosil 5C18-MS
**Column temperature:** 50
**Mobile phase:** Gradient. MeCN:10 mM pH 4.2 potassium phosphate buffer 0:100 for 2 min, to 75:25 over 7.5 min, maintain at 75:25 for 5.5 min. Isocratic. MeCN:buffer 50:50
**Flow rate:** 1.5
**Detector:** UV (wavelength not specified)

## CHROMATOGRAM

**Retention time:** 13.4 (gradient) or 6.7 (isocratic)

## OTHER SUBSTANCES

**Simultaneous:** acetazolamide (7.9), acyclovir (7.0), allopurinol (7.3), aniracetam (8.6), benazepril (11.5), betamethasone (10.8), camostat mesylate (8.8), carbamazepine (10.8), cilazapril (10.7), cimetidine (7.5), clofibrate (15.0), clonazepam (11.3), cyclophosphamide (9.9), delapril (11.9), dexamethasone (10.9), diazepam (12.5), digoxin (9.0), dilazep (10.6), diltiazem (10.2), docarpamine (11.4), eperisone (8.7), ethosuximide (9.0), fenbufen (11.8), fluconazole (9.2), flutamide (12.7), fominoben (12.6), hydrocortisone acetate (12.1), imidapril (10.1), indomethacin (12.6), irinotecan (9.2), maprotiline (10.5), methotrexate (8.0), nefiracetam (9.7), nifedipine (11.9), nitrazepam (11.2), pentobarbital (10.9), phenobarbital (10.0), phenytoin (10.8), pindolol (8.3), pranoprofen (10.4), prednisolone (10.3), primidone (8.9), quinapril (10.5), spironolactone (12.4), sulpiride (7.6), sulthiame (9.3), tolbutamide (11.6), tranilast (10.5), triamcinolone (11.2), warfarin (12.0), zonisamide (9.5) (gradient retention times; isocratic conditions may differ)

## REFERENCE

Sugiyama, T.; Matsuyama, R.; Usui, S.; Katagiri, Y.; Hirano, K. Selection of mobile phases in high-performance liquid chromatographic determination for medicines, *Biol.Pharm.Bull.*, **2000**, *23*, 274–278.

# Prednicarbate

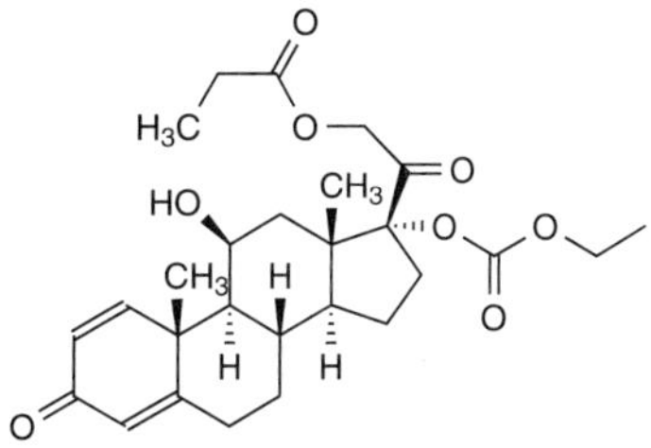


**Molecular formula:** $C_{27}H_{36}O_8$
**Molecular weight:** 488.57
**CAS Registry No:** 73771-04-7
**Merck Index:** 13, 7805

## SAMPLE

**Matrix:** cell incubations
**Sample preparation:** Sonicate cells, extract twice with 3 mL portions of ethyl acetate by vortexing for 1 min, centrifuge at 1000 rpm for 5 min. Evaporate the combined organic layers to dryness under a stream of nitrogen, reconstitute the residue with 1 mL MeOH, vortex for 1 min, evaporate to dryness, reconstitute the residue with 100 μL MeOH, centrifuge, inject a 20 μL aliquot.

## HPLC VARIABLES

**Column:** 125 × 4 5 μm LiChrospher 100 RP-18
**Mobile phase:** Gradient. MeCN:water from 20:80 to 100:0 over 20 min.
**Flow rate:** 1
**Injection volume:** 20
**Detector:** UV 254

## CHROMATOGRAM

**Retention time:** 15.1
**Internal standard:** betamethasone (9.4)
**Limit of detection:** 10 ng/mL
**Limit of quantitation:** 100 ng/mL

## OTHER SUBSTANCES

**Extracted:** prednisolone (7.8), prednisolone 17-ethylcarbonate (11.1), prednisolone 21-ethylcarbonate (12.5)

## REFERENCE

Gysler, A.; Lange, K.; Korting, H.C.; Schäfer-Korting, M. Prednicarbate biotransformation in human foreskin keratinocytes and fibroblasts, *Pharm.Res.*, **1997**, *14*, 793–797.

# Propionylpromazine

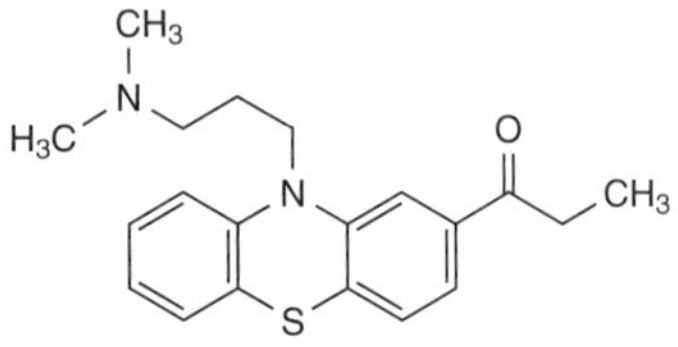

**Molecular formula:** $C_{20}H_{24}N_2OS$
**Molecular weight:** 340.49
**CAS Registry No:** 3568-24-9
**Merck Index:** 13, 7921

## SAMPLE

**Matrix:** formulations
**Sample preparation:** Vortex 100 mg formulation with 10 mL hexane, sonicate for 5 min, add 10 mL 50 mM HCl, vortex, centrifuge at 835 g for 2 min. Repeat the extraction of the hexane layer with 10 mL 50 mM HCl. Combine the aqueous layers and make up to 50 mL with 50 mM HCl. Dilute a 2 mL aliquot to 10 mL with MeCN:mobile-phase buffer 70:30, filter (0.45 μm PTFE), inject a 1 μL aliquot. (Protect from light. Use of hexane may not be necessary, depending on the nature of the formulation.)

## HPLC VARIABLES

**Column:** 150 × 2 5 μm ODS/H C18 (Keystone)
**Column temperature:** 40
**Mobile phase:** MeCN:buffer 85:15 (Prepare the buffer by diluting 25 mL of a 200 mM heptanesulfonic acid premixed solution (Alltech) to 1 L with water.)
**Flow rate:** 0.3
**Injection volume:** 1
**Detector:** UV 280

## CHROMATOGRAM

**Retention time:** 3
**Limit of detection:** 0.09%

## REFERENCE

Hurlbut, D.B.; Primus, T.M.; Goodall, M.J.; Volz, S.A.; Johnston, J.J. Determination of propionylpromazine hydrochloride in formulation matrices using reversed-phase ion-pair small bore liquid chromatography, *J.AOAC Int.*, **1999**, *82*, 1321–1328.

## SAMPLE

**Matrix:** tissue
**Sample preparation:** Condition a Sep-Pak C18 SPE cartridge with 5 mL MeOH and 5 mL water. Homogenize kidney with a kitchen grinder. Weigh out a 5 g sample and add 20 mL MeCN with continuous gentle mixing, mix vigorously on a vibromixer at 1500 rpm for 30 s, sonicate for 2 min, centrifuge at 4000 g for 5 min. Mix 7.5 mL sample extract and 40 mL 10% NaCl and add to SPE cartridge, wash with 1 mL 10 mM sulfuric acid, wash with 2 mL air, elute with 2 mL acidic MeCN. Place eluate in a washed tube and evaporate to 300 μL at 70° under a stream of nitrogen, mix gently, add 1 mL *n*-hexane, mix on a vibromixer for 30 s, centrifuge at 2000 g, inject a 50 μL aliquot of the aqueous phase. (Acidic MeCN was 1 mL 50 mM sulfuric acid and 100 mL MeCN. The washed tube was prepared by rinsing with concentrated ammonia, water, and acetone and drying under a stream of nitrogen.)

## HPLC VARIABLES

**Guard column:** 10 × 2.1 37–50 μm Bondapak C18
**Column:** 300 × 3.9 Bondapak C18
**Mobile phase:** MeCN:water 55:45 containing 2.46 g/L anhydrous sodium acetate, pH adjusted to 6.5 with acetic acid
**Flow rate:** 1.2

**Injection volume:** 50
**Detector:** UV 240

## CHROMATOGRAM
**Retention time:** 18
**Limit of detection:** 4 ng/g

## OTHER SUBSTANCES
**Extracted:** acepromazine (13), azaperol (Fex 246 em 351) (5), azaperone (8), carazolol (5), haloperidol (8.5), chlorpromazine (25), xylazine (7)

## KEY WORDS
kidney; pig; SPE

## REFERENCE
Keukens, H.J.; Aerts, M.M.L. Determination of residues of carazolol and a number of tranquilizers in swine kidney by high-performance liquid chromatography with ultraviolet and fluorescence detection, *J.Chromatogr.*, **1989**, *464*, 149–161.

---

## SAMPLE
**Matrix:** tissue
**Sample preparation:** Condition a Bond-Elut C18 SPE cartridge with 5 mL MeOH and 5 mL water. Cut pig kidney or liver into small pieces and homogenize. 5 g Homogenate + 10 mL MeCN, shake, vortex for 30 s, sonicate for 3 min, vortex for 30 s, sonicate for 3 min, centrifuge at 10000 g for 20 min. Add 7.5 mL supernatant and 40 mL 10% NaCl to the SPE cartridge at about 1 mL/min, do not allow cartridge to dry out, wash with 850 μL 10 mM sulfuric acid, dry with air, elute with 3.5 mL acidic MeCN. Evaporate the eluate to dryness under a stream of nitrogen at 50°, reconstitute the residue in 300 μL 10 mM sulfuric acid, vortex briefly, add 1 mL hexane, vortex for 30 s, centrifuge at 2000 g for 5 min, inject an aliquot of the aqueous layer. (Acidic MeCN was 1 mL 50 mM sulfuric acid in 100 mL MeCN.)

## HPLC VARIABLES
**Guard column:** Hypersil 5 μm SAS C1
**Column:** 250 mm long 5 μm Hypersil SAS C1
**Mobile phase:** MeCN:water 50:50 containing 0.77 g/L ammonium acetate
**Flow rate:** 2
**Detector:** E, ESA Model 5100A Coulochem, first electrode +0.4 V, second electrode (which was monitored) +0.7 V, Model 5020 guard cell after pump but before injector at +0.75 V

## CHROMATOGRAM
**Retention time:** 25
**Limit of detection:** 2 ng/g

## OTHER SUBSTANCES
**Extracted:** acepromazine (18), azaperol (5), azaperone (6.5), carazolol (9), chlorpromazine (32), haloperidol (12.5), xylazine (10)

## KEY WORDS
kidney; liver; pig; SPE

## REFERENCE
Rose, M.D.; Shearer, G. Determination of tranquilisers and carazolol residues in animal tissue using high-performance liquid chromatography with electrochemical detection, *J.Chromatogr.*, **1992**, *624*, 471–477.

## ANNOTATED BIBLIOGRAPHY

Delahaut, P.; Levaux, C.; Eloy, P.; Dubois, M. Validation of a method for detecting and quantifying tranquillisers and a β-blocker in pig tissues by liquid chromatography-tandem mass spectrometry, *Anal.Chim.Acta*, **2003**, *483*, 335–340. [SPE; LOD 1–10 ng/g; propionylpromazine; acepromazine; chlorpromazine; xylazine; carazolol; azaperone; azaperol; haloperidol; isobutcar; levamisole is internal standard]

Quintana, M.C.; Blanco, M.H.; Lacal, J.; Hernández, L. Analysis of pharmaceutical residues in bovine liver by HPLC, *J.Liq.Chromatogr.Rel.Technol.*, **2001**, *24*, 735–745. [LOD 21–298 ng/mL; dexamethasone; betamethasone; betamethasone acetate; xylazine; haloperidol; acepromazine; propionylpromazine; chlorpromazine]

# Propoxycaine hydrochloride

**Molecular formula:** $C_{16}H_{27}ClN_2O_3$
**Molecular weight:** 330.86
**CAS Registry No:** 550-83-4
**Merck Index:** 13, 7930

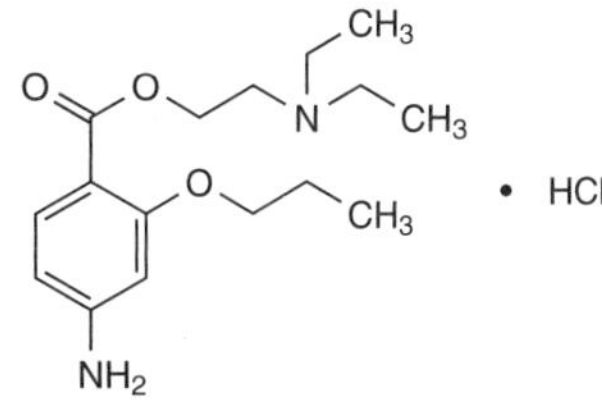


## SAMPLE
**Matrix:** formulations
**Sample preparation:** Dilute with mobile phase, inject a 20 μL aliquot.

## HPLC VARIABLES
**Column:** 300 × 3.9 μBondapak phenyl
**Mobile phase:** MeCN:buffer 30:70 (The buffer was 25 mM pH 3.0 potassium dihydrogen phosphate containing 50 mM sodium heptanesulfonate.)
**Flow rate:** 1
**Injection volume:** 20
**Detector:** UV 254

## CHROMATOGRAM
**Retention time:** 18.9

## OTHER SUBSTANCES
**Simultaneous:** levonordefrin (3.6), norepinephrine (3.9), procaine (9.8)

## KEY WORDS
injections; stability-indicating

## REFERENCE
Storms, M.L.; Stewart, J.T. Stability-indicating HPLC assays for the determination of prilocaine and procaine drug combinations, *J.Pharm.Biomed.Anal.*, **2002**, *30*, 49–58.

# Propylene glycol

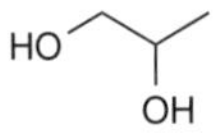


**Molecular formula:** $C_3H_8O_2$
**Molecular weight:** 76.09
**CAS Registry No:** 57-55-6
**Merck Index:** 13, 7947

## SAMPLE

**Matrix:** blood, tissue

**Sample preparation:** Plasma. Mix 50 μL plasma with 10 μL 50 μg/mL IS and 50 μL 4 M NaOH, add 1 mL benzoyl chloride solution, vortex for 30 min, add 25 μL 1% glycine, vortex for 15 min, let stand at room temperature for 45 min, centrifuge at 4000 rpm for 5 min. Remove a 100 μL aliquot of the pentane layer and evaporate it to dryness under a stream of nitrogen at 40°, reconstitute the residue with 1 mL MeOH, inject a 10 μL aliquot. Tissue. Homogenize rat lung tissue with 4 vol of water. Mix 100 μL homogenate with 10 μL 100 μg/mL IS, let stand at room temperature for 30 min, vortex thoroughly, centrifuge at 4000 rpm for 10 min. Mix 50 μL of the supernatant with 50 μL 4 M NaOH, add 1 mL benzoyl chloride solution, vortex for 30 min, add 25 μL 1% glycine, vortex for 15 min, let stand at room temperature for 45 min, centrifuge at 4000 rpm for 5 min. Remove a 100 μL aliquot of the pentane layer and evaporate it to dryness under a stream of nitrogen at 40°, reconstitute the residue with 200 μL MeOH, inject a 10 μL aliquot. (Prepare benzoyl chloride solution by mixing 125 μL benzoyl chloride with 10 mL pentane.)

## HPLC VARIABLES

**Column:** 100 × 2.1 3.5 μm Symmetry Shield RP-18 (Waters)

**Mobile phase:** MeOH:water:formic acid 78:22:0.1

**Flow rate:** 0.25

**Injection volume:** 10

**Detector:** MS, Applied Biosystems API365, APCI, ionspray 4500 V, declustering potential 1 V, focusing potential 120 V, entrance potential -2.5 V, collision entrance potential 8.57 V, collision energy 13 V, collision exit potential 16 V, probe 450°, turbo gas flow 7 L/min, m/z 285.1–163.2

## CHROMATOGRAM

**Retention time:** 3.5

**Internal standard:** 1,4-butanediol (m/z 299.0–177.2) (4.4)

**Limit of detection:** 269 ng/mL (plasma), 1.12 μg/g (lung)

**Limit of quantitation:** 448 ng/mL (plasma), 1.62 μg/g (lung)

## KEY WORDS

derivatization; lung; plasma; rat

## REFERENCE

Gao, S.; Wilson, D.M.; Edinboro, L.E.; McGuire, G.M.; Williams, S.G.P.; Karnes, H.T. Improvement of sensitivity for the determination of propylene glycol in rat plasma and lung tissue using HPLC/tandem MS and derivatization with benzoyl chloride, *J.Liq.Chromatogr.Rel.Technol.*, **2003**, *26*, 3413–3431.

# Propylhexedrine

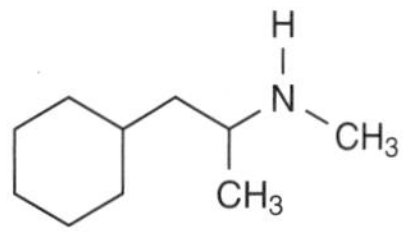

**Molecular formula:** $C_{10}H_{21}N$
**Molecular weight:** 155.28
**CAS Registry No:** 101-40-6
**Merck Index:** 13, 7952

## SAMPLE

**Matrix:** blood
**Sample preparation:** Condition a 200 mg Bond-Elut C18 SPE cartridge with 1 mL MeOH, 1 mL water, and 2 mL 10 mM pH 9.3 ammonium carbonate buffer. Centrifuge whole blood or serum at 14 000 g for 5 min. Vortex 0.2–1 mL of the supernatant with IS and 2 mL 10 mM pH 9.3 ammonium carbonate buffer, centrifuge at 5000 g for 10 min, add 2 mL of the supernatant to the SPE cartridge and allow to pass through over 5 min, wash with 2 mL 10 mM pH 9.3 ammonium carbonate buffer, dry with vacuum for 5 min, elute with 500 μL MeOH:500 mM acetic acid 90:10. Add 10 μL 1 mM HCl to the eluate and evaporate to dryness under a stream of nitrogen, reconstitute the residue with 100 μL mobile phase, centrifuge at 14 000 g for 4 min, inject a 5–20 μL aliquot of the supernatant.

## HPLC VARIABLES

**Column:** 125 × 3 4 μm Superspher 100 RP 18
**Mobile phase:** MeCN:50 mM pH 3.0 ammonium formate buffer 25:75
**Flow rate:** 0.8
**Injection volume:** 5–20
**Detector:** MS, Finnigan MAT SSQ 7000, APCI, positive ion mode, sheath gas nitrogen 70 psi, auxiliary gas nitrogen 20 mL/min, capillary 190°, vaporizer 450°, corona current 5 μA, m/z 204–156–157

## CHROMATOGRAM

**Retention time:** 3.9
**Internal standard:** norfenfluramine (m/z 212–204–187–159 (4.4)
**Limit of detection:** 2 ng/mL

## OTHER SUBSTANCES

**Extracted:** amphetamine (m/z 147–136–119–91) (3.7), cathinone (m/z 150–147–132) (3), ephedrine (m/z 160-166–148) (3.1), fenfluramine (m/z 232–212–159) (7), MDA (m/z 180–163–147) (3.8), MDEA (m/z 215–208–163) (5.3), MDMA (m/z 199–194–163–133) (4.3), methamphetamine (m/z 160–150–119–91) (4.2), norfenfluramine (m/z 212–204–187–159) (4.4), phentermine (m/z 160–150–133–91) (4.4), phenylethylamine (m/z 147–122–105) (3.1), phenylpropanolamine (m/z 152–147–134) (2.8)

## KEY WORDS

serum; SPE; whole blood

## REFERENCE

Bogusz, M.J.; Krüger, K.-D.; Maier, R.-D. Analysis of underivatized amphetamines and related phenethylamines with high-performance liquid chromatography-atmospheric pressure chemical ionization mass spectrometry, *J.Anal.Toxicol.*, **2000**, *24*, 77–84.

# Protirelin

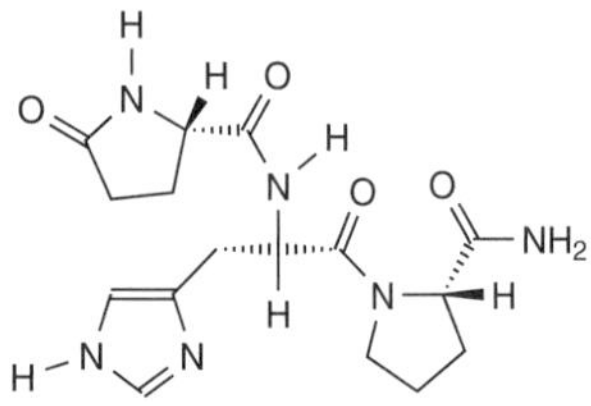

**Molecular formula:** $C_{16}H_{22}N_6O_4$
**Molecular weight:** 362.38
**CAS Registry No:** 24305-27-9
**Merck Index:** 13, 9663

## SAMPLE
**Matrix:** blood
**Sample preparation:** Mix 1 mL serum with 2 mL water and 7.5 mL cold EtOH/HCl, let stand at 4° for 12 h, centrifuge at 2800 rpm at 4° for 30 min, adjust the pH of the supernatant to 8.3 with concentrated ammonium hydroxide, let stand at 4° for 15 min, centrifuge at 2800 rpm at 4° for 20 min, adjust the pH of the supernatant to 5.3 with 4 M HCl. For each 1 mL of extract, add 25 μL 2 M ammonium acetate, adjust pH to 5.3; to each 10 mL of extract, slowly add 15 mL cold EtOH and 25 mL diethyl ether, let stand at 4° for 12 h, centrifuge at 2800 rpm at 4° for 30 min. Collect the precipitate and dry it under nitrogen, reconstitute with 100 mM pH 3.10 sodium dihydrogen phosphate, inject an aliquot containing 1.1–3.3 nmol. (Prepare EtOH/HCl by adding 7.5 mL concentrated HCl to 375 mL 95% EtOH. Sample preparation from Oyama,H.; Tenku,A.; Kakita,K.; Matsumura,S.; Nishida,S.; Horino,M. Recovery of human C-peptide by acid-ethanol extraction. *Endocrinol.Japon.* **1978**, *25*, 493–498.)

## HPLC VARIABLES
**Column:** 250 × 4.5 Spherisorb S5 ODS2
**Column temperature:** 37
**Mobile phase:** Gradient. MeCN: 100 mM pH 3.10 sodium dihydrogen phosphate from 0:100 to 28:72 over 14 min, to 31.2:68.8 over 8 min, to 32.8:67.2 over 8 min, to 42.8 over 5 min, to 60:40 over 3.5 min, maintain at 60:40.
**Flow rate:** 1
**Injection volume:** 200
**Detector:** UV 220

## CHROMATOGRAM
**Retention time:** 7.5

## OTHER SUBSTANCES
**Extracted:** porcine C-peptide (21), porcine glucagon (24), porcine monocomponent insulin (25), porcine pancreatic polypeptide (36), porcine proinsulin (30), somatostatin (24), vasoactive intestinal polypeptide (19)

## KEY WORDS
cord; plasma

## REFERENCE
Knip, M. Analysis of pancreatic peptide hormones by reversed-phase high-performance liquid chromatography, *Horm.Metab.Res.*, **1984**, *16*, 487–491.

## SAMPLE
**Matrix:** blood
**Sample preparation:** Mix whole blood with ice-cold EtOH, centrifuge (?), evaporate to dryness under reduced pressure or a stream of nitrogen at 35°, reconstitute with 0.1% trifluoroacetic acid, centrifuge briefly, inject an aliquot of the supernatant.

## HPLC VARIABLES

**Column:** 250 × 4.6 Ultrasphere ODS
**Mobile phase:** Gradient. MeCN:water:trifluoroacetic acid from 0:100:0.1 to 100:0:0.1 over 50 min
**Flow rate:** 1
**Injection volume:** 1000
**Detector:** RIA (of fractions)

## CHROMATOGRAM

**Retention time:** 16

## KEY WORDS

rat

## REFERENCE

Sheward, W.J.; Harmar, A.J.; Fraser, H.M.; Fink, G. Thyrotropin-releasing hormone in rat pituitary stalk blood and hypothalamus: studies with high performance liquid chromatography, *Endocrinology*, **1983**, *113*, 1865–1869.

# Prulifloxacin

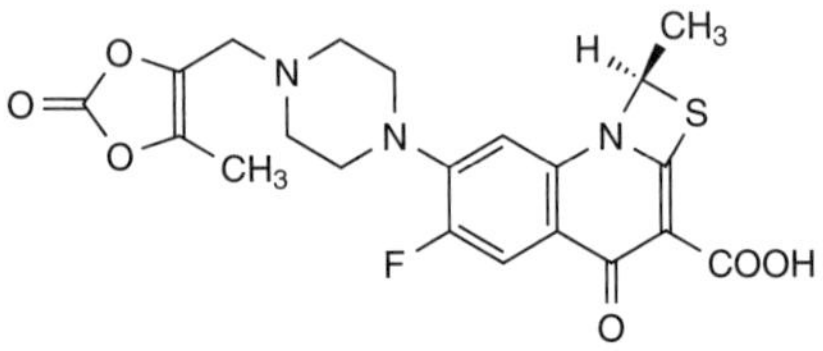

**Molecular formula:** $C_{21}H_{20}FN_3O_6S$
**Molecular weight:** 461.46
**CAS Registry No:** 123447-62-1

## SAMPLE

**Matrix:** blood, urine

**Sample preparation:** Inject urine directly. Condition a 200 mg Sep-Pak Vac PS-1 SPE cartridge with 3 mL MeOH and 3 mL 1% acetic acid. Mix 500 μL plasma with 1 mL 1% acetic acid, add to the SPE cartridge, wash with 1 mL 1% acetic acid, elute with 4 mL MeCN:1% acetic acid 90:10. Evaporate the eluate to dryness under reduced pressure, reconstitute the residue with MeOH, inject an aliquot.

## HPLC VARIABLES

**Column:** 250 × 4.6 Capcell Pak C18 SG120 (Shiseido)
**Column temperature:** 40
**Mobile phase:** MeCN:MeOH:50 mM pH 2 sodium phosphate buffer 10:20:60
**Flow rate:** 1.0–1.7
**Detector:** UV 275; Radioactivity ($^{14}C$)

## CHROMATOGRAM

**Retention time:** 15

## OTHER SUBSTANCES

**Extracted:** metabolites

## KEY WORDS

dog; monkey; plasma; rat

## REFERENCE

Okuyama, Y.; Morino, A. Pharmacokinetics of prulifloxacin. 3rd communication: metabolism in rats, dogs and monkeys, *Arzneimittelforschung*, **1997**, *47*, 293–298.

# Pyrethrins

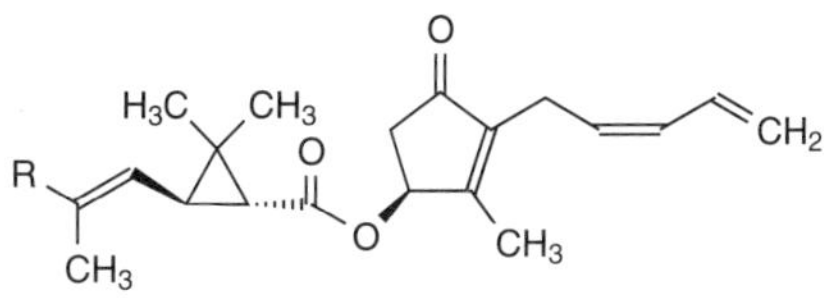

R = $CH_3$ (I) or $COOCH_3$ (II)

**Molecular formula:** $C_{21}H_{28}O_3$ (pyrethrin I), $C_{22}H_{28}O_5$ (pyrethrin II)

**Molecular weight:** 328.44 (pyrethrin I), 372.45 (pyrethrin II)

**CAS Registry No:** 121-21-1 (pyrethrin I), 121-29-9 (pyrethrin II)

**Merck Index:** 13, 8054

## SAMPLE

**Matrix:** formulations

**Sample preparation:** Dissolve 5 g formulation in 100 mL isopropanol that contains 5 mL IS, dilute with isopropanol, filter (0.45 μm PTFE), inject a 10 μL aliquot.

## HPLC VARIABLES

**Column:** 200 × 4.6 3 μm Pinnacle C8 (Restek)

**Column temperature:** 50

**Mobile phase:** Gradient. MeCN:MeOH:water 40:15:45 for 10 min, to 45:15:40 over 5 min, maintain at 45:15:40 for 10 min, to 50:15:35 over 10 min, to 55:15:30 over 10 min. (For MS, all components contained 0.1% trifluoroacetic acid.)

**Flow rate:** 0.8

**Injection volume:** 10

**Detector:** UV 240; MS Finnigan TSQ 7000, electrospray, triple-stage quadrupole, positive ion mode, capillary 5.0 kV, electrospray vaporizer 400°, capillary 210°, collision gas argon 1.5 mtorr

## CHROMATOGRAM

**Retention time:** 21 (pyrethrin II), 37 (pyrethrin I)

**Internal standard:** 2,2-dimethylpropiophenone (9)

**Limit of detection:** 160 ng/mL (pyrethrin I, UV), 60 ng/mL (pyrethrin II, UV)

**Limit of quantitation:** 430 ng/mL (pyrethrin I, UV), 120 ng/mL (pyrethrin II, UV)

## OTHER SUBSTANCES

**Simultaneous:** cinerin I (35), cinerin II (20), jasmolin I (42), jasmolin II (26), (*S*)-methoprene (45), piperonyl butoxide (24)

## KEY WORDS

validation details

## REFERENCE

Wang, I.-H.; Subramanian, V.; Moorman, R.; Ko, J.; Johnson, D. A validated reversed-phase HPLC method for analyzing pyrethrins, piperonyl butoxide, and (*S*)-methoprene in pesticide formulations, *LC.GC*, **1999**, *17*, 260–275.

## SAMPLE

**Matrix:** solutions

## HPLC VARIABLES

**Column:** 250 × 4 5 μm Polygosil C18 (MetaChem)

**Mobile phase:** MeOH:isopropanol:EtOH 35:15:50

**Flow rate:** 0.5

**Detector:** UV 225

## CHROMATOGRAM
**Retention time:** 15 (pyrethrin II), 31 (pyrethrin I)
MetaChem Catalog; **1995**; page 46.

## SAMPLE
**Matrix:** solutions

## HPLC VARIABLES
**Column:** 150 × 8 Nucleosil 5 NO2
**Mobile phase:** *n*-Hexane:THF 96:4
**Flow rate:** 2
**Detector:** UV 220

## CHROMATOGRAM
**Retention time:** 18 (pyrethrin I)

## OTHER SUBSTANCES
**Simultaneous:** bioallethrin (15, 16), cinerin (14), jasmolin (12.5)

## KEY WORDS
normal phase

## REFERENCE
Ando, T.; Kurotsu, Y.; Uchiyama, M. High performance liquid chromatographic separation of the stereoisomers of natural pyrethrins and related compounds, *Agric.Biol.Chem.*, **1986**, *50*, 491–493.

## SAMPLE
**Matrix:** solutions

## HPLC VARIABLES
**Column:** two 300 × 4 10 μm μPorasil columns in series
**Mobile phase:** Dichloromethane:water-saturated dichloromethane 50:50
**Flow rate:** 0.6
**Injection volume:** 10
**Detector:** UV 254; Refractive Index

## CHROMATOGRAM
**Retention time:** 18–57 (numerous isomers)

## KEY WORDS
normal phase

## REFERENCE
Otieno, D.A.; Jondiko, I.J.; McDowell, P.G.; Kezdy, F.J. Quantitative.analysis of the pyrethrins by HPLC, *J.Chromatogr.Sci.*, **1982**, *20*, 566–570.

## SAMPLE
**Matrix:** urine
**Sample preparation:** Add 4 g solid NaCl, 3.5 mL MeCN, and 5 mL saturated NaCl solution to 5 mL MeCN, shake for 1 min. Remove the MeCN layer and extract the aqueous layer with 1 mL MeCN. Combine the MeCN layers and adjust to a known volume (0.5–1 mL), mix, filter (0.45 μm), inject a 40 μL aliquot.

## HPLC VARIABLES

**Column:** 150 × 3 3 μm Luna C18(2) (Phenomenex)
**Column temperature:** 30
**Mobile phase:** Gradient. MeCN:water 10:90 for 1 min, to 90:10 over 30 min, maintain at 90:10 for 4 min, to 100:0 over 1 min, maintain at 100:0 for 10 min, return to initial conditions over 1 min.
**Flow rate:** 0.5
**Injection volume:** 40
**Detector:** UV 235

## CHROMATOGRAM

**Retention time:** 29.6 (pyrethrin I), 33.7 (pyrethrin II)
**Limit of detection:** 4 ng/mL (pyrethrin I), 40 ng/mL (pyrethrin II)

## OTHER SUBSTANCES

**Extracted:** allethrin (31.8, LOD 5 ng/mL), bifenthrin (37, LOD 5 ng/mL), cyfluthrin (34.3, LOD 5 ng/mL), fenvalerate (35.3, LOD 2 ng/mL), *cis*-permethrin (35.7, LO 5 ng/mL), *trans*-permethrin (36.3, LOD 5 ng/mL), phenothrin (36.4, LOD 5 ng/mL), *m*-phenoxybenzyl alcohol (21, LOD 5 ng/mL), resmethrin (35.2, LOD 5 ng/mL), tetramethrin (31.4, LOD 5 ng/mL)

## REFERENCE

Loper, B.L.; Anderson, K.A. Determination of pyrethrin and pyrethroid pesticides in urine and water matrices by liquid chromatography with diode array detection, *J.AOAC Int.*, **2003**, *86*, 1236–1240.

## ANNOTATED BIBLIOGRAPHY

Essig, K.; Zhao, Z. Method development and validation of a high-performance liquid chromatographic method for pyrethrum extract, *J.Chromatogr.Sci.*, **2001**, *39*, 473–480. [pyrethrins; cinerin; jasmolin; normal phase]

Wang, I.-H.; Moorman, R.; Burleson, J. Simultaneous determination of dipropyl pyridine-2,5-dicarboxylate, N-octyl bicycloheptene dicarboximide, piperonyl butoxide, and pyrethrins in pet shampoo by reversed phase high-performance liquid chromatography, *J.Liq.Chromatogr.Rel.Technol.*, **1996**, *19*, 3293–3304.

# Quetiapine

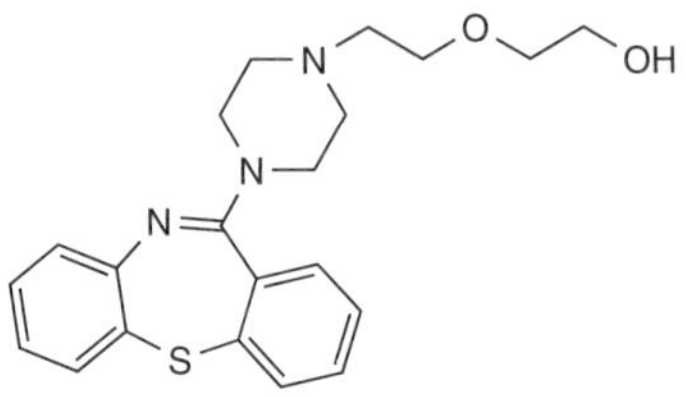

**Molecular formula:** $C_{21}H_{25}N_3O_2S$
**Molecular weight:** 383.52
**CAS Registry No:** 111974-69-7, 111974-72-2 (hemifumarate)
**Merck Index:** 13, 8127

## SAMPLE
**Matrix:** blood
**Sample preparation:** Mix 400 µL plasma and 10 µL 2.5 µg/mL IS in MeCN:MeOH 50:50, add 25 µL 15% ammonium hydroxide, add 3 mL ethyl acetate, vortex for 30 s, centrifuge for 5 min. Add the upper organic layer to 1 mL 200 mM HCl, vortex for 30 s, centrifuge for 5 min, discard the organic layer. Basify the aqueous layer with 500 µL 15% ammonium hydroxide, add 3 mL ethyl acetate, vortex for 30 s, centrifuge for 5 min. Evaporate the organic layer to dryness under a stream of nitrogen at 35°, reconstitute the residue with 30 µL MeCN:MeOH:20 mM pH 7.4 phosphate buffer containing 30 mM KCl 10:30:60, inject a 25 µL aliquot.

## HPLC VARIABLES
**Column:** 150 × 2.1 5 µm Zorbax Stablebond phenyl SB-Ph
**Mobile phase:** MeCN:MeOH:20 mM pH 7.4 phosphate buffer containing 30 mM KCl 10:50:40
**Flow rate:** 0.25
**Injection volume:** 25
**Detector:** UV 225; E, Bioanalytical Systems LC-44, precell 0.25 V, analytical cell 0.55 V, Ag/AgCl reference cell

## CHROMATOGRAM
**Retention time:** 9 (UV 225)
**Internal standard:** M 214,652 (2-(2-[4-(8-chlorodibenzo[b,f][1,4]thiazepin-11-yl)piperazin-1-yl]ethoxy)ethanol) (UV 225) (16)
**Limit of quantitation:** 2.5 ng/mL

## OTHER SUBSTANCES
**Extracted:** metabolite ICI 214,227 (7-hydroxylated) (E) (5), metabolite M 236,303 (7-hydroxylated, *N*-dealkylated) (E) (14)

## KEY WORDS
pharmacokinetics; plasma

## REFERENCE
Davis, P.C.; Wong, J.; Gefvert, O. Analysis and pharmacokinetics of quetiapine and two metabolites in human plasma using reversed-phase HPLC with ultraviolet and electrochemical detection, *J.Pharm.Biomed.Anal.*, **1999**, *20*, 271–282.

## SAMPLE
**Matrix:** blood
**Sample preparation:** Condition a 3 mL 3M-Empore mixed-phase SPE extraction disc with 1 mL MeOH and 1 mL water. Mix 900 µL serum with 100 µL 3 µg/mL IS in serum and 2 mL 100 mM pH 6.0 potassium dihydrogen phosphate buffer, add to the SPE disc, wash with 1 mL water, wash with 1 mL 1 M acetic acid, wash with 1 mL *n*-hexane, wash with 2 mL *n*-hexane:ethyl acetate 50:50, wash with 1 mL MeOH, elute with 1 mL

isopropanol:25% ammonia:dichloromethane 20:2:78. Evaporate the eluate to dryness, reconstitute the residue with 250 μL MeCN:water 30:70, inject a 100 μL aliquot.

## HPLC VARIABLES

**Column:** 250 × 4.6 5 μm Nucleosil 100-5-Protect 1 (endcapped)
**Column temperature:** 25
**Mobile phase:** MeCN:25 mM pH 7.0 potassium dihydrogen phosphate 40:60
**Flow rate:** 1
**Injection volume:** 100
**Detector:** UV 230

## CHROMATOGRAM

**Retention time:** 11.7
**Internal standard:** melperone (8.8)
**Limit of quantitation:** 5 ng/mL

## OTHER SUBSTANCES

**Extracted:** amisulpride (6.1), amitriptyline (23.4), *m*-chlorophenylpiperazine (8.0), citalopram (13.3), clomipramine (30.8), clozapine (30.9), desipramine (12.8), *O*-desmethylvenlafaxine (4.8), diazepam (11.0), doxepin (18.3), fluoxetine (17.8), 9-hydroxyrisperidone (6.6), imipramine (20.6), maprotiline (15.3), melperone (8.8), mianserin (29.0), mirtazapine (16.6), moclobemide (5.6), nefazodone (32.5), norclomipramine (19.2), norclozapine (14.4), nordoxepin (10.9), norfluoxetine (13.4), nortriptyline (14.5), paroxetine (15.3), quetiapine (11.7), reboxetine (10.2), risperidone (11.1), sertraline (33.6), sulpiride (4.1), trimipramine (21.5), venlafaxine (7.3), ziprasidone (26.4)
**Simultaneous:** benperidol (11.5), chlorprothixene (36.4), fluphenazine (31.0), haloperidol (15.3), normirtazapine (8.3), olanzapine (21.0), pimozide (44.1), promethazine (28.1), thioridazine (43.2), trifluperidol (20.8), zolpidem (10.2)
**Noninterfering:** biperiden, buspirone, carbamazepine, lorazepam, perazine, valproic acid, zopiclone, zotepine
**Interfering:** dibenzepin (11.5), fluvoxamine (11.6), normaprotiline (11.5), opipramol (11.6)

## KEY WORDS

serum; SPE

## REFERENCE

Frahnert, C.; Rao, M.L.; Grasmäder, K. Analysis of eighteen antidepressants, four atypical antipsychotics and active metabolites in serum by liquid chromatography: a simple tool for therapeutic drug monitoring, *J.Chromatogr.B*, **2003**, *794*, 35–47.

## ANNOTATED BIBLIOGRAPHY

Hasselstrom, J.; Linnet, K. Fully automated on-line quantification of quetiapine in human serum by solid phase extraction and liquid chromatography, *J.Chromatogr.B*, **2003**, *798*, 9–16. [LOD 10.3 nM; trifluoperazine is internal standard]

Mandrioli, R.; Fanali, S.; Ferranti, A.; Raggi, M.A. HPLC analysis of the novel antipsychotic drug quetiapine in human plasma, *J.Pharm.Biomed.Anal.*, **2002**, *30*, 969–977. [triprolidine is internal standard; SPE; LOQ 4 ng/mL]

Thieme, D.; Sachs, H. Improved screening capabilities in forensic toxicology by application of liquid chromatography-tandem mass spectrometry, *Anal.Chim.Acta*, **2003**, *492*, 171–186. [hair; LC-MS; alprazolam; dothiepin; piritramide; cocaine; lorazepam; lormetazepam; clonazepam; flunitrazepam; bromazepam; midazolam; flurazepam; nitrazepam; temazepam; medazepam; nordazepam; diazepam; methylclonazepam; triazolam; oxazepam; haloperidol; benperidol; sulpiride; amisulpride; mirtazapine; citalopram; olanzapine; paroxetine; fluoxetine; sertraline; zopiclone; zolpidem; risperidone; quetiapine; fentanyl; pipamperone; meperidine; buprenorphine; propoxyphene; pentazocine; phenazocine; EDDP;

tilidine; methadone; morphine; codeine; dihydrocodeine; acetylmorphine; amphetamine; ephedrine; norephedrine; pseudoephedrine; methylephedrine; amphetaminil; benzphetamine; methylphenidate; nikethamide; amfeparone; clobenzorex; atropine; scopolamine; ajmaline; aconitine; colchicine; strychnine; metoprolol; acebutolol; propranolol; sotalol; atenolol; bisoprolol; amiloride; triamterene; warfarin; brodifacoum; coumatetralyl; phenprocoumon; methaqualone; clomethiazole; acetaminophen; methoxamine; vecuronium; neostigmine; LSD]

# Quinfamide

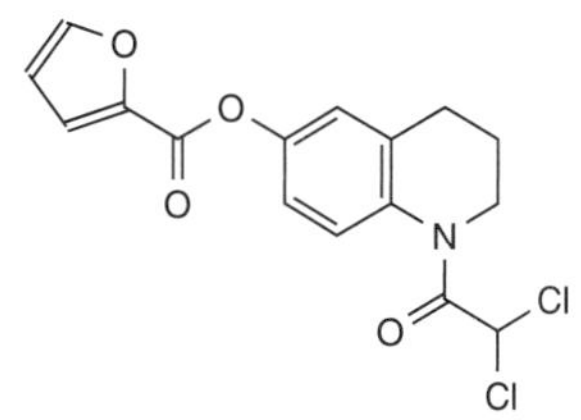

**Molecular formula:** $C_{16}H_{13}Cl_2NO_4$
**Molecular weight:** 354.19
**CAS Registry No:** 62265-68-3
**Merck Index:** 13, 8147

## SAMPLE

**Matrix:** blood, feces, urine
**Sample preparation:** Condition a Sep-Pak C18 SPE cartridge with 2 mL MeOH, 1 mL 5 μg/mL diazepam in MeOH (sic), and 2 mL water. Dissolve 10 mg lyophilized feces in 10 mL MeCN. Add 1 mL plasma, urine, or feces solution to the SPE cartridge, wash with 12 mL water, elute with 1 mL MeCN, vortex the eluate, inject a 20 μL aliquot.

## HPLC VARIABLES

**Column:** 250 × 4.6 5 μm Zorbax SB-CN
**Mobile phase:** MeCN:MeOH:water 10:50:40
**Flow rate:** 0.7
**Injection volume:** 20
**Detector:** UV 269

## CHROMATOGRAM

**Retention time:** 9.7
**Internal standard:** diazepam (8.5)
**Limit of quantitation:** 80 ng/mL

## OTHER SUBSTANCES

**Extracted:** 1-(dichloroacetyl)-1,2,3,4-tetrahydro-6-quinolol (metabolite) (6.5)

## KEY WORDS

plasma; SPE

## REFERENCE

Morales, J.M.; Jung, C.H.; Alarcón, A.; Barreda, A. Solid-phase extraction and liquid chromatographic quantitation of quinfamide in biological samples, *J.Chromatogr.B*, **2000**, *746*, 133–139.

# Quinupristin

**Molecular formula:** $C_{53}H_{67}N_9O_{10}S$
**Molecular weight:** 1022.24
**CAS Registry No:** 120138-50-3
**Merck Index:** 13, 8176

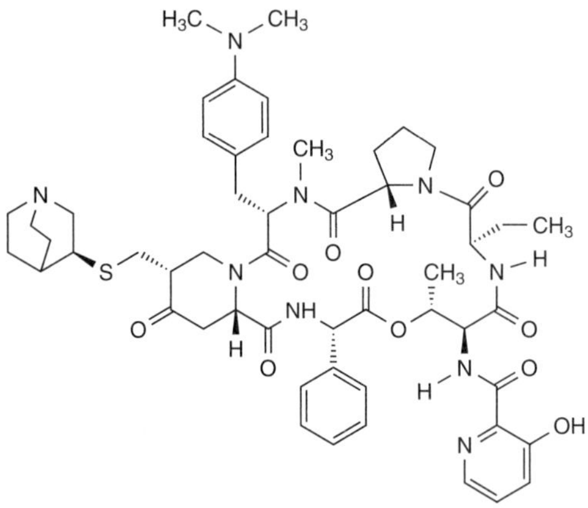

## SAMPLE

**Matrix:** blood
**Sample preparation:** Condition a CN SPE cartridge (Lida-Interchim) with 1 mL MeOH, 1 mL water, and 1 mL buffer. Add 1 mL 3.8% sodium citrate and 2.5 mL 250 mM HCl to 10 mL whole blood. Shake gently by hand and centrifuge at 2000 g at 4° for 15 min. Add 1 mL buffer and 50 μL 100 μg/mL IS in MeOH to 1.35 mL acidified plasma. Vortex for a few seconds, centrifuge at 4000 g at 4° for 5 min. Add either supernatant to the SPE cartridge and dry the SPE cartridge with 3 mL air. Elute with 500 μL MeOH:water 70:30 containing 3.5 mM pentane sulfonic acid, inject an aliquot. (The buffer was a mixture of 85 mM pH 3.0 citric acid monohydrate containing 81 mM NaOH and 60 mM HCl.)

## HPLC VARIABLES

**Guard column:** 10 μm μBondapak C18
**Column:** 125 × 4.6 5 μm Kromasil C18 (Higgins Analytical)
**Mobile phase:** Gradient. MeCN:buffer 30:70 for 11 min, then 32:68 for 4 min (step gradient), to 40:60 over 0.6 min, maintain at 40:60 for 0.4 min, at 38:62 for 18 min (step gradient), at 80:20 for 2 min (step gradient), re-equilibrate at 30:70 for 9 min. (Prepare buffer by adding 800 μL 70% perchloric acid to 1 L water.)
**Flow rate:** 0.5 for 11 min, 1 for 25 min, 0.5 for 9 min
**Injection volume:** 500
**Detector:** UV 235

## CHROMATOGRAM

**Retention time:** 22.1
**Internal standard:** dimethylamino-3-propyl thiomethylene-5 virginiamycin S (31.0)
**Limit of quantitation:** 25 pg/mL

## OTHER SUBSTANCES

**Extracted:** metabolites, dalfopristin (12.7), pristinamycin II A (24.1)

## KEY WORDS

plasma; SPE; whole blood

## REFERENCE

Le Liboux, A.; Pasquier, O.; Montay, G. Simultaneous high-performance liquid chromatographic determination of quinupristin, dalfopristin and their main metabolites in human plasma, *J.Chromatogr.B*, **1998**, *708*, 161–168.

## SAMPLE
**Matrix:** formulations
**Sample preparation:** Inject a 5–10 μL aliquot of the infusion solution.

## HPLC VARIABLES
**Column:** 125 × 4 5 μm LiChrospher-100 RP18
**Column Temperature:** 40 ± 1
**Mobile phase:** Gradient. A:B from 0:100 to 66:34 over 42.5 min, return to initial conditions over 1.5 min, re-equilibrate for 5 min. A was MeCN:buffer 65:35. B was MeCN:buffer 20:80. The buffer was 30 mM potassium dihydrogen phosphate adjusted to pH 2.9 with phosphoric acid.
**Flow rate:** 1.1
**Injection volume:** 5–10
**Detector:** UV 254

## CHROMATOGRAM
**Retention time:** 23.9, 23.1, 27.0
**Limit of quantitation:** 0.12%

## OTHER SUBSTANCES
**Simultaneous:** impurities, dalfopristin (LOQ 0.05%) (8.5)

## KEY WORDS
infusion; injection; stability-indicating

## REFERENCE
Vasselle, B.; Gousset, G.; Bounine, J.-P. Development and validation of a high-performance liquid chromatographic stability-indicating method for the analysis of Synercid in quality control, stability and compatibility studies, *J.Pharm.Biomed.Anal.*, **1999**, *19*, 641–657.

## ANNOTATED BIBLIOGRAPHY
Abdel-Hamid, M.E.; Phillips, O.A. LC-MS/MS determination of Synercid injections, *J.Pharm.Biomed. Anal.*, **2003**, *32*, 1167–1174. [pristinamycin; quinupristin; dalfopristin]

# Rabeprazole

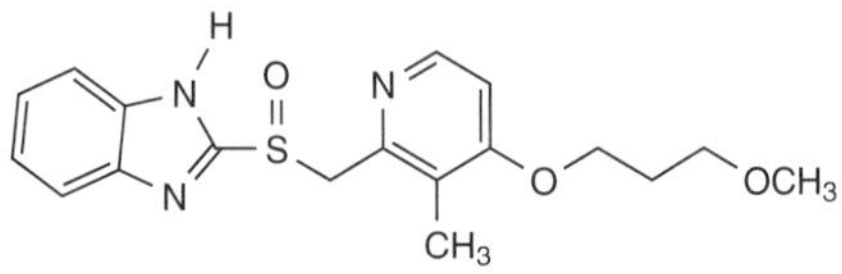

**Molecular formula:** $C_{18}H_{21}N_3O_3S$
**Molecular weight:** 359.45
**CAS Registry No:** 117976-89-3, 117976-90-6 (Na salt)
**Merck Index:** 13, 8181

## SAMPLE

**Matrix:** blood
**Sample preparation:** Mix 1 mL plasma with 100 μL 1% diethylamine in water, add 100 μL 0.1% diethylamine in MeOH, add 1 mL pH 10.38 Britton–Robinson buffer, add 4 mL ethyl acetate, shake for 10 min, centrifuge at 1500 g for 5 min, repeat extraction twice. Combine the organic layers and evaporate to dryness under a stream of nitrogen at 40°, reconstitute the residue with 100 μL 0.1% diethylamine in MeOH, inject a 30 μL aliquot.

## HPLC VARIABLES

**Column:** 150 × 4.6 5 μm Inertsil C8
**Column temperature:** 40
**Mobile phase:** MeCN:100 mM pH 7.00 phosphate buffer 28:72, adjusted to pH 7.00 with phosphoric acid
**Flow rate:** 1.4
**Injection volume:** 30
**Detector:** UV 288

## CHROMATOGRAM

**Retention time:** 6.5
**Internal standard:** IS735 (5-methyl-2-[{4-(3-methoxypropoxy)-3-methylpyridin-2-yl}methylsulfinyl]-1H-benzimidazole sodium salt) (10.1)
**Limit of quantitation:** 5 ng/mL

## OTHER SUBSTANCES

**Extracted:** metabolites

## KEY WORDS

pharmacokinetics; plasma

## REFERENCE

Nakai, H.; Shimamura, Y.; Kanazawa, T.; Yasuda, S.; Kayano, M. Determination of a new $H^+$-$K^+$ ATPase inhibitor (E3810) and its four metabolites in human plasma by high-performance liquid chromatography, *J.Chromatogr.B*, **1994**, *660*, 211–220.

## SAMPLE

**Matrix:** formulations
**Sample preparation:** Weigh out powdered tablet containing 15 mg rabeprazole, dissolve in 100 mL water, filter, inject a 20 μL aliquot.

## HPLC VARIABLES

**Column:** 250 × 4.6 4.6 μm VP-ODS Shim-pack
**Mobile phase:** MeOH:water 70:30
**Flow rate:** 2
**Injection volume:** 20
**Detector:** UV 284

## CHROMATOGRAM

**Retention time:** 5.2
**Limit of detection:** 25 ng/mL
**Limit of quantitation:** 76 ng/mL

## OTHER SUBSTANCES

**Simultaneous:** degradants

## KEY WORDS

comparison with HPTLC; validation data

## REFERENCE

el-Gindy, A.; El-Yazby, F.; Maher, M.M. Spectrophotometric and chromatographic determination of rabeprazole in presence of its degradation products, *J.Pharm.Biomed.Anal.*, **2003**, *31*, 229–242.

## ANNOTATED BIBLIOGRAPHY

Mano, N.; Oda, Y.; Takakuwa, S.; Chiku, S.; Nakata, H.; Asakawa, N. Plasma direct injection high-performance liquid chromatographic method for simultaneously determining E3810 enantiomers and their metabolites by using flavoprotein-conjugated column, *J.Pharm.Sci.*, **1996**, *85*, 903–907. [chiral; dog; column-switching]

Takakuwa, S.; Chiku, S.; Nakata, H.; Yuzuriha, T.; Mano, N.; Asakawa, N. Enantioselective high-performance liquid chromatographic assay for determination of the enantiomers of a new anti-ulcer agent, E3810, in Beagle dog plasma and rat plasma, *J.Chromatogr.B*, **1995**, *673*, 113–122. [chiral; LOQ 30 ng/mL]

# Ractopamine

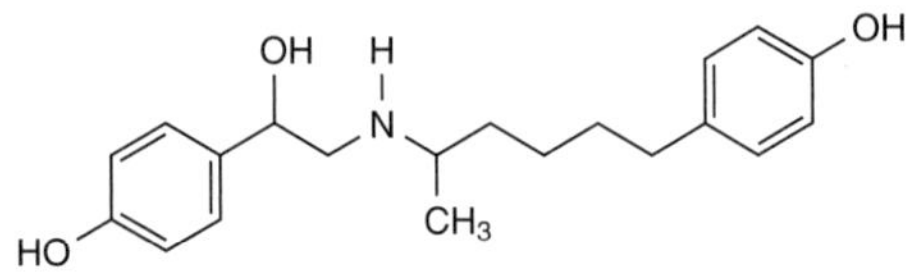


**Molecular formula:** $C_{18}H_{23}NO_3$
**Molecular weight:** 301.38
**CAS Registry No:** 97825-25-7, 90274-24-1 (HCl)
**Merck Index:** 13, 8184

## SAMPLE

**Matrix:** tissue, urine

**Sample preparation:** Condition a ChromP SPE cartridge (Supelco) with 6 mL ethyl acetate, 6 mL MeOH, and 6 mL water. Condition a Screen DAU SPE cartridge (Supelco) with 2 mL MeOH, 2 mL water, and 2 mL 100 mM pH 6 phosphate buffer. Freeze-dry and grind 15 g tissue or 200–500 mg retina, add 12 mL MeOH, add 15 mL 2 M pH 5.2 acetate buffer, stir for 30 min, centrifuge at 2000 g for 15 min. Remove the supernatant and evaporate the MeOH. Mix 10 mL urine with 2 mL 2 M pH 5.2 acetate buffer. Add the tissue or urine preparations to 400 μL Helix pomatia preparation containing 25 U/μL (Sigma), heat at 60° for 15 h, centrifuge, add the supernatant to the ChromP SPE cartridge, wash with 5 mL hexane, wash with 12 mL hexane:diethyl ether 70:30, elute with 24 mL diethyl ether. Evaporate the eluate to dryness, reconstitute with 6 mL 100 mM pH 6 phosphate buffer, add to the DAU SPE cartridge, wash with 1 mL 1 M acetic acid, wash with 6 mL MeOH, elute with 6 mL ethyl acetate:32% ammonium hydroxide 97:3. Evaporate the eluate to dryness, reconstitute the residue with 50 μL MeOH:water:acetic acid 5:95:0.5, inject a 10 μL aliquot.

## HPLC VARIABLES

**Guard column:** 10 × 2 5 μm Nucleosil C18AB
**Column:** 50 × 2 5 μm Nucleosil C18AB
**Mobile phase:** Gradient. MeOH:0.5% acetic acid from 95:5 to 50:50 over 10 min, to 10:90 over 10 min, maintain at 10:90 for 10 min (?), re-equilibrate at initial conditions for 10 min. (sic).
**Flow rate:** 0.22
**Injection volume:** 10
**Detector:** MS, Micromass QuattroLC, triple quadrupole, nebulizing gas nitrogen 90 L/h, desolvation gas nitrogen 600 L/h, source 120°, desolvation 350°, capillary 3.5 kV, sampling cone 25 V, collision gas argon 0.4 μbar, collision energy 15–25 V, m/z 302–284–150

## CHROMATOGRAM

**Retention time:** 8.58
**Internal standard:** isoxsuprine (m/z 302–284–164–136–121–107–91) (10.35)
**Limit of detection:** 10 ng/kg
**Limit of quantitation:** 30 ng/kg

## KEY WORDS

kidney; liver; lung; meat; pig; retina; SPE

## REFERENCE

Antignac, J.-P.; Marchand, P.; Le Bizec, B.; Andre, F. Identification of ractopamine residues in tissue and urine samples at ultra-trace level using liquid chromatography-positive electrospray tandem mass spectrometry, *J.Chromatogr.B*, **2002**, *774*, 59–66.

## SAMPLE

**Matrix:** tissue

**Sample preparation:** Condition a Sep-Pak Alumina A SPE cartridge with ethyl acetate. Add IS to tissue, extract 10 g homogenized tissue three times with 20 mL portions of MeOH, and combine the extracts. Evaporate an 8 mL aliquot to dryness, reconstitute with 1 mL 25 mM pH 5.0 sodium acetate buffer, add 20 μL β-glucuronidase (Helix pomatia) (Sigma), heat at 65° for 2 h, mix with 2 mL 25 mM pH 10.3 sodium borate buffer, extract three times with 7 mL portions of ethyl acetate, add the combined extracts to the alumina SPE cartridge, wash with ethyl acetate, dry under vacuum, elute with 10 mL MeOH. Evaporate the eluate to dryness, reconstitute the residue with 4 mL 1 M acetic acid, add to a 6 mL 500 mg Oasis MCX SPE cartridge, elute with 4 mL 2% ammonia in MeOH. Evaporate the eluate to dryness, reconstitute the residue with 500 μL 2% acetic acid, inject a 100 μL aliquot.

---

## HPLC VARIABLES

**Column:** 250 × 4.6 5 μm Supelcosil LC-18-DB
**Mobile phase:** MeCN:water:glacial acetic acid 32:68:2 containing 0.87 g/L 1-pentanesulfonic acid
**Flow rate:** 1
**Injection volume:** 100
**Detector:** F ex 226 em 306

---

## CHROMATOGRAM

**Retention time:** 5.4
**Internal standard:** ritodrine (4.7)
**Limit of quantitation:** 1 ng/g

---

## KEY WORDS

cow; pig; muscle; SPE

---

## REFERENCE

Shishani, E.; Chai, S.C.; Jamokha, S.; Aznar, G.; Hoffman, M.K. Determination of ractopamine in animal tissues by liquid chromatography-fluorescence and liquid chromatography/tandem mass spectrometry, *Anal.Chim.Acta*, **2003**, *483*, 137–145.

---

## SAMPLE

**Matrix:** tissue
**Sample preparation:** Condition a Sep-Pak Alumina A SPE cartridge with ethyl acetate. Add IS to tissue, extract 10 g homogenized tissue three times with 20 mL portions of MeOH, and combine the extracts. Evaporate an 8 mL aliquot to dryness, reconstitute with 1 mL 25 mM pH 5.0 sodium acetate buffer, add 20 μL β-glucuronidase (Helix pomatia) (Sigma), heat at 65° for 2 h, mix with 2 mL 25 mM pH 10.3 sodium borate buffer, extract three times with 7 mL portions of ethyl acetate, add the combined extracts to the alumina SPE cartridge, wash with ethyl acetate, dry under vacuum, elute with 10 mL MeOH. Evaporate the eluate to dryness, reconstitute the residue with 4 mL 1 M acetic acid, add to a 6 mL 500 mg Oasis MCX SPE cartridge, elute with 4 mL 2% ammonia in MeOH. Evaporate the eluate to dryness, reconstitute the residue with 200 μL MeOH for MS detection, inject an aliquot.

---

## HPLC VARIABLES

**Column:** 50 × 2.1 5 μm Discovery RP-Amide C-16 (Supelco)
**Mobile phase:** Gradient. MeCN:pH 4.5 ammonium acetate buffer from 5:95 to 100:0 over 2 min.
**Flow rate:** 0.5
**Detector:** MS, PE Sciex API-III, positive mode, nebulizer 500°, interface 55°, nebulizer gas 0.6 L/min, auxiliary flow 2.8 L/min, collision gas argon, m/z 302–164, 302–107, 302–121

---

## CHROMATOGRAM

**Internal standard:** ritodrine (m/z 288–150)
**Limit of quantitation:** 1 ng/g

## KEY WORDS

cow; pig; muscle; SPE

## REFERENCE

Shishani, E.; Chai, S.C.; Jamokha, S.; Aznar, G.; Hoffman, M.K. Determination of ractopamine in animal tissues by liquid chromatography-fluorescence and liquid chromatography/tandem mass spectrometry, *Anal.Chim.Acta*, **2003**, *483*, 137–145.

## ANNOTATED BIBLIOGRAPHY

Turberg, M.P.; Macy, T.D.; Lewis, J.J.; Coleman, M.R. Determination of ractopamine hydrochloride in swine, cattle, and turkey feeds by liquid chromatography with coulometric detection, *J.AOAC Int.*, **1994**, *77*, 840–847. [SPE; LOQ 1.25 ppm]

Turberg, M.P.; Macy, T.D.; Lewis, J.J.; Coleman, M.R. Determination of ractopamine hydrochloride in swine and turkey tissues by liquid chromatography with coulometric detection, *J.AOAC Int.*, **1995**, *78*, 1394–1402. [liver; kidney; muscle; fat; SPE; LOD 0.5 ppb]

Turberg, M.P.; Rodewald, J.M.; Coleman, M.R. Determination of ractopamine in monkey plasma and swine serum by high-performance liquid chromatography with electrochemical detection, *J.Chromatogr.B*, **1996**, *675*, 279–285. [SPE; LOQ 2 ng/mL]

# Raloxifene

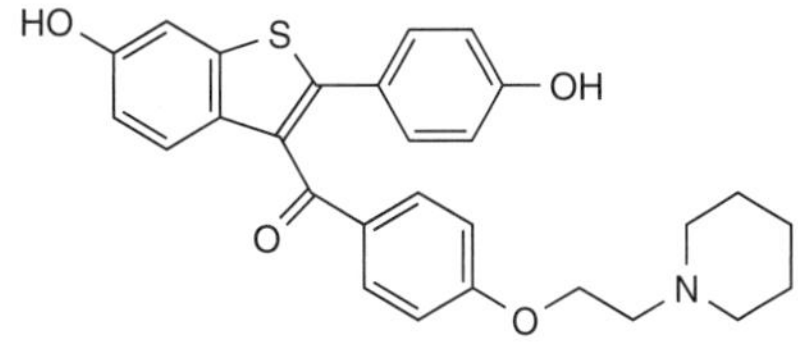


**Molecular formula:** $C_{28}H_{27}NO_4S$
**Molecular weight:** 473.59
**CAS Registry No:** 84449-90-1, 82640-04-8 (HCl)
**Merck Index:** 13, 8190

## SAMPLE
**Matrix:** blood
**Sample preparation:** Mix 100 ng IS with 300 μL plasma, add 600 μL acetone, mix, let stand on ice for 10 min, centrifuge. Evaporate the supernatant to dryness under reduced pressure, reconstitute the residue with 100 μL 500 mM pH 9.0 sodium carbonate and 100 μL dichloromethane, mix thoroughly, centrifuge, inject a 40 μL aliquot of the organic layer.

## HPLC VARIABLES
**Column:** 10 μm Zorbax C8 or LiChrosorb C8
**Mobile phase:** MeCN:100 mM pH 4.0 sodium acetate 50:50
**Flow rate:** 2
**Injection volume:** 40
**Detector:** E, BAS LC-3, glassy carbon electrode +0.65 V

## CHROMATOGRAM
**Retention time:** 5.8
**Internal standard:** (4-(2-(dimethylamino)ethoxy)phenyl)(6-hydroxy-2-(4-hydroxyphenyl)benzo[b]thien-3-yl)methanone (3.9)

## KEY WORDS
dog; monkey; plasma; rat

## REFERENCE
Lindstrom, T.D.; Whitaker, N.G.; Whitaker, G.W. Disposition and metabolism of a new benzothiophene antiestrogen in rats, dogs and monkeys, *Xenobiotica*, **1984**, *14*, 841–847.

## SAMPLE
**Matrix:** solutions
**Sample preparation:** Inject a 1 μL aliquot of a solution in MeOH:water 10:90.

## HPLC VARIABLES
**Column:** 20 × 2 5 μm DASH BetaBasic C8 (ThermoHypersil Keystone)
**Mobile phase:** Gradient. A was MeCN:water:formic acid 5:95:0.1. B was MeCN:water:formic acid 95:5:0.1. A:B 100:0 for 0.2 min, to 0:100 over 1.5 min.
**Flow rate:** 1.5
**Injection volume:** 1
**Detector:** MS, PE Sciex API-3000, TurboIonspray, electrospray 4500 V, ring 290 V, orifice 60 V, drying gas 400°, 20% of column effluent entered the detector, m/z 474.1–112

## CHROMATOGRAM
**Retention time:** 0.97

## OTHER SUBSTANCES
**Simultaneous:** amitriptyline (m/z 278.3–233) (1.1), aprepitant (MK-869) (m/z 535.3–277) (1.4), diclofenac (m/z 296.1–215) (1.35), enoxacin (m/z 321.2–234) (0.7),

fenofibrate (m/z 360.9–233) (1.6), finasteride (m/z 373.2–317) (1.2), indinavir (m/z 614.4–421) (0.93), pioglitazone (357.2–134) (0.87)

## REFERENCE

Romanyshyn, L.A.; Tiller, P.R. Ultra-short columns and ballistic gradients: considerations for ultra-fast chromatographic liquid chromatographic-tandem mass spectrometric analysis, *J.Chromatogr.A*, **2001**, *928*, 41–51.

# Ramosetron

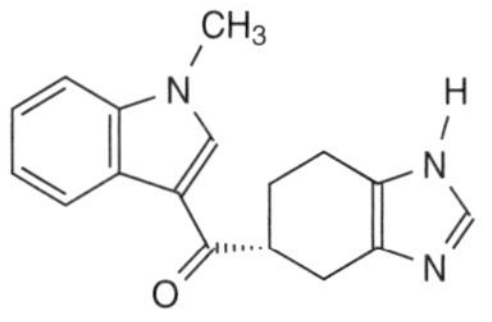

**Molecular formula:** $C_{17}H_{17}N_3O$
**Molecular weight:** 279.34
**CAS Registry No:** 132036-88-5
**Merck Index:** 13, 8195

---

## SAMPLE

**Matrix:** blood, urine
**Sample preparation:** Shake 1 (animal) or 2 (human) mL plasma or 1 mL urine with 200 μL 100 μg/mL IS in water, 1 mL saturated sodium bicarbonate, and 5 mL ethyl acetate for 15 min, centrifuge at 1200 g for 10 min. Remove the organic layer and add it to 2.5 mL 400 mM HCl, shake for 15 min, centrifuge at 800 g for 5 min. Remove the aqueous layer, add 2 mL saturated sodium bicarbonate solution to the aqueous layer, stir, let stand at room temperature for 20 min, add 4.5 mL ethyl acetate, shake for 15 min, centrifuge at 800 g for 5 min. Evaporate the organic layer to dryness under reduced pressure, reconstitute the residue with 100 μL mobile phase, inject a 60 μL aliquot.

## HPLC VARIABLES

**Column:** 250 × 4.6 TSK-gel ODS-80-Tm (Tosoh) (plasma) or 250 × 4.6 Cosmosil 5C18AR (Nacalai Tesque) (urine)
**Mobile phase:** MeCN:100 mM potassium dihydrogen phosphate:100 mM phosphoric acid 25:37.5:37.5 (plasma) or MeCN:100 mM potassium dihydrogen phosphate:100 mM phosphoric acid 33:33:33 containing 400 μM sodium dodecyl sulfate (urine)
**Flow rate:** 1
**Injection volume:** 60
**Detector:** UV 311

## CHROMATOGRAM

**Retention time:** 10
**Internal standard:** GSA 110 (9-methyl-3-[(5-methylimidazole-4-yl)methyl]-2,3,4,9-tetrahydrothiopyrano[2,3-b]indole-4-one fumarate) (14)
**Limit of detection:** 200 pg/mL (S/N 3, human plasma), 500 pg/mL (rat, dog plasma), 1 ng/mL (human urine)

## KEY WORDS

dog; human; plasma; rat

## REFERENCE

Miura, H.; Takeshige, T.; Kobayashi, S.-i.; Higuchi, S. A simple method for the determination of YM060 in plasma and urine by high performance liquid chromatography, *Biomed.Chromatogr.*, **1994**, *8*, 103–104.

---

## SAMPLE

**Matrix:** solutions

## HPLC VARIABLES

**Column:** 150 × 4 Crestpak C18S-10
**Mobile phase:** MeCN:buffer 15:85 (The buffer was 3% triethylamine adjusted to pH 2.0 with phosphoric acid.)
**Flow rate:** 2
**Detector:** Radioactivity ($^{11}C$)

## CHROMATOGRAM

**Retention time:** 7.3

## REFERENCE

Ishiwata, K.; Ishii, K.; Ishii, S.-I.; Senda, M. Synthesis of 5-HT3 receptor antagonists, [$^{11}$C]Y-25130 and [$^{11}$C]YM060, *Appl.Radiat.Isot.*, **1995**, *46*, 907–910.

# Rapacuronium bromide

**Molecular formula:** $C_{37}H_{61}BrN_2O_4$
**Molecular weight:** 677.81
**CAS Registry No:** 156137-99-4
**Merck Index:** 13, 8201

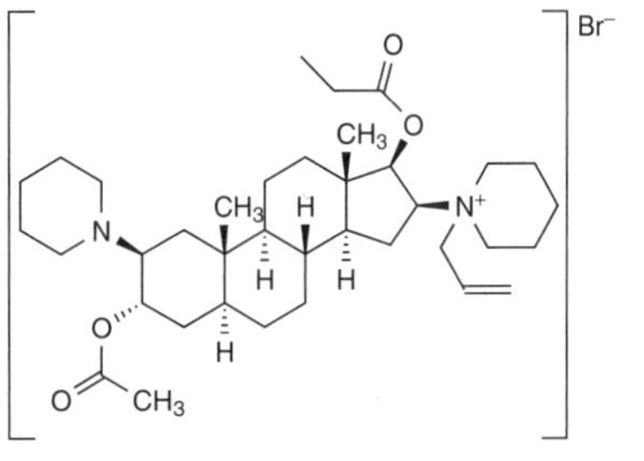

## SAMPLE

**Matrix:** bile, blood, stoma fluid, tissue, urine
**Sample preparation:** Caution! The method as described is applied to and validated for rocuronium. However, Wierda et al. (Wierda,J.M.K.H.; Beaufort,A.M.; Kleef,U.W.; Smeulers,N.J.; Agoston,S. Preliminary investigations of the clinical pharmacology of three short-acting non-depolarizing neuromuscular blocking agents, Org 9453, Org 9489 and Org 9487. *Can.J.Anaesth.* **1994**, *41*, 213–220.) state that "the compounds were determined by means of the HPLC method described for rocuronium bromide, which was modified and validated for ... Org 9487" [rapacuronium]. Homogenize (Ultra-Turrax) 1 g tissue with 9 mL 1 M sodium dihydrogen phosphate for 10 min. Acidify 1 mL plasma, urine, or bile with 200 μL 1 M sodium dihydrogen phosphate. Homogenize 1 mL stoma fluid with 200 μL 1 M sodium dihydrogen phosphate. Make up 50–1000 μL plasma, 200–1000 μL urine, 5–200 μL bile, 1000 μL stoma fluid, or 100–1000 μL tissue homogenate to 2 mL with water, add 1 mL buffer, add 150 ng IS, add 7 mL dichloromethane, vortex for 15 s, centrifuge at 740 g for 5 min. Remove the lower organic layer and evaporate it to dryness under a stream of air at 37°, reconstitute the residue in 200 μL mobile phase, inject a 100 μL aliquot (or less). (The buffer was prepared by mixing 6 mL of an aqueous solution containing 7.505 mg/mL glycine and 5.85 mg/mL NaCl, with 4 mL 100 mM NaOH and 6.2 g KI.)

## HPLC VARIABLES

**Guard column:** 4 × 6 μBondapak C18
**Column:** 150 × 3.9 5 μm Lichrospher 100-RP18
**Mobile phase:** Dioxane:buffer 16:84 (Caution! Dioxane is a carcinogen!) (The buffer was 100 mM sodium dihydrogen phosphate containing 0.11 mM 9,10-dimethoxyanthracene-2-sulfonate and 0.11 mM 1-heptanesulfonic acid, pH adjusted to 3.0 with orthophosphoric acid. After each series of analyses, flush column with 15 mL water and 75 mL MeOH.)
**Flow rate:** 1
**Injection volume:** 100
**Detector:** F ex 385 em 452 following post-column extraction. The column effluent mixed with dichloroethane pumped at 1 mL/min and the mixture flowed through a 1 m × 0.25 mm ID stainless steel coil to a phase separator (Organon International) and then the organic phase flowed through the detector (*J.Chromatogr.* **1987**, *421*, 327; *Anal.Chim.Acta* **1987**, *192*, 267).

## CHROMATOGRAM

**Retention time:** 9 (for rocuronium)
**Internal standard:** 1-(3α,17β-dihydroxy-2β-morpholino-5α-androstan-16β-yl)-1-methylpiperidinium bromide (Org 7402, Organon) (21)
**Limit of detection:** 3 ng (plasma), 4 ng (urine, bile), 5 ng (tissue)
**Limit of quantitation:** 10 ng/mL (plasma), 25 ng/mL (urine), 100 ng/mL (bile), 20 ng/mL (stoma fluid)

## OTHER SUBSTANCES

**Simultaneous:** cefotaxime, ceftazidime, ceftriaxone, cefuroxime, cefamandole, cephradine, dixyrazine, meperidine, metocurine, metoprolol, sulfamethoxazole, trimethoprim

**Interfering:** alizapride, atracurium, ketamine, ketogan, lidocaine, metoclopramide, nimodipine, prochlorperazine, tubocurarine

**Noninterfering:** alfentanil, aprotinin, atropine, bupivacaine, chlorpromazine, dalteparin, dexamethasone, diazepam, dopamine, droperidol, etomidate, fentanyl, furosemide, gallamine, haloperidol, midazolam, morphine, neostigmine, nitroglycerin, nitroprusside, oxytocin, pancuronium, pentobarbital, phenylephrine, phenytoin, pipecuronium, piperacillin, promethazine, propofol, ranitidine, succinylcholine, sufentanil, terbutaline, thiopental, vecuronium, verapamil

## KEY WORDS

human; dog; plasma; liver; lung

## REFERENCE

Kleef, U.W.; Proost, J.H.; Roggeveld, J.; Wierda, J.M.K.H. Determination of rocuronium and its putative metabolites in body fluids and tissue homogenates, *J.Chromatogr.*, **1993**, *621*, 65–76.

# Remifentanil

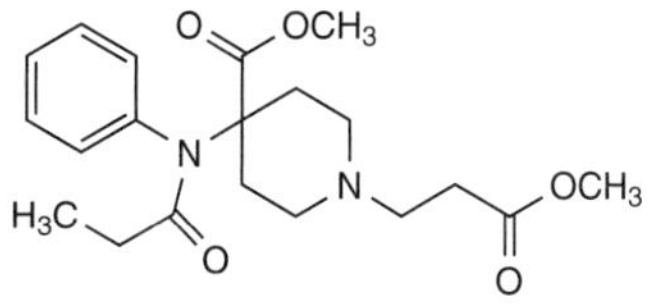

**Molecular formula:** $C_{20}H_{28}N_2O_5$
**Molecular weight:** 376.45
**CAS Registry No:** 132875-61-7
**Merck Index:** 13, 8216

## SAMPLE

**Matrix:** blood
**Sample preparation:** Mix 500 μL whole blood with 10 μL 50% citric acid and 25 μL 500 ng/mL IS in 1 mM HCl, add 500 μL 100 mM pH 7.4 phosphate buffer, vortex until homogeneous, add 2 mL dichloromethane, shake mechanically for 10 min, centrifuge at 13 000 rpm for 10 min. Evaporate the lower dichloromethane layer to dryness under a stream of nitrogen at room temperature, reconstitute the residue with 125 μL mobile phase, inject a 20 μL aliquot.

## HPLC VARIABLES

**Column:** 33 × 4.6 Pecosphere C18 (Perkin-Elmer)
**Mobile phase:** MeCN:chloroform 50:50 containing 2 mM ammonium acetate (Caution! Chloroform is a carcinogen!)
**Flow rate:** 0.3
**Injection volume:** 20
**Detector:** MS, PE Sciex API-III Plus triple quadrupole, turbo ionspray, positive ion mode, turbo 200°, orifice 60 V, auxiliary gas nitrogen at 4 L/min, curtain gas at 1.2 L/min, m/z 377–228

## CHROMATOGRAM

**Retention time:** 1.3
**Internal standard:** $d_4$-remifentanil (381–232)
**Limit of quantitation:** 0.1 ng/mL

## KEY WORDS

whole blood

## REFERENCE

Bender, J.; van den Elshout, J.; Selinger, K.; Broeders, G.; Dankers, J.; van der Heiden, C. Determination of remifentanil in human heparinised whole blood by tandem mass spectrometry with short-column separation, *J.Pharm.Biomed.Anal.*, **1999**, *21*, 559–567.

## SAMPLE

**Matrix:** blood
**Sample preparation:** Condition a Bond Elut phenyl SPE cartridge with 1 mL MeOH and 1 mL 50 mM pH 3 potassium phosphate buffer. Add 500 μL plasma and 750 μL 800 ng/mL IS in 50 mM pH 3 potassium phosphate buffer to the SPE cartridge, wash with 1 mL 50 mM pH 3 potassium phosphate buffer, wash with 1 mL MeCN, elute with 1 mL MeOH. Evaporate the eluate to dryness under reduced pressure, reconstitute the residue with 200 μL mobile phase, inject a 50–100 μL aliquot.

## HPLC VARIABLES

**Guard column:** 4 × 3 C1
**Column:** 150 × 4.6 5 μm Spherisorb C1
**Mobile phase:** MeCN:MeOH:50 mM pH 3 potassium phosphate buffer:water 18:12:4.8:65.2
**Flow rate:** 1.5

**Injection volume:** 50–100
**Detector:** UV 210

## CHROMATOGRAM

**Retention time:** 9
**Internal standard:** GI97559 (ethyl 3-{4-methoxycarbonyl-4-[(1-oxopropyl)-phenylamino]-1-piperidine}propanoate) (10.5)
**Limit of detection:** 0.5 ng/mL
**Limit of quantitation:** 8 ng/mL

## OTHER SUBSTANCES

**Extracted:** demethoxyremifentanil (metabolite) (4)

## KEY WORDS

dog; pharmacokinetics; plasma; SPE

## REFERENCE

Kabbaj, M.; Varin, F. Simultaneous solid-phase extraction combined with liquid chromatography with ultraviolet absorbance detection for the determination of remifentanil and its metabolite in dog plasma, *J.Chromatogr.B*, **2003**, *783*, 103–111.

## ANNOTATED BIBLIOGRAPHY

Haidar, S.H.; Liang, Z.; Selinger, K.; Hamlett, L.; Eddington, N.D. Determination of remifentanil, an ultra-short-acting opioid anesthetic, in rat blood by high-performance liquid chromatography with ultraviolet detection, *J.Pharm.Biomed.Anal.*, **1996**, *14*, 1727–1732. [LOQ 2.5 ng/mL]

Selinger, K.; Lanzo, C.; Sekut, A. Determination of remifentanil in human and dog blood by HPLC with UV detection, *J.Pharm.Biomed.Anal.*, **1994**, *12*, 243–248. [LOQ 1 ng/mL]

Vishwanathan, K.; Stewart, J.T. HPLC determination of a propofol and remifentanil mixture, *J.Liq.Chromatogr.Rel.Technol.*, **1999**, *22*, 923–931. [LOD 200 ng/mL]

# Repaglinide

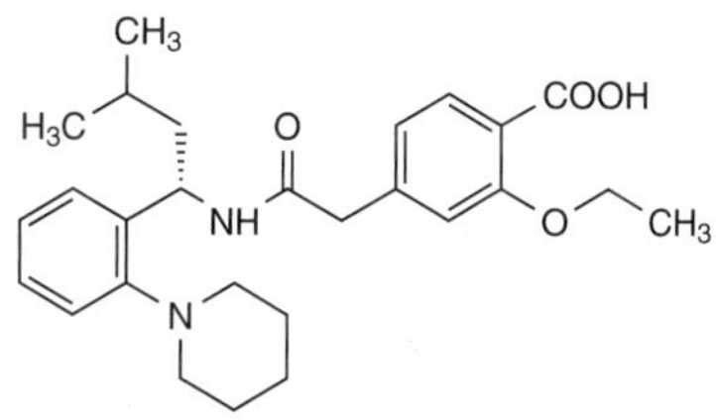

**Molecular formula:** $C_{27}H_{36}N_2O_4$
**Molecular weight:** 452.58
**CAS Registry No:** 135062-02-1
**Merck Index:** 13, 8220

## SAMPLE
**Matrix:** blood
**Sample preparation:** Centrifuge plasma at 13 000 g for 2 min and dilute with 2 vol of 200 mM HCl, inject a 160 μL aliquot onto column A and elute to waste with water; after 3 min, backflush the contents of column A onto column B with mobile phase, monitor the effluent from column B. At the end of the separation, wash column A with MeCN:MeOH:dioxane 24:68:8. (Caution! Dioxane is a carcinogen!)

## HPLC VARIABLES
**Column:** A 17 × 2.9 30–40 μm Perisorb RP-2 (Merck); B 17 × 4.6 5 μm ODS-Hypersil + 125 × 4 5 μm LiChrospher 100 RP-18
**Mobile phase:** MeCN:MeOH:dioxane:buffer 18.24:51.68:6.08:24 (Caution! Dioxane is a carcinogen!) (The buffer was 3 g potassium dihydrogen phosphate and 0.5 g lithium perchlorate in 1 L water, adjusted to pH 2.7 with orthophosphoric acid.)
**Flow rate:** 1
**Injection volume:** 160
**Detector:** E, Waters 460, 1.04 V (to prevent contamination, column effluent passes through the detector only from 7.5–13.8 min)

## CHROMATOGRAM
**Retention time:** 9
**Limit of detection:** 5 ng/mL

## KEY WORDS
column-switching; plasma

## REFERENCE
Greischel, A.; Beschke, K.; Rapp, H.; Roth, W. Quantitation of the new hypoglycaemic agent AG-EE 388 ZW in human plasma by automated high-performance liquid chromatography with electrochemical detection, *J.Chromatogr.*, **1991**, *568*, 246–252.

## SAMPLE
**Matrix:** formulations
**Sample preparation:** Dissolve powdered tablet containing 25 mg repaglinide and 25 mg nimesulide in 25 mL MeOH, shake well, filter (0.45 μm), dilute filtrate with mobile phase to a concentration of 1 μg/mL, inject a 20 μL aliquot.

## HPLC VARIABLES
**Guard column:** 20 mm long Shim-pack ODS (Shimadzu)
**Column:** 150 × 4.6 5 μm Shim-pack RP-C18
**Column temperature:** 30
**Mobile phase:** MeOH:buffer 50:50 (The buffer was 0.1% triethylamine adjusted to pH 7 with 1% orthophosphoric acid.)
**Flow rate:** 1
**Injection volume:** 20
**Detector:** UV 235

## CHROMATOGRAM

**Retention time:** 3.4
**Internal standard:** nimesulide (2.04)
**Limit of quantitation:** 100 ng/mL

## KEY WORDS

tablets

## REFERENCE

Gandhimathi, M.; Ravi, T.K.; Renu, S.K. Determination of repaglinide in pharmaceutical formulations by HPLC with UV detection, *Anal.Sci.*, **2003**, *19*, 1675–1677.

## ANNOTATED BIBLIOGRAPHY

Niemi, M.; Neuvonen, P.J.; Kivistö, K.T. The cytochrome P4503A4 inhibitor clarithromycin increases the plasma concentrations and effects of repaglinide, *Clin.Pharmacol.Ther.*, **2001**, *70*, 58–65. [LC-MS; LOQ 50 pg/mL; indomethacin is internal standard]

Reddy, K.V.S.R.K.; Babu, J.M.; Mathad, V.T.; Eswaraiah, S.; Reddy, M.S.; Dubey, P.K.; Vyas, K. Impurity profile study of repaglinide, *J.Pharm.Biomed.Anal.*, **2003**, *32*, 461–467.

Thomsen, M.S.; Chassard, D.; Evène, E.; Nielsen, K.K.; Jorgensen, M. Pharmacokinetics of repaglinide in healthy Caucasian and Japanese subjects, *J.Clin.Pharmacol.*, **2003**, *43*, 23–28. [LC-MS; LOQ 200 pg/mL; SPE]

# Ricinoleic acid

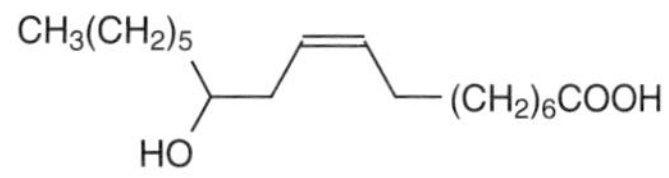


**Molecular formula:** $C_{18}H_{34}O_3$
**Molecular weight:** 289.46
**CAS Registry No:** 141-22-0
**Merck Index:** 13, 8295

## SAMPLE

**Matrix:** blood

**Sample preparation:** Vortex 20 μL plasma, 10 μL 2 mg/mL margaric acid in MeOH, and 200 μL 500 mM KOH in EtOH for 30 s, centrifuge at 2500 g for 10 min. Remove the supernatant and heat it at 100° for 30 min, cool, add 200 μL 1 M HCl, add 2 mL chloroform (Caution! Chloroform is a carcinogen!), mix vigorously for 5 min, centrifuge at 460 g for 5 min. Remove the organic layer and evaporate it to dryness under reduced pressure at 43°, reconstitute the residue in 500 μL benzene (Caution! Benzene is a carcinogen!), sonicate for 1 min, add 500 μL 2% oxalyl chloride in benzene, heat at 70° for 30 min, evaporate to dryness under reduced pressure, reconstitute with 100 μL benzene, add 100 μL 40 mM 1-naphthylamine (Caution! 1-Naphthylamine is a carcinogen!) in benzene, vortex for 30 s, heat at 37° for 30 min, evaporate to dryness, reconstitute with MeOH:MeCN:water 72:13:15, vortex for 20 s, inject a 20 μL aliquot. (This method was developed for Cremophor EL, which is saponified to ricinoleic acid and then derivatized with 1-naphthylamine. However, the method should work for ricinoleic acid in plasma.)

## HPLC VARIABLES

**Column:** two 100 × 3 5 μm Spherisorb ODS-I glass columns in series
**Mobile phase:** MeCN:MeOH:10 mM pH 7.0 potassium phosphate buffer 13:72:15
**Flow rate:** 0.4
**Injection volume:** 20
**Detector:** UV 280

## CHROMATOGRAM

**Retention time:** 10.0
**Internal standard:** margaric acid (27.8)
**Limit of detection:** 0.005%
**Limit of quantitation:** 0.01%

## OTHER SUBSTANCES

**Simultaneous:** arachidonic acid, linoleic acid, linolenic acid, myristic acid, oleic acid, palmitic acid, palmitoleic acid, stearic acid

## KEY WORDS

derivatization; human; mouse; plasma

## REFERENCE

Sparreboom, A.; van Tellingen, O.; Huizing, M.T.; Nooijen, W.J.; Beijnen, J.H. Determination of polyoxyethyleneglycerol triricinolate 35 (Cremophor EL) in plasma by pre-column derivatization and reversed-phase high-performance liquid chromatography, *J.Chromatogr.B*, **1996**, *681*, 355–362.

# Rifaximin

**Molecular formula:** $C_{43}H_{51}N_3O_{11}$
**Molecular weight:** 785.88
**CAS Registry No:** 80621-81-4
**Merck Index:** 13, 8304

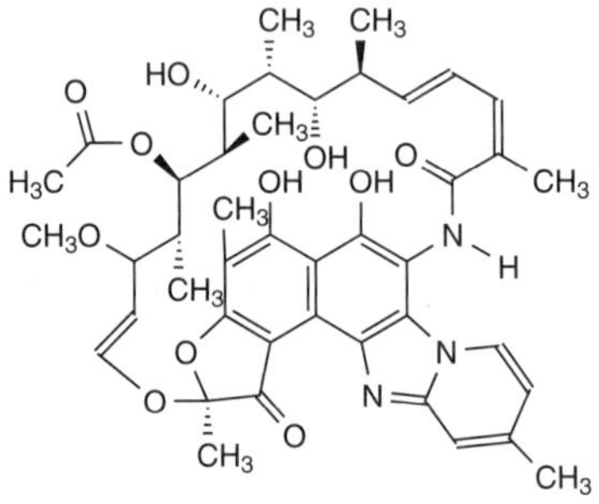

## SAMPLE

**Matrix:** blood, urine

**Sample preparation:** Mix 1 mL plasma or 2 mL urine with 100 μL 1 μg/mL IS solution, add 6 mL ethyl acetate, shake horizontally for 10 min, centrifuge. Remove the organic layer and, for urine samples only, wash with 1 mL 200 mM NaOH. Evaporate a 5 mL aliquot of the organic layer to dryness under reduced pressure, reconstitute the residue with 200 μL MeCN:mobile phase 20:80, vortex, sonicate for 10 min, inject a 40 μL aliquot.

## HPLC VARIABLES

**Column:** 150 × 4.6 5 μm Kromasil C18

**Mobile phase:** MeCN:water:500 mM pH 7.2 phosphate buffer 47:48:5 containing 1 g/L tetrabutylammonium phosphate

**Injection volume:** 40

**Detector:** E, ESA Coulochem Model 5100A, guard cell Model 5020 + 0.7 V, analytical cell Model 5010, channel 1 + 0.28 V, channel 2 + 0.6 V

## CHROMATOGRAM

**Retention time:** 7.5

**Internal standard:** 17α-estradiol (6.0)

**Limit of quantitation:** 2 ng/mL

## KEY WORDS

plasma

## REFERENCE

Descombe, J.J.; Dubourg, D.; Picard, M.; Palazzini, E. Pharmacokinetic study of rifaximin after oral administration in healthy volunteers, *Int.J.Clin.Pharmacol.Res.*, **1994**, *14*, 51–56.

# Rilmazafone

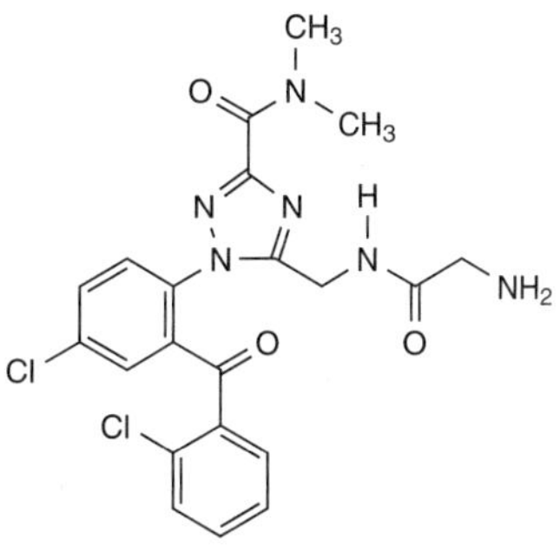


**Molecular formula:** $C_{21}H_{20}Cl_2N_6O_3$
**Molecular weight:** 475.34
**CAS Registry No:** 99593-25-6
**Merck Index:** 13, 8305

## SAMPLE

**Matrix:** blood
**Sample preparation:** Mix 1 mL plasma with 100 μL 1 M pH 5.0 acetate buffer and 5 mL EtOH, centrifuge at 1000 rpm for 15 min. Evaporate 5 mL of the supernatant to dryness, reconstitute the residue with 5 mL 100 mM pH 5.0 acetate buffer, add 4 mL of this solution to a Sep-Pak C18 SPE cartridge, elute with 5 mL MeOH, evaporate the eluate to dryness, reconstitute the residue with 200 μL 100 mM pH 11.6 sodium carbonate buffer, inject a 5–10 μL aliquot.

## HPLC VARIABLES

**Guard column:** 50 × 4 Nucleosil 10C18
**Column:** 150 × 4.6 Nucleosil 5C18
**Mobile phase:** MeCN:buffer 35:65 (Prepare the buffer by dissolving one vial of Waters PIC reagent A in 650 mL water.)
**Flow rate:** 1.5
**Injection volume:** 5–10
**Detector:** UV 254

## CHROMATOGRAM

**Retention time:** 7

## OTHER SUBSTANCES

**Extracted:** metabolites

## KEY WORDS

plasma; rat

## REFERENCE

Matsubara, T.; Touchi, A.; Yamada, N.; Sugeno, K. Induction of rat liver microsomal drug-metabolizing enzymes by a new sleep inducer 450191-S and plasma levels of 450191-S-metabolites, *J.Pharmacobiodyn.*, **1986**, *9*, 249–256.

# Risedronate sodium

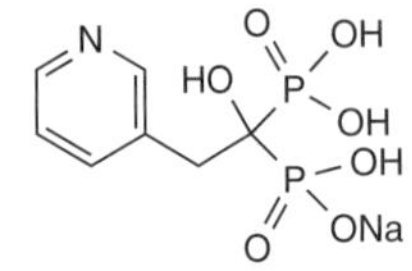


**Molecular formula:** $C_7H_{10}NNaO_7P_2$
**Molecular weight:** 305.09
**CAS Registry No:** 115436-72-1, 105462-24-6 (free acid)
**Merck Index:** 13, 8315

## SAMPLE

**Matrix:** urine

**Sample preparation:** Condition a 1 mL 30 mg HLB SPE cartridge (Waters) with 2 mL MeOH, 1 mL water, and 1 mL 10 mM mM 1-octyltriethylammonium phosphate. Immediately upon collection, add 25 μL 6 M HCl to each 1 mL urine, freeze at – 20° until ready to analyze. Mix 5 mL urine with 25 μL 125 μg/mL IS in water, add 50 μL 1.25 M calcium chloride solution, add 65 μL 30% NaOH, vortex; if no precipitate is observed, add 10 μL aliquots of 30% NaOH until a visible precipitate forms, centrifuge at 5020 g for 10 min, discard the supernatant, dissolve the precipitate in 50 μL 1 M HCl; if necessary, add 25 μL aliquots of 1 M HCl until the precipitate is completely dissolved, dilute with 5 mL water, add 50 μL NaOH to produce a second precipitate, centrifuge, discard the aqueous layer. Repeat the process. Dissolve the third precipitate in 500 μL 50 mM ethylene glycol-bis(*β*-aminoethyl ether)-*N*,*N*,*N'*,*N'*-tetraacetic acid (EGTA), sonicate for 5 min, vortex, add 4.5 mL water, add 100 μL 500 mM 1-octyltriethylammonium phosphate, vortex for 10 s, add to the SPE cartridge at 0.3–0.5 mL/min, wash with 1 mL water, wash with 1 mL MeOH:water 5:95, elute with 1 mL MeOH by centrifuging at 70 g for 5 min. Evaporate the eluate to dryness under a stream of nitrogen at 50°, reconstitute the residue with 500 μL 10 mM pH 6.25 sodium phosphate containing 1 mM etidronate, inject a 100 μL aliquot onto column A and elute to waste with mobile phase A; after 3.5 min, divert the effluent from column A onto column B and continue to elute with mobile phase A; after another 3 min, remove column A from the circuit, elute column B with mobile phase B, monitor the effluent from column B.

## HPLC VARIABLES

**Column:** A 50 × 4.6 3.5 μm X-Terra RP 18 (Waters); B 150 × 4.6 4 μm Synergi Polar RP (Phenomenex)

**Column temperature:** 30 (A and B)

**Mobile phase:** A MeCN:10 mM sodium phosphate containing 5 mM 1-octyltriethylammonium phosphate and 1 mM etidronate 8:92, apparent pH 6.25; B MeCN:11 mM sodium phosphate containing 5 mM 1-octyltriethylammonium phosphate and 1.1 mM etidronate 13:87, apparent pH 6.25

**Flow rate:** 1

**Injection volume:** 100

**Detector:** UV 262

## CHROMATOGRAM

**Retention time:** 18.5

**Internal standard:** 2-(3-pyridinyl)oxy-1-hydroxymethane diphosphonic acid tetraammonium salt (23)

**Limit of quantitation:** 7.5 ng/mL

## KEY WORDS

column-switching; SPE

## REFERENCE

Vallano, P.T.; Shugarts, S.B.; Kline, W.F.; Woolf, E.J.; Matuszewski, B.K. Determination of risedronate in human urine by column-switching ion-pair high-performance liquid chromatography with ultraviolet detection, *J.Chromatogr.B*, **2003**, *794*, 23–33.

# Rizatriptan

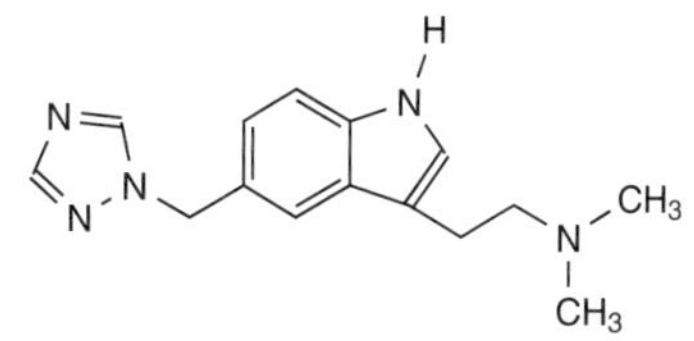

**Molecular formula:** $C_{15}H_{19}N_5$
**Molecular weight:** 269.34
**CAS Registry No:** 144034-80-0, 145202-66-0 (benzoate)
**Merck Index:** 13, 8324

## SAMPLE

**Matrix:** blood
**Sample preparation:** Condition a 1 mL 100 mg Bond Elut C2 SPE cartridge with two 1 mL portions of MeOH and two 1 mL portions of water. Mix 1 mL plasma with 100 μL 100 ng/mL IS in water, add to the SPE cartridge, wash with 1 mL water, wash with two 1 mL portions of MeOH:water 30:70, elute with 1 mL MeOH:10 mM pH 5.0 ammonium acetate 60:40. Evaporate the eluate to dryness at 50°, reconstitute the residue with 200 μL mobile phase, sonicate for 10 min, vortex, inject a 25 μL aliquot.

## HPLC VARIABLES

**Column:** 250 × 4.6 5 μm Spherisorb CN
**Mobile phase:** MeCN:MeOH:water:trifluoroacetic acid 54:4:42:0.1
**Flow rate:** 1
**Injection volume:** 25
**Detector:** MS, PE Sciex API-III triple quadrupole, nebulizer probe 500°, positive ionization, nebulizer gas at 80 psi, auxiliary gas at 2 L/min, corona discharge +3 μA, 0.1143 mm orifice, orifice 45 V, collision gas argon, m/z 270–201

## CHROMATOGRAM

**Retention time:** 5
**Internal standard:** L-743,214 (*N,N*-diethyl analogue) (m/z 298–229) (5.5)
**Limit of quantitation:** 0.5 ng/mL

## KEY WORDS

pharmacokinetics; plasma; SPE

## REFERENCE

McLoughlin, D.A.; Olah, T.V.; Ellis, J.D.; Gilbert, J.D.; Halpin, R.A. Quantitation of the $5HT_{1D}$ agonists MK-462 and sumatriptan in plasma by liquid chromatography-atmospheric pressure chemical ionization mass spectrometry, *J.Chromatogr.A*, **1996**, *726*, 115–124.

# Rofecoxib

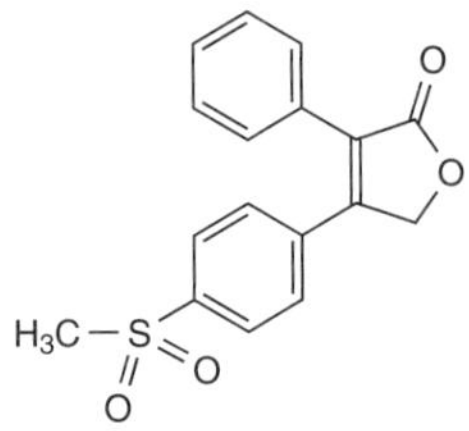

**Molecular formula:** $C_{17}H_{14}O_4S$
**Molecular weight:** 314.36
**CAS Registry No:** 162011-90-7
**Merck Index:** 13, 8330

## SAMPLE

**Matrix:** blood
**Sample preparation:** Add 50 μL 100 mM NaOH to 100 μL serum dropwise while gently vortexing, mix thoroughly, add 1 mL 600 ng/mL IS in ethyl acetate, vortex for 30 s, centrifuge at 1200 g for 10 min. Evaporate the organic layer to dryness under a stream of nitrogen at 37°, reconstitute the residue with 100 μL mobile phase, inject a 40 μL aliquot.

## HPLC VARIABLES

**Column:** 150 × 3.9 5 μm Novapak C18
**Mobile phase:** MeCN:50 mM sodium acetate 40:60, pH 6.4
**Flow rate:** 1
**Injection volume:** 40
**Detector:** UV 273

## CHROMATOGRAM

**Retention time:** 3.23
**Internal standard:** 5-ethyl-5-tolyl barbituric acid (2.17)
**Limit of quantitation:** 20 ng/mL

## OTHER SUBSTANCES

**Noninterfering:** EDTA, heparin, ibuprofen, indomethacin, naproxen

## KEY WORDS

plasma

## REFERENCE

Aravind, M.K.; Prescilla, R.; Ofenstein, J.P. A rapid and sensitive high-performance liquid chromatography assay for rofecoxib in human serum, *J.Chromatogr.Sci.*, **2002**, *40*, 26–28.

## SAMPLE

**Matrix:** blood
**Sample preparation:** Mix 1 mL plasma with 100 μL 10 μg/mL IS in MeCN and 1 mL pH 9.8 carbonate buffer, add 8 mL MTBE, rotate for 15 min. Evaporate the organic layer to dryness, reconstitute the residue with 500 μL MeCN, vortex for 1 min, add 50 μL water, vortex and sonicate for 15 min, inject a 25 μL aliquot.

## HPLC VARIABLES

**Guard column:** 20 × 3 YMC ODS AQ
**Column:** 100 × 3 3 μm YMC ODS AQ
**Mobile phase:** MeCN:water 50:50
**Flow rate:** 0.4
**Injection volume:** 25
**Detector:** MS, PE Sciex API-III Plus triple quadrupole, ionspray, negative ionization, orifice – 40 V, corona – 40 μA, nebulizer gas air at 80 psi, collision gas argon, m/z 313–257

## CHROMATOGRAM

**Retention time:** 5

**Internal standard:** 4-(4-methanesulfonylphenyl)-3-(4-methylphenyl)-5H-furan-2-one (m/z 327–271) (6.5)

**Limit of quantitation:** 100 pg/mL

## KEY WORDS

pharmacokinetics; plasma

## REFERENCE

Chavez-Eng, C.M.; Constanzer, M.L.; Matuszewski, B.K. Determination of Rofecoxib (MK-0966), a cyclooxygenase-2 inhibitor, in human plasma by high-performance liquid chromatography with tandem mass spectrometric detection, *J.Chromatogr.B*, **2000**, *748*, 31–39.

## SAMPLE

**Matrix:** blood

**Sample preparation:** Vortex 1 mL plasma with 50 μL MeCN and 25 μL 400 ng/mL IS in MeCN, add 1 mL 100 mM pH 5 acetate buffer, vortex, add 8 mL hexane:dichloromethane 50:50, mix on a flat-bed shaker for 5 min, centrifuge at 1500 g for 5 min, freeze in dry ice/acetone. Evaporate the organic layer to dryness under a stream of nitrogen at 50°, reconstitute the residue with 1 mL mobile phase, inject a 150 μL aliquot.

## HPLC VARIABLES

**Guard column:** 20 × 4 5 μm BDS-Hypersil C18

**Column:** 100 × 4.6 5 μm BDS-Hypersil C18

**Mobile phase:** MeCN:water 35:65

**Flow rate:** 1.2

**Injection volume:** 150

**Detector:** F ex 250 em 375 following post-column reaction. The column effluent flowed through a 10 m ×0.3 mm ID reaction coil irradiated at 254 nm (Astec Beam Boost) to the detector.

## CHROMATOGRAM

**Retention time:** 6

**Internal standard:** 4-(4-methanesulfonylphenyl)-3-(4-methylphenyl)-5H-furan-2-one (9)

**Limit of quantitation:** 0.5 ng/mL

## KEY WORDS

plasma; post-column photochemical derivatization

## REFERENCE

Woolf, E.; Fu, I.; Matuszewski, B. Determination of rofecoxib, a cyclooxygenase-2 specific inhibitor, in human plasma using high-performance liquid chromatography with post-column photochemical derivatization and fluorescence detection, *J.Chromatogr.B*, **1999**, *730*, 221–227.

## ANNOTATED BIBLIOGRAPHY

Abdel-Hamid, M.E. LC-MS analysis of selected sulfur-containing non-steroid antiinflammatory agents: applications to pharmaceutical products, *J.Liq.Chromatogr.Rel.Technol.*, **2000**, *23*, 3095–3107. [LC-MS; LOQ 100 ng/mL; rofecoxib; sulindac; celecoxib; piroxicam; tenoxicam]

Chavez-Eng, C.M.; Constanzer, M.L.; Matuszewski, B.K. High-performance liquid chromatographic-tandem mass spectrometric evaluation and determination of stable isotope labeled analogs of rofecoxib in human plasma samples from oral bioavailability studies, *J.Chromatogr.B*, **2002**, *767*, 117–129. [LOQ 100 pg/mL]

Ehrich, E.W.; Dallob, A.; De Lepeleire, I.; Van Hecken, A.; Riendeau, D.; Yuan, W.; Porras, A.; Wittreich, J.; Seibold, J.R.; De Schepper, P.; Mehlisch, D.R.; Gertz, B.J. Characterization of rofecoxib as a cyclooxygenase-2 isoform inhibitor and demonstration of analgesia in the dental pain model, *Clin.Pharmacol.Ther.*, **1999**, *65*, 336–347. [fluorescence detection]

Halpin, R.A.; Geer, L.A.; Zhang, K.E.; Marks, T.M.; Dean, D.C.; Jones, A.N.; Melillo, D.; Doss, G.; Vyas, K.P. The absorption, distribution, metabolism and excretion of rofecoxib, a potent and selective cyclooxygenase-2 inhibitor, in rats and dogs, *Drug Metab.Dispos.*, **2000**, *28*, 1244–1254. [LOQ 1 ng/mL; plasma; post-column photochemical derivatization; fluorescence detection]

Halpin, R.A.; Porras, A.G.; Geer, L.A.; Davis, M.R.; Cui, D.; Doss, G.A.; Woolf, E.; Musson, D.; Matthews, C.; Mazenko, R.; Schwartz, J.I.; Lasseter, K.C.; Vyas, K.P.; Baillie, T.A. The disposition and metabolism of rofecoxib, a potent and selective cyclooxygenase-2 inhibitor, in human subjects, *Drug Metab.Dispos.*, **2002**, *30*, 684–693. [plasma; urine; bile; feces; LOQ 0.5 ng/mL; post-column photochemical derivatization; fluorescence detection]

Hsieh, J.Y.-K.; Lin, L.; Matuszewski, B.K. High-throughput liquid chromatographic determination of rofecoxib in human plasma using a fully automated on-line solid-phase extraction system, *J.Liq.Chromatogr.Rel.Technol.*, **2001**, *24*, 799–812. [LOQ 0.5 ng/mL; plasma; post-column photochemical derivatization; fluorescence detection; SPE]

Krishna Reddy, K.V.S.R.; Babu, J.M.; Dubey, P.K.; Chandra Sekhar, B.; Om Reddy, G.; Vyas, K. Isolation and characterisation of process-related impurities in rofecoxib, *J.Pharm.Biomed.Anal.*, **2002**, *29*, 355–360.

Mao, B.; Abrahim, A.; Ge, Z.; Ellison, D.K.; Hartman, R.; Prabhu, S.V.; Reamer, R.A.; Wyvratt, J. Examination of rofecoxib solution decomposition under alkaline and photolytic stress conditions, *J.Pharm.Biomed.Anal.*, **2002**, *28*, 1101–1113. [stability-indicating]

Matthews, C.Z.; Woolf, E.J.; Matuszewski, B.K. Improved procedure for the determination of rofecoxib in human plasma involving 96-well solid-phase extraction and fluorescence detection, *J.Chromatogr.A*, **2002**, *949*, 83–89. [LOQ 0.5 ng/mL; plasma; post-column photochemical derivatization; fluorescence detection; SPE]

Niederberger, E.; Tegeder, I.; Schäfer, C.; Seegel, M.; Grösch, S.; Geisslinger, G. Opposite effects of rofecoxib on nuclear factor-kappaB and activating protein-1 activation, *J.Pharmacol.Exp.Ther.*, **2003**, *304*, 1153–1160. [plasma; post-column photochemical derivatization; fluorescence detection]

Radhakrishna, T.; Rao, D.S.; Reddy, G.O. LC determination of rofecoxib in bulk and pharmaceutical formulations, *J.Pharm.Biomed.Anal.*, **2001**, *26*, 617–628.

Slaughter, D.; Takenaga, N.; Lu, P.; Assang, C.; Walsh, D.J.; Arison, B.H.; Cui, D.; Halpin, R.A.; Geer, L.A.; Vyas, K.P.; Baillie, T.A. Metabolism of rofecoxib in vitro using human liver subcellular fractions, *Drug Metab.Dispos.*, **2003**, *31*, 1398–1408. [LC-MS]

Vallano, P.T.; Mazenko, R.S.; Woolf, E.J.; Matuszewski, B.K. Monolithic silica liquid chromatography columns for the determination of cyclooxygenase II inhibitors in human plasma, *J.Chromatogr.B*, **2002**, *779*, 249–257. [LOQ 0.5 ng/mL; plasma; post-column photochemical derivatization; fluorescence detection; SPE]

Werner, U.; Werner, D.; Mundkowski, R.; Gillich, M.; Brune, K. Selective and rapid liquid chromatography-mass spectrometry method for the quantification of rofecoxib in pharmacokinetic studies with humans, *J.Chromatogr.B*, **2001**, *760*, 83–90. [LC-MS; plasma; LOQ 1 ng/mL; celecoxib is internal standard]

Werner, U.; Werner, D.; Pahl, A.; Mundkowski, R.; Gillich, M.; Brune, K. Investigation of the pharmacokinetics of celecoxib by liquid chromatography-mass spectrometry, *Biomed.Chromatogr.*, **2002**, *16*, 56–60. [LC-MS; rofecoxib is internal standard]

# Ropinirole

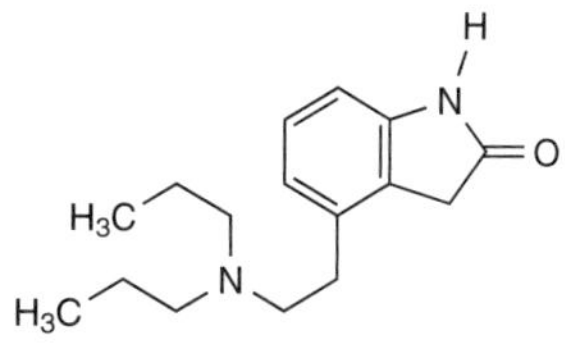


**Molecular formula:** $C_{16}H_{24}N_2O$
**Molecular weight:** 260.37
**CAS Registry No:** 91374-21-9, 91374-20-8 (HCl)
**Merck Index:** 13, 8338

## SAMPLE

**Matrix:** blood
**Sample preparation:** Condition a 3 mL low displacement C18 SPE cartridge (Baker) with 3 column vol of MeOH and 3 column vol of water. Mix 1 mL plasma with 50 μL 2 μg/mL IS in water, add to the SPE cartridge, wash with 10 mL water, wash with 10 mL MeCN, elute with 3.5 mL MeCN:water:ammonium hydroxide 100:2:0.5. Evaporate the eluate to dryness under a stream of nitrogen at 35°, immediately reconstitute the residue with 300 μL mobile phase, mix vigorously, inject a 10–100 μL aliquot.

## HPLC VARIABLES

**Column:** 250 × 4.6 5 μm Ultrasphere ODS
**Mobile phase:** MeCN:70 mM pH 3.8 ammonium formate buffer 25:75 containing 0.3% EDTA and 0.005% sodium octyl sulfate
**Flow rate:** 1
**Injection volume:** 10–100
**Detector:** UV 250

## CHROMATOGRAM

**Retention time:** 9.4
**Internal standard:** 4-(2-di-*N*,*N*-propylaminoethyl)-7-methoxy-2-(3H)-indoline HCl (11.5)
**Limit of detection:** 5 ng/mL
**Limit of quantitation:** 10 ng/mL

## KEY WORDS

dog; human; pharmacokinetics; plasma; rat; SPE

## REFERENCE

Swagzdis, J.E.; Mico, B.A. Liquid chromatographic determination of 4-(2-di-*N*,*N*-propylaminoethyl)-2-(3H)-indolone in rat, dog, and human plasma with ultraviolet detection, *J.Pharm.Sci.*, **1986**, *75*, 90–91.

# Rosiglitazone

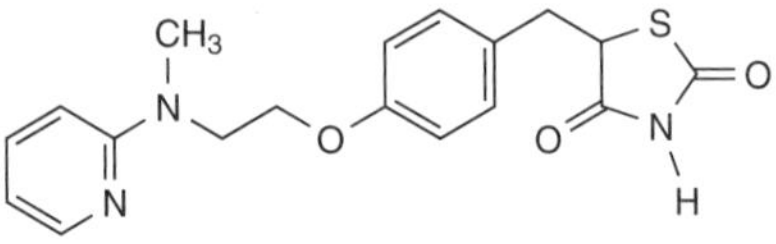

**Molecular formula:** $C_{18}H_{19}N_3O_3S$
**Molecular weight:** 357.43
**CAS Registry No:** 122320-73-4, 155141-29-0 (maleate)
**Merck Index:** 13, 8346

## SAMPLE

**Matrix:** blood
**Sample preparation:** Vortex 1 mL plasma with 50 μL 10 μg/mL IS in pH 2.3 buffer for 1 min, add 200 μL 20 mM pH 9.3 disodium tetraborate solution, vortex for 1 min, add 5 mL dichloromethane, shake horizontally for 10 min, centrifuge at 735 g for 10 min. Evaporate the organic layer to dryness under a stream of nitrogen at room temperature, reconstitute the residue with 5 mL *n*-hexane:dichloromethane 80:20, vortex for 30 s, add 350 μL pH 2.3 buffer, vortex for 2 min, centrifuge at 735 g for 10 min, inject a 100 μL aliquot. (Prepare pH 2.3 buffer by dissolving 0.136 g of potassium dihydrogen phosphate in 800 ml of water, adjusting to pH 2.3 with 10% HCl, and diluting to 1 L with water.)

## HPLC VARIABLES

**Column:** 150 × 4.6 5 μm Zorbax SB C18
**Column temperature:** 30
**Mobile phase:** MeOH:buffer 30:70 (Prepare the buffer by dissolving 1.41 g of disodium hydrogen phosphate and 1.56 g of sodium dihydrogen phosphate in 800 ml water, adjusting to pH 2.6 with orthophosphoric acid and diluting to 1 L with water.)
**Flow rate:** 1.2
**Injection volume:** 100
**Detector:** UV 245

## CHROMATOGRAM

**Retention time:** 8.3
**Internal standard:** pioglitazone (18.0)
**Limit of quantitation:** 5 ng/mL

## KEY WORDS

plasma

## REFERENCE

Kolte, B.L.; Raut, B.B.; Deo, A.A.; Bagool, M.A.; Shinde, D.B. Liquid chromatographic method for the determination of rosiglitazone in human plasma, *J.Chromatogr.B*, **2003**, *788*, 37–44.

## SAMPLE

**Matrix:** blood
**Sample preparation:** Mix 250 μL plasma with 10 μg IS for 15 s, add 3 mL ethyl acetate, vortex for 1 min, centrifuge at 2000 rpm for 10 min. Evaporate the organic layer to dryness under a stream of nitrogen at 40°, reconstitute the residue with 750 μL mobile phase, inject a 50 μL aliquot.

## HPLC VARIABLES

**Column:** 250 × 4.6 5 μm Hichrom KR100-5C18-250A
**Mobile phase:** MeCN:MeOH:10 mM potassium dihydrogen phosphate 50:10:40, adjusted to pH 6.5 with triethylamine

**Flow rate:** 1
**Injection volume:** 50
**Detector:** F ex 247 em 367

## CHROMATOGRAM
**Retention time:** 8
**Internal standard:** celecoxib (16)
**Limit of quantitation:** 5 ng/mL

## KEY WORDS
plasma

## REFERENCE
Mamidi, R.N.V.S.; Benjamin, B.; Ramesh, M.; Srinivas, N.R. Simple method for the determination of rosiglitazone in human plasma using a commercially available internal standard, *Biomed.Chromatogr.*, **2003**, *17*, 417–420.

## ANNOTATED BIBLIOGRAPHY
Cox, P.J.; Ryan, D.A.; Hollis, F.J.; Harris, A.-M.; Miller, A.K.; Vousden, M.; Cowley, H. Absorption, disposition, and metabolism of rosiglitazone, a potent thiazolidinedione insulin sensitizer, in humans, *Drug Metab.Dispos.*, **2000**, *28*, 772–780. [plasma; urine; feces; UV detection; LC-MS; SPE]

Di Cicco, R.A.; Allen, A.; Carr, A.; Fowles, S.; Jorkasky, D.K.; Freed, M.I. Rosiglitazone does not alter the pharmacokinetics of metformin, *J.Clin.Pharmacol.*, **2000**, *40*, 1280–1285. [dialysis; fluorescence detection; LOQ 2.5 ng/mL]

Mamidi, R.N.V.S.; Chaluvadi, M.R.; Benjamin, B.; Ramesh, M.; Katneni, K.; Babu, A.P.; Bhanduri, J.; Rao, N.M.U.; Rajagopalan, R. HPLC method for the determination of rosiglitazone in human plasma and its application in a clinical pharmacokinetic study, *Arzneimittelforschung*, **2002**, *52*, 560–564. [fluorescence detection; LOQ 5 ng/mL]

Muxlow, A.-M.; Fowles, S.; Russell, P. Automated high-performance liquid chromatography method for the determination of rosiglitazone in human plasma, *J.Chromatogr.B*, **2001**, *752*, 77–84. [dialysis; fluorescence detection; LOQ 3 ng/mL]

Radhakrishna, T.; Satyanarayana, J.; Satyanarayana, A. LC determination of rosiglitazone in bulk and pharmaceutical formulation, *J.Pharm.Biomed.Anal.*, **2002**, *29*, 873–880.

Rao, M.N.V.S.; Mullangi, R.; Katneni, K.; Ravikanth, B.; Babu, A.P.; Rani, U.P.; Naidu, M.U.R.; Srinivas, N.R.; Rajagopalan, R. Lack of effect of sucralfate on the absorption and pharmacokinetics of rosiglitazone, *J.Clin.Pharmacol.*, **2002**, *42*, 670–675. [fluorescence detection; LOQ 5 ng/mL]

# Rosuvastatin calcium

**Molecular formula:** $2C_{22}H_{27}FN_3O_6S.Ca$
**Molecular weight:** 1001.14
**CAS Registry No:** 147098-20-2

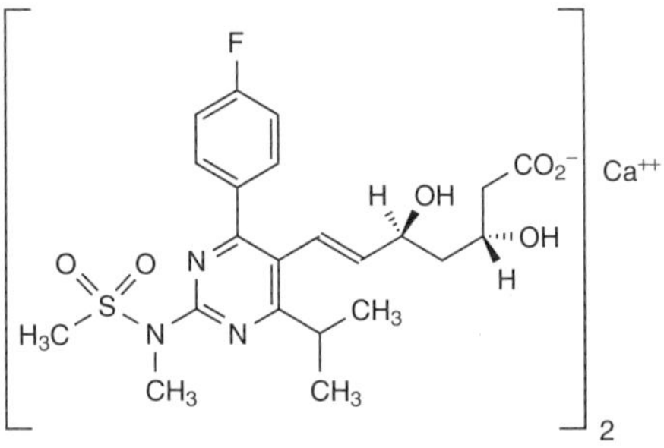


## SAMPLE

**Matrix:** blood
**Sample preparation:** Condition a 30 mg Oasis HLB SPE cartridge (in 96 well plate) with 1 mL MeOH and 1 mL 0.5% acetic acid. Mix 500 μL plasma with 500 μL 100 mM pH 4 acetate buffer, 50 μL 150 ng/mL IS in MeOH:1 M acetic acid 50:50, and 750 μL 1 M acetic acid, vortex for 2 s, centrifuge at 738 g for 7 min, add 1.7 mL of the supernatant to the SPE cartridge, wash with 1 mL MeOH:0.5% acetic acid 30:70, elute with 1 mL 0.5% acetic acid in MeOH. Evaporate the eluate to dryness under a stream of nitrogen at 40°, reconstitute the residue with 130 μL 0.5% acetic acid, centrifuge at 1700 or 9500 g for 10 min, inject a 100 μL aliquot.

## HPLC VARIABLES

**Column:** 150 × 4.6 5 μm Luna C18(2)
**Mobile phase:** MeOH:0.2% formic acid 70:30 (divert first 2 min of run to waste)
**Flow rate:** 1
**Injection volume:** 100
**Detector:** MS, PE Sciex API 365, TurboIonspray 450°, 0.2 mL/min entered detector, positive ion mode, ionspray 3000 V, ring 160 V, orifice 44 V, nebulizer gas nitrogen at 14 units, TurboIonspray gas nitrogen at 7 L/min, collision gas nitrogen 4 units, curtain gas nitrogen 8 units, deflector −250 V, electron multiplier 2900 V, collision energy −47.5 V, m/z 482.2−258.2

## CHROMATOGRAM

**Retention time:** 3.6
**Internal standard:** $d_6$-rosuvastatin (m/z 488.2−264.2)
**Limit of quantitation:** 100 pg/mL

## KEY WORDS

plasma; SPE

## REFERENCE

Hull, C.K.; Penman, A.D.; Smith, C.K.; Martin, P.D. Quantification of rosuvastatin in human plasma by automated solid-phase extraction using tandem mass spectrometric detection, *J.Chromatogr.B*, **2002**, *772*, 219−228.

# Sarafloxacin

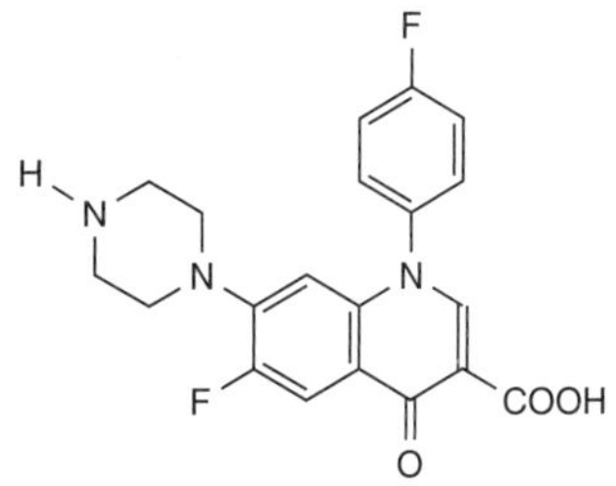


**Molecular formula:** $C_{20}H_{17}F_2N_3O_3$
**Molecular weight:** 385.36
**CAS Registry No:** 98105-99-8, 91296-87-6 (HCl)
**Merck Index:** 13, 8447

## SAMPLE
**Matrix:** blood
**Sample preparation:** Mix 200 μL plasma with 800 μL 750 ng/mL IS in 100 mM pH 7.4 phosphate buffer, add 6 mL chloroform (Caution! Chloroform is a carcinogen!), shake at 200 oscillations/min for 30 min, centrifuge at 13 000 g for 6 min. Evaporate the organic layer to dryness under a stream of nitrogen at 40°, reconstitute the residue with 200 μL phosphate-buffered saline, inject a 10–80 μL aliquot.

## HPLC VARIABLES
**Guard column:** Guard-Pak Novapak C18
**Column:** 150 × 3.9 5 μm Novapak C18
**Mobile phase:** MeCN:buffer 18:82 (The buffer was 20 mM potassium dihydrogen phosphate containing 6 mM phosphoric acid and 12 mM tetraethylammonium bromide adjusted to pH 3.0 with 2 M NaOH.)
**Flow rate:** 1
**Injection volume:** 10–80
**Detector:** F ex 338 em 425

## CHROMATOGRAM
**Retention time:** 5.6
**Internal standard:** norfloxacin (2.2)
**Limit of detection:** 6 ng/mL
**Limit of quantitation:** 12 ng/mL

## OTHER SUBSTANCES
**Extracted:** difloxacin (6.2)

## KEY WORDS
plasma; rabbit

## REFERENCE
Garcia, M.A.; Solans, C.; Aramayona, J.J.; Rueda, S.; Bregante, M.A. Simultaneous determination of difloxacin and its primary metabolite sarafloxacin in rabbit plasma, *Chromatographia*, **2000**, *51*, 487–490.

## SAMPLE
**Matrix:** blood
**Sample preparation:** Condition a 1 mL Bond Elut C2 SPE cartridge with two 1 mL portions of MeOH and three 1 mL portions of 2 mM phosphoric acid. Add 300 μL 500 mM phosphoric acid to the SPE cartridge, add 250 μL serum, add 50 μL 500 ng/mL IS in mobile phase, add 250 μL water, wash with 100 μL water, wash with 100 μL 100 mM phosphoric acid, elute with five 100 μL aliquots of MeOH:500 mM phosphoric acid 70:30, dilute the eluate with 500 μL water, inject a 20 μL aliquot.

## HPLC VARIABLES
**Guard column:** 10 × 3 5 μm PLRP-S (Polymer Laboratories)
**Column:** 150 × 4.6 5 μm PLRP-S (Polymer Laboratories)
**Mobile phase:** MeCN:MeOH:2 mM phosphoric acid 20:8:72
**Flow rate:** 0.9
**Injection volume:** 20
**Detector:** F ex 278 em 440

## CHROMATOGRAM
**Retention time:** 5
**Internal standard:** enrofloxacin (3.5)
**Limit of detection:** 5 ng/g

## KEY WORDS
fish; serum; SPE

## REFERENCE
Steffenak, I.; Hormazabal, V.; Yndestad, M. A rapid assay for the determination of sarafloxacin (A-55620) in fish serum by high performance liquid chromatography, *J.Liq.Chromatogr.*, **1991**, *14*, 1983–1988.

## ANNOTATED BIBLIOGRAPHY
Holtzapple, C.K.; Pishko, E.J.; Stanker, L.H. Separation and quantification of two fluoroquinolones in serum by on-line high-performance immunoaffinity chromatography, *Anal.Chem.*, **2000**, *72*, 4148–4153. [LOD 1.7 ng/mL; fluorescence detection]

Holtzapple, C.K.; Buckley, S.A.; Stanker, L.H. Determination of fluoroquinolones in serum using an on-line clean-up column coupled to high-performance immunoaffinity-reversed-phase liquid chromatography, *J.Chromatogr.B*, **2001**, *754*, 1–9. [LOD 1.7 ng/mL; fluorescence detection]

# Selamectin

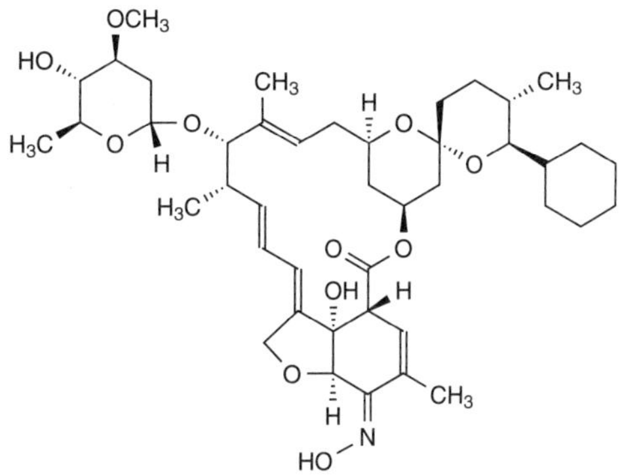

**Molecular formula:** $C_{43}H_{63}NO_{11}$
**Molecular weight:** 769.96
**CAS Registry No:** 220119-17-5
**Merck Index:** 13, 8500

## SAMPLE

**Matrix:** blood
**Sample preparation:** Condition a 1 mL 100 mg Isolute C18 SPE cartridge (Jones Chromatography) with 1 mL MeOH and 1 mL water. Vortex 0.2 (cat) or 1 (dog) mL plasma with 10 μL 1 μg/mL IS in MeOH and 1 mL MeCN:water 30:70, add to the SPE cartridge, wash with 1 mL water, dry under vacuum, elute with two 500 μL portions of MeOH. Evaporate the eluate to dryness under a stream of nitrogen at 37°, make sure sample is completely dry, reconstitute the residue with 100 μL triethylamine:MeCN 50:50, mix thoroughly, add 150 μL trifluoroacetic acid:MeCN 33:67, mix thoroughly. Evaporate the sample to 75 μL under a stream of nitrogen at 40° (ca. 20 min), add 250 μL 2.0 M ammonia in MeOH, mix gently. Evaporate the sample to 75 μL under a stream of nitrogen at 40° (ca. 10 min), add 200 μL MeCN, mix thoroughly, inject a 100 μL aliquot.

## HPLC VARIABLES

**Column:** 250 × 3.2 Spherisorb RPB
**Mobile phase:** MeCN:THF:water 68:15:17
**Flow rate:** 0.5
**Injection volume:** 100
**Detector:** F ex 360 em 450

## CHROMATOGRAM

**Retention time:** 14
**Internal standard:** UK-127,053 (methoxyselamectin) (11.5)
**Limit of quantitation:** 0.2 ng/mL (dog), 1 ng/mL (cat)

## KEY WORDS

cat; derivatization; dog; plasma; SPE

## REFERENCE

Walker, D.K.; Fenner, K.S. A sensitive method for the measurement of the novel pet endectocide, selamectin (UK-124,114), in dog and cat plasma by chemical derivatisation and high-performance liquid chromatography with fluorescence detection, *J.Pharm.Biomed.Anal.*, **2000**, *24*, 105–111.

# Sermorelin

**Molecular formula:** $C_{149}H_{246}N_{44}O_{42}S$
**Molecular weight:** 3357.94
**CAS Registry No:** 86168-78-7
**Merck Index:** 13, 8535

## SAMPLE
**Matrix:** blood
**Sample preparation:** Condition a Sep-Pak C18 SPE cartridge with MeOH, 0.01% trifluoroacetic acid in MeCN, and 0.01% trifluoroacetic acid in water. Mix 200 μL serum with 800 μL cold 50 mM pH 0.8 (sic) phosphate buffer, add to the SPE cartridge, elute with 2 mL MeCN:water:trifluoroacetic acid 40:60:0.01, lyophilize the eluate, reconstitute with initial mobile phase, inject an aliquot.

## HPLC VARIABLES
**Column:** 150 × 3.9 10 μm μBondapak C18
**Mobile phase:** Gradient. MeCN:water:trifluoroacetic acid from 20:80:0.01 to 36:64:0.01 over 40 min, wash at 40:60:0.01 for 10 min.
**Flow rate:** 1.5
**Detector:** UV 214

## CHROMATOGRAM
**Retention time:** 33

## OTHER SUBSTANCES
**Extracted:** metabolites

## KEY WORDS
rat; serum; SPE

## REFERENCE
Boulanger, L.; Roughly, P.; Gaudreau, P. Catabolism of rat growth hormone-releasing factor(1–29) amide in rat serum and liver, *Peptides*, **1992**, *13*, 681–689.

## SAMPLE
**Matrix:** blood
**Sample preparation:** Condition a Sep-Pak C18 SPE cartridge with 2 mL 80% acetic acid and 4 mL 10 mM trifluoroacetic acid. Mix 1 mL plasma with 200 μL 1 M trifluoroacetic acid, place on ice, add to the SPE cartridge, wash with 3 mL 100 mM trifluoroacetic acid, pass 20 mL air through the cartridge, elute with 3 mL 80% acetic acid, inject an aliquot.

## HPLC VARIABLES
**Column:** 150 × 3.9 5 μm Delta-Pak C18 (Waters)
**Mobile phase:** Gradient. A was MeCN:water:trifluoroacetic acid 95:5:0.1. B was 0.1% trifluoroacetic acid. A:B from 34:66 to 50:50 over 60 min.
**Flow rate:** 1
**Detector:** UV 215

## CHROMATOGRAM
**Retention time:** 20

## OTHER SUBSTANCES
**Extracted:** metabolites

## KEY WORDS
pig; plasma; SPE

## REFERENCE
Su, C.M.; Jensen, L.R.; Heimer, E.P.; Felix, A.M.; Pan, Y.C.; Mowles, T.F. In vitro stability of growth hormone releasing factor (GRF) analogs in porcine plasma, *Horm.Metab.Res.*, **1991**, *23*, 15–21.

## ANNOTATED BIBLIOGRAPHY
Zarandi, M.; Serfozo, P.; Zsigo, J.; Deutch, A.H.; Janaky, T.; Olsen, D.B.; Bajusz, S.; Schally, A.V. Potent agonists of growth hormone-releasing hormone. II, *Pept.Res.*, **1992**, *5*, 190–193.

# Sibutramine

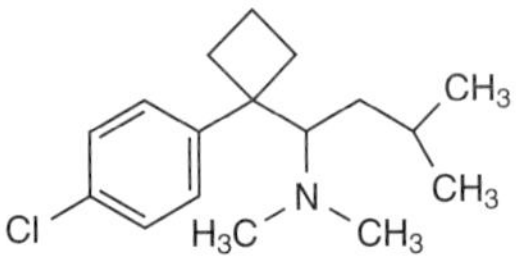


**Molecular formula:** $C_{17}H_{26}ClN$
**Molecular weight:** 279.86
**CAS Registry No:** 106650-56-0, 125494-59-9 (HCl monohydrate)
**Merck Index:** 13, 8559

## SAMPLE
**Matrix:** blood
**Sample preparation:** Mix 1 mL plasma, 50 μL 80 μg/mL IS in MeOH, 1 mL saturated sodium bicarbonate, and 5 mL cyclohexane, centrifuge. Evaporate the organic layer to dryness under a stream of nitrogen, reconstitute the residue with 100 μL mobile phase, inject a 50 μL aliquot.

## HPLC VARIABLES
**Column:** 250 × 4.6 5 μm Hypersil ODS-2 C18
**Column temperature:** 25
**Mobile phase:** MeOH:10 mM pH 3.5 ammonium acetate buffer 75:25 (divert first 3 min to waste)
**Flow rate:** 1
**Injection volume:** 50
**Detector:** MS, quadrupole, electrospray, drying gas nitrogen 10.5 L/min, nebulizer 45 psi, drying gas 350°, capillary 4 kV, positive ion mode, fragmenter 70 V, m/z 280

## CHROMATOGRAM
**Retention time:** 5
**Internal standard:** phenoprolamine HCl (*N*-(2-(3,4-dimethoxyphenyl)ethyl)-*N*-(2-(2,6-dimethylphenoxy)-1-methylethyl)amine) (m/z 344) (3.5)
**Limit of quantitation:** 50 ng/mL

## OTHER SUBSTANCES
**Extracted:** metabolites

## KEY WORDS
plasma

## REFERENCE
Ding, L.; Hao, X.; Huang, X.; Zhang, S. Simultaneous determination of sibutramine and its *N*-desmethyl metabolites in human plasma by liquid chromatography-electrospray ionization-mass spectrometry: method and clinical applications, *Anal.Chim.Acta*, **2003**, *492*, 241–248.

## SAMPLE
**Matrix:** solutions
**Sample preparation:** Inject a 10 μL aliquot of a 1 mg/mL solution.

## HPLC VARIABLES
**Guard column:** 50 mm long Chiralcel OD
**Column:** 250 × 4.6 10 μm Chiralcel OD
**Column temperature:** 30
**Mobile phase:** Hexane:EtOH:trifluoroacetic acid 93:7:0.05
**Flow rate:** 1
**Injection volume:** 10
**Detector:** UV 225

## CHROMATOGRAM

**Retention time:** 10 (−), 14 (+)

## KEY WORDS

chiral

## REFERENCE

Radhakrishna, T.; Narayana, C.L.; Rao, D.S.; Vyas, K.; Reddy, G.O. LC method for the determination of assay and purity of sibutramine hydrochloride and its enantiomers by chiral chromatography, *J.Pharm.Biomed.Anal.*, **2000**, *22*, 627–639.

## SAMPLE

**Matrix:** solutions

## HPLC VARIABLES

**Column:** 250 × 4.6 5 μm Microsorb-MV (Varian)
**Mobile phase:** MeOH:water:triethylamine 80:20:0.3, pH adjusted to 4.5 with 85% phosphoric acid
**Flow rate:** 1.1
**Injection volume:** 20
**Detector:** UV 225

## CHROMATOGRAM

**Retention time:** 4

## KEY WORDS

stability-indicating

## REFERENCE

Segall, A.I.; Collado, E.A.; Ricci, R.A.; Pizzorno, M.T. Reversed-phase HPLC determination of sibutramine hydrochloride in the presence of its oxidatively-induced degradation products, *J.Liq.Chromatogr.Rel.Technol.*, **2003**, *26*, 977–986.

# Sildenafil

**Molecular formula:** $C_{22}H_{30}N_6O_4S$
**Molecular weight:** 474.59
**CAS Registry No:** 139755-83-2, 171599-83-0 (citrate)
**Merck Index:** 13, 8563

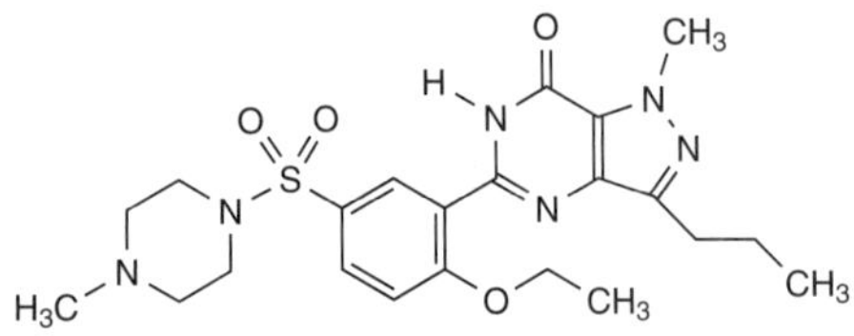


## SAMPLE

**Matrix:** blood

**Sample preparation:** Condition a 25 mg Certify mixed-mode SPE cartridge in a 96 well plate (Varian) with 500 µL MeOH and 500 µL 5% acetic acid. Mix 350 µL plasma with 20 µL 500 ng/mL IS in MeOH:water 50:50 and 350 µL 5% acetic acid, add to the SPE cartridge, wash with 500 µL 5% acetic acid, wash with 500 µL MeOH, dry for 3 min, elute with two 350 µL portions of MeCN:ammonium hydroxide 98:2. Evaporate the eluate to dryness, reconstitute the residue with 200 µL 0.05% trifluoroacetic acid in MeCN, inject a 10 µL aliquot.

## HPLC VARIABLES

**Column:** 50 × 3 5 µm Betasil silica (Keystone)

**Mobile phase:** Unspecified. However, in similar analyses, the authors have reported MeCN:water:trifluoroacetic acid 90:10:0.1 (*J.Pharm.Biomed.Anal.* **2003**, *32*, 609), MeCN: water:formic acid 90:10:0.1 (*J.Chromatogr.B* **1999**, *735*, 255), and MeCN:water:trifluoroacetic acid 95:5:0.05 (*J.Chromatogr.B* **2001**, *754*, 387)

**Flow rate:** 0.4

**Injection volume:** 10

**Detector:** MS, PE Sciex API 3000, electrospray, positive ion mode, ionspray needle 5000 V, turbo gas 400°, auxiliary gas 8 L/min, nebulizer gas 12 units, curtain gas 8 units, collision gas 4 units, declustering 46 V, focusing 200 V, collision energy 77 V, m/z 475–283

## CHROMATOGRAM

**Retention time:** 1.7

**Internal standard:** 1-[[3-(7-methoxy-1-methyl-3-propyl-1*H*-pyrazolo[4,3-d]pyrimidin-5-yl)-4-ethoxyphenyl]sulfonyl]-4-methylpiperazine (m/z 489–297) (1.6)

**Limit of quantitation:** 1 ng/mL

## OTHER SUBSTANCES

**Extracted:** desmethylsildenafil (metabolite) (m/z 461–283) (1.6)

## KEY WORDS

plasma; SPE

## REFERENCE

Eerkes, A.; Addison, T.; Naidong, W. Simultaneous assay of sildenafil and desmethylsildenafil in human plasma using liquid chromatography-tandem mass spectrometry on silica column with aqueous-organic mobile phase, *J.Chromatogr.B*, **2002**, *768*, 277–284.

## SAMPLE

**Matrix:** blood

**Sample preparation:** Vortex 1 mL plasma with 100 µL 20 µg/mL IS in MeOH and 100 µL 1 M NaOH for 5 s, add 3 mL ethyl acetate, vortex for 5 min, centrifuge at 2950 g for 10 min. Evaporate the organic layer to dryness under a stream of nitrogen at 40°, reconstitute the residue with 200 μL mobile phase, inject a 100 μL aliquot.

### HPLC VARIABLES
**Column:** 150 × 4.6 5 μm Inertsil 5 ODS-2
**Mobile phase:** MeCN:buffer 55:45 (The buffer was 30 mM potassium dihydrogen phosphate adjusted to pH 6.0 with 1 M NaOH.)
**Flow rate:** 0.5
**Injection volume:** 100
**Detector:** UV 230

### CHROMATOGRAM
**Retention time:** 7.5
**Internal standard:** butylparaben (12)
**Limit of quantitation:** 10 ng/mL

### KEY WORDS
pharmacokinetics; plasma

### REFERENCE
Sheu, M.-T.; Wu, A.-B.; Yeh, G.-C.; Hsia, A.; Ho, H.-O. Development of a liquid chromatographic method for bioanalytical applications with sildenafil, *J.Chromatogr.B*, **2003**, *791*, 255–262.

### SAMPLE
**Matrix:** formulations
**Sample preparation:** Weigh out powdered tablets containing 100 mg of active, dissolve in 100 mL MeCN:water 50:50, add IS, dilute with mobile phase, inject a 20 μL aliquot.

### HPLC VARIABLES
**Column:** 250 × 4.6 5 μm LiChrospher C18
**Mobile phase:** MeCN:water 52:48
**Flow rate:** 1
**Injection volume:** 20
**Detector:** UV 245

### CHROMATOGRAM
**Retention time:** 7.2
**Internal standard:** piroxicam (2.2)
**Limit of detection:** 15 ng/mL
**Limit of quantitation:** 50 ng/mL

### KEY WORDS
stability-indicating; tablets

### REFERENCE
Dinesh, N.D.; Vishukumar, B.K.; Nagaraja, P.; Gowda, N.M.M.; Rangappa, K.S. Stability indicating RP-LC determination of sildenafil citrate (Viagra) in pure form and in pharmaceutical samples, *J.Pharm.Biomed.Anal.*, **2002**, *29*, 743–748.

## ANNOTATED BIBLIOGRAPHY
Aboul-Enein, H.Y.; Hefnawy, M.M. Rapid determination of sildenafil citrate in pharmaceutical preparations using monolithic silica HPLC column, *J.Liq.Chromatogr.Rel.Technol.*, **2003**, *26*, 2897–2908. [LOD 25 ng/mL]

Cooper, J.D.H.; Muirhead, D.C.; Taylor, J.E.; Baker, P.R. Development of an assay for the simultaneous determination of sildenafil (Viagra) and its metabolite (UK-103,320) using automated sequential trace enrichment of dialysates and high-performance liquid chromatography, *J.Chromatogr.B*, **1997**, *701*, 87–95.

Daraghmeh, N.; Al-Omari, M.; Badwan, A.A.; Jaber, A.M.Y. Determination of sildenafil citrate and related substances in the commercial products and tablet dosage form using HPLC, *J.Pharm.Biomed.Anal.*, **2001**, *25*, 483–492.

Jeong, C.K.; Lee, H.-Y.; Jang, M.-S.; Kim, W.B.; Lee, H.S. Narrowbore high-performance liquid chromatography for the simultaneous determination of sildenafil and its metabolite UK-103,320 in human plasma using column switching, *J.Chromatogr.B*, **2001**, *752*, 141–147. [LOQ 10 ng/mL]

Kim, J.; Ji, H.Y.; Kim, S.J.; Lee, H.W.; Lee, S.-S.; Kim, D.S.; Yoo, M.; Kim, W.B.; Lee, H.S. Simultaneous determination of sildenafil and its active metabolite UK-103,320 in human plasma using liquid chromatography-tandem mass spectrometry, *J.Pharm.Biomed.Anal.*, **2003**, *32*, 317–322. [LOQ 2 ng/mL]

Lee, M.; Min, D.I. Determination of sildenafil citrate in plasma by high-performance liquid chromatography and a case for the potential interaction of grapefruit juice with sildenafil citrate, *Ther.Drug Monit.*, **2001**, *23*, 21–26. [LOD 10 ng/mL]

Liaw, J.; Chang, T.-W. Determination of transdermal sildenafil in nude mouse skin by reversed-phase high-performance liquid chromatography, *J.Chromatogr.B*, **2001**, *765*, 161–166. [LOQ 5 ng/mL]

Nagaraju, V.; Sreenath, D.; Rao, J.T.; Rao, R.N. Separation and determination of synthetic impurities of sildenafil (Viagra) by reversed-phase high-performance liquid chromatography, *Anal.Sci.*, **2003**, *19*, 1007–1011.

Segall, A.I.; Vitale, M.F.; Perez, V.L.; Palacios, M.L.; Pizzorno, M.T. Reversed-phase HPLC determination of sildenafil citrate in the presence of its oxidative-induced degradation products, *J.Liq.Chromatogr.Rel.Technol.*, **2000**, *23*, 1377–1386. [stability-indicating]

Tracqui, A.; Ludes, B. HPLC-MS for the determination of sildenafil citrate (Viagra) in biological fluids. Application to the salivary excretion of sildenafil after oral intake, *J.Anal.Toxicol.*, **2003**, *27*, 88–94. [saliva; plasma; LOD 0.2 ng/mL; LOQ 0.5 ng/mL]

Walker, D.K.; Ackland, M.J.; James, G.C.; Muirhead, G.J.; Rance, D.J.; Wastall, P.; Wright, P.A. Pharmacokinetics and metabolism of sildenafil in mouse, rat, rabbit, dog and man, *Xenobiotica*, **1999**, *29*, 297–310. [SPE; LOQ 1 ng/mL]

Warrington, J.S.; Shader, R.I.; von Moltke, L.L.; Greenblatt, D.J. In vitro biotransformation of sildenafil (Viagra): identification of human cytochromes and potential drug interactions, *Drug Metab.Dispos.*, **2000**, *28*, 392–397. [buspirone is internal standard]

Warrington, J.S.; von Moltke, L.L.; Harmatz, J.S.; Shader, R.I.; Greenblatt, D.J. The effect of age on sildenafil biotransformation in rat and mouse liver microsomes, *Drug Metab.Dispos.*, **2003**, *31*, 1306–1309.

# Simethicone

**CAS Registry No:** 8050-81-5
**Merck Index:** 13, 3241

## SAMPLE

**Matrix:** formulations
**Sample preparation:** Add formulation (oral liquid or crushed tablet) to 25 mL dichloromethane and 25 mL diluted HCl (2:1), shake vigorously for 5 min, let stand for 1 h. Dry 5 mL of the lower organic layer over 500 mg anhydrous sodium sulfate, filter (0.45 μm nylon), inject an aliquot of the filtrate.

## HPLC VARIABLES

**Column:** 250 × 4.6 5 μm Alltima C8
**Mobile phase:** Gradient. MeCN:chloroform from 45:55 to 15:85 over 5 min, return to initial conditions over 5 min, re-equilibrate for 5 min. (Caution! Chloroform is a carcinogen!)
**Flow rate:** 1
**Detector:** Evaporative Light Scattering Detector, Alltech model 500, drift tube 95°, nebulizer 2 L/min

## CHROMATOGRAM

**Retention time:** 9.2

## KEY WORDS

crushed tablet; oral liquid

## REFERENCE

Moore, D.E.; Liu, T.X.; Miao, W.G.; Edwards, A.; Elliss, R. A RP-LC method with evaporative light scattering detection for the assay of simethicone in pharmaceutical formulations, *J.Pharm.Biomed.Anal.*, **2002**, *30*, 273–278.

# Sivelestat

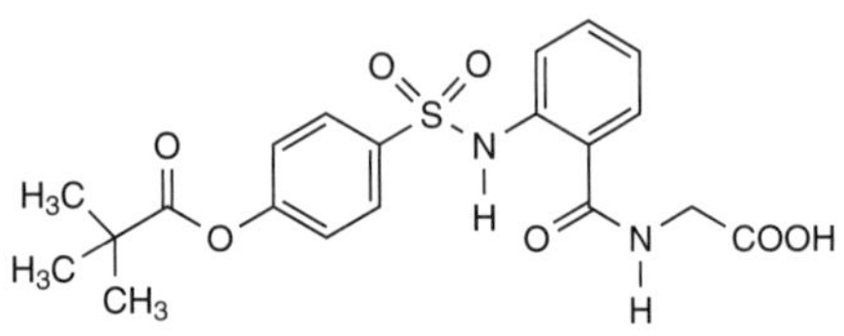

**Molecular formula:** $C_{20}H_{22}N_2O_7S$
**Molecular weight:** 434.47
**CAS Registry No:** 127373-66-4, 201677-61-4 (Na salt tetrahydrate)
**Merck Index:** 13, 8629

## SAMPLE
**Matrix:** blood, urine
**Sample preparation:** Mix plasma or urine with IS, filter (0.22 μm), inject a 40 (plasma) or 10 (urine) μL aliquot onto column A and elute to waste with mobile phase A; after 7 min, backflush the contents of column A onto column B with mobile phase B; after 3 min, remove column A from the circuit and backflush the contents of column B onto column C with mobile phase C; after 3 min, remove column B from the circuit, continue to elute column C with mobile phase C, monitor the effluent from column C.

## HPLC VARIABLES
**Column:** A 10 × 4.6 5 μm SPS-C18 + 150 × 4.6 5 μm SPS-C18 Semipermeable-Surface (Regis); B 150 × 4.6 5 μm YMC-A602; C 150 × 4.6 5 μm Capcell Pak C18 (Shiseido)
**Mobile phase:** A MeCN:50 mM pH 7 phosphate buffer 10:90; B MeCN:water 10:90; C Gradient. MeCN:20 mM pH 3.8 potassium dihydrogen phosphate buffer 20:80 for 10 min, to 40:50 over 20 min, maintain at 40:50 for 15 min.
**Flow rate:** 1
**Injection volume:** 40 (plasma), 10 (urine)
**Detector:** UV 240

## CHROMATOGRAM
**Retention time:** 32
**Internal standard:** ONO-EI-547 (*N*-[2′-[4-(2,2-dimethylpropionyloxy)-3-methylphenylsulfonylamino]-5′-methylbenzoyl]aminoacetic acid) (38), ONO-EI-537 (*N*-[2′-[4-hydroxy-3-methylphenylsulfonylamino]benzoyl]aminoacetic acid) (20)
**Limit of quantitation:** 156 ng/mL (plasma), 1.56 μg/mL (urine)

## KEY WORDS
column-switching; plasma

## REFERENCE
Shintani, T.; Takamoto, M.; Sawada, M.; Aishita, H.; Nakagawa, T. Simultaneous determination of human neutrophil elastase inhibitor (ONO-5046) and its metabolite in plasma and urine by direct injection column-switching HPLC, *J.Pharm.Biomed.Anal.*, **1994**, *12*, 397–405.

## SAMPLE
**Matrix:** blood, urine
**Sample preparation:** Mix plasma or urine with IS, dilute 500 μL plasma or 100 μL urine with water and 1 M HCl, add to a Sep-Pak C18 SPE cartridge, elute with MeOH. Add the eluate to a Bond Elut SAX SPE cartridge, elute with MeOH:100 mM HCl 60:40, extract the eluate with ethyl acetate. Evaporate the organic layer to dryness, reconstitute the residue with 150 μL mobile phase, inject a 50 μL aliquot.

## HPLC VARIABLES
**Column:** 150 × 4.6 3 μm Chemcosorb 3Dph
**Mobile phase:** Gradient. MeCN:20 mM pH 3.8 potassium dihydrogen phosphate buffer from 20:80 to 40:50 over 40 min, maintain at 40:50 for 15 min.

**Flow rate:** 1
**Injection volume:** 50
**Detector:** UV 240

## CHROMATOGRAM

**Internal standard:** ONO-EI-547 (*N*-[2′-[4-(2,2-dimethylpropionyloxy)-3-methylphenylsulfonylamino]-5′-methylbenzoyl]aminoacetic acid), ONO-EI-537 (*N*-[2′-[4-hydroxy-3-methylphenylsulfonylamino]benzoyl]aminoacetic acid)
**Limit of quantitation:** 125 ng/mL (plasma), 1.25 μg/mL (urine)

## KEY WORDS

comparison with column-switching (above); plasma; SPE

## REFERENCE

Shintani, T.; Takamoto, M.; Sawada, M.; Aishita, H.; Nakagawa, T. Simultaneous determination of human neutrophil elastase inhibitor (ONO-5046) and its metabolite in plasma and urine by direct injection column-switching HPLC, *J.Pharm.Biomed.Anal.*, **1994**, *12*, 397–405.

# Sodium oxybate

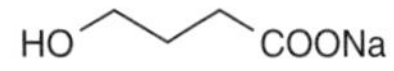

**Molecular formula:** $C_4H_7NaO_3$
**Molecular weight:** 126.09
**CAS Registry No:** 502-85-2, 591-81-1 (free acid)
**Merck Index:** 13, 4840

## SAMPLE

**Matrix:** solutions

**Sample preparation:** Mix 200 μL of a solution in MeOH:water 10:90 with 150 μL 20 mM 0.5 mM tetrakis(decyl)ammonium bromide in 5 mM pH 7.0 phosphate buffer and 100 μL 4.2 mg/mL 2-bromoacetyl-6-methoxynaphthalene in acetone, stir at 70° for 65 min, add 150 μL 80 μg/mL IS in MeCN, sonicate for 1 min, inject a 50 μL aliquot. (Preparation of 2-bromoacetyl-6-methoxynaphthalene is given in *Organic Syntheses*, Collective Volume 6, page 175; available free of cost at www.orgsyn.org.)

## HPLC VARIABLES

**Column:** 250 × 4.5 Hypersil 5 ODS
**Column temperature:** 35
**Mobile phase:** MeCN:MeOH:25 mM pH 4.5 phosphate buffer 18.7:15.3:66
**Flow rate:** 1.7
**Injection volume:** 50
**Detector:** F ex 300 em 460

## CHROMATOGRAM

**Retention time:** 35

**Internal standard:** 4-(6-methoxy-2-naphthyl)-4-oxobutanoic acid (Synthesis of 4-(6-methoxy-2-naphthyl)-4-oxobutanoic acid is as follows. Dissolve 5 g 2-acetyl-6-methoxynaphthalene (6′-methoxy-2′-acetonaphthone) in the minimum amount of warm glacial acetic acid, add 2.5 g glyoxylic acid, reflux for 24 h, evaporate to dryness under reduced pressure, dissolve the residue in chloroform, extract three times with 5% sodium carbonate solution (Caution! Chloroform is a carcinogen!). Combine the extracts and acidify them with concentrated HCl, filter to recover the product, recrystallize from MeOH/water to obtain 4-(6-methoxy-2-naphthyl)-4-oxo-2-butenoic acid (mp 165–168°). Dissolve 4 g 4-(6-methoxy-2-naphthyl)-4-oxo-2-butenoic acid in the minimum amount of THF, add palladium on charcoal, hydrogenate until 450 mL hydrogen are absorbed, filter, evaporate to dryness under reduced pressure, recrystallize from acetic acid to obtain 4-(6-methoxy-2-naphthyl)-4-oxobutanoic acid as a white solid (mp 148°) (*Farmaco Ed.Sci.* **1982**, *37*, 171).) (25)

## KEY WORDS

derivatization

## REFERENCE

Gatti, R.; Bousquet, E.; Bonazzi, D.; Cavrini, V. Determination of carboxylic acid salts in pharmaceuticals by high-performance liquid chromatography after pre-column fluorogenic labelling, *Biomed.Chromatogr.*, **1996**, *10*, 19–24.

# Somatropin

**Molecular formula:** $C_{990}H_{1529}N_{263}O_{299}S_7$
**Molecular weight:** 22124.12
**CAS Registry No:** 9002-72-6,
12629-01-5 (human)
**Merck Index:** 13, 8789

## SAMPLE

**Matrix:** blood

**Sample preparation:** Prepare an immunoaffinity chromatography column by circulating 5 mL 500 μg/mL polyclonal antibodies raised against porcine somatotropin in 200 mM pH 8.3 sodium bicarbonate containing 500 mM NaCl through a 1 mL Hitrap NHS-activated immunoaffinity chromatography column (Amersham Pharmacia) for 12 h. Wash and deactivate the column by passing in sequence three times 6 mL 500 mM pH 8.3 ethanolamine containing 500 mM NaCl and 6 mL 100 mM pH 4 sodium acetate containing 500 mM NaCl. Slowly pass 10 mL serum (0.5 drop/s) through the column, wash with 5 mL 200 mM pH 8.3 sodium bicarbonate containing 500 mM NaCl, elute with 5 mL EtOH:water 70:30, concentrate with a diafiltration device (MW cut-off 10 000 Da, Vivasciences) while centrifuging at 6000 g for 30 min, inject a 25 μL aliquot of the residue. Alternatively, dissolve 1 g urea in 10 mL plasma, filter with a diafiltration device (MW cut-off 30 000 Da, Vivasciences) while centrifuging at 6000 g for 30 min, pass 5 mL through a Hiload 26/60 Superdex 75 pg GPC column (Amersham Pharmacia), concentrate the appropriate fraction with a diafiltration device (MW cut-off 10 000 Da, Vivasciences) while centrifuging at 6000 g for 30 min, inject a 25 μL aliquot of the residue.

## HPLC VARIABLES

**Column:** 50 × 1 3.5 μm Zorbax Extend-C18

**Mobile phase:** Gradient. MeCN:1% acetic acid from 10:90 to 90:10 over 5 min

**Flow rate:** 0.06

**Injection volume:** 25

**Detector:** MS, Agilent 1100 MSD, electrospray, positive ion, nebulizer 20 psi, capillary 5000 V, drying gas 7 L/min at 350°, voltage applied to capillary 190 V, m/z 1279.2–1359.1 (natural porcine), m/z 1270.5–1349.5 (recombinant porcine)

## CHROMATOGRAM

**Limit of detection:** 10 ng/mL

## KEY WORDS

immunoaffinity; pig; plasma; serum; ultrafiltrate

## REFERENCE

Blokland, M.H.; Sterk, S.S.; van Ginkel, L.A.; Stephany, R.W.; Heck, A.J.R. Analysis for endogenous and recombinant porcine somatotropine in serum, *Anal.Chim.Acta*, **2003**, *483*, 201–206.

# Squalane

**Molecular formula:** $C_{30}H_{62}$
**Molecular weight:** 422.81
**CAS Registry No:** 111-01-3
**Merck Index:** 13, 8846

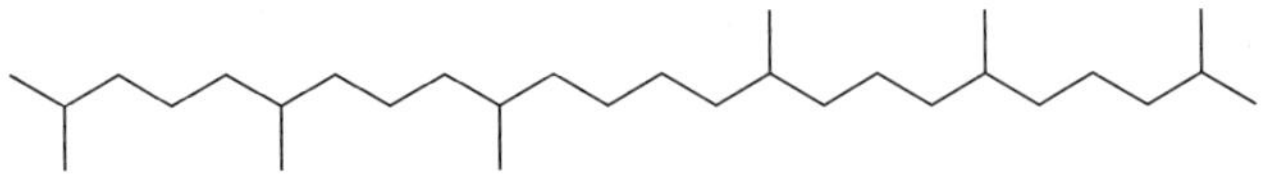

## SAMPLE

**Matrix:** cell cultures

**Sample preparation:** Centrifuge cell culture at 2772 g at 4° for 20 min, extract the supernatant with diethyl ether. Evaporate the extracts to dryness, reconstitute the residue with *n*-propanol, inject an aliquot.

## HPLC VARIABLES

**Column:** Nucleosil-100 C18
**Mobile phase:** *n*-Propanol
**Flow rate:** 0.5
**Detector:** Refractive Index

## REFERENCE

Berekaa, M.M.; Steinbüchel, A. Microbial degradation of the multiply branched alkane 2,6,10,15,19,23-hexamethyltetracosane (squalane) by *Mycobacterium fortuitum* and *Mycobacterium ratisbonense*, *Appl.Environ.Microbiol.*, **2000**, *66*, 4462–4467.

# Squalene

**Molecular formula:** $C_{30}H_{50}$
**Molecular weight:** 410.72
**CAS Registry No:** 111-02-4
**Merck Index:** 13, 8847

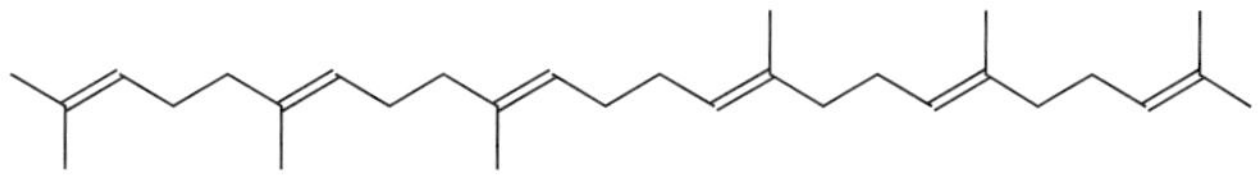

## SAMPLE

**Matrix:** cell cultures
**Sample preparation:** Centrifuge cell culture at 2772 g at 4° for 20 min, extract the supernatant with diethyl ether. Evaporate the extracts to dryness, reconstitute the residue with *n*-propanol, inject an aliquot.

## HPLC VARIABLES

**Column:** Nucleosil-100 C18
**Mobile phase:** *n*-Propanol
**Flow rate:** 0.5
**Detector:** UV (wavelength not specified)

## REFERENCE

Berekaa, M.M.; Steinbüchel, A. Microbial degradation of the multiply branched alkane 2,6,10,15,19,23-hexamethyltetracosane (squalane) by *Mycobacterium fortuitum* and *Mycobacterium ratisbonense*, *Appl.Environ.Microbiol.*, **2000**, *66*, 4462–4467.

## SAMPLE

**Matrix:** food
**Sample preparation:** Mix 500 mg solid food or milk powder, 5 g liquid milk, or 100–200 mg oil or fat with 10 mL EtOH, add 2 mL 50% KOH solution, heat at 70° for 8 min with periodic agitation, cool, add 20 mL hexane:diisopropyl ether 75:25, shake mechanically for 5 min, add 30 mL water, invert 10 times, centrifuge at 180 g for 10 min. Evaporate 10 mL of the organic layer to dryness at <45°, reconstitute the residue with 1 mL EtOH or hexane, inject a 20–50 μL aliquot.

## HPLC VARIABLES

**Guard column:** Guard-Pak C18
**Column:** 5 μm Rad-Pak C18
**Mobile phase:** MeOH
**Flow rate:** 1
**Injection volume:** 20–50
**Detector:** UV 212

## CHROMATOGRAM

**Retention time:** 32

## OTHER SUBSTANCES

**Simultaneous:** campesterol (28), cholesterol (24), β-sitosterol (30), α-tocopherol (15), δ-tocopherol (12), γ-tocopherol (14)

## REFERENCE

Indyk, H.E. Simultaneous liquid chromatographic determination of cholesterol, phytosterols and tocopherols in foods, *Analyst*, **1990**, *115*, 1525–1530.

## ANNOTATED BIBLIOGRAPHY

Domnas, A.; Biswas, S.S.; Gallagher, P.A. Squalene metabolism in two species of *Lagenidium*, *Can.J. Microbiol.*, **1994**, *40*, 523–531.

Piretti, M.V.; Pagliuca, G.; Tarozzi, G. Simultaneous reversed-phase high-performance liquid chromatographic separation of non-polar isoprenoid lipids and their determination, *J.Chromatogr.B*, **1995**, *674*, 177–185.

Sulpice, J.C.; Ferezou, J. Squalene isolation by HPLC and quantitative comparison by HPLC and GLC, *Lipids*, **1984**, *19*, 631–635.

# Stanozolol

**Molecular formula:** $C_{21}H_{32}N_2O$
**Molecular weight:** 328.49
**CAS Registry No:** 10418-03-8
**Merck Index:** 13, 8873

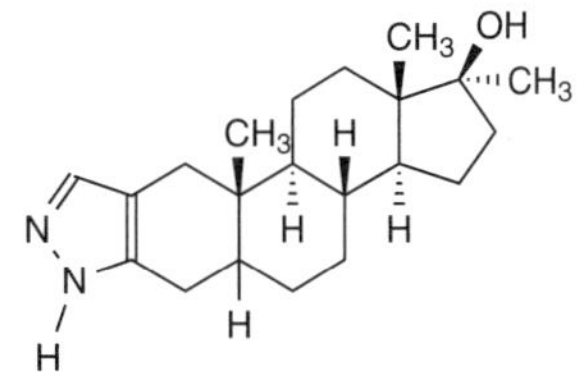

## SAMPLE

**Matrix:** blood
**Sample preparation:** Condition a 1 mL 100 mg Bond Elut C1 SPE cartridge with 2 column vol of MeCN and 2 column vol of water. Mix 500 μL serum with 25 μL 5 μg/mL IS in MeCN, add 500 μL water, vortex for 5 s, add to the SPE cartridge, wash with 2 column vol of water, wash with 1 column vol of MeCN:water 10:90, elute with 1 mL MeCN:water 45:55, concentrate the eluate to 250 μL under a stream of nitrogen at 45°, inject a 50 μL aliquot.

## HPLC VARIABLES

**Guard column:** 15 × 4.6 7 μm silica (Brownlee)
**Column:** 220 × 4.6 5 μm silica (Brownlee)
**Column temperature:** 60
**Mobile phase:** MeCN:100 mM pH 2.5 sodium phosphate buffer 15:85
**Flow rate:** 1
**Injection volume:** 50
**Detector:** UV 230

## CHROMATOGRAM

**Retention time:** 12.7
**Internal standard:** spironolactone (7.8)
**Limit of detection:** 7 ng/mL

## OTHER SUBSTANCES

**Extracted:** fluoxymesterone (UV 247) (5.4), methandrostenolone (UV 247) (9.3), methyltestosterone (UV 247) (9.6), nandrolone (UV 247) (8.2), testosterone (UV 247) (8.6), zeranol (UV 263) (3.6)

## KEY WORDS

serum: SPE

## REFERENCE

Lampert, B.L.; Stewart, J.T. Determination of anabolic steroids and zeranol in human serum by isocratic reverse phase HPLC on silica, *J.Liq.Chromatogr.*, **1989**, *12*, 3231–3249.

# Succimer

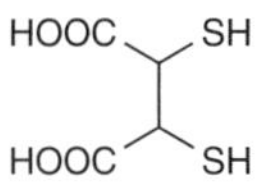

**Molecular formula:** $C_4H_6O_4S_2$
**Molecular weight:** 182.22
**CAS Registry No:** 304-55-2
**Merck Index:** 13, 8949

## SAMPLE

**Matrix:** blood
**Sample preparation:** Mix 100–250 μL whole blood or plasma with 50 μL 50–125 mM dithiothreitol and 100 mM pH 8.3 ammonium bicarbonate buffer containing IS to a final volume of 2 mL, purge headspace with nitrogen, mix vigorously for 1 min, let stand in the dark at room temperature for 30 min, filter (2 mL Centricon 30) while centrifuging at 6000 rpm for 1 h. Add 200 μL 80 mM monobromobimane in 100 mM ammonium bicarbonate buffer to the ultrafiltrate, purge with nitrogen, mix vigorously for 1 min, let stand in the dark at room temperature for 10 min, wash with 2 vol of dichloromethane, shake for 15 s, centrifuge at 2500 rpm for 2 min, repeat the wash. Add 17 μL 6 M HCl to the aqueous layer, inject an aliquot.

## HPLC VARIABLES

**Column:** 250 × 4.6 5 μm Spherisorb ODS-2
**Mobile phase:** Gradient. A was 10 mM tetrabutylammonium bromide in MeOH containing 10 mM sodium acetate. B was 10 mM tetrabutylammonium bromide in 10 mM pH 4.1 acetate buffer. A:B 52.5:47.5 for 10 min, to 90:10 over 2 min, maintain at 90:10 for 5 min, return to initial conditions over 2 min, re-equilibrate at initial conditions for 11 min.
**Flow rate:** 1
**Injection volume:** 10
**Detector:** F ex 356 em 450 (see also Maiorino,R.M.; Barry,T.J.; Aposhian,H.V. Determination and metabolism of dithiol chelating agents: Electrolytic and chemical reduction of oxidized dithiols in urine. *Anal.Biochem.* **1987**, *160*, 217–226.)

## CHROMATOGRAM

**Retention time:** 7.1
**Internal standard:** 2,3-dimercaptopropane-1-sulfonic acid (10.2)
**Limit of quantitation:** 5 μM

## OTHER SUBSTANCES

**Extracted:** dithiothreitol (5.5)

## KEY WORDS

derivatization; plasma; whole blood

## REFERENCE

Maiorino, R.M.; Akins, J.M.; Blaha, K.; Carter, D.E.; Aposhian, H.V. Determination and metabolism of dithiol chelating agents: X. In humans, *meso*-2,3-dimercaptosuccinic acid is bound to plasma proteins *via* mixed disulfide formation, *J.Pharmacol.Exp.Ther.*, **1990**, *254*, 570–577.

# Succinylcholine chloride

**Molecular formula:** $C_{14}H_{30}Cl_2N_2O_4$
**Molecular weight:** 361.31
**CAS Registry No:** 71-27-2, 55-94-7 (bromide), 541-19-5 (iodide)
**Merck Index:** 13, 8959 (chloride), 8958 (bromide), 8960 (iodide)

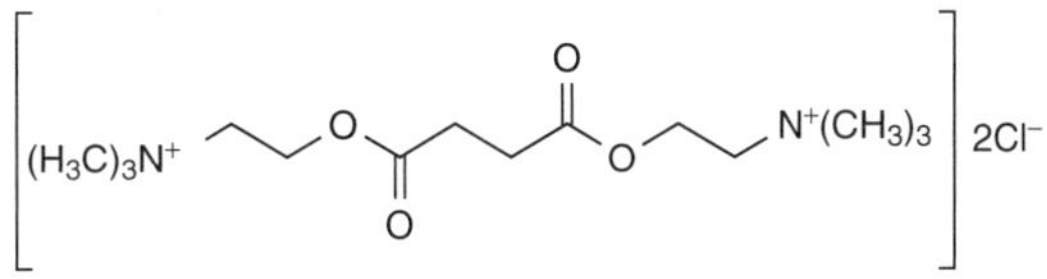

## SAMPLE
**Matrix:** blood
**Sample preparation:** Condition a Bond Elut C2 SPE cartridge with 3 mL TMAH buffer and 3 mL water. Mix 1 mL plasma with 1 mL water and 100 μL 100 ng/mL IS in 100 mM pH 5.0 phosphate buffer, add to the SPE cartridge, wash with 3 mL water, wash with 3 mL MeCN, wash with 3 mL MeOH, elute with two 250 μL portions of TMAH buffer, inject a 150 μL aliquot of the eluate. (Prepare TMAH buffer by dissolving 275.6 mg tetramethylammonium chloride in 1 mL water and adding 240 mL MeOH, adjust to apparent pH 3.0 with 100 mM HCl, make up to 250 mL with MeOH.)

## HPLC VARIABLES
**Column:** 125 × 4.6 5 μm Spherisorb CN
**Column temperature:** 35
**Mobile phase:** MeCN:MeOH:30 mM phosphoric acid 35:25:45, adjusted to apparent pH 5.00 with concentrated ammonium hydroxide
**Flow rate:** 2
**Injection volume:** 150
**Detector:** E, ESA Coulochem II, 5010 analytical cell, detector 1 450 mV (screen), detector 2 750 mV (monitored)

## CHROMATOGRAM
**Retention time:** 9.2
**Internal standard:** pipecuronium (19.8)
**Limit of quantitation:** 250 ng/mL

## OTHER SUBSTANCES
**Simultaneous:** mivacurium (21), rocuronium (18), vecuronium (21)

## KEY WORDS
plasma; SPE

## REFERENCE
Gao, H.; Roy, S.; Donati, F.; Varin, F. Determination of succinylcholine in human plasma by high-performance liquid chromatography with electrochemical detection, *J.Chromatogr.B*, **1998**, *718*, 129–134.

## SAMPLE
**Matrix:** blood
**Sample preparation:** Vortex thoroughly 400 μL plasma with 40 μL 500 mg/mL trichloroacetic acid, centrifuge at 12 500 g for 5 min, inject a 100 μL aliquot.

## HPLC VARIABLES
**Column:** 250 × 4.6 8 μm Cp-tm-Spher C8 (Chrompack)
**Mobile phase:** MeCN:MeOH:50 mM pH 5.0 potassium phosphate buffer 35:5:60
**Flow rate:** 1.2

**Injection volume:** 100
**Detector:** F ex 257 em 282

## CHROMATOGRAM

**Retention time:** 6.3
**Limit of quantitation:** 100 ng/mL

## KEY WORDS

pharmacokinetics; plasma

## REFERENCE

Lagerwerf, A.J.; Vanlinthout, L.E.; Vree, T.B. Rapid determination of succinylcholine in human plasma by high-performance liquid chromatography with fluorescence detection, *J.Chromatogr.*, **1991**, *570*, 390–395.

# Sulfabromomethazine

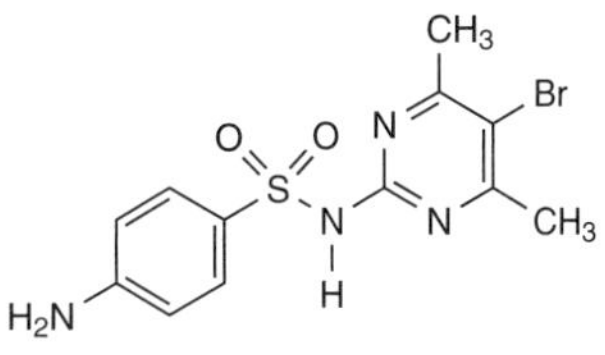

**Molecular formula:** $C_{12}H_{13}BrN_4O_2S$
**Molecular weight:** 357.23
**CAS Registry No:** 116-45-0
**Merck Index:** 13, 8983

## SAMPLE

**Matrix:** tissue
**Sample preparation:** Cut 20 g tissue into small pieces and add to 25 mL dichloromethane, let stand for 20–30 min with frequent stirring with a glass rod, filter through glass wool, repeat extraction twice more. Combine the dichloromethane layers and shake with 10 mL 1.5 M sulfuric acid for 15 min. Remove the aqueous phase and add it to 10 mL dichloromethane, shake for 10 min, discard the organic phase, repeat the wash, add 1 mL 10% dipotassium hydrogen phosphate to the aqueous phase, add 3 mL 40% NaOH, mix well, allow to cool, adjust pH to 5–6 with 1.5 M sulfuric acid or 40% NaOH, add 5 mL dichloromethane, extract for 15 min, repeat the extraction. Pass the organic layers through anhydrous sodium sulfate, evaporate to 1 g on a hot plate at 60–70° (in a hood!), inject a 10 μL aliquot. For confirmation of sulfabromomethazine, add 1 mL MeOH to the extract, add 2 drops acetic anhydride, heat at 80–90° on a hot plate until acetic acid fumes are no longer seen, reconstitute with 1 g dichloromethane, inject a 10 μL aliquot. (Note: Extraction from the sample matrix is only demonstrated for sulfamethazine, but it should work for sulfabromomethazine.)

## HPLC VARIABLES

**Column:** 250 mm long MicroPak CN-10
**Mobile phase:** Isooctane:chloroform:MeOH:acetic acid 30.5:65:4:0.5 (Caution! Chloroform is a carcinogen!)
**Flow rate:** 0.33
**Injection volume:** 10
**Detector:** UV 254

## CHROMATOGRAM

**Retention time:** 3.56 (4.16 for acetyl sulfabromomethazine)
**Limit of detection:** 20 ppb (for sulfamethazine)

## OTHER SUBSTANCES

**Extracted:** acetylsulfachlorpyridazine (6.68), acetylsulfadiazine (6.78), acetylsulfadimethoxine (4.70), acetylsulfaethoxypyridazine (5.23), acetylsulfamethazine (5.34), sulfachlorpyridazine (5.60), sulfadiazine (5.55), sulfadimethoxine (3.99), sulfaethoxypyridazine (4.60), sulfamethazine (4.34)
**Noninterfering:** sulfamerazine, sulfathiazole, sulfanilamide, sulfapyridine, sulfaquinoline

## KEY WORDS

cow; liver; fat; kidney; muscle

## REFERENCE

Seymour, D.; Rupe, B.D. High-pressure liquid chromatographic determination of sulfamethazine residues in beef tissues, *J.Pharm.Sci.*, **1980**, *69*, 701–703.

# Sulfachlorpyridazine

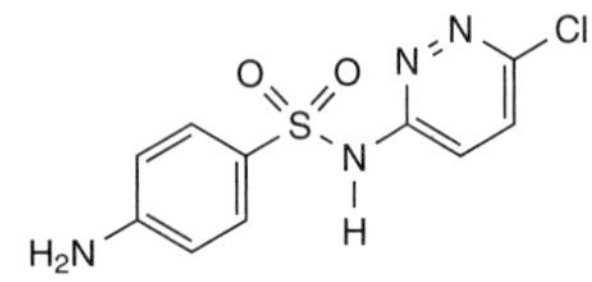


**Molecular formula:** $C_{10}H_9ClN_4O_2S$
**Molecular weight:** 284.73
**CAS Registry No:** 80-32-0, 23282-55-5 (Na salt)
**Merck Index:** 13, 8985

## SAMPLE

**Matrix:** honey
**Sample preparation:** Condition a 200 mg Oasis HLB SPE cartridge with 3 mL MeCN and two 2 mL portions of water. Dissolve 7.5 g honey in 15 mL 2 M HCl, let stand at room temperature for 30 min, add 30 mL 300 mM citric acid, mix, filter. Adjust the pH of a 20 mL aliquot of the filtrate to 3.5–4.5 with 25% ammonia, immediately add to the SPE cartridge and pass through within 10–15 min, wash with three 3 mL portions of water, let dry for 4 min, elute with 3 mL MeCN. Evaporate the eluate to a small volume under reduced pressure at 40°, add 500 μL mobile phase, vortex, inject a 10 μL aliquot.

## HPLC VARIABLES

**Guard column:** 2 mm long 5 μm Nucleosil 100 C18 HD
**Column:** 50 × 2 5 μm Nucleosil 100-5 C18 HD
**Mobile phase:** Gradient. A was MeCN:water:formic acid 5:94.7:0.3. B was MeCN:formic acid 99.7:0.3. A:B from 100:0 to 70:30 over 10 min, maintain at 70:30 for 2 min, return to initial conditions over 0.1 min, re-equilibrate for 6.9 min.
**Flow rate:** 0.2
**Injection volume:** 10
**Detector:** MS, Micromass Quattro LCZ, triple quadrupole, electrospray, capillary 3.25 kV, cone 25 V, extractor 3 V, RF lens 0.3 V, source 90°, desolvation 250°, cone gas flow 50 L/h, desolvation gas nitrogen 560 L/h, nebulizer gas flow is factory preset, LM 1 resolution 12, HM 1 resolution 12, ion energy 1 V, entrance voltage −5 V, exit voltage 0 V, collision cell pressure 2 μbar argon, LM 2 resolution 12, HM 2 resolution 12, ion energy 2 V, multiplier voltage 650 V, m/z 285–156

## CHROMATOGRAM

**Retention time:** 12.9
**Limit of detection:** 1 ppb

## OTHER SUBSTANCES

**Extracted:** chlortetracycline (m/z 479–444) (LOD 1 ppb) (13.1), flumequine (m/z 262–244) (LOD 1 ppb) (17.9), oxytetracycline (m/z 461–426) (LOD 4 ppb) (11.1), sulfacetamide (m/z 215–156) (LOD 4 ppb) (5.6), sulfachloropyrazine (m/z 285–156) (LOD 1 ppb) (15.8), sulfadiazine (m/z 251–156) (LOD 2 ppb) (6.5), sulfadimethoxine (m/z 311–156) (LOD 1 ppb) (16.0), sulfadoxine (m/z 311–156) (LOD 0.4 ppb) (13.5), sulfaguanidine (m/z 215–156) (LOD 2 ppb) (2.0), sulfamerazine (m/z 265–156) (LOD 1 ppb) (8.9), sulfamethazine (m/z 279–186) (LOD 1 ppb) (10.4), sulfamethizole (m/z 271–156) (LOD 0.5 ppb) (11.5), sulfamethoxazole (m/z 254–156) (LOD 1 ppb) (14.0), sulfamethoxypyridazine (m/z 281–156) (LOD 0.4 ppb) (11.5), sulfanilamide (m/z 173–156) (LOD 4 ppb) (2.4), sulfapyridine (m/z 250–156) (LOD 11 ppb) (8.3), sulfathiazole (m/z 256>) (LOD 156 ppb) (1 8.6), sulfisoxazole (m/z 268–156) (LOD 1 ppb) (14.9), tetracycline (m/z 445–410) (LOD 4 ppb) (10.5)

## KEY WORDS

SPE

## REFERENCE

Kaufmann, A.; Roth, S.; Ryser, B.; Widmer, M.; Guggisberg, D. Quantitative LC/MS-MS determination of sulfonamides and some other antibiotics in honey, *J.AOAC Int.*, **2002**, *85*, 853–860.

## SAMPLE
**Matrix:** solutions
**Sample preparation:** Inject an aliquot of a solution in MeOH.

## HPLC VARIABLES
**Column:** 300 × 3.9 μBondapak C18
**Mobile phase:** MeCN:water:acetic acid 12.5:86.5:1
**Flow rate:** 1.6
**Injection volume:** 10
**Detector:** UV 254

## CHROMATOGRAM
**Retention time:** 13

## OTHER SUBSTANCES
**Simultaneous:** sulfabenzamide (25), sulfacetamide (4), sulfadiazine (5), sulfadimethoxine (40), sulfamerazine (7), sulfamethazine (8), sulfamethizole (9), sulfamethoxazole (16), sulfanilamide (2.5), sulfapyridine (6), sulfisoxazole (20)

## REFERENCE
Roos, R.W. High pressure liquid chromatographic determination of sulfisoxazole in pharmaceuticals and separation patterns of other sulfonamides, *J.Assoc.Off.Anal.Chem.*, **1981**, *64*, 851–854.

## SAMPLE
**Matrix:** tissue
**Sample preparation:** Condition a 500 mg Chromabond SA cation-exchange SPE cartridge (Macherey-Nagel) with 6 mL hexane, dry under vacuum for 10 min, condition with 6 mL dichloromethane:acetone:acetic acid 50:50:2, do not allow to go dry. Homogenize (Polytron) 10 g sample with 60 mL dichloromethane:acetone 50:50 for 30 s, rinse the apparatus with 2–3 mL dichloromethane:acetone 50:50, centrifuge the mixture at 2500 rpm for 10 min. Filter (cotton wool) the supernatant and wash it through with a little dichloromethane:acetone 50:50, add 5 mL acetic acid to the filtrate, mix, remove one-tenth of this mixture and add it to the SPE cartridge at 2 mL/min, do not allow the SPE cartridge to run dry, wash with 5 mL water, wash with 5 mL MeOH, dry under vacuum for 10 min, pass gaseous ammonia through the SPE cartridge until the acid is neutralized (when air is passed through the cartridge, moist pH paper should turn blue), elute with 3 mL MeOH at 1–2 mL/min, carefully evaporate to dryness under reduced pressure (100 mbar) at 40°, reconstitute with 500 μL initial mobile phase, centrifuge, inject a 50 μL aliquot of the supernatant.

## HPLC VARIABLES
**Column:** 125 × 4 5 μm LiChrospher 100 RP-18
**Mobile phase:** Gradient. A was MeCN:20 mM pH 5 sodium acetate buffer 5.5:94.5. B was MeCN:EtOH:20 mM pH 5 sodium acetate buffer 50:10:40. A:B from 100:0 to 0:100 over 32 min (concave gradient), return to initial conditions over 4 min, re-equilibrate at initial conditions for 10 min.
**Flow rate:** 0.8
**Injection volume:** 50
**Detector:** UV 270; F ex 395 em 495 following post-column reaction. The column effluent mixed with ice-cold reagent pumped at 0.3 mL/min and this mixture flowed through a 2.3 m × 0.5 mm ID coil in a cooled ultrasonic bath to the detector. (Prepare reagent by dissolving 25 mg fluorescamine in 25 mL MeCN and adding 75 mL buffer and 200 μL mercaptoethanol. The buffer was 20 mM sodium dihydrogen phosphate adjusted to pH 3 with 1 M phosphoric acid.)

## CHROMATOGRAM

**Retention time:** 22

**Limit of detection:** 0.5–5 ppb

## OTHER SUBSTANCES

**Extracted:** sulfadiazine (10), sulfadimethoxine (28), sulfadoxine (24), sulfaguanidine (3), sulfamerazine (16.5), sulfamethazine (20), sulfamethizole (18), sulfamethoxypyridazine (21), sulfanilamide (3.7), sulfapyridine (15.7), sulfathiazole (15)

## KEY WORDS

kidney; muscle; post-column reaction; SPE

## REFERENCE

Pacciarelli, B.; Reber, S.; Douglas, C.; Dietrich, S.; Etter, R. Bestimmung von 12 Sulfonamiden in Fleisch und Nieren mittels HPLC und Nachsäulenderivatisierung [Determination of 12 sulfonamides in meat and kidney by HPLC and post column derivatization], *Mitt.geb.Lebensmittelunters.Hyg.*, **1991**, *82*, 45–55.

## SAMPLE

**Matrix:** tissue

**Sample preparation:** Cut 20 g tissue into small pieces and add to 25 mL dichloromethane, let stand for 20–30 min with frequent stirring with a glass rod, filter through glass wool, repeat extraction twice more. Combine the dichloromethane layers and shake with 10 mL 1.5 M sulfuric acid for 15 min. Remove the aqueous phase and add it to 10 mL dichloromethane, shake for 10 min, discard the organic phase, repeat the wash, add 1 mL 10% dipotassium hydrogen phosphate to the aqueous phase, add 3 mL 40% NaOH, mix well, allow to cool, adjust pH to 5–6 with 1.5 M sulfuric acid or 40% NaOH, add 5 mL dichloromethane, extract for 15 min, repeat the extraction. Pass the organic layers through anhydrous sodium sulfate, evaporate to 1 g on a hot plate at 60–70° (in a hood!), inject a 10 µL aliquot. For confirmation of sulfachlorpyridazine, add 1 mL MeOH to the extract, add 2 drops acetic anhydride, heat at 80–90° on a hot plate until acetic acid fumes are no longer seen, reconstitute with 1 g dichloromethane, inject a 10 µL aliquot. (Note: Extraction from the sample matrix is only demonstrated for sulfamethazine, but it should work for sulfachlorpyridazine.)

## HPLC VARIABLES

**Column:** 250 mm long MicroPak CN-10

**Mobile phase:** Isooctane:chloroform:MeOH:acetic acid 30.5:65:4:0.5 (Caution! Chloroform is a carcinogen!)

**Flow rate:** 0.33

**Injection volume:** 10

**Detector:** UV 254

## CHROMATOGRAM

**Retention time:** 5.60 (6.68 for acetylsulfachlorpyridazine)

**Limit of detection:** 20 ppb (for sulfamethazine)

## OTHER SUBSTANCES

**Extracted:** acetylsulfabromomethazine (4.16), acetylsulfadiazine (6.78), acetylsulfadimethoxine (4.70), acetylsulfaethoxypyridazine (5.23), acetylsulfamethazine (5.34), sulfabromomethazine (3.56), sulfadiazine (5.55), sulfadimethoxine (3.99), sulfaethoxypyridazine (4.60), sulfamethazine (4.34)

**Noninterfering:** sulfamerazine, sulfathiazole, sulfanilamide, sulfapyridine, sulfaquinoline

## KEY WORDS

cow; liver; fat; kidney; muscle

## REFERENCE

Seymour, D.; Rupe, B.D. High-pressure liquid chromatographic determination of sulfamethazine residues in beef tissues, *J.Pharm.Sci.*, **1980**, *69*, 701–703.

## ANNOTATED BIBLIOGRAPHY

Combs, M.T.; Ashraf-Khorassani, M.; Taylor, L.T. Method development for the separation of sulfonamides by supercritical fluid chromatography, *J.Chromatogr.Sci.*, **1997**, *35*, 176–180. [sulfamethazine; sulfamerazine; sulfapyridine; sulfadimethoxine; sulfadiazine; sulfaquinoxaline; sulfachlorpyridazine; sulfathiazole]

Combs, M.T.; Ashraf-Khorassani, M.; Taylor, L.T. HPLC/atmospheric pressure chemical ionization-mass spectroscopy of eight regulated sulfonamides, *J.Pharm.Biomed.Anal.*, **1999**, *19*, 301–308. [LOD 0.05–6 ng; sulfathiazole; sulfamethazine; sulfamerazine; sulfapyridine; sulfadimethoxine; sulfadiazine; sulfaquinoxaline; sulfachlorpyridazine]

McGrane, M.; O'Keeffe, M.; Smyth, M.R. The analysis of sulphonamide drug residues in pork muscle using automated dialysis, *Anal.Lett.*, **1999**, *32*, 481–495. [LOD 40 ng/g; sulfadiazine; sulfathiazole; sulfapyridine; sulfamerazine; sulfamethizole; sulfamethazine; sulfamethoxypyridazine; sulfachlorpyridazine; sulfisoxazole]

Thomas, G.K.; Millar, R.G.; Anstis, P.W. Stability of sulfonamide antibiotics in spiked pig liver tissue during frozen storage, *J.AOAC Int.*, **1997**, *80*, 988–995. [sulfamethazine; sulfadimethoxine; sulfachlorpyridazine; sulfathiazole; sulfaquinoxaline]

# Sulfaethoxypyridazine

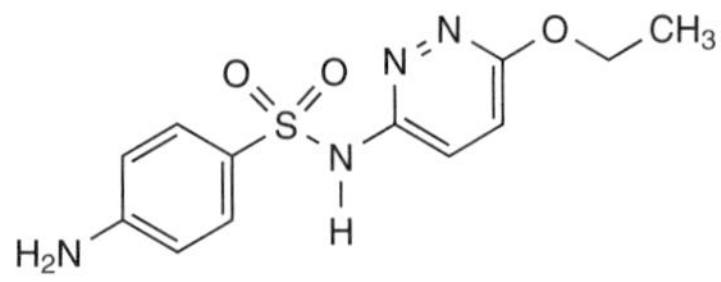

**Molecular formula:** $C_{12}H_{14}N_4O_3S$
**Molecular weight:** 294.32
**CAS Registry No:** 963-14-4

## SAMPLE

**Matrix:** tissue
**Sample preparation:** Condition a 3 mL 500 mg Sep-Pak C18 SPE cartridge with 20 mL MeOH and 20 mL water. Shake 5 g homogenized tissue and 25 mL chloroform mechanically for 2 min (Caution! Chloroform is a carcinogen!), centrifuge at 3000 g for 5 min, repeat the extraction. Combine the lower organic layers, add 10 mL 10% NaCl in 100 mM NaOH, shake vigorously for 1 min. Remove the upper aqueous layer and centrifuge it at 1500 g for 10 min. Remove 8 mL of the upper aqueous layer and add it to 10 mL 1 M pH 6 sodium dihydrogen phosphate, vortex for 20 s, add to the SPE cartridge, wash with 20 mL water, elute with 1 mL MeOH. Add 2 mL 50 mM sodium dihydrogen phosphate to the eluate, filter (0.45 μm), inject a 20–50 μL aliquot. (SPE cartridge must have a carbon loading of 14%.)

## HPLC VARIABLES

**Column:** 250 × 4.6 5 μm Spherisorb C18 ODS
**Mobile phase:** MeOH:50 mM sodium dihydrogen phosphate 30:70
**Flow rate:** 0.8
**Injection volume:** 20–50
**Detector:** UV 265

## CHROMATOGRAM

**Retention time:** 24.5
**Limit of detection:** 2 ng/g (for sulfamethazine)

## OTHER SUBSTANCES

**Extracted:** sulfamethazine (10.5)
**Simultaneous:** sulfachlorpyridazine (12.1), sulfadiazine (3.8), sulfadimethoxine (45), sulfadoxine (17.2), sulfamerazine (6.4), sulfamethoxazole (14), sulfamethoxypyridazine (11.7), sulfanilamide (8.0), sulfathiazole (4.2)

## KEY WORDS

kidney; liver; muscle; pig; SPE; the method is validated for sulfamethazine and sulfaethoxypyridazine is the internal standard

## REFERENCE

Boison, J.O.K.; Keng, L.J.-Y. Determination of sulfamethazine in bovine and porcine tissues by reversed-phase liquid chromatography, *J.AOAC Int.*, **1994**, *77*, 558–564.

## SAMPLE

**Matrix:** tissue
**Sample preparation:** Cut 20 g tissue into small pieces and add to 25 mL dichloromethane, let stand for 20–30 min with frequent stirring with a glass rod, filter through glass wool, repeat extraction twice more. Combine the dichloromethane layers and shake with 10 mL 1.5 M sulfuric acid for 15 min. Remove the aqueous phase and add it to 10 mL dichloromethane, shake for 10 min, discard the organic phase, repeat the wash, add 1 mL 10% dipotassium hydrogen phosphate to the aqueous phase, add 3 mL 40% NaOH, mix well, allow to cool, adjust pH to 5–6 with 1.5 M sulfuric acid or 40% NaOH, add 5 mL dichloromethane, extract for 15 min, repeat the extraction. Pass the

organic layers through anhydrous sodium sulfate, evaporate to 1 g on a hot plate at 60–70° (in a hood!), inject a 10 μL aliquot. For confirmation of sulfaethoxypyridazine, add 1 mL MeOH to the extract, add 2 drops acetic anhydride, heat at 80–90° on a hot plate until acetic acid fumes are no longer seen, reconstitute with 1 g dichloromethane, inject a 10 μL aliquot. (Note: Extraction from the sample matrix is only demonstrated for sulfamethazine, but it should work for sulfaethoxypyridazine.)

## HPLC VARIABLES

**Column:** 250 mm long MicroPak CN-10
**Mobile phase:** Isooctane:chloroform:MeOH:acetic acid 30.5:65:4:0.5 (Caution! Chloroform is a carcinogen!)
**Flow rate:** 0.33
**Injection volume:** 10
**Detector:** UV 254

## CHROMATOGRAM

**Retention time:** 4.60 (5.23 for acetylsulfaethoxypyridazine)
**Limit of detection:** 20 ppb (for sulfamethazine)

## OTHER SUBSTANCES

**Extracted:** acetylsulfabromomethazine (4.16), acetylsulfachlorpyridazine (6.68), acetylsulfadiazine (6.78), acetylsulfadimethoxine (4.70), acetylsulfamethazine (5.34), sulfabromomethazine (3.56), sulfachlorpyridazine (5.60), sulfadiazine (5.55), sulfadimethoxine (3.99), sulfamethazine (4.34)
**Noninterfering:** sulfamerazine, sulfathiazole, sulfanilamide, sulfapyridine, sulfaquinoline

## KEY WORDS

cow; liver; fat; kidney; muscle

## REFERENCE

Seymour, D.; Rupe, B.D. High-pressure liquid chromatographic determination of sulfamethazine residues in beef tissues, *J.Pharm.Sci.*, **1980**, *69*, 701–703.

# Sulfamerazine

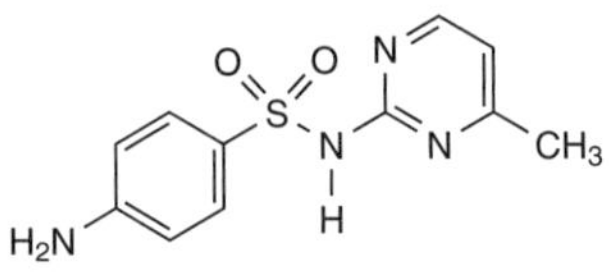

**Molecular formula:** $C_{11}H_{12}N_4O_2S$
**Molecular weight:** 264.31
**CAS Registry No:** 127-79-7, 127-58-2 (Na salt)
**Merck Index:** 13, 8998

## SAMPLE
**Matrix:** blood
**Sample preparation:** Vortex 1 mL plasma with 5–20 μg IS, 1 mL 10 mM pH 6.8 dipotassium hydrogen phosphate buffer, and 10 mL ethyl acetate:isopropanol 98:2 for 10 s, centrifuge at 1000 g for 10 min. Evaporate the organic layer to dryness under a stream of nitrogen at 60°, reconstitute the residue with 100 μL MeOH, inject a 5–50 μL aliquot.

## HPLC VARIABLES
**Column:** 250 × 4.6 5 μm Spherisorb ODS1
**Mobile phase:** MeCN:MeOH:20 mM sodium dihydrogen phosphate 5:4:91
**Flow rate:** 1.5
**Injection volume:** 5–50
**Detector:** E, EDT Research LCA 15, +1.0 V

## CHROMATOGRAM
**Retention time:** 21.5
**Internal standard:** sulfamethazine (36)
**Limit of quantitation:** 1 μg/mL

## OTHER SUBSTANCES
**Extracted:** sulfadiazine (13.0), sulfapyridine (18.5)

## KEY WORDS
pharmacokinetics; plasma; sheep

## REFERENCE
Mallett, D.N.; Gulaid, A.A.; Dennis, M.J. High-performance liquid chromatographic method with electrochemical detection for the concomitant assay of sulphadiazine, sulphamerazine and sulphapyridine in plasma, *J.Chromatogr.*, **1988**, *428*, 190–195.

## SAMPLE
**Matrix:** blood
**Sample preparation:** Mix 1 mL plasma with 1 mL MeOH, centrifuge at 2000 g for 10 min, inject a 20 μL aliquot.

## HPLC VARIABLES
**Guard column:** Corasil/C18
**Column:** μBondapak C18
**Mobile phase:** MeOH:water 40:60
**Flow rate:** 0.8
**Injection volume:** 20
**Detector:** UV 254

## CHROMATOGRAM
**Retention time:** 5.40
**Limit of quantitation:** 10 ng/mL

## OTHER SUBSTANCES

**Extracted:** acetylsulfisoxazole (4.40), sulfaguanidine (3.90), sulfisoxazole (4.20), sulfamethazine (6.10), sulfanilamide (3.05), sulfapyridine (5.00), sulfathiazole (3.30)

## KEY WORDS

plasma

## REFERENCE

Suber, R.L.; Edds, G.T. High performance liquid chromatographic determinations of sulfonamides by ionic suppression, *J.Liq.Chromatogr.*, **1980**, *3*, 257–268.

## ANNOTATED BIBLIOGRAPHY

Goehl, T.J.; Mathur, L.K.; Strum, J.D.; Jaffe, J.M.; Pitlick, W.H.; Shah, V.P.; Poust, R.I.; Colaizzi, J.L. Simple high-pressure liquid chromatographic determination of trisulfapyrimidines in human serum, *J.Pharm.Sci.*, **1978**, *67*, 404–406. [sulfamerazine; sulfadiazine; sulfamethazine]

# Sulfanitran

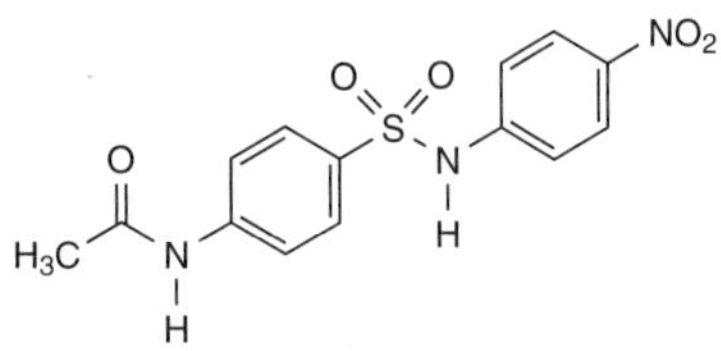

**Molecular formula:** $C_{14}H_{13}N_3O_5S$
**Molecular weight:** 335.34
**CAS Registry No:** 122-16-7
**Merck Index:** 13, 9019

## SAMPLE
**Matrix:** feed, premix
**Sample preparation:** Shake ground feed containing 1.5 mg sulfanitran with 80 mL MeOH at 60° for 20 min, cool, let stand for 2 h, make up to 100 mL with MeOH, mix well, filter (Whatman No. 2 paper), filter (0.50 μm), inject a 20 μL aliquot. Shake premix containing 150 mg sulfanitran with 100 mL DMF on a rotary shaker for 1 h, filter (Whatman No. 2 paper), filter (0.50 μm), dilute 1 mL filtrate to 50 mL with DMF, inject a 20 μL aliquot.

## HPLC VARIABLES
**Guard column:** 100 × 3 (OD) Bondapak C18/Corasil
**Column:** 300 × 4 μBondapak C18
**Mobile phase:** MeCN:water 45:55
**Flow rate:** 1
**Injection volume:** 20
**Detector:** UV 254

## CHROMATOGRAM
**Retention time:** 6

## OTHER SUBSTANCES
**Extracted:** dinsed (9)

## REFERENCE
Eaves, K.L.; Colvin, B.M.; Hanks, A.R.; Bushway, R.J. High pressure liquid chromatographic determination of sulfanitran and dinsed in medicated feeds and premixes, *J.Assoc.Off.Anal.Chem.*, **1977**, *60*, 1064–1066.

## SAMPLE
**Matrix:** tissue
**Sample preparation:** Prepare a 30 × 6 column of 80–200 mesh neutral alumina (Brockman Activity I) and wash with two 2 mL portions of chloroform:ethyl acetate 50:50 (Caution! Chloroform is a carcinogen!). Blend (Polytron) 2.5 g partially frozen tissue with 20 mL chloroform:ethyl acetate:DMSO 50:50:0.8 at medium speed for 45 s, centrifuge at 3500 rpm for 5 min (10 min for liver), pass the organic layer through a plug of glass wool. Pass 15 mL filtrate through the alumina column, wash with three 1 mL portions of chloroform, wash with 3 mL chloroform, force remaining chloroform out with air pressure until dry, pass air through the column for an additional 5 min, elute with MeOH:50 mM pH 6.0 potassium phosphate buffer 50:50, collect first 2 mL eluate, inject a 50 μL aliquot. (Protect from light.)

## HPLC VARIABLES
**Column:** 250 × 4.6 5 μm Supelcosil LC-18
**Mobile phase:** MeOH:buffer 42.5:57.5 (The buffer was 50 mM potassium dihydrogen phosphate containing 1 mM EDTA, adjusted to pH 6.0 with 1 M NaOH.)
**Flow rate:** 1
**Injection volume:** 50

**Detector:** E, Bioanalytical Systems BAS Model LC-4B, glassy carbon electrode −0.8V, Ag/AgCl reference electrode

## CHROMATOGRAM

**Retention time:** 13.2
**Limit of detection:** <6 ng/g

## OTHER SUBSTANCES

**Extracted:** aklomide (5.3), furazolidone (4.0), nitrofurazone (4.1), nitromide (6.7), zoalene (6.2)

## KEY WORDS

chicken; liver; muscle; SPE

## REFERENCE

Parks, O.W. Liquid chromatographic-electrochemical detection screening procedure for six nitro-containing drugs in chicken tissues at low ppb level, *J.Assoc.Off.Anal.Chem.*, **1989**, *72*, 567–569.

# Sultamicillin

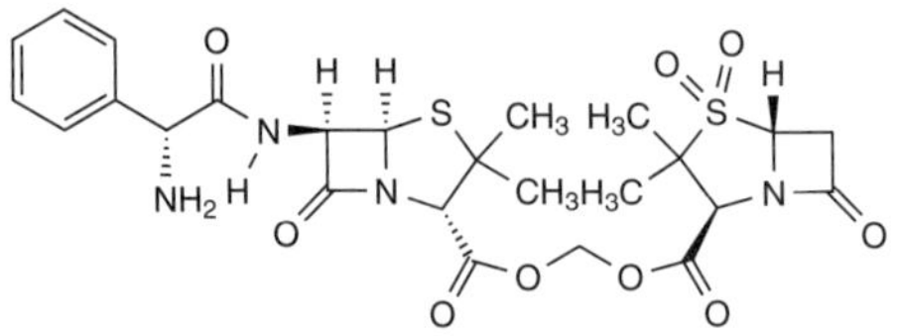

**Molecular formula:** $C_{25}H_{30}N_4O_9S_2$
**Molecular weight:** 594.66
**CAS Registry No:** 76497-13-7, 83105-70-8 (tosylate)
**Merck Index:** 13, 9083

## SAMPLE

**Matrix:** bulk
**Sample preparation:** Inject a 20 μL aliquot of a solution in MeCN:25 mM pH 7.0 phosphate buffer 70:30.

## HPLC VARIABLES

**Column:** 150 × 4.6 5 μm Kromasil C18
**Column temperature:** 20
**Mobile phase:** MeCN:25 mM pH 7.0 phosphate buffer 48:52
**Flow rate:** 1
**Injection volume:** 20
**Detector:** UV 215

## CHROMATOGRAM

**Retention time:** 4.92
**Limit of detection:** 200 ng/mL

## OTHER SUBSTANCES

**Extracted:** degradants, impurities

## REFERENCE

Laviana, L.; Fernández-Marí, F.; Bayod, M.; Blanco, D. HPLC for in-process control in the production of sultamicillin, *J.Pharm.Biomed.Anal.*, **2003**, *26*, 321–328.

## SAMPLE

**Matrix:** formulations
**Sample preparation:** Sonicate a finely powdered tablet with 50 mL MeCN for 10 min, make up to 100 mL with MeCN, filter (Whatman No. 42 paper), dilute a 1 mL aliquot to 100 mL with MeCN, inject a 20 μL aliquot.

## HPLC VARIABLES

**Column:** 250 × 3.9 5 μm Shodex C18
**Mobile phase:** MeCN:20 mM sodium dihydrogen phosphate 20:60, pH adjusted to 3.0 with dilute phosphoric acid
**Injection volume:** 20
**Detector:** UV 215

## CHROMATOGRAM

**Retention time:** 7.8
**Limit of detection:** 10 μg/mL
**Limit of quantitation:** 40 μg/mL

## OTHER SUBSTANCES

**Extracted:** *p*-toluenesulfonic acid (3.7)

## KEY WORDS
tablets

## REFERENCE
Argekar, A.P.; Kunjir, S.S. Quantitative estimation of sultamicillin p-toluenesulfonate in pharmaceutical preparations by reverse-phase high performance liquid chromatography, *J.Pharm.Biomed.Anal.*, **1996**, *15*, 423–427.

## SAMPLE
**Matrix:** solutions

## HPLC VARIABLES
**Column:** 250 × 4.6 10 μm Cosmosil 5C18-MS
**Column temperature:** 50
**Mobile phase:** Gradient. MeCN:10 mM pH 2.5 potassium phosphate buffer 0:100 for 2 min, to 75:25 over 7.5 min, maintain at 75:25 for 5.5 min. Isocratic. MeCN:buffer 15:85.
**Flow rate:** 1.5
**Detector:** UV (wavelength not specified)

## CHROMATOGRAM
**Retention time:** 8.3 (gradient) or 4.3 (isocratic)

## OTHER SUBSTANCES
**Simultaneous:** acetaminophen (7.9), alacepril (10.9), ampicillin (7.9), aspirin (10.0), caffeine (8.5), carbenicillin (9.5), cefotiam (7.2), chlorpromazine (10.8), cromolyn (8.9), enalapril (9.9), loperamide (11.6), ofloxacin (8.3), procainamide (7.4), procaine (7.9), propranolol (9.6), tegafur (8.4), temocapril (12.3), theophylline (8.0), tulobuterol (8.9) (gradient retention times; isocratic conditions may differ)

## REFERENCE
Sugiyama, T.; Matsuyama, R.; Usui, S.; Katagiri, Y.; Hirano, K. Selection of mobile phases in high-performance liquid chromatographic determination for medicines, *Biol.Pharm.Bull.*, **2000**, *23*, 274–278.

# Tacalcitol

**Molecular formula:** $C_{27}H_{44}O_3$
**Molecular weight:** 416.63
**CAS Registry No:** 57333-96-7
**Merck Index:** 13, 9114

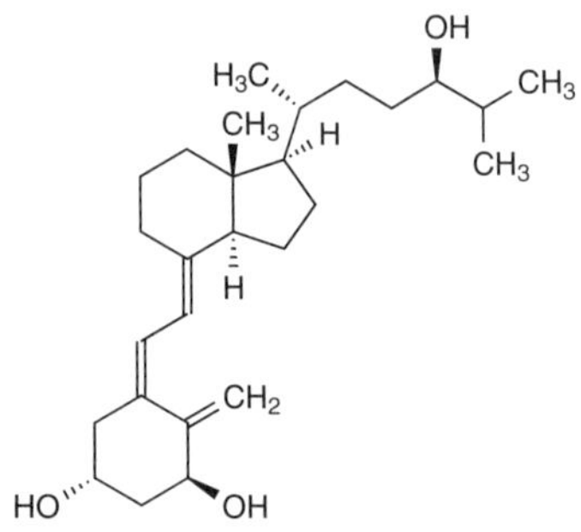


## SAMPLE
**Matrix:** solutions

## HPLC VARIABLES
**Column:** 25 × 6.2 Zorbax SIL
**Mobile phase:** *n*-Hexane:isopropanol 90:10
**Flow rate:** 1.5
**Detector:** Radioreceptor assay

## CHROMATOGRAM
**Retention time:** 20

## REFERENCE
Shigeno, C.; Yamamoto, I.; Dokoh, S.; Hino, M.; Aoki, J.; Yamada, K.; Morita, R.; Kameyama, M.; Torizuka, K. Identification of 1,24(*R*)-dihydroxyvitamin $D_3$-like bone-resorbing lipid in a patient with cancer-associated hypercalcemia, *J.Clin.Endocrinol.Metab.*, **1985**, *61*, 761–768.

## SAMPLE
**Matrix:** solutions

## HPLC VARIABLES
**Column:** 25 × 4.6 Zorbax ODS
**Mobile phase:** MeOH:water 80:20
**Flow rate:** 1
**Detector:** Radioreceptor assay

## CHROMATOGRAM
**Retention time:** 19

## REFERENCE
Shigeno, C.; Yamamoto, I.; Dokoh, S.; Hino, M.; Aoki, J.; Yamada, K.; Morita, R.; Kameyama, M.; Torizuka, K. Identification of 1,24(*R*)-dihydroxyvitamin $D_3$-like bone-resorbing lipid in a patient with cancer-associated hypercalcemia, *J.Clin.Endocrinol.Metab.*, **1985**, *61*, 761–768.

# Talipexole

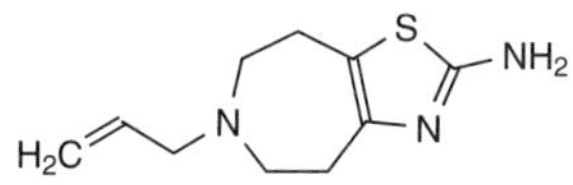

**Molecular formula:** $C_{10}H_{15}N_3S$
**Molecular weight:** 209.32
**CAS Registry No:** 101626-70-4, 36085-73-1 (di HCl)
**Merck Index:** 13, 9129

## SAMPLE
**Matrix:** blood, urine
**Sample preparation:** Vortex 1 mL urine or plasma with 50 μL 1% acetic acid, 100 μL 1 M NaOH, and 6 mL diethyl ether for 5 min, centrifuge at 2500 g for 5 min, freeze in dry ice/acetone. Add the organic layer to 100 (plasma) or 200 (urine) μL buffer, mix, freeze in dry ice/acetone, discard the organic layer, remove traces of organic solvent with a stream of nitrogen, thaw, inject a 50 μL aliquot. (Prepare the buffer by dissolving 1.7 g potassium dihydrogen phosphate, 1.7 g sodium acetate, and 0.5 g sodium heptanesulfonate in 500 mL water, adjust pH to 3.5 with acetic acid. Use only polypropylene containers.)

## HPLC VARIABLES
**Guard column:** 15 × 3.2 7 μm Brownlee RP-8
**Column:** 250 × 4.6 5 μm Zorbax Rx C8
**Mobile phase:** MeCN:buffer 13:87 (Prepare mobile phase by dissolving 10.2 g potassium dihydrogen phosphate, 10.2 g sodium acetate, and 4.5 g sodium heptanesulfonate in 3 L water, adjust pH to 3.5 with acetic acid, add 450 mL MeCN.)
**Flow rate:** 1.2
**Injection volume:** 50
**Detector:** E, ESA 5100A detector, 5020 guard cell 0.65 V, 5011 analytical cell, electrode 1 0.2 V (impurity screening), electrode 2 0.6 V (monitored); UV 286

## CHROMATOGRAM
**Retention time:** 10
**Limit of quantitation:** 50 pg/mL (plasma, E), 10 ng/mL (urine, UV) (for pramipexole)

## OTHER SUBSTANCES
**Extracted:** pramipexole (15)

## KEY WORDS
plasma; method is validated for pramipexole and talipexole is IS; recovery of talipexole from plasma is 98.2% and from urine is 95.1%

## REFERENCE
Lau, Y.Y.; Hanson, G.D.; Ichhpurani, N. Determination of pramipexole (U-98,528) in human plasma and urine by high-performance liquid chromatography with electrochemical and ultraviolet detection. *J.Chromatogr.B*, **1996**, *683*, 217–223.

## SAMPLE
**Matrix:** blood
**Sample preparation:** Vortex 1 mL plasma with 50 μL 1% acetic acid, 100 μL 1 M NaOH, and 6 mL MTBE for 5 min, centrifuge at 630 g for 5 min, freeze in dry ice/acetone. Evaporate the organic layer to dryness under a stream of nitrogen, reconstitute the residue with 100 μL MeOH:water 95:5, inject a 70 μL aliquot. (Use only polypropylene containers.)

## HPLC VARIABLES

**Column:** 150 × 4.6 5 μm Zorbax SB CN
**Mobile phase:** MeOH:water:100 mM ammonium acetate 80:15:5
**Flow rate:** 1.2
**Injection volume:** 70
**Detector:** MS, PE Sciex API-III triple quadrupole, nebulizer 470°, gas-phase chemical ionization, positive ion mode, Corona discharge 3 μA, nebulizer gas at 550 kPa, auxiliary gas 1.5 L/min, orifice 45 V, m/z 210–141

## CHROMATOGRAM

**Retention time:** 2.2
**Limit of quantitation:** 50 pg/mL (for pramipexole)

## OTHER SUBSTANCES

**Extracted:** pramipexole (m/z 212–153) (3.4)

## KEY WORDS

plasma; method is validated for pramipexole, and talipexole is IS; recovery of talipexole is 85.4%

## REFERENCE

Lau, Y.Y.; Hanson, G.D.; Ichhpurani, N. Determination of pramipexole (U-98,528) in human plasma and urine by high-performance liquid chromatography with electrochemical and ultraviolet detection, *J.Chromatogr.B*, **1996**, *683*, 217–223.

# Taltirelin

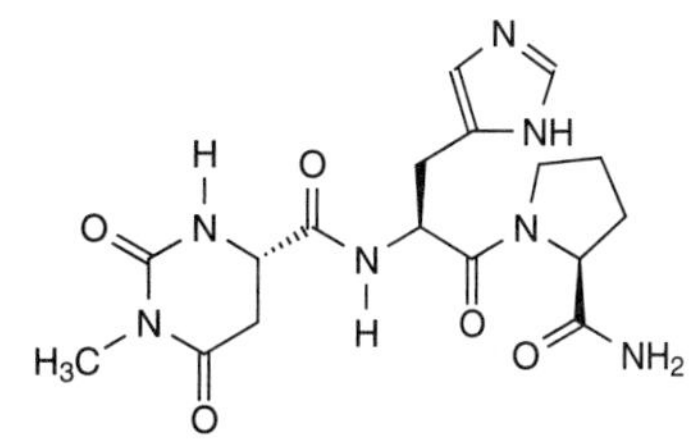


**Molecular formula:** $C_{17}H_{23}N_7O_5$
**Molecular weight:** 405.41
**CAS Registry No:** 103300-74-9, 201677-75-0 (tetrahydrate)
**Merck Index:** 13, 9135

## SAMPLE

**Matrix:** blood

**Sample preparation:** Condition a 3 mL 60 mg Oasis HLB divinylbenzene-*N*-vinylpyrrolidone SPE cartridge with 1 mL MeOH, 1 mL water, and 1 mL 200 mM sodium carbonate. Mix 500 μL plasma with 50 μL mobile phase A and 1 mL 200 mM sodium carbonate, add to the SPE cartridge, wash with 2 mL water, elute with 1 mL MeOH. Evaporate the eluate to dryness under a stream of nitrogen at 40°, reconstitute the residue with 250 μL 50 ng/mL IS in mobile phase A, inject a 25 μL aliquot.

## HPLC VARIABLES

**Column:** 150 × 2 5 μm Daisopak SP-120-30DS-BP (Daiso, Osaka)

**Column temperature:** 40

**Mobile phase:** Gradient. A was MeOH:0.1% formic acid 7.5:92.5. B was MeOH:0.1% formic acid 90:10. A:B 100:0 for 3 min, to 0:100 over 1 min, maintain at 0:100 for 3 min, return to initial conditions over 0.1 min, re-equilibrate for 12.9 min.

**Flow rate:** 0.2

**Injection volume:** 25

**Detector:** MS, PE Sciex API 3000, TurboIonSpray, positive ionization, source 475°, source flow rate 7 L/min, ionspray 5000 V, nebulizer gas air at 14 units, curtain gas nitrogen at 12 units, collision gas nitrogen at 4 units, collision energy – 80 eV, m/z 406–264

## CHROMATOGRAM

**Retention time:** 3.1

**Internal standard:** *N*-[(*S*)-hexahydro-1-methyl-2,6-dioxo-4-pyrimidinylcarbonyl]-L-histidine (m/z 310–264) (2.7)

**Limit of quantitation:** 17 pg/mL

## OTHER SUBSTANCES

**Extracted:** taltirelin acid-type (4.9)

## KEY WORDS

plasma; SPE

## REFERENCE

Horimoto, S.; Mayumi, T.; Tagawa, K.; Yamakita, H.; Yoshikawa, M. Determination of taltirelin, a new stable thyrotropin-releasing hormone analogue, in human plasma by high-performance liquid chromatography turbo-ionspray ionization tandem mass spectrometry, *J.Pharm.Biomed.Anal.*, **2002**, *30*, 1361–1369.

# Technetium Tc 99m bicisate

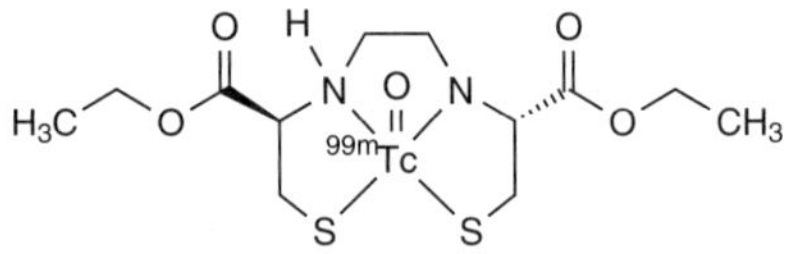


**Molecular formula:** $C_{12}H_{21}N_2O_5S_2^{99m}Tc$
**Molecular weight:** 436.44
**CAS Registry No:** 121281-41-2
**Merck Index:** 13, 9181

## SAMPLE
**Matrix:** solutions

## HPLC VARIABLES
**Column:** 250 × 4.6 Dynamax C8 (Rainin)
**Mobile phase:** Gradient. MeCN:50 mM ammonium acetate (unspecified)
**Flow rate:** 1
**Detector:** UV 254; Radioactivity ($^{99}$Tc)

## CHROMATOGRAM
**Retention time:** 7.1

## REFERENCE
Walovitch, R.C.; Franceschi, M.; Picard, M.; Cheesman, E.H.; Hall, K.M.; Makuch, J.; Watson, M.W.; Zimmerman, R.E.; Watson, A.D.; Ganey, M.V.; Williams, S.J.; Holman, B.L. Metabolism of $^{99m}$Tc-L,L-ethyl cysteinate dimer in healthy volunteers, *Neuropharmacology*, **1991**, *30*, 283–292.

# Tegaserod

**Molecular formula:** $C_{16}H_{23}N_5O$
**Molecular weight:** 301.39
**CAS Registry No:** 145158-71-0, 189188-57-6 (maleate)
**Merck Index:** 13, 9194

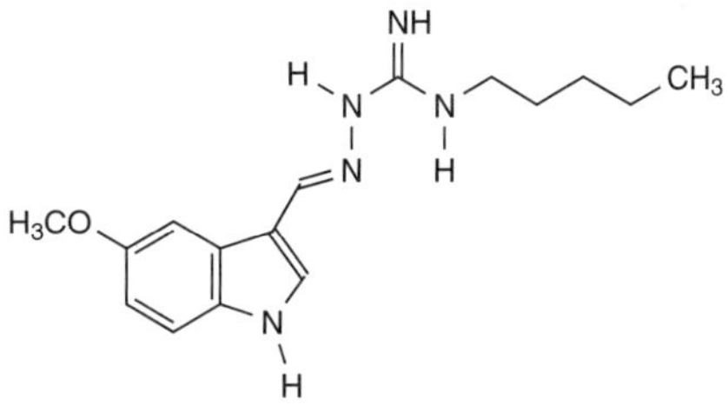


## SAMPLE
**Matrix:** microsomal incubations
**Sample preparation:** Add an equal volume of cold MeOH, 70% perchloric acid, or 10% trichloroacetic acid, cool on dry ice, centrifuge at 100 000 g for 10 min, inject an aliquot of the supernatant.

## HPLC VARIABLES
**Guard column:** 20 × 4.6 5 μm Supelco LC-18
**Column:** 150 × 4.6 5 μm Supelco LC-18
**Column temperature:** 40
**Mobile phase:** Gradient. MeCN:10 mM pH 5.4 ammonium acetate 0:100 for 10 min, to 10:90 over 20 min, to 90:10 over 41 min, to 100:0 over 19 min.
**Flow rate:** 1
**Detector:** Radioactivity ($^{14}C$) (scintillator flow 3 mL/min); MS, Finnigan MAT TSQ 700 triple-stage quadrupole, electrospray, 0.25 mL/min of column effluent mixed with 0.1 mL/min MeCN and entered the detector, sheath liquid methanol at 0.1 mL/min, sheath gas nitrogen 45 psi, auxiliary gas nitrogen 5 units, spray 4.5 kV, transfer capillary 240°, collision offset 30 V

## CHROMATOGRAM
**Retention time:** 67

## OTHER SUBSTANCES
**Extracted:** metabolites

## KEY WORDS
human; liver

## REFERENCE

Vickers, A.E.M.; Zollinger, M.; Dannecker, R.; Tynes, R.; Heitz, F.; Fischer, V. In vitro metabolism of tegaserod in human liver and intestine: assessment of drug interactions, *Drug Metab.Dispos.*, **2001**, *29*, 1269–1276.

# Telithromycin

**Molecular formula:** $C_{43}H_{65}N_5O_{10}$
**Molecular weight:** 812.00
**CAS Registry No:** 191114-48-4
**Merck Index:** 13, 9199

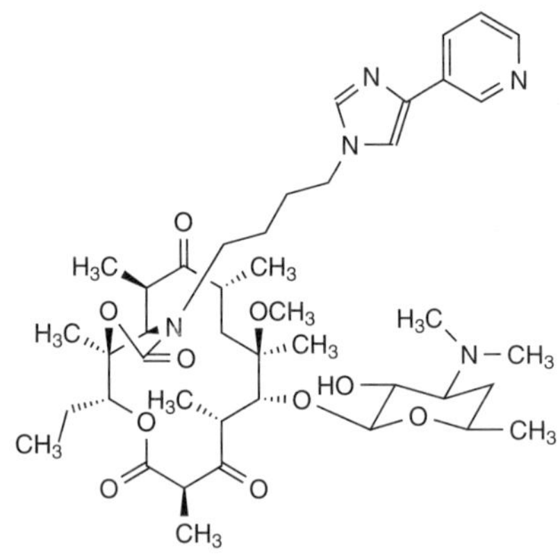

## SAMPLE

**Matrix:** blood
**Sample preparation:** Mix plasma with MeCN, centrifuge. Evaporate the supernatant to dryness under a stream of nitrogen, reconstitute the residue with MeCN:MeOH:50 mM ammonium acetate 18.1:21.9:60 containing IS, inject an aliquot.

## HPLC VARIABLES

**Column:** 125 × 4 5 μm Purospher RP-18e
**Mobile phase:** MeCN:MeOH:50 mM ammonium acetate 24:29:52
**Flow rate:** 1
**Detector:** F ex 263 em 460

## CHROMATOGRAM

**Internal standard:** RU 66260
**Limit of quantitation:** 5 ng/mL

## KEY WORDS

guinea pig; pharmacokinetics; plasma

## REFERENCE

Edelstein, P.H.; Edelstein, M.A. In vitro activity of the ketolide HMR 3647 (RU 6647) for *Legionella* spp., its pharmacokinetics in guinea pigs, and use of the drug to treat guinea pigs with *Legionella pneumophila* pneumonia, *Antimicrob.Agents Chemother.*, **1999**, *43*, 90–95.

# Telmesteine

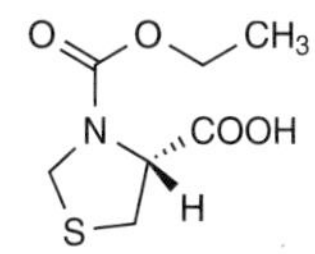


**Molecular formula:** $C_7H_{11}NO_4S$
**Molecular weight:** 205.23
**CAS Registry No:** 122946-43-4

## SAMPLE

**Matrix:** bile, blood, urine
**Sample preparation:** Plasma. Mix 300 μL plasma with 600 μL MeCN, centrifuge at 1000 g for 10 min. Evaporate the supernatant to dryness under a stream of nitrogen, reconstitute the residue with 50 μL initial mobile phase, inject a 30 μL aliquot. Urine, bile. Add 1 mL urine or bile to an activated Sep-Pak C18 SPE cartridge, wash with two 1 mL portions of water, elute with 1 mL MeOH. Evaporate the eluate to dryness under a stream of nitrogen, reconstitute the residue with 100 μL initial mobile phase, inject a 30 μL aliquot.

## HPLC VARIABLES

**Column:** 150 × 4.6 5 μm YMC-Pak C8
**Mobile phase:** Gradient. MeCN:0.1% acetic acid from 20:80 to 40:60 over 25 min, maintain at 40:60 for 5 min
**Flow rate:** 1
**Injection volume:** 30
**Detector:** Radioactivity ($^{14}C$); MS, Agilent LC-MSD Trap, 10% of column effluent entered MS detector

## CHROMATOGRAM

**Retention time:** 21.1

## OTHER SUBSTANCES

**Extracted:** glucuronide conjugate (15.8)

## KEY WORDS

pharmacokinetics; plasma; rat; SPE

## REFERENCE

Lee, J.; Son, J.; Rhee, S.W.; Kim, D.H. Absorption, distribution, metabolism and excretion of telmesteine, a mucolitic agent, in rat, *Xenobiotica*, **2003**, *33*, 755–765.

# Telmisartan

**Molecular formula:** $C_{33}H_{30}N_4O_2$
**Molecular weight:** 514.62
**CAS Registry No:** 144701-48-4
**Merck Index:** 13, 9209

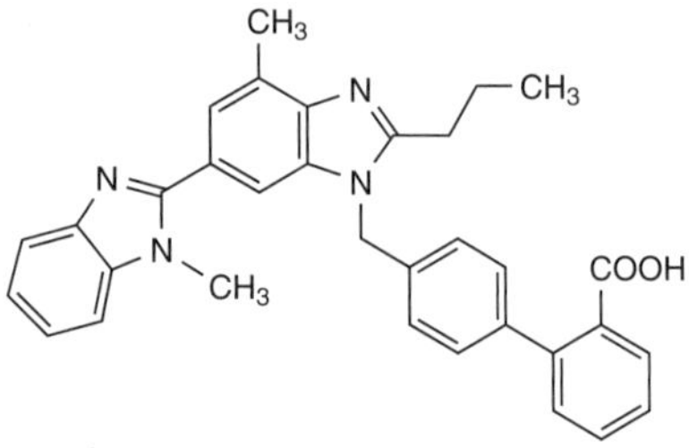


## SAMPLE

**Matrix:** blood
**Sample preparation:** Inject a 100 μL aliquot of plasma onto column A and elute to waste with mobile phase A; after 3 min, backflush the contents of column A onto column B using mobile phase B; after 3 min, elute column B with mobile phase B, monitor the effluent from column B.

## HPLC VARIABLES

**Column:** A 17 × 4.6 unspecified enrichment column; B 125 × 4.6 Nucleosil 100-5 C18
**Mobile phase:** A 50 mM pH 4.5 ammonium acetate; B MeCN:50 mM pH 4.5 ammonium acetate 10:90; C Gradient. MeCN:50 mM pH 4.5 ammonium acetate 30:70 for 3 min, to 60:40 over 12 min
**Flow rate:** 1
**Injection volume:** 100
**Detector:** F ex 305 em 365

## CHROMATOGRAM

**Retention time:** 18
**Limit of quantitation:** 2.5 ng/mL

## OTHER SUBSTANCES

**Extracted:** metabolites

## KEY WORDS

column-switching; plasma; rat

## REFERENCE

Ebner, T.; Heinzel, G.; Prox, A.; Beschke, K.; Wachsmuth, H. Disposition and chemical stability of telmisartan 1-*O*-acylglucuronide, *Drug Metab.Dispos.*, **1999**, *27*, 1143–1149.

## SAMPLE

**Matrix:** urine
**Sample preparation:** Condition a 1 mL 100 mg Bond Elut C8 SPE cartridge with 2 mL MeOH and 1 mL 100 mM pH 6 phosphate buffer. Shake 1 mL urine with 500 μL 100 mM pH 6 phosphate buffer, centrifuge at 3500 rpm (16.1 g) at 4° for 5 min, add a 1 mL aliquot to the SPE cartridge, wash with 1 mL MeOH:100 mM pH 6 phosphate buffer 30:70, dry under full vacuum for 20 min, elute with 1 mL MeOH. Evaporate the eluate to dryness under a stream of nitrogen at 40°, reconstitute the residue with 500 μL mobile phase, inject a 20 μL aliquot.

## HPLC VARIABLES

**Guard column:** 20 × 3.9 4 μm Novapak C18
**Column:** 150 × 3.9 4 μm Novapak C18
**Mobile phase:** MeCN:5 mM pH 6.0 phosphate buffer 45:55
**Flow rate:** 0.5

**Injection volume:** 20
**Detector:** F ex 305 em 365

## CHROMATOGRAM
**Retention time:** 2.5
**Limit of quantitation:** 1 ng/mL

## KEY WORDS
SPE

## REFERENCE
Torrealday, N.; González, L.; Alonso, R.M.; Jiménez, R.M.; Ortiz Lastra, E. Experimental design approach for the optimisation of a HPLC-fluorimetric method for the quantitation of the angiotensin II receptor antagonist telmisartan in urine, *J.Pharm.Biomed.Anal.*, **2003**, *32*, 847–857.

# Temocapril

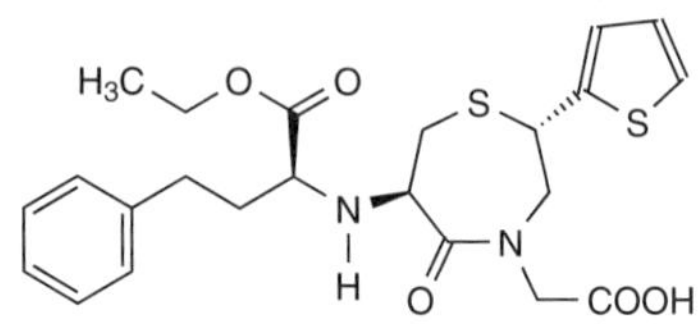


**Molecular formula:** $C_{23}H_{28}N_2O_5S_2$
**Molecular weight:** 476.62
**CAS Registry No:** 111902-57-9
**Merck Index:** 13, 9215

## SAMPLE
**Matrix:** solutions

## HPLC VARIABLES
**Column:** 250 × 4.6 Mightysil RP-18
**Mobile phase:** MeOH:water:phosphoric acid 50:50:0.05
**Flow rate:** 1
**Detector:** UV 220

## OTHER SUBSTANCES
**Simultaneous:** enalapril

## REFERENCE
Kitagawa, S.; Takeda, J.; Sato, S. Uptake of enalapril by rabbit small intestinal brush-border membrane vesicles, *Biol.Pharm.Bull.*, **1999**, *22*, 762–764.

## SAMPLE
**Matrix:** solutions

## HPLC VARIABLES
**Column:** 250 × 4.6 10 μm Cosmosil 5C18-MS
**Column temperature:** 50
**Mobile phase:** Gradient. MeCN:10 mM pH 2.5 potassium phosphate buffer 0:100 for 2 min, to 75:25 over 7.5 min, maintain at 75:25 for 5.5 min. Isocratic. MeCN:buffer 50:50.
**Flow rate:** 1.5
**Detector:** UV (wavelength not specified)

## CHROMATOGRAM
**Retention time:** 12.3 (gradient) or 5.7 (isocratic)

## OTHER SUBSTANCES
**Simultaneous:** acetaminophen (7.9), alacepril (10.9), ampicillin (7.9), aspirin (10.0), caffeine (8.5), carbenicillin (9.5), cefotiam (7.2), chlorpromazine (10.8), cromolyn (8.9), enalapril (9.9), loperamide (11.6), ofloxacin (8.3), procainamide (7.4), procaine (7.9), propranolol (9.6), sultamicillin tosylate (8.3), tegafur (8.4), theophylline (8.0), tulobuterol (8.9) (gradient retention times; isocratic conditions may differ)

## REFERENCE
Sugiyama, T.; Matsuyama, R.; Usui, S.; Katagiri, Y.; Hirano, K. Selection of mobile phases in high-performance liquid chromatographic determination for medicines, *Biol.Pharm.Bull.*, **2000**, *23*, 274–278.

# Temozolomide

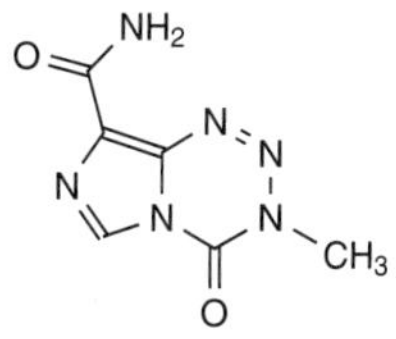

**Molecular formula:** $C_6H_6N_6O_2$
**Molecular weight:** 194.15
**CAS Registry No:** 85622-93-1
**Merck Index:** 13, 9218

## SAMPLE
**Matrix:** blood
**Sample preparation:** Vortex 500 μL plasma with 500 μL 1 μg/mL IS in water, 50 μL 1 M HCl, and 5 mL ethyl acetate for 10 min, centrifuge at 4500 g for 5 min. Evaporate the organic layer to dryness under a stream of nitrogen at 45°, reconstitute the residue with 300 μL mobile phase, inject a 20 μL aliquot.

## HPLC VARIABLES
**Column:** 150 × 4.6 5 μm Ultrasphere ODS
**Mobile phase:** MeCN:0.1% acetic acid 10:90
**Flow rate:** 1
**Injection volume:** 20
**Detector:** UV 316

## CHROMATOGRAM
**Retention time:** 2.7
**Internal standard:** ethazolastone (5.0)
**Limit of quantitation:** 100 ng/mL

## KEY WORDS
pharmacokinetics; plasma

## REFERENCE
Kim, H.; Likhari, P.; Parker, D.; Statkevich, P.; Marco, A.; Lin, C.-C.; Nomeir, A.A. High-performance liquid chromatographic analysis and stability of anti-tumor agent temozolomide in human plasma, *J.Pharm.Biomed.Anal.*, **2001**, *24*, 461–468.

## SAMPLE
**Matrix:** blood, urine
**Sample preparation:** Condition a 100 mg Chromabond C18 SPE cartridge (Macherey-Nagel) with two 1 mL portions of MeOH and two 1 mL portions of 0.5% acetic acid. Mix 1 mL plasma with 100 μL 1 M HCl. Mix 10 mL urine with 1 mL 1 M HCl. Mix 235 μL acidified plasma or urine with 115 μL 20 (plasma) or 160 (urine) μg/mL IS in 100 mM HCl. Add a 160 μL aliquot of this solution to the SPE cartridge, pull through under light vacuum, let stand for 1 min (important!), wash with 750 μL 0.5% acetic acid, dry under vacuum for 5 min, elute with 1.25 mL MeOH. Evaporate the eluate to dryness under a stream of nitrogen at room temperature, reconstitute the residue with 200 μL 0.5% acetic acid, centrifuge, inject a 30 μL aliquot.

## HPLC VARIABLES
**Guard column:** 20 × 4 5 μm HP ODS-Hypersil
**Column:** 100 × 4.6 5 μm HP ODS-Hypersil
**Mobile phase:** MeOH:0.5% acetic acid 10:90
**Flow rate:** 1
**Injection volume:** 30
**Detector:** UV 330

## CHROMATOGRAM

**Retention time:** 3.2
**Internal standard:** ethazolastone (7.4)
**Limit of quantitation:** 200 ng/mL (plasma), 2 μg/mL (urine)

## KEY WORDS

pharmacokinetics; plasma; SPE

## REFERENCE

Shen, F.; Decosterd, L.A.; Gander, M.; Leyvraz, S.; Biollax, J.; Lejeune, F. Determination of temozolomide in human plasma and urine by high-performance liquid chromatography after solid-phase extraction, *J.Chromatogr.B*, **1995**, *667*, 291–300.

## ANNOTATED BIBLIOGRAPHY

Chowdhury, S.K.; Laudicina, D.; Blumenkrantz, N.; Wirth, M.; Alton, K.B. An LC/MS/MS method for the quantitation of MTIC (5-(3-N-methyltriazen-1-yl)-imidazole-4-carboxamide), a bioconversion product of temozolomide, in rat and dog plasma, *J.Pharm.Biomed.Anal.*, **1999**, *19*, 659–668.

Kim, H.K.; Lin, C.-C.; Parker, D.; Veals, J.; Lim, J.; Likhari, P.; Statkevich, P.; Marco, A.; Nomeir, A.A. High-performance liquid chromatographic determination and stability of 5-(3-methyltriazen-1-yl)-imidazo-4-carboximide, the biologically active product of the antitumor agent temozolomide, in human plasma, *J.Chromatogr.B*, **1997**, *703*, 225–233.

Ma, J.; Li, S.; Reed, K.; Guo, P.; Gallo, J.M. Pharmacodynamic-mediated effects of the angiogenesis inhibitor SU5416 on the tumor disposition of temozolomide in subcutaneous and intracerebral glioma xenograft models, *J.Pharmacol.Exp.Ther.*, **2003**, *305*, 833–839. [plasma; rat]

# Tenofovir disoproxil fumarate

**Molecular formula:** $C_{19}H_{30}N_5O_{10}P.C_4H_4O_4$
**Molecular weight:** 635.51
**CAS Registry No:** 202138-50-9, 147127-20-6 (tenofovir only)
**Merck Index:** 13, 9223

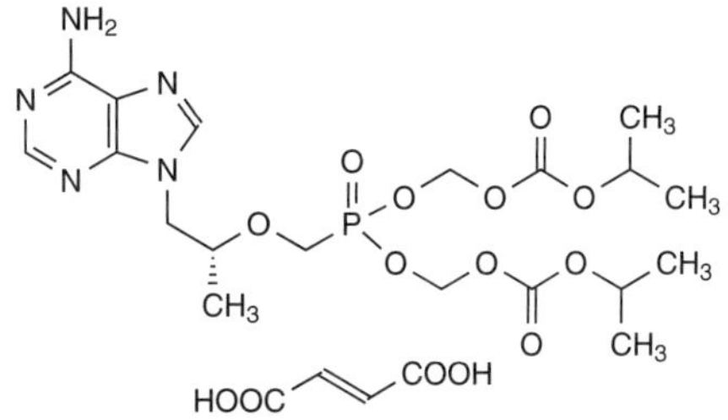

## SAMPLE

**Matrix:** blood
**Sample preparation:** Mix 200 μL plasma with 20 μL 2.5 μg/mL IS in water, add 600 μL MeOH, vortex for 30 s, centrifuge at 2200 g for 10 min. Evaporate the supernatant to dryness under a stream of nitrogen at 60°, reconstitute the residue with 200 μL reagent, heat at 80° for 40 min, cool at −20° for 10 min, inject a 40–80 μL aliquot. (The reagent was 0.34% chloroacetaldehyde in 100 mM pH 4.5 acetate buffer, prepared just before use.)

## HPLC VARIABLES

**Guard column:** 15 × 3 (Cluzeau)
**Column:** 250 × 3 5 μm C8 plus satisfaction (Cluzeau, France)
**Column temperature:** 35
**Mobile phase:** MeCN:5 mM pH 6 phosphate buffer containing 5 mM tetrabutylammonium chloride 15:85
**Flow rate:** 0.5
**Injection volume:** 40–80
**Detector:** F ex 236 em 420

## CHROMATOGRAM

**Retention time:** 11.7 (of tenofovir)
**Internal standard:** adefovir (9.5)
**Limit of quantitation:** 5 ng/mL

## KEY WORDS

derivatization; plasma

## REFERENCE

Jullien, V.; Tréluyer, J.-M.; Pons, G.; Rey, E. Determination of tenofovir in human plasma by high-performance liquid chromatography with spectrofluorimetric detection, *J.Chromatogr.B*, **2003**, *785*, 377–381.

## SAMPLE

**Matrix:** blood
**Sample preparation:** Mix 200 μL plasma with 1 μg/mL IS and 400 μL 0.1% trifluoroacetic acid in MeCN, centrifuge. Evaporate the supernatant to dryness under reduced pressure at room temperature, reconstitute the residue with 200 μL 0.34% chloroacetaldehyde in 100 mM pH 4.5 sodium acetate buffer, vortex, centrifuge. Heat the supernatant at 95° for 40 min, evaporate to dryness, reconstitute with 200 μL mobile phase A, inject a 20 μL aliquot.

## HPLC VARIABLES

**Guard column:** 15 × 3.2 Brownlee RP-18 Newguard
**Column:** 150 × 4.6 Zorbax RX-C18
**Column temperature:** 35
**Mobile phase:** Gradient. A was MeCN:buffer 5:95. B was MeCN:buffer 65:35. A:B 100:0 for 8 min, to 0:100 for 10 min, return to initial conditions (step gradient), re-equilibrate for 8 min. (The buffer was 20 mM pH 6.0 phosphate buffer containing 5 mM tetrabutylammonium hydrogen phosphate.)
**Flow rate:** 1.5
**Injection volume:** 20
**Detector:** F ex 236 em 420

## CHROMATOGRAM

**Internal standard:** 9-[2-(phosphonomethoxy)ethyl]adenine (PMEA)

## KEY WORDS

derivatization; dog; plasma; method validated for tenofovir only

## REFERENCE

Shaw, J.-P.; Sueoka, C.M.; Oliyai, R.; Lee, W.A.; Arimilli, M.N.; Kim, C.U.; Cundy, K.C. Metabolism and pharmacokinetics of novel oral prodrugs of 9-[(R)-2-(phosphonomethoxy)propyl]adenine (PMPA) in dogs, *Pharm.Res.*, **1997**, *14*, 1824–1829.

## SAMPLE

**Matrix:** blood, microsomal incubations
**Sample preparation:** Mix 50 μL plasma or microsomal incubation with 100 μL 0.1% trifluoroacetic acid in MeCN, centrifuge for 5 min at 14 000 rpm, inject a 50 μL aliquot of the supernatant.

## HPLC VARIABLES

**Guard column:** 15 × 3.2 Brownlee RP-18 Newguard
**Column:** 250 × 4.6 5 μm Zorbax RX-C18
**Column temperature:** 35
**Mobile phase:** Gradient. MeCN:20 mM pH 6.0 phosphate buffer from 0:100 to 65:35 over 15 min, return to initial conditions (step gradient), re-equilibrate for 7 min.
**Flow rate:** 1
**Injection volume:** 50
**Detector:** UV 262

## KEY WORDS

dog; intestine; liver; plasma; method validated for tenofovir and tenofovir disoproxil

## REFERENCE

Shaw, J.-P.; Sueoka, C.M.; Oliyai, R.; Lee, W.A.; Arimilli, M.N.; Kim, C.U.; Cundy, K.C. Metabolism and pharmacokinetics of novel oral prodrugs of 9-[(R)-2-(phosphonomethoxy)propyl]adenine (PMPA) in dogs, *Pharm.Res.*, **1997**, *14*, 1824–1829.

## ANNOTATED BIBLIOGRAPHY

Cundy, K.C.; Sueoka, C.; Lynch, G.R.; Griffin, L.; Lee, W.A.; Shaw, J.-P. Pharmacokinetics and bioavailability of the anti-human immunodeficiency virus nucleotide analog 9-[(R)-2-(phosphonomethoxy)-propyl]adenine (PMPA) in dogs, *Antimicrob.Agents Chemother.*, **1998**, *42*, 687–690. [derivatization; adefovir is internal standard; fluorescence detection; LOQ 25 ng/mL; tenofovir only]

Sentenac, S.; Fernandez, C.; Thuillier, A.; Lechat, P.; Aymard, G. Sensitive determination of tenofovir in human plasma samples using reversed-phase liquid chromatography, *J.Chromatogr.B*, **2003**, *793*, 317–324. [SPE; tenofovir only; LOQ 10 ng/mL]

Sparidans, R.W.; Crommentuyn, K.M.L.; Schellens, J.H.M.; Beijnen, J.H. Liquid chromatographic assay for the antiviral nucleotide analogue tenofovir in plasma using derivatization with chloroacetaldehyde, *J.Chromatogr.B*, **2003**, *791*, 227–233. [tenofovir only; derivatization; adefovir is internal standard; LOQ 20 ng/mL; fluorescence detection]

Van Gelder, J.; Witvrouw, M.; Pannecouque, C.; Henson, G.; Bridger, G.; Naesens, L.; De Clercq, E.; Annaert, P.; Shafiee, M.; Van den Mooter, G.; Kinget, R.; Augustijns, P. Evaluation of the potential of ion pair formation to improve the oral absorption of two potent antiviral compounds, AMD3100 and PMPA, *Int.J.Pharm.*, **1999**, *186*, 127–136. [tenofovir only]

# Teprenone

**Molecular formula:** $C_{23}H_{38}O$
**Molecular weight:** 330.55
**CAS Registry No:** 6809-52-5
**Merck Index:** 13, 9228

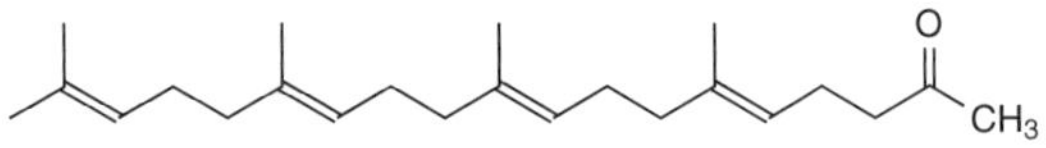


## SAMPLE

**Matrix:** blood

**Sample preparation:** Condition a 200 mg Bond Elut C18 SPE cartridge with 1 column vol of MeOH and 2 column vol of water. Vortex 1 mL plasma, 100 μL 10 μg/mL IS in MeOH, and 1 mL water, add 1 mL EtOH, add 100 μL 50% KOH solution, heat in a boiling water bath for 10 min, cool, add 3 mL hexane, shake for 10 min, centrifuge at 1670 g for 5 min, repeat extraction twice more. Combine the organic layers, evaporate to dryness under a stream of nitrogen at 45°, reconstitute the residue with 200 μL 0.5% HCl in isopropanol, add 200 μL 0.2% dansylhydrazine in isopropanol, heat at 35° for 60 min, add 1 mL water, add to the SPE cartridge, wash twice with 3 mL portions of water, wash with 1 mL MeCN, elute with three 500 μL portions of MeOH:chloroform 50:50 (Caution! Chloroform is a carcinogen!). Evaporate the eluate to dryness under a stream of nitrogen at 45°, reconstitute the residue with 200 μL EtOH, sonicate, centrifuge at 1670 g for 5 min, inject a 50 μL aliquot.

## HPLC VARIABLES

**Column:** 150 × 6 5 μm YMC-A312 C18
**Column temperature:** 30
**Mobile phase:** MeCN:water:triethylamine 92.5:7.5:0.015
**Flow rate:** 1.8
**Injection volume:** 50
**Detector:** F ex 365 em 515

## CHROMATOGRAM

**Retention time:** 9 (cis-5, syn), 10 (trans-5, syn), 13 (cis-5, anti), 14 (trans-5, anti)
**Internal standard:** 7,11,15,19-tetramethyl-6,10,14,18-eicosatetraen-3-one (15, 16 (syn/anti))
**Limit of quantitation:** 8 ng/mL (cis), 12 ng/mL (trans)

## KEY WORDS

derivatization; pharmacokinetics; plasma; SPE; teprenone is a cis:trans (39.8:60.2) mixture at the 5-double bond and produces syn and anti derivatives at the ketone

## REFERENCE

Seki, T.; Hashida, N.; Kanazawa, T. Determination of tetraprenylacetone in human plasma by high-performance liquid chromatography with fluorescence derivatization using dansylhydrazine, *J. Chromatogr.*, **1988**, *424*, 410–415.

# Teriparatide

**Molecular formula:** $C_{181}H_{291}N_{55}O_{51}S_2$
**Molecular weight:** 4117.72
**CAS Registry No:** 52232-67-4, 99294-94-7 (acetate)
**Merck Index:** 13, 9241

## SAMPLE

**Matrix:** solutions

## HPLC VARIABLES

**Column:** 250 × 10 Vydac C-18
**Mobile phase:** Gradient. MeCN:0.1% trifluoroacetic acid from 20:80 to 60:40 over 40 min.
**Detector:** UV (wavelength not specified)

## REFERENCE

Oldenburg, K.R.; D'Orfani, A.L.; Selick, H.E. A method for the high-level expression of a parathyroid hormone analog in *Escherichia coli*, *Protein Expr.Purif.*, **1994**, *5*, 278–284.

# Tetrachlorvinphos

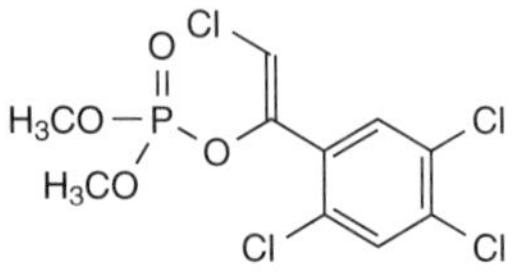


**Molecular formula:** $C_{10}H_9Cl_4O_4P$
**Molecular weight:** 365.96
**CAS Registry No:** 22248-79-9
**Merck Index:** 13, 9267

## SAMPLE

**Matrix:** solutions

## HPLC VARIABLES

**Column:** 250 × 2.2 6–8 μm silica
**Mobile phase:** *n*-Hexane:EtOH 98.5:1.5
**Flow rate:** 0.55
**Injection volume:** 5
**Detector:** Transport flame ionization

## CHROMATOGRAM

**Retention time:** 3 ($\beta$ isomer), 4 ($\alpha$ isomer)

## KEY WORDS

normal phase

## REFERENCE

Szalontai, G. High-performance liquid chromatography of organophosphorus insecticides, *J.Chromatogr.*, **1976**, *124*, 9–16.

# Tetrahydrozoline

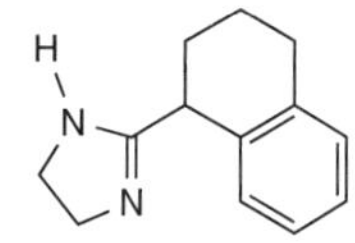

**Molecular formula:** $C_{13}H_{16}N_2$
**Molecular weight:** 200.28
**CAS Registry No:** 84-22-0, 522-48-5 (HCl)
**Merck Index:** 13, 9292

## SAMPLE

**Matrix:** blood
**Sample preparation:** Mix 1 mL whole blood with 20 μL 1 μg/mL dibenzepin in MeOH:water 50:50, add 300 μL pH 11 Tris buffer, mix, add 500 μL butyl acetate, vortex for 2 min, centrifuge. Preserve the aqueous layer (A). Remove the organic layer and add it to 75 μL 10 mM ammonium acetate buffer containing 0.1% formic acid (pH 3.2), evaporate (?). Add 75 μL MeCN, sonicate for 5 min, centrifuge at 5000 rpm for 5 min, keep the extract as B. Add 20 μL 1 μg/mL enalapril in MeOH:water 50:50 and 120 mg NaCl to the aqueous layer (A), mix, add 500 μL pH 3 phosphate buffer, add 600 μL 8.5% phosphoric acid, add 5 mL dichloromethane:isopropanol 95:5, shake at 250 cycles/min in a bench-top shaker for 30 min, centrifuge at 5000 rpm for 5 min. Remove the lower organic layer and evaporate it to dryness under a stream of air at 45°. Reconstitute the residue with 150 μL initial mobile phase, sonicate, centrifuge, combine with extract B, inject a 30 μL aliquot. (Sample preparation from Gergov,M.; Robson,J.N.; Ojanperä,I.; Heinonen,O.P.; Vuori,E. Simultaneous screening and quantitation of 18 antihistamine drugs in blood by liquid chromatography ionspray tandem mass spectrometry. *Forensic Sci.Inter.* **2001**, 121, 108–115.)

## HPLC VARIABLES

**Guard column:** 40 mm long 4 μm Purospher RP-18 LiChro Cart 4-4
**Column:** 100 × 2.1 4 μm Genesis C18 (Jones Chromatography)
**Column temperature:** 35
**Mobile phase:** Gradient. MeCN:buffer from 20:80 to 100:0 over 10 min, maintain at 0:100 for 3 min, re-equilibrate at initial conditions for 5 min. (The buffer was 10 mM ammonium acetate containing 0.1% formic acid (pH 3.2).)
**Flow rate:** 0.2
**Injection volume:** 30
**Detector:** MS, PE Sciex API 365 triple-stage quadrupole LC-MS-MS, PE Sciex TurboIonSpray interface, positive ion mode, needle voltage 5.2 kV, nebulizer gas air at 60 psi, curtain gas nitrogen at 40 psi, collision cell gas nitrogen at 40 psi, turbo ionspray heater 375°, heater gas flow 7 L/min

## CHROMATOGRAM

**Retention time:** 3.6
**Internal standard:** dibenzepin, enalapril
**Limit of detection:** 100 ng/mL

## OTHER SUBSTANCES

**Extracted:** acebutolol (3.8, LOD 0.1 μg/mL), acetaminophen (2.5, LOD <5 μg/mL), acrivastine (5.7, LOD <0.02 μg/mL), alprazolam (6.1, LOD <0.02 μg/mL), alprenolol (5.4, LOD 0.01 μg/mL), amantadine (3.4, LOD 0.1 μg/mL), amiloride (2.0, LOD 0.1 μg/mL), aminophenazone (2.8, LOD <5 μg/mL), amiodarone (10.2, LOD 0.05 μg/mL), amitriptyline (6.6, LOD <0.02 μg/mL), astemizole (5.8, LOD <0.02 μg/mL), atenolol (1.7, LOD 0.30 μg/mL), azacyclonol (5.1, LOD 0.02 μg/mL), benzhexol (6.6, LOD <0.02 μg/mL), benzoylecgonine (3.3, LOD 0.01 μg/mL), betaxolol (5.5, LOD 0.01 μg/mL), biperidine (6.2, LOD <0.02 μg/mL), bisoprolol (5.0, LOD <0.02 μg/mL), brompheniramine (5.3, LOD 0.002 μg/mL), bupivacaine (5.1, LOD <0.02 μg/mL), buprenorphine (5.9, LOD 0.01 μg/mL), buspirone (5.1, LOD 0.002 μg/mL), caffeine (2.8, LOD 1 μg/mL), carbamazepine

(6.1, LOD <0.02 μg/mL), carbinoxamine (5.1, LOD 0.002 μg/mL), carisoprodol (6.7, LOD <5 μg/mL), carvedilol (6.2, LOD <0.02 μg/mL), celiprolol (4.3, LOD 0.05 μg/mL), cetirizine (6.3, LOD 0.05 μg/mL), chlorcyclizine (6.6, LOD <0.02 μg/mL), chlordiazepoxide (5.7, LOD <0.02 μg/mL), chlormezanone (5.8, LOD <5 μg/mL), chloroquine (2.7, LOD 0.02 μg/mL), chlorpheniramine (5.1, LOD 0.002 μg/mL), chlorpromazine (7.0, LOD 0.02 μg/mL), chlorpropamide (6.7, LOD <5 μg/mL), chlorprothixene (7.0, LOD <0.02 μg/mL), cinnarizine (7.9, LOD <0.02 μg/mL), citalopram (5.7, LOD <0.02 μg/mL), clemastine (7.7, LOD 0.02 μg/mL), clobazam (7.3, LOD <0.02 μg/mL), clobutinol (5.3, LOD 0.02 μg/mL), clomethiazole (6.2, LOD 0.5 μg/mL), clomipramine (7.1, LOD <0.02 μg/mL), clonazepam (6.6, LOD <0.02 μg/mL), clonidine (2.8, LOD 0.1 μg/mL), clozapine (5.6, LOD <0.02 μg/mL), cocaine (4.6, LOD <0.02 μg/mL), codeine (2.5, LOD 0.1 μg/mL), coumatetralyl (8.4, LOD 0.05 μg/mL), cyclizine (5.8, LOD <0.02 μg/mL), dextropropoxyphene (6.6, LOD 0.05 μg/mL), demoxepam (5.8, LOD 0.02 μg/mL), dextromethorphan (5.5, LOD <0.02 μg/mL), diazepam (8.1, LOD 0.02 μg/mL), diltiazem (5.8, LOD <0.02 μg/mL), diphenhydramine (5.7, LOD <0.02 μg/mL), dipyridamole (5.4, LOD 0.005 μg/mL), disopyramine (4.4, LOD <0.02 μg/mL), dixyrazine (6.8, LOD 0.005 μg/mL), doxapram (4.8, LOD <0.02 μg/mL), doxepin (5.9, LOD <0.02 μg/mL), dronabinol (12.3, LOD 0.05 μg/mL), ebastine (9.6, LOD 0.005 μg/mL), embutramide (6.7, LOD 0.005 μg/mL), ergotamine (5.5, LOD 0.005 μg/mL), ethenzamide (5.0, LOD 0.05 μg/mL), ethylmorphine (3.2, LOD 0.05 μg/mL), ethylparathion (9.7, LOD <5 μg/mL), etodroxizine (6.4, LOD <0.02 μg/mL), felodipine (9.6, LOD 0.02 μg/mL), fenazepam (7.5, LOD <0.02 μg/mL), fenfluramine (5.3, LOD <0.02 μg/mL), fenkamfamine (5.1, LOD <0.02 μg/mL), fentanyl (5.5, LOD <0.02 μg/mL), fexofenadine (6.3, LOD <0.02 μg/mL), flecainide (5.9, LOD <0.02 μg/mL), fluconazole (4.0, LOD 0.1 μg/mL), flumazenil (5.2, LOD <0.02 μg/mL), flunitrazepam (7.1, LOD 0.002 μg/mL), fluoxetine (6.8, LOD 0.1 μg/mL), flupentixol (7.5, LOD 0.18 μg/mL), fluvoxamine (6.3, LOD 0.02 μg/mL), glibenclamide (8.5, LOD <0.02 μg/mL), glipizide (6.8, LOD <0.05 μg/mL), haloperidol (6.1, LOD <0.02 μg/mL), histapyrrodine (6.3, LOD 0.02 μg/mL), hydrocodone (3.0, LOD 0.05 μg/mL), hydroxychloroquine (2.4, LOD <0.3 μg/mL), hydroxyzine (6.3, LOD <0.02 μg/mL), imipramine (6.4, LOD 0.05 μg/mL), indomethacin (8.6, LOD 0.05 μg/mL), isoniazid (2.2, LOD 3 μg/mL), isradipine (8.6, LOD 0.05 μg/mL), ketamine (3.6, LOD <0.05 μg/mL), ketobemidone (3.3, LOD <0.05 μg/mL), ketoprofen (7.3, LOD 0.1 μg/mL), ketorolac (6.2, LOD 0.05 μg/mL), labetalol (4.9, LOD 0.05 μg/mL), lamotrigine (4.0, LOD 0.1 μg/mL), levocabastine (5.8, LOD 0.01 μg/mL), levomepromazine (6.5, LOD 0.02 μg/mL), lidocaine (3.7, LOD <0.05 μg/mL), loratadine (9.3, LOD 0.002 μg/mL), lorazepam (6.6, LOD 0.02 μg/mL), lormetazepam (7.4, LOD <0.02 μg/mL), LSD (4.7, LOD <0.02 μg/mL), malathion (8.9, LOD 10 μg/mL), maprotiline (6.4, LOD <0.02 μg/mL), MDMA (3.3, LOD 0.02 μg/mL), meclozine (8.5, LOD <0.02 μg/mL), medazepam (6.3, LOD <0.02 μg/mL), meloxicam (7.1, LOD 0.01 μg/mL), melperone (5.0, LOD <0.02 μg/mL), meperidine (4.7, LOD <0.02 μg/mL), mepivacaine (3.7, LOD <0.02 μg/mL), meprobamate (4.9, LOD 0.1 μg/mL), mesoridazine (5.4, LOD <0.02 μg/mL), methamphetamine (3.3, LOD 0.05 μg/mL), methadone (6.7, LOD <0.02 μg/mL), methylparathion (8.6, LOD 10 μg/mL), methylphenidate (4.2, LOD <0.02 μg/mL), metoclopramide (3.8, LOD <0.02 μg/mL), metoprolol (4.1, LOD 0.02 μg/mL), metronidazole (2.6, LOD 1 μg/mL), mexiletine (4.4, LOD 0.05 μg/mL), mianserin (5.7, LOD <0.02 μg/mL), midazolam (5.9, LOD <0.02 μg/mL), mirtazapine (4.4, LOD <0.02 μg/mL), mizolastine (5.5, LOD 0.01 μg/mL), moclobemide (3.7, LOD 0.05 μg/mL), molindone (4.0, LOD <0.02 μg/mL), monoacetylmorphine (2.7, LOD 0.1 μg/mL), morphine (2.0, LOD 0.1 μg/mL), nicotine (2.2, LOD 0.05 μg/mL), nifedipine (7.5, LOD 0.02 μg/mL), nikethamide (3.6, LOD <0.02 μg/mL), nitrazepam (6.5, LOD <0.02 μg/mL), nizatidine (1.7, LOD 1 μg/mL), nomifensine (4.6, LOD <0.02 μg/mL), nortriptyline (6.4, LOD <0.02 μg/mL), norverapamil (6.2, LOD 1 μg/mL), noscapine (5.0, LOD <0.02 μg/mL), olanzapine (3.0, LOD 0.05 μg/mL), ondansetron (4.6, LOD <0.02 μg/mL), orphenadrine (6.1, LOD <0.02 μg/mL), oxazepam (6.3, LOD <0.02 μg/mL), oxcarbazepine (5.3, LOD 0.02 μg/mL), oxprenolol (4.7, LOD 0.02 μg/mL), oxycodone (2.8, LOD 0.05 μg/mL), papaverine (4.8, LOD <0.02 μg/mL), paroxetine (6.2, LOD 0.02 μg/mL), pemoline (3.3, LOD 0.05 μg/mL), pentazocine (5.0, LOD <0.02 μg/mL), pentifylline (7.3, LOD <5 μg/mL), pentoxyverine (6.6, LOD <0.02 μg/mL), perphenazine (6.9, LOD 0.002 μg/mL), phenazone (3.9, LOD 0.05 μg/mL), phencyclidine (5.3, LOD 0.05 μg/mL), pheniramine (4.1, LOD 0.02 μg/mL), phenylbutazone (9.0, LOD <5 μg/mL), phenylpropanolamine (2.5, LOD 0.3 μg/mL), phenytoin (6.1,

LOD 0.05 μg/mL), pindolol (3.3, LOD 0.05 μg/mL), piroxicam (6.6, LOD 0.02 μg/mL), pitofenone (5.4, LOD <0.02 μg/mL), pizotifen (6.5, LOD <0.02 μg/mL), practolol (1.8, LOD 0.1 μg/mL), prazosin (4.1, LOD 0.05 μg/mL), prilocaine (3.8, LOD <0.02 μg/mL), primidone (4.0, LOD <5 μg/mL), procainamide (2.2, LOD 0.05 μg/mL), prochlorperazine (7.5, LOD 0.02 μg/mL), promazine (6.2, LOD <0.02 μg/mL), promethazine (6.0, LOD 0.05 μg/mL), propafenone (6.3, LOD <0.02 μg/mL), propranolol (5.4, LOD 0.02 μg/mL), propyphenazone (6.6, LOD 0.50 μg/mL), pseudoephedrine (2.6, LOD 1 μg/mL), quinine (4.2, LOD 0.02 μg/mL), ranitidine (1.8, LOD 0.1 μg/mL), risperidone (4.9, LOD <0.02 μg/mL), rocurone (3.8, LOD 0.1 μg/mL), ropivacaine (4.6, LOD <0.02 μg/mL), salicylamide (4.2, LOD <5 μg/mL), selegiline (4.1, LOD 0.05 μg/mL), sertindole (7.2, LOD <0.02 μg/mL), sertraline (6.8, LOD 0.02 μg/mL), sulindac (6.5, LOD 0.02 μg/mL), simazine (6.0, LOD 0.1 μg/mL), sincocaine (6.5, LOD <0.02 μg/mL), sisapride (5.9, LOD <0.02 μg/mL), sotalol (2.1, LOD 0.1 μg/mL), strychnine (5.3, LOD 0.05 μg/mL), sulpiride (1.9, LOD 0.1 μg/mL), sulthiame (4.1, LOD 0.05 μg/mL), temazepam (7.2, LOD <0.02 μg/mL), terbutaline (2.3, LOD 0.1 μg/mL), terfenadine (8.1, LOD 0.002 μg/mL), terodiline (6.7, LOD <0.02 μg/mL), tetracaine (5.7, LOD <0.02 μg/mL), theobromine (2.3, LOD <5 μg/mL), theophylline (2.4, LOD <5 μg/mL), thioridazine (7.5, LOD 0.02 μg/mL), timolol (3.8, LOD 0.05 μg/mL), thiothixene (6.7, LOD 0.02 μg/mL), tolbutamide (7.1, LOD <5 μg/mL), toremifene (8.7, LOD 0.02 μg/mL), tramadol (4.2, LOD 0.02 μg/mL), trazodone (5.2, LOD <0.02 μg/mL), triamterene (3.2, LOD 0.1 μg/mL), triazolam (6.7, LOD 0.002 μg/mL), trimeprazine (6.4, LOD <0.02 μg/mL), trimethoprim (3.1, LOD 0.05 μg/mL), trimipramine (6.7, LOD <0.02 μg/mL), venlafaxine (4.9, LOD 0.02 μg/mL), verapamil (6.5, LOD <0.02 μg/mL), warfarin (7.9, LOD <0.02 μg/mL), yohimbine (4.5, LOD <0.02 μg/mL), zolpidem (4.7, LOD <0.02 μg/mL), zopiclone (4.0, LOD 0.1 μg/mL)

## KEY WORDS

whole blood

## REFERENCE

Gergov, M.; Ojanperä, I.; Vuori, E. Simultaneous screening for 238 drugs in blood by liquid chromatography-ionspray tandem mass spectrometry with multiple-reaction monitoring, *J.Chromatogr.B*, **2003**, *795*, 41–53.

# Thalidomide

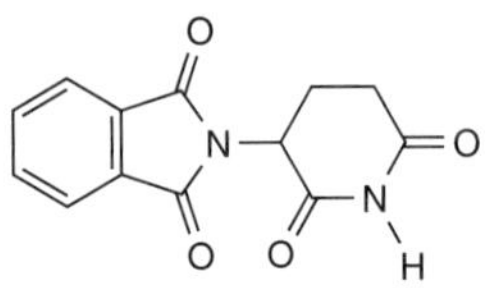

**Molecular formula:** $C_{13}H_{10}N_2O_4$
**Molecular weight:** 258.23
**CAS Registry No:** 50-35-1
**Merck Index:** 13, 9326

## SAMPLE
**Matrix:** blood
**Sample preparation:** Condition a 1 mL 100 mg C18 SPE cartridge (Varian) with 1 column vol of MeOH and one column volume of water. Vortex 1 mL 25 mM pH 2.5 phosphate buffer with 1 mL plasma, add 100 μL 100 μg/mL labetalol in MeOH, vortex, add to the SPE cartridge, wash with 2 column vol of water, dry under vacuum, elute with 1 mL anhydrous diethyl ether, evaporate the eluate to dryness under a stream of nitrogen, reconstitute the residue with 130 μL mobile phase, inject a 50 μL aliquot.

## HPLC VARIABLES
**Guard column:** Bioptic AV-1 (Meta Chem Technologies, Torrance CA)
**Column:** 150 × 4.5 Bioptic AV-1 (Meta Chem Technologies, Torrance CA)
**Mobile phase:** Isopropanol:100 mM pH 4 phosphate buffer 2:98
**Flow rate:** 0.6
**Injection volume:** 50
**Detector:** UV 300

## CHROMATOGRAM
**Retention time:** 8.4 (*S*-(−)), 10.5 (*R*-(+))
**Internal standard:** labetalol (6.7)
**Limit of detection:** 50 ng/mL
**Limit of quantitation:** 100 ng/mL

## OTHER SUBSTANCES
**Extracted:** metabolites

## KEY WORDS
chiral; plasma; SPE

## REFERENCE
Haque, A.; Stewart, J.T. Determination of racemic thalidomide in human plasma by use of an avidin column and solid phase extraction, *J.Liq.Chromatogr.Rel.Technol.*, **1998**, *21*, 2151–2163.

## SAMPLE
**Matrix:** blood
**Sample preparation:** Mix 500 μL serum with 20 μL 250 μg/mL IS in MeOH, add 500 μL 10% trichloroacetic acid solution with constant vortexing for 30 s, centrifuge at 2700 g for 5 min, inject a 20 μL aliquot.

## HPLC VARIABLES
**Guard column:** 20 × 4.6 Discovery C18 (Supelco)
**Column:** 125 × 4.6 5 μm Discovery C18 (Supelco)
**Mobile phase:** MeOH:buffer 75:25 (The buffer was 10 mM potassium dihydrogen phosphate adjusted to pH 3.0 with 43% phosphoric acid.)
**Flow rate:** 1.5
**Injection volume:** 20
**Detector:** UV 220

## CHROMATOGRAM

**Retention time:** 7.9
**Internal standard:** phenacetin (15.0)
**Limit of quantitation:** 222 ng/mL

## OTHER SUBSTANCES

**Simultaneous:** clofazimine (8.3), dapsone (3.5), indinavir (4.0), isoniazid (2.1), lamivudine (2.2), nevirapine (15.0), rifampin (2.5), saquinavir (2.5), stavudine (2.1), trimethoprim (4.8), ritonavir (2.2), zidovudine (4.0)
**Noninterfering:** cyclosporine, rifabutin, sulfamethoxazole, prednisolone

## KEY WORDS

pharmacokinetics; serum

## REFERENCE

Toraño, J.S.; Verbon, A.; Guchelaar, H.-J. Quantitative determination of thalidomide in human serum with high-performance liquid chromatography using protein precipitation with trichloroacetic acid and ultraviolet detection, *J.Chromatogr.B*, **1999**, *734*, 203–210.

## ANNOTATED BIBLIOGRAPHY

Boughton, B.J.; Sheehan, T.M.; Wood, J.; O'Brien, D.; Butler, M.; Simpson, A.; Hale, K.A. High-performance liquid chromatographic assay of plasma thalidomide: stabilization of specimens and determination of a tentative therapeutic range for chronic graft-versus-host disease, *Ann.Clin.Biochem.*, **1995**, *32*, 79–83.

Delon, A.; Favreliere, S.; Couet, W.; Courtois, P.; Bouquet, S. Rapid and sensitive determination of thalidomide in human plasma by high- performance liquid chromatography, *J.Liq.Chromatogr.*, **1995**, *18*, 297–309. [SPE; pharmacokinetics; ciprofloxacin is internal standard; LOQ 62.5 ng/mL; simultaneous acyclovir, flucytosine, metronidazole, azathioprine, ceftazidime, cefotaxime; non-interfering cyclosporine, cyclophosphamide, prednisolone, hydroxyzine, nifedipine, diltiazem, amphotericin, clonazepam, clobazam, diazepam]

Eriksson, T.; Bjorkman, S.; Fyge, A.; Ekberg, H. Determination of thalidomide in plasma and blood by high-performance liquid chromatography: avoiding hydrolytic degradation, *J.Chromatogr.*, **1992**, *582*, 211–216.

Eriksson, T.; Bjorkman, S.; Roth, B.; Fyge, A.; Hoglund, P. Enantiomers of thalidomide: blood distribution and the influence of serum albumin on chiral inversion and hydrolysis, *Chirality*, **1998**, *10*, 223–228. [chiral]

Haque, A.; Stewart, J.T. Determination of racemic thalidomide in human plasma by use of an avidin column and solid phase extraction (Abstract 4163), *Pharm.Res.*, **1997**, *14*, S684–S685.

Huupponen, R.; Pykkö, K. Stability of thalidomide in human plasma, *Clin.Chem.*, **1995**, *41*, 1199.

Teo, S.K.; Chandula, R.S.; Harden, J.L.; Stirling, D.I.; Thomas, S.D. Sensitive and rapid method for the determination of thalidomide in human plasma and semen using solid-phase extraction and liquid chromatography-tandem mass spectrometry, *J.Chromatogr.B*, **2002**, *767*, 145–151. [LOQ 2 ng/mL]

# Thialbarbital

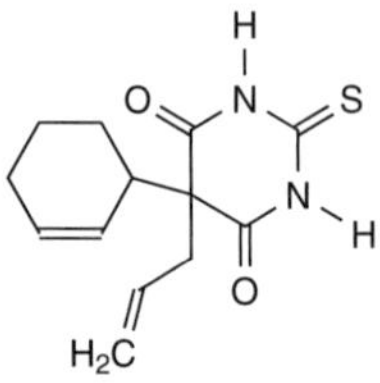


**Molecular formula:** $C_{13}H_{16}N_2O_2S$
**Molecular weight:** 264.35
**CAS Registry No:** 467-36-7
**Merck Index:** 13, 9363

## SAMPLE
**Matrix:** solutions

## HPLC VARIABLES
**Column:** 150 × 2 Chiralcel OJ-R
**Mobile phase:** MeCN:water 30:70
**Flow rate:** 0.2
**Injection volume:** 8
**Detector:** UV 235

## CHROMATOGRAM
**Retention time:** 21, 27 (enantiomers)

## KEY WORDS
chiral

## REFERENCE
Aboul-Enein, H.Y.; Schmid, M.G.; Tuscher, C.; Gübitz, G.; Laffranchini, M.; Bojarski, J. Chiral separation of several thiobarbiturates on a cellulose tris(4-methylbenzoate) phase under reversed-phase conditions, *J.Liq.Chromatogr.Rel.Technol.*, **2001**, *24*, 69–77.

# Thyrotropin

**Molecular formula:** $C_{1039}H_{1602}N_{274}O_{307}S_{27}$
**Molecular weight:** 23709.28
**CAS Registry No:** 9002-71-5
**Merck Index:** 13, 9870

## SAMPLE
**Matrix:** solutions

## HPLC VARIABLES
**Guard column:** Vydac 214 FSK 54
**Column:** 250 × 4.6 5 μm C4 Vydac 214 TP 54
**Column temperature:** 25
**Mobile phase:** Gradient. A was 50 mM pH 7.0 phosphate buffer. B was MeCN:50 mM pH 7.0 phosphate buffer 50:50. A:B from 75:25 to 0:100 over 40 min. (Place a silica column packed with 7.9–12.4 μm LiChrosorb Si-60 silica between pump and injector.)
**Flow rate:** 0.5
**Injection volume:** 5–200
**Detector:** UV 280

## CHROMATOGRAM
**Retention time:** 33–36
**Limit of detection:** 200 ng

## REFERENCE
Ezequiel de Oliveira, J.; de Mendonça, F.; Peroni, C.N.; Bartolini, P.; Ribela, M.T.C.P. Determination of Chinese hamster ovary cell-derived recombinant thyrotropin by reversed-phase liquid chromatography, *J.Chromatogr.B*, **2003**, *787*, 345–355.

## SAMPLE
**Matrix:** solutions

## HPLC VARIABLES
**Guard column:** 75 × 7.5 SW
**Column:** 600 × 7.5 10 μm G2000 SW
**Mobile phase:** 20 mM pH 7.0 phosphate buffer containing 150 mM NaCl
**Flow rate:** 1
**Injection volume:** 5–50
**Detector:** UV 280 (?)

## CHROMATOGRAM
**Retention time:** 17

## REFERENCE
Ezequiel de Oliveira, J.; de Mendonça, F.; Peroni, C.N.; Bartolini, P.; Ribela, M.T.C.P. Determination of Chinese hamster ovary cell-derived recombinant thyrotropin by reversed-phase liquid chromatography, *J.Chromatogr.B*, **2003**, *787*, 345–355.

# Tiagabine

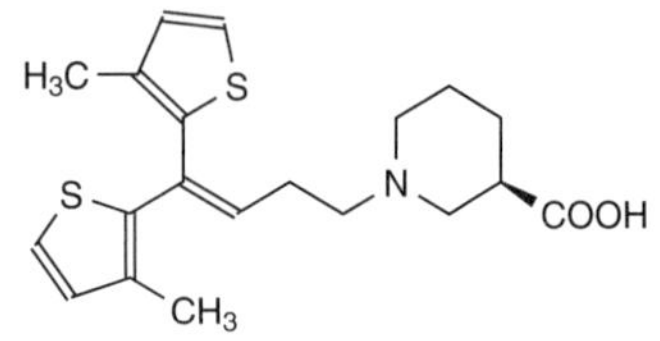


**Molecular formula:** $C_{20}H_{25}NO_2S_2$
**Molecular weight:** 375.56
**CAS Registry No:** 115103-54-3, 145821-59-6 (HCl)
**Merck Index:** 13, 9493

## SAMPLE
**Matrix:** blood
**Sample preparation:** Condition a 3 mL Bond Elut C8 SPE cartridge with 5 mL MeOH and 2 mL water. Mix 1 mL plasma with 200 μL water and 200 μL 1 μg/mL IS in MeOH:water 50:50, centrifuge at 1000 g for 15 min, add the supernatant to the SPE cartridge, wash with 2 mL water, wash with 2 mL MeOH:water 5:95, elute with 1 mL MeOH. Evaporate the eluate to dryness under a stream of air at 40°, reconstitute the residue with 200 μL mobile phase, inject a 50–60 μL aliquot.

## HPLC VARIABLES
**Column:** 250 × 4.6 5 μm Ultrasphere ODS C18
**Mobile phase:** MeCN:MeOH:water 37:10:53 containing 10 mM sodium dihydrogen phosphate and 5 mM sodium octanesulfonate
**Flow rate:** 1.2
**Injection volume:** 50–60
**Detector:** E, Environmental Sciences Coulochem Model 5100A, porous graphite electrode – 0.50 V (screen), porous graphite electrode +0.76 V (monitored), guard cell +0.95 V (before injector)

## CHROMATOGRAM
**Retention time:** 16
**Internal standard:** (R)-*N*-(4-(3-methyl-2-thienyl)-4-(2-thienyl)but-3-enyl)nipecotic acid (12)
**Limit of quantitation:** 2 ng/mL

## KEY WORDS
plasma; SPE

## REFERENCE
Gustavson, L.E.; Chu, S.Y. High-performance liquid chromatographic procedure for the determination of tiagabine concentrations in human plasma using electrochemical detection, *J.Chromatogr.*, **1992**, *574*, 313–318.

# Tiletamine hydrochloride

**Molecular formula:** $C_{12}H_{17}NOS.HCl$
**Molecular weight:** 259.80
**CAS Registry No:** 14176-50-2

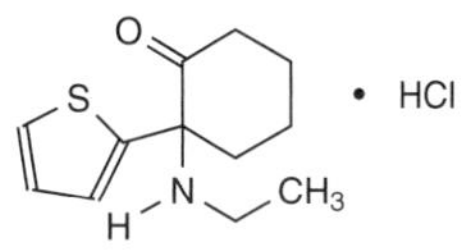

## SAMPLE

**Matrix:** blood, tissue

**Sample preparation:** Add IS to serum or tissue, make alkaline with pH 9.5 borate buffer, extract with ethyl acetate. Extract the organic layer with 1 mL 100 mM HCl. Basify the aqueous layer with 100 mg sodium borate, extract with ethyl acetate. Evaporate the organic layer to dryness under a stream of nitrogen, reconstitute the residue with MeCN:water 25:75 (serum) or mobile phase (tissue), inject a 70 μL aliquot.

## HPLC VARIABLES

**Column:** 250 × 4 5 μm LiChrospher 60 RP-select B C8
**Mobile phase:** MeCN:50 mM pH 6.8 phosphate buffer 26:74
**Flow rate:** 1
**Injection volume:** 70
**Detector:** UV 233

## CHROMATOGRAM

**Retention time:** 18.9
**Internal standard:** pindolol (9.7)
**Limit of quantitation:** 10 ng/mL

## OTHER SUBSTANCES

**Extracted:** zolazepam (LOQ 2 ng/mL) (12.5)

## KEY WORDS

bear; muscle; serum

## REFERENCE

Semple, H.A.; Gorecki, D.K.J.; Farley, S.D.; Ramsay, M.A. Pharmacokinetics and tissue residues of Telazol in free-ranging polar bears, *J.Wildlife Dis.*, **2000**, *36*, 653–662.

## SAMPLE

**Matrix:** blood, urine

**Sample preparation:** Mix 1 mL pH 7 buffer with 5 mL whole blood or urine, add 1 μg IS1, add 6 μg IS2, add 7 mL hexane:toluene:isoamyl alcohol 90:5:5, vortex, centrifuge, remove the organic layer. Add 1 mL pH 9.5 buffer to the aqueous layer, extract with 6 mL hexane:isoamyl alcohol 99:1. Combine the organic layers, evaporate to dryness under a stream of nitrogen, reconstitute the residue with 250 μL MeCN:water:trifluoroacetic acid 45:55:0.05, evaporate to half volume to remove MeCN, inject a 50 μL aliquot.

## HPLC VARIABLES

**Guard column:** 5 μm Altima C18
**Column:** 20 × 2.1 5 μm Altima C18
**Column temperature:** 40
**Mobile phase:** Gradient. A was MeCN:water:trifluoroacetic acid 10:90:0.05. B was MeCN:water:trifluoroacetic acid 80:20:0.05. A:B 100:0 for 5 min, to 0:100 over 50 min, re-equilibrate at initial conditions for 15 min.

**Flow rate:** 0.25
**Injection volume:** 50
**Detector:** MS, HP 5989B, electron impact; UV 205; UV 290

---

## CHROMATOGRAM

**Internal standard:** SKF-525 A (IS1), 5-ethyl-5-*p*-tolylbarbituric acid (IS2)

---

## OTHER SUBSTANCES

**Extracted:** zolazepam

---

## KEY WORDS

whole blood

---

## REFERENCE

Cording, C.J.; Deluca, R.; Camporese, T.; Spratt, E. A fatality related to the veterinary anesthetic Telazol, *J.Anal.Toxicol.*, **1999**, *23*, 552–555.

---

## SAMPLE

**Matrix:** tissue
**Sample preparation:** Homogenize 1.5 g tissue with 300 μL 1 μg/mL IS in water and 6 mL 500 mM pH 9.5 sodium borate buffer, add 10 mL ethyl acetate, vortex for 20 min, repeat the extraction. Combine the organic layers and extract twice with 2 mL portions of 100 mM HCl. Combine the aqueous layers and make basic with 200 mg sodium borate, extract with 12 mL ethyl acetate. Evaporate the organic layer to dryness under a stream of nitrogen, reconstitute the residue with 200 μL MeCN:water 20:80, inject a 70 μL aliquot.

---

## HPLC VARIABLES

**Column:** 250 × 4 5 μm LiChrospher 60 RP-select B C8
**Mobile phase:** MeCN:50 mM pH 5.5 sodium phosphate buffer 16:84
**Flow rate:** 1
**Injection volume:** 70
**Detector:** UV 233

---

## CHROMATOGRAM

**Retention time:** 45
**Internal standard:** ripazapam (51.5)

---

## OTHER SUBSTANCES

**Extracted:** tiletamine metabolite CI-398 (23.5), zolazepam (58), zolazepam metabolites

---

## KEY WORDS

bear; fat; kidney; muscle

---

## REFERENCE

Semple, H.A.; Gorecki, D.K.J.; Farley, S.D.; Ramsay, M.A. Pharmacokinetics and tissue residues of Telazol in free-ranging polar bears, *J.Wildlife Dis.*, **2000**, *36*, 653–662.

# Tilmicosin

**Molecular formula:** $C_{46}H_{80}N_2O_{13}$
**Molecular weight:** 869.13
**CAS Registry No:** 108050-54-0
**Merck Index:** 13, 9516

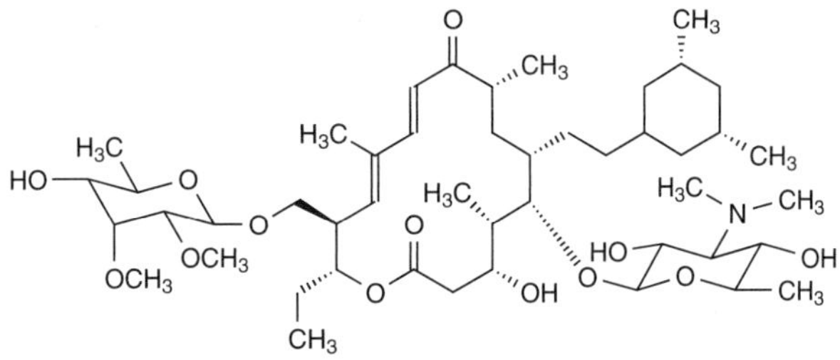


## SAMPLE

**Matrix:** blood

**Sample preparation:** Condition a 500 mg Bond Elut C18 SPE cartridge with MeOH and water. Add 2 mL serum to the SPE cartridge, wash with water, wash with 5% ammonium hydroxide, wash with water, elute with 2 mL MeOH:acetic acid 5:95. Evaporate the eluate to dryness under a stream of nitrogen, reconstitute the residue with 1 mL diluent, inject an aliquot. (The diluent was MeOH:water:1 M dibutylammonium phosphate 50:47.5:2.5.)

## HPLC VARIABLES

**Column:** 250 × 4.6 5 μm Hi-Chrom Reversible phenyl (Regis)

**Mobile phase:** Gradient. A was MeCN:water 50:50 adjusted to pH 2.5 with orthophosphoric acid. B was water adjusted to pH 2.5 with orthophosphoric acid. C was 168 mL dibutylamine in 700 mL water, adjust to pH 2.5 with orthophosphoric acid, make up to 1 L with water. A:B:C 100:0:0 for 3 min, to 55:30:15 over 1 min, maintain at 55:30:15 for 7 min, return to initial conditions over 1 min, re-equilibrate for 8 min.

**Flow rate:** 1.5

**Injection volume:** 50

**Detector:** UV 280

## CHROMATOGRAM

**Retention time:** 11

**Limit of detection:** 2.6 ng/mL

**Limit of quantitation:** 50 mg/mL

## KEY WORDS

cow; pharmacokinetics; serum; sheep

## REFERENCE


Modric, S.; Webb, A.I.; Derendorf, H. Pharmacokinetics and pharmacodynamics of tilmicosin in sheep and cattle, *J.vet.Pharmacol.Therap.*, **1998**, *21*, 444–452.

# Tiludronic acid

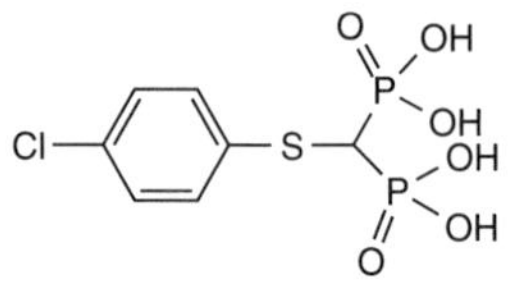


**Molecular formula:** $C_7H_9ClO_6P_2S$

**Molecular weight:** 318.61

**CAS Registry No:** 89987-06-4, 149845-07-8 (Na salt), 155453-10-4 (Na salt hemihydrate)

**Merck Index:** 13, 9518

## SAMPLE

**Matrix:** blood, urine

**Sample preparation:** Mix 200 μL plasma or urine with 400 μL 5 μg/mL IS in water, add 100 μL 1 M NaOH, add 100 μL 180 mM calcium chloride, add 1 mL water, vortex, centrifuge at 11 600 g. Discard the liquid phase and dissolve the residue in 100 μL 1 M HCl, add 200 μL 1 M NaOH, add 100 μL 180 mM calcium chloride (only for plasma samples), add 1 mL water, vortex, centrifuge at 11 600 g. Discard the liquid phase and vortex the residue with 200 μL mobile phase containing 100 mM disodium EDTA until dissolution is complete, inject a 20–100 μL aliquot.

## HPLC VARIABLES

**Column:** 150 × 4.1 5 μm PRP-1 (Hamilton)

**Column temperature:** 22

**Mobile phase:** MeCN:buffer 13:87 (The buffer was 50 mM pH 11.8 disodium hydrogen phosphate containing 5 mM tetrabutylammonium phosphate.)

**Flow rate:** 1

**Injection volume:** 20–100

**Detector:** UV 280

## CHROMATOGRAM

**Retention time:** 3

**Internal standard:** (3-trifluoromethyl)-thiomethylene bisphosphonic acid (4)

**Limit of quantitation:** 50 ng/mL

## KEY WORDS

baboon; human; monkey; mouse; plasma

## REFERENCE

Fels, J.-P.; Guyonnet, J.; Berger, Y.; Cautreels, W. Determination of (4-chlorophenyl)thiomethylene bisphosphonic acid, a new bisphosphonate, in biological fluids by high-performance liquid chromatography, *J.Chromatogr.*, **1988**, *430*, 73–79.

# Tirilazad

**Molecular formula:** $C_{38}H_{52}N_6O_2$
**Molecular weight:** 624.86
**CAS Registry No:** 110101-66-1
**Merck Index:** 13, 9540

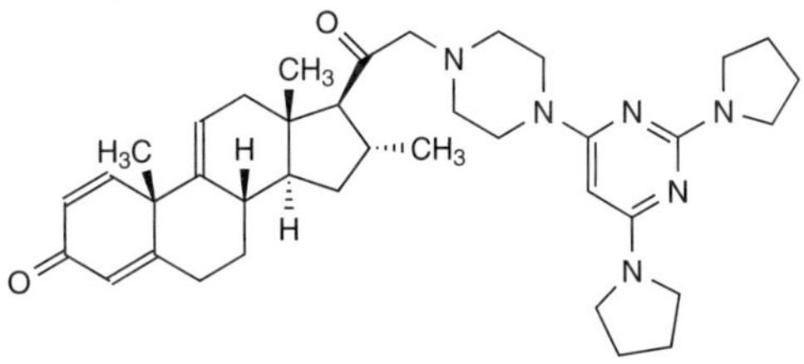

## SAMPLE

**Matrix:** blood
**Sample preparation:** Condition a 1 mL 100 mg Bond Elut C18 SPE cartridge with 2 mL MeOH and 2 mL water. Vortex 500 μL plasma with 500 μL 5 μg/mL IS in MeCN for 1 min, centrifuge at 1500 g at 4°, add 800 μL of the supernatant to the SPE cartridge, wash with two 1 mL portions of MeCN:water 50:50, elute with two 1 mL portions of MeOH. Evaporate the eluate to dryness under a stream of nitrogen at 40°, reconstitute the residue with 350 μL MeCN, vortex for 1 min, dilute with 150 μL water, inject an aliquot.

## HPLC VARIABLES

**Guard column:** 15 × 3.2 7 μm NewGuard RP-8 (Brownlee)
**Column:** 250 × 4.6 5 μm Suplecosil LC-8
**Mobile phase:** MeCN:buffer 75:25 (The buffer was 22 mM triethylamine adjusted to pH 5.0 with glacial acetic acid.)
**Flow rate:** 1
**Injection volume:** 100
**Detector:** UV 254

## CHROMATOGRAM

**Retention time:** 10.5
**Internal standard:** 21-[4-(2,6-di-1-pyrrolidinyl-4-pyrimidinyl)-1-piperazinyl]-3α-hydroxy-16α-methyl-5α-pregna-20-one (16)
**Limit of quantitation:** 12 ng/mL (S/N 10)

## KEY WORDS

plasma; rat

## REFERENCE

Cox, J.W.; Pullen, R.H. High-performance liquid chromatographic determination of a 21-aminosteroid antioxidant in plasma, *J.Chromatogr.*, **1988**, *424*, 293–302.

## SAMPLE

**Matrix:** blood
**Sample preparation:** Add 150 μL water and 300 μL IS in MeCN to 100 μL plasma, centrifuge, add 450 μL of the supernatant to a conditioned (unspecified) Advanced Automated Sample Processor C8 SPE cartridge containing 0.5 mL water (?), wash with 500 μL MeCN:water 25:75, elute the contents directly onto column A in series with column B using mobile phase, remove column A from the circuit, continue to elute column B with mobile phase, monitor the effluent from column B.

## HPLC VARIABLES

**Column:** A 15 × 3.2 7 μm Brownlee New Guard C18; B 250 × 4.6 Supelco LC8
**Mobile phase:** MeCN:water:triethylamine:acetic acid 80:20:0.1:0.2
**Flow rate:** 1.5
**Detector:** UV 254

## CHROMATOGRAM

**Retention time:** 8
**Internal standard:** (16$\alpha$)-21-[4-(2,6-di-1-pyrrolidinyl-4-pyrimidinyl)-1-piperazinyl]-16-methylpregna-3-ol-20-one (PNU-76824) (15)
**Limit of quantitation:** 5.4 ng/mL

## KEY WORDS

column-switching; pharmacokinetics; plasma; rat

## REFERENCE

Wang, Y.; Mesfin, G.-M.; Rodriguez, C.A.; Slater, J.G.; Schuette, M.R.; Cory, A.L.; Higgins, M.J. Venous irritation, pharmacokinetics, and tissue distribution of tirilazad in rats following intravenous administration of a novel supersaturated submicron lipid emulsion, *Pharm.Res.*, **1999**, *16*, 930–938.

## SAMPLE

**Matrix:** microsomal incubations
**Sample preparation:** Vortex 500 µL microsomal incubation with 500 µL MeCN for 20 s, centrifuge at 2100 g at 4° for 25 min, inject an aliquot of the supernatant.

## HPLC VARIABLES

**Column:** 250 × 4.6 5 µm Kromasil C18
**Mobile phase:** Gradient. MeCN:100 mM ammonium acetate from 10:90 to 100:0 in 60 min, maintain at 100:0 for 40 min
**Flow rate:** 0.5
**Detector:** MS, Finnigan 4021 quadrupole, thermal pneumatic nebulizer with momentum separator, positive ion CI mode with ammonia as moderating gas, tip 180°, expansion region 80°, helium 150 psi with a 360 × 75 µm fused silica capillary; Radioactivity ($^{14}C$)

## CHROMATOGRAM

**Retention time:** 80

## OTHER SUBSTANCES

**Extracted:** metabolites

## KEY WORDS

human; liver

## REFERENCE

Wienkers, L.C.; Steenwyk, R.C.; Sanders, P.E.; Pearson, P.G. Biotransformation of tirilazad in human: 1. Cytochrome P450 3A-mediated hydroxylation of tirilazad mesylate in human liver microsomes, *J.Pharmacol.Exp.Ther.*, **1996**, *277*, 982–990.

## ANNOTATED BIBLIOGRAPHY

Cox, J.W.; Pullen, R.H. Irregular retention properties of 21-aminosteroid antioxidants in octylsilane bonded-phase high-performance liquid chromatography, *J.Chromatogr.*, **1988**, *424*, 285–292.

Fleishaker, J.C.; Straw, R.N.; Cross, C.J. Pharmacokinetics of tirilazad and U-89678, an active, reduced metabolite, following acute head trauma in adults, *J.Pharm.Sci.*, **1997**, *86*, 434–437. [LOQ 5 ng/mL]

Hoerle, S.L.; Snider, B.G. Determination of degradation products occurring in acidic solutions of a 21-aminosteroid (tirilazad) using a gradient HPLC method, *J.Liq.Chromatogr.*, **1995**, *18*, 3269–3282.

Wienkers, L.C.; Steenwyk, R.C.; Mizsak, S.A.; Pearson, P.G. In vitro metabolism of tirilazad mesylate in male and female rats. Contribution of cytochrome P4502C11 and $\Delta^4$-5$\alpha$-reductase, *Drug Metab.Dispos.*, **1995**, *23*, 383–392.

# Tirofiban

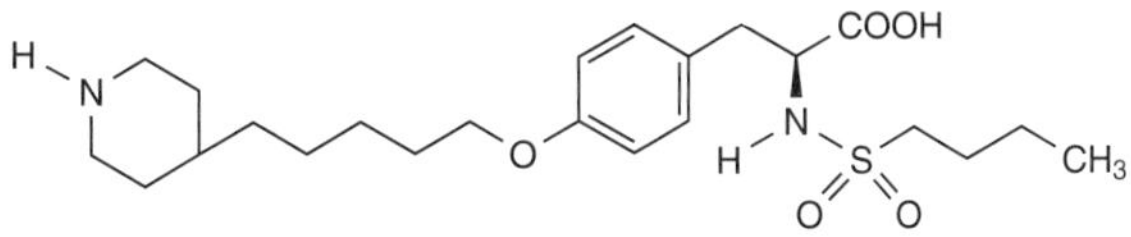

**Molecular formula:** $C_{22}H_{36}N_2O_5S$
**Molecular weight:** 440.60
**CAS Registry No:** 144494-65-5, 150915-40-5 (HCl monohydrate)
**Merck Index:** 13, 9541

## SAMPLE
**Matrix:** blood
**Sample preparation:** Vortex 500 μL plasma with 50 μL 100 ng/mL IS and 500 μL 100 mM perchloric acid briefly, add 5 mL 1-chlorobutane, shake for 10 min, centrifuge, extract the aqueous layer twice more with 5 mL portions of ethyl acetate. Combine the organic layers and evaporate them to dryness under a stream of nitrogen at 40°, reconstitute the residue with 250 μL EtOH:toluene 50:50, evaporate to dryness under a stream of nitrogen at 40°, add 400 μL 10 mM dimethylaminopyridine in MeCN, add 40 μL trifluoroacetic anhydride, heat at 40° for 30 min, remove excess reagents with a stream of nitrogen, reconstitute the residue with 150 μL mobile phase, inject a 25 μL aliquot.

## HPLC VARIABLES
**Column:** 250 × 4.6 5 μm Zorbax RX-C18
**Mobile phase:** MeCN:1 mM ammonium acetate 41.5:58.5, pH 6.0
**Injection volume:** 25
**Detector:** MS, Sciex Model API-III triple-quadrupole, API, negative ion mode, orifice potential – 60 V, electron multiplier +4.4 kV, interface 50°, nebulizer 500°, nebulizer air 1 L/min, curtain gas 700 mL/min, m/z 535

## CHROMATOGRAM
**Retention time:** 5.2
**Internal standard:** *N*-phenylsulfonyl-*O*-[4-(butane-4-piperidinyl]-L-tyrosine (L-702, 128) (5.1)
**Limit of quantitation:** 400 pg/mL

## KEY WORDS
derivatization; pharmacokinetics; plasma

## REFERENCE
Ellis, J.D.; Hand, E.L.; Gilbert, J.D. Use of LC-MS/MS to cross-validate a radioimmunoassay for the fibrinogen receptor antagonist, Aggrastat (tirofiban hydrochloride) in human plasma, *J. Pharm. Biomed. Anal.*, **1997**, *15*, 561–569.

## SAMPLE
**Matrix:** solutions

## HPLC VARIABLES
**Column:** 250 × 4.6 5 μm Symmetry C18 (Waters)
**Column temperature:** 40
**Mobile phase:** MeCN:10 mM pH 2.3 potassium phosphate buffer 22:78
**Flow rate:** 1.5
**Injection volume:** 50
**Detector:** UV 227

## KEY WORDS
stability-indicating

## REFERENCE
Bergquist, P.A.; Manas, D.; Hunke, W.A.; Reed, R.A. Stability and compatibility of tirofiban hydrochloride during simulated Y-site administration with other drugs, *Amer.J.Health-Syst.Pharm.*, **2001**, *58*, 1218–1223.

## ANNOTATED BIBLIOGRAPHY
Bergquist, P.A.; Zimmerman, J.; Kenney, R.R.; Han, R.Y.-H.; Hunke, W.A. Stability of tirofiban hydrochloride in three commonly used i.v. solutions and polyvinyl chloride administration sets, *Am.J.Health-Syst.Pharm.*, **1999**, *56*, 1627–1629. [stability-indicating]

Garabito, M.J.; Jimenez, L.; Bautista, F.J.; Santos-Rubio, M.D.; Perez-Rodrigo, I. Stability of tirofiban hydrochloride in 0.9% sodium chloride injection for 30 days, *Am.J.Health-Syst.Pharm.*, **2001**, *58*, 1850–1851. [stability-indicating]

Vickers, S.; Theoharides, A.D.; Arison, B.; Balani, S.K.; Cui, D.; Duncan, C.A.; Ellis, J.D.; Gorham, L.M.; Polsky, S.L.; Prueksaritanont, T.; Ramjit, H.G.; Slaughter, D.E.; Vyas, K.P. In vitro and in vivo studies on the metabolism of tirofiban, *Drug Metab.Dispos.*, **1999**, *27*, 1360–1366. [rat; dog; bile; urine; feces; radiolabeled; metabolites]

# Tiropramide

**Molecular formula:** $C_{28}H_{41}N_3O_3$
**Molecular weight:** 467.64
**CAS Registry No:** 55837-29-1, 53567-47-8 (HCl)
**Merck Index:** 13, 9543

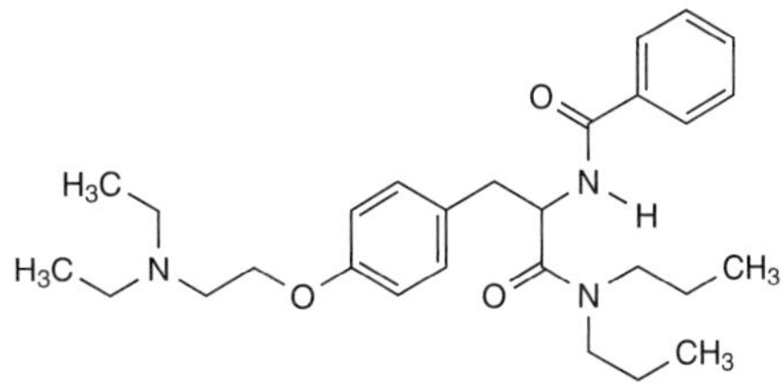

## SAMPLE

**Matrix:** blood
**Sample preparation:** Filter (0.22 μm PVDF) plasma, inject a 100 μL aliquot onto column A and elute to waste with mobile phase A at 0.5 mL/min; after 4.5 min, direct the effluent from column A onto column B and then to waste (continue to use mobile phase A); after another 3 min, backflush the contents of column B onto column C with mobile phase B at 0.1 mL/min, monitor the effluent from column C for 10.5 min.

## HPLC VARIABLES

**Column:** A 20 × 4 5 μm Capcell Pak MF Ph-1 polymer-coated mixed function (change after 35 samples); B 35 × 2 5 μm Capcell Pak C18 UG 120; C 250 × 1.5 Capcell Pak C18 UG 120
**Column temperature:** 30 (columns A and C only)
**Mobile phase:** A MeCN:50 mM pH 7.0 potassium phosphate buffer 12:88; B MeCN:10 mM pH 7.0 potassium phosphate buffer 50:50
**Flow rate:** A 0.5; B 0.1
**Injection volume:** 100
**Detector:** UV 230

## CHROMATOGRAM

**Retention time:** 14
**Limit of quantitation:** 5 ng/mL

## KEY WORDS

column-switching; pharmacokinetics; plasma

## REFERENCE

Baek, S.K.; Lee, S.S.; Park, E.J.; Sohn, D.H.; Lee, H.S. Semi-micro high-performance liquid chromatographic analysis of tiropramide in human plasma using column-switching, *J.Pharm.Biomed.Anal.*, **2003**, *31*, 185–189.

## SAMPLE

**Matrix:** blood
**Sample preparation:** Mix 100 μL plasma with 10 μL 50 ng/mL IS in MeOH and 100 μL 50 μM NaOH, add 800 μL MTBE, vortex at high speed for 5 min, centrifuge at 5000 g for 5 min. Evaporate the organic layer to dryness under a stream of nitrogen at 35°, reconstitute the residue with 40 μL MeCN:water 50:50, vortex for 2 min, inject a 10 μL aliquot.

## HPLC VARIABLES

**Column:** 100 × 2 3 μm Luna C8
**Column temperature:** 30
**Mobile phase:** MeCN:10 mM pH 4.5 ammonium formate buffer 50:50
**Flow rate:** 0.2
**Injection volume:** 10

**Detector:** MS, Micromass Quattro LC, tandem quadrupole, positive ionization electrospray, ion source 120°, desolvation 250°, cone 42 V, collision energy 24 eV, collision gas argon, m/z 469–367

## CHROMATOGRAM

**Retention time:** 2.0
**Internal standard:** cisapride (cone 45 V, m/z 467–184) (1.9)
**Limit of quantitation:** 2 ng/mL

## KEY WORDS

pharmacokinetics; plasma

## REFERENCE

Lee, H.W.; Ji, H.Y.; Kim, H.H.; Cho, H.-Y.; Lee, Y.-B.; Lee, H.S. Determination of tiropramide in human plasma by liquid chromatography-tandem mass spectrometry, *J.Chromatogr.B*, **2003**, *796*, 395–400.

# Tizanidine

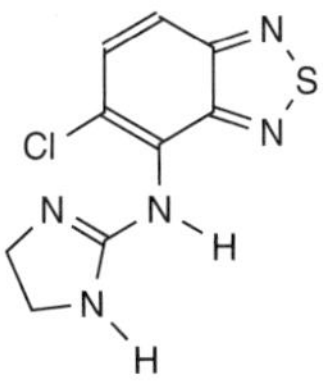

**Molecular formula:** $C_9H_8ClN_5S$
**Molecular weight:** 253.72
**CAS Registry No:** 51322-75-9, 64461-82-1 (HCl)
**Merck Index:** 13, 9561

## SAMPLE

**Matrix:** formulations
**Sample preparation:** Weigh out ground tablets containing 2 mg tizanidine, add 20 mL mobile phase, sonicate for 5 min, make up to 50 mL with mobile phase, filter, discard first 10 mL filtrate. Dilute 2 mL of the subsequent filtrate to 10 mL with mobile phase, inject a 20 μL aliquot.

## HPLC VARIABLES

**Column:** 150 × 5 5 μm Hypersil CN
**Mobile phase:** MeCN:MeOH:buffer 18:57:50 (Prepare the buffer by dissolving 2.5 mL 50 mM sodium 1-heptanesulfonate in water and 800 μL triethylamine in 800 mL water, adjust pH to 3.3 with glacial acetic acid, make up to 1 L with water.)
**Flow rate:** 1
**Injection volume:** 20
**Detector:** UV 227

## CHROMATOGRAM

**Retention time:** 7
**Limit of quantitation:** 51 ng/mL

## OTHER SUBSTANCES

**Simultaneous:** degradants, impurities

## KEY WORDS

stability-indicating; tablets

## REFERENCE

Qi, M.-L.; Wang, P.; Wang, L. Validated liquid chromatography method for assay of tizanidine in drug substance and formulated products, *Anal.Chim.Acta*, **2003**, *478*, 171–177.

## SAMPLE

**Matrix:** solutions
**Sample preparation:** Inject an aliquot of a 100 μg/mL solution in MeOH.

## HPLC VARIABLES

**Column:** 150 × 4.6 5 μm JASCO-RP-C18
**Column temperature:** 50
**Mobile phase:** Carbon dioxide:MeOH:acetic acid 90.9:8.918:0.182
**Flow rate:** 1.5
**Injection volume:** 20
**Detector:** UV 235

## CHROMATOGRAM

**Retention time:** 8.81
**Internal standard:** guaifenesin (5.40)
**Limit of detection:** 600 ng/mL

## OTHER SUBSTANCES
**Simultaneous:** baclofen (13.64), chlorzoxazone (3.03), methocarbamol (3.92)

## KEY WORDS
outlet pressure 12.75 Mpa; SFC

## REFERENCE
Bhoir, I.C.; Raman, B.; Sundaresan, M.; Bhagwat, A.M. Development of an isocratic SFC method for four centrally active muscle relaxant drugs, *Anal.Lett.*, **1998**, *31*, 1533–1542.

# Tolcapone

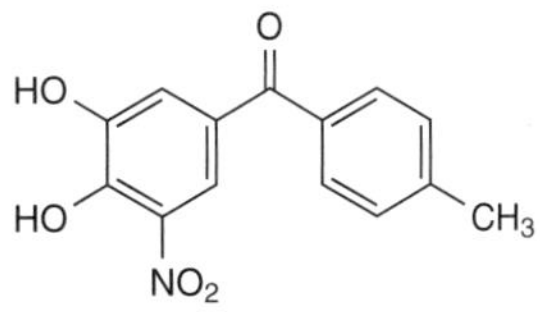

**Molecular formula:** $C_{14}H_{11}NO_5$
**Molecular weight:** 273.24
**CAS Registry No:** 134308-13-7
**Merck Index:** 13, 9585

## SAMPLE
**Matrix:** blood
**Sample preparation:** Mix 200 μL plasma with 25 μL IS in MeCN, add 300 μL MeCN, let stand at 4° for 15 min, centrifuge for 5 min. Mix 100 μL of the supernatant with 100 μL 50 mM pH 2 sodium dihydrogen phosphate buffer, inject an 80 μL aliquot.

## HPLC VARIABLES
**Guard column:** 10 × 4 5 μm Hypersil ODS
**Column:** 125 × 4 5 μm Hypersil ODS
**Mobile phase:** MeOH:THF:buffer 55:45:5 (Prepare the buffer by mixing 550 mL MeOH containing 11.5 g/L *N*-hexylmethylamine with 450 mL 50 mM sodium dihydrogen phosphate, adjust apparent pH to 2.1 with phosphoric acid, add 50 mL THF.)
**Flow rate:** 1
**Injection volume:** 80
**Detector:** UV 270

## CHROMATOGRAM
**Retention time:** 7
**Internal standard:** 3-hydroxy-4-methoxy-4′-methyl-5-nitrobenzophenone (10.5)
**Limit of quantitation:** 50 ng/mL

## OTHER SUBSTANCES
**Extracted:** 3-*O*-methyl metabolite (9)

## KEY WORDS
dog; human; monkey; mouse; plasma; rat

## REFERENCE
Heizmann, P.; Schmitt, M.; Leube, J.; Martin, H.; Saner, A. Determination of the catechol-*O*-methyltransferase inhibitor tolcapone and three of its metabolites in animal and human plasma and urine by reversed-phase high-performance liquid chromatography with ultraviolet detection, *J.Chromatogr.B*, **1999**, *730*, 153–160.

## SAMPLE
**Matrix:** urine
**Sample preparation:** Mix 25 μL urine with 25 μL IS in MeCN and 200 μL MeCN:10% phosphoric acid 12.5:87.5, vortex, centrifuge, inject a 50 μL aliquot.

## HPLC VARIABLES
**Guard column:** 10 × 4 3 μm Spherisorb ODS-1
**Column:** 125 × 4 3 μm Spherisorb ODS-1
**Column temperature:** 25
**Mobile phase:** MeCN:THF:50 mM sodium dihydrogen phosphate 17:13:70, adjusted to apparent pH 2.1 with phosphoric acid
**Flow rate:** 1
**Injection volume:** 50
**Detector:** UV 270

## CHROMATOGRAM

**Retention time:** 18

**Internal standard:** 3-hydroxy-4-methoxy-4′-methyl-5-nitrobenzophenone (27)

**Limit of quantitation:** 200 ng/mL

## OTHER SUBSTANCES

**Extracted:** 4′-carboxylic acid metabolite (8), glucuronide metabolite (5), 3-*O*-methyl metabolite (22)

## REFERENCE

Heizmann, P.; Schmitt, M.; Leube, J.; Martin, H.; Saner, A. Determination of the catechol-*O*-methyl transferase inhibitor tolcapone and three of its metabolites in animal and human plasma and urine by reversed-phase high-performance liquid chromatography with ultraviolet detection, *J.Chromatogr.B*, **1999**, *730*, 153–160.

## ANNOTATED BIBLIOGRAPHY

Dingemanse, J.; Jorga, K.M.; Schmitt, M.; Gieschke, R.; Fotteler, B.; Zürcher, G.; Da Prada, M.; van Brummelen, P. Integrated pharmacokinetics and pharmacodynamics of the novel-O-methyltransferase inhibitor tolcapone during first administration to humans, *Clin.Pharmacol.Ther.*, **1995**, *57*, 508–517. [LOQ 50 ng/mL]

Lave, T.; Dupin, S.; Schmitt, M.; Kapps, M.; Meyer, J.; Morgenroth, B.; Chou, R.C.; Jaeck, D.; Coassolo, P. Interspecies scaling of tolcapone, a new inhibitor of catechol-O-methyltransferase (COMT). Use of in vitro data from hepatocytes to predict metabolic clearance in animals and humans, *Xenobiotica*, **1996**, *26*, 839–851.

# Topiramate

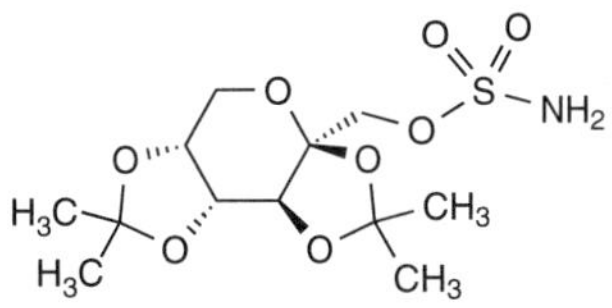


**Molecular formula:** $C_{12}H_{21}NO_8S$
**Molecular weight:** 339.36
**CAS Registry No:** 97240-79-4
**Merck Index:** 13, 9625

## SAMPLE

**Matrix:** blood, CSF
**Sample preparation:** Vortex 200 μL plasma or CSF with 25 μL 20 μg/mL IS in MeCN:water 50:50 and 400 μL MeCN, centrifuge at 10 900 rpm for 5 min. Mix a 250 μL aliquot of the supernatant with 100 μL 0.1% formic acid, inject a 10 μL aliquot.

## HPLC VARIABLES

**Column:** 50 × 2 3 μm Luna C18(2)
**Column temperature:** 40
**Mobile phase:** MeOH:200 μM ammonium acetate:formic acid 40:60:0.1
**Flow rate:** 0.25
**Injection volume:** 10
**Detector:** MS, PE Sciex API 3000, nebulizer 375°, nebulizer gas nitrogen at 75 psi, corona needle − 3 mA, auxiliary gas nitrogen at 10 units, curtain gas nitrogen at 8 units, negative ion mode, collision gas nitrogen at 5 units, collision energy 60 eV, m/z 338.2–78.1

## CHROMATOGRAM

**Retention time:** 2.95
**Internal standard:** 1,2:3,4-bis-*O*-(1-methylethylidene-α-D-galactopyranose sulfamate) (collision energy 30 eV, m/z 338.2–96.1) (3.50)
**Limit of quantitation:** 40 ng/mL

## KEY WORDS

comparison with FPIA; plasma

## REFERENCE

Christensen, J.; Hojskov, C.S.; Poulsen, J.H. Liquid chromatography tandem mass spectrometry assay for topiramate analysis in plasma and cerebrospinal fluid: Validation and comparison with fluorescence-polarization immunoassay, *Ther.Drug Monit.*, **2002**, *24*, 658–664.

## SAMPLE

**Matrix:** blood, urine
**Sample preparation:** Plasma. Mix 500 μL plasma with 30 μL 100 μg/mL IS in MeOH, extract with 10 mL diethyl ether. Evaporate the organic layer to dryness, reconstitute the residue with 75 μL MeOH:10 mM ammonium acetate 50:50, inject a 10 μL aliquot. Urine. Mix 100 μL urine with 900 μL water and 30 μL 100 μg/mL IS in MeOH, extract with 10 mL diethyl ether. Evaporate the organic layer to dryness, reconstitute the residue with 150 μL MeOH:10 mM ammonium acetate 50:50, inject a 10 μL aliquot.

## HPLC VARIABLES

**Column:** 100 × 2.1 3.5 μm Symmetry C18 (Waters)
**Mobile phase:** Gradient. MeOH:10 mM ammonium acetate 5:95 for 2 min, to 50:50 over 4 min.
**Flow rate:** 0.25
**Injection volume:** 10

**Detector:** MS, Bruker Esquire-LC, electrospray, positive ion mode, nebulizer 20 psi, capillary 4300 V, drying gas 300° and 8 L/min, m/z 357

## CHROMATOGRAM
**Retention time:** 14
**Internal standard:** $d_{12}$-topiramate (m/z 369)
**Limit of quantitation:** 625 ng/mL

## OTHER SUBSTANCES
**Extracted:** metabolites

## KEY WORDS
pharmacokinetics; plasma

## REFERENCE
Britzi, M.; Soback, S.; Isoherranen, N.; Levy, R.H.; Perucca, E.; Doose, D.R.; Maryanoff, B.E.; Bialer, M. Analysis of topiramate and its metabolites in plasma and urine of healthy subjects and patients with epilepsy by use of a novel liquid chromatography-mass spectrometry assay, *Ther.Drug Monit.*, **2003**, *25*, 314–322.

## ANNOTATED BIBLIOGRAPHY
Christensen, J.; Hojskov, C.S.; Dam, M.; Poulsen, J.H. Plasma concentration of topiramate correlates with cerebrospinal fluid concentration, *Ther.Drug Monit.*, **2001**, *23*, 529–535.

Contin, M.; Riva, R.; Albani, F.; Baruzzi, A. Simple and rapid liquid chromatographic-turbo ion spray mass spectrometric determination of topiramate in human plasma, *J.Chromatogr.B*, **2001**, *761*, 133–137. [LOQ 250 ng/mL]

Duong, H.T.; Guh, H.Y.; Ko, C.Y.; Micheel, A.P.; Thakur, M.S. A HPLC assay for coating integrity of topiramate sprinkle formulation, *J.Pharm.Biomed.Anal.*, **2002**, *29*, 69–74. [sprinkles; LOD 0.07%; refractive index detector]

Langman, L.J.; Kaliciak, H.A.; Boone, S.A. Fatal acute topiramate toxicity, *J.Anal.Toxicol.*, **2003**, *27*, 323–324. [topiramate; LC-MS; gabapentin; vigabatrin; droperidol is internal standard]

# Topotecan

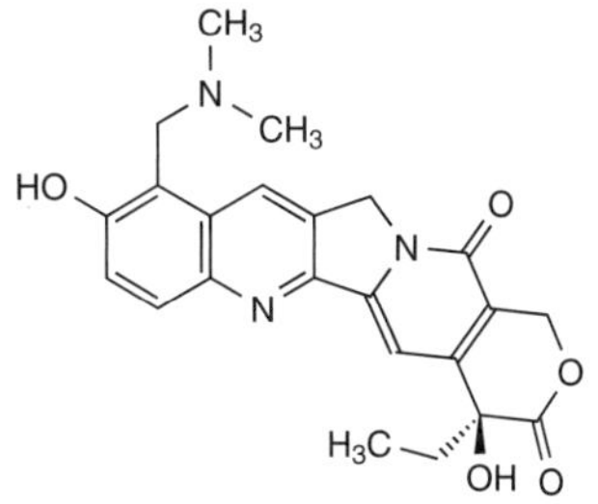


**Molecular formula:** $C_{23}H_{23}N_3O_5$
**Molecular weight:** 421.44
**CAS Registry No:** 123948-87-8, 119413-54-6 (HCl)
**Merck Index:** 13, 9626

## SAMPLE
**Matrix:** blood
**Sample preparation:** Vortex 200 μL plasma with 800 μL cold (dry ice) MeOH for 10 s, centrifuge at 7000 g for 2 min. Dilute 2 vol of supernatant with 1 vol of water (to measure lactone form) or 1.5% phosphoric acid (to measure total amount), inject a 20 μL aliquot.

## HPLC VARIABLES
**Guard column:** 10 × 3 ChromGuard RP SS (Varian)
**Column:** 150 × 3 3.5 μm SB-C18 (Agilent)
**Column temperature:** 50
**Mobile phase:** MeOH:buffer 27:73 (The buffer was 75 mM potassium dihydrogen phosphate containing 0.2% triethylamine, adjusted to pH 6.5 with KOH.)
**Flow rate:** 0.8
**Injection volume:** 20
**Detector:** F ex 376 em 530

## CHROMATOGRAM
**Retention time:** 13.7
**Limit of quantitation:** 250 pg/mL

## OTHER SUBSTANCES
**Extracted:** *N*-desmethyltopotecan (LOQ 50 pg/mL) (7.6)

## KEY WORDS
plasma

## REFERENCE
Bai, F.; Kirstein, M.N.; Hanna, S.K.; Iacono, L.C.; Johnston, B.; Stewart, C.F. Determination of plasma topotecan and its metabolite *N*-desmethyl topotecan as both lactone and total form by reversed-phase liquid chromatography with fluorescence detection, *J.Chromatogr.B*, **2003**, *784*, 225–232.

## SAMPLE
**Matrix:** blood
**Sample preparation:** Vortex 900 μL plasma with 100 μL buffer for 20 s. Add a 200 μL aliquot to 800 μL cold (−40°) MeOH, vortex for 20 s, centrifuge for 2 min. Mix a 500 μL aliquot of the supernatant with 500 μL buffer, inject a 200 μL aliquot. (The buffer was 8 mM disodium hydrogen phosphate containing 1 mM potassium dihydrogen phosphate, 137 mM NaCl, and 3 mM KCl and was adjusted to pH 2.0 for lactone or pH 11.0 for carboxylate forms.)

## HPLC VARIABLES
**Guard column:** Guard-Pak C18 NovaPak
**Column:** 150 × 3.9 4 μm NovaPak C18

**Mobile phase:** MeCN:buffer 12:88 (The buffer was 3% triethylamine adjusted to pH 5.5 with glacial acetic acid.)
**Flow rate:** 1
**Injection volume:** 200
**Detector:** F (tunable) ex 390 em 520, F (filter) ex 350–470 em 510–650

---

## CHROMATOGRAM

**Retention time:** 2.5 (carboxylate), 5.2 (lactone)
**Limit of detection:** 300 pg/mL (lactone, filter F), 260 pg/mL (lactone, tunable F), 150 pg/mL (carboxylate, filter F), 100 pg/mL (carboxylate, tunable F)
**Limit of quantitation:** 750 pg/mL (lactone, filter F), 500 pg/mL (lactone, tunable F), 500 pg/mL (carboxylate, filter F), 250 pg/mL (carboxylate, tunable F)

---

## KEY WORDS

plasma

---

## REFERENCE

Warner, D.L.; Burke, T.G. Comparison of filter and tunable fluorescence detection for the HPLC simultaneous quantitation of lactone and carboxylate forms of topotecan in plasma, *J.Liq.Chromatogr.Rel. Technol.*, **1997**, *20*, 1523–1537.

---

## ANNOTATED BIBLIOGRAPHY

Beijnen, J.H.; Smith, B.R.; Keijer, W.J.; van Gijn, R.; ten Bokkel Huinink, W.W.; Vlasveld, L.T.; Rodenhuis, S.; Underberg, W.J. High-performance liquid chromatographic analysis of the new antitumour drug SK&F 104864-A (NSC 609699) in plasma, *J.Pharm.Biomed.Anal.*, **1990**, *8*, 789–794.

Boucaud, M.; Pinguet, F.; Poujol, S.; Astre, C.; Bressolle, F. Sensitive high performance liquid chromatographic fluorescence determination of topotecan in human plasma and parotid saliva, *J.Liq.Chromatogr.Rel.Technol.*, **2000**, *23*, 2373–2390. [LOQ 50 pg/mL]

Loos, W.J.; Stoter, G.; Verweij, J.; Schellens, J.H.M. Sensitive high-performance liquid chromatographic fluorescence assay for the quantitation of topotecan (SKF 104864-A) and its lactone ring-opened product (hydroxy acid) in human plasma and urine, *J.Chromatogr.B*, **1996**, *678*, 309–315. [LOQ 0.1 ng/mL; fluorescence detection]

Loos, W.J.; van Zomeren, D.M.; Gelderblom, H.; Verweij, J.; Nooter, K.; Stoter, G.; Sparreboom, A. Determination of topotecan in human whole blood and unwashed erythrocytes by high-performance liquid chromatography, *J.Chromatogr.B*, **2002**, *766*, 99–105. [LOQ 200 pg/mL; fluorescence detection]

Rosing, H.; Doyle, E.; Davies, B.E.; Beijnen, J.H. High-performance liquid chromatographic determination of the novel antitumour drug topotecan as the total of the lactone plus carboxylate forms in human plasma, *J.Chromatogr.B*, **1995**, *668*, 107–115. [LOQ 0.05 ng/mL; fluorescence detection]

Rosing, H.; Doyle, E.; Beijnen, J.H. The impact of column temperature in the high performance liquid chromatographic analysis of topotecan in rat and dog plasma, *J.Pharm.Biomed.Anal.*, **1996**, *15*, 279–286. [LOQ 0.1 ng/mL; fluorescence detection]

Rosing, H.; van Zomeren, D.M.; Doyle, E.; ten Bokkel Huinink, W.W.; Schellens, J.H.M.; Bult, A.; Beijnen, J.H. Quantification of topotecan and its metabolite *N*-desmethyltopotecan in human plasma, urine and faeces by high-performance liquid chromatographic methods, *J.Chromatogr.B*, **1999**, *727*, 191–203. [LOQ 100 pg/mL; fluorescence detection]

# Tosufloxacin

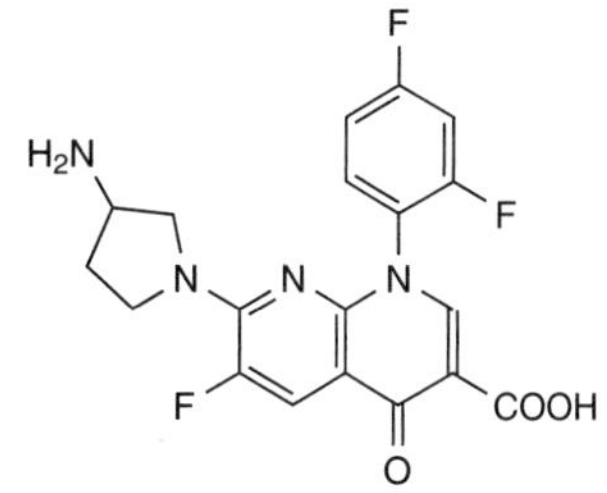


**Molecular formula:** $C_{19}H_{15}F_3N_4O_3$
**Molecular weight:** 404.34
**CAS Registry No:** 100490-36-6. 104051-69-6 (HCl), 115964-29-9 (tosylate)
**Merck Index:** 13, 9632

## SAMPLE

**Matrix:** bile, blood, urine
**Sample preparation:** Dilute urine 1:20. Dilute bile 1:10. Vortex 500 μL serum, diluted urine, or diluted bile with 3.2 mL dichloromethane, rotate at 20 rpm for 10 min, centrifuge at 1000 g for 10 min. Remove 3 mL of the lower organic phase and add it to 200 μL pH 2.5 acetic acid, rotate at 20 rpm for 10 min, centrifuge at 1000 g for 10 min, inject a 20 μL aliquot of the aqueous layer.

## HPLC VARIABLES

**Column:** 150 × 4.6 5 μm Ultrasphere C18
**Mobile phase:** MeCN:buffer 18:82, pH adjusted to 2 with 14.6 M phosphoric acid (The buffer was 10 mM sodium dihydrogen phosphate containing 5 mM tetrabutylammonium bromide.)
**Flow rate:** 2
**Injection volume:** 20
**Detector:** F ex 265 em 433

## CHROMATOGRAM

**Retention time:** 2
**Limit of detection:** 20 ng/mL (serum), 500 ng/mL (urine), 250 ng/mL (bile)

## OTHER SUBSTANCES

**Noninterfering:** amikacin, aztreonam, carbamazepine, cephalosporins, ciprofloxacin, clavulanic acid, difloxacin, digitoxin, digoxin, fleroxacin, fosfomycin, furosemide, gentamicin, imipenem, lidocaine, netilmicin, norfloxacin, ofloxacin, pefloxacin, penicillins, phenobarbital, phenytoin, primidone, procainamide, quinidine, rifampin, salicylic acid, teicoplanin, theophylline, tobramycin, valproic acid, vancomycin

## KEY WORDS

human; pig; serum

## REFERENCE

Koechlin, C.; Jehl, F.; Linger, L.; Monteil, H. High-performance liquid chromatography for the determination of three new fluoroquinolones, fleroxacin, temafloxacin and A-64730, in biological fluids, *J.Chromatogr.*, **1989**, *491*, 379–387.

## SAMPLE

**Matrix:** wastewater
**Sample preparation:** Condition a mixed-phase (C8 and benzenesulfonate) cation-exchange SPE disc (Varian) with 2 mL MeOH and 2 mL water at pH 3. Adjust pH of wastewater to 3, pass 50–150 mL through the SPE disc at 1 mL/min, dry under vacuum for 5 min, elute with 2.5 mL 5% ammonia in MeOH:water 15:85. Neutralize the eluate with 500 μL 85% phosphoric acid, inject a 200 μL aliquot.

## HPLC VARIABLES

**Guard column:** 20 × 3 5 μm Discovery RP-AmideC16 (Supelco)
**Column:** 250 × 3 5 μm Discovery RP-AmideC16 (Supelco)
**Column temperature:** 50
**Mobile phase:** Gradient. MeCN:25 mM pH 2.4 orthophosphoric acid from 5:95 to 7:93 over 17 min, maintain at 7:93 for 5 min, to 17:83 over 13 min, to 85:15 (step gradient), maintain at 85:15 for 5 min, return to initial conditions over 2 min, re-equilibrate for 10 min.
**Flow rate:** 0.7
**Injection volume:** 200
**Detector:** F ex 278 em 445

## CHROMATOGRAM

**Retention time:** 35

## OTHER SUBSTANCES

**Extracted:** ciprofloxacin (19.5), danofloxacin (23), difloxacin (28), enrofloxacin (22.5), fleroxacin (14), lomefloxacin (21), norfloxacin (18), ofloxacin (15.5), pipemidic acid (6),

## KEY WORDS

SPE

## REFERENCE

Golet, E.M.; Alder, A.C.; Hartmann, A.; Ternes, T.A.; Giger, W. Trace determination of fluoroquinolone antibacterial agents in urban wastewater by solid-phase extraction and liquid chromatography with fluorescence detection, *Anal.Chem.*, **2001**, *73*, 3632–3638.

## ANNOTATED BIBLIOGRAPHY

Lyon, D.J.; Cheung, S.W.; Chan, C.Y.; Cheng, A.F. Rapid HPLC assay of clinafloxacin, fleroxacin, levofloxacin, sparfloxacin and tosufloxacin, *J.Antimicrob.Chemother.*, **1994**, *34*, 446–448. [in conjunction with Chan,C.Y.; Lam,A.W.; French,G.L. Rapid HPLC assay of fluoroquinolones in clinical specimens. *J.Antimicrob.Chemother.* **1989**, 23, 597–604.]

# Travoprost

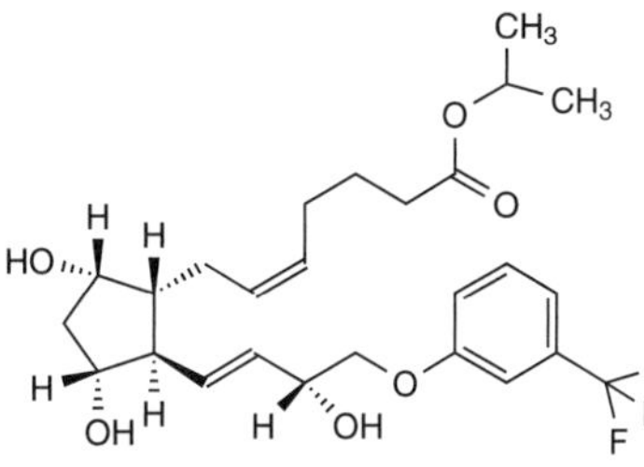

**Molecular formula:** $C_{26}H_{35}F_3O_6$
**Molecular weight:** 500.55
**CAS Registry No:** 157283-68-6

## SAMPLE

**Matrix:** blood

**Sample preparation:** Condition a SPEC 3 mL 15 mg MP1 nonpolar reversed-phase/ strong cation-exchange SPE Cartridge (Ansys) with 500 μL MeOH and 500 μL 40 mM formic acid. Mix 1 mL plasma with 50 μL 400 IU/mL rabbit liver esterase (Sigma) in buffered saline, heat at 37° for 45 min. (This hydrolyzes travoprost to the free acid.) Mix 1 mL plasma with 15 μL 20 ng/mL IS and 1 mL 100 mM formic acid, add to the SPE cartridge, rinse tube with 0.5–1 mL water, add rinse to the SPE cartridge, wash with 500 μL water, dry under vacuum, wash with two 500 μL portions of toluene:dichloromethane 60:40, dry under vacuum, elute with 600 μL toluene:methylformate 20:80. Evaporate the eluate to dryness under a stream of nitrogen at 30°, reconstitute the residue with 125 μL MeOH:water 50:50, inject a 35 μL aliquot.

## HPLC VARIABLES

**Column:** 150 × 2 5 μm Phenomenex Columbus C18

**Mobile phase:** MeOH:5 mM pH 6.3 ammonium formate 70:30

**Flow rate:** 0.2

**Injection volume:** 35

**Detector:** MS, Micromass Quattro LC, electrospray, capillary 3.0 kV, sample cone 40 V, extraction cone 2 V, RF lens 0.3 V, source temperature 125°, drying gas 250°, MS1 parameters LM resolution 14, HM resolution 14, ion energy 1.2 V, entrance and exit set to 0 and 1, collision energy 30 eV, MS2 parameters LM resolution 15.0, HM resolution 15.0, ion energy 1.2 V, multiplier 650 V, nebulizing gas 75 L/h, drying gas 570 L/h, collision gas argon, m/z 457–161

## CHROMATOGRAM

**Retention time:** 5.3 (as travoprost free acid)

**Internal standard:** AL-5848X (tetradeutero travoprost free acid) (m/z 461–161)

**Limit of quantitation:** 10 pg/mL

## KEY WORDS

plasma; SPE

## REFERENCE

McCue, B.A.; Cason, M.M.; Curtis, M.A.; Faulkner, R.D.; Dahlin, D.C. Determination of travoprost and travoprost free acid in human plasma by electrospray HPLC/MS/MS, *J.Pharm.Biomed.Anal.*, **2002**, *28*, 199–208.

# Trenbolone

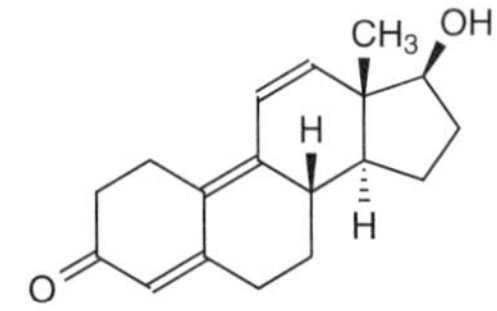


**Molecular formula:** $C_{18}H_{22}O_2$

**Molecular weight:** 270.37

**CAS Registry No:** 10161-33-8, 10161-34-9 (acetate), 23454-33-3 (cyclohexylmethylcarbonate)

**Merck Index:** 13, 9659

## SAMPLE

**Matrix:** blood

**Sample preparation:** Serum. Condition a 3 mL Oasis HLB SPE cartridge with 3 mL MeOH and 3 mL water. Add 1 mL serum containing 100 ng IS to the SPE cartridge, wash with 3 mL water, elute with 2 mL MeOH. Evaporate the eluate to dryness under a stream of nitrogen at 45°, reconstitute the residue with 100 μL MeOH:water 50:50, inject a 5 μL aliquot. Urine. Condition a 360 mg Sep-Pak C18 SPE cartridge with 5 mL MeOH and 10 mL water. Adjust the pH of 5 mL urine containing 100 ng IS to 5.5 with acetate buffer, add 50 μL β-glucuronidase-aryl sulfatase (Type H-2 Helix Pomatia, Boehringer Mannheim), heat at 37° overnight, cool, add to the SPE cartridge, wash with 5 mL water, elute with 3 mL MeOH. Evaporate the eluate to dryness under a stream of nitrogen at 45°, reconstitute the residue with 100 μL MeOH:water 50:50, inject a 5 μL aliquot.

## HPLC VARIABLES

**Guard column:** 4 × 2 Phenomenex C18

**Column:** 250 × 2.1 5 μm Hypersil C18

**Mobile phase:** MeOH:2.88 mM ammonium formate 65:35

**Flow rate:** 0.15

**Injection volume:** 5

**Detector:** MS, PE Sciex API 365, turbo ionspray, positive mode, source 4000 V, orifice 30 V, ring 250 V, nebulizer gas nitrogen at 8 units, curtain gas nitrogen at 8 units, collision gas nitrogen at 3 units, collision energy – 42 eV, vaporizer 450°, m/z 271.2–253.3–227.2–199.2

## CHROMATOGRAM

**Retention time:** 13.8

**Internal standard:** methyltestosterone (m/z 303.1–109.1–97.0) (22.9)

**Limit of detection:** 350 pg/mL

**Limit of quantitation:** 1 ng/mL

## KEY WORDS

cow; serum; SPE

## REFERENCE

Buiarelli, F.; Cartoni, G.P.; Coccioli, F.; De Rossi, A.; Neri, B. Determination of trenbolone and its metabolite in bovine fluids by liquid chromatography-tandem mass spectrometry, *J.Chromatogr.B*, **2003**, *784*, 1–15.

# Treprostinil

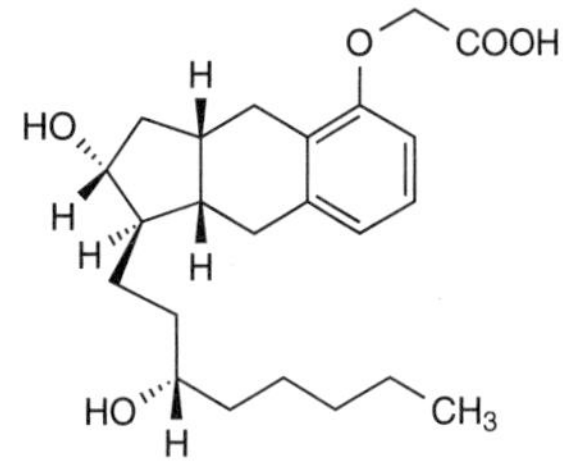


**Molecular formula:** $C_{23}H_{34}O_5$
**Molecular weight:** 390.51
**CAS Registry No:** 81846-19-7

## SAMPLE

**Matrix:** blood
**Sample preparation:** Extract 1 mL plasma with hexane:ethyl acetate 70:30. Evaporate the organic layer to dryness under a stream of nitrogen, reconstitute the residue with MeOH:water:100 mM ammonium formate:formic acid 50:47.5:2.5:0.05, inject a 25 μL aliquot.

## HPLC VARIABLES

**Column:** 100 × 2 Betasil C18 (Keystone)
**Mobile phase:** MeCN:water:100 mM ammonium formate:formic acid 33.25:61.75:5:0.1
**Flow rate:** 0.3
**Injection volume:** 25
**Detector:** MS, PE Sciex API-III or API 365, ionspray, atmospheric pressure ionization, negative ionization

## CHROMATOGRAM

**Retention time:** 2.5
**Internal standard:** LRXA-97 JO2 (a dimethylene homolog) (3.5)
**Limit of quantitation:** 25 pg/mL

## KEY WORDS

pharmacokinetics; plasma

## REFERENCE

Wade, M.; Baker, F.J.; Roscigno, R.; DellaMaestra, W.; Hunt, T.L.; Lai, A.A. Absolute bioavailability and pharmacokinetics of treprostinil sodium administered by acute subcutaneous infusion, *J.Clin.Pharmacol.*, **2004**, *44*, 83–88.

## SAMPLE

**Matrix:** formulations
**Sample preparation:** Inject a 20 μL aliquot of a 4–130 μg/mL solution in water, 0.9% NaCl, or 5% dextrose.

## HPLC VARIABLES

**Column:** 250 × 4.6 5 μm Hypersil ODS
**Mobile phase:** Gradient. MeCN:water:trifluoroacetic acid 40:60:0.1 for 17 min, 78:22:0.1 for 18 min (step gradient (?)), initial conditions for 7 min (step gradient (?)).
**Flow rate:** 2
**Injection volume:** 20
**Detector:** UV 217

## CHROMATOGRAM

**Retention time:** 15.0
**Limit of detection:** 0.025%
**Limit of quantitation:** 0.05%

## OTHER SUBSTANCES

**Simultaneous:** *m*-cresol (preservative) (3.4), degradants

## KEY WORDS

injections; stability-indicating

## REFERENCE

Phares, K.R.; Weiser, W.E.; Miller, S.P.; Myers, M.A.; Wade, M. Stability and preservative effectiveness of treprostinil sodium after dilution in common intravenous diluents, *Am.J.Health-Syst.Pharm.*, **2003**, *60*, 916–922.

# Trichlorfon

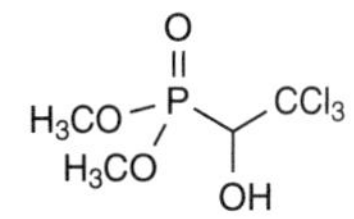


**Molecular formula:** $C_4H_8Cl_3O_4P$
**Molecular weight:** 257.44
**CAS Registry No:** 52-68-6
**Merck Index:** 13, 9696

## SAMPLE

**Matrix:** blood

**Sample preparation:** Vortex 1 mL plasma and 5 mL chloroform for 15 s, centrifuge at 800 g at 0–5° for 15 min (Caution! Chloroform is a carcinogen!). Remove a 4.5 mL aliquot of the organic layer and add it to 3.5 mL 5 M HCl, vortex for 15 s, centrifuge at 800 g at 0–5° for 15 min. Remove a 4 mL aliquot of the chloroform layer and add it to 200 mg anhydrous calcium sulfate, vortex, centrifuge at 800 g at 0–5° for 5 min. Evaporate the chloroform layer to dryness in an acid-washed vial under a stream of nitrogen at room temperature, reconstitute the residue with 120 μL water, vortex, centrifuge at 11 000 g at 0–5° for 5 min, inject a 100 μL aliquot. (However, Aden Abdi et al. claim that substantial base-catalyzed conversion of trichlorfon to dichlorvos can occur under these conditions and plasma should be acidified with phosphoric acid before extraction. (Aden Abdi,Y.; Villén,T.; Gustafsson,L.L.; Ericsson,O.; Sjöqvist,F. Methodological commentary on the analysis of metrifonate and dichlorvos in biological samples. *J.Chromatogr.* **1993**, *612*, 336–337.) These workers recommend mixing blood with an equal volume of 740 mM phosphoric acid immediately upon collection. Trichlorfon can still be extracted from the acidified sample. (Aden Abdi,Y.; Villén,T. Pharmacokinetics of metrifonate and its rearrangement product dichlorvos in whole blood. *Pharmacol.Toxicol.* **1991**, *68*, 137–139.))

## HPLC VARIABLES

**Guard column:** 37–50 μm C18/Corasil
**Column:** 300 × 3.9 10 μm C18
**Mobile phase:** MeOH:THF:1 mM pH 3.0 sodium octanesulfonate 10:0.1:89.9
**Flow rate:** 2
**Injection volume:** 200
**Detector:** UV 210

## CHROMATOGRAM

**Retention time:** 10.8
**Limit of detection:** 1 μg/mL

## KEY WORDS

plasma

## REFERENCE

Unni, L.K.; Hannant, M.E.; Becker, R.E. High-performance liquid chromatographic method using ultraviolet detection for measuring metrifonate and dichlorvos levels in human plasma, *J.Chromatogr.*, **1992**, *573*, 99–103.

# Triethanolamine

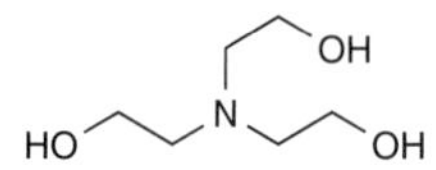


**Molecular formula:** $C_6H_{15}NO_3$
**Molecular weight:** 149.19
**CAS Registry No:** 102-71-6
**Merck Index:** 13, 9739

## SAMPLE
**Matrix:** solutions

## HPLC VARIABLES
**Column:** 100 × 4 Alltech Universal Cation
**Mobile phase:** 3 mM methanesulfonic acid
**Flow rate:** 1
**Injection volume:** 100
**Detector:** Conductivity (nonsuppressed mode, indirect conductivity detection)

## CHROMATOGRAM
**Retention time:** 6

## OTHER SUBSTANCES
**Simultaneous:** ammonium (4), calcium (12), ethanolamine (4.2), lithium (3), magnesium (10), potassium (5.5), sodium (3.5),

## REFERENCE
Saari-Nordhaus, R.; Anderson, J.M. Jr. Alternative approach to enhancing cation selectivity in ion chromatography, *J.Chromatogr.A*, **2004**, *1039*, 123–127.

# Trifluridine

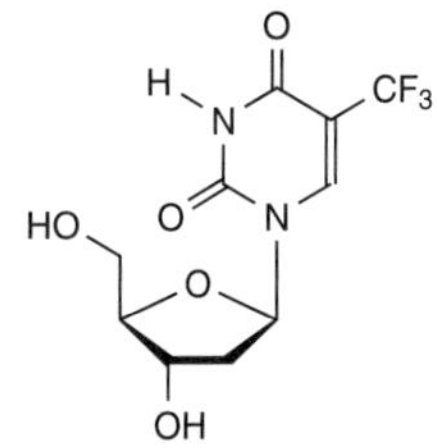


**Molecular formula:** $C_{10}H_{11}F_3N_2O_5$
**Molecular weight:** 296.20
**CAS Registry No:** 70-00-8
**Merck Index:** 13, 9758

## SAMPLE

**Matrix:** blood, tissue
**Sample preparation:** Mix 1 mL blood with 2 mL ice-cold MeCN, centrifuge, filter, inject a 5–20 μL aliquot. Homogenize brain tissue with 0.5 mL water, add 2 mL MeCN, centrifuge, filter, inject a 5–20 μL aliquot.

## HPLC VARIABLES

**Guard column:** C18 Corasil (Whatman)
**Column:** 100 × 3 Chrompack C18
**Mobile phase:** MeCN:50 mM sodium dihydrogen phosphate 15:85
**Flow rate:** 0.3–0.8
**Injection volume:** 5–20
**Detector:** UV 254, UV 280

## KEY WORDS

brain; rat; whole blood

## REFERENCE

Rand, K.H.; Bodor, N.; el Koussi, A.A.; Raad, I.; Miyake, A.; Houck, H.; Gildersleeve, N. Potential treatment of herpes simplex virus encephalitis by brain-specific delivery of trifluorothymidine using a dihydropyridine in equilibrium pyridinium salt type redox delivery system, *J.Med.Virol.*, **1986**, *20*, 1–8.

# Triptorelin

**Molecular formula:** $C_{64}H_{82}N_{18}O_{13}$
**Molecular weight:** 1311.45
**CAS Registry No:** 57773-63-4, 140194-24-7 (acetate)
**Merck Index:** 13, 9818

## SAMPLE

**Matrix:** blood
**Sample preparation:** Mix 500 μL plasma with 2 mL EtOH for 30 min, centrifuge at 3000 g for 15 min. Evaporate the supernatant to ca. 500 μL under a stream of nitrogen at 40° for 1 h, lyophilize, reconstitute with mobile phase, inject a 600 μL aliquot.

## HPLC VARIABLES

**Column:** Partisil PXS 10/25 SCX
**Mobile phase:** EtOH:200 mM pH 4.6 ammonium acetate 10:90
**Flow rate:** 1.6
**Injection volume:** 600
**Detector:** Radioimmunoassay of fractions

## CHROMATOGRAM

**Retention time:** 14

## KEY WORDS

plasma

## REFERENCE

Barron, J.L.; Millar, R.P.; Searle, D.I. Metabolic clearance and plasma half-disappearance time of D-TRP[6] and exogenous luteinizing hormone-releasing hormone, *J.Clin.Endocrinol.Metab.*, **1982**, *54*, 1169–1173.

## SAMPLE

**Matrix:** blood
**Sample preparation:** Condition a Sep-Pak C18 SPE cartridge with 3 mL MeOH and 5 mL water. Add 500 μL plasma to the SPE cartridge, wash with 5 mL 0.1% trifluoroacetic acid, elute with 5 mL MeOH:water:trifluoroacetic acid 80:19:1. Evaporate the eluate, freeze the residual aqueous solution, freeze, lyophilize, reconstitute with 300 μL 0.1% trifluoroacetic acid, inject a 100 μL aliquot.

## HPLC VARIABLES

**Column:** μBondapak C18
**Mobile phase:** Gradient. A was 11 mM trifluoroacetic acid in MeCN:water 70:30. B was 11 mM trifluoroacetic acid in 3 mM acetic acid. A:B from 5:95 to 95:5 over 20 min.
**Flow rate:** 1
**Injection volume:** 100
**Detector:** Radioimmunoassay of fractions; UV 210

## CHROMATOGRAM

**Retention time:** 18

## KEY WORDS

dog; pharmacokinetics; plasma; rat; SPE

## REFERENCE

Ezan, E.; Drieu, K.; Chapelat, M.; Rougeot, C.; Dray, F. Radioimmunoassay of [D-Trp$^6$]-luteinizing hormone-releasing hormone: its application to animal pharmacokinetic studies after single injection and long-acting formulation administration, *Regul.Pept.*, **1986**, *14*, 155–167.

# Troleandomycin

**Molecular formula:** $C_{41}H_{67}NO_{15}$
**Molecular weight:** 813.97
**CAS Registry No:** 2751-09-9
**Merck Index:** 13, 9839

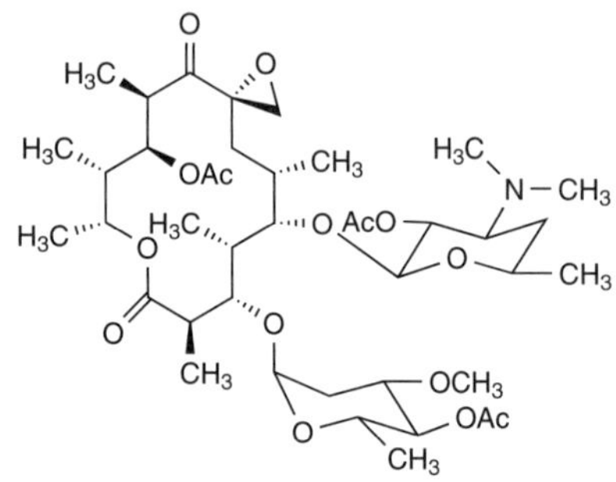

## SAMPLE
**Matrix:** solutions
**Sample preparation:** Inject a 10 μL aliquot of a 40 mg/mL solution in MeCN.

## HPLC VARIABLES
**Column:** 250 × 4.6 5 μm Xterra RP18
**Column temperature:** 30
**Mobile phase:** MeCN:200 mM pH 6.0 ammonium acetate:water 45:5:50
**Flow rate:** 1
**Injection volume:** 10
**Detector:** UV 205

## CHROMATOGRAM
**Retention time:** 22
**Limit of detection:** 0.02% (S/N 3)
**Limit of quantitation:** 0.05%

## REFERENCE
Chepkwony, H.K.; Roets, E.; Hoogmartens, J. Liquid chromatography of troleandomycin, *J.Chromatogr.A*, **2001**, *914*, 53–58.

# Troxipide

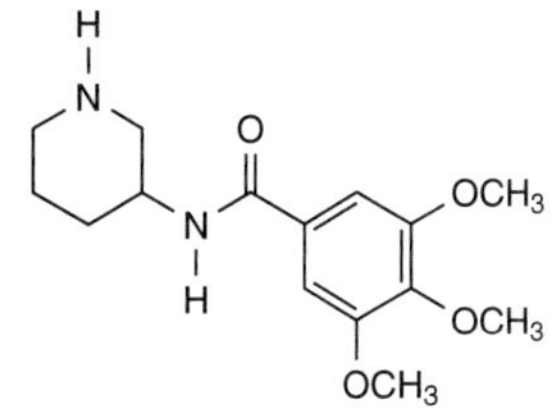

**Molecular formula:** $C_{15}H_{22}N_2O_4$
**Molecular weight:** 294.35
**CAS Registry No:** 99777-81-8
**Merck Index:** 13, 9860

## SAMPLE

**Matrix:** tissue
**Sample preparation:** Homogenize 5–30 mg gastric mucosal tissue with 1.2 mL 20 mM pH 7.4 phosphate buffer, centrifuge at 3000 rpm for 10 min. Mix 1 mL of the supernatant with 100 μL IS solution and 1.5 mL MeCN, let stand at 5–6° for 1 h, centrifuge at 3000 rpm at 4° for 10 min. Dry 2 mL of the supernatant, reconstitute with 1 mL EtOH, dry, reconstitute with 120 μL injection solvent (unspecified), inject a 60 μL aliquot.

## HPLC VARIABLES

**Guard column:** 10 × 4 5 μm Capcell Pak C18 UG 120
**Column:** 50 × 4.6 3 μm Capcell Pak C18 UG 120
**Mobile phase:** MeCN:50 mM pH 5.0 acetate buffer containing 0.08% sodium 1-hexanesulfonate (ratio not given)
**Flow rate:** 0.6
**Injection volume:** 60
**Detector:** UV 260

## CHROMATOGRAM

**Internal standard:** 3-amino-1-[3,4,5-trimethoxybenzoyl]piperidine HCl

## KEY WORDS

gastric mucosa

## REFERENCE

Kusugami, K.; Ina, K.; Hosokawa, T.; Kobayashi, F.; Kusajima, H.; Momo, K.; Nishio, Y. Troxipide, a novel antiulcer compound, has inhibitory effects on human neutrophil migration and activation induced by various stimulants, *Dig.Liver Dis.*, **2000**, *32*, 305–311.

# Ubenimex

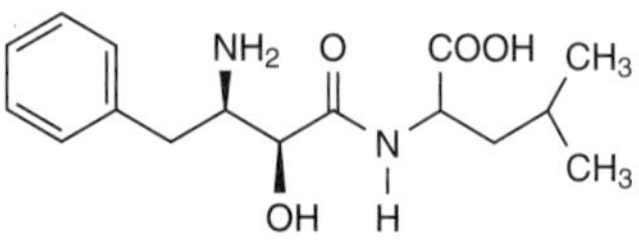

**Molecular formula:** $C_{16}H_{24}N_2O_4$
**Molecular weight:** 308.37
**CAS Registry No:** 58970-76-6
**Merck Index:** 13, 9910

## SAMPLE

**Matrix:** blood
**Sample preparation:** Mix 20 μL serum with 20 μL water and 200 μL 4.2 mM acetic acid, heat in a boiling water bath for 5 min, centrifuge at 1000 g for 5 min. Mix 100 μL of the supernatant with 50 μL 1.5 M ammonium hydroxide and 25 μL 3 mM sodium periodate, let stand at room temperature for 20 min, add 25 μL 12 mM sodium sulfite to decompose excess periodate, add 200 μL DDB solution, heat at 37° for 50 min, add 50 μL 1 M NaOH, inject a 100 μL aliquot. (Prepare DDB solution by dissolving 26.5 mg 4,5-dimethoxy-1,2-diaminobenzene monohydrochloride in 100 mL 300 mM HCl, use within 3 h. Dihydrochloride available from Molecular Probes or Dojindo Laboratories)

## HPLC VARIABLES

**Column:** 150 × 4 5 μm LiChrosorb RP-18
**Mobile phase:** MeCN:100 mM pH 8.7 tris-HCl buffer 25:75
**Flow rate:** 0.8
**Injection volume:** 100
**Detector:** F ex 320 em 390

## CHROMATOGRAM

**Retention time:** 4.5
**Limit of detection:** 200 ng/mL

## OTHER SUBSTANCES

**Extracted:** *p*-hydroxybestatin (10.5)

## KEY WORDS

derivatization; serum

## REFERENCE

Ishida, J.; Yamaguchi, M.; Kai, M.; Ohkura, Y.; Nakamura, M. Determination of bestatin in serum by high-performance liquid chromatography with fluorescence detection, *J.Chromatogr.*, **1984**, *305*, 381–389.

# Unoprostone isopropyl ester

**Molecular formula:** $C_{25}H_{44}O_5$
**Molecular weight:** 424.61
**CAS Registry No:** 120373-24-2, 120373-36-6 (free acid)
**Merck Index:** 13, 9917

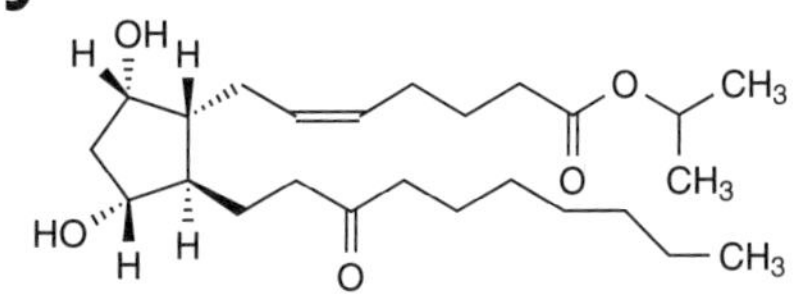

## SAMPLE
**Matrix:** aqueous humor, tissue
**Sample preparation:** Inject a 50 μL aliquot of aqueous humor directly. Homogenize eye tissue with MeCN:0.1% acetate 20:80 at 4°, centrifuge at 15000 rpm at 4° for 30 min. Filter (Millipore Ultra-Free CL, 300 000 cut-off) while centrifuging at 15 000 rpm at 4° for 30 min, inject an aliquot of the ultrafiltrate

## HPLC VARIABLES
**Column:** 150 × 4.6 C18-5A (Shodex)
**Mobile phase:** Gradient. MeCN:water:acetic acid from 20:80:0.1 to 100:0:0.1 over 40 min
**Injection volume:** 50
**Detector:** Radioactivity ($^3$H)

## CHROMATOGRAM
**Retention time:** 30

## OTHER SUBSTANCES
**Extracted:** metabolites

## KEY WORDS
eye; rabbit

## REFERENCE
Kashiwagi, K.; Iizuka, Y.; Tsukahara, S. Metabolites of isopropyl unoprostone as potential ophthalmic solutions to reduce intraocular pressure in pigmented rabbits, *Jpn.J.Pharmacol.*, **1999**, *81*, 56–62.

# Valacyclovir

**Molecular formula:** $C_{13}H_{20}N_6O_4$
**Molecular weight:** 324.33
**CAS Registry No:** 124832-26-4, 124832-27-5 (HCl)
**Merck Index:** 13, 9966

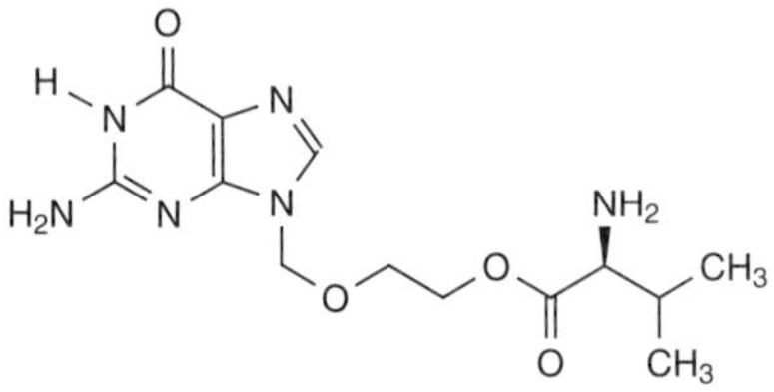

## SAMPLE

**Matrix:** blood
**Sample preparation:** Acidify plasma with trichloroacetic acid to a final trichloroacetic acid concentration of 3%, centrifuge at 9000 g at 4° for 10 min, inject an aliquot of the ultrafiltrate.

## HPLC VARIABLES

**Column:** 250 × 4.6 Adsorbosphere C18 (Alltech)
**Mobile phase:** MeCN:100 mM pH 3.5 ammonium formate buffer 10:90
**Flow rate:** 1
**Detector:** UV 254

## CHROMATOGRAM

**Retention time:** 8
**Limit of quantitation:** 80 ng/mL

## KEY WORDS

plasma; ultrafiltrate

## REFERENCE

Weller, S.; Blum, M.R.; Doucette, M.; Burnette, T.; Cederberg, D.M.; de Miranda, P.; Smiley, M.L. Pharmacokinetics of the acyclovir pro-drug valaciclovir after escalating single- and multiple-dose administration to normal volunteers, *Clin.Pharmacol.Ther.*, **1993**, *54*, 595–605.

## SAMPLE

**Matrix:** urine
**Sample preparation:** Acidify urine with trichloroacetic acid to a final trichloroacetic acid concentration of 1%, filter (Centrifree) while centrifuging at 2000 g at 4° for 10–20 min, inject an aliquot of the ultrafiltrate.

## HPLC VARIABLES

**Column:** 250 × 4.6 Adsorbosphere C18 (Alltech)
**Mobile phase:** Gradient. A was 50 mM pH 3.5 ammonium formate buffer. B was MeCN:50 mM pH 7.2 ammonium phosphate buffer 50:50. A:B ratio not given.
**Flow rate:** 1
**Detector:** UV 254, UV 280

## CHROMATOGRAM

**Retention time:** 27
**Limit of quantitation:** 160 ng/mL

## OTHER SUBSTANCES

**Extracted:** acyclovir (20), CMMG (14)

## KEY WORDS

ultrafiltrate

## REFERENCE

Weller, S.; Blum, M.R.; Doucette, M.; Burnette, T.; Cederberg, D.M.; de Miranda, P.; Smiley, M.L. Pharmacokinetics of the acyclovir pro-drug valaciclovir after escalating single- and multiple-dose administration to normal volunteers, *Clin.Pharmacol.Ther.*, **1993**, *54*, 595–605.

# Valdecoxib

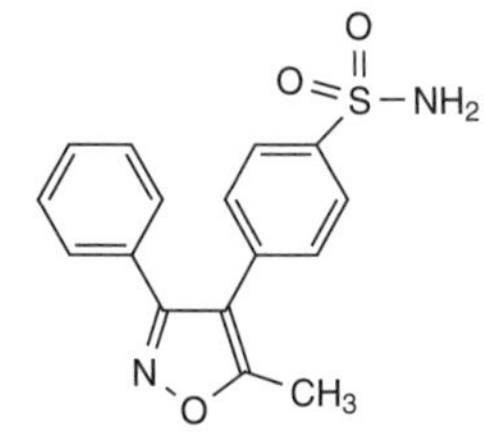

**Molecular formula:** $C_{16}H_{14}N_2O_3S$
**Molecular weight:** 314.36
**CAS Registry No:** 181695-72-7

## SAMPLE

**Matrix:** blood
**Sample preparation:** Condition a 1 mL 100 mg Bond Elut C18 SPE cartridge with 2 mL MeOH and 2 mL water. Centrifuge plasma at 2000 g at 4° for 5 min, vortex 400 μL of the supernatant with 400 μL 100 ng/mL IS in water, add to the SPE cartridge, wash with 2 mL water, elute with 250 μL MeCN. Evaporate the eluate to dryness under a stream of nitrogen, reconstitute the residue with 100 μL mobile phase, inject a 20 μL aliquot.

## HPLC VARIABLES

**Column:** 50 × 2.1 5 μm Zorbax XDB-C8
**Mobile phase:** MeCN:water 50:50 containing 10 mM ammonium acetate
**Flow rate:** 0.1
**Injection volume:** 20
**Detector:** MS, PE Sciex API-III Plus quadrupole, electrospray, negative ion mode, electrospray interface – 3700 V, orifice – 62 V, nebulizer gas nitrogen at 50 psi, curtain gas nitrogen at 1.8 L/min, collision gas argon, collision offset energy 25 eV, m/z 313–118

## CHROMATOGRAM

**Retention time:** 3
**Internal standard:** 4-(5-methyl-3-(3-fluorophenyl)-isoxazol-4-yl)benzenesulfonamide (m/z 331–118) (3.5)
**Limit of quantitation:** 0.5 ng/mL

## OTHER SUBSTANCES

**Extracted:** metabolite (4-(5-hydroxymethyl-3-phenylisoxazol-4-yl)benzenesulfonamide) (m/z 329–196) (2)

## KEY WORDS

plasma; SPE

## REFERENCE

Zhang, J.Y.; Fast, D.M.; Breau, A.P. Development and validation of an automated SPE-LC-MS/MS assay for valdecoxib and its hydroxylated metabolite in human plasma, *J.Pharm.Biomed.Anal.*, **2003**, *33*, 61–72.

## SAMPLE

**Matrix:** blood, feces, urine
**Sample preparation:** Blood. Extract plasma or red blood cells three times with MeCN. Combine the supernatants and evaporate to dryness under a stream of nitrogen, reconstitute the residue with mobile phase A, inject an aliquot. Urine. Centrifuge urine at 2000 g for 5 min, inject an aliquot. Feces. Homogenize feces three times with MeCN and twice with mobile phase A.

## HPLC VARIABLES

**Column:** 150 × 3.9 4 μm Novapak C8

**Mobile phase:** Gradient. A was MeCN:MeOH:25 mM pH 4 ammonium acetate buffer 3:6:81. B was MeCN:MeOH:25 mM pH 4 ammonium acetate buffer 20:40:60. A:B from 100:0 to 0:100 over 25 min, maintain at 0:100 for 5 min, return to initial conditions over 2 min, re-equilibrate for 8 min.
**Flow rate:** 1
**Detector:** UV 240; Radioactivity ($^{14}C$); MS, PE Sciex API-III Plus triple quadrupole, negative electrospray, capillary – 3700 V, orifice – 65 V, nebulizer gas at 50 psi, curtain gas nitrogen at 1.8 L/min, auxiliary gas nitrogen at 2 L/min, collision gas argon, collision energy 30 eV, turbo ionspray 400°, 0.2 mL/min of column effluent entered the detector

## CHROMATOGRAM
**Retention time:** 26

## OTHER SUBSTANCES
**Extracted:** metabolites

## KEY WORDS
mouse; pharmacokinetics; plasma; red blood cells

## REFERENCE
Zhang, J.Y.; Yuan, J.J.; Wang, Y.-F.; Bible, R.H. Jr.; Breau, A.P. Pharmacokinetics and metabolism of a COX-2 inhibitor, valdecoxib, in mice, *Drug Metab.Dispos.*, **2003**, *31*, 491–501.

## ANNOTATED BIBLIOGRAPHY
Yuan, J.J.; Yang, D.-C.; Zhang, J.Y.; Bible, R.; Karim, A.; Findlay, J.W.A. Disposition of a specific cyclooxygenase-2 inhibitor, valdecoxib, in human, *Drug Metab.Dispos.*, **2002**, *30*, 1013–1021. [radioactivity detection ($^{14}C$); LC-MS; metabolites]

Zhang, J.Y.; Fast, D.M.; Breau, A.P. Determination of valdecoxib and its metabolites in human urine by automated solid-phase extraction-liquid chromatography-tandem mass spectrometry, *J.Chromatogr.B*, **2003**, *785*, 123–134. [LOQ 1 ng/mL]

# Valganciclovir

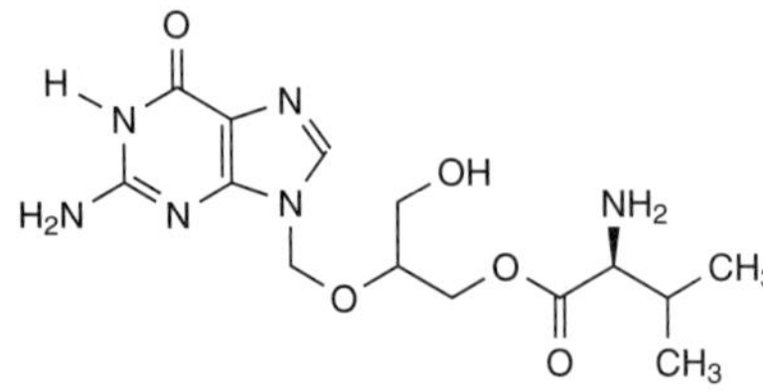


**Molecular formula:** $C_{14}H_{22}N_6O_5$
**Molecular weight:** 354.36
**CAS Registry No:** 175865-60-8
**Merck Index:** 13, 9973

## SAMPLE

**Matrix:** blood
**Sample preparation:** Vortex 250 μL plasma with 100 μL 1 M HCl and 100 μL cold 15% trichloroacetic acid, centrifuge, inject an aliquot of the supernatant. (A column-switching system is used so that only the fraction of interest is diverted from the initial column on to the analytical column. Details are not provided.)

## HPLC VARIABLES

**Column:** BDS Hypersil C18
**Mobile phase:** MeCN:0.0425% phosphoric acid 5:95 (diastereomers are separated)
**Detector:** UV 254

## CHROMATOGRAM

**Limit of quantitation:** 40 ng/mL

## KEY WORDS

pharmacokinetics; plasma

## REFERENCE

Brown, F.; Banken, L.; Saywell, K.; Arum, I. Pharmacokinetics of valganciclovir and ganciclovir following multiple oral dosages of valganciclovir in HIV- and CMV-seropositive volunteers, *Clin.Pharmacokinet.*, **1999**, *37*, 167–176.

## SAMPLE

**Matrix:** formulations
**Sample preparation:** Dilute oral solution with 25 mM pH 2.5 phosphate buffer containing 5 μg/mL IS, filter (0.2 μm), inject a 10 μL aliquot.

## HPLC VARIABLES

**Guard column:** 15 × 4.6 5 μm MetaGuard Inertsil ODS-3
**Column:** 100 × 4.6 5 μm Inertsil ODS-3
**Mobile phase:** MeCN:25 mM pH 2.5 phosphate buffer 2.5:97.5
**Flow rate:** 1.6
**Injection volume:** 10
**Detector:** UV 254

## CHROMATOGRAM

**Retention time:** 5.9, 6.9 (diastereomers)
**Internal standard:** hypoxanthine (1.8)

## OTHER SUBSTANCES

**Simultaneous:** ganciclovir (2.5)

## KEY WORDS

oral solution; stability-indicating

## REFERENCE

Anaizi, N.H.; Dentinger, P.J.; Swenson, C.F. Stability of valganciclovir in an extemporaneously compounded oral liquid, *Am.J.Health-Syst.Pharm.*, **2002**, *59*, 1267–1270.

## ANNOTATED BIBLIOGRAPHY

Henkin, C.C.; Griener, J.C.; Ten Eick, A.P. Stability of valganciclovir in extemporaneously compounded liquid formulations, *Am.J.Health-Syst.Pharm.*, **2003**, *60*, 687–690. [stability-indicating]

# Valrubicin

**Molecular formula:** $C_{34}H_{36}F_3NO_{13}$
**Molecular weight:** 723.64
**CAS Registry No:** 56124-62-0
**Merck Index:** 13, 9981

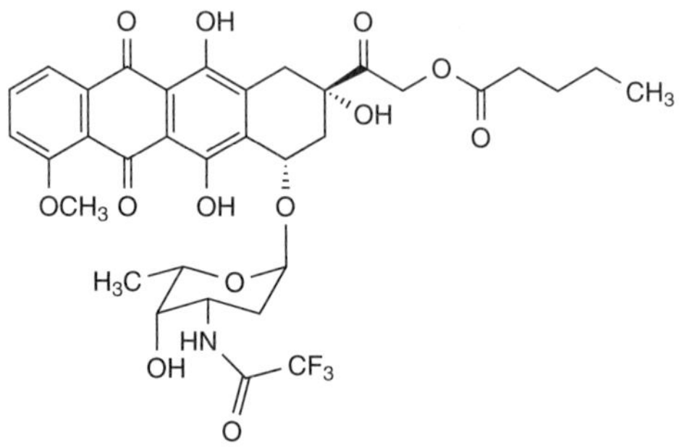


## SAMPLE

**Matrix:** tissue
**Sample preparation:** Homogenize (Duall ground glass grinder) 1 g of tissue with 9 mL 50 mM pH 8.5 Tris HCl buffer containing 3% sodium dodecyl sulfate, extract three times with 2 vol of ethyl acetate:*n*-propanol 90:10. Evaporate the combined extracts to dryness at 45°, reconstitute the residue with MeOH, inject an aliquot.

## HPLC VARIABLES

**Column:** μBondapak phenyl
**Mobile phase:** Gradient. MeCN:pH 4.00 ammonium formate buffer from 32:68 to 65:35 over 7 min.
**Flow rate:** 5
**Detector:** F ex 482 em Schoeffel no. 2–73 filter

## CHROMATOGRAM

**Retention time:** 7

## OTHER SUBSTANCES

**Extracted:** metabolites

## KEY WORDS

liver; mouse; small intestine

## REFERENCE

Israel, M.; Karkowsky, A.M.; Khetarpal, V.K. Distribution of radioactivity and anthracycline-fluorescence in tissues of mice one hour after [$^{14}$C]-labeled AD 32 administration. Evidence for tissue aglycone formation, *Cancer Chemother.Pharmacol.*, **1981**, *6*, 25–30.

# Valsartan

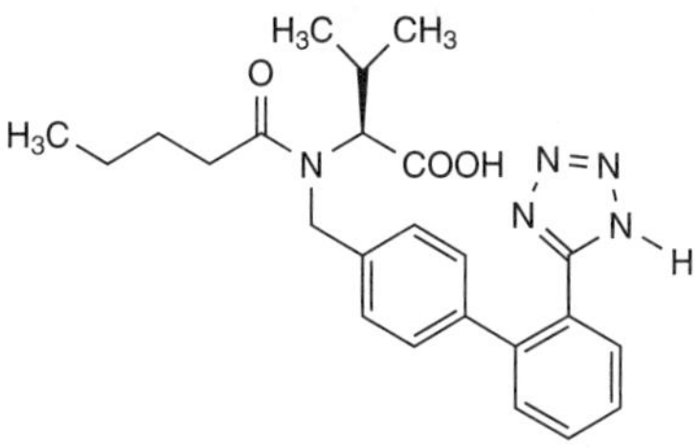


**Molecular formula:** $C_{24}H_{29}N_5O_3$
**Molecular weight:** 435.52
**CAS Registry No:** 137862-53-4
**Merck Index:** 13, 9982

### SAMPLE

**Matrix:** blood
**Sample preparation:** Condition a 1 mL 100 mg Bond Elut C8 SPE cartridge with 2 mL MeOH and 1 mL 100 mM pH 2 phosphate buffer. Mix 250 μL plasma with IS, add 250 μL 1 M phosphoric acid, shake, centrifuge at 10 000 g at 4° for 5 min, add the supernatant to the SPE cartridge, wash with 500 μL MeOH:100 mM pH 2 phosphate buffer 50:50, dry at full vacuum for 20 min, elute with 500 μL MeOH. Add 100 μL MeOH:ethylene glycol 90:10 to the eluate (to prevent adsorption of the drug), vortex, evaporate to dryness under a stream of nitrogen at 40°C, reconstitute the residue with 250 μL of the initial mobile phase, inject a 20 μL aliquot.

### HPLC VARIABLES

**Guard column:** 20 × 3.9 4 μm Novapak C18 (Waters)
**Column:** 300 × 3.9 10 μm μBondapak C18
**Mobile phase:** Gradient. MeCN:5 mM pH 4 acetate buffer from 30:70 to 60:40 over 15 min, to 95:5 over 6 min, return to initial conditions over 3 min, re-equilibrate at initial conditions for 1 min.
**Flow rate:** from 1 to 1.2 over 15 min, maintain at 1.2 for 6 min, to 1 over 3 min, maintain at 1 for 1 min
**Injection volume:** 20
**Detector:** F ex 250 em 375

### CHROMATOGRAM

**Retention time:** 14.4
**Internal standard:** bumetanide (13.5)
**Limit of quantitation:** 50 ng/mL

### OTHER SUBSTANCES

**Extracted:** metabolites, candesartan (22.6, LOQ 3 ng/mL), irbesartan (12.6, LOQ 50 ng/mL), losartan (11.5, LOQ 16 ng/mL)

### KEY WORDS

plasma; SPE

### REFERENCE

González, L.; López, J.A.; Alonso, R.M.; Jiménez, R.M. Fast screening method for the determination of angiotensin II receptor antagonists in human plasma by high-performance liquid chromatography with fluorimetric detection, *J.Chromatogr.A*, **2002**, *949*, 49–60.

### SAMPLE

**Matrix:** formulations
**Sample preparation:** Sonicate finely ground tablets in MeOH for 5 min, centrifuge, dilute with MeOH containing 290 ng/mL IS, inject a 20 μL aliquot.

## HPLC VARIABLES

**Column:** 150 × 4.6 5 μm Supelcosil LC 18
**Mobile phase:** MeCN:20 mM pH 3.2 phosphate buffer 45:55
**Flow rate:** 0.9
**Injection volume:** 20
**Detector:** UV 225

## CHROMATOGRAM

**Retention time:** 6.99
**Internal standard:** trimethoprim (2.96)
**Limit of detection:** 17 ng/mL
**Limit of quantitation:** 58 ng/mL

## OTHER SUBSTANCES

**Extracted:** hydrochlorothiazide (2.20)

## KEY WORDS

comparison with derivative spectrophotometry; tablets

## REFERENCE

Satana, E.; Altinay, S.; Göger, N.G.; Ozkan, S.A.; Sentürk, Z. Simultaneous determination of valsartan and hydrochlorothiazide in tablets by first-derivative ultraviolet spectrophotometry and LC, *J.Pharm.Biomed.Anal.*, **2001**, *25*, 1009–1013.

## ANNOTATED BIBLIOGRAPHY

Carlucci, G.; Di Carlo, V.; Mazzeo, P. Simultaneous determination of valsartan and hydrochlorothiazide in tablets by high-performance liquid chromatography, *Anal.Lett.*, **2000**, *33*, 2491–2500.

Daneshtalab, N.; Lewanczuk, R.Z.; Jamali, F. High-performance liquid chromatographic analysis of angiotensin II receptor antagonist valsartan using a liquid extraction method, *J.Chromatogr.B*, **2002**, *766*, 345–349. [fluorescence detection; LOQ 10 ng/mL; losartan is internal standard]

Francotte, E.; Davatz, A.; Richert, P. Development and validation of chiral high-performance liquid chromatographic methods for the quantitation of valsartan and of the tosylate of valinebenzyl ester, *J.Chromatogr.B*, **1996**, *686*, 77–83. [LOQ 0.1%; LOD 0.04%]

González, L.; Alonso, R.M.; Jiménez, R.M. A high-performance liquid chromatographic method for screening angiotensin II receptor antagonists in human urine, *Chromatographia*, **2000**, *52*, 735–740. [SPE; LOQ 400 ng/mL; losartan; irbesartan; valsartan; candesartan]

Sioufi, A.; Marfil, F.; Jaouen, A.; Cardot, J.M.; Godbillon, J.; Ezzet, F.; Lloyd, P. The effect of age on the pharmacokinetics of valsartan, *Biopharm.Drug Dispos.*, **1998**, *19*, 237–244. [LOQ 5 ng/mL; fluorescence detection]

Tatar, S.; Saglik, S. Comparison of UV- and second derivative-spectrophotometric and LC methods for the determination of valsartan in pharmaceutical formulation, *J.Pharm.Biomed.Anal.*, **2002**, *30*, 371–375. [capsules; losartan is internal standard; LOD 200 ng/mL; LOQ 1 μg/mL]

Waldmeier, F.; Flesch, G.; Müller, P.; Winkler, T.; Kriemler, H.-P.; Bühlmayer, P.; de Gasparo, M. Pharmacokinetics, disposition and biotransformation of [$^{14}$C]-radiolabelled valsartan in healthy male volunteers after a single oral dose, *Xenobiotica*, **1997**, *27*, 59–71.

# Zaleplon

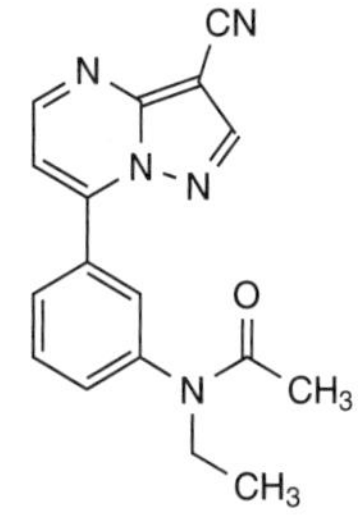

**Molecular formula:** $C_{17}H_{15}N_5O$
**Molecular weight:** 305.33
**CAS Registry No:** 151319-34-5
**Merck Index:** 13, 10165

## SAMPLE

**Matrix:** blood

**Sample preparation:** Evaporate 100 μL 1 μg/mL IS in MeOH in the bottom of a tube using a stream of nitrogen, add 1 mL whole blood, add 1 mL buffer, vortex for 1 min, add 5 mL dichloromethane:hexane:ethyl acetate 50:40:10, shake horizontally at 200 cycles/min for 30 min, centrifuge at 2000 rpm for 30 min. Evaporate the organic layer to dryness under a stream of nitrogen, reconstitute the residue with 100 μL initial mobile phase, inject an aliquot. (Prepare the buffer by adjusting the pH of 1 L saturated ammonium chloride solution to 9.5 with 25% ammonium hydroxide.)

## HPLC VARIABLES

**Guard column:** Inertsil ODS-3

**Column:** 150 × 2 3 μm Inertsil ODS-3

**Mobile phase:** Gradient. MeCN:1 mM pH 4.0 ammonium formate buffer 10:90 for 2 min, to 60:40 over 15 min, maintain at 60:40 for 3 min, return to initial conditions over 1 min, re-equilibrate for 10 min.

**Flow rate:** 0.2

**Detector:** MS, PE Sciex API 150EX, single quadrupole, turbo ionspray, atmospheric pressure ionization, positive mode, heater gas nitrogen, ionspray 5000 V, nebulizer gas nitrogen at 1.2 L/min, curtain gas nitrogen at 1.0 L/min, turbo probe 475°, orifice 35 V, ring 175 V, m/z 306.21

## CHROMATOGRAM

**Retention time:** 14.35

**Internal standard:** methaqualone (m/z 251.11, orifice 20 V, ring 150 V) (16.12)

**Limit of detection:** 0.1 ng/mL (S/N 5)

**Limit of quantitation:** 1 ng/mL

## OTHER SUBSTANCES

**Extracted:** zolpidem (m/z 308.21, orifice 40 V, ring 175 V) (9.78)

## KEY WORDS

whole blood

## REFERENCE


Giroud, C.; Augsburger, M.; Menetrey, A.; Mangin, P. Determination of zaleplon and zolpidem by liquid chromatography-turbo-ionspray mass spectrometry: application to forensic cases, *J.Chromatogr.B*, **2003**, *789*, 131–138.


## SAMPLE

**Matrix:** blood

**Sample preparation:** Add IS to 200 μL plasma, add 400 μL MeCN, vortex, centrifuge, inject an aliquot of the supernatant.

### HPLC VARIABLES

**Column:** 150 × 4.6 5 μm CSC-S-Octyl
**Mobile phase:** MeCN:100 mM sodium dihydrogen phosphate:100 mM disodium hydrogen phosphate 32:32.5:32.5
**Flow rate:** 1
**Detector:** F ex 345 em 460

### CHROMATOGRAM

**Internal standard:** CL-218872
**Limit of quantitation:** 0.5 ng/mL

### OTHER SUBSTANCES

**Extracted:** deethylzaleplon

### KEY WORDS

pharmacokinetics; plasma

### REFERENCE

Greenblatt, D.J.; Harmatz, J.S.; von Moltke, L.L.; Ehrenberg, B.L.; Harrel, L.; Corbett, K.; Counihan, M.; Graf, J.A.; Darwish, M.; Mertzanis, P.; Martin, P.T.; Cevallos, W.H.; Shader, R.I. Comparative kinetics and dynamics of zaleplon, zolpidem, and placebo, *Clin.Pharmacol.Ther.*, **1998**, *64*, 553–561.

---

### SAMPLE

**Matrix:** microsomal incubations
**Sample preparation:** Add 2 vol of ice-cold MeCN to the microsomal incubation, stir vigorously, centrifuge at 1500 g for 15 min. Evaporate 500 μL of the supernatant to dryness under a stream of nitrogen, reconstitute the residue with 200–400 μL initial mobile phase, inject a 50 μL aliquot.

### HPLC VARIABLES

**Column:** 150 × 4.6 Develosil ODS-UG-5
**Mobile phase:** Gradient. MeCN:50 mM pH 6.8 potassium phosphate buffer from 20:80 to 60:40 over 18 min.
**Flow rate:** 1
**Injection volume:** 50
**Detector:** UV 245

### CHROMATOGRAM

**Retention time:** 17.6
**Internal standard:** CL 218,872 (20.8)
**Limit of quantitation:** 50 nM

### OTHER SUBSTANCES

**Extracted:** *N*-desethyl-5-oxozaleplon (3.4), *N*-desethylzaleplon (14.5), 5-oxozaleplon (8.5)

### KEY WORDS

liver; monkey; rat

### REFERENCE

Kawashima, K.; Hosoi, K.; Naruke, T.; Shiba, T.; Kitamura, M.; Watabe, T. Aldehyde oxidase-dependent marked species difference in hepatic metabolism of the sedative-hypnotic, zaleplon, between monkeys and rats, *Drug Metab.Dispos.*, **1999**, *27*, 422–428.

## ANNOTATED BIBLIOGRAPHY

Feng, F.; Jiang, J.; Dai, H.; Wu, J. Development and validation of a high-performance liquid chromatography-electrospray ionization-mass spectrometry assay for the determination of zaleplon in human plasma, *J.Chromatogr.Sci.*, **2003**, *41*, 17–21. [LOD 100 pg/mL; triazolam is internal standard]

# Zaltoprofen

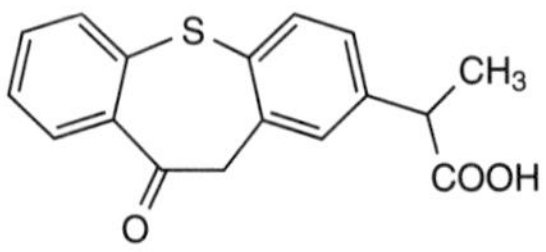

**Molecular formula:** $C_{17}H_{14}O_3S$
**Molecular weight:** 298.36
**CAS Registry No:** 74711-43-6, 89482-01-9 ((S)-form)
**Merck Index:** 13, 10166

## SAMPLE
**Matrix:** microsomal incubations
**Sample preparation:** Mix microsomal incubation with MeCN containing IS, centrifuge at 9000 g for 10 min, inject an aliquot of the supernatant

## HPLC VARIABLES
**Guard column:** 10 × 4 Capcell Pak C18 UG120 (Shiseido)
**Column:** 250 × 4.6 L-Column ODS (Chemicals Evaluation and Research Institute, Tokyo)
**Column temperature:** 35
**Mobile phase:** MeCN:water:acetic acid 60:40:1
**Flow rate:** 0.7
**Detector:** UV 330

## CHROMATOGRAM
**Internal standard:** mefenamic acid

## KEY WORDS
human; liver

## REFERENCE
Furuta, S.; Akagawa, N.; Kamada, E.; Hiyama, A.; Kawabata, Y.; Kowata, N.; Inaba, A.; Matthews, A.; Hall, M.; Kurimoto, T. Involvement of CYP2C9 and UGT2B7 in the metabolism of zaltoprofen, a nonsteroidal anti-inflammatory drug, and its lack of clinically significant CYP inhibition potential, *Br.J.Clin.Pharmacol.*, **2002**, *54*, 295–303.

## SAMPLE
**Matrix:** microsomal incubations
**Sample preparation:** Mix microsomal incubation with 1 M pH 4.5 acetate buffer and dichloromethane, add IS, centrifuge at 2000 g for 10 min. Evaporate the organic layer to dryness under a stream of nitrogen, reconstitute the residue with mobile phase, inject an aliquot.

## HPLC VARIABLES
**Guard column:** 10 × 4 Capcell Pak C18 UG120 (Shiseido)
**Column:** 250 × 4.6 Capcell Pak C18 UG120 (Shiseido)
**Column temperature:** 35
**Mobile phase:** MeCN:buffer 30:70 (The buffer was 65 mM sodium dihydrogen phosphate containing 5 mM tetrabutylammonium chloride, adjusted to pH 6.0 with triethylamine.)
**Flow rate:** 1
**Detector:** UV 270

## CHROMATOGRAM
**Retention time:** 24
**Internal standard:** methyl paraben (8)

## OTHER SUBSTANCES

**Extracted:** metabolites

## KEY WORDS

human; liver

## REFERENCE

Furuta, S.; Akagawa, N.; Kamada, E.; Hiyama, A.; Kawabata, Y.; Kowata, N.; Inaba, A.; Matthews, A.; Hall, M.; Kurimoto, T. Involvement of CYP2C9 and UGT2B7 in the metabolism of zaltoprofen, a nonsteroidal anti-inflammatory drug, and its lack of clinically significant CYP inhibition potential, *Br.J.Clin.Pharmacol.*, **2002**, *54*, 295–303.

# Zanamivir

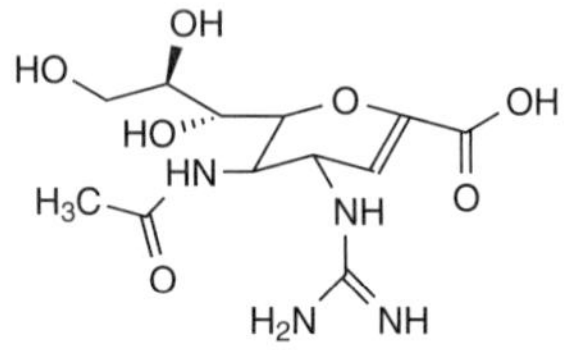


**Molecular formula:** $C_{12}H_{20}N_4O_7$
**Molecular weight:** 332.31
**CAS Registry No:** 139110-80-8
**Merck Index:** 13, 10167

## SAMPLE

**Matrix:** blood
**Sample preparation:** Condition a 1 mL 50 mg isolute SCX SPE cartridge with 500 μL MeOH and 500 μL 10% acetic acid water. Vortex 200 μL serum with 100 μL 5 μg/mL IS in water, 500 μL MeCN, and 100 μL 3% acetic acid in water, let stand at room temperature for 5 min, centrifuge at 1400 g for 10 min, add the supernatant to the SPE cartridge, elute with two 500 μL portions of 10% triethylamine in MeOH:water 50:50. Evaporate the eluate to dryness under a stream of nitrogen at 60°, reconstitute the residue with 200 μL mobile phase, vortex for 10 s, centrifuge at 5000 g for 5 min, inject a 20–200 μL aliquot.

## HPLC VARIABLES

**Column:** 100 × 4.6 5 μm Hypersil SAS C1
**Mobile phase:** MeCN:water:acetic acid 50:50:1
**Flow rate:** 1
**Injection volume:** 20–200
**Detector:** MS, PE Sciex API 300 triple quadrupole, TurboIonSpray, positive ion mode, 20% of column effluent entered the detector, collision gas nitrogen, collision energy 28 eV, m/z 333–60

## CHROMATOGRAM

**Retention time:** 4
**Internal standard:** $^{13}C,^{15}N_2$-zanamivir (m/z 336–63)
**Limit of quantitation:** 10 ng/mL

## KEY WORDS

serum; SPE

## REFERENCE

Allen, G.D.; Brookes, S.T.; Barrow, A.; Dunn, J.A.; Grosse, C.M. Liquid chromatographic-tandem mass spectrometric method for the determination of the neuraminidase inhibitor zanamivir (GG167) in human serum, *J.Chromatogr.B*, **1999**, *732*, 383–393.

## SAMPLE

**Matrix:** blood
**Sample preparation:** Condition a Bond Elut SCX SPE cartridge with 2 mL MeOH and 2 mL 10% formic acid. Mix 1 mL serum with 1 mL 10% formic acid, add to the SPE cartridge, wash with 2 mL 1% trifluoroacetic acid in MeOH, wash with 2 mL water, elute with four 500 μL portions of 10% triethylamine in MeOH:water 50:50. Evaporate the eluate to dryness under a stream of nitrogen at 70°, reconstitute the residue with 75 μL 4 mM benzoin in ethylene glycol, add 150 μL reagent, vortex, heat at 100° for 3 min, cool, inject a 100 μL aliquot. (Prepare the reagent by diluting 750 μL β-mercaptoethanol, 2.52 g sodium sulfite, and 10 mL 5 M KOH to 100 mL with water.)

## HPLC VARIABLES

**Column:** 100 × 4.6 Hypersil ODS

**Mobile phase:** Gradient. MeCN:buffer 20:80 for 12 min, to 80:20 (step gradient), maintain at 80:20 for 5 min, re-equilibrate at initial conditions for 10 min. (Prepare the buffer by adding 100 mL 1 M tris(hydroxymethyl)methylamine (Tris) and 29.4 mL 1 M HCl to water and making up to 2 L with water (50 mM pH 8.5).)
**Flow rate:** 1
**Injection volume:** 100
**Detector:** F ex 325 em 442

---

## CHROMATOGRAM
**Retention time:** 14
**Limit of quantitation:** 10 ng/mL

---

## KEY WORDS
derivatization; serum; SPE

---

## REFERENCE
Stubbs, R.J.; Harker, A.J. Automated high-performance liquid chromatographic method for the determination of a neuraminidase inhibitor, *J.Chromatogr.B*, **1995**, *670*, 279–285.

# Zinostatin

**CAS Registry No:** 9014-02-2, 123760-07-6
**Merck Index:** 13, 10222

## SAMPLE

**Matrix:** solutions

## HPLC VARIABLES

**Column:** TSK G-3000SW
**Mobile phase:** 10 mM pH 7.9 ammonium bicarbonate containing 30 mM NaCl
**Flow rate:** 1
**Detector:** UV 254, UV 280

## REFERENCE

Maeda, H.; Ueda, M.; Morinaga, T.; Matsumoto, T. Conjugation of poly(styrene-co-maleic acid) derivatives to the antitumor protein neocarzinostatin: pronounced improvements in pharmacological properties, *J.Med.Chem.*, **1985**, *28*, 455–461.

# Zofenopril calcium

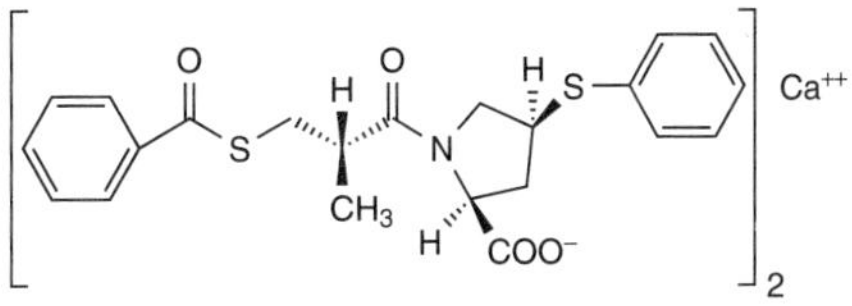

**Molecular formula:** $C_{44}H_{44}CaN_2O_8S_4$
**Molecular weight:** 897.17
**CAS Registry No:** 81938-43-4, 81872-10-8 (free acid)

## SAMPLE

**Matrix:** blood

**Sample preparation:** Collect 10 mL whole blood in a heparinized tube containing 20 mg *N*-ethylmaleimide, centrifuge. Shake 1 mL plasma with 20 μL MeOH containing 10 μg/mL IS1 and 15 μg/mL IS2, 1 mL 2 M phosphoric acid containing 2% tetrabutylammonium hydrogen sulfate, and 7 mL toluene in a PTFE-lined tube on a rotating mixer at 32 rpm for 15 min, centrifuge at 1500 g for 5 min. Evaporate the organic layer to dryness under a stream of nitrogen at 60°, reconstitute the residue with 200 μL MeOH:water 50:50, inject an aliquot.

## HPLC VARIABLES

**Guard column:** 10 × 3.2 5 μm C18 (Hichrom)

**Column:** 75 × 4.6 3 μm Luna C18

**Mobile phase:** Gradient. MeCN:26 mM pH 4.5 ammonium acetate buffer 20:80 for 0.5 min, to 80:20 over 2.5 min, maintain at 80:20 for 3 min, re-equilibrate at initial conditions for 3 min.

**Flow rate:** 0.4

**Detector:** MS, PE Sciex API 365 triple quadrupole, TurboIonSpray, probe 450°, negative ion mode, orifice – 35 V, ring – 200 V, nebulizer gas nitrogen at 10 units, curtain gas nitrogen at 12 units, auxiliary gas air, collision gas nitrogen at 2 units, m/z 428–137

## CHROMATOGRAM

**Retention time:** 6

**Internal standard:** IS1 (*N*-[3-mercapto-2-methylpropionyl]-4-(4-fluorophenylthio)-L-proline benzoate) (m/z 446–137) (6), IS2 (m/z 467–308) (Prepare a solution of IS2 by reacting 7 mg *N*-[3-mercapto-2-methylpropionyl]-4-(4-fluorophenylthio)-L-proline with 1.75 mL of a 25 mg/mL solution of *N*-ethylmaleimide in 60 mM pH 7 buffer in the dark at room temperature for 1 h, make up to 10 mL with acetone.) (5.3)

**Limit of detection:** 50 pg/mL (S/N 3)

**Limit of quantitation:** 1 ng/mL

## OTHER SUBSTANCES

**Extracted:** zofenoprilat (*N*-[3-mercapto-2-methylpropionyl]-4-(phenylthio)-L-proline) (m/z 449–290 (as derivative)) (LOQ 2 ng/mL, LOD 100 pg/mL) (5.3)

## KEY WORDS

derivatization; pharmacokinetics; plasma; whole blood

## REFERENCE

Dal Bo, L.; Mazzucchelli, P.; Marzo, A. Assay of zofenopril and its active metabolite zofenoprilat by liquid chromatography coupled with tandem mass spectrometry, *J.Chromatogr.B*, **2000**, *749*, 287–294.

## SAMPLE

**Matrix:** solutions

## HPLC VARIABLES

**Column:** 250 × 4.6 5 μm Econosil

**Mobile phase:** MeCN:water:triethylamine:phosphoric acid 65:35:0.02:0.13, pH 2.4
**Flow rate:** 1
**Detector:** UV 220

---

## CHROMATOGRAM
**Internal standard:** cephalothin

---

## OTHER SUBSTANCES
**Also analyzed:** captopril, enalaprilat, fosinopril, quinapril, quinaprilat, ramipril

---

## REFERENCE
Lin, C.J.; Akarawut, W.; Smith, D.E. Competitive inhibition of glycylsarcosine transport by enalapril in rabbit renal brush border membrane vesicles: interaction of ACE inhibitors with high-affinity $H^+$/peptide symporter, *Pharm.Res.*, **1999**, *16*, 609–615.

---

## SAMPLE
**Matrix:** solutions

---

## HPLC VARIABLES
**Column:** 150 × 3.9 μBondapak phenyl
**Column temperature:** 30–40
**Mobile phase:** MeOH:water:85% phosphoric acid 68:32:0.2
**Detector:** UV 215–220

---

## OTHER SUBSTANCES
**Simultaneous:** zofenoprilat

---

## REFERENCE
Ranadive, S.A.; Chen, A.X.; Serajuddin, A.T. Relative lipophilicities and structural-pharmacological considerations of various angiotensin-converting enzyme (ACE) inhibitors, *Pharm.Res.*, **1992**, *9*, 1480–1486.

---

## ANNOTATED BIBLIOGRAPHY
Marzo, A.; Dal Bo, L.; Mazzucchelli, P.; Monti, N.C.; Crivelli, F.; Ismaili, S.; Giusti, A.; Uhr, M.R. Pharmacokinetic and pharmacodynamic comparative study of zofenopril and enalapril in healthy volunteers, *Arzneimittelforschung*, **2002**, *52*, 233–242. [LC-MS; LOQ 5 ng/mL]

# Zolazepam hydrochloride

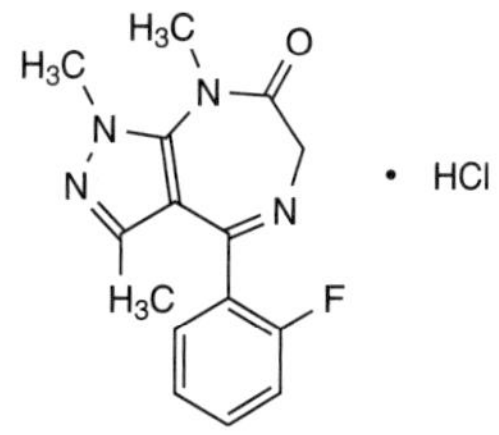

**Molecular formula:** $C_{15}H_{15}FN_4O.HCl$
**Molecular weight:** 322.77
**CAS Registry No:** 33754-49-3, 31352-82-6 (free base)

## SAMPLE

**Matrix:** blood, tissue
**Sample preparation:** Add IS to serum or tissue, make alkaline with pH 9.5 borate buffer, extract with ethyl acetate. Extract the organic layer with 1 mL 100 mM HCl. Basify the aqueous layer with 100 mg sodium borate, extract with ethyl acetate. Evaporate the organic layer to dryness under a stream of nitrogen, reconstitute the residue with MeCN:water 25:75 (serum) or mobile phase (tissue), inject a 70 μL aliquot.

## HPLC VARIABLES

**Column:** 250 × 4 5 μm LiChrospher 60 RP-select B C8
**Mobile phase:** MeCN:50 mM pH 6.8 phosphate buffer 26:74
**Flow rate:** 1
**Injection volume:** 70
**Detector:** UV 233

## CHROMATOGRAM

**Retention time:** 12.5
**Internal standard:** pindolol (9.7)
**Limit of quantitation:** 2 ng/mL

## OTHER SUBSTANCES

**Extracted:** tiletamine (LOQ 10 ng/mL) (18.9)

## KEY WORDS

bear; muscle; serum

## REFERENCE

Semple, H.A.; Gorecki, D.K.J.; Farley, S.D.; Ramsay, M.A. Pharmacokinetics and tissue residues of Telazol in free-ranging polar bears, *J. Wildlife Dis.*, **2000**, *36*, 653–662.

## SAMPLE

**Matrix:** blood, urine
**Sample preparation:** Mix 1 mL pH 7 buffer with 5 mL whole blood or urine, add 1 μg IS1, add 6 μg IS2, add 7 mL hexane:toluene:isoamyl alcohol 90:5:5, vortex, centrifuge, remove the organic layer. Add 1 mL pH 9.5 buffer to the aqueous layer, extract with 6 mL hexane:isoamyl alcohol 99:1. Combine the organic layers, evaporate to dryness under a stream of nitrogen, reconstitute the residue with 250 μL MeCN:water:trifluoroacetic acid 45:55:0.05, evaporate to half volume to remove MeCN, inject a 50 μL aliquot.

## HPLC VARIABLES

**Guard column:** 5 μm Altima C18
**Column:** 20 × 2.1 5 μm Altima C18
**Column temperature:** 40

**Mobile phase:** Gradient. A was MeCN:water:trifluoroacetic acid 10:90:0.05. B was MeCN:water:trifluoroacetic acid 80:20:0.05. A:B 100:0 for 5 min, to 0:100 over 50 min, re-equilibrate at initial conditions for 15 min.
**Flow rate:** 0.25
**Injection volume:** 50
**Detector:** MS, HP 5989B, electron impact; UV 205; UV 290

---

## CHROMATOGRAM
**Internal standard:** SKF-525 A (IS1), 5-ethyl-5-*p*-tolylbarbituric acid (IS2)

---

## OTHER SUBSTANCES
**Extracted:** tiletamine

---

## KEY WORDS
whole blood

---

## REFERENCE
Cording, C.J.; Deluca, R.; Camporese, T.; Spratt, E. A fatality related to the veterinary anesthetic Telazol, *J.Anal.Toxicol.*, **1999**, *23*, 552–555.

---

## SAMPLE
**Matrix:** tissue
**Sample preparation:** Homogenize 1.5 g tissue with 300 μL 1 μg/mL IS in water and 6 mL 500 mM pH 9.5 sodium borate buffer, add 10 mL ethyl acetate, vortex for 20 min, repeat the extraction. Combine the organic layers and extract twice with 2 mL portions of 100 mM HCl. Combine the aqueous layers and make basic with 200 mg sodium borate, extract with 12 mL ethyl acetate. Evaporate the organic layer to dryness under a stream of nitrogen, reconstitute the residue with 200 μL MeCN:water 20:80, inject a 70 μL aliquot.

---

## HPLC VARIABLES
**Column:** 250 × 4 5 μm LiChrospher 60 RP-select B C8
**Mobile phase:** MeCN:50 mM pH 5.5 sodium phosphate buffer 16:84
**Flow rate:** 1
**Injection volume:** 70
**Detector:** UV 233

---

## CHROMATOGRAM
**Retention time:** 58
**Internal standard:** ripazapam (51.5)

---

## OTHER SUBSTANCES
**Extracted:** tiletamine (45), metabolites

---

## KEY WORDS
bear; fat; kidney; muscle

---

## REFERENCE
Semple, H.A.; Gorecki, D.K.J.; Farley, S.D.; Ramsay, M.A. Pharmacokinetics and tissue residues of Telazol in free-ranging polar bears, *J.Wildlife Dis.*, **2000**, *36*, 653–662.

# CUMULATIVE INDEX

This index provides a comprehensive listing of the drugs for which analytical methods have been described in this volume and in earlier volumes in the series (available from John Wiley & Sons, Inc.). These volumes are designated as follows:

I HPLC Methods for Pharmaceutical Analysis
II HPLC Methods for Pharmaceutical Analysis, Volume 2
III HPLC Methods for Pharmaceutical Analysis, Volume 3
IV HPLC Methods for Pharmaceutical Analysis, Volume 4
V HPLC Methods for Recently Approved Pharmaceuticals (this volume)
CE Capillary Electrophoresis Methods for Pharmaceutical Analysis

(For obvious reasons non-interfering compounds are not listed in this index.)

Abacavir, V 1
Acamprosate, II 1
Acarbose, CE 1, V 5
Acebutolol, CE 1, II 2
Aceclofenac, II 17
Acefylline, II 19
Acenocoumarol, CE 9, II 20
Acepromazine, CE 11, II 22
Aceprometazine, II 29
Acesulfame, II 31
Acetaminophen, CE 12, I 1, II 32
Acetazolamide, CE 27, II 37
Acetiamine, II 46
Acetohexamide, CE 28, II 47
Acetophenazine, II 49
Acetorphan, II 54
Acetyl sulfisoxazole, V 6
Acetylcholine, II 56
Acetylcysteine, CE 30, II 66
Acipimox, II 84
Acitretin, II 85
Aconitine, II 93
Acrivastine, II 95, V 7
Actinoquinol, II 96
Acyclovir, CE 33, I 27, II 97
Adapalene, V 10
Adefovir, II 101
Adefovir dipivoxil, V 11
Adenosine triphosphate, CE 39, II 111
Adenosine, CE 35, II 103
Adiphenine, I 33, II 118
Adrafinil, II 119
Adrenocorticotropic Hormone, V 13
Afloqualone, V 15
Ajmaline, II 120
Alacepril, V 16
Albendazole, II 122
Albuterol, CE 41, I 35, II 129
Alclometasone 17,21-Dipropionate, V 18
Aldesleukin, CE 51, II 131
Alendronate sodium, II 132
Alfentanil, II 136
Alfuzosin, II 138
Alitretinoin, V 21
Allantoin, CE 52, II 141
Allethrin, V 24
Allobarbital, CE 52
Allopurinol, CE 53, II 145
Alminoprofen, II 149
Almitrine, II 152
Almotriptan, V 27
Alosetron, CE 54, V 29
Alpidem, II 153
Alpiropride, II 156
Alprazolam, CE 55, I 48, II 158
Alprenolol, CE 57, II 160
Alprostadil, CE 62, II 166
Alteplase, I 56

Altretamine, II 181
Amantadine, CE 62, II 183
Ambenonium chloride, II 187
Amcinonide, V 30
Amdinocillin, II 189
Amfenac, II 190
Amidox, II 191
Amifostine, II 193
Amikacin, CE 64, II 195
Amiloride, CE 67, II 198
Amineptine, II 205
Amino acids, CE 68, II 206
Aminobenzoic acid, CE 88, II 306
Aminocaproic acid, II 315
Aminoglutethimide, CE 91, II 318
Aminohippuric acid, CE 93, II 321
Aminolevulinic acid, V 33
Aminophylline, II 328
Aminopyrine, CE 94
Amiodarone, CE 95, II 331
Amiprilose, II 340
Amisulpride, II 341
Amitraz, II 348
Amitriptyline, CE 97, I 58, II 349
Amlexanox, II 355
Amlodipine, CE 101, I 87, II 356
Amobarbital, CE 103, II 359
Amodiaquin, II 368
Amorolfine, CE 107
Amosulalol, II 370
Amoxapine, II 372
Amoxicillin, CE 109, I 91, II 382
Amphetamine, CE 112, II 388
Amphotericin, II 424
Ampicillin, CE 124, I 107, II 434
Amprenavir, V 36
Amprolium, II 441
Amrinone, II 446
Amsacrine, II 449
Amylase, II 451
Amylocaine, II 452
Anagrelide, V 42
Anakinra, V 43
Angiotensin II, CE 127, II 454
Aniracetam, II 465
Anistreplase, II 467
Antazoline, II 468
Anthralin, II 470
Antipyrine, CE 130, II 472
Antitrypsin, II 487
Apigenin, II 489
Apomorphine, II 490
Apraclonidine, V 45
Apramycin, II 491
Aprepitant, V 46
Aprindine, II 493
Aprobarbital, CE 133, II 494
Aprotinin, CE 135, II 498
Aranidipine, V 48
Arbaprostil, II 500
Arbekacin, CE 136
Arbidol, II 502
Arecoline, II 503
Argatroban, II 504
Arotinolol, CE 137, V 49
Arteether, V 52
Artemisinin, II 505
Artesunate, CE 138
Articaine, V 54
Ascorbic acid, CE 139, II 509
Asparaginase, V 57
Aspartame, CE 147, II 526
Aspirin, CE 149, I 122, II 529
Aspoxicillin, CE 154, II 532
Astemizole, I 137, II 534
Atazanavir sulfate, V 58
Atenolol, CE 155, I 140, II 535
Atipamezole, V 60
Atomoxetine hydrochloride, V 62
Atorvastatin, V 64
Atosiban, CE 170, V 66
Atovaquone, II 543
Atracurium besylate, II 545
Atropine, CE 171, I 164, II 547
Auranofin, II 553
Avermectin, II 555
Avitriptan, II 557
Azacyclonol, II 558
Azaperone, CE 176, II 560
Azasetron, II 563
Azatadine, II 564
Azathioprine, CE 177, I 177, II 566
Azelaic acid, II 568
Azelastine, CE 177, II 572
Azithromycin, I 181, II 575
Azlocillin, II 578
Azosemide, II 580
Aztreonam, II 582

Bacampicillin, II 586
Bacitracin, CE 180, II 588
Baclofen, CE 180, II 590
Balofloxacin, V 67
Bambermycins, V 69
Bambuterol, CE 183, II 596
Bamethan, CE 185, II 598
Bamifylline, II 600
Barbital, CE 187, II 601
Barnidipine, II 604
Beclobrate, II 607
Beclomethasone, CE 190, II 609
Beclomethasone dipropionate, I 184
Befunolol, II 614, V 70
Benactyzine, II 615
Benazepril, I 188, II 617
Bendroflumethiazide, CE 191, II 619
Benfluorex, II 627
Benidipine, II 628
Benoxaprofen, II 630
Benperidol, II 632

Benproperine, CE 194
Benserazide, CE 196
Benzalkonium chloride, CE 197, II 635, V 71
Benzbromarone, II 640
Benzethidine, II 641
Benzethonium chloride, II 643
Benzocaine, CE 199, II 645
Benzoctamine, II 655
Benzoic acid, CE 201, II 657
Benzonatate, II 676
Benzoyl peroxide, II 677
Benzphetamine, CE 208, II 679
Benzquinamide, CE 208, II 682
Benzthiazide, CE 209, II 684
Benztropine mesylate, II 688
Benzydamine, II 692
Benzyl alcohol, II 693
Benzyl benzoate, II 694
Bepridil, II 698
Berberine, II 702
Besipirdine, II 704
Beta-carotene, II 705
Betahistine, II 717
Betaine, II 718, V 72
Betamethasone, CE 210, I 191, II 720
Betaxolol, CE 212, II 723
Bethanechol chloride, V 74
Betotastine, II 729
Bevantolol, II 730
Bexarotene, V 75
Bezafibrate, II 732
BHT, CE 213
Biapenem, V 77
Bicalutamide, II 733
Bidisomide, II 734
Bifonazole, CE 214, II 735
Bimatoprost, V 79
Bioresmethrin, II 737, V 80
Biotin, CE 214, II 738
Biperiden, CE 217, II 744
Bisacodyl, II 746
Bisantrene, II 748
Bisoprolol, CE 219, II 752
Bitolterol, II 757
Bivalirudin, V 81
Bleomycin sulfate, II 759
Boldenone, II 763, V 82
Boldine, II 767
Bopindolol, II 768
Bosentan, V 83
β-Boswellic acid, V 86
Bretylium tosylate, CE 221, II 771
Brimonidine, V 88
Brodifacoum, II 772
Brodimoprim, CE 222, II 773
Bromadiolone, II 775
Bromazepam, CE 223, II 776
Bromfenac, V 90
Bromhexine, II 781
Bromocriptine, I 201, II 783
Bromodiphenhydramine, II 784
Bromperidol, II 786
Brompheniramine, CE 226, II 788
Brotizolam, CE 230, II 798
Brovincamine, V 92
Bucillamine, II 801, V 93
Bucladesine, II 802
Buclizine, CE 230, II 803
Bucumolol, II 807
Budesonide, II 809
Budipine, V 94
Buflomedil, II 816
Bufotenine, II 817
Bufuralol, II 819
Bulaquine, V 95
Bumadizon, II 820
Bumetanide, CE 231, II 822
Bunazosin, II 832
Bupivacaine, CE 233, II 833
Bupranolol, CE 244, II 851
Buprenorphine, II 854
Bupropion, CE 246, II 865
Buserelin, CE 247, II 868
Buspirone, CE 247, I 204, II 871
Busulfan, II 873
Butabarbital, CE 248, II 877
Butacaine, II 884, V 97
Butalbital, CE 250, I 211
Butamben, II 887, V 99
Butamirate, CE 253
Butethal, II 888
Butethamate, CE 254, II 890
Buthionine sulfoximine, CE 256, II 892
Butibufen, II 894
Butoconazole, V 100
Butorphanol, II 895
Butyl flufenamate, V 101
Butylscopolammonium bromide, CE 257

Cabergoline, II 897
Cadralazine, II 899
Caffeine, CE 257, II 901
Calcifediol, II 943
Calcipotriene, II 946
Calcitonin, CE 274, II 948
Calcitriol, II 952
Cambendazole, V 102
Camostat, II 955
Camphor, II 956
Camptothecin, II 958
Canadine, CE 274
Candesartan cilexetil, V 104
Candoxatril, II 960
Capecitabine, V 106
Capreomycin, CE 275
Capsaicin, II 962
Captan, CE 276, II 964
Captodiamine, II 966
Captopril, CE 276, I 215, II 967
Carazolol, CE 278, II 976

Carbadox, II 977
Carbamazepine, CE 280, I 224, II 984
Carbaryl, CE 284, II 991
Carbenicillin, CE 285, II 1006
Carbetapentane, II 1010
Carbidopa, CE 286, II 1011
Carbimazole, II 1016
Carbinoxamine, CE 286, II 1017
Carbisocaine, II 1021
Carboplatin, CE 288, I 252, II 1022
Carbromal, II 1025
Carbutamide, II 1027
Carbuterol, CE 288
Carisoprodol, I 255
Carmustine, II 1028
Carnitine, CE 290, II 1031
Carpipramine, II 1042
Carprofen, CE 291, II 1044
Carteolol, CE 295, II 1046
Carumonam, II 1053
Carvedilol, CE 298, II 1055
Casanthranol, V 108
Caspofungin, V 109
Castor oil, V 112
Cefaclor, I 257, II 1061
Cefadroxil, I 265, II 1062
Cefamandole, CE 302, II 1065
Cefatrizine, II 1069
Cefazolin, II 1070
Cefbuperazone, II 1081, V 113
Cefdinir, II 1082
Cefditoren, V 114
Cefepime, II 1084
Cefetamet, II 1089
Cefixime, CE 303, I 274, II 1090
Cefmenoxime, CE 304, II 1091
Cefmetazole, II 1095
Cefminox, CE 304, II 1098
Cefoaxime, I 278
Cefodizime, CE 305, II 1099
Cefonicid, II 1102
Cefoperazone, CE 305, II 1104
Ceforanide, II 1115
Cefoselis, V 116
Cefotaxime, CE 306, II 1116
Cefotetan, II 1118
Cefotiam, II 1121
Cefoxitin, II 1125
Cefozopran, V 117
Cefpimizole, CE 308, II 1133
Cefpiramide, CE 309, II 1135
Cefpirome, CE 310, II 1139
Cefpodoxime proxetil, II 1142
Cefprozil, I 289, II 1146
Ceftazidime, CE 311, I 291, II 1147
Cefteram, II 1151
Ceftibuten, II 1153
Ceftiofur, CE 313, II 1160
Ceftizoxime, CE 313, II 1165
Ceftriaxone, CE 314, I 303, II 1170
Cefuroxime, CE 315, I 311, II 1173
Cefuzonam, V 118
Celecoxib, V 119
Celiprolol, CE 317, II 1176
Centchroman, II 1183
Cephalexin, CE 320, I 317, II 1185
Cephaloridine, II 1189
Cephalosporin C, II 1193
Cephalothin, CE 321, II 1194
Cephapirin sodium, CE 322, II 1200
Cephradine, CE 323, II 1208
Cerivastatin, V 123
Ceronapril, CE 324
Cetirizine, II 1214
Cetrorelix, V 125
Cetyl alcohol, V 128
Cevimeline hydrochloride, V 130
Chelidonine, II 1219
Chenodiol, CE 325, II 1221
Chlophedianol, CE 326
Chloral hydrate, II 1237
Chlorambucil, CE 327, II 1239
Chloramphenicol, CE 328, II 1244
Chlorcyclizine, CE 329, II 1257
Chlordiazepoxide, CE 331, II 1261
Chlorguanide, II 1276
Chlorhexidine, II 1278
Chlormadinone, II 1283
Chlormezanone, CE, II 1284
Chlorobutanol, V 132
Chloroprocaine, II 1288, V 133
Chloropyramine, II 1290
Chloroquine, CE 333, II 1291
Chlorothiazide, CE 338, II 1303
Chloroxine, II 1310
Chloroxylenol, II 1311
Chlorphenesin, II 1314
Chlorpheniramine, CE 340, II 1317
Chlorphenoxamine, CE 352, II 1334
Chlorpromazine, CE 353, II 1336
Chlorpropamide, CE 355, II 1356
Chlorprothixene, CE 357, II 1364
Chlorpyrifos, CE 358, II 1370
Chlortetracycline, CE 358, II 1375
Chlorthalidone, CE 362, II 1392
Chlorzoxazone, II 1402
Cholic acid, CE 365, II 1408
Choline, CE 366, II 1411
Chorionic gonadotropin, V 134
Chymopapain, II 1423
Chymotrypsin, CE 368, II 1424
Cicletanine, CE 368, II 1425
Ciclopirox, II 1429
Cicloprofen, CE 371
Cicloprolol, II 1431
Ciclotropium bromide, II 1432
Cidofovir, II 1434
Cifenline, II 1436
Cilastatin, I 330
Cilazapril, II 1441

Cilnidipine, V 135
Cilostazol, II 1443
Cimetidine, CE 372, I 334, II 1445
Cimetropium bromide, V 136
Cinchonidine, II 1449
Cinchonine, II 1452
Cinnarizine, II 1454
Cinolazepam, II 1457
Cinoxacin, II 1458
Ciprofibrate, II 1460
Ciprofloxacin, CE 375, I 347, II 1463
Ciprostene, II 1474
Cirazoline, II 1475
Cisapride, CE 376, I 364, II 1476
Cisatracurium besylate, V 137
Cisplatin, CE 376, II 1478
Citalopram, II 1487
Citric acid, V 139
Cladribine, II 1492
Clarithromycin, CE 377, I 368, II 1495
Clavulanic acid, CE 378, I 371, II 1497
Clemastine, II 1498
Clenbuterol, CE 378, II 1501
Clentiazem, CE 383
Clidinium bromide, CE 384, II 1508
Clinafloxacin, II 1510
Clindamycin, CE 388, I 379, II 1512
Clioquinol, II 1514, V 142
Clobazam, CE 388, II 1516
Clobenzorex, II 1519
Clobetasol, CE 390
Clobetasol 17-propionate, V 143
Clobetasone, CE 391
Clobutinol, CE 391
Clocinizine, II 1521
Clodronic acid, II 1522
Clofazimine, II 1525
Clofibrate, II 1527
Clofibric acid, II 1529
Clomethiazole, II 1532
Clomiphene, CE 393, II 1533
Clomipramine, CE 394, II 1536
Clonazepam, CE 395, I 383, II 1556
Clonidine, II 1559
Clonixin, II 1568
Clopenthixol, CE 399
Clopidogrel, V 147
Clopidol, V 149
Clorazepate, CE 399, II 1570
Cloricromen, V 151
Clorprenaline, CE 400
Clorsulon, V 152
Closantel, II 1577
Clothiapine, II 1578
Clotiazepam, CE 401, II 1579
Clotrimazole, I 395
Cloxacillin, CE 402, II 1580
Cloxazolam, CE 404
Cloxyquin, II 1596
Clozapine, CE 406, I 400, II 1597
Cocaine, CE 407, II 1602
Codeine, CE 413, I 412, II 1628
Colchicine, II 1633
Colistin, II 1640, V 153
Copovithane, II 1642
Corticosterone, CE 424, II 1643
Corticotropin, CE 426, II 1644
Cortisone, CE 427, II 1646
Cortivazol, II 1665
Cotinine, II 1666
Coumachlor, CE 433
Coumaphos, CE 434, II 1668
Coumarin, II 1671
Cromolyn, I 436, II 1673
Crotamiton, II 1675
Cyamemazine, II 1677
Cyclamic acid, II 1679
Cyclandelate, II 1680
Cyclazocine, II 1682
Cyclizine, CE 435, II 1685
Cyclobarbital, CE 436
Cyclobenzaprine, I 439, II 1691
Cyclodrine, CE 437
Cycloguanil, II 1692
Cyclopentamine, II 1694
Cyclopentolate, CE 438, II 1696
Cyclophosphamide, II 1697
Cycloserine, II 1702
Cyclosporine, I 446, II 1704
Cyclothiazide, II 1708
Cyfluthrin, II 1712
Cyheptamide, II 1714
Cymarin, II 1716
Cypermethrin, II 1718, V 155
Cyproheptadine, II 1723
Cyromazine, II 1729
Cysteamine, II 1731
Cytarabine, CE 439, II 1736

Dacarbazine, II 1741
Dactinomycin, II 1746
Dalfopristin, V 156
Dalteparin, V 158
Danazol, II 1747
Danofloxacin, II 1752
Danthron, II 1754
Dantrolene, II 1756
Dapsone, CE 440, II 1759
Daptomycin, V 159
Daunorubicin, CE 441, II 1768
Debrisoquin, CE 442, II 1776
Decoquinate, II 1780
Deferiprone, V 161
Deferoxamine, II 1781
Defibrotide, II 1786
Deflazacort, II 1787, V 162
Dehydrocholic acid, II 1788
Delapril, II 1790
Delavirdine, II 1791
Demecarium bromide, II 1793

Demeclocycline, CE 443, II 1794
Demexiptiline, II 1802
Demoxepam, II 1804
Denopamine, CE 445, II 1805
Deoxycholic acid, CE 447, II 1807
Deserpidine, CE 448, II 1821
Desipramine, CE 448, II 1823
Desloratadine, V 164
Deslorelin, II 1860
Desmopressin, II 1861
Desogestrel, I 457, V 166
Desonide, II 1862
Desoximetasone, V 167
Desoxycorticosterone, V 169
Detomidine, II 1863
Dexamethasone, CE 451, II 1865
Dexpanthenol, II 1887
Dexrazoxane, V 172
Dextran, II 1889, V 174
Dextromethorphan, CE 456, II 1890
Dextromoramide, II 1907
Dextrose, CE 457, II 1910
Dextrothyroxine, II 1942
Dezocine, II 1944
Diacerein, V 176
Diamorphine, CE 469, II 1945
Diatrizoate sodium, II 1951
Diaveridine, II 1953
Diazepam, CE 471, I 458, II 1954
Diazinon, II 1960
Diazoxide, II 1965
Dibekacin, II 1968
Dibucaine, CE 478, II 1970
Dichloroacetic acid, II 1973, V 177
Dichlorophen, V 178
Dichlorphenamide, CE 478, II 1974
Dichlorvos, CE 479, II 1976
Diclazuril, V 179
Diclofenac sodium, CE 480, I 486, II 1979
Dicloxacillin, CE 481, II 1993
Dicumarol, II 2006
Dicyclomine, I 500
Didanosine, CE 483, II 2007
Didox, II 2016
Dienestrol, II 2017
Diethylcarbamazine, II 2019
Diethylpropion, II 2021
Diethylstilbestrol, II 2026
Difemerine, II 2033
Difenoxin, II 2034
Diflorasone, II 2035
Difloxacin, II 2036
Diflunisal, CE 483, II 2038
Digitoxin, CE 484, II 2045
Digoxin, CE 485, I 503, II 2052
Dihydralazine, II 2055
Dihydroartemisinin, II 2057
Dihydrocodeine, CE 486, II 2058
Dihydroergotamine, II 2062
Dihydrostreptomycin, CE 488, II 2066
Dihydrotachysterol, V 181
Dihydroxyacetone, II 2071
Diitiazem, I 517
Dilevalol, CE 489, II 2073
Diltiazem, CE 490, II 2075
Dimenhydrinate, CE 495, II 2079
Dimethindene, CE 496, II 2080
Dimethyl sulfoxide, V 183
Dimethylaminobenzoic acid, II 2085
Dimetridazole, II 2087
Dinitolmide, V 185
Dinoprost, CE 500, II 2088
Dinoprostone, CE 501, II 2110
Diphenadione, II 2140
Diphenhydramine, CE 502, II 2141
Diphenidol, II 2159
Diphenoxylate, CE 503, II 2161
Dipipanone, II 2165
Dipivefrin, CE 504, V 186
Diprenorphine, II 2167
Dipyridamole, II 2170
Dipyrone, CE 506, II 2178
Dirithromycin, II 2184
Disopyramide, CE 507, II 2186
Disulfiram, II 2194
Dithiazanine iodide, V 187
Diuron, II 2198
Dobutamine, CE 513, II 2199
Docarpamine, V 188
Docetaxel, CE 516, II 2203
Docusate sodium, II 2205
Dofetilide, V 189
Dolasetron, V 191
Domperidone, II 2206
Donepezil, V 193
Dopamine, CE 517, II 2207
Dopexamine, II 2235
Doramectin, II 2236
Dorzolamide hydrochloride, II 2238
Dothiepin, II 2242
Doxacurium chloride, II 2248
Doxapram, CE 523, II 2249
Doxazosin, I 531, II 2254
Doxefazepam, V 195
Doxepin, CE 524, II 2255
Doxercalciferol, V 196
Doxifluridine, II 2276
Doxofylline, II 2277
Doxorubicin, CE 526, II 2279
Doxycycline, CE 527, I 534, II 2291
Doxylamine, CE 531, II 2295
Drofenine, II 2302
Droloxifene, II 2303
Dronabinol, II 2305
Droperidol, II 2312
Dropropizine, II 2320, V 198
Drospirenone, V 199
Droxicam, V 200
Droxidopa, V 201
Duloxetine, II 2321

Dyclonine, II 2323
Dyphylline, CE 537, II 2324

Ebastine, III 1
Ebiratide, III 2
Ebrotidine, V 202
Econazole, CE 537, III 4
Edaravone, V 204
Edrophonium chloride, III 6
EDTA, CE 538, III 7, V 206
Efavirenz, V 208
Eflornithine, CE 540, III 10
Efonidipine, III 12
Efrotomycin, V 212
Egualen, V 213
Eletriptan, V 214
Eliprodil, III 13
Emedastine, III 14
Emetine, III 14
Emtricitabine, V 215
Enalapril, CE 541, I 543, III 15
Enalaprilat, CE 542, III 17
Encainide, III 19
Enilconazole, III 23
Enocitabine, III 24
Enoxacin, CE 542, III 24
Enoxaparin sodium, V 217
Enoximone, III 33
Enprostil, III 35
Enrofloxacin, III 36
Entacapone, V 218
Eperisone, CE 543, V 220
Ephedrine, CE 544, III 42
Epicillin, III 57
Epidermal growth factor, III 57
Epinastine, CE 560, III 60
Epinephrine, CE 562, III 61
Epinine, III 83
Epirizole, III 84
Epirubicin, CE 572, III 84
Epithiazide, CE 572
Eplerenone, V 222
Epoetin, CE 573, I 546, III 90
Epoprostenol, CE 574, III 90, V 224
Eprinomectin, III 91
Eprosartan, V 225
Eptastigmine, III 92
Eptazocine, V 227
Eptifibatide, V 228
Equilenin, CE 575
Equilin, III 93
Erdosteine, V 229
Ergocornine, III 94
Ergocristine, III 95
Ergocristinine, III 96
Ergocryptine, III 97
Ergoloid mesylates, III 98
Ergonovine, CE 576, III 99
Ergosine, III 104
Ergotamine, CE 576, III 105, V 230
Ertapenem, V 234
Erythrityl tetranitrate, III 108
Erythromycin, CE 578, I 549, III 109
Esculin, III 111
Esmolol, III 112
Estazolam, CE 580, III 114
Estradiol, CE 583, I 557, III 118
Estramustine, III 121
Estriol, III 123
Estrogens, conjugated, CE 586, I 569, III 124
Estrone, CE 587, III 128
Ethacrynic acid, CE 589, III 135
Ethambutol, III 141
Ethaverine, III 144
Ethchlorvynol, III 146
Ethenzamide, CE 590
Ethinyl estradiol, CE 591, I 579, III 147
Ethionamide, III 150
Ethoheptazine, CE 592
Ethopabate, III 151, V 236
Ethopropazine, CE 593, III 152
Ethosuximide, III 155, CE 594
Ethotoin, III 164
Ethyl icosapentate, V 237
Ethylmorphine, III 166
Ethynodiol diacetate, I 588
Ethynodiol, CE 596, III 166
Etidocaine, III 167
Etidronic acid, III 169
Etilefrin, CE 597
Etizolam, CE 599, III 170
Etodolac, CE 601, I 590, III 172
Etonitazene, III 173
Etonogestrel, V 238
Etoposide, I 597, III 174
Etoricoxib, V 240
Etorphine, III 177, V 242
Etoxeridine, III 179
Etretinate, III 181
Eugenol, III 184
Exemestane, V 243
Exifone, III 185
Ezetimibe, V 245

Factor VIII, III 186
Fadrozole, V 247
Falecalcitriol, V 248
Famciclovir, III 187
Famotidine, CE 603, I 605, III 189
Febantel, III 190
Felbamate, CE 603, III 191
Felbinac, III 196
Felodipine, III 199
Felypressin, III 203
Fenbendazole, III 204
Fenbufen, III 212
Fenbutrazate, III 216
Fencamfamine, III 217
Fenchone, III 219
Fendiline, CE 604

Fenethazine, III 219
Fenfluramine, CE 606, III 220
Fenofibrate, III 228
Fenoldopam, III 229
Fenoprofen, CE 608, III 230
Fenoterol, CE 614, III 242
Fenoverine, III 243
Fenoxycarb, V 250
Fenozolone, III 244
Fenproporex, III 245
Fenprostalene, III 246
Fentanyl, CE 615, III 248
Fenthion, III 256
Fentiazac, III 258
Fenticonazole, III 259, V 251
Feprazone, III 260
Fexofenadine, V 253
Fibrinolysin, III 261
Finasteride, I 611, III 262
Flavoxate, III 263
Flecainide, CE 616, III 265
Fleroxacin, CE 618, III 271
Flezelastine, III 278
Floctafenine, III 278
Flomoxef, V 257
Florfenicol, V 258
Flosequinan, III 280
Floxacillin, III 282
Floxuridine, III 283
Flubendazole, III 289
Fluconazole, CE 618, I 615, III 291
Flucytosine, CE 619, III 293
Fludarabine, III 297
Fludeoxyglucose F18, III 299
Fludiazepam, CE 620
Fludrocortisone, CE 622, III 301, V 260
Flufenamic acid, CE 623, III 303
Fluindione, III 305
Flumazenil, III 306
Flumequine, CE 624, III 310
Flumethasone, CE 624, III 312
Flunarizine, III 315
Flunisolide, III 316
Flunitrazepam, CE 625, III 317
Flunixin, CE 628, III 321
Flunoxaprofen, III 325
Fluocinolone acetonide, CE 628, III 325
Fluocinonide, CE 630, III 329
Fluocortolone, CE 631, III 330
Fluorometholone, III 330
Fluorouracil, III 333
Fluoxetine, CE 632, I 621, III 347
Fluoxymesterone, III 351
Flupentixol, CE 633, III 357
Fluphenazine, CE 634, III 358
Flupirtine, III 366
Fluprostenol, V 262
Flurandrenolide, V 264
Flurazepam, CE 634, III 367
Flurbiprofen, CE 638, III 375
Flurithromycin, V 267
Flurogestone acetate, V 268
Flutamide, I 633, III 395
Flutazolam, III 396
Fluticasone propionate, CE 648, V 270
Flutoprazepam, III 397
Flutrimazole, III 398, V 273
Fluvastatin, I 634, III 398
Fluvoxamine, CE 649, III 399
Folic acid, CE 650, III 409
Fomepizole, V 274
Fomivirsen, V 276
Fonazine, III 414
Fondaparinux, V 277
Formestane, V 278
Formoterol, CE 650, III 415, V 279
Fosamprenavir calcium, V 281
Foscarnet sodium, III 417
Fosfosal, III 418
Fosinopril, CE 652, III 419, V 283
Fosphenytoin, III 420, V 284
Fotemustine, III 421
Frovatriptan, V 286
Fructose, III 422
Fumagillin, V 288
Furaltadone, III 423
Furazolidone, III 424
Furosemide, CE 652, I 637, III 434
Furprofen, III 440
Fusidic acid, III 440

Gabapentin, CE 655, III 442
Gadodiamide, III 446
Gadopentetic acid, III 447
Galantamine, V 290
Gallamine triethiodide, III 447
Gallopamil, CE 656, III 449
Ganciclovir, III 452
Ganirelix, V 292
Gatifloxacin, V 293
Gefitinib, V 295
Gemcitabine, V 296
Gemeprost, III 454
Gemfibrozil, I 654
Gemifloxacin, V 298
Gentamicin, CE 659, III 455
Gentian violet, III 462
Gentisic acid, III 466
Gestodene, V 300
Gestrinone, V 301
Glafenine, III 467
Glibornuride, III 468
Gliclazide, III 469
Glimepiride, III 470
Glipizide, CE 660, I 659, III 471
Glucagon, III 472
Gluconolactone, III 472
Glutethimide, CE 663, III 473
Glyburide, CE 664, I 665, III 480
Glycerin, CE 667, III 483, V 302

Glycopyrrolate, III 487
Gonadorelin, CE 668, III 488
Gonadotropin, III 494
Goserelin, CE 669, I 672, III 495
Gossypol, III 495
Gramicidin, III 496
Granisetron, III 497
Grepafloxacin, III 503
Griseofulvin, III 503
Guaiacol, III 506
Guaifenesin, CE 669, I 673, III 508
Guanabenz, III 509, V 304
Guanadrel, V 305
Guanethidine, III 511
Guanfacine, III 512
Gusperimus, III 513

Halazepam, CE 672, III 515
Halcinonide, III 518
Halobetasol propionate, V 306
Halofantrine, CE 673, III 518
Halofuginone, V 308
Haloperidol, CE 674, III 525
Halothane, III 543
Haloxazolam, CE 677, III 543
Heparin, CE 678, III 544
Heptaminol, CE 680
Hesperidin, III 547
Hetacillin, III 551
Hetastarch, V 310
Hexachlorophene, III 553
Hexobarbital, CE 681, III 553
Hirudin, CE 682
Histamine, CE 683, III 554
Histrelin, III 571
Homatropine, CE 686, III 572
Hyaluronic acid, CE 691, III 575
Hydralazine, III 577
Hydrochiorothiazide, I 682, CE 693, III 583
Hydrocodone, I 699, III 586
Hydrocortisone, CE 695, I 705, III 587
Hydroflumethiazide, CE 707, III 592
Hydrogen peroxide, III 597
Hydromorphone, CE 708, III 598
Hydroquinone, CE 708, III 605, V 311
Hydroxychloroquine, CE 709, III 609
Hydroxyprogesterone, CE 710, III 615
Hydroxypropyl methylcellulose, III 623
Hydroxyquinoline, III 624
Hydroxyurea, III 625
Hydroxyzine, CE 713, III 626
Hygromycin B, V 312

Ibafloxacin, III 633
Ibogaine, III 634
Ibopamine, III 636
Ibufenac, III 638
Ibuprofen, CE 714, I 741, III 638
Ibutilide, III 646
Idarubicin, CE 724, III 647
Idazoxan, CE 724
Idebenone, III 648
Idoxuridine, III 649
Ifosfamide, III 652
Iloprost, III 654, V 313
Imatinib, V 314
Imidapril, III 655
Imidocarb, V 316
Imipenem, I 769, III 657
Imipramine, CE 725, III 658
Indalpine, III 689
Indapamide, CE 729, I 774, III 691
Indeloxazine hydrochloride, III 692
Indinavir, III 692
Indobufen, III 695
Indocyanine green, III 698
Indomethacin, CE 730, III 699
Indoprofen, CE 734
Indoramin, III 718
Inositol, III 719
Insulin, CE 738, I 778, III 719
Interferon, CE 741, I 787, III 721
Iobenguane, V 318
Iodinated glycerol, III 723
Iodixanol, V 320
Iodochlorhydroxyquin, III 723
Iodoquinol, III 726
Iofratol, III 727
Iohexol, CE 743, III 728
Iopamidol, CE 744, III 729
Iopanoic acid, CE 746, V 322
Iopromide, V 324
Iothalamate, III 730
Iothalamic acid, CE 746
Ioversol, V 326
Ipecac, III 735
Ipratropium bromide, CE 749, I 791, V 327
Ipriflavone, III 738, V 328
Iprindole, III 739
Iproniazid, III 740
Irbesartan, III 741
Irinotecan, III 742
Isepamicin, III 747
Isocarboxazid, CE 751, III 750
Isoetharine, III 752
Isofezolac, III 753
Isoflupredone, III 753, V 329
Isoflurane, III 754
Isoniazid, CE 752, III 755
Isopropamide iodide, V 330
Isoproterenol, CE 753, III 763
Isosorbide dinitrate, III 772
Isosorbide mononitrate, III 776
Isothipendyl, CE 763, III 778
Isoxicam, III 780
Isoxsuprine, CE 765, III 785
Isradipine, III 789
Itopride, V 332
Itraconazole, CE 767, III 792
Ivermectin, III 799

Josamycin, CE 768, III 809

Kanamycin, CE 768, III 812
Ketamine, CE 769, III 815
Ketanserin, III 822
Ketoconazole, CE 776, I 792, III 824
Ketoprofen, CE 777, I 798, III 826
Ketorolac, I 819, III 833
Ketotifen, III 834
Kinetin, V 333

Labetalol, CE 787, III 835
Lacidipine, III 843
Lactulose, III 844
Lafutidine, V 334
Lamivudine, CE 792, III 848, V 335
Lamotrigine, CE 794, III 849
Lansoprazole, CE 794, III 855
Lasalocid A, III 858
Latanoprost, V 339
Lazabemide, III 862
Leflunomide, V 341
Lenampicillin, III 863
Lercanidipine, V 343
Letrozole, III 864, V 345
Leucovorin, CE 795, III 864
Leuprolide, I 829
Levallorphan, CE 795, III 869
Levamisole, CE 796, III 870
Levetiracetam, V 346
Levobunolol, CE 797, III 874
Levodopa, CE 798, III 875
Levonordefrin, III 887, V 348
Levorphanol, CE 802, III 888
Levosimendan, V 349
Levothyroxine sodium, CE 804, III 896
Levothyroxine, I 831
Lidamidine, V 351
Lidocaine, CE 805, III 901
Lidoflazine, III 920
Limaprost, III 921
Lincomycin, CE 809, III 922, V 352
Lindane, V 354
Linezolid, III 924, V 355
Linoleic acid, III 924
Linuron, III 955
Liothyronine, CE 810, III 956, V 357
Lisinopril, CE 811, I 836, III 961
Lisofylline, III 962
Lisuride, III 962
Litracen, CE 812
Lobeline, III 963
Lobenzarit, III 964
Lodamide, III 1722
Lodoxamide, III 965
Lofepramine, III 966
Lomefloxacin, CE 813, III 967
Lomerizine, V 358
Lometrexol, III 971
Lomustine, III 972
Lonidamine, III 973
Loperamide, III 975
Lopinavir, V 359
Loprazolam, III 977
Loracarbef, I 839
Loratadine, I 841, III 979
Lorazepam, CE 813, I 844, III 979
Lormetazepam, III 982
Lornoxicam, III 983
Losartan, III 984
Loteprednol etabonate, V 362
Lovastatin, I 858
Loxapine, III 990
Loxoprofen, III 999
Lufenuron, III 1004
Lypressin, CE 816, III 1005
Lysozyme, CE 816, III 1007

Mabuterol, III 1013
Maduramicin, III 1014
Mafenide, III 1014
Malachite green, III 1015
Malotilate, III 1017
Manidipine, III 1018
Mannitol, CE 821, III 1019
Maprotiline, CE 821, III 1025
Marbofloxacin, V 364
Masoprocol, V 367
Maxacalcitol, V 368
Mazindol, III 1039
Mebendazole, III 1042
Mecamylamine, III 1050
Mechlorethamine, III 1051
Meclizine, CE 823, III 1053
Meclocycline, CE 825, III 1057
Meclofenamic acid, CE 825, III 1058
Meclofenoxate, III 1061
Medazepam, CE 826, III 1063
Medetomidine, V 369
Medifoxamine, III 1066
Medroxyprogesterone, III 1069
Medroxyprogesterone acetate, I 862
Mefenamic acid, CE 828, III 1072
Mefenorex, III 1083
Mefloquine hydrochloride, CE 828, III 1083
Megazol, III 1089
Megestrol acetate, III 1090
Meglutol, III 1093, V 371
Melatonin, V 372
Melengestrol acetate, V 375
Melitracen, III 1094
Meloxicam, III 1094
Melphalan, CE 833, III 1095
Memantine, V 378
Menadione, III 1102
Menotropins, III 1106
Menthol, III 1107, V 380
Mepazine, III 1108
Mepenzolate bromide, CE 833, III 1109, V 381

Meperidine, CE 835, III 1110
Mephenesin, III 1119
Mephentermine, III 1121
Mephenytoin, CE 837, III 1125
Mephobarbital, CE 837, III 1131
Mepindolol, CE 841
Mepivacaine, CE 843, III 1138
Mepixanox, V 383
Meprobamate, III 1145
Meptazinol, III 1146
Mequinol, V 384
Mequitazine, CE 848
Mercaptobenzothiazole, CE 850, III 1149
Mercaptopurine, III 1151
Meropenem, III 1159
Mesalamine, CE 851, III 1163
Mescaline, III 1168
Mesna, III 1169
Mesoridazine, CE 853, III 1172
Mestranol, CE 854, III 1181
Metaclazepam, CE 854
Metapramine, III 1184
Metaproterenol, CE 856, III 1188
Metaraminol, III 1192
Metformin, CE 861, III 1194
Methacholine chloride, CE 862, III 1200
Methacycline, CE 863, III 1201
Methadone, CE 863, III 1201
Methamphetamine, CE 868, III 1216
Methaphenilene, CE 876
Methapyrilene, III 1241
Methaqualone, CE 877, III 1243
Metharbital, CE 877
Methazolamide, III 1245
Methdilazine, III 1247
Methenamine, V 385
Methicillin sodium, III 1249
Methimazole, III 1255
Methocarbamol, III 1256
Methohexital, III 1262
Methoprene, III 1266, V 386
Methotrexate, CE 879, III 1266
Methotrimeprazine, CE 881, III 1278
Methoxamine, CE 882, III 1285
Methoxsalen, III 1290
Methoxychlor, V 387
Methoxyphenamine, CE 882, III 1292
Methoxypromazine, III 1293
Methscopolamine bromide, III 1294
Methsuximide, III 1295
Methyclothiazide, CE 883, III 1299
Methyl salicylate, III 1301
Methyldopa, CE 884, III 1304
Methyldopate, III 1309
Methylene blue, CE 884, III 1310
Methylephedrine, III 1313
Methylergonovine, III 1314
Methylphenidate, CE 885, I 867
Methylprednisolone, CE 886, I 874, III 1317
Methyltestosterone, CE 888, III 1319, V 392
Methylthiouracil, CE 889
Methyprylon, III 1327
Methysergide, III 1328
Metiamide, III 1330
Metipranolol, CE 890, III 1331
Metoclopramide, III 1332
Metocurine iodide, III 1341
Metolazone, CE 892, III 1342
Metopimazine, III 1346
Metoprolol, CE 893, I 887, III 1347
Metrizamide, V 393
Metronidazole, III 1354
Metyrapone, CE 904, III 1363
Metyrosine, V 394
Mexiletine, CE 904, III 1365
Mezlocillin, III 1376
Mianserin, CE 906, III 1380
Mibefradil, III 1384
Mibolerone, III 1384
Micafungin, V 396
Miconazole, CE 909, III 1386
Midazolam, I 911, III 1390
Midodrine, CE 909
Mifepristone, III 1395
Milnacipran, V 398
Miloxacin, III 1396
Milrinone, III 1397
Miltefosine, III 1399
Minaprine, III 1400
Minocycline, CE 910, III 1401
Minoxidil, III 1407
Miokamycin, III 1409
Mirtazapine, V 400, III 1410
Misoprostol, III 1411, V 405
Mitoguazone, III 1412
Mitolactol, III 1412
Mitomycin, III 1413
Mitotane, III 1419
Mitoxantrone, III 1421
Mivacurium chloride, III 1424
Mizolastine, V 406
Mizoribine, III 1426
Moclobemide, CE 912, III 1427
Modafinil, III 1432
Moexipril, V 411
Mofezolac, V 412
Molindone, CE 913, III 1433
Molsidomine, III 1433
Mometasone furoate, I 924, V 413
Monensin, III 1434, V 415
Montelukast, III 1441
Moperone, III 1443
Moprolol, III 1445
Morantel, III 1445, V 416
Morazone, III 1447
Moricizine, III 1448
Moroxydine, III 1449
Morphine, CE 913, III 1450
Morsuximide, III 1483
Mosapride, V 417

Moxidectin, III 1484
Moxifloxacin, V 420
Moxonidine, III 1485, V 423
Mupirocin, I 925
Muzolimine, III 1485
Mycophenolic acid, III 1486

Nabumetone, I 927, III 1488
Nadifloxacin, V 424
Nadolol, CE 923, III 1488
Nadoxolol, III 1501
Nafarelin, CE 928, III 1502
Nafcillin, CE 929, III 1503
Nafronyl, CE 930, III 1511
Naftifine, III 1512
Naftopidil, V 425
Nalbuphine, CE 932, III 1513
Nalidixic acid, CE 933, III 1519
Nalmefene, III 1525
Nalorphine, CE 933, III 1526
Naloxone, III 1537
Naltrexone, III 1546
Nandrolone, CE 936, III 1552, V 427
Naphazoline, CE 937, III 1557
Naproxen, CE 938, I 929, III 1565
Narasin, III 1570, V 429
Naratriptan, III 1573
Nartograstim, V 430
Natamycin, III 1574
Nateglinide, V 431
Nebivolol, V 433
Nedocromil, CE 947, III 1576
Nefazodone, III 1577
Nefopam, CE 948, III 1579
Nelfinavir, III 1581, V 435
Nemonapride, III 1582
Neomycin, CE 951, I 951, III 1583
Neostigmine bromide, CE 952, III 1584
Nequinate, V 440
Neridronic acid, V 441
Netilmicin, CE 953, III 1587
Nevirapine, V 443
Niacin, CE 954, III 1591
Niacinamide, CE 959, III 1598
Nialamide, III 1605
Niaprazine, III 1606
Nicarbazin, III 1607, V 447
Nicardipine, CE 965, III 1610
Nicergoline, III 1616
Niclosamide, III 1616
Nicorandil, III 1618
Nicotine, CE 968, III 1620
Nifedipine, CE 971, I 958, III 1630
Nifenalol, CE 972, III 1633
Niflumic acid, III 1634
Nifuroxazide, III 1637
Nikethamide, CE 973, III 1637
Nilutamide, III 1638, V 448
Nilvadipine, III 1639
Nimesulide, III 1640
Nimetazepam, CE 947
Nimodipine, CE 975, III 1641
Nipradilol, V 449
Nisin, CE 976
Nisoldipine, CE 976, III 1646
Nitazoxanide, V 450
Nitenpyram, V 452
Nitrazepam, CE 977, III 1649
Nitrendipine, CE 981, III 1654
Nitrofurantoin, I 968, III 1660
Nitrofurazone, III 1661
Nitroglycerin, I 973
Nizatidine, I 979, III 1667
Nomegestrol, V 453
Nomifensine, III 1668
Nonoxynol-9, III 1671, V 454
Nordazepam, CE 982
Norepinephrine, CE 982, III 1672
Norethindrone, CE 992, I 984, III 1697
Norethynodrel, CE 994, III 1698
Norfenefrine, CE 994
Norfloxacin, CE 996, III 1698
Norgestimate, CE 997, III 1711
Norgestrel, CE 998, I 993, III 1713
Norpseudoephedrine, CE 999
Nortriptyline, CE 1000, I 1000, III 1715
Noscapine, CE 1003, III 1720
Novobiocin, III 1723
Nylidrin, CE 1004, III 1726
Nystatin, CE 1005, III 1728, V 455

Octhilinone, III 1730
Octocrylene, V 456
Octopamine, CE 1006, III 1731
Octreotide, III 1732
Octyl methoxycinnamate, III 1734
Odapipam, III 1735
Ofloxacin, CE 1007, I 1029, III 1735
Olanzapine, III 1741
Olaquindox, III 1743
Oleandomycin, CE 1012, III 1744
Oleic acid, V 457
Olmesartan, V 469
Olopatadine, V 470
Olsalazine, III 1744
Omeprazole, CE 1012, I 1044, III 1745
Ondansetron, CE 1013, I 1052, III 1747
Ontazolast, III 1749
Opipramol, CE 1014, III 1749
Orbifloxacin, V 472
Orlistat, V 475
Ormetoprim, CE 1015, III 1752
Ornidazole, CE 1016
Orphenadrine, CE 1018, III 1753
Oryzalin, III 1756
Oseltamivir, V 477
Oxacillin, CE 1021, III 1757
Oxaliplatin, V 479
Oxamniquine, CE 1023, III 1768
Oxandrolone, III 1769

Oxaprozin, I 1062
Oxatomide, III 1770
Oxazepam, CE 1023, III 1771
Oxazolam, III 1793
Oxcarbazepine, III 1793
Oxeladin, III 1798
Oxfendazole, III 1799
Oxibendazole, III 1805
Oxiconazole, V 481
Oxiracetam, III 1806
Oxolinic acid, CE 1027, III 1808
Oxomemazine, CE 1028
Oxprenolol, CE 1029, III 1811
Oxybenzone, III 1817
Oxybutynin chloride, CE 1034, III 1819
Oxyclozanide, III 1819
Oxycodone, I 1065, III 1820
Oxymetazoline, CE 1035, III 1821
Oxymetholone, III 1825
Oxymorphone, III 1826
Oxyphenbutazone, III 1828
Oxyphencyclimine, CE 1036
Oxyquinoline, III 1829
Oxytetracycline, CE 1037, III 1830
Oxytocin, CE 1043, III 1850
Ozagrel, III 1855

Paclitaxel, CE 1045, I 1072, IV 1
Padimate O, IV 6
Palinavir, IV 6
Pamidronate, IV 7
Pamoic acid, CE 1047
Pancreatin, IV 12
Pancuronium bromide, IV 13
Panipenem, V 483
Panomifene, IV 14
Pantoprazole, CE 1047, IV 15
Pantothenic acid, CE 1048, IV 17
Papaverine, CE 1050, IV 20
Paramethadione, IV 29
Paramethasone, IV 30
Parathyroid hormone, IV 31
Parbendazole, IV 32
Parconazole, IV 33
Parecoxib, V 484
Pargyline, IV 34
Paricalcitol, V 485
Paromomycin, CE 1054, IV 36
Paroxetine, IV 39
Pazufloxacin, V 487
Pefloxacin, CE 1056, IV 44
Pemirolast, IV 51
Pemoline, CE 1056, IV 51
Penbutolol, CE 1057, IV 55
Penciclovir, CE 1057, IV 62, V 488
Penclomedine, IV 63
Penfluridol, IV 64
Penicillamine, IV 66
Penicillin G, CE 1058, IV 76
Penicillin V, CE 1061, I 1085, IV 95
Pentaerythritol tetranitrate, IV 98, V 490
Pentamidine, CE 1062, IV 100
Pentazocine, CE 1062, IV 104
Penthienate bromide, IV 115
Pentobarbital, CE 1064, IV 116
Pentosan polysulfate, V 491
Pentostatin, IV 124
Pentoxifylline, CE 1071, I 1096, IV 130
Perflubron, V 492
Pergolide, IV 131
Perhexiline, IV 132
Pericyazine, IV 133
Perindopril, IV 134
Permethrin, IV 135
Perospirone, V 493
Perphenazine, CE 1071, IV 139
Phenacemide, IV 146
Phenacetin, CE 1073, IV 147
Phenadoxone, IV 149
Phenampromide, IV 150
Phenazocine, IV 151
Phenazopyridine hydrochloride, CE 1075, IV 154, V 494
Phencyclidine, IV 156
Phendimetrazine, CE 1075, IV 159
Phenelzine, IV 161
Phenformin, CE 1076, IV 166
Phenglutarimide, IV 166
Phenindamine, IV 167
Phenindione, IV 169
Pheniramine, CE 1077, IV 170
Phenmetrazine, CE 1080, IV 178
Phenobarbital, CE 1081, IV 181
Phenol, CE 1091, IV 205
Phenolphthalein, IV 224
Phenomorphan, IV 228
Phenoperidine, IV 229
Phenothiazine, CE 1094, IV 230
Phenoxybenzamine, CE 1095, IV 235
Phenprocoumon, CE 1098
Phensuximide, CE 1098, IV 237
Phentermine, CE 1099, IV 239, V 497
Phentolamine, IV 245
Phenyl salicylate, IV 248
Phenylbutazone, IV 249
Phenylephrine, CE 1100, IV 258
Phenylpropanolamine, CE 1102, I 1103, IV 266
Phenyltoloxamine, IV 267
Phenytoin, CE 1108, I 1117, IV 270
Phloroglucinol, IV 274
Pholcodine, IV 275
Pholedrine, CE 1111
Phosmet, IV 276
Phosphatidylcholine, IV 277, V 501
Phosphatidylglycerol, V 504
Physostigmine, CE 1114, IV 284
Picotamide, IV 289
Picroside, IV 290
Picumeterol, CE 1114
Pidotimod, IV 290

Piketoprofen, V 508
Pilocarpine, CE 1115, IV 291
Pilsicainide, V 509
Piminodine, IV 295
Pimobendan, IV 296
Pimozide, CE 1116, IV 298
Pinacidil, IV 303
Pinaverium bromide, IV 306
Pindolol, CE 1117, IV 307
Pioglitazone, V 511
Pipamazine, IV 323
Pipamperone, IV 324
Pipazethate, IV 326
Pipecuronium bromide, IV 327
Pipemidic acid, CE 1134, IV 328
Piperacetazine, IV 330
Piperacillin, CE 1134, IV 331
Piperazine, IV 340
Pipercuronium bromide, V 514
Piperidolate, IV 341
Piperine, IV 342
Piperocaine, IV 343
Piperonyl butoxide, IV 344
Piperoxan, CE 1136
Pipotiazine, IV 347
Pipradrol, IV 348
Piracetam, CE 1137, IV 349
Pirarubicin, IV 350
Pirbuterol, CE 1137
Pirenzepine, IV 352
Piretanide, CE 1140, IV 353
Piritramide, IV 354
Pirlimycin, V 515
Pirmenol, IV 355
Piromidic acid, CE 1141, IV 358
Piroxicam, CE 1141, I 1140, IV 359
Pivampicillin, IV 361
Pizotyline, IV 362
Podofilox, IV 364
Poloxalene, V 518
Polyethylene glycol, CE 1144
Polymyxin, I 1150, IV 365
Polythiazide, IV 367
Porfiromycin, IV 369
Practolol, IV 370
Pralidoxime chloride, IV 371
Pramipexole, V 519
Pramiracetam, IV 373
Pramlintide, IV 374
Pramoxine, IV 375
Pranlukast, V 521
Pranoprofen, CE 1144
Pravastatin sodium, I 1153, IV 378
Prazepam, CE 1145, IV 380
Praziquantel, CE 1146, IV 390
Prazosin, CE 1147, IV 393
Prednicarbate, IV 402, V 523
Prednisolone, CE 1147, IV 403
Prednisone, CE 1150, I 1156, IV 421
Premafloxacin, IV 424
Prenalterol, CE 1151
Prenylamine, IV 424
Prifinium bromide, IV 425
Prilocaine, CE 1152, IV 426
Primaquine, CE 1157, IV 432
Primidone, CE 1161, IV 437
Pristinamycin, IV 448
Proadifen, IV 449
Probenecid, CE 1164, IV 450
Probucol, IV 459
Procainamide, CE 1165, IV 460
Procaine, CE 1168, IV 469
Procarbazine, IV 478
Prochlorperazine, CE 1172, IV 481
Procyclidine, CE 1173, IV 488
Progabide, IV 490
Progesterone, CE 1175, IV 492
Proglumide, CE 1179
Proheptazine, IV 505
Prolintane, CE 1180, IV 506
Promazine, CE 1180, IV 507
Promethazine, CE 1182, I 1171, IV 515
Pronethalol, IV 517
Propafenone, CE 1187, IV 518
Propantheline bromide, IV 529
Proparacaine, IV 532
Propentofylline, CE 1190, IV 533
Properidine, IV 534
Propicillin, IV 535
Propiomazine, CE 1190, IV 536
Propionylpromazine, V 524
Propiverine, IV 541
Propofol, I 1185, IV 542
Propoxur, CE 1191, IV 545
Propoxycaine hydrochloride, V 527
Propoxyphene, CE 1191, I 1189, IV 551
Propranolol, CE 1193, IV 555
Propylene glycol, IV 602, V 528
Propylhexedrine, IV 605, V 529
Propylparaben, IV 606
Propylthiouracil, CE 1213, IV 607
Propyphenazone, CE 1214
Proscillaridin, IV 609
Prospidium chloride, CE 1214
Protamine sulfate, IV 609
Prothipendyl, IV 611
Protionamide, IV 612
Protirelin, CE 1215, IV 613, V 530
Protriptyline, CE 1216, IV 614
Prulifloxacin, V 532
Pseudoephedrine, CE 1219, IV 627
Psilocybin, IV 636
Pumactant, CE 1226, IV 637
Puromycin, IV 639
Pyrantel, IV 640
Pyrazinamide, IV 644
Pyrazoloacridine, IV 647
Pyrethrins, V 533
Pyridostigmine bromide, IV 648
Pyridoxal, CE 1227, IV 651

Pyridoxamine dihydrochloride, CE 1228, IV 656
Pyridoxine, CE 1230, IV 658
Pyrilamine, CE 1235, IV 667
Pyrimethamine, CE 1236, IV 675
Pyrisuccideanol, IV 683
Pyrithione, IV 684
Pyrithyldione, IV 684
Pyrocatechol, IV 685
Pyrrobutamine, CE 1238

Quazepam, IV 686
Quercetin, IV 689
Quetiapine, V 536
Quinacrine, CE 1238, IV 690
Quinagolide, CE 1239
Quinaldic acid, IV 692
Quinapril, I 1202, IV 693
Quinestrol, IV 694
Quinethazone, IV 695
Quinfamide, V 539
Quinidine, CE 1240, IV 696
Quinine, CE 1243, IV 710
Quinupramine, IV 723
Quinupristin, V 540

Rabeprazole, V 542
Ractopamine, V 544
Raloxifene, V 547
Ramatroban, IV 725
Ramipril, I 1205, IV 726
Ramosetron, V 549
Ranitidine, CE 1246, I 1209, IV 727
Rapacuronium bromide, V 551
Rebamipide, IV 731
Reboxetine, IV 731
Remifentanil, V 553
Remikiren, IV 733
Remoxipride, CE 1251, IV 733
Repaglinide, V 555
Repirinast, IV 736
Reproterol, CE 1253, IV 737
Rescinnamine, IV 738
Reserpine, CE 1255, IV 740
Resorcinol, CE 1256, IV 748
Retinoic acid, CE 1256, I 1223, IV 753
Ribavirin, IV 757
Riboflavin, CE 1258, IV 760
Ricinoleic acid, IV 769, V 557
Rifabutin, IV 769
Rifampin, IV 773
Rifapentine, IV 782
Rifaximin, IV 784, V 558
Rilmazafone, IV 785, V 559
Rilmenidine, IV 786
Riluzole, IV 786
Rimantadine, CE 1262, IV 787
Risedronate sodium, V 560
Risperidone, IV 789
Ritodrine, CE 1263, IV 793
Ritonavir, IV 795
Rizatriptan, IV 798, V 561
Robenidine, IV 799
Rocuronium bromide, IV 800
Rofecoxib, V 562
Rokitamycin, IV 802
Romurtide, IV 803
Ronidazole, IV 803
Ropinirole, IV 804, V 565
Ropivacaine, CE 1264, IV 805
Rosiglitazone, V 566
Rosoxacin, CE 1265, IV 806
Rosuvastatin calcium, V 568
Rotenone, IV 807
Roxarsone, IV 809
Roxatidine acetate, IV 810
Roxithromycin, IV 811
Rufloxacin, IV 814
Rutin, CE 1266, IV 817

Saccharin, CE 1267, IV 818
Salicylamide, CE 1268, IV 825
Salicylic acid, CE 1270, IV 830
Salinomycin, IV 856
Salmeterol, I 1234
Salsalate, IV 859
Sampatrilat, IV 860
Saperconazole, IV 861
Saquinavir, IV 862
Sarafloxacin, IV 863, V 569
Scopolamine, CE 1279, I 1235, IV 865
Scopoletin, IV 867
Secnidazole, IV 868
Secobarbital, CE 1281, IV 869
Secretin, IV 880
Selamectin, V 571
Selegiline, CE 1286, IV 881
Selfotel, IV 883
Semduramicin, IV 884
Seratrodast, IV 885
Sermorelin, V 572
Sertaconazole, IV 886
Sertindole, IV 887
Sertraline, I 1244, IV 888
Sibutramine, V 574
Sildenafil, V 576
Simazine, IV 888
Simethicone, V 579
Simvastatin, I 1249, IV 889
Sincalide, IV 890
Sirolimus, IV 892
Sisomicin, IV 894
Sivelestat, V 580
Sobrerol, IV 895
Sodium nitroprusside, IV 895
Sodium oxybate, V 582
Somatomedin, IV 897
Somatrem, CE 1288, IV 899
Somatropin, CE 1289, I 1252, IV 900, V 583
Sorivudine, IV 902

Sotalol, CE 1291, IV 904
Sparfloxacin, CE 1295, IV 917
Spectinomycin, IV 922
Spiperone, IV 926
Spiramycin, IV 927
Spirapril, CE 1296, IV 928
Spironolactone, CE 1296, IV 929
Squalane, V 584
Squalene, V 585
Stanozolol, IV 935, V 587
Stavudine, CE 1297, IV 937
Stearic acid, IV 941
Stiripentol, IV 942
Streptokinase, IV 943
Streptomycin, CE 1298, IV 944
Streptozocin, IV 948
Strychnine, IV 950
Succimer, IV 951, V 588
Succinylcholine chloride, IV 953, V 589
Sucralfate, I 1256
Sucrose, IV 954
Sufentanil, CE 1299, IV 954
Sulbactam, I 1257
Sulbenicillin, IV 956
Sulbentine, IV 957
Sulconazole, CE 1300, IV 958
Sulfabenzamide, CE 1301, IV 959
Sulfabromomethazine, V 591
Sulfacetamide, CE 1302, IV 962
Sulfachlorpyridazine, CE 1303, V 592
Sulfadiazine, CE 1305, IV 970
Sulfadimethoxine, CE 1309, IV 991
Sulfadoxine, CE 1312, IV 1010
Sulfaethidole, IV 1018
Sulfaethoxypyridazine, V 596
Sulfaguanidine, CE 1313, IV 1019
Sulfalene, CE 1314, IV 1020
Sulfamerazine, CE 1316, IV 1020, V 598
Sulfameter, CE 1318
Sulfamethazine, CE 1320, IV 1022
Sulfamethizole, CE 1324, IV 1047
Sulfamethoxazole, CE 1324, I 1263, IV 1054
Sulfamethoxypyridazine, IV 1059
Sulfamonomethoxine, CE 1330, IV 1059
Sulfanilamide, CE 1331, IV 1061
Sulfanilic acid, CE 1333, IV 1075
Sulfanitran, V 600
Sulfapyridine, CE 1334, IV 1076
Sulfaquinoxaline, CE 1335, IV 1079
Sulfasalazine, IV 1087
Sulfathiazole, CE 1339, IV 1090
Sulfinpyrazone, IV 1105
Sulfisomidine, CE 1342
Sulfisoxazole, CE 1344, IV 1109
Sulfur, IV 1114
Sulindac, CE 1347, IV 1115
Sulpiride, CE 1349, IV 1128
Sultamicillin, V 602
Sultopride, IV 1130
Sumatriptan, CE 1352, I 1284, IV 1132
Suprofen, CE 1353, IV 1134
Suramin, CE 1357
Suriclone, IV 1136
Synephrine, CE 1358, IV 1138

Tacalcitol, V 604
Tacrine, CE 1362, IV 1138
Tacrolimus, IV 1145
Talinolol, CE 1362
Talipexole, V 605
Taltirelin, V 607
Tamoxifen, CE 1364, IV 1150
Tamsulosin, IV 1160
Taurine, CE 1366, IV 1161
Tazarotene, IV 1176
Tazobactam, IV 1176
Tazofelone, IV 1180
Technetium Tc 99m bicisate, V 608
Tegafur, IV 1181
Tegaserod, V 609
Teicoplanin, IV 1182
Telithromycin, V 610
Telmesteine, V 611
Telmisartan, V 612
Temafloxacin, IV 1185
Temazepam, CE 1368, I 1286, IV 1189
Temocapril, V 614
Temocillin, IV 1191
Temozolomide, V 615
Teniposide, IV 1192
Tenofovir disoproxil fumarate, V 617
Tenoxicam, CE 1369, IV 1198
Teprenone, V 620
Terazosin, I 1302, IV 1203
Terbinafine, I 1306, IV 1204
Terbutaline, CE 1370, IV 1205
Terconazole, I 1310
Terfenadine, I 1312, IV 1214
Teriparatide, V 621
Teroxirone, IV 1217
Terpin, IV 1218
Tertatolol, IV 1218
Testolactone, IV 1220
Testosterone, CE 1383, IV 1222
Tetracaine, CE 1387, IV 1240
Tetrachlorvinphos, V 622
Tetracycline, CE 1390, I 1321, IV 1250
Tetrahydrozoline, CE 1395, IV 1256, V 623
Tetramethrin, IV 1259
Tetrazepam, IV 1262
Thalidomide, CE 1399, IV 1264, V 626
Thebaine, CE 1401, IV 1264
Thenyldiamine, IV 1265
Theobromine, IV 1266
Theodrenaline, CE 1402
Theophylline, CE 1404, I 1336, IV 1268
Thiabendazole, IV 1272
Thiacetarsamide, IV 1282
Thialbarbital, V 628
Thiambutene, IV 1282

Thiamine, CE 1413, IV 1283
Thiamphenicol, CE 1418, IV 1296
Thiamylal, IV 1296
Thiethylperazine, IV 1301
Thimerosal, IV 1304
Thiobutabarbital, IV 1307
Thioctic acid, IV 1308
Thioguanine, IV 1309
Thiopental, CE 1419, IV 1313
Thiopropazate, IV 1329
Thioproperazine, IV 1330
Thioridazine, CE 1422, IV 1333
Thiosalicylic acid, IV 1346
Thiotepa, IV 1349
Thiothixene, CE 1424, IV 1351
Thonzylamine, IV 1360
Thymol, IV 1361
Thymopentin, IV 1363
Thyrotropin, V 629
Tiagabine, IV 1363, V 630
Tiamenidine, IV 1364
Tiamulin, IV 1364
Tianeptine, IV 1366
Tiapride, IV 1368
Tiaprofenic acid, CE 1425, IV 1370
Ticarcillin, CE 1427, IV 1373
Ticlopidine, CE 1428, IV 1377
Tiemonium iodide, IV 1380
Tiletamine hydrochloride, V 631
Tilisolol, IV 1381
Tilmicosin, IV 1383, V 633
Tiludronic acid, V 634
Timepidium bromide, CE 1429
Timiperone, IV 1384
Timolol, CE 1429, I 1371, IV 1384
Tinidazole, IV 1388
Tioclomarol, IV 1389
Tioconazole, CE 1434, IV 1391
Tiopronin, CE 1437, IV 1393
Tiracizine, IV 1397
Tirilazad, V 635
Tirofiban, V 637
Tiropramide, V 639
Tizanidine, V 641
Toborinone, IV 1398
Tobramycin, CE 1437, IV 1399
Tocainide, CE 1438, IV 1404
Tofisopam, IV 1411
Tolazamide, CE 1440, IV 1414
Tolazoline, IV 1416
Tolbutamide, CE 1443, IV 1419
Tolcapone, V 643
Tolfenamic acid, IV 1429
Tolmetin, CE 1446, IV 1430
Tolnaftate, IV 1435
Toloxatone, IV 1436
Tolperisone, CE 1447
Tolpropamine, IV 1439
Tolrestat, IV 1440
Tolterodine, IV 1442
Tolycaine, IV 1443
Topiramate, V 645
Topotecan, IV 1445, V 647
Toremifene, IV 1446
Torsemide, IV 1449
Tosufloxacin, V 649
Tramadol, CE 1448, I 1383, IV 1453
Tramazoline, IV 1455
Trandolapril, IV 1455
Tranexamic acid, IV 1456
Tranilast, IV 1458
Tranylcypromine, CE 1449, IV 1459
Travoprost, V 651
Trazodone, IV 1467
Trenbolone, IV 1476, V 652
Treprostinil, V 653
Tretoquinol, CE 1450
Triamcinolone, CE 1453, I 1385, IV 1483
Triamterene, CE 1456, I 1398, IV 1486
Triazolam, CE 1457, IV 1489
Trichlorfon, IV 1499, V 655
Trichlormethiazide, CE 1460, IV 1501
Triclocarban, IV 1505
Triclosan, IV 1506
Trientine, IV 1507
Triethanolamine, V 656
Trifluoperazine, CE 1461, IV 1507
Trifluperidol, IV 1518
Triflupromazine, CE 1463, IV 1521
Trifluridine, IV 1527, V 657
Trihexyphenidyl, CE 1463, IV 1528
Trilostane, IV 1532
Trimazosin, IV 1533
Trimebutine, CE 1468, IV 1534
Trimeperidine, IV 1534
Trimeprazine, CE 1469, IV 1535
Trimetazidine, IV 1545
Trimethadione, IV 1546
Trimethobenzamide, IV 1546
Trimethoprim, CE 1472, I 1409, IV 1548
Trimetrexate, IV 1551
Trimidox, IV 1554
Trimipramine, CE 1475, IV 1554
Trioxsalen, IV 1568
Tripelennamine, CE 1480, IV 1569
Triprolidine, CE 1481, IV 1574
Triptorelin, IV 1583, V 658
Tritoqualine, IV 1584
Troglitazone, CE 1482, IV 1584
Troleandomycin, CE 1483, V 660
Tromethamine, CE 1484, IV 1585
Tropacocaine, IV 1589
Tropic acid, IV 1590
Tropicamide, CE 1484, IV 1590
Tropisetron, IV 1591
Trospium chloride, IV 1593
Trovafloxacin, IV 1594
Troxipide, V 661
Trypsin, IV 1594
Tryptophan, IV 1596

Tubocurarine chloride, IV 1596
Tylosin, IV 1600
Tyloxapol, IV 1606
Tymazoline, IV 1607

Ubenimex, V 662
Undecylenic acid, IV 1607
Unoprostone isopropyl ester, V 663
Urea, IV 1609
Urokinase, IV 1613
Ursodiol, CE 1489, IV 1614

Valacyclovir, IV 1626, V 664
Valdecoxib, V 666
Valganciclovir, V 668
Valproic acid, CE 1490, I 1429, IV 1627
Valrubicin, V 670
Valsartan, V 671
Valspodar, IV 1629
Vancomycin, CE 1491, I 1438, IV 1630
Vanillin, IV 1632
Vasopressin, IV 1633
Vecuronium bromide, IV 1636
Venlafaxine, CE 1491, I 1445, IV 1639
Verapamil, CE 1492, I 1448, IV 1640
Vesnarinone, IV 1647
Vidarabine, IV 1648
Vigabatrin, IV 1654
Viloxazine, IV 1657
Vinblastine, CE 1502, IV 1659
Vincamine, IV 1665
Vincristine, CE 1503, IV 1667
Vindesine, IV 1673
Vinorelbine, IV 1674
Vinylbital, IV 1678
Virginiamycin, IV 1679
Vitamin A, CE 1504, IV 1680
Vitamin $B_{12}$, CE 1506, IV 1705
Vitamin $D_2$, CE 1508, IV 1710
Vitamin $D_3$, CE 1509, IV 1714
Vitamin E, CE 1511, IV 1719
Vitamin K, IV 1741
Vitamin $K_1$, IV 1742
Voriconazole, IV 1747

Warfarin, CE 1512, I 1466, IV 1748

Xamoterol, IV 1752
Xanomeline, IV 1754
Xipamide, IV 1755
Xylazine, IV 1757
Xylometazoline, CE 1520, IV 1760
Xylose, CE 1521, IV 1764

Yohimbine, IV 1777

Zafirlukast, IV 1783
Zalcitabine, CE 1529, IV 1784
Zaleplon, V 673
Zaltoprofen, V 676
Zanamivir, V 678
Zearalenone, IV 1788
Zeranol, IV 1789
Zidovudine, CE 1529, I 1480, IV 1794
Zileuton, IV 1795
Zinostatin, IV 1797, V 680
Zipeprol, IV 1798
Ziprasidone, IV 1799
Zofenopril calcium, CE 1530, V 681
Zolazepam hydrochloride, V683
Zolpidem, CE 1531, I 1491, IV 1799
Zonisamide, IV 1801
Zopiclone, CE 1532, IV 1804
Zorubicin, IV 1812
Zoxazolamine, IV 1813
Zuclopenthixol, IV 1815

# CROSS-INDEX TO OTHER SUBSTANCES

In many cases other compounds may be chromatographed under the conditions described in the monographs in this book. Using the information in the monographs it may be possible to develop chromatographic procedures for these compounds. The compounds fall into the following categories:

| | | |
|---|---|---|
| AL | Also chromatographed | Compounds which can be analyzed at the same time. It is not specified whether or not they interfere but they can be extracted. |
| EX | Extracted | Compounds which can be extracted from the matrix in question and analyzed at the same time and do not interfere. |
| IN | Interfering | Compounds which interfere with the analysis of the target compound. |
| IS | Internal Standard | A compound added at some time during the sample preparation procedure to act as a standard. |
| SI | Simultaneous | Compounds which can be analyzed at the same time and do not interfere. Note that the compound may or may not be extracted from the matrix in question. |

(For obvious reasons non-interfering compounds are not listed in this index.)


Abacavir (EX) 335, 336
Abacavir 5'-carboxylate (EX) 3
Abacavir 5'-glucuronide (EX) 3
Acamprosate (EX) 387
Acebutolol (EX) 7, 230, 253, 387, 400, 406, 434, 495, 623; (IN) 499
Acefylline heptaminol (EX) 387
Acenocoumarol (EX) 387
Acepromazine (EX) 387, 525
Aceprometazine (EX) 387
Acetaminophen (EX) 7, 232, 254, 384, 387, 401, 407, 495, 498, 623; (IS) 315; (SI) 16, 173, 178, 202, 603, 614
Acetamiprid (EX) 452
Acetazolamide (EX) 387; (SI) 188, 220, 522
Acetiamin (EX) 387
Acetophenetidine (SI) 178
Acetorphan (EX) 387
Acetylcodeine (AL) 98
Acetylleucine (EX) 387
Acetylmethionine (EX) 387
6-Acetylmorphine (EX) 387
Acetylsulfabromomethazine (EX) 591, 594, 597
Acetylsulfachlorpyridazine (EX) 591, 594, 597
Acetylsulfadiazine (EX) 591, 594, 597
Acetylsulfadimethoxine (EX) 591, 594, 597
Acetylsulfaethoxypridazine (EX) 591, 594, 597
Acetylsulfamethazine (EX) 591, 594, 597
Acetylsulfisoxazole (EX) 599
*N*-Acetylprocainamide (EX) 495, 498
*N*-Acetyl-L-tyrosine ethyl ester (EX) 389
Acrivastine (EX) 230, 253, 400, 406, 623
Actinoquinol (EX) 387
Acyclovir (EX) 387, 488, 664; (SI) 188, 220, 522
Adefovir (EX) 12; (IS) 617
Adenosine (EX) 387; (IS) 372
Adipic acid (EX) 139, 371
Adrafinil (EX) 387
Aklomide (EX) 601
Alacepril (SI) 603, 614

Albendazole (EX) 102, 103, 387
Alclometasone 17,21-dipropionate (SI) 30, 32, 143, 145, 167, 170, 264, 271, 306
Aldicarb (EX) 387
Alendronate (SI) 441
Alfuzosin (EX) 387
Alimemazine (EX) 387
Alizapride (IN) 552
Allethrin (EX) 80, 535
Allopurinol (SI) 188, 220, 522
Alminoprofen (EX) 387
Almitrine (EX) 387
Alpiropride (IS) 50
Alprazolam (EX) 7, 230, 253, 387, 400, 406, 498, 623; (IN) 496; (SI) 194
Alprenolol (EX) 7, 230, 253, 400, 406, 434, 623
Alprostadil (EX) 460
Altazide (EX) 387
Altretamine (EX) 387
Amantadine (EX) 7, 230, 253, 400, 406, 623; (IS) 378
Ambemonium (EX) 387; (IS) 137
Amcinonide (SI) 18, 20, 143, 145, 167, 170, 264, 271, 306
Amfepramone (EX) 387
Amiloride (EX) 7, 230, 253, 387, 400, 406, 623
Amineptine (EX) 387, 495, 498
*p*-Aminobenzoic acid (EX) 97; (SI) 456
4-Amino-3-nitroanisole (IS) 334
Aminophenazone (EX) 7, 230, 253, 400, 406, 623
*p*-Aminosalicylic acid (EX) 390
Aminophylline (IN) 204; (SI) 178
Amiodarone (EX) 7, 230, 253, 387, 400, 406, 623; (IS) 343
Amisulpride (EX) 387, 537
Amitriptyline (EX) 7, 230, 253, 387, 400, 406, 495, 498, 537, 623; (SI) 47, 383, 512, 547
Amlodipine (EX) 387
Ammonium (SI) 656
Amobarbital (EX) 387, 495, 498; (SI) 178
Amoxapine (EX) 378, 387, 495, 498
Amoxicillin (EX) 387; (SI) 202
Amphetamine (AL) 98; (EX) 387, 495, 497, 498, 529
Amphotericin B (EX) 387
Ampicillin (EX) 387; (SI) 16, 603, 614
Amprenavir (EX) 209, 281, 282, 359, 360, 436, 444
Androstenedione (EX) 169
Anethole (EX) 387
Angiotensin I (SI) 13
Angiotensin II (SI) 13
Angiotensin III (SI) 13
Anhydrodaptomycin (SI) 159
Anhydroglucitol (EX) 302
Aniracetam (SI) 188, 220, 522
Anthraquinone (IS) 357
Antipyrine (EX) 204; (SI) 178
Antrenyl (SI) 331, 382
Apomorphine (EX) 387
Aprepitant (SI) 512, 547
Aprindine (EX) 387
Aprobarbital (EX) 495, 498; (SI) 178
Arachidonic acid (EX) 457, 458, 459, 460, 461, 462; (SI) 557
Arecoline (EX) 387
Artemisinin (IS) 52
Articaine (AL) 98
Articainic acid (EX) 55
Ascorbic acid (SI) 140
Aspartame (EX) 387
$\beta$-Asp daptomycin (SI) 159
Aspirin (SI) 16, 178, 202, 603, 614
Astemizole (AL) 38; (EX) 7, 231, 253, 387, 400, 406, 623
Atenolol (EX) 7, 231, 253, 387, 400, 406, 434, 495, 498, 623
Atipamazole (EX) 369, 370
Atracurium (IN) 552
Atrazine (EX) 387
Atrial natriuretic peptide (SI) 13
Atropine (EX) 327, 387, 495, 498
Azacyclonol (EX) 7, 231, 253, 400, 406, 623
Azaperol (EX) 525
Azaperone (EX) 525
Azithromycin (SI) 267

Bacampicillin (EX) 387
Baclofen (SI) 642
Bamethan (EX) 387
Bamifylline (EX) 387
Barbital (EX) 387, 495, 499; (SI) 178
Baythroid (cyfluthrin) (AL) 25; (EX) 26, 80, 535
Beclomethasone (EX) 168, 265
Beclomethasone 17,21-dipropionate (SI) 20, 32, 145, 167, 170, 264, 271, 306, 414
Befunolol (EX) 434
Benazepril (EX) 387; (SI) 104, 188, 220, 522
Bendroflumethiazide (EX) 387
Benfluorex (EX) 387
Benperidol (SI) 537
Bentazone (EX) 387
Benzanilide (IS) 6
Benzbromarone (EX) 387
Benzene (EX) 387
Benzhexol (EX) 7, 231, 253, 400, 406, 495, 499, 623
Benzoic acid (SI) 178
Benzophenone (EX) 387
Benzophenone-3 (SI) 456
Benzophenone-4 (SI) 456
Benzoylecgonine (EX) 7, 231, 253, 387, 400, 406, 495, 499, 623
Benztropine (EX) 495, 499
Benzydamine (EX) 387
Benzyl alcohol (IS) 136; (SI) 516
3-Benzylidene-*d*, *l*-camphor (SI) 456
Benzylidene camphorsulfonic acid (SI) 456
1-Benzylimidazole (IS) 100
Bepridil (EX) 387

Betahistine (EX) 388
Betamethasone (EX) 168, 265, 388; (IS) 523; (SI) 18, 20, 30, 32, 143, 145, 167, 170, 188, 220, 264, 271, 306, 522
Betamethasone-17-acetate (SI) 18, 30, 143
Betamethasone-21-acetate (SI) 18, 20, 30, 32, 143, 145, 167, 170, 264, 271, 306
Betamethasone 17-benzoate (SI) 18, 20, 30, 32, 143, 145, 167, 170, 271, 306
Betamethasone-17,21-diacetate (SI) 18, 30, 143
Betamethasone 17,21-dipropionate (IS) 18, 20, 30, 32, 143, 145, 167, 170, 264, 271, 306, 413
Betamethasone-17,21-divalerate (SI) 18, 30, 143
Betamethasone propionate (SI) 18, 30, 143
Betamethasone-17-propionate-21-butyrate (SI) 18, 30, 143
Betamethasone-17-propionate-21-stearate (SI) 18, 30, 143
Betamethasone 17-valerate (SI) 18, 20, 30, 32, 143, 145, 167, 170, 264, 271, 306
Betamethasone-17-valerate-21-acetate (SI) 18, 30, 143
Betaxolol (EX) 7, 231, 253, 388, 400, 406, 434, 623
Bezafibrate (EX) 388
BHT (EX) 388
Bidesethylchloroquine (EX) 388
Bifenthrin (EX) 26, 80, 535
Bifonazole (SI) 482
Bioallethrin (SI) 534
Biotin (Vitamin H) (EX) 391
Biperiden (EX) 388
Biperidine (EX) 7, 231, 253, 400, 406, 623
Biscoumacetate (EX) 388
Bisoprolol (EX) 7, 231, 253, 388, 400, 406, 434, 623
Boldine (EX) 388
Bombesin (SI) 13
Bradykinin (SI) 13
Bromadiolone (EX) 388
Bromazepam (EX) 388, 398, 495, 499
Bromocriptine (EX) 233, 388
4-Bromo-2,5-dimethoxyphenylethylamine (EX) 498
3-Bromoprimaquine diphosphate (IS) 96
5-Bromouracil (SI) 173
Brompheniramine (EX) 7, 231, 253, 388, 400, 406, 499, 623; (IN) 496
Buclizide (EX) 388
Budesonide (SI) 20, 32, 145, 167, 170, 264, 271, 306, 414
Buflomedil (EX) 388
Bumetanide (EX) 388; (IS) 105, 671
Bunitrolol (EX) 434
Bupivacaine (AL) 98; (EX) 7, 231, 254, 400, 406, 623
Bupranolol (IS) 433
Buprenorphine (EX) 7, 231, 254, 388, 400, 406, 495, 499, 623
Buspirone (EX) 7, 231, 254, 388, 400, 406, 495, 499, 623
Butabarbital (EX) 495, 499; (IN) 178
1,4-Butanediol (IS) 528
Butethal (SI) 178
Butobarbital (EX) 388
Butofilolol (EX) 434
2-*n*-Butylmercaptobenzimidazole (IS) 103
Butyl methoxydibenzoylmethane (SI) 456
Butyl paraben (EX) 388; (IS) 577
*N*-Butylscopolammonium bromide (EX) 388

Cafedrine (EX) 388
Caffeic acid (EX) 388
Caffeine (AL) 98; (EX) 7, 231, 254, 384, 388, 400, 406, 495, 499, 623; (SI) 16, 173, 178, 202, 603, 614, 656
Camostat mesylate (SI) 188, 220, 522
Campesterol (SI) 585
Camphor benzalkonium methosulfate (SI) 456
Candesartan (EX) 671
Canrenoate (EX) 388
Capric acid (EX) 461; (SI) 463
Caproic acid (EX) 461
Caprylic acid (EX) 461
Captodiamine (EX) 388
Captopril (AL) 682; (EX) 388; (IS) 229
Carazolol (EX) 434, 525
Carbamazepine (EX) 7, 231, 254, 388, 400, 406, 495, 499, 623; (IN) 349; (IS) 431; (SI) 188, 220, 522
Carbaryl (EX) 250, 388
Carbenicillin (SI) 16, 603, 614
Carbimazole (EX) 388
Carbinoxamine (EX) 8, 231, 254, 388, 401, 407, 624
Carbofuran (EX) 250; (SI) 354
Carboplatin (EX) 480
Carbovir (EX) 3
Carbutamide (EX) 388
Carisoprodol (EX) 8, 231, 254, 401, 407, 624
Carnitine (EX) 388
$\beta$-Carotene (EX) 22
Carpipramine (EX) 388
Carteolol (EX) 388, 434
Carvedilol (EX) 8, 231, 254, 401, 407, 426, 434, 624; (IS) 425
Cathinone (EX) 497, 529
Cefaclor (EX) 113, 118, 257
Cefadroxil (EX) 388
Cefamandole (EX) 113, 118, 257; (SI) 552
Cefatrizine (EX) 388
Cefazolin (EX) 113, 118, 257, 388
Cefbuperazone (EX) 118, 257
Cefixime (EX) 113, 118, 257, 388
Cefmenoxime (EX) 113, 118, 257
Cefmetazole (EX) 113, 118, 257
Cefoperazone (EX) 113, 118, 257
Cefotaxime (EX) 113, 118, 257; (IS) 114; (SI) 552
Cefotetan (EX) 113, 118, 257

Cefotiam (EX) 113, 118, 257; (SI) 16, 603, 614
Cefpiramide (EX) 113, 118, 257
Cefpirome sulfate (IS) 116
Cefpodoxime (EX) 388
Cefsulodin (EX) 113, 118, 257
Ceftazidime (EX) 113, 118, 257; (SI) 552
Ceftizoxime (EX) 113, 118, 257
Ceftriaxone (EX) 113, 118, 257, 388; (SI) 552
Cefuroxime (EX) 113, 118, 257, 388; (SI) 552
Cefuzonam (EX) 113, 257
Celecoxib (IS) 567
Celiprolol (EX) 8, 231, 254, 388, 401, 407, 434, 624
Cephalexin (EX) 113, 118, 257, 388
Cephaloglycine (EX) 113, 118, 257
Cephaloridine (EX) 113, 118, 257
Cephalothin (EX) 113, 118, 257; (IS) 682
Cephradine (SI) 552
Cetirizine (EX) 8, 231, 254, 388, 401, 407, 624
Chlorambucil (EX) 388
Chloramphenicol (EX) 388, 495, 499; (IS) 258; (SI) 178
Chlorcyclizine (EX) 8, 231, 254, 401, 407, 624
Chlordiazepoxide (EX) 8, 231, 254, 388, 401, 407, 495, 499, 624; (SI) 383
Chlorhexidine (EX) 388
Chloridazone (EX) 388
Chlormadinone (EX) 375, 388
Chlormethiazole (EX) 495; (IN) 499
Chlormezanone (EX) 8, 231, 254, 388, 401, 407, 624
Chlorobenzoic acid (EX) 388
Chloromadinone acetate (EX) 269, 376
Chlorophacinone (EX) 388
4-Chlorophenylbiguanide (EX) 388
*m*-Chlorophenylpiperazine (EX) 378, 537
Chloropicrin (EX) 388
Chloroquine (EX) 8, 231, 254, 388, 401, 407, 624; (SI) 95
Chlorotestosterone acetate (EX) 269, 376
8-Chlorotheophylline (IS) 173
Chlorothiazide (SI) 178
5-Chlorouracil (SI) 173
Chlorpheniramine (EX) 8, 231, 254, 388, 401, 407, 499, 624; (IN) 496; (SI) 331
Chlorpromazine (EX) 8, 231, 254, 388, 401, 407, 495, 499, 525, 624; (IS) 383; (SI) 16, 194, 603, 614
Chlorpropamide (EX) 8, 231, 254, 388, 401, 407, 624; (SI) 178
Chlorprothixene (EX) 8, 231, 254, 401, 407, 624; (SI) 537
Chlorpyrifos (SI) 354
Chlortetracycline (EX) 592
Chlortoluron (EX) 388
Chlorzoxazone (SI) 642
Cholesterol (SI) 585
Cibenzoline (EX) 388
Cicletanine (EX) 388
Cilazapril (EX) 388; (SI) 104, 188, 220, 522
Cimetidine (EX) 388, 495, 499; (SI) 188, 202, 220, 522
Cinchonine (EX) 388
Cinerin (SI) 534
Cinerin I (SI) 533
Cinerin II (SI) 533
Cinnarizine (EX) 8, 231, 254, 388, 401, 407, 624; (IS) 358
Cinoxacin (EX) 365
Ciprofibrate (EX) 388
Ciprofloxacin (EX) 365, 388, 473, 650; (IN) 364; (IS) 294; (SI) 293, 472
Cisapride (EX) 388; (IS) 194, 640
Citalopram (EX) 8, 231, 254, 401, 407, 495, 499, 537, 624; (SI) 399
Citric acid (EX) 371
Clarithromycin (SI) 267
Clemastine (EX) 8, 231, 254, 401, 407, 624
Clenbuterol (EX) 388; (IS) 304, 473
Clidinium bromide (EX) 330, 381, 388; (SI) 331, 382
Clindamycin (EX) 352, 353, 388; (SI) 251, 252, 516
Clobazam (EX) 8, 231, 254, 388, 401, 407, 495, 499, 624
Clobenzorex (EX) 388
Clobetasol 17-propionate (SI) 18, 20, 30, 32, 167, 170, 264, 271, 306
Clobetasone butyrate (SI) 18, 30, 143
Clobutinol (EX) 8, 231, 254, 401, 407, 624
Clocinizine (EX) 388
Clodronate (SI) 441
Clofazimine (SI) 627
Clofibrate (EX) 388; (SI) 188, 220, 522
Clomethiazole (EX) 8, 231, 254, 388, 401, 407, 624
Clomipramine (EX) 8, 231, 254, 388, 401, 407, 495, 499, 537, 624
Clonazepam (EX) 8, 231, 254, 388, 401, 407, 495, 499, 624; (SI) 188, 220, 522
Clonidine (EX) 8, 231, 254, 388, 401, 407, 495, 499, 624; (IS) 423; (SI) 45
Cloperastine (EX) 495, 499
Cloranolol (EX) 434
Clorazepate (EX) 388
Clotiazepam (EX) 388
Clotrimazole (EX) 273
Clovoxamine (EX) 378
Cloxacillin (EX) 388
Cloxazolam (EX) 495, 499
Clozapine (EX) 8, 231, 254, 401, 407, 499, 537, 624; (IN) 496
Cocaine (EX) 8, 231, 254, 388, 401, 407, 495, 499, 624
Codeine (AL) 98; (EX) 8, 231, 254, 388, 401, 407, 495, 499, 624; (IS) 290; (SI) 202
Codethyline (EX) 388
Colchicine (EX) 388; (SI) 178
Colchicoside (EX) 388
Colchicosine (EX) 388

Corticosterone (EX) 169, 171
Cortisone (EX) 169; (SI) 18, 20, 30, 32, 143, 145, 163, 167, 170, 178, 261, 264, 271, 306
Cortisone 21-acetate (SI) 19, 20, 30, 32, 143, 145, 167, 170, 264, 271, 306
Cortivazol (EX) 388
Cotinine (EX) 388
Coumachlor (EX) 388
Coumafen (EX) 388
Coumarin (SI) 178
Coumatetralyl (EX) 8, 231, 254, 388, 401, 407, 624
*m*-Cresol (SI) 654
Crimidine (EX) 388
Cromolyn (SI) 16, 603, 614
Cyamemazine (EX) 388
Cyclandelate (EX) 388
Cyclizine (EX) 8, 231, 254, 401, 407, 624
Cycloguanil (EX) 388
Cyclohexymethylguanidine (IS) 305
Cyclophosphamide (SI) 188, 220, 522
Cyclothiazide (SI) 178
Cyfluthrin (AL) 25; (EX) 26, 80, 535
Cyhalothrin (EX) 155
Cyheptamide (SI) 178
Cypermethrin (AL) 26; (SI) 24
Cyproheptadine (EX) 388
Cyromazine (EX) 388

Dacarbazine (EX) 388
Dalfopristin (EX) 540; (SI) 541
Danazol (SI) 178
Danofloxacin (EX) 365, 473, 650; (SI) 364; 472
Danthron (SI) 178
Dapsone (EX) 388; (IN) 304; (SI) 178, 627
Deacetylalacepril (EX) 17
Decanol (EX) 380
Deethylzaleplon 674
Deflazacort (EX) 261
Dehydrocorticosterone (SI) 19, 30, 143
Delapril (SI) 188, 220, 522
Delavirdine (EX) 208, 435, 443; (IN) 36
Delmadinone acetate (EX) 269, 376
Deltamethrin (EX) 155
Demoxepam (EX) 8, 231, 254, 401, 407, 624
3′-Deoxy-2′,3′-didehydrothymidine (IS) 215
11-Deoxycortisol (EX) 169, 260
Deoxycortone (EX) 168, 265
21-Deoxydexamethasone (EX) 168, 265
Deoxymethasone (SI) 19, 30, 144
Desethylatrazine (EX) 388
*N*-Desethyl-5-oxozaleplon (EX) 674
*N*-Desethylzaleplon (EX) 674
Desfluoroaprepitant (IS) 46, 47
Desipramine (EX) 378, 388, 495, 499, 537
Desmethylcitalopram (EX) 378; (SI) 399
Desmethylclomipramine (EX) 378
Desmethyldoxepin (EX) 378
Desmethylmaprotiline (EX) 378
Desmethylmianserin (EX) 378
Desmethylmirtazapine (SI) 399
Desmethylsertraline (EX) 378
Desmethyltrimipramine (EX) 378
Desmethylvenlafaxine (EX) 537; (SI) 399
Desonide (SI) 20, 32, 145, 167, 170, 265, 271, 306
Desoximetasone (EX) 265; (SI) 20, 32, 145, 167, 170, 264, 271, 306
Desoxy-2-phenobarbital (EX) 388
Desoxycorticosterone acetate (SI) 20, 32, 145, 167, 264, 271, 306
Desoxycorticosterone pivalate (SI) 20, 32, 145, 167, 264, 271, 306
Detomidine (IS) 60, 369
Dexamethasone (EX) 168, 260, 265, 388; (SI) 19, 20, 30, 32, 144, 145, 167, 170, 188, 220, 264, 271, 306, 522
Dexamethasone 21-acetate (EX) 169; (SI) 19, 20, 31, 32, 144, 145, 167, 170; 264, 271, 306; (IS) 414
Dexamethasone isonicotinate (SI) 19, 31, 144
Dexamethasone pivalate (SI) 19, 31, 144
Dexamethasone valerate (SI) 19, 31, 144
Dextromethorphan (EX) 8, 231, 254, 401, 407, 388, 624
Dextromoramide (EX) 388
Dextropropoxyphene (EX) 8, 231, 254, 388, 401, 407, 624; see also propoxyphene (EX) 495, 499
Dextrose (EX) 302, 303
DFdU (2′,2′-difluorodeoxyuridine) (EX) 296
Diamorphine (AL) 98; (EX) 388
Diarachidoylphosphatidylcholine (EX) 502, 505
Diatrizoate (EX) 322, 325
Diaveridine (EX) 388
Diazepam (EX) 8, 231, 254, 388, 401, 407, 495, 499, 537, 624; (IS) 195, 539; (SI) 188, 194, 202, 220, 522
Diazinon (EX) 388
Dibekacin (EX) 388
Dibenzepin (IN) 537; (IS) 7, 230, 253, 400, 406, 623
Dibucaine (AL) 98
Dichlorisone (EX) 168, 265
Dichlorprop (EX) 388
Diclofenac (EX) 388; (SI) 47, 512, 547
Didanosine (EX) 1, 2, 335, 336; (SI) 2
Diethofencarb (EX) 250
Diethylstilbestrol (EX) 388; (SI) 178
Difemerine (EX) 388
Diflorasone 17,21-diacetate (SI) 20, 32, 145, 167, 170, 264, 271, 306
Difloxacin (EX) 569, 650; (SI) 364, 472
Diflucortolone valerate (SI) 19, 31, 144
2′,2′-Difluorodeoxyuridine (dFdU) (EX) 296
Digitoxin (EX) 388
Digoxin (EX) 388; (SI) 188, 220, 522
Dihomo-gamma-linolenic acid (EX) 458
Dihydralazine (EX) 388
Dihydrobefunolol (EX) 70
Dihydrocodeine (EX) 388, 495, 499

Dihydrostreptomycin (EX) 312
3,4-Dihydroxybenzylamine (IS) 348
Dilazep (SI) 188, 220, 522
Dilinoleylphosphatidylcholine (EX) 502, 505
Diltiazem (EX) 8, 231, 254, 388, 401, 407, 495, 499, 624; (IS) 494; (SI) 188, 220, 522
Dimefline (SI) 383
Dimerazole (EX) 388
3-Dimethylaminomethyleneimino-2,4,6-triiodobenzoic acid (IS) 393
*N,N*-Dimethylglycine (EX) 72
Dimethyl phthalate (EX) 388
2,2′-Dimethylpropiophenone (IS) 533
Dimetridazole (EX) 388
Diminazene (EX) 316
2,4-Dinitrochlorobenzene (IS) 237
Dinoprost (EX) 460
Dinoprostone (EX) 460
Dinoseb (EX) 388
Dinsed (EX) 600
Dipalmitolyphosphatidylcholine 506
Diphacinone (EX) 388
Diphenhydramine (EX) 8, 231, 254, 401, 407, 495, 499, 624
Dipheniramine (EX) 388
Diphenylphosphate (IS) 284
Diphosphatidylglycerol (EX) 502, 505
Diprophylline (EX) 388
Dipropyline (EX) 388
Dipyridamole (EX) 8, 231, 254, 388, 401, 407, 624
Dipyrone (SI) 383
Disopyramine (EX) 8, 231, 254, 388, 401, 407, 495, 499, 624
Disulfiram (EX) 388
Di-syston (EX) 388
Dithiothreitol (EX) 588
Diuron (EX) 388
Dixyrazine (EX) 8, 231, 254, 401, 407, 624; (SI) 552
Docarpamine (SI) 220, 522
Docosahexaenoic acid (EX) 458, 459
Docosanoic acid (EX) 457
Dodecyl alcohol (SI) 129
Domperidone (EX) 388
Dopamine (IS) 394
Dosulepin (EX) 388
Dothiepin (EX) 495, 499
Doxapram (EX) 8, 231, 254, 401, 407, 624; (SI) 383
Doxazosin (EX) 189
Doxepin (EX) 8, 231, 254, 388, 401, 407, 495, 499, 537, 624
Doxorubicin (EX) 388
Doxylamine (EX) 388
Dronabinol (EX) 8, 232, 255, 402, 408, 624; (SI) 178
Droperidol (EX) 388
Dropropizine (EX) 388

Ebastine (EX) 8, 231, 254, 401, 407, 624
Econazole (EX) 388; (IS) 251
EDDP (EX) 388
Efavirenz (EX) 36, 38, 360, 435, 436, 443, 444
Eicosanoic acid (EX) 457
Eicosapentaenoic acid (EX) 458, 459
Eicosapentanoic acid (EX) 457
Eicosatrienoic acid (EX) 457
Eicosyl alcohol (SI) 129
Elaidic acid (EX) 459
Embutramide (EX) 8, 231, 254, 388, 401, 407, 624
Emetine (EX) 388
Emtricitabine diphosphate (EX) 216
Emtricitabine monophosphate (EX) 216
Emtricitabine triphosphate (EX) 216
Enalapril (EX) 388; (IS) 7, 230, 253, 400, 406, 623; (SI) 16, 603, 614
Enalaprilat (AL) 682
Endorphins (SI) 14
Enoxacin (EX) 365, 388; (SI) 47, 512, 547
Enrofloxacin (EX) 365, 473, 650; (SI) 364, 421, 472, 570
Eperisone (SI) 188, 522
Ephedrine (AL) 98; (EX) 388, 495, 497, 499, 529
Epigalantamine (EX) 291
Epinephrine (AL) 98; (EX) 348, 389, 498
Ergotamine (EX) 8, 254, 401, 407, 624
Erythritol (EX) 302
Erythromycin (EX) 495, 499; (SI) 267
Esculin (EX) 389
Eserine (EX) 389
Esmolol (EX) 434
Estazolam (EX) 389
Estradiol (EX) 169; (SI) 178
17$\alpha$-Estradiol (IS) 558
$\beta$-Estradiol (EX) 389
Estriol (EX) 169, 389; (SI) 178
Estrone (EX) 169; (SI) 178
Ethanolamine (SI) 656
Ethazolastone (IS) 615, 616
Ethenzamide (EX) 8, 231, 254, 401, 407, 624
Ethinyl estradiol (EX) 300; (SI) 239
Ethiofencarb (EX) 250
Ethosuximide (EX) 389; (SI) 178, 188, 220, 522
4-Ethoxyphenol (IS) 384
3-(4′-Ethylbenzylidene)-*d,l*-camphor (SI) 456
Ethylene glycol (EX) 302
Ethylmorphine (AL) 98; (EX) 8, 231, 254, 401, 407, 624
Ethyl paraben (EX) 389; (IS) 160
Ethylparathion (EX) 8, 231, 254, 401, 407, 624
Etidocaine (AL) 98; (IS) 54
Etidronate (SI) 441
Etodolac (EX) 389; (IS) 162
Etodroxizine (EX) 8, 231, 254, 401, 407, 624
Eugenol (SI) 178

Famciclovir (EX) 489
Famotidine (EX) 389

Febantel (EX) 103
Felodipine (EX) 8, 231, 254, 389, 401, 407, 624
Fenazepam (EX) 8, 231, 254, 401, 407, 624
Fenbendazole (EX) 102, 103
Fenbufen (EX) 389; (SI) 188, 220, 522
Fenfluramine (EX) 8, 231, 254, 378, 389, 401, 407, 497, 498, 499, 529, 624; (IN) 496
Fenkamfamine (EX) 8, 231, 254, 401, 407, 624
Fenobucarb (EX) 250
Fenofibrate (EX) 389; (SI) 47, 512, 548
Fenoprofen (EX) 389; (SI) 178
Fenoverine (EX) 389
Fenozolone (EX) 389
Fenpiverinium bromide (IS) 331
Fenpropathrin (AL) 26
Fenproporex (EX) 389
Fentanyl (EX) 8, 231, 254, 389, 401, 407, 495, 499, 624
Fenvalerate (AL) 26; (EX) 26, 80, 535
Fexofenadine (EX) 8, 231, 401, 407, 624
Finasteride (SI) 47, 512, 548
Flavone (EX) 389; (IS) 480
Flecaine (EX) 389
Flecainide (EX) 8, 231, 254, 401, 407, 495, 499, 624
Fleroxacin (EX) 650
Floctafenine (EX) 389
Flomoxef (EX) 113, 118
Flubendazole (EX) 103, 389
Fluclorolone acetonide (EX) 168, 265
Fluconazole (EX) 8, 231, 254, 389, 401, 407, 624; (SI) 188, 220, 522
Flucytosine (EX) 389
Fludrocortisone (EX) 168, 265; (IS) 260
Fludrocortisone 21-acetate (IS) 163, 261; (SI) 19, 20, 31, 32, 144, 145, 167, 170, 264, 271, 306
Flumazenil (EX) 8, 231, 254, 401, 407, 495, 499, 624
Flumequine (EX) 365, 473, 592
Flumethasone (EX) 168, 265
Flumethasone 21-pivalate (SI) 19, 20, 31, 32, 144, 145, 167, 170, 264, 271, 306
Flumethrin (EX) 155
Flunarizine (EX) 358, 389
Flunindione (EX) 389
Flunisolide (SI) 20, 32, 145, 167, 170, 264, 271, 306
Flunitrazepam (EX) 8, 231, 254, 389, 401, 407, 499, 624; (IN) 36, 496
Fluocinolone acetonide (EX) 168, 265; (SI) 19, 20, 31, 32, 144, 145, 167, 170, 264, 271, 306
Fluocinonide (EX) 168, 265; (SI) 19, 20, 31, 32, 144, 145, 167, 170, 264, 271, 306
Fluocortin butyl ester (SI) 19, 31, 144
Fluocortolone (EX) 168, 265
Fluocortolone caproate (SI) 19, 31, 144
Fluocortolone pivalate (SI) 19, 31, 144
6-Fluoroalosetron (IS) 29
5-Fluorocytosine (SI) 173
Fluorometholone (EX) 168, 265; (SI) 19, 20, 31, 32, 144, 145, 167, 170, 264, 271, 306
9-α-Fluoroprednisolone (SI) 19, 31, 144
9-α-Fluoroprednisolone acetate (SI) 19, 31, 144
5-Fluorouracil (EX) 389; (SI) 173, 178
Fluoxetine (EX) 8, 231, 254, 378, 389, 401, 407, 495, 499, 537, 624; (IS) 498; (SI) 399
Fluoxymesterone (EX) 427, 587; (SI) 178
Flupentixol (EX) 8, 231, 254, 389, 401, 407, 495, 499, 624
Fluphenazine (EX) 389, 495, 499; (SI) 383, 537
Fluprednisolone (EX) 168, 265; (SI) 20, 32, 145, 167, 170, 264, 271, 306
Flurandrenolide (EX) 168; (SI) 19, 20, 31, 32, 144, 145, 167, 170, 271, 306
Flurazepam (EX) 495, 499
Flurbiprofen (EX) 389
Flurogestone acetate (EX) 376
Flutamide (EX) 389; (SI) 188, 220, 522
Fluticasone 17-propionate (SI) 20, 32, 145, 167, 170, 264, 306, 414
Fluvoxamine (EX) 8, 231, 254, 378, 389, 401, 407, 495, 499, 624; (IN) 537; (SI) 399
Folic acid (EX) 389; (SI) 2
Fominoben (SI) 188, 220, 522
Fosinopril (AL) 682
Fosinoprilat (EX) 283
FTU (EX) 215
Fumaric acid (EX) 389
Furaltadone (EX) 389
Furazolidone (EX) 389, 601
Furosemide (EX) 389, 495, 499; (SI) 178
Fusidic acid (EX) 389

Ganciclovir (SI) 2, 668
Gentamicin (EX) 312, 389
Gitoxigenin (SI) 178
Glibenclamide (EX) 8, 231, 254, 389, 401, 407, 624
Gliclazide (EX) 389
Glipizide (EX) 8, 231, 254, 389, 401, 407, 624
Glutathione (EX) 389
Glutethimide (EX) 495, 499; (SI) 178
Glycopyrrolate (IS) 136; (SI) 331
Gonadorelin (LHRH) (SI) 13
Grepafloxacin (SI) 293
Griseofulvin (EX) 389
Guaiacol (EX) 389; (SI) 178
Guaifenesin (EX) 389, 495, 499; (IS) 641
Guanfacine (EX) 389

Halcinonide (SI) 19, 20, 31, 32, 144, 145, 167, 170, 264, 271, 306
Halobetasol propionate (SI) 20, 32, 145, 167, 170, 264, 271
Halofantrine (EX) 389
Haloperidol (EX) 8, 231, 254, 389, 401, 407, 495, 499, 525, 624; (SI) 537
Harpagoside (EX) 389

Hematoporphyrin (EX) 389
Heptadecanoic acid (IS) 462
Heptanoic acid (EX) 461
Hexachlorobenzene (SI) 354
Hexobarbital (IS) 335; (SI) 178
Hippuric acid (SI) 178
Histapyrrodine (EX) 8, 231, 254, 401, 407, 624
Histidine (EX) 389
Homosalate (SI) 456
Hydralazine (EX) 495, 499
Hydrochlorothiazide (EX) 104, 389, 672
Hydrocodone (EX) 8, 231, 254, 401, 407, 624
Hydrocortisone (EX) 168, 169, 260, 265, 389; (SI) 19, 20, 31, 32, 144, 145, 163, 167, 170, 178, 261, 264, 271, 306, 414
Hydrocortisone 21-acetate (SI) 19, 20, 31, 32, 144, 145, 167, 170, 188, 264, 271, 306, 522
Hydrocortisone-17-butyrate (SI) 19, 20, 31, 32, 144, 145, 167, 170, 264, 271, 306
Hydrocortisone 21-cypionate (SI) 20, 32, 145, 167, 170, 264, 271, 306
Hydrocortisone pivalate (SI) 19, 31, 144
Hydrocortisone 17-valerate (SI) 20, 32, 145, 167, 170, 264, 272, 306
Hydroflumethiazide (SI) 104
Hydroquinone (SI) 178
*p*-Hydroxybestatin (EX) 662
4-Hydroxybutyrate (EX) 389
Hydroxychloroquine (EX) 8, 231, 254, 401, 407, 624
21-Hydroxydeflazacort (EX) 163, 261
18-Hydroxydeoxycorticosterone (EX) 171
5-Hydroxyindole-3-acetic acid (IS) 77
10-Hydroxynortriptyline (EX) 378
Hydroxyprogesterone (EX) 169
9-Hydroxyrisperidone (EX) 537
3-Hydroxytyramine (EX) 389
Hydroxyzine (EX) 8, 231, 254, 389, 401, 407, 495, 499, 624
Hyoscyamine (EX) 389
Hypoxanthine (IS) 668

Ibuprofen (EX) 389, 495, 499; (SI) 178
Iloprost (EX) 405
Imidacloprid (EX) 452
Imidapril (SI) 188, 220, 522
Imidazole (EX) 273, 389
Imipramine (EX) 8, 231, 254, 389, 401, 407, 495, 499, 537, 624; (IN) 202; (IS) 403; (SI) 383
Inchonidine (EX) 389
Indenolol (EX) 434
Indinavir (AL) 38; (EX) 36, 37, 38, 208, 209, 359, 360, 435, 436, 443, 444; (SI) 47, 512, 548, 627
Indomethacin (EX) 8, 231, 254, 389, 401, 407, 495, 499, 624; (IS) 341; (SI) 178, 188, 221, 522
Indoramine (EX) 389
Iodoform (EX) 389
Iohexol (EX) 324, 325, 326
Iothalamic acid (EX) 322
Iotrolan (EX) 325
Ioxaglic acid (EX) 322
Irbesartan (EX) 105, 671
Irinotecan (SI) 188, 221, 522
Isoamyl *p*-methoxycinnamate (SI) 456
Isoflupredone (EX) 168, 265
Isoniazid (EX) 8, 231, 254, 401, 407, 624; (SI) 627
Iso-pirlimycin (IS) 516
Isoprocarb (EX) 250
Isopropamide iodide (EX) 381; (SI) 382
Isopropyl dibenzoylmethane (SI) 456
Isoquercitrin (EX) 389
Isosorbide (EX) 389
Isosorbide dinitrate (EX) 490
Isothipendyl (EX) 389
Isotretinoin (EX) 21, 22; (SI) 23
Isoxicam (SI) 200
Isoxsuprine (IS) 544
Isradipine (EX) 8, 231, 254, 389, 401, 407, 624

Jasmolin (SI) 534
Jasmolin I (SI) 533
Jasmolin II (SI) 533
Josamycine (EX) 389

Ketamine (EX) 8, 231, 254, 389, 401, 407, 495, 624; (IN) 499, 552
Ketobemidone (EX) 8, 231, 254, 401, 407, 624
Ketoconazole (AL) 38; (EX) 389, 495, 499
Ketogan (IN) 552
Ketoprofen (EX) 8, 231, 254, 389, 401, 407, 624
Ketorolac (EX) 8, 231, 254, 401, 407, 624

Labetalol (EX) 8, 231, 254, 401, 407, 434, 495, 499, 624; (IS) 49, 626
Lacidipine (EX) 389
Lactic acid (EX) 389
Lamivudine (EX) 1, 2; (SI) 2, 627
Lamotrigine (EX) 8, 231, 254, 401, 407, 624
Lanolin (SI) 112
Lanzoprazole (EX) 389
Lasalocid (EX) 415, 429
Latamoxef (EX) 113, 118, 257
Laudanosine (EX) 137; (SI) 138
Lauric acid (EX) 457, 458, 459, 460, 461; (SI) 463
Levamisole (EX) 389
Levobunolol (EX) 434; (IS) 70
Levocabastine (EX) 8, 231, 254, 401, 407, 624
Levodopa (EX) 389
Levomepromazine (EX) 8, 231, 254, 389, 401, 407, 624; (SI) 194, 383
Levonordefrin (SI) 527
Levopenbutolol (EX) 389
Levothyroxine (EX) 357
Levrazoxane (EX) 172
LHRH (gonadorelin) (SI) 13
Lidocaine (AL) 98; (EX) 8, 231, 254, 389, 401, 407, 495, 499, 624; (IN) 552; (IS) 133

Lincomycin (SI) 516
Linoleic acid (EX) 457, 458, 459, 460, 461, 462; (SI) 463, 557
Linolenic acid (EX) 457, 458, 459, 460, 461, 462; (SI) 463, 557
Gamma-linolenic acid (EX) 457
Linuron (EX) 389
Lipotropic hormone (SI) 14
Lisinopril (SI) 104
Lisuride (EX) 233, 389
Lithium (SI) 656
Lobeline (EX) 389
Loflazepate (EX) 389
Lomefloxacin (EX) 650; (IS) 420, 487
Lomustine (EX) 389
Loperamide (SI) 16, 603, 614
Lopinavir (EX) 37, 38, 209, 436, 444
Loprazolam (EX) 389
Loratadine (EX) 8, 164, 165, 231, 254, 389, 401, 407, 624
Lorazepam (EX) 8, 231, 254, 389, 401, 407, 495, 499, 624
Lormetazepam (EX) 8, 231, 254, 401, 407, 495, 499, 624
Losartan (EX) 105, 671; (IS) 124
Loxapine (EX) 389, 495, 499
LSD (EX) 8, 231, 254, 389, 401, 407, 624
Lufenuron (EX) 389
Luteinizing hormone (SI) 134
Lysophosphatidylcholine (EX) 502, 505

Magnesium (SI) 656
Malathion (EX) 8, 231, 254, 401, 407, 624
Malic acid (SI) 140
Maprotiline (EX) 8, 231, 254, 378, 389, 401, 407, 495, 499, 537, 624; (SI) 188, 221, 522
Marbofloxacin (SI) 472
Margaric acid (EX) 458, 460; (IS) 457, 459, 461, 557
MCPA (EX) 389
MDA (EX) 389, 497, 529; (IN) 499
MDEA (EX) 497, 529
MDMA (EX) 8, 231, 254, 389, 401, 407, 495, 499, 529, 624; (IN) 499
MDPA (EX) 389
Mebendazole (EX) 102, 103, 389
Mebezonium (EX) 389
Meclozine (EX) 8, 231, 254, 389, 401, 407, 624
Mecoprop (EX) 389
Medazepam (EX) 8, 231, 254, 389, 401, 407, 624
Medetomidine (EX) 60, 61, 389
Medroxyprogesterone (EX) 389
Medroxyprogesterone acetate (EX) 269, 376
Mefenamic acid (IS) 676; (SI) 178
Mefenorex (EX) 389
Mefloquine (EX) 389
Megasol (SI) 456
Megestrol (EX) 375, 389
Megestrol acetate (EX) 269, 376
Meglutol (EX) 139
Melanotropin (SI) 14
Melatonin (EX) 389
Melengestrol acetate (EX) 269
Meloxicam (EX) 8, 231, 254, 401, 407, 624
Melperone (EX) 8, 231, 254, 401, 407, 537, 624; (IS) 537
Melphalan (EX) 389
Menotropins (SI) 14
Mepenzolate bromide (EX) 330; (SI) 331
Meperidine (pethidine) (EX) 8, 231, 254, 389, 401, 407, 495, 499, 624; (SI) 552
Mepindolol (EX) 434
Mepivacaine (EX) 8, 54, 231, 254, 401, 407, 624
Meprednisone (SI) 20, 32, 145, 167, 170, 264, 272, 306
Meprobamate (EX) 8, 231, 254, 401, 407, 624
Meropenem (IS) 234
Mesalamine (EX) 389
Mesoridazine (EX) 8, 231, 254, 401, 407, 496, 499, 624
Mestranol (SI) 239
Metamitron (EX) 389
Metaproterenol (EX) 389
Metformin (EX) 389, 512
Methacycline (EX) 389
Methadone (AL) 38; (EX) 8, 231, 254, 389, 401, 407, 496, 499, 624
Methamphetamine (EX) 8, 231, 254, 389, 401, 407, 495, 497, 529, 624; (IN) 499
Methandienone (IS) 82
Methandrostenolone (EX) 427, 587
Methaqualone (EX) 495, 499; (IS) 673
Methiocarb (EX) 250
Methionine (EX) 389
Methocarbamal (SI) 178, 642
Metholcarb (EX) 250
Methoprene (SI) 533
Methoprene acid (EX) 386
Methotrexate (SI) 188, 221, 522
*p*-Methoxylevodropropizine (IS) 198
Methoxyselamectin (IS) 571
Methyclothiazide (EX) 389
Methylbenacyzine (SI) 331
Methyldopa (SI) 178, 389, 394
3-Methyldopamine (EX) 389
4-Methyldopamine (EX) 389
Methylene blue (SI) 187
Methylephedrine (EX) 495, 499
7-Methyleptazocine (IS) 227
Methylhydroxyprogesterone (EX) 389
3-Methylleflunomide (EX) 341
5-Methyl-2-nitrophenol (IS) 255
Methyl paraben (EX) 389; (IS) 512, 676; (SI) 178, 331
Methylparathion (EX) 8, 231, 254, 401, 407, 624
Methylphenidate (EX) 8, 231, 254, 401, 407, 496, 499, 624
Methylprednisolone (EX) 168, 260, 265, 389; (SI) 19, 20, 31, 32, 144, 145, 167, 170, 264, 270, 306

Methylprednisolone 21-acetate (SI) 20, 32, 145, 167, 170, 264, 272, 306
3-Methylpyrazole (IS) 274
Methyl salicylate (SI) 178
Methyltestosterone (EX) 427, 587; (IS) 652; (SI) 178
Metipranolol (EX) 434
Metoclopramide (EX) 8, 231, 254, 318, 384, 389, 401, 407, 495, 624; (IN) 499, 552
Metocurine (SI) 552
Metoprolol (EX) 8, 231, 254, 389, 401, 407, 434, 495, 499, 624; (SI) 552
Metronidazole (EX) 8, 231, 254, 389, 401, 407, 624
Mexiletine (EX) 8, 231, 254, 389, 401, 407, 624; (IS) 509
Mianserin (EX) 8, 231, 254, 389, 401, 407, 495, 499, 537, 624
Miconazole (EX) 389; (IS) 252
Midazolam (EX) 8, 60, 231, 254, 369, 389, 401, 407, 495, 499, 624
Mifepristone (IS) 301
Minaprin (EX) 389
Minocycline (EX) 389
Minoxidil (EX) 389
Mirtazapine (EX) 8, 231, 254, 407, 537, 624; (SI) 399
Misoprostol (EX) 313
Mitomycin C (EX) 89
Mivacurium (EX) 137; (SI) 589
Mizolastine (EX) 8, 231, 254, 401, 624
Moclobemide (EX) 8, 231, 254, 389, 401, 407, 495, 499, 537, 624
Molindone (EX) 8, 231, 254, 401, 407, 495, 499, 624
Molsidomine (EX) 389
Mometasone (EX) 413
Mometasone 17-furoate (SI) 19, 20, 31, 32, 144, 145, 167, 170, 264, 272, 306
Monensin (EX) 429
Monoacetylmorphine (AL) 98; (EX) 8, 231, 254, 401, 407, 624
Monodesbutylhalofantrine (EX) 389
Monodesethylchloroquine (EX) 389
Moprolol (EX) 434
Moroxydine (EX) 389
Morphine (AL) 98; (EX) 8, 231, 254, 389, 401, 407, 495, 499, 624
Moxifloxacin (SI) 293
Myristic acid (EX) 457, 458, 459, 460, 461, 462; (SI) 463, 557
Myristoleic acid (EX) 458, 459, 460, 461

Nadolol (EX) 389, 434, 495, 499
Nadoxolol (EX) 389
Naftidrofuryl (EX) 389
Naftopidil (IS) 426
Nalidixic acid (EX) 365, 389, 473
Nalorphine (EX) 389
Naloxone (EX) 389, 495, 499
Naltrexone (EX) 389
Nandrolone (EX) 587
2-Naphthoic acid (IS) 313, 405
Naphthol (SI) 354
Naproxen (EX) 495, 499; (IS) 147
Narasin (EX) 415
Nefazodone (EX) 537
Nefiracetam (SI) 188, 221, 522
Nefopam (EX) 389
Nelfinavir (AL) 38; (EX) 36, 37, 38, 208, 209, 281, 359, 360, 435, 443, 444
Nemonapride (EX) 389
Neomycin (EX) 312
Neopynamine (EX) 389
Neostigmine (EX) 389; (SI) 331
Netilmicin (EX) 389; (IS) 153
Nevirapine (EX) 1, 38, 208, 209, 335, 360, 435, 436; (SI) 2, 627
Niacin (EX) 389; (SI) 178
Niacinamide (EX) 389
Niaprazine (EX) 389
Nicarbazin (EX) 149
Nicardipine (EX) 390
Nicergoline (EX) 390
Niclosamide (EX) 390
Nicotine (EX) 8, 231, 254, 401, 407, 624
Nifedipine (EX) 8, 99, 231, 254, 390, 401, 407, 495, 499, 624; (IS) 48; (SI) 188, 221, 522
Nifenalol (EX) 434
Niflumic acid (EX) 390
Nifuroxazide (EX) 390; (IS) 450
Nikethamide (EX) 8, 231, 254, 390, 401, 407, 624; (SI) 383
Nimesulide (IS) 556
Nimodipine (IN) 552
Nitrazepam (EX) 8, 231, 254, 390, 401, 407, 495, 499, 624; (SI) 188, 194, 221, 522
Nitrendipine (EX) 390
*o*-Nitroacetanilide (IS) 77
Nitrofural (EX) 390
Nitrofurantoin (SI) 178
Nitrofurazone (EX) 601
Nitroglycerin (EX) 490
Nitromide (EX) 601
Nizatidine (EX) 8, 231, 254, 390, 401, 407, 624
Nomifensine (EX) 8, 231, 254, 401, 407, 624
Nonanoic acid (EX) 461
Nonanol (EX) 380
Noramidopyrine (EX) 390
Norclomipramine (EX) 390, 537
Norclozapine (EX) 378, 537
Nordiazepam (EX) 398, 495, 499
Nordoxepin (EX) 537
Norephedrine (EX) 390
Norepinephrine (EX) 348, 390, 498; (SI) 527
Norethisterone (EX) 390
Norfenfluramine (EX) 378, 497, 529; (S) 529
Norfloxacin (EX) 365, 650; (IN) 364; (IS) 472, 569
Norfluoxetine (EX) 378, 495, 499, 537; (SI) 399
Norgalantamine (EX) 290

Pyrantel tartrate (IS) 416
Pyrazinamide (EX) 390; (SI) 2
Pyrethrin I (EX) 26, 80
Pyrethrin II (EX) 26, 80
Pyridostigmine (EX) 390
Pyridoxine (Vitamin $B_6$) (EX) 391
Pyrimethamine (EX) 390
Pyrithydione (SI) 178
Pyrocatechol (SI) 178
Pyroglutamic acid (EX) 390

Quercetin (EX) 390
Quetiapine (EX) 537
Quinapril (AL) 682; (EX) 390; (SI) 188, 221, 522
Quinaprilat (AL) 682
Quinidine (EX) 390, 495, 499
Quinine (EX) 9, 232, 255, 390, 402, 408, 625; (IS) 365
Quinupramine (EX) 390
Quinupristin (EX) 156; (SI) 157

Raloxifene (SI) 47, 512
Ramipril (AL) 682; (EX) 390
Ranitidine (EX) 9, 232, 255, 390, 402, 408, 495, 499, 625; (SI) 2, 202
Reboxetine (EX) 537
Reserpine (EX) 390; (SI) 178
Resmethrin (AL) 25; (EX) 26, 535
Resorcinol (EX) 390; (SI) 178
all-trans Retinal (EX) 22
Retinyl acetate (IS) 22
all-trans Retinyl stearate (EX) 22
all-trans Retinyl palmitate (EX) 22
Rhein (EX) 108
Riboflavin (Vitamin $B_2$) (EX) 391
Rifabutin (AL) 38; (EX) 390
Rifampin (AL) 38; (EX) 390; (SI) 2, 627
Rifamycin (EX) 390
Rilmenidine (EX) 390
Rimantadine (EX) 378
Ripazapam (IS) 632, 684
Risperidone (EX) 9, 232, 255, 402, 408, 499, 537, 625; (IN) 496; (IS) 418
Ritodrine (IS) 545
Ritonavir (AL) 38; (EX) 36, 37, 38, 208, 209, 359, 360, 435, 436, 443, 444; (SI) 627
Rocurone (EX) 9, 232, 255, 402, 408, 625
Rocuronium (EX) 137; (SI) 589
Ronidazole (EX) 390
Ropivacaine (EX) 9, 232, 255, 402, 408, 625
Rosmarinic acid (EX) 390
Rosoxacin (EX) 390
Roxithromycin (EX) 390; (SI) 267
Rutine (EX) 390

Saccharin (EX) 390; (SI) 178
Salicylamide (EX) 9, 232, 255, 402, 408, 625; (SI) 178
Salicylic acid (EX) 384, 390, 495, 499; (SI) 178, 293
Salinomycin (EX) 415, 429
Saquinavir (AL) 38; (EX) 36, 37, 38, 208, 209, 359, 360, 435, 436, 443, 444, 627
Sarafloxacin (EX) 473; (SI) 364, 472
Scopolamine (EX) 390
Secnidazole (EX) 390
Secobarbital (EX) 390, 495, 499; (SI) 178
Selegiline (EX) 9, 232, 255, 390, 402, 408, 625
Sennidin A (EX) 108
Sennidin B (EX) 108
Sertindole (EX) 9, 232, 255, 402, 408, 625
Sertraline (EX) 9, 232, 255, 378, 402, 408, 495, 499, 537, 625; (SI) 399
Simazine (EX) 9, 232, 255, 390, 402, 408, 625
Sincocaine (EX) 9, 232, 255, 402, 408, 625
Sisapride (EX) 9, 232, 255, 402, 408, 625
Sisomicin (EX) 390
$\beta$-Sitosterol (SI) 585
Sodium (SI) 656
Somatoliberin (SI) 13
Somatostatin (EX) 530
Somatropin (SI) 14
Sotalol (EX) 9, 232, 255, 390, 402, 408, 434, 625
Spectinomycin (EX) 312
Sphingomyelin (EX) 501, 502, 505
Spirenone (EX) 199
Spironolactone (EX) 390, 495, 499; (IS) 427, 587; (SI) 188, 221, 522
Stanozolol (EX) 427; (SI) 178
Stavudine (EX) 1, 2, 335, 336; (SI) 2, 627
Stearic acid (EX) 457, 458, 459, 460, 461, 462; (SI) 463, 557
Stearyl alcohol (SI) 129
Streptomycin (EX) 312
Strychnine (EX) 9, 232, 255, 390, 402, 408, 625
Succinic acid (SI) 140
Sucrose (EX) 303; (SI) 5
Sulbutiamine (EX) 390
Sulfabenzamide (SI) 593
Sulfabromomethazine (EX) 594, 597
Sulfacetamide (EX) 592; (SI) 178, 593
Sulfachloropyrazine (EX) 390, 592
Sulfachlorpyridazine (EX) 591, 597; (SI) 596
Sulfadiazine (EX) 390, 591, 592, 594, 597, 598; (SI) 593, 596
Sulfadimethoxine (EX) 390, 591, 592, 594, 597; (SI) 178, 593, 596
Sulfadoxine (EX) 390, 592, 594; (SI) 596
Sulfaethidole (SI) 178
Sulfaethoxypridazine (EX) 591, 594
Sulfaguanidine (EX) 390, 592, 594, 599
Sulfamerazine (EX) 592, 594; (SI) 178, 593, 596
Sulfamethazine (EX) 591, 592, 593, 594, 597, 599; (IS) 598; (SI) 178, 593
Sulfamethizole (EX) 592, 594; (SI) 178, 593
Sulfamethoxazole (EX) 390, 495, 499, 592; (SI) 2, 178, 552, 593, 596

Sulfamethoxypyridazine (EX) 390, 592, 594; (SI) 596
Sulfanilamide (EX) 390, 592, 594, 599; (SI) 6, 178, 593, 596
Sulfapyridine (EX) 592, 594, 598, 599; (SI) 178, 593
Sulfaquinoxaline (IN) 185
Sulfathiazole (EX) 390, 592, 594, 599; (SI) 596
Sulfisoxazole (EX) 592, 599; (S) 6, 178, 593
Sulindac (EX) 9, 232, 255, 390, 402, 408, 625; (SI) 178
Sulpiride (EX) 9, 232, 255, 390, 402, 408, 495, 499, 537, 625; (SI) 188, 221, 522
Sultamicillin tosylate (SI) 16, 614
Sulthiame (EX) 9, 232, 255, 402, 408, 625; (SI) 188, 221, 522
Sultopride (EX) 390
Synephrine (EX) 390

Talinolol (EX) 434
Tamoxifen (EX) 390
Tartaric acid (SI) 140
Tegafur (SI) 16, 603, 614
Temazepam (EX) 9, 232, 255, 390, 402, 408, 495, 499, 625
Temocapril (SI) 16, 603
Tenoxicam (EX) 390
Terbutaline (EX) 9, 232, 255, 390, 402, 408, 625
Terephthalydiene dicamphor sulfonic acid (SI) 456
Terfenadine (AL) 38; (EX) 9, 232, 255, 390, 402, 408, 495, 499, 625
Terodiline (EX) 9, 232, 255, 402, 408, 625
Terpine (EX) 390
Testosterone (EX) 82, 169, 427, 428, 587; (IS) 268, 428
Testosterone acetate (SI) 178
Testosterone 17$\beta$-cypionate (SI) 178
Testosterone enanthate (SI) 178
Testosterone propionate (IS) 392; (SI) 178
Tetracaine (EX) 9, 232, 255, 390, 402, 408, 625; (IS) 97
Tetracosanoic acid (EX) 457
Tetracycline (EX) 390, 592
Tetradecyl alcohol (SI) 129
Tetrahydrozoline (EX) 9, 232, 255, 402, 408
Tetramethrin (AL) 25, 26; (EX) 26, 80, 535; (SI) 24
Tetramisole (EX) 390
Tetraplatin (EX) 480
Tetrazepam (EX) 390; (IS) 398
Theobromine (EX) 9, 232, 255, 390, 402, 408, 625; (SI) 178
Theodrenaline (EX) 390
Theophylline (EX) 9, 232, 255, 384, 390, 402, 408, 495, 499, 625; (SI) 16, 173, 202, 603, 614
Thiabendazole (EX) 102, 103
Thiamine (Vitamin $B_1$) (EX) 391
Thiamphenicol (EX) 390; (SI) 259
Thiobarbituric acid (SI) 178
Thiobencarb (EX) 250
Thiopental (thiopentone) (EX) 390, 495, 499
Thioproperazine (EX) 390
Thioridazine (EX) 9, 232, 255, 390, 402, 408, 495, 499, 625; (SI) 383, 537
Thiosalicylic acid (SI) 178
Thiothixene (EX) 9, 232, 255, 402, 408, 495, 499, 625
Thyroid-stimulating hormone (SI) 134
Tianeptine (EX) 390
Tiapride (EX) 390
Tiaprofenic acid (EX) 391
Ticlopidine (EX) 391
Tiemonium iodide (EX) 391; (SI) 331
Tiletamine (EX) 683, 684
Timolol (EX) 9, 89, 232, 255, 391, 402, 408, 434, 625; (IS) 151
Tinidazole (EX) 391
Tioclomarol (EX) 391
Tiospirone (IS) 493
Tobramycin (EX) 312, 391
Tocainide (EX) 495, 499
Tocopherol (EX) 22; (SI) 585
Tofisopam (EX) 391
Tolbutamide (EX) 9, 232, 255, 402, 408, 495, 499, 625; (SI) 178, 188, 221, 522
Toliprolol (EX) 434
Tolmetin (SI) 178
Toloxatone (EX) 391
Tolperisone (IS) 220
Toluene (EX) 391
*p*-Toluenesulfonic acid (EX) 602
*o*-Toluidine (AL) 98
Toremifene (EX) 9, 232, 255, 402, 408, 625
Tramadol (EX) 9, 232, 255, 402, 408, 625
Trandolapril (EX) 391
Tranilast (SI) 188, 221, 522
Trazodone (EX) 9, 232, 255, 391, 402, 408, 495, 499, 625
Tretinoin (EX) 21, 22; (SI) 23
Triamcinolone (EX) 168, 265; (SI) 19, 20, 31, 32, 144, 145, 167, 170, 178, 188, 221, 264, 272, 306, 522
Triamcinolone acetonide (EX) 168, 265; (SI) 19, 20, 31, 32, 144, 145, 167, 170, 178, 264, 272, 306
Triamcinolone 16,21-diacetate (SI) 19, 20, 31, 32, 144, 145, 167, 170, 264, 272, 306
Triamcinolone hexacetonide (SI) 20, 32, 145, 167, 170, 264, 272, 306
Triamterene (EX) 9, 232, 255, 391, 402, 408, 625
Triazolam (EX) 9, 232, 255, 391, 402, 408, 495, 499, 625
Trichlorocarbanilide (EX) 391
Trichloromethiazide (EX) 391
Triclabendazole (EX) 103
Tridecanoic acid (EX) 461
Trifluoperazine (EX) 391
Trifluperidol (EX) 391; (SI) 537

Triflupromazine (EX) 495, 499
Trihexyphenidyl (EX) 391
2,4,6-Triiodobenzoic acid (IS) 322
Trimebutine (EX) 391
Trimeprazine (EX) 9, 232, 255, 402, 408, 496, 499, 625
Trimetazidine (EX) 391
Trimethoprim (AL) 38; (EX) 9, 232, 255, 391, 402, 408, 496, 499, 625; (IS) 672; (SI) 2, 552, 627
Trimipramine (EX) 9, 232, 255, 391, 402, 408, 496, 499, 537, 625
Triprolidine (EX) 391
Tritoqualine (EX) 391
Tropatepine (EX) 391
Trovafloxacin (IS) 293
Tryptophan (SI) 293
Tubocurarine (IN) 552
Tulobuterol (SI) 16, 603, 614

Undecanoic acid (EX) 461
Urocanic acid (SI) 456

Valdecoxib (SI) 484
Valethamate (SI) 331, 382
Valsartan (EX) 105
Vanillin (EX) 391
Vasoactive intestinal polypeptide (EX) 530
Vasopressin (SI) 13
Vecuronium (EX) 137, 514; (SI) 589
Venlafaxine (EX) 9, 232, 255, 402, 408, 537, 625; (SI) 399
Verapamil (EX) 9, 232, 255, 391, 402, 408, 496, 499, 625
Veratrine (EX) 391
Veratrole (EX) 391
Viloxazine (EX) 391
Vinblastine (EX) 391
Vincamine (EX) 391; (IS) 92
Vincristine (EX) 391
Vinylbital (EX) 391
Virginiamycin (EX) 391
Vitamin A (EX) 21, 22
Vitamin $B_1$ (thiamine) (EX) 391
Vitamin $B_2$ (riboflavin) (EX) 391
Vitamin $B_5$ (pantothenic acid) (EX) 391
Vitamin $B_6$ (pyridoxine) (EX) 391
Vitamin $B_{12}$ (EX) 391
Vitamin C (EX) 391
Vitamin $D_2$ (IS) 181
Vitamin E (EX) 22; (SI) 585
Vitamin H (biotin) (EX) 391

Warfarin (EX) 9, 232, 255, 391, 402, 408, 625; (IS) 342; (SI) 178, 188, 221, 522

Xanthydrol (EX) 391
Xipamide (EX) 391
Xylazine (EX) 525
Xylene (EX) 391

Yohimbine (EX) 9, 232, 255, 391, 402, 408, 625

Zalcitabine (EX) 1, 2, 335, 336
Zeranol (EX) 427, 587
Zidovudine (EX) 1, 2, 335, 336; (SI) 2, 627
Zipeprol (EX) 391
Ziprasidone (EX) 537
Zoalene (EX) 601
Zofenoprilat (EX) 681; (SI) 682
Zolazepam (EX) 631, 632
Zolpidem (EX) 9, 232, 255, 391, 402, 408, 625, 673; (SI) 537
Zonisamide (SI) 188, 221, 522
Zopiclone (EX) 9, 232, 255, 402, 408, 496, 499, 625
Zuclopenthixol (EX) 391, 496, 499